AF389325

Animal Perinatology: E
Health of the Dam and

Animal Perinatology: Behavior and Health of the Dam and Her Offspring

Editors

Daniel Mota-Rojas
Julio Martínez-Burnes
Agustín Orihuela

Basel • Beijing • Wuhan • Barcelona • Belgrade • Novi Sad • Cluj • Manchester

Editors
Daniel Mota-Rojas
Universidad Autónoma
Metropolitana
Mexico City, Mexico

Julio Martínez-Burnes
Universidad Autónoma de
Tamaulipas
Ciudad Victoria, Mexico

Agustín Orihuela
Universidad Autónoma del
Estado de Morelos
Cuernavaca, Mexico

Editorial Office
MDPI
St. Alban-Anlage 66
4052 Basel, Switzerland

This is a reprint of articles from the Special Issue published online in the open access journal *Animals* (ISSN 2076-2615) (available at: https://www.mdpi.com/journal/animals/special_issues/animal_perinatology).

For citation purposes, cite each article independently as indicated on the article page online and as indicated below:

Lastname, A.A.; Lastname, B.B. Article Title. *Journal Name* **Year**, *Volume Number*, Page Range.

ISBN 978-3-0365-9196-4 (Hbk)
ISBN 978-3-0365-9197-1 (PDF)
doi.org/10.3390/books978-3-0365-9197-1

© 2023 by the authors. Articles in this book are Open Access and distributed under the Creative Commons Attribution (CC BY) license. The book as a whole is distributed by MDPI under the terms and conditions of the Creative Commons Attribution-NonCommercial-NoDerivs (CC BY-NC-ND) license.

Contents

About the Editors

Daniel Mota-Rojas

Daniel Mota-Rojas has significant research expertise in applied animal behavior and welfare. He obtained a Bachelor of Science degree in Veterinary Medicine and Animal Husbandry at the Metropolitan Autonomous University (UAM) in Mexico City. He earned a Master's in Production and Animal Health Sciences from the National Autonomous University of Mexico (UNAM) and a Ph.D. in Biological Sciences from the UAM in Mexico City. Currently, he serves as the Commissioner for the Ph.D. program in Biological and Health Sciences at the UAM. Dr. Mota-Rojas served as the Leader in Biomedical Science, focusing on the Pathophysiology of Stress, Behavior, and Welfare of Domestic Animals. His research centers on evaluating the role of cognitive bias in animal emotions, the neurological modulation of facial expressions, quality of life, quality of death, mother–young bonding, euthanasia, thermal biology, and the behavior and welfare of domestic and wild animals. He is a guest lecturer and international speaker on Animal Behavior and Welfare. Dr. Mota-Rojas is a member of the National System of Researchers of CONACYT in Mexico (level III) and has authored over 220 scientific articles indexed in SCOPUS with more than 3400 international citations. He is a member of the Mexican Network of Animal Welfare, the Mexican Academy of Sciences, and the Mexican Veterinary Academy. He is the editor of 11 books, encompassing topics related to animal behavior, cognition, emotional states, attachment and welfare, thermoregulation, global warming, pain assessment, facial expression, neonatology, and maternal behavior in domestic animals, including the water buffalo. Additionally, he serves on the Editorial Boards of several journals, including "Frontiers in Veterinary Science" (Switzerland), "Journal of Buffalo Science" (Lifescience Global, Canada), "Dog Behavior" (Italy), "CABI Reviews" (UK), and the journal "Animals" (MDPI, Basel, Switzerland).

Julio Martínez-Burnes

Julio Martínez-Burnes is a Veterinarian and Zootechnician with a degree from the Autonomous University of Tamaulipas, Mexico. He earned a Master's in Veterinary Sciences (Animal Pathology) from the National Autonomous University of Mexico (UNAM) and holds a Ph.D. in Pathology from the University of Prince Edward Island, Canada. He was a full-time Professor at the Faculty of Veterinary Medicine and Zootechnics (FMVZ) and holds the title of Professor Emeritus at the Autonomous University of Tamaulipas. Dr. Martínez-Burnes has taught General Pathology and Systemic Pathology at the undergraduate and postgraduate levels for over 40 years. He is a member of the Mexican Veterinary Academy and the Mexican Association of Veterinary Pathologists. He is the author of more than 90 scientific articles indexed in SCOPUS with over 1400 international citations. He is a National System of Researchers member. Dr. Martínez-Burnes served as the Leader of the Academic Body in Animal Health, with PROMEP certifications, and is a National Research member. His research focuses on perinatology, pathology, and diseases of the respiratory system in various species of domestic animals and wildlife. He is the author and co-author of articles in national and international journals and has contributed to books on animal pathology and perinatology.

Agustín Orihuela

Agustín Orihuela is an agronomist and animal scientist with a degree from the Autonomous University of Chapingo. He holds a Master's and a Ph.D. in Animal Production from the National Autonomous University of Mexico (UNAM). Dr. Orihuela completed his postdoctoral studies in

Animal Behavior at the University of California, Davis. He is the author of over 150 scientific articles indexed in SCOPUS with more than 1900 international citations. Currently, he holds the position of Full Professor in Animal Welfare and Animal Behavior at the Autonomous University of the State of Morelos. Dr. Orihuela is recognized as a Professor Emeritus by the National System of Researchers of CONACYT in Mexico. His research interests encompass reproduction, animal behavior, and animal welfare. He is the author and co-author of articles in national and international journals. He has contributed to books on animal behavior, with a focus on the mother–offspring bond, allosuckling behavior, and maternal behavior.

Preface

It is a privilege and a pleasure to introduce this scientific publication that compiles original and review manuscripts in Animal Perinatology.

Veterinary medicine is advancing rapidly with the expansion of knowledge and emerging technologies. New medical specialization areas provide more precise understandings of numerous aspects of animal health. One of these fields is animal perinatology, a branch of obstetrics and gynecology concerned primarily with fetal anatomy and physiology analyses in the late stages of pregnancy, labor, the postpartum period, and lactation. Studies in animal perinatology span not only the broader field of obstetrics, essentially focused on the fetus and neonate, but also, significantly, aspects related to the welfare, behavioral characteristics, and health status of the mother. Thus, the phases of the birth process examined in this specialized field include gestation, parturition, puerperium, and lactation. For these reasons, animal perinatology has gained enormous importance. It continues to generate invaluable knowledge and tools to improve the monitoring of the *in-utero* health of fetuses, neonates, and dams into the postpartum period. Today, obstetrics and perinatology are guided by one fundamental goal: to ensure the optimum health and well-being of fetuses and dams.

In veterinary science, parturition refers to the entire birthing process, the typical event that marks the end of pregnancy (or gestation) in domesticated animals. It is of pre-eminent importance in the animal production cycle, as guaranteeing normal, problem-free births helps ensure the economic success of breeding operations, and supports the health and welfare of all associated production animals. While it is true that the vast majority of parturitions proceed with no complications, there is no doubt that livestock production units and their personnel must be well-prepared in advance to deal with the potential problems and difficulties that can arise during the birthing process.

The term 'parturition', then, encompasses a multifaceted physiological event involving a daunting array of physiological, morphological, hormonal, and behavioral changes. For many animal species, the birthing process is a pivotal moment, one that, for dams, is often the life event associated with the most intense pain they will suffer. Although there is an immense corpus of research on pain during labor in human mothers, reports on this phenomenon in domesticated animals are scarce. Fortunately, recent studies on topics related to parturition reveal an increased awareness of this concern in the broader society, among animal owners, and in academia. Today, this growing body of scientific research contains abundant information on this topic for numerous species of farm animals.

Compared to other phases of the reproductive cycle, perinatal processes in farm animals have been little studied in recent decades, even though birthings can (i) become complicated and require intervention by the owners of production units, their personnel, or veterinarians and (ii) cause significant losses in the perinatal period in many domesticated species, whose females (bitches, cows, sows, buffaloes, and ewes, among others) frequently require attention during parturition to prevent the repercussions—direct or indirect—of perinatal asphyxia.

The consequences of the complications arising during birth can threaten neonates' and dams' welfare, and even their survival. This is especially true of primiparous females with narrow pelvises and in cases of size disproportion between the fetus and the mother's birth canal. In this regard, the term dystocia refers broadly to anomalies that can complicate labor, especially certain factors that can significantly intensify pain. These phenomena are thought to occur less often in polytocous species, for it is widely believed that giving birth to multiple newborns of smaller size can lessen labor-induced pain compared to the conditions of humans and other monotocous species.

Another critical topic of animal perinatology addressed herein is maternal behavior, including dams' active and passive responses in two key areas: willingness to feed their offspring and displays of protective behavior toward them. Some species are characterized by particular forms of conduct toward their own offspring that may endure long periods. Others, however, appear less selective and protect their young for less time. We know that expressions of maternal behavior are mediated by several processes in the brain: acceptance, social recognition, the inhibition of rejection/fear, and increased motivation to provide care. What triggers these forms of maternal conduct? The answer is exposure to adequate natural stimuli, especially for pregnant females or in the stages of farrowing and lactation. However, it is interesting that even virgin females and males may exhibit such behaviors when sensitized to neonates through cohabitation or cross-sensitization (via mating).

One hotly debated theme in this regard is whether it is best to ease the pain dams suffer when giving birth, especially given that labor-related pain serves a primary physiological function, namely, maintaining frequent, robust myometrial contractions. Here, again, the biological effects of pain can impact the welfare of the dam and her fetus(es). Some experts support using analgesic therapy to alleviate pain, but current evidence indicates that some analgesics can interfere with the natural labor process. Studies show that opioids and non-steroidal anti-inflammatory drugs can indirectly hinder uterine contractions by lowering oxytocin release. At the same time, local pain relievers reduce the number of contractions but intensify their strength, perhaps enhancing maternal performance during parturition.

Regarding the perinatal phase, research shows that this has a significant effect on newborn survival. The range of phenomena that can impact the welfare of a litter is diverse and varies during gestation, parturition, and the first weeks of post-uterine life. Hence, they require careful consideration by breeders and veterinarians alike. It is clear today that assessing fetal and neonatal welfare is vital for lowering the rates of stillbirths. For this reason, careful evaluations of fetal health before performing any obstetric intervention and of the health of newborns are essential to guarantee that when medical intervention is indicated, it can be provided in a timely, accurate manner. In cases of severe oxygen deprivation, for example, performing a cesarean section may be a more feasible, less life-threatening option for the fetus than other obstetric procedures. When this condition is diagnosed postpartum, various resuscitation techniques can be performed to attempt to save the newborn's life. What is essential, however, is to recognize that the range of possible postpartum interventions in animal production units is limited, emphasizing the importance of preventing stillbirths.

Offspring survival is of paramount importance. We now know this is directly related to two main factors: vitality scores at birth and environmental conditions. The two main challenges that confront a newborn are maintaining thermal stability and beginning to nurse as soon as possible so as to ensure ingestion of colostrum, an element of the mother's milk that augments its energy reserves and improves its probability of survival. Vitality scores measure the strength and energy that neonates show immediately post-birth. This often entails applying a numerical scale based on monitoring several factors: heartbeat, respiratory frequency, the color of mucous membranes, the time required to stand up, and signs of meconium staining. There are numerous factors that can influence vitality, as there are treatments available to support it.

Mother–Young Bonding: This key element in neonate survival requires the development of mutual recognition between the newborn and its mother. This bonding process enhances neonate survival by stimulating the dam to perform nurturing behaviors. This is stimulated through several sensory inputs, mainly the series of cues (auditory, tactile, visual, olfactory) emitted by the neonate to stimulate specific brain structures in the dam that help establish the emotional recognition of her offspring. In species that give birth to immature young—altricial animals, among others—mothers must provide comprehensive care. The latter's postpartum maturation determines the intensity of mother–offspring interaction.

Nursing behavior is another crucial theme elucidated in this Special Issue. In social animal species, cementing solid bonds with their young allows dams to offer several kinds of maternal care: protection from predators, strengthening the immune system, and providing nutrition. Even though lactation imposes high energy demands on mothers, social behaviors known as "allonursing" (when a female nurses another dam's infant) and "allosuckling" (when a newborn"feeds from any available female) have been documented in wild and domestic terrestrial and aquatic species. These behaviors are elements of a multifaceted strategy, as published studies suggest they are species-specific and depend on factors that include living conditions, social dynamics, and kinship. While allonursing and allosuckling have potential advantages, they may also endanger the health and welfare of unrelated dams and newborns, thereby increasing the risk of spreading pathogens.

These important topics are among the many intriguing themes and proposals explored in depth and breadth of this publication. Readers will benefit enormously from perusing all the articles in this product of international scientific collaboration.

Words fail us to adequately express our thanks to over 60 researchers from 16 countries who contributed to this scientific publication. We feel immensely fortunate to have benefitted from their collaboration. We are grateful for the trusting and unwavering support of the authors in sharing the experiments and scientific findings in this Special Issue on Animal Perinatology.

Daniel Mota-Rojas, Julio Martínez-Burnes, and Agustín Orihuela
Editors

Article

Placental Development and Physiological Changes in Pregnant Ewes in Silvopastoral and Open Pasture Systems during the Summer

Julia Morgana Vieira Dada [1], Matheus Luquirini Penteado dos Santos [1], Ana Paula Schneiders Dani [2], Cecília Paulina Johann Dammann [2], Letícia Pinto [2], Frederico Márcio Corrêa Vieira [1,*] and Flávia Regina Oliveira de Barros [1]

[1] Biometeorology Study Group (GEBIOMET), Federal University of Technology—Paraná (UTFPR), Dois Vizinhos 85660-000, Brazil
[2] Coordination of the Bioprocess and Biotechnology Engineering, Federal University of Technology—Paraná (UTFPR), Dois Vizinhos 85660-000, Brazil
* Correspondence: fredericovieira@utfpr.edu.br

Citation: Dada, J.M.V.; Santos, M.L.P.d.; Dani, A.P.S.; Dammann, C.P.J.; Pinto, L.; Vieira, F.M.C.; Barros, F.R.O.d. Placental Development and Physiological Changes in Pregnant Ewes in Silvopastoral and Open Pasture Systems during the Summer. *Animals* **2023**, *13*, 478. https://doi.org/10.3390/ani13030478

Academic Editors: Daniel Mota-Rojas, Julio Martínez-Burnes and Agustín Orihuela

Received: 28 December 2022
Revised: 20 January 2023
Accepted: 27 January 2023
Published: 30 January 2023

Copyright: © 2023 by the authors. Licensee MDPI, Basel, Switzerland. This article is an open access article distributed under the terms and conditions of the Creative Commons Attribution (CC BY) license (https://creativecommons.org/licenses/by/4.0/).

Simple Summary: Heat stress is a physiological condition where an animal fails to adequately dissipate body heat; this results in increased blood flow in the animal's core and negatively affects its physiological system. Considering this problem, this study aimed to analyze the reproductive and physiological changes in ewes subjected to heat stress during pregnancy. Twenty-four pregnant crossbred ewes were kept at UTFPR-DV's (Brazil) silvopastoral (SP) or open pasture (OP) systems throughout their pregnancy. During the experiment, microclimatic variables, sheep's blood samples, and physiological variables were collected every two weeks. After the birth of the lambs, the placentas were also collected. Our results showed that both systems were stressful for the sheep, but the SP system had lower air and grass temperatures than the OP system. The respiratory and heart rates of animals from the OP system were higher than those from the SP system. As regards the animals' immune cells, their mobilization was not affected by the systems, while the neutrophil count was only affected by time. Regarding placental biometry, it was observed that placentas in twin pregnancies had a greater membrane area. We concluded that the type of production system used affects the thermal comfort of pregnant ewes; an SP system can offer more amenable microclimatic conditions, which result in greater comfort for the ewes.

Abstract: This study aimed to analyze the reproductive and physiological changes in ewes subjected to heat stress during pregnancy at UTFPR-Brazil. Twenty-four pregnant crossbred ewes were kept in a silvopastoral system (SP) or an open pasture system (OP) throughout the final trimester of pregnancy. Both systems were stressful, but the SP system had lower air temperature than the OP system (26.0 ± 0.38 and 26.9 ± 0.41 °C, respectively; $p = 0.0288$). Moreover, the radiant thermal load of the two groups presented a difference of 34 Wm^{-2} ($p = 0.0288$), and the grass temperature was also lower in the SP system compared to that in the OP system (23.4 ± 0.37 and 25.6 ± 0.44 °C, respectively; $p = 0.0043$). The respiratory and heart rates of animals from the OP group were higher than those from the SP group ($p < 0.001$), but no difference was observed in the mobilization of white blood cells ($p = 0.4777$), and the neutrophil count was only affected by time ($p < 0.0001$). As regards placental biometry, placentas in twin pregnancies had a greater membrane area ($p = 0.0223$), but no differences between the systems were observed in placental weight ($p = 0.1522$) and the number of cotyledons ($p = 0.5457$). We concluded that the type of rearing system used affects the thermal comfort of pregnant ewes, and that an SP system can offer more amenable microclimatic conditions, which result in greater comfort for the ewes.

Keywords: heat stress; sheep production system; sheep pregnancy; placental development

---

1. Introduction

An animal's environment is defined as comfortable when the animal is in its thermoneutral zone; that is, the heat produced by metabolism (thermogenesis) is lost (thermolysis) to the environment without prejudice to the animal's performance [1]. When this process does not occur properly, and a higher body temperature is maintained, thermal stress due to heat is created, and it is, therefore, necessary for the animal to use devices capable of restoring its thermal balance with the environment. For sheep, the thermoneutral zone, which is its ideal ambient temperature, is established between 15 and 20 °C [2].

Heat stress is caused by a combination of high temperature and relative humidity. This causes animal behavioural and physiological changes, affecting food and water intake, growth, reproduction, milk production, haematological changes, and concentrations of cortisol and thyroid hormones in plasma [3–5].

In warm-blooded animals, behavioural change is the first mechanism for achieving body heat loss when the ambient temperature is high, followed by increased respiratory and heart rates [6]. As noted above, a high ambient temperature results in behavioural and other physiological changes in animals, affecting food and water intake, growth, reproduction, and milk production, in addition to haematological changes and changes in plasma concentrations of cortisol and thyroid hormones [5,7,8].

During pregnancy, the placenta nourishes the developing fetus and transports oxygen to meet its demands. In sheep, the larger the maternal—fetal surface, the greater the placental transport capacity; this is associated with the thickness of the placental barrier and can allow greater exchange of metabolic substrates [9]. As the placenta is the organ through which respiratory gases, nutrients, and wastes are exchanged between the maternal and fetal systems, placental vascular development is thought to play a vital role in ensuring the adequate exchange of nutrients and oxygen and ultimately in determining the fetal weight at birth [10,11]. However, few studies examine how placental and fetal development are affected by rearing systems, especially regarding thermal stress, and how an environment with a pleasant microclimate can be beneficial.

The benefits of SP systems in regulating physiological systems in ewes during pregnancy compared with open pasture systems are not extensively discussed in the literature. We hypothesized that physiological and placental parameters are negatively affected when sheep are exposed directly to solar radiation. Thus, this study aimed to analyze the reproductive aspects and physiological changes of ewes kept in different rearing systems during pregnancy, to monitor the birth and the weight of their lambs, and to characterize the microclimatic variables of the systems in which the ewes were kept.

2. Materials and Methods

2.1. Study Area, Animals, and Experimental Design

The research was conducted at the Teaching, Research, and Extension of Sheep and Goat Farming Unit of the Federal University of Technology—Paraná, Campus Dois Vizinhos (UTFPR-DV). This project was approved by the Comissão de Ética no Uso de Animais (Animal Use Ethics Committee) (CEUA) of the UTFPR-DV, under protocol 2020-18, CEUA UTFPR.

Twenty-four Dorper × Santa Inês crossbred ewes, pregnant during a breeding season from 2 November to 16 December 2020 (with an average pregnancy duration of 50.4 ± 9.17 days), aged between 3 and 4.5 years, and with an average weight of 60 kg, were used in this study. They were divided into groups of 12 animals each and subjected to two treatments according to the rearing systems: the silvopastoral system (SP) and the open pasture system (OP). In the OP, the females remained in the open air without shade, and in the SP, the females had shade provided by trees at their disposal. In both treatments, the animals were kept in their respective paddocks during pregnancy; that is, rotation was not performed.

All the animals were kept on a pasture (*Panicum maximum Jacq* cv. Aruana) and supplemented with a concentrate of 1% of body weight (corn and soybeans); mineral salt and water were also offered ad libitum.

Blood sample collections were carried out every 15 days from January to March 2021, with 6 collections in this period. The births took place in April and May (Figure 1).

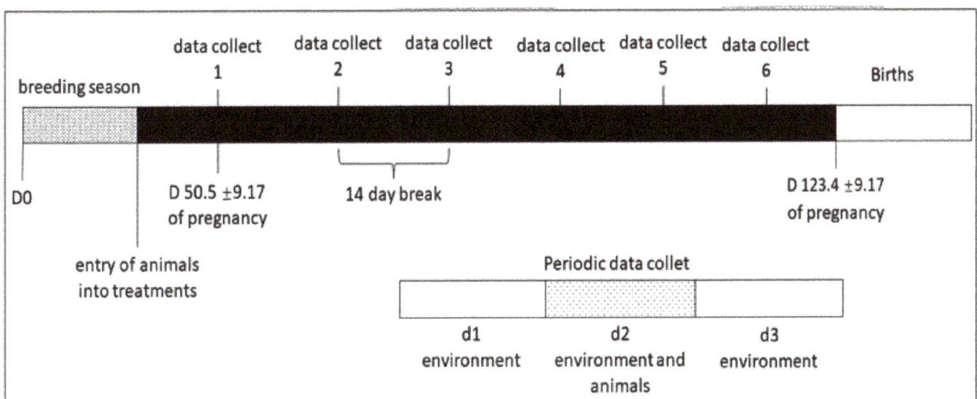

Figure 1. Representation of the breeding season scheme, periodic collections (main and microclimatic), and calving period.

It is important to note that, even in the breeding season, the people responsible for the blood sample collections were "introduced" to the sheep so that the animals could become accustomed to their presence. Therefore, no additional stress was created during the collections.

2.2. Microclimatic Variables

The microclimate of the two environments was evaluated by studying the environmental variables: air temperature (AT, °C), black globe temperature (BGT, °C), dew point temperature (DPT, °C), wind speed (WS, m s^{-1}), and relative humidity (RH, %). These variables were collected for 12 h (7:00 am to 7:00 pm) on days when physiological variables and the animals' blood were collected (Figure 1).

The temperature variables (AT, BGT, DPT) and RH were measured using Onset® data loggers (HOBO U12-013), with a temperature measurement range between −20 and 70 °C, with an accuracy of ±0.35 °C and a relative humidity measurement range between 5 and 95%, with an accuracy of ±2.5%.

These devices were installed at 1.5 m in the SP (5 devices) and OP (1 device) areas. They had one internal channel and two external channels. The internal channel was used to couple a thermocouple cable (sensor) to the black globe (15 cm diameter hollow polyethene ball painted with matte black paint). WS was measured using a Mastech® digital anemometer (MS6252A).

2.3. Thermal Comfort Indices

The radiant heat load (RHL) of the microclimatic variables of both the OP and the SP production systems was evaluated according to the equation proposed by Dixon [12]:

$$RHL = \sigma(MRT)^4 \qquad (1)$$

where *RHL* was the radiant heat load (W m^{-2}), σ was the constant of Stefan-Boltzmann (5.67 × 10^{-8}, W m^{-2} K^{-4}), and *MRT* was the mean radiant temperature (K), as described below:

$$MRT = 100 \times \left\{ [2.51 \times (WS^{\frac{1}{2}}) \times (BGT - AT)] + \left[\left(BGT \times 100^{-1} \right)^{4} \right] \right\}^{\frac{1}{4}} \qquad (2)$$

where WS was the wind speed (m s^{-1}), BGT was the black globe temperature (K), and AT was the air temperature (K).

2.4. Evaluation of Physiological Variables

The physiological variables analyzed were mean surface temperature, heart and respiratory rates, and rectal temperature. Measurements were taken from all the animals during the midday period. To collect these data, the animals were contained to minimize the stress caused by handling and not interfere with the data collection. The measurements were performed in the following order: heart and respiratory rates (HR and RR, respectively), surface temperature (ST), rectal temperature (RT), and finally, blood collection.

With the animals from the SP system, the collections were carried out in the paddock where they were kept. In the OP system, the animals were taken in pairs to a sheepfold and isolated in pens.

Two people were required to carry out the containment of the animals, one of whom was positioned on the right side of the animal while the other collected the data. HR and RR were measured by auscultation with a stethoscope. In HR, the stethoscope was positioned in the left thoracic region at the height of the aortic arch. The heartbeats were counted for 30 s, and then the value obtained was multiplied by 2 to obtain the number of beats per minute (beats. min^{-1}). In RR, the stethoscope was positioned over the trachea to count the air passage activity for 30 s. The value obtained was multiplied by 2 to obtain the respiration rate per minute.

The average surface temperature of the animals was measured with a Flir TG165 infrared thermometer (with a measurement range of -25 to 380 °C, a precision of $\pm 1.5\%$ or 1.5 °C, and an emissivity of 0.90). Animals under the SP system were measured in the shade, and animals kept on the OP system were measured in direct sunlight. A person positioned 1 m from the animal measured the temperature at five points on the animal: head, neck, back, flank, and hind limb at thigh height. Then, the simple arithmetic average of the measured temperatures was calculated, as in previous studies [5,7].

Rectal temperature was measured using a mercury thermometer (5198.10, Incoterm, Brazil), which was introduced into the animal's rectum so that the metallic tip remained in contact with the mucosa for one minute.

2.5. Peripheral Blood Sampling and Characterization of Leukocytes

Blood samples were collected between 11 a.m. and 1 p.m. by performing jugular vein punctures. They were stored in test tubes containing 3.2% sodium citrate (m/v), maintaining a ratio of 1:10 (citrate: blood) for later leukocyte characterization. The morphological analysis was carried out by adapting the methodology Thrall (2014) described. The smears were prepared from a drop of blood (approximately 20 µL) on a microscope slide, stained with panoptic dye (Newprov, Pinhais, Brazil). The morphological analysis of leukocytes was performed by optical microscopy, using a 100× objective lens together with immersion oil (Laborclin, Pinhais, Brazil). The quantitative analysis was carried out through the Neubauer Chamber using 4% Türk's liquid (m/v) at a ratio of 1:20 (blood: Türk's solution) in light microscopy ($n = 15$ per group).

2.6. Births and the Collection of Placentas

The ewes were kept in a suspended fold one week before the expected calving date. They were fed corn silage ad libitum, and the same supplementation conditions were applied (corn and soybean concentrate, mineral salt, and water). At delivery, the placentas were collected and taken to the laboratory. The lambs were weighed at birth and 10 days old. At the end of the pregnancies, 17 placentas were collected from the birth of 18 ewes. Pregnancies were classified as a single (1 lamb) or twins (2 lambs).

2.7. Placental Biometry

At the laboratory, the placentas were opened on a bench and photographed, and their cotyledons were counted. ImageJ software was used to define the area (membrane and cotyledons). The areas of interest were manually outlined using the "freehand" and "brush" tools, identified, and stored in the program. The data were processed, and the measurement was carried out, providing the areas delimited in cm^2. The rulers visible in the photographs (Figure 2) created a scale in centimetres on ImageJ.

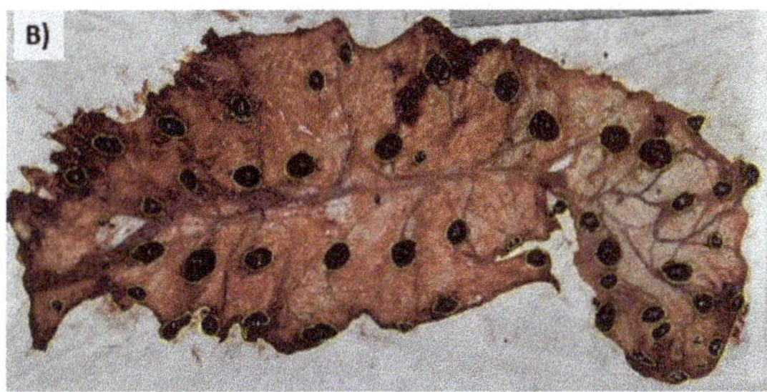

Figure 2. Representation of area measurement. (**A**) Placenta of single-gestation ewe. (**B**) Markings (yellow lines) and area and cotyledon numbering in ImageJ.

2.8. Statistical Analysis

2.8.1. Environmental and Physiological Variables

The experimental design was completely randomized, with 24 animals as biological replications (with 12 paddocks as experimental units and 2 animals per paddock as observational units). For the confirmatory analysis, mixed models were used, with time and treatment as fixed effects and date and paddock as random effects. The models were tested using the statistical software R [13] and the lme4 package [14]. The data were adjusted using ordinary least squares to enable an examination of the accuracy of the transformation of the response variables due to possible deviations from the assumptions of a linear model. In case of the need for transformations, the method of maximum likelihood was used. With the model adjusted, the data were analyzed using analysis of variance, and the type III F test was used for the model's fixed effects. When significant, the Tukey test was used for multiple comparisons of means, with the level of significance declared when the *p*-value was greater than 0.05.

2.8.2. Placental Biometry

To perform the statistical analysis, all the data were tested for normality and homogeneity (Shapiro-Wilk), considering a 2×6 factorial system (independent variable Production System × independent variable Time). The effects of Production System, Time, and System × Time interaction were verified. Normally distributed data were analyzed using Student's t-test. Data that presented non-normal distribution were analyzed using the Mann-Whitney test. Dependent variables were considered: placental mass, placental area, and the number of cotyledons.

2.8.3. Leukocyte Characterization

All the data were tested for normality and homogeneity of variances before being analyzed by two-way ANOVA, considering a 2 × 6 factorial system (independent variable Production System × independent variable Time). Effects of Production System, Time, and System × Time interaction were verified. The groups were compared using Tukey's post-test with a significance level of 5%. Non-parametric data were transformed using the square root function and submitted to two-way ANOVA. Transformed data that did not meet the assumptions for analysis of variance were analyzed using the non-parametric Kruskal-Wallis's test, followed by Dunn's post-test using a significance level of 5%. Dependent variables were considered: total leukocytes, lymphocytes, neutrophils, monocytes, eosinophils, and basophils. All statistical analyses and graph productions were performed using Prism GraphPad version 7.0 software for macOS.

3. Results

3.1. Microclimate Characterization

Although variation was observed between the morning and afternoon shifts for all microclimatic variables, it was decided to focus on treatments since these were the primary goal of this experiment. The average AT for the SP was 0.9 °C below the average AT for the OP, as described in Table 1. As expected, a difference was observed in the AT from the shaded system compared to the OP ($p = 0.0288$), but no interaction was observed between these and the shifts. For RH, there was no variation between production systems and their interaction. As regards WS, no difference between treatments and their interactions with shifts was found. However, an effect on the GT was detected for the production systems ($p = 0.0043$) since the mean temperature of the OP was 2.2 °C above the temperature of the shaded pasture. For RHL, the only treatment effect observed was $p = 0.0288$ (Table 1), with an average variation of 34 W m^{-2} between the production systems. The lowest temperatures for both OP and SP were observed at 7 a.m. on 11 March 2021, while the higher temperatures for both systems were registered at 3 p.m. two days later (13 March 2021).

Table 1. Microclimatic variables between rearing systems throughout the summer (mean ± SD; median, maximum, minimum).

Variables		Open Pasture (OP)	Silvopastoral System (SP)	*p* Value
Air temperature (AT, °C)	Mean ± SD	26.9 ± 0.41 A	26.0 ± 0.38 B	0.0288
	Minimum	17.0	17.4	
	Median	26.2	26.2	
	Maximum	34.3	33.3	
Relative humidity (RH, %)	Mean ± SD	67.2 ± 3.42 A	68.4 ± 3.37 A	0.4011
	Minimum	28.1	28.3	
	Median	69.5	69.5	
	Maximum	93.7	93.9	
Wind speed (WS, m s^{-1})	Mean ± SD	1.25 ± 0.20 A	1.02 ± 0.13 A	0.0939
	Minimum	0	0	
	Median	0.86	0.86	
	Maximum	4.04	5.69	
Grass temperature (GT, °C)	Mean ± SD	25.6 ± 0.44 A	23.4 ± 0.37 B	0.0043
	Minimum	11.0	9.6	
	Median	23.8	23.8	
	Maximum	37.6	31.4	

Table 1. *Cont.*

Variables		Open Pasture (OP)	Silvopastoral System (SP)	*p* Value
Radiant heat load (RHL, W m^{-2})	Mean $\pm$ SD	610 $\pm$ 12.3 A	576 $\pm$ 10.9 B	0.288
	Minimum			
	Median			
	Maximum			

Means with different letters within each factor differ among themselves (Tukey's test; $p < 0.05$). AT (air temperature); RH (relative humidity); WS (wind speed); GT (grass temperature); and RHL (radiant heat load).

Finally, some microclimatic variables were correlated, proving AT's dependence proportional to GT and RHL and inversely proportional to RH ($p < 0.001$). Also, through this correlation, it was noted that the WS had little influence on the other variables, having significant importance only in calculating the RHL ($p < 0.001$).

3.2. Thermoregulatory Variables

By observing the movement of the ewes' flanks, the RR of ewes exposed to the sun was higher than that of ewes that remained in the shade ($p < 0.001$), since direct sun exposure increased the number of movements per minute by 38.5%. Furthermore, the HR of the sheep in the shaded system was considerably lower than the HR of the sheep in the open pasture ($p < 0.001$) since the blood-pumping rate is proportional to the heat dissipation rate (Table 2). The direct exposure of the animal to the sun significantly increased its MST by 2.8 °C compared to animals from the SP ($p < 0.001$). However, the RT was not affected by the presence of shade ($p = 0.6742$; Table 2). Additionally, regarding the RT, both systems had a normal distribution and were similar.

Table 2. Physiological variables between rearing systems throughout the summer (mean $\pm$ SD; median, maximum, minimum).

Variables		Open Pasture (OP)	Silvopastoral System (SP)	*p* Value
Heart rate (HR, beats min.$^{-1}$)	Mean $\pm$ SD	116 $\pm$ 4.85 A	100 $\pm$ 4.84 B	<0.001
	Minimum	64	48	
	Median	117	97	
	Maximum	178	166	
Respiratory rate (RR, mov. min.$^{-1}$)	Mean $\pm$ SD	104.1 $\pm$ 10.0 A	76.4 $\pm$ 10.1 B	<0.001
	Minimum	24	36	
	Median	102	68	
	Maximum	198	182	
Rectal temperature (RT, °C)	Mean $\pm$ SD	39.35 $\pm$ 0.06 A	39.33 $\pm$ 0.06 A	0.06742
	Minimum	38.4	38.4	
	Median	39.4	39.3	
	Maximum	40.0	40.4	
Mean surface temperature (MST, °C)	Mean $\pm$ SD	32.8 $\pm$ 0.76 A	30.0 $\pm$ 0.76 B	<0.001
	Minimum	26.0	25.3	
	Median	33.4	29.7	
	Maximum	44.4	38.5	

Means with different letters, within each factor, differ among themselves (Tukey's test; $p < 0.05$). HR (heart rate); RR (respiratory rate); RT (rectal temperature); and MST (mean surface temperature).

3.3. Leukocyte Characterization

The mobilization of white blood cells (WBCs) was not affected by the production system or by time and their interaction (Figure 3a). As regards lymphocytes, there was a difference in the cell count between days 14 and 70 for the SP ($p = 0.0024$) and between 14 and 24 and 84 for the OP ($p = 0.0259$; $p = 0.0365$, respectively). Even so, it was observed that there was a significant effect of time ($p = 0.0055$) and interaction ($p = 0.0374$) between the production systems (Figure 3b). The neutrophil cell count started near the lower reference limit and showed constant growth until day 84 for both systems. For the OP, however, a difference was spotted between days 14 and 56 ($p = 0.0290$). Finally, an effect of time ($p < 0.0001$) and interaction was observed ($p = 0.0001$; Figure 3c).

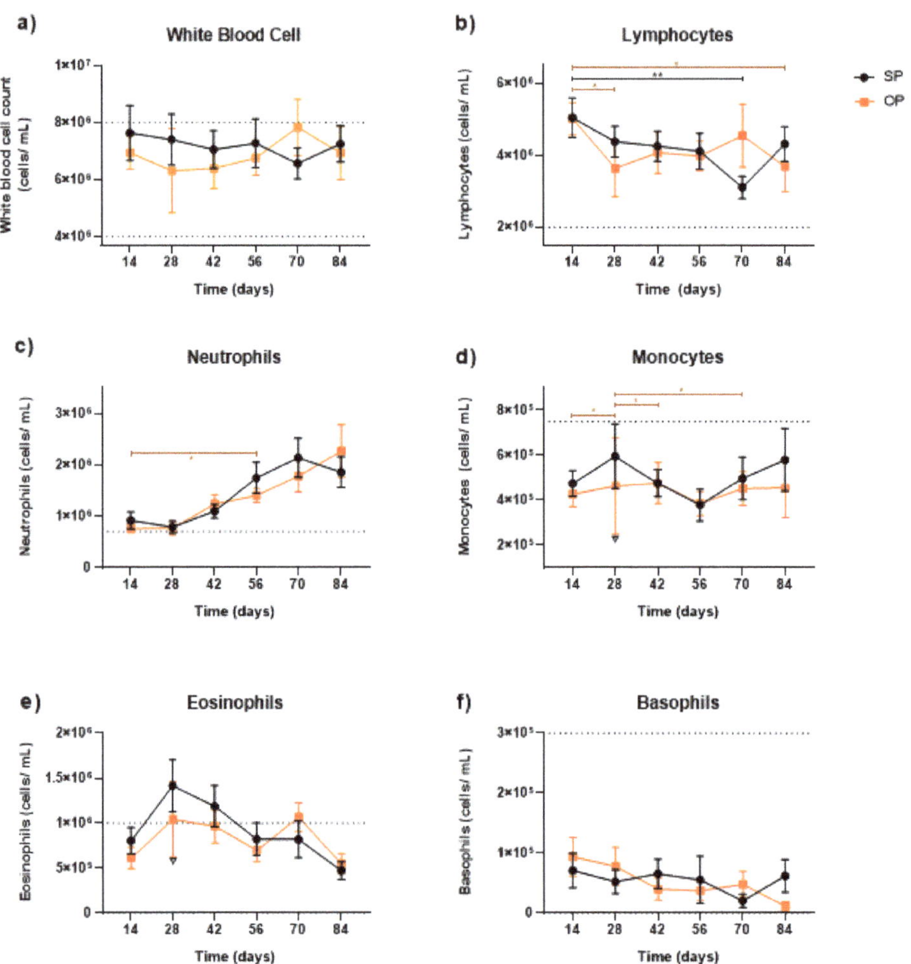

Figure 3. Mobilization of (**a**) total leukocytes; (**b**) lymphocytes; (**c**) neutrophils; (**d**) monocytes; (**e**) eosinophils; and (**f**) basophils (cells/mL of blood) during exposure to heat stress over the summer. * represents statistical variation over time ($p < 0.005$) of the OP system. ** represents statistical variation over time ($p < 0.005$) of the SP system. All values are represented as Mean ± SD.

The OP system presented lower monocyte recruitment than the SP system on day 28 ($p = 0.0119$; Figure 3d). Nevertheless, this day contrasted significantly with days 14, 42, and 70 ($p = 0.0476$; $p = 0.0387$; $p = 0.0206$, respectively). As for the eosinophils, an effect of interaction between systems and time was observed ($p = 0.0398$; Figure 3e). The mobilization of the SP cells on days 28 and 42 were slightly above the upper reference limit, having a statistical distinction from the OP on day 28 ($p = 0.0424$). Finally, the only effect of interaction was detected for the basophils ($p = 0.0325$; Figure 3f).

3.4. Reproduction Variables

3.4.1. Gestation and Lambs

The number of lambs born, the sex of the lambs, and their weight and weight gain ten days after birth were analyzed, but no system effect was observed ($p = 0.6239$; $p = 0.9403$; $p = 0.5218$; $p = 0.3079$, respectively). Although the median indicates a longer gestation for the SP system (145.5 days) as compared to the gestation for the OP system (140 days), there was no system effect on the duration of pregnancy and birth weight ($p = 0.7954$; $p = 0.9441$, respectively) (Table 3).

Table 3. Mean values (mean $\pm$ SEM) of the reproductive variables and lamb's weight between rearing systems.

Variables	Open Pasture (OP)	Silvopastoral System (SP)	*p* Value
Duration of pregnancy (days)	138.6 $\pm$ 12.79	137.6 $\pm$ 16.39	0.7954
Birth weight (lambs) (kg)	3.937 $\pm$ 0.4506	3.921 $\pm$ 0.5943	0.9441
Ten days weight (lambs) (kg)	5.788 $\pm$ 0.8883	6.124 $\pm$ 1.1257	0.5218
Weight gain (lambs) (kg)	1.774 $\pm$ 1.010	2.186 $\pm$ 0.7481	0.3079

Tukey's test; $p < 0.05$.

There was no effect of pregnancy on the duration of pregnancy and birth weight ($p = 0.7715$; $p = 0.8519$, respectively). When considering the weight and weight gain of the lambs ten days after birth, there was an effect of gestation ($p = 0.0163$; $p = 0.0026$, respectively) favoring simple pregnancy (Table 3). Interestingly, the production systems showed different types of placentation. While 22% of pregnancies in the OP group were twins and all were dichorionic, in the SP group 30% were twins, and all were monochorionic. There was no effect of the type of pregnancy in respect of the placental area, the placental mass, and the number of cotyledons ($p = 0.2161$; $p = 0.3866$; $p = 0.2859$, respectively). Numerically, monochorionic placentas had more cotyledons.

3.4.2. Placental Biometry

There was no system effect (OP $\times$ SP) regarding the total area of cotyledons per placenta, the area, and the mass of the chorionic membrane ($p = 0.8456$; $p = 0.7178$; $p = 0.9092$, respectively). However, there was a significant correlation between the cotyledon area and the placental mass, but only for the OP group ($p = 0.0148$, $R^2 = 0.8954$). A positive correlation was also observed when the same analysis was performed, disregarding the rearing system ($p = 0.0055$, $r^2 = 0.5934$). However, there was no significant difference ($p = 0.2541$) between the OP and SP systems regarding the number of cotyledons, as shown in Table 4.

Table 4. Mean values (mean $\pm$ SEM) of the placental biometry of ewes kept in the silvopastoral system (SP) and open pasture (OP).

Variables	Open Pasture (OP)	Silvopastoral System (SP)	*p* Value
Membrane area (simple pregnancy) (cm^2)	2296 $\pm$ 473.6	2298 $\pm$ 490.4	0.9578
Cotyledon area (simple pregnancy) (cm^2)	234.7 $\pm$ 69.59	226.4 $\pm$ 68.62	0.8456
Number of cotyledons	60.22 $\pm$ 22.06	70.75 $\pm$ 12.61	0.2541
Placental mass (kg)	0.3423 $\pm$ 0.08057	0.3370 $\pm$ 0.1084	0.9092

(Mann-Whitney test).

4. Discussion

4.1. Microclimate Characterization

Because of their wool, sheep suffer more from the effects of HS. Owing to the physico-chemical properties of wool, sheep in high-temperature environments have a lower rate of loss of sensible heat by conduction and convection (body surface—environment), which makes it difficult for heat to be dissipated and for water to evaporate [15].

The SP system, characterized by a combination of trees, pasture, and animals in the same space, aims to improve the thermal comfort of these animals through the shade provided by trees [16]. The microclimate generated has a positive influence on the grazing time [17] and the rumination of sheep [7], reduces the air temperature and increases the air humidity [18], and reduces the RHL thanks to the radiative interception provided by the canopy of trees [19,20]. The OP, which is the most common rearing system in the subtropical areas, does not present any trees (i.e., there is no shade), and this leads to higher air and grass temperatures alongside high RHL values.

Both systems remained within the parameters of thermal comfort for sheep, with temperatures ranging between 20 and 30 °C [21] and relative humidity close to 60% [22]. However, our relative humidity values were slightly higher than the reference, while the air temperature was lower than in our previous study [7]. Recent studies [23,24] also showed that the OP had higher AT and RH values when compared to the SP, which corroborates our findings. As regards WS, although our results did not show any difference, other studies found that WS in the OP was higher than in the SP [25].

According to Silva [26], RHL values above 570 W m^{-2} indicate a harsh environment for female animals. As regards high air temperature and humidity, higher RHL values were expected in the OP than in the SP; this was confirmed by the variance of 34 W m^{-2} between both systems. From this, the microclimate of both production systems was stressful for the animals, despite the OP having significantly higher values than the SP. Therefore, this exposure may cause discomfort to animals in direct contact with the pasture. Recent studies [7,20,27] analyzed the behaviour of adult females and their lambs and related direct sun exposure with increased GT and CTR; this affects them negatively and corroborates the premise that the OP provides a more stressful environment for the animals that are kept there. Our findings regarding RHL also concur with those previously described in the literature [19].

4.2. Physiological Variables

The physiological variables of sheep that can be used to assess whether they are under heat stress are respiratory rate, heart rate, and rectal temperature [28]. Thus, the ewes will trigger physiological mechanisms to improve homeostasis as the ambient temperature rises. In the present study, sheep from the OP system have a higher HR and RR, which is an apparent sign of heat stress. Similar results were found by Nejad and Sung [29], who, when analyzing the physiological parameters of sheep subjected to heat stress and limited water supply, observed that sheep without heat stress had lower RR values and gasping scores (*p*-value 0.02 and 0.03, respectively).

Despite the differences between the other physiological variables, TR is the main variable used to identify the state of the thermal stress of the animal. Therefore, even though the sheep were in challenging microclimatic conditions, they could maintain their body temperature within the normal range of this variable, which may be linked to the increase in HR and RR. The RT of both systems was similar, in contrast to that presented in the literature, where ewes under heat stress present a high RT [30–32].

The surface temperature can be used as a parameter for evaluating the thermal comfort of sheep, as there is a correlation between environmental indices and rectal and skin surface temperature [33]. Thus, the highest value of MST was detected for the OP system, where the highest temperatures were observed.

4.3. Leukocytes

We also hypothesized that heat stress (HS) negatively affects the sheep's immune system. However, our results showed that the recruitment of WBCs was not affected by direct exposure to solar radiation (Table 1). Although some studies analyze the physiological [5], behavioural [7], and biochemical [34] response to heat stress in a silvopastoral system, the literature lacks information regarding the mobilization of leukocytes in adult sheep, whether heat-stressed or in a shaded production system. HS is commonly known to affect leukocyte concentration in broiler chickens [35], dairy cattle [36], human workers [37], and even Australian abalone [38].

Our results in respect of WBC are in accordance with what is described in the literature for heat-stressed and non-heat-stressed sheep. However,, we observed higher eosinophil numbers along with higher monocyte concentration for a time. Karthik [32] observed higher WBC populations in sheep in an extensive heat-stressed system compared to an intensive one. Yet, the harsh environment did not affect the subpopulations of WBCs (neutrophils, lymphocytes, eosinophils, monocytes, and basophils). Contrary to our results, Wojtas et al. [34] discovered that total leukocyte mobilization in sheep was influenced by the movement and temperature of the air, being in lower concentrations when under high temperatures. Caroprese et al. [39] presented an alternative to combat the negative effects of HS in sheep based on supplementation with polyunsaturated fatty acids. Their results showed that supplementing flaxseed may improve the cell-mediated immune response in sheep exposed to solar radiation. Swanson et al. [40] found that heat-stressed lambs presented higher WBC, monocyte, and granulocyte concentrations in blood plasma compared to non-heat-stressed lambs. At the same time, Liu et al. [41] showed that lymphocyte count was reduced in the stressed environment. Cortisol is a hormone produced by the pituitary gland in response to HS exposure. It is one of the most important factors leading to reduced leukocyte mobilization [42]. It is possible that in the present study, despite the low but significant difference between the systems' RHLs, the increase in the cortisol hormone was not enough to significantly affect the WBC concentration in the plasma. However, more studies involving plasma cortisol analysis are needed to confirm this hypothesis.

4.4. Gestation and Lambs

In several studies, the effects of heat stress on the body weight at birth of ewes and other ruminants are observed [32,43–45]. When comparing the birth weight of lambs in a controlled environment with the birth weight of lambs from ewes exposed to heat stress, a lower weight is noted in the second group; however, this article does not present this system effectively. According to Marai et al. [43], this inferiority in weight is due to a disturbance in placental growth generated by heat, which affects fetal development and, consequently, the birth weight of the lambs.

There was no system effect on the number of lambs born. In contrast, Van Wettere et al. [45] found fewer lambs born to mated ewes exposed to heat stress in their work. Their studies show harmful alterations in the fertility and fecundity of ewes and the production and quality of semen in the rams. Other studies indicate that non-heat-stressed ewes are 2.43 times more likely to become pregnant than heat-stressed ewes [44].

The average weight gain in goats eight days after birth is lower than in sheep raised in heat-stress systems; this may be associated with indications of low placental efficiency [46]. However, the present study does not show any statistically significant system effect on weight and weight gain ten days after birth. In cows, Casamassima et al. [47] find that heat stress significantly reduces milk production and quality, which negatively impacts lamb growth.

Heat stress can reduce the duration of pregnancy in cows and goats, directly impacting the reduction in birth weight [48,49]. However, the present study does not show a system effect on the duration of the gestation of ewes, even with a lower mean of the OP system (140 days) compared to the SP system (145.5 days).

Interestingly, the production systems showed different types of twin placentation. While the pregnancies in the OP group were all dichorionic, those in the SP group were all monochorionic. Although studies such as those conducted by Dwyer et al. [50] and Özyürek and Türkyilmaz [51] state that twin gestation lambs have an increase in placental weight and the number of cotyledons, the type of pregnancy showed no effect on the area, the mass of the placenta, and the number of cotyledons.

In humans and ewes, monochorionic twin pregnancies have complications arising from vascular anastomoses, which is not observed in dichorionic pregnancies. Additionally, in twin pregnancies, there is unequal sharing of the placenta by the fetuses [52,53]. In ewes, this can lead to the birth of unequal-sized lambs, which impacts their survival and development.

4.5. Placental Biometry

There was no system effect (OP $\times$ SP) for the total area of cotyledons per placenta and the area of the chorionic membrane ($p = 0.8456$; $p = 0.7178$, respectively). However, there was a significant correlation between the cotyledon area and the placental mass, but only for the SP group ($p = 0.0148$, $r^2 = 0.8954$). When the same analysis was performed disregarding the rearing system, a positive correlation was also observed ($p = 0.0055$, $r^2 = 0.5934$), allowing the inference that the cotyledon area and the placental mass are correlated.

Placental efficiency can be defined as the placenta's capacity to support the fetus's growth: the greater the weight of the fetus in relation to the placental weight, the greater the placental efficiency [54]. Since placental vascular architecture is closely related to placental efficiency, it is essential to evaluate placental biometry to obtain correlations.

Decreased placental development also limits the supply of oxygen and nutrients, and this leads to decreased fetal plasma glucose and fructose rates and reduced fetal growth rates [55].

Such results corroborate what is described in the literature, since there is proportionality between the placental mass and the number of cotyledons [50]. A greater number of cotyledons can be related to a greater passage of nutrients between the fetus and the mother [56]). However, there was no significant difference ($p = 0.2541$) between the OP and SP systems regarding the number of cotyledons (Figure 3).

The increase in placental weight in multiparous mothers may be due to the expansion and vascularization of the uterus after multiple pregnancies and advancing maternal age [50]. The age of the ewe and its reproductive maturity can significantly influence placentation [57]. Finally, drawing on our findings, we recommend using the silvopastoral system to raise ewes and lambs to minimize the effects of heat stress on these animals.

5. Conclusions

No significant differences between rearing systems (OP and SP) regarding gestation variables and lambing were observed. We observed some interesting points concerning the types of gestation (twin pregnancy and simple pregnancy); each system presented just one example of twin pregnancy; however, this did not affect the remaining variables.

Author Contributions: Conceptualization, formal analysis, writing, review—editing, supervision, F.M.C.V. and F.R.O.d.B.; Conceptualization, data curation, investigation, writing—original draft, J.M.V.D. and M.L.P.d.S.; Investigation, writing—original draft, A.P.S.D., C.P.J.D. and L.P. All authors have read and agreed to the published version of the manuscript.

Funding: The Coordination for the Improvement of Higher Education Personnel (CAPES) granted a scholarship to the first author.

Institutional Review Board Statement: This project was approved by the Comissão de Ética no Uso de Animais (Animal Use Ethics Committee) (CEUA) of the UTFPR-DV, under protocol 2020-18, CEUA-UTFPR.

Informed Consent Statement: Not applicable.

Data Availability Statement: Not applicable.

Acknowledgments: We thank the Molecular Biology Laboratory—BioMol and Cell Biology Laboratory—LACEL from UTFPR. We also thank Vicente de Paulo Macedo for the partnership with UNEPE Ovinocaprinocultura of UTFPR and the research groups Biometeorology Study Group (Gebiomet) and LabBarros from UTFPR.

Conflicts of Interest: The authors declare no conflict of interest. The funders had no role in the design of the study; in the collection, analyses, or interpretation of data; in the writing of the manuscript; or in the decision to publish the results.

References

1. ALVES, F.; de Almeida, R.G. Diretrizes Técnicas Para Produção de Carne Com Baixa Emissão de Carbono Certificada Em Pastagens Tropicais: Carne Baixo Carbono (CBC). 2020. Available online: http://www.infoteca.cnptia.embrapa.br/infoteca/handle/doc/1120985 (accessed on 24 November 2022).
2. Perez-Garcia, V.; Turco, M.Y. Keep Calm and the Placenta Will Carry On. *Dev. Cell* **2020**, *54*, 295–296. [CrossRef] [PubMed]
3. da Silva, E.M.N.; de Souza, B.B.; de Silva, G.A.; Cezar, M.F.; de Souza, W.H.; Benício, T.M.A.; Freitas, M.M.S. Avaliação Da Adaptabilidade de Caprinos Exóticos e Nativos No Semi-Árido Paraibano. *Ciência E Agrotecnologia* **2006**, *30*, 516–521. [CrossRef]
4. Coelho, L.A.; Sasa, A.; Bicudo, S.D.; Balieiro, J.C.C. Concentrações Plasmáticas de Testosterona, Triiodotironina (T3) e Tiroxina (T4) Em Bodes Submetidos Ao Estresse Calórico. *Arq. Bras. Med. Vet. Zootec* **2008**, *60*, 1338–1345. [CrossRef]
5. Santos, M.L.P.; Dada, J.M.V.; Muniz, P.C.; Nunes-Zotti, M.L.A.; Barros, F.R.O.D.; Vieira, F.M.C. Physiological Responses of Santa Inês x Dorper Ewes and Lambs to Thermal Environment of Silvopasture and Open Pasture Systems. *Small Rumin. Res.* **2021**, *205*. [CrossRef]
6. Rathwa, S.D.; Vasava, A.A.; Pathan, M.M.; Madhira, S.P.; Patel, Y.G.; Pande, A.M. Effect of Season on Physiological, Biochemical, Hormonal, and Oxidative Stress Parameters of Indigenous Sheep. *Vet. World* **2017**, *10*, 650–654. [CrossRef]
7. Dada, J.M.V.; Santos, M.L.P.D.; Muniz, P.C.; Nunes-Zotti, M.L.A.; de Barros, F.R.O.; Vieira, F.M.C. Postpartum Behavioural Response of Santa Inês x Dorper Ewes and Lambs in a Silvopastoral System. *Small Rumin. Res.* **2021**, *203*. [CrossRef]
8. McManus, C.M.; Faria, D.A.; Lucci, C.M.; Louvandini, H.; Pereira, S.A.; Paiva, S.R. Heat Stress Effects on Sheep: Are Hair Sheep More Heat Resistant? *Theriogenology* **2020**, *155*, 157–167. [CrossRef]
9. Limesand, S.W.; Camacho, L.E.; Kelly, A.C.; Antolic, A.T. Impact of Thermal Stress on Placental Function and Fetal Physiology. *Anim. Reprod* **2018**, *15*, 886–898. [CrossRef]
10. Kaufmann, P.; Mayhew, T.M.; Charnock-Jones, D.S. Aspects of Human Fetoplacental Vasculogenesis and Angiogenesis. II. Changes During Normal Pregnancy. *Placenta* **2004**, *25*, 114–126. [CrossRef] [PubMed]
11. Reynolds, L.P.; Borowicz, P.P.; Vonnahme, K.A.; Johnson, M.L.; Grazul-Bilska, A.T.; Redmer, D.A.; Caton, J.S. Placental Angiogenesis in Sheep Models of Compromised Pregnancy. *J. Physiol.* **2005**, *565*, 43–58. [CrossRef] [PubMed]
12. Dixon, J.E.; Wells, G.D.; Esmay, M.L. Convective Heat Transfer Coefficient for Poultry Excreta. *Trans. ASAE* **1978**, *21*, 0534–0536. [CrossRef]
13. R Core Team. R: A Language and Environment for Statistical Computing. 2013. Available online: http://www.R-project.org (accessed on 27 December 2022).
14. Bates, D.; Mächler, M.; Bolker, B.; Walker, S. Fitting Linear Mixed-Effects Models Using Lme4. *J. Stat. Softw.* **2015**, *67*. [CrossRef]
15. Sejian, V.; Bhatta, R.; Gaughan, J.; Malik, P.K.; Naqvi, S.M.K.; Lal, R. Adapting Sheep Production to Climate Change. In *Sheep Production Adapting to Climate Change*; Springer: Singapore, 2017; pp. 1–29.
16. Fey, R.; Contro Malavasi, U.; de Matos Malavasi, M. Silvopastoral system: A review regarding the family agriculture. *Rev. De Agric. Neotrop.* **2015**, *2*. [CrossRef]
17. Alvarado-Canché, A.; Martínez, B.; Castillo, L.E.; Piñeiro-Vázquez, A.; Canul-Soli, J.R. Comportamiento Productivo y Alimenticio de Ovinos En Pastoreo En Sistemas Silvopastoriles Con Leucaena Leucocephala y Cynodon Plectostachyus. *Rev. Bio Cienc.* **2017**, *4*, 1–11.
18. Karvatte, N.; Klosowski, E.S.; de Almeida, R.G.; Mesquita, E.E.; de Oliveira, C.C.; Alves, F.V. Shading Effect on Microclimate and Thermal Comfort Indexes in Integrated Crop-Livestock-Forest Systems in the Brazilian Midwest. *Int. J. Biometeorol.* **2016**, *60*, 1933–1941. [CrossRef]
19. Pezzopane, J.R.M.; Nicodemo, M.L.F.; Bosi, C.; Garcia, A.R.; Lulu, J. Animal Thermal Comfort Indexes in Silvopastoral Systems with Different Tree Arrangements. *J. Biol.* **2019**, *79*, 103–111. [CrossRef]
20. Vieira, F.M.C.; Pilatti, J.A.; Czekoski, Z.M.W.; Fonsêca, V.F.C.; Herbut, P.; Angrecka, S.; de Souza Vismara, E.; de Paulo Macedo, V.; dos Santos, M.C.R.; Paśmionka, I. Effect of the Silvopastoral System on the Thermal Comfort of Lambs in a Subtropical Climate: A Preliminary Study. *Agriculture* **2021**, *11*, 790. [CrossRef]
21. Baêta, F.C.; Souza, C.F. *Ambiência Em Edificações Rurais: Conforto Animal*, 2nd ed.; UFV: Viçosa, Brazil, 2010.
22. Eustáquio Filho, A.; Teodoro, S.M.; Chaves, M.A.; dos Santos, P.E.F.; da Silva, M.W.R.; Murta, R.M.; de Carvalho, G.G.P.; de Souza, L.E.B. Zona de Conforto Térmico de Ovinos Da Raça Santa Inês Com Base Nas Respostas Fisiológicas. *Rev. Bras. De Zootec.* **2011**, *40*, 1807–1814. [CrossRef]
23. Lopes, L.B.; Eckstein, C.; Pina, D.S.; Carnevalli, R.A. The Influence of Trees on the Thermal Environment and Behaviour of Grazing Heifers in Brazilian Midwest. *Trop Anim. Health Prod.* **2016**, *48*, 755–761. [CrossRef]
24. de Sousa, K.T.; Deniz, M.; do Vale, M.M.; Dittrich, J.R.; Hötzel, M.J. Influence of Microclimate on Dairy Cows' Behavior in Three Pasture Systems during the Winter in South Brazil. *J. Biol.* **2021**, *97*, 102873. [CrossRef]
25. Deniz, M.; Schmitt Filho, A.L.; Farley, J.; de Quadros, S.F.; Hötzel, M.J. High Biodiversity Silvopastoral System as an Alternative to Improve the Thermal Environment in the Dairy Farms. *Int. J. Biometeorol.* **2019**, *63*, 83–92. [CrossRef] [PubMed]

26. da Silva, W.E.; Leite, J.H.G.M.; de Sousa, J.E.R.; Costa, W.P.; da Silva, W.S.T.; Guilhermino, M.M.; Asensio, L.A.B.; Façanha, D.A.E. Daily Rhythmicity of the Thermoregulatory Responses of Locally Adapted Brazilian Sheep in a Semiarid Environment. *Int. J. Biometeorol.* **2017**, *61*, 1221–1231. [CrossRef] [PubMed]

27. Pent, G.J.; Greiner, S.P.; Munsell, J.F.; Tracy, B.F.; Fike, J.H. Lamb Performance in Hardwood Silvopastures, II: Animal Behavior in Summer1. *Transl. Anim. Sci.* **2020**, *4*, 363–375. [CrossRef]

28. Hyder, I.; Pasumarti, M.; Reddy, P.R.; Prasad, C.S.; Kumar, K.A.; Sejian, V. Thermotolerance in Domestic Ruminants: A HSP_{70} Perspective. In *Heat Shock Proteins in Veterinary Medicine and Sciences*; Springer: Berlin/Heidelberg, Germany, 2017; pp. 3–35.

29. Ghassemi Nejad, J.; Sung, K.-I. Behavioral and Physiological Changes during Heat Stress in Corriedale Ewes Exposed to Water Deprivation. *J. Anim. Sci. Technol.* **2017**, *59*, 13. [CrossRef] [PubMed]

30. Pehlivan, E.; Kaliber, M.; Konca, Y.; Dellal, G. Effect of Shearing on Some Physiological and Hormonal Parameters in Akkaraman Sheep. *Asian-Australas J. Anim. Sci.* **2020**, *33*, 848–855. [CrossRef] [PubMed]

31. Dias e Silva, T.P.; da Costa Torreão, J.N.; Torreão Marques, C.A.; de Araújo, M.J.; Bezerra, L.R.; kumar Dhanasekaran, D.; Sejian, V. Effect of Multiple Stress Factors (Thermal, Nutritional and Pregnancy Type) on Adaptive Capability of Native Ewes under Semi-Arid Environment. *J. Biol.* **2016**, *59*, 39–46. [CrossRef]

32. Karthik, D.; Suresh, J.; Reddy, Y.R.; Sharma, G.R.K.; Ramana, J.V.; Gangaraju, G.; Reddy, P.P.R.; Reddy, Y.P.K.; Yasaswini, D.; Adegbeye, M.J.; et al. Adaptive Profiles of Nellore Sheep with Reference to Farming System and Season: Physiological, Hemato-Biochemical, Hormonal, Oxidative-Enzymatic and Reproductive Standpoint. *Heliyon* **2021**, *7*, e07117. [CrossRef]

33. Seixas, L.; de Melo, C.B.; Menezes, A.M.; Ramos, A.F.; Paludo, G.R.; Peripolli, V.; Tanure, C.B.; Costa Junior, J.B.G.; McManus, C. Study on Environmental Indices and Heat Tolerance Tests in Hair Sheep. *Trop. Anim. Health Prod.* **2017**, *49*, 975–982. [CrossRef]

34. Wojtas, K.; Cwynar, P.; Kołacz, R. Effect of Thermal Stress on Physiological and Blood Parameters in Merino Sheep. *Bull. Vet. Inst. Pulawy* **2014**, *58*, 283–288. [CrossRef]

35. Kamel, N.N.; Ahmed, A.M.H.; Mehaisen, G.M.K.; Mashaly, M.M.; Abass, A.O. Depression of Leukocyte Protein Synthesis, Immune Function and Growth Performance Induced by High Environmental Temperature in Broiler Chickens. *Int. J. Biometeorol.* **2017**, *61*, 1637–1645. [CrossRef]

36. Bagath, M.; Krishnan, G.; Devaraj, C.; Rashamol, V.P.; Pragna, P.; Lees, A.M.; Sejian, V. The Impact of Heat Stress on the Immune System in Dairy Cattle: A Review. *Res. Vet. Sci.* **2019**, *126*, 94–102. [CrossRef] [PubMed]

37. Gharibi, V.; Khanjani, N.; Heidari, H.; Ebrahimi, M.H.; Hosseinabadi, M.B. The Effect of Heat Stress on Hematological Parameters and Oxidative Stress among Bakery Workers. *Toxicol. Ind. Health* **2020**, *36*, 1–10. [CrossRef] [PubMed]

38. Hooper, C.; Day, R.; Slocombe, R.; Benkendorff, K.; Handlinger, J.; Goulias, J. Effects of Severe Heat Stress on Immune Function, Biochemistry and Histopathology in Farmed Australian Abalone (Hybrid Haliotis Laevigata × Haliotis Rubra). *Aquaculture* **2014**, *432*, 26–37. [CrossRef]

39. Caroprese, M.; Ciliberti, M.G.; Annicchiarico, G.; Albenzio, M.; Muscio, A.; Sevi, A. Hypothalamic-Pituitary-Adrenal Axis Activation and Immune Regulation in Heat-Stressed Sheep after Supplementation with Polyunsaturated Fatty Acids. *J. Dairy Sci.* **2014**, *97*, 4247–4258. [CrossRef] [PubMed]

40. Swanson, R.M.; Tait, R.G.; Galles, B.M.; Duffy, E.M.; Schmidt, T.B.; Petersen, J.L.; Yates, D.T. Heat Stress-Induced Deficits in Growth, Metabolic Efficiency, and Cardiovascular Function Coincided with Chronic Systemic Inflammation and Hypercatecholaminemia in Ractopamine-Supplemented Feedlot Lambs. *J. Anim. Sci.* **2020**, *98*. [CrossRef]

41. Liu, H.; Li, K.; Mingbin, L.; Zhao, J.; Xiong, B. Effects of Chestnut Tannins on the Meat Quality, Welfare, and Antioxidant Status of Heat-Stressed Lambs. *Meat Sci.* **2016**, *116*, 236–242. [CrossRef]

42. Seiler, A.; Fagundes, C.P.; Christian, L.M. The Impact of Everyday Stressors on the Immune System and Health. In *Stress Challenges and Immunity in Space*; Springer International Publishing: Cham, Germany, 2020; pp. 71–92.

43. Marai, I.F.M.; El-Darawany, A.A.; Fadiel, A.; Abdel-Hafez, M.A.M. Physiological Traits as Affected by Heat Stress in Sheep—A Review. *Small Rumin. Res.* **2007**, *71*, 1–12. [CrossRef]

44. Romo-Barron, C.B.; Diaz, D.; Portillo-Loera, J.J.; Romo-Rubio, J.A.; Jimenez-Trejo, F.; Montero-Pardo, A. Impact of Heat Stress on the Reproductive Performance and Physiology of Ewes: A Systematic Review and Meta-Analyses. *Int. J. Biometeorol.* **2019**. [CrossRef]

45. van Wettere, W.H.E.J.; Kind, K.L.; Gatford, K.L.; Swinbourne, A.M.; Leu, S.T.; Hayman, P.T.; Kelly, J.M.; Weaver, A.C.; Kleemann, D.O.; Walker, S.K. Review of the Impact of Heat Stress on Reproductive Performance of Sheep. *J. Anim. Sci. Biotechnol.* **2021**, *12*, 26. [CrossRef]

46. Silva, P.S.; Hooper, H.B.; Manica, E.; Merighe, G.K.F.; Oliveira, S.A.; Traldi, A.S.; Negrão, J.A. Heat Stress Affects the Expression of Key Genes in the Placenta, Placental Characteristics, and Efficiency of Saanen Goats and the Survival and Growth of Their Kids. *J. Dairy Sci.* **2021**, *104*, 4970–4979. [CrossRef]

47. Casamassima, D.; Sevi, A.; Palazzo, M.; Ramacciato, R.; Colella, G.E.; Bellitti, A. Effects of Two Different Housing Systems on Behavior, Physiology and Milk Yield of Comisana Ewes. *Small Rumin. Res.* **2001**, *41*, 151–161. [CrossRef] [PubMed]

48. Tao, S.; Dahl, G.E. Invited Review: Heat Stress Effects during Late Gestation on Dry Cows and Their Calves. *J. Dairy Sci.* **2013**, *96*, 4079–4093. [CrossRef] [PubMed]

49. Coloma-García, W.; Mehaba, N.; Llonch, P.; Caja, G.; Such, X.; Salama, A.A.K. Prenatal Heat Stress Effects on Gestation and Postnatal Behavior in Kid Goats. *PLoS ONE* **2020**, *15*, e0220221. [CrossRef] [PubMed]

50. Dwyer, C.M.; Calvert, S.K.; Farish, M.; Donbavand, J.; Pickup, H.E. Breed, Litter and Parity Effects on Placental Weight and Placentome Number, and Consequences for the Neonatal Behaviour of the Lamb. *Theriogenology* **2005**, *63*, 1092–1110. [CrossRef] [PubMed]
51. Özyürek, S.; Türkyilmaz, D. Determination of Relationships between Placental Characteristics and Birth Weight in Morkaraman Sheep. *Arch. Anim. Breed* **2020**, *63*, 39–44. [CrossRef] [PubMed]
52. Zhao, D.; Lipa, M.; Wielgos, M.; Cohen, D.; Middeldorp, J.M.; Oepkes, D.; Lopriore, E. Comparison between Monochorionic and Dichorionic Placentas with Special Attention to Vascular Anastomoses and Placental Share. *Twin Res. Hum. Genet.* **2016**, *19*, 191–196. [CrossRef]
53. Brzozowska, A.; Wojtasiak, N.; Błaszczyk, B.; Stankiewicz, T.; Udała, J.; Wieczorek-Dąbrowska, M. The effects of non-genetic factors on the morphometric parameters of sheep placenta and the birth weight of lambs. *Large Anim. Rev.* **2020**, *26*, 119–126.
54. Wilson, M.E.; Ford, S.P. Comparative Aspects of Placental Efficiency. *Reprod Suppl.* **2001**, *58*, 223–232. [CrossRef]
55. Mellor, D.J. Nutritional and Placental Determinants of Foetal Growth Rate in Sheep and Consequences for the Newborn Lamb. *Br. Vet. J.* **1983**, *139*, 307–324. [CrossRef]
56. Alexander, G. Studies on the placenta of the sheep (*Ovis aries* L.). *Reproduction* **1964**, *7*, 307–322. [CrossRef]
57. Ocak, S.; Ogun, S.; Onder, H. Relationship between Placental Traits and Maternal Intrinsic Factors in Sheep. *Anim. Reprod. Sci.* **2013**, *139*, 31–37. [CrossRef] [PubMed]

Disclaimer/Publisher's Note: The statements, opinions and data contained in all publications are solely those of the individual author(s) and contributor(s) and not of MDPI and/or the editor(s). MDPI and/or the editor(s) disclaim responsibility for any injury to people or property resulting from any ideas, methods, instructions or products referred to in the content.

Article

Changes of Acid-Base Variables in Dairy Cows with Chronically Implanted Fetal and Maternal Catheters during Late Gestation and Calving

Ottó Szenci [1,*], Gijsbert Cornelis Van Der Weyden [2], Lea Lénárt [1] and Marcel Antoine Marie Taverne [2]

[1] Department of Obstetrics and Food Animal Medicine Clinic, University of Veterinary Medicine Budapest, H-2225 Üllő, Hungary
[2] Department of Farm Animal Health, Foetal and Perinatal Biology, Faculty of Veterinary Medicine, Utrecht University, 3584 CL Utrecht, The Netherlands
* Correspondence: szenci.otto@univet.hu

Simple Summary: A fetal catheterization is an efficient tool allowing longitudinal in vivo studies on hormonal and metabolic changes, including fetal blood gases and acid-base changes. These surgical techniques made it possible to take blood samples daily under aseptic conditions to determine arterial and/or venous blood samples for acid-base variables (like pH, blood gas tensions: partial pressure of carbon dioxide and partial pressure of oxygen, oxygen saturation, bicarbonate concentration, total carbon dioxide, and base excess). All these examinations may contribute to a better understanding of the physiological changes that occur during calving, which may help reach a significant reduction in losses caused by perinatal mortality, which is still high today.

Abstract: The objective of the present study was to evaluate the changes in maternal and fetal arterial acid-base variables withdrawn from catheterized dams and fetuses during the last days before and during calving. The average gestation length in nine cows with chronically catheterized fetuses was 285 ± 10 (SD) days. The arterial acid-base variables of a catheterized dam and fetus were very stable during late gestation. Four newborn calves showed small differences between prenatal and postnatal pH values (-0.035). At the same time, pCO_2 values started to increase significantly ($p = 0.02$), indicating a shift towards physiological respiratory acidosis during calving. The partial pressure of oxygen and oxygen saturation values showed some non-significant improvements immediately after birth, while the other acid-base parameters did not differ. The remaining five newborn calves showed a significant decrease in arterial blood pH ($p < 0.01$) and BE ($p = 0.01$), while pCO_2 tended to be higher ($p = 0.06$), indicating a shift towards physiological respiratory and metabolic acidosis, while the other acid-base parameters hardly differed. It is essential to mention that physiological ($n = 2$) and mild metabolic acidosis ($n = 2$) developed gradually in four newborn calves during the second stage of calving, lasting about ≤ 2 h. In contrast, in the remaining newborn calf the physiological metabolic acidosis developed during the last 3 min of birth because immediately before birth, the BE value was 0.4 mmol/L. After birth, it was -5.4 mmol/L. The results indicate that the acid-base variables may start to move gradually in the direction of expressed respiratory and metabolic acidosis only after appearing the amniotic sac and fetal feet in the vulva during the second stage of labor; therefore, it is essential to complete obstetrical assistance in time.

Keywords: dairy cow; fetal cannulation; maternal cannulation; calving; blood gases; acid-base parameters

Citation: Szenci, O.; Van Der Weyden, G.C.; Lénárt, L.; Taverne, M.A.M. Changes of Acid-Base Variables in Dairy Cows with Chronically Implanted Fetal and Maternal Catheters during Late Gestation and Calving. *Animals* **2022**, *12*, 2448. https://doi.org/10.3390/ani12182448

Academic Editors: Daniel Mota-Rojas, Julio Martínez-Burnes and Agustín Orihuela

Received: 31 August 2022
Accepted: 11 September 2022
Published: 16 September 2022

Publisher's Note: MDPI stays neutral with regard to jurisdictional claims in published maps and institutional affiliations.

Copyright: © 2022 by the authors. Licensee MDPI, Basel, Switzerland. This article is an open access article distributed under the terms and conditions of the Creative Commons Attribution (CC BY) license (https://creativecommons.org/licenses/by/4.0/).

1. Introduction

The prevalence of perinatal mortality (death of a mature fetal calf after at least 260 days of gestation during calving or in the first 24 to 48 h of postnatal life [1,2] is still very high (3.5 to 8%) and contributes to considerable economic losses, especially in Holstein-Friesian dairy farms [3,4]. The perinatal mortality rate in Holstein-Friesian heifers reached a loss of

11 to 13.2% in the last decades [5–9], which nowadays shows a static or declining trend [10]. All of these emphasize the importance of examining the causal factors of perinatal mortality because asphyxia plays a critical role in perinatal mortality (58.3% [11] and 44.7% [12]).

It has been known in human practice for a long time that all fetuses develop more or less severe hypoxia due to the rupture of chorioallantoic and amniotic sacs and uterine contractions consequent metabolic acidosis [13]. Under normoxic conditions, glucose as the primary energy source is reduced to pyruvate via the citric acid cycle to the final products CO_2 and H_2O. During oxygen shortage, glucose can only be metabolized anaerobically to pyruvate and then reduced to lactic acid. Anaerobic glycolysis, however, has a significant disadvantage because energy production is reduced, the carbohydrate reserves are rapidly exhausted, and metabolic acidosis develops by the accumulation of acid metabolites (lactic acid). At calving, all fetuses, therefore, suffer from respiratory as well as metabolic acidosis. The degree of respiratory and metabolic acidosis in the perinatal period can be assigned into three groups according to their pH values [14]:

Group 1: blood pH > 7.2–physiological acidosis = slight, combined respiratory and metabolic acidosis.

Group 2: blood pH 7.2–7.0–moderate acidosis = mild to expressed, combined respiratory and metabolic acidosis.

Group 3: blood pH < 7.0–severe acidosis = severe, combined respiratory and metabolic acidosis.

The BE values can also be used to evaluate the degree of acidosis in the perinatal period: physiological acidosis BE <−6.0 mmol/L, moderate acidosis: BE: −6.0 and −12.9 mmol/L, and severe acidosis: ≥−13 mmol/L [15]. The degree of acidosis finally determines whether the fetus lives or dies. Before that, the organism's regulatory system of chemical buffering in the blood operates to keep the offspring alive. Bicarbonate is the essential buffer. The other ones are hemoglobin, plasma proteins, and phosphate buffers, as reviewed by Szenci [16]. Fetal cannulation may help us understand when respiratory and metabolic acidosis in the fetus to be born can develop to decrease the prevalence of stillbirth on dairy farms.

A significant advance was reached when the method for insertion and maintenance of catheters in umbilical and uterine vessels under chronic conditions was introduced for the sheep and goats by Meschia et al. [17]. Since then, the technique has been widely used for various studies on placental transfer in conscious sheep [18,19] and other large domestic animals such as the cow [20], mare [21,22], and pig [22,23].

The bovine catheter was usually inserted into the umbilical artery and/or vein [20,22,24–26] through a placentome selected near the junction of the uterine horns. The anterior tibial artery [24–29], saphenous artery [30], or saphenous vein [24–27,29–32] were also used to insert a catheter into the caudal fetal artery, aorta and/or fetal caudal vena cava. A fetal medial or metatarsal vein was also tried for cannulation; however, it was often dislodged [20]. The main uterine artery [24–26,29,32] and/or vein [22], or a circumflex iliac artery and/or vein [20,27], were usually catheterized on the maternal side of the cow. The maternal jugular vein was also used for the insertion of a catheter [30]. These surgical techniques made it possible to take blood samples daily under aseptic conditions to determine arterial and/or venous blood samples for acid-base variables (like pH, partial pressure of carbon dioxide /pCO_2/, partial pressure of oxygen /pO_2/, oxygen saturation /SaO_2/, base excess /BE/, bicarbonate concentration /HCO_3^-/ and total carbon dioxide /TCO_2/, however, most of the cases only the pH and the blood gas tensions were determined [20,22,30]. Schmidt et al. [26] reported the changes in pH, hemoglobin, pCO_2, pO_2, SaO_2, SO_2-content, ionized sodium, potassium, chloride, and calcium concentration, and glucose in whole blood in pregnant cows operated under general and local anesthesia while Sangild et al. [29] measured lactate and cortisol concentrations additionally in dams pregnant with in-vivo and AI produced embryos.

Independently of the duration of the surgery, a well-trained surgical team is needed for fetal cannulation to decrease the risk factors during operations [28]. At the same time, Smith et al. [26] also received a good result when pregnant dams for fetal catheterization were performed upon a standing cow in local anesthesia (total duration of the surgery: 2 to 2.5 h) compared with general anesthesia in the dorsal position (total duration of the surgery: 3 to 3.5 h).

Because of the extended duration of obstetrical traction (more than 2 min) [33] or in case of the delayed second stage of labor in primiparous dams [30], severe metabolic acidosis may develop, which endangers the chances of survival of newborn calves [15,33]. In contrast, Vannouchi et al. [34] have recently reported that calves born after 2 h of calving showed decreased vitality, hypercapnia, hypoxia, and increased antioxidant status (glutathione peroxidase) due to respiratory acidosis while severe metabolic acidosis did not develop. All these results call attention to further studies to determine precisely the reasons that lead to a shift in the blood gases and acid-base parameters during calving. Therefore, it was hypothesized that by measuring the metabolic parameters like BE, HCO_3^-, and TCO_2 besides pH and blood gases, respiratory and/or metabolic acidosis in the fetus to be born could be accurately differentiated.

The present study aimed to determine the changes in maternal and fetal arterial acid-base variables during the last days before and during calving in chronically catheterized Dutch-Friesian dams and their fetuses.

2. Materials and Methods

2.1. Animals and Surgery

The Veterinary Faculty Council (the State University of Utrecht, The Netherlands) approved the use and treatment of animals in this study. After buying healthy pregnant animals, fetal and maternal catheterization of nine pluriparous Dutch-Friesian pregnant cows (body condition score between 3.2 to 3.5) was performed at the Department of Veterinary Obstetrics, Gynaecology, and A.I. (Utrecht) as previously described by Taverne et al. [27]. Briefly, after withholding food for 48 h and water for 24 h operation was performed under general anesthesia in dorsal recumbency. The animals were premedicated intravenously with acepromazine (10 mg/100 kg bodyweight: Vetranquil R, Clin-Midy, France) and atropine (2 mg/100 kg bodyweight). Anesthesia was induced by intravenous infusion of a mixture of guaifenesin (10 g/100 kg bodyweight; Gujatal R, Aesculaap, The Netherlands) and thiopental sodium (0.7 g/100 kg bodyweight; Nesdonal R, Rhone Poulenc, France). The cow was intubated and ventilated with 2–3% fluothane in oxygen/nitrous oxide.

Following surgical preparation, the uterus was approached through a ventral midline incision. The right or left fetal hind limb was identified by intraabdominal palpation and moved so that the foot lay in the abdominal incision and under an intercotyledonary area of the uterus. The fetal leg was withdrawn from the uterus until the anterior surface of the hock joint was easily accessible. The fetal anterior tibial artery was catheterized with a polyvinyl catheter (0.75 mm I.D. × 1.45 mm O.D., Dural Plastics, Minto, Australia) which was advanced approximately 50 cm to lie with its tip in the dorsal aorta. The uterus was closed with one row of continuous sutures, which included all fetal membranes (chorioallantoic and amniotic membranes) and the uterine wall. The second row of continuous Lembert sutures was used to cover the first row. The catheter was exteriorized through the lateral abdominal wall and temporarily wrapped in a sterile towel. The abdominal midline incision was then closed using standard procedures.

Subsequently, the cow was positioned on her right side and prepared for cannulation of the left circumflex iliac artery. A polyvinyl catheter (1.02 mm I.D. × 1.78 mm O.D., Norton Plastics, Hayward, CA, USA) was advanced approximately 40 cm to lie with its tip in the dorsal aorta of the cow. Finally, both catheters were tunneled subcutaneously to the most dorsal area of the sub-lumbar fossa. They were fixed to the skin and protected within gauze swabs soaked in alcohol within a plastic bag. A more detailed description of this operation illustrated with several pictures has been published recently [28]. On the morning following surgery, the cow was placed in a pen, where she remained until after calving. The outer ends of the catheters were transferred to a hood containing a small container of alcohol. Hypodermic needles capped with Luer-lock injection caps were inserted into the ends of the catheters, which were filled with heparinized saline (200 IU/mL) between samplings.

The mean (±SD) gestational age on the day of surgery was 266 ± 13 (between 254 and 279) days (Table 1) and the cows had their second to fourth gestation (3.2 ± 0.7).

Table 1. The data of catheterized cows and fetuses.

Cow	Duration of Gestation (Days)	Duration between Operation and Calving (Days)	Period of Sampling (Post-Surgery)	Number of Maternal Blood Samples before Calving	Number of Fetal Blood Samples before Calving	Retained Fetal Membranes (Y/N)
A	254	31	D9–D31	10	16	Y
B	269	25	D6–D25	14	12	N
C	262	11	D6–D11	6	8	Y
D	258	16	D6–D16	10	11	N
E	265	14	D4–D14	11	*	Y
F	261	17	D5–D17	7	7	Y
G	NK	12	D5–D12	8	8	N
I	277	20	D5–D20	17	17	N
J	279	21	D5–D21	17	17	N
Mean (±SD)	266 ± 9	19 ± 6	6 ± 1 **	11 ± 4 N = 100	12 ± 4 N = 96	4/5

NK: not known, * fetal catheter was blocked, ** First blood samples were withdrawn after operation.

2.2. Evaluation of Neonatal Vitality

Neonatal vitality was evaluated immediately after delivery using the vitality score recommended by Szenci [35]. The following categories were used: a score of 3 indicated normal tonicity, head erect, and regular reflectory movements; a score of 2 indicated low tonicity, sternal recumbency with fetal head requiring support, reduced number and intensity of reflectory movements; and a score of 1 indicated toneless, head dropping, limbs extended, and cardiac activity present and a score of 0 indicated toneless, head dropping, limbs extended, and cardiac activity absent (dead).

2.3. Blood Sampling Protocol

Arterial blood samples were taken daily, usually between 8.00 and 12.00 a.m., starting on the sixth day (between 4 and 9 days) after surgery. When cows showed signs of impending parturition, sampling was repeated. Immediately after calving, blood samples were withdrawn from the newborn calves. On several occasions, samples could not be obtained (blocked catheters, unobserved start of calving), therefore, a strict protocol could not be followed in each of the cows. In two cows, the clinical signs of impending parturition were not expressed, therefore, we could withdraw the last blood samples only 17 and 23 h before calving, respectively. As the acid-base parameters did not differ from the other animals they were not removed from the evaluation.

Before arterial blood sampling, heparinized saline solution was removed from each catheter, and 1 mL of blood was withdrawn for acid-base measurements as described previously [36]. Briefly, the dead space of a 2-mL plastic syringe and an attached needle was filled with heparin solution (1000 IU/mL), and blood samples were taken anaerobically. Air bubbles observed in the sample were removed immediately, and the sample was mixed by rolling. The syringe was capped with a rubber stopper and placed on a bed of crushed ice. In each case, arterial blood samples were analyzed for acid-base variables (pH, pCO_2, pO_2, SaO_2, BE, HCO_3^- and TCO_2 using an acid-base analyzer (AVL Gas-Check 938, AVL Medical Instruments, Graz, Austria) at 37 °C within 15 min. After sampling, heparinized saline solution was used again to remove the blood from the catheter. After closing the outer end of the catheter with a Luer-lock injection cap, it was put back into a small container filled with alcohol. After calving, the fetal catheter was cut. A hypodermic needle capped with a Luer-lock injection cap was inserted into the end of the catheter to continue the withdrawal of additional arterial blood samples.

2.4. Statistical Analysis

Data were analyzed using R3.5.2. Statistical software [37]. Acid-base data were checked for normality with the Shapiro–Wilk test. In prepartum samples, the mean, the standard deviation (SD), and the 5th and 95th percentiles were calculated for both the dams and the fetuses. We used a generalized linear model to determine the effect of sample number for each acid-base variable, using sample number and animal ID as factors. Pearson's product-moment correlation was used to determine the prepartum correlation between fetal and maternal acid-base values. Each parameter from both maternal and fetal samples taken around the onset of calving was compared to the prepartum and the immediate prepartum samples in the dam and the immediate fetal prepartum and the immediate neonatal post-partum values using two-sided paired t-tests. In the case of newborn calves, according to the BE values (0.7 to -1.7 mmol/L vs. -3.7 to -7.2 mmol/L), two neonatal groups were established and analyzed separately. A probability of $p < 0.05$ was considered statistically significant.

3. Results

All cows recovered well after surgery, and the mean interval between operation and calving was 19 ± 6 (SD) days, resulting in mean gestational age of 285 ± 8 (between 273 and 300) days at calving. We were able to take an average of 11 ± 4 (between 6 to 17) blood samples from the cows and 12 ± 4 (between 7 to 17) blood samples from eight fetuses because one catheter was blocked before calving (Table 1). All single calves without obstetrical assistance were born alive within one h after the hooves appeared in the vagina in anterior presentations, and each newborn calves had a normal vitality score (Score 3). Neonatal blood samplings were continued after birth through the same catheter. After calving, 4 of 9 cows had retained fetal membranes.

The mean (SD) and 5th and 95th percentiles of the maternal blood gases and acid-base parameters before calving ($n = 100$) and about ≤ 2 h before calving ($n = 9$) are given in Table 2. The prepartum maternal acid-base values showed no significant differences over time until about ≤ 2 h before calving, although some individual maternal differences were observed between animals. Regarding the maternal arterial acid-base variables, only the pH values increased significantly ($p < 0.001$). Conversely, pCO_2 and pO_2 values decreased significantly ($p < 0.01$ and $p = 0.01$, respectively) until about ≤ 2 h before calving, demonstrating the effect of imminent calving on pH and blood gases. At the same time, maternal prepartum acid-base values did not show any differences between cows with and without retained fetal membranes after calving.

Table 2. Changes in maternal arterial blood gases and acid-base parameters in catheterized cows (n = 9) before calving and about ≤ 2-h before calving. Due to daily sampling from dams, 100 arterial blood samples could be withdrawn.

Acid-Base Parameters	Maternal Arterial Blood			Maternal Arterial Blood about ≤ 2-h before Calving	*p*-Value
	Mean $\pm$ SD	5th Percentile	95th Percentile	Mean $\pm$ SD	
pH	7.427 ± 0.029	7.381	7.467	7.447 ± 0.029	0.001
pCO_2 (kPa)	5.02 ± 0.41	4.30	5.60	4.68 ± 0.31	0.01
pO_2 (kPa)	13.43 ± 1.63	11.30	16.87	12.64 ± 1.77	0.01
SaO_2 (%)	96.9 ± 0.8	95.6	98.1	96.7 ± 1.1	0.09
BE (mmol/L)	0.56 ± 1.92	-2.72	3.72	1.3 ± 2.1	0.22
HCO_3^- (mmol/L)	24.5 ± 2.0	20.6	27.4	24.6 ± 2.0	0.82
TCO_2 (mmol/L)	25.6 ± 1.9	22.3	28.4	25.1 ± 1.5	0.58

SD: standard deviation.

The mean (SD) and 5th and 95th percentiles of the fetal/neonatal blood gases and acid-base parameters before calving ($n = 96$) and about ≤ 2 h before calving ($n = 8$) as well as immediately after calving ($n = 9$) are given in Table 3. We could withdraw the arterial blood sample from the missing calf after birth, and it showed similar acid-base values (BE: 0.0 mmol/L) to those for the three newborn calves; therefore, these values were not removed. The prepartum fetal acid-base values showed no significant differences over time until about ≤ 2 h before calving, although some individual differences were observed between fetuses. At the same time, the neonatal pH and BE values ($n = 9$) were significantly lower ($p < 0.01$ and $p = 0.03$, respectively), while pCO_2 was significantly higher ($p = 0.03$) in the immediate post-partum samples compared to prepartum samples (data not shown).

Table 3. Changes in fetal arterial blood gases and acid-base parameters before calving and about <2-h before calving in catheterized fetuses and immediately after calving in newborn calves.

Acid-Base Parameters	Fetal Arterial Blood			Fetal Arterial Blood about $\leq$2-h before Birth ($n = 8$ *)	Neonatal Arterial Blood (Group A: $n = 4$)	Neonatal Arterial Blood (Group B: $n = 5$)	*p*-Value	
	Mean $\pm$ SD	5th Percentile	95th Percentile	Mean $\pm$ SD	(Mean $\pm$ SD)	(Mean $\pm$ SD)	Fetal $\leq$2-h vs. Neonatal (Group A)	Fetal $\leq$2-h vs. Neonatal (Group B)
pH	7.351 $\pm$ 0.019	7.326	7.384	7.347 $\pm$ 0.019	7.312 $\pm$ 0.019	7.209 $\pm$ 0.073	0.08	0.01
pCO_2 (kPa)	6.52 $\pm$ 0.48	5.89	7.26	6.35 $\pm$ 0.70	7.24 $\pm$ 0.26	8.26 $\pm$ 1.81	0.02	0.06
pO_2 (kPa)	3.61 $\pm$ 0.65	2.43	4.45	2.83 $\pm$ 0.66	4.11 $\pm$ 0.86	3.19 $\pm$ 0.61	0.49	0.16
SaO_2 (%)	47.9 $\pm$ 11.9	23.2	62.6	31.7 $\pm$ 13.2	51.9 $\pm$ 14.7	30.9 $\pm$ 11.0	0.63	0.64
BE (mmol/L)	0.51 $\pm$ 1.28	-1.66	2.72	-0.4 ± 2.0	-0.3 ± 0.9	-5.4 ± 1.3	0.43	0.01
HCO_3^- (mmol/L)	26.5 $\pm$ 1.7	24.1	29.8	25.6 $\pm$ 2.6	26.8 $\pm$ 0.7	23.4 $\pm$ 1.8	0.77	0.43
TCO_2 (mmol/L)	27.6 $\pm$ 1.4	25.4	29.6	26.2 $\pm$ 2.0	28.4 $\pm$ 0.7	24.5 $\pm$ 1.5	0.63	0.83

* Due to the blocking of a fetal catheter, blood sample could not be withdrawn before calving.

Except for SaO_2, all acid-base parameters between maternal and fetal arterial blood showed a significant positive correlation before calving. In contrast, from about ≤ 2 h before calving, only pCO_2 and acid-base parameters (BE, HCO_3^- and TCO_2) showed a significant positive correlation albeit to a reduced extent (Table 4).

Table 4. Correlation between maternal and fetal acid-base parameters before calving.

Acid-Base Parameters	Before Caving			About $\leq$2-h before Calving		
	Pearson's Correlation	Confidence Intervals	*p*-Value	Pearson's Correlation	Confidence Intervals	*p*-Value
pH	0.47	0.29 to 0.62	0.001	0.57	-0.22 to 0.91	0.14
pCO_2 (kPa)	0.48	0.30 to 0.63	0.001	0.87	0.42 to 0.98	0.01
pO_2 (kPa)	0.28	0.07 to 0.46	0.01	0.44	-0.38 to 0.87	0.27
SaO_2 (%)	-0.01	-0.23 to 0.21	0.93	0.28	-0.53 to 0.82	0.51
BE (mmol/L)	0.44	0.26 to 0.6	0.001	0.81	0.24 to 0.96	0.02
HCO_3^- (mmol/L)	0.43	0.24 to 0.59	0.001	0.77	0.03 to 0.96	0.04
TCO_2 (mmol/L)	0.50	0.30 to 0.65	0.001	0.88	0.47 to 0.98	0.01

Among the nine newborn calves, there were four newborn calves with similar BE values (0.7 to -1.7 mmol/L) immediately after birth than those before delivery. Only pCO_2 showed a significant increase ($p = 0.02$) compared to samples taken 23 ± 13 min (except for one case when it was 23 h) before calving (Table 3). Concurrently, post-partum pH tended to be lower ($p = 0.08$). In these four calves, after birth, physiological respiratory acidosis dominated. In the remaining five newborn calves, the mean pH and BE values were significantly lower ($p < 0.01$) after birth, while pCO_2 tended to be higher ($p = 0.06$)

than in samples taken 124 ± 43 min (except for one case when it was 17 h) before calving. Three newborn calves were born with physiological respiratory and metabolic acidosis, while the remaining two had moderate respiratory and metabolic acidosis (pH: 7.167 and 7.089, while BE: -6.6 and -7.2 mmol/L, respectively). None of the newborn calves were lost in the post-parturient period.

4. Discussion

In agreement with previous findings [22,29,30], catheterized dams and fetuses' arterial blood pH and blood gases are very stable during late gestation. Our results also apply to the other acid-base variables like BE, HCO_3^-, and TCO_2 (Tables 2 and 3). The daily maternal and fetal blood pH and blood gases also changed a little up to the day of birth. Differences between maternal and fetal blood gases ensured the placental transfer of blood gases, while there was a significant pH gradient between the fetal and maternal arterial blood. According to Comline and Silver [22], the differences between maternal and fetal arterial samples were 0.052 for pH, 7.8 mmHg (1.04 kPa) for pCO_2, and 60.8 mmHg (8.1 kPa) for pO_2, respectively. In our case ($n = 86$, data not shown), the mean differences calculated from the blood withdrawn from the aorta were very similar: 0.075 for pH, 1.46 kPa for pCO_2, and 9.92 kPa for pO_2, respectively. Sangild et al. [29] also reported similar pH value differences (0.080 vs. 0.074, respectively) for dams with in-vitro produced (IVP) embryos and dams with AI embryos between 2–5 days after surgery during late pregnancy. In comparison, there were somewhat higher differences regarding the maternal and fetal pCO_2 values (IVP dam vs. fetus: 11.9 mmHg /1.58 kPa/ and AI dam vs. fetus: 26.1 mmHg /3.48 kPa/) between 2 to 5 days after surgery. According to our results, the other acid-base parameters (BE, HCO_3^- and TCO_2) were hardly different.

Except for SaO_2, all acid-base parameters between maternal and fetal arterial blood showed a significant positive correlation before calving. In contrast, from about ≤ 2 h before calving, only pCO_2 and acid-base parameters (BE, HCO_3^- and TCO_2) showed a significant positive correlation, albeit to a reduced extent, indicating the changes caused by uterine contraction at the beginning of calving (Table 4).

It was also found by Comline et al. [20] that fetal blood pH remained stable both immediately before and during normal parturition, and it was only after rupture of the umbilical cord that there was a fall in blood pH associated with a high pCO_2 level. In contrast, Wilson et al. [30] indicated that fetal pCO_2 values might elevate in the last 24 h before birth. According to our results, we had four newborn calves that showed little difference between prenatal and postnatal pH values (-0.035). In contrast, pCO_2 values started to increase significantly ($p = 0.02$), indicating a shift towards physiological respiratory acidosis. In contrast, the pO_2 and SaO_2 showed some non-significant improvements (1.28 kPa, 20.2 %) immediately after birth. At the same time, the other acid-base parameters did not differ. The remaining five newborn calves showed a significant decrease in arterial blood pH ($p < 0.01$) and BE ($p = 0.01$), while pCO_2 tended to be higher ($p = 0.06$), indicating a shift towards physiological respiratory and metabolic acidosis. At the same time, the other acid-base parameters hardly differed. It is essential to mention that metabolic acidosis developed gradually in four newborn calves during about ≤ 2 h of calving, while in the remaining newborn calf during the last 3 min of birth because immediately before birth, the BE value was 0.4 mmol/L and immediately after birth it was -5.4 mmol/L. Metabolic acidosis appears to be due to electrolyte imbalance and L-lactate accumulation due to anaerobic glycolysis for no tissue oxygenation.

In the field, the prevalence of normal fetuses (physiological acidosis: pH > 7.2) before spontaneous delivery and obstetrical assistance withdrawing venous blood from *v. metacarpalis volaris superficialis* [38,39] or capillary blood [40] can vary between 74.2 and 85% [38–40], while before Caesarean section withdrawing blood from *v. umbilicalis* or *v. digitalis dorsalis communis III* [14,41,42] before extraction from the uterus between 51.2 and 63.6%, respectively. A close correlation could be found between the duration of the second stage of calving and the duration of obstetrical assistance because the prevalence of severely acidotic calves (pH < 7.0) can be increased if the duration of the second stage of calving is extended (<2 h: 0%, 2 to 4 h: 19%, 4 to 7 h: 44%: [15]). In contrast, besides respiratory acidosis, Vannouchi et al. [34] reported normal mean metabolic parameters (BE: −5.4 mmol/L, HCO_3^-: 20.5 mmol/L, TCO_2: 21.9 mmol/L) even if the duration of calving was longer than >4 h. At the same time, a significant respiratory depression (altered lung gas exchange and delayed lung clearance) was also observed [43]. Therefore, it is important to emphasize for the dairy practice that, in agreement with Comline and Siver [22], not only the pH value and the blood gases but the other acid-base variables will start to move in the direction of expressed respiratory and metabolic acidosis only after appearing the amniotic sac and fetal feet in the vulva during the second stage of labor. Therefore, it is crucial to start obstetrical assistance within 70 min after amniotic sac appearance or 65 min after the appearance of feet in the vulva [44]. If we start obstetrical assistance too early, we may increase not only the prevalence of stillbirth rate but the prevalence of injuries to the birth canal and retained fetal membranes [45]. At the same time, Villettaz Robichaud et al. [46] could not confirm the negative effect of early intervention (within 15 min after the first sight of both front hooves in the vulva) on the stillbirth rate; however, sterile obstetrical lubricant was applied liberally to the dam's birth canal around the fetus before performing the examination and providing obstetrical assistance. The effect on the prevalence of injuries of the birth canal and retained fetal membranes was not reported. At the same time, Vannouchi et al. [47] could not confirm any connection between retained fetal membranes and extended duration of calving (>4 h). In our case, the prepartum maternal acid-base parameters did not differ if retained fetal membranes occurred after calving. In contrast, dystocia [15,16,35,40] or extended duration of obstetrical traction (more than 2 min) [33] may also contribute to the development of expressed respiratory and metabolic acidosis in the fetus being born.

5. Conclusions

To decrease the prevalence of dystocia and stillbirth and to improve animal welfare, one of our most important management activities during the periparturient period is to provide obstetrical assistance at an appropriate time after detecting the onset of the second stage of labor. In addition, maternal and fetal cannulation can help us understand those physiological and pathological changes that may endanger the fetuses' life to be born. This is all the more important because apart from buffering the metabolic acidosis in a dairy farm, instruments for effectively clearing the airways from amniotic mucus and providing artificial respiration are still not widely available; therefore, we must emphasize the importance of prevention of neonatal asphyxia.

Author Contributions: Investigation and writing: O.S., G.C.V.D.W. and M.A.M.T.; statistical analysis: L.L. All authors have read and agreed to the published version of the manuscript.

Funding: This research received no external funding.

Institutional Review Board Statement: The animal study protocol was approved by the Veterinary Faculty Council, the State University of Utrecht, Utrecht, The Netherlands.

Informed Consent Statement: Not applicable.

Data Availability Statement: The data presented in this study are available on request from the corresponding author.

Conflicts of Interest: The authors declare no conflict of interest.

References

1. Szenci, O.; Kiss, M.B. Perinatal calf losses in large cattle production units. *Acta Vet. Hung.* **1982**, *30*, 85–95.
2. Mee, J.F. Stillbirths—What can you do? *Cattle Pract.* **1999**, *7*, 277–281.
3. Mee, J.F. Managing the dairy cow at calving time. *Vet. Clin. Food Anim.* **2004**, *20*, 521–546. [CrossRef] [PubMed]
4. Mee, J.F. Newborn dairy calf management. *Vet. Clin. Food Anim.* **2008**, *24*, 1–17. [CrossRef] [PubMed]

5. Meyer, C.L.; Berger, P.J.; Koehler, K.J. Interactions among factors affecting stillbirths in Holstein cattle in the United States. *J. Dairy Sci.* **2000**, *83*, 2657–2663. [CrossRef]
6. Harbers, A.; Segeren, L.; De Jong, G. Genetic parameters for stillbirth in the Netherlands. *Interbull Bull.* **2000**, *25*, 117–122.
7. Berglund, B.; Steinbock, L.; Elvander, M. Causes of stillbirth and time of death in Swedish Holstein calves examined post mortem. *Acta Vet. Scand.* **2003**, *44*, 111–120. [CrossRef]
8. Steinbock, L.; Näsholm, A.; Berglund, B.; Johansson, K.; Philipsson, J. Genetic effects on stillbirth and calving difficulty in Swedish Holsteins at first and second calving. *J. Dairy Sci.* **2003**, *86*, 2228–2235. [CrossRef]
9. Gustafsson, H.; Kindahl, H.; Berglund, B. Stillbirths in Holstein heifers-some results from Swedish research. *Acta Vet. Scand.* **2007**, *49* (Suppl. 1), S17. [CrossRef]
10. Mee, J.F. Epidemiology of bovine perinatal mortality. In *Bovine Prenatal, Perinatal and Neonatal Medicine*; Szenci, O., Mee, J.F., Bleul, U., Taverne, M.A.M., Eds.; Hungarian Association for Buiatrics: Budapest, Hungary, 2021; pp. 224–232.
11. Schuijt, G. Iatrogenic fractures of ribs and vertebrae during delivery in perinataly dying calves: 235 cases (1978–1988). *J. Am. Vet. Med. Assoc.* **1990**, *197*, 1196–1202.
12. Mock, T.; Mee, J.F.; Dettwiler, M.; Rodriguez-Campos, S.; Hüsler, J.; Michel, B.; Häfliger, I.; Drögemüller, C.; Bodmer, M.; Hirsbrunner, G. Evaluation of an investigative model in dairy herds with high calf perinatal mortality rates in Switzerland. *Theriogenology* **2020**, *148*, 48–59. [CrossRef] [PubMed]
13. Saling, E. *Das Kind Im Bereich Der Geburtshilfe*; George Thieme Verlag: Stuttgart, Germany, 1966.
14. Eigenmann, U.J.E.; Grunert, E.; Born, E. Untersuchungen über den Einfluss der Schnittentbindung auf den Säurenbasenhaushalt sowie die Plasmaglukosekonzentration neugeborener Kälbern. *Dtsch. Tierärztl. Wschr.* **1981**, *88*, 433–437.
15. Held, T. Klinische und blutgasanalytische Untersuchungen bei kalbenden Rindern und deren Feten. Inaugural Dissertation, Tierärzliche Hochschule Hannover, Hannover, Germany, 1983.
16. Szenci, O. Role of acid-base disturbances in perinatal mortality of calves: Review. *Vet. Bull.* **2003**, *73*, 7R–14R. [CrossRef]
17. Meschia, G.; Cotter, J.R.; Breathnach, C.S.; Barron, D.H. The hemoglobin, oxygen, carbon dioxide and hydrogen ion concentrations in the umbilical bloods of sheep and goats as sampled via indwelling plastic catheters. *Q. J. Exp. Physiol. Cogn. Med. Sci.* **1965**, *50*, 185–195. [CrossRef] [PubMed]
18. Comline, R.S.; Silver, M. Daily changes in foetal and maternal blood of conscious pregnant ewes, with catheters in umbilical and uterine vessels. *J. Physiol.* **1970**, *209*, 567–586. [CrossRef]
19. Comline, R.S.; Silver, M. The composition of foetal and maternal blood during parturition in the ewe. *J. Physiol.* **1972**, *222*, 233–256. [CrossRef]
20. Comline, R.S.; Hall, L.W.; Lavelle, R.; Nathanielsz, P.W.; Silver, M. Parturition in the cow: Endocrine changes in animals with chronically implanted catheters in the foetal and maternal circulations. *J. Endocrinol.* **1974**, *63*, 451–472. [CrossRef]
21. Comline, R.S.; Hall, L.W.; Lavelle, R.; Silver, M. The use of intravascular catheters for long-term studies on the mare and fetus. *J. Reprod. Fertil. Suppl.* **1975**, *23*, 583–588.
22. Comline, R.S.; Silver, M. A comparative study of blood gas tensions, oxygen affinity and red cell 2,3 DPG concentrations in foetal and maternal blood in the mare, cow and sow. *J. Physiol.* **1974**, *242*, 805–826. [CrossRef]
23. Randall, G.C. Daily changes in the blood of conscious pigs with catheters in foetal and uterine vessels during late gestation. *J. Physiol.* **1977**, *270*, 719–731. [CrossRef]
24. Ferrell, C.L. Maternal and fetal influences on uterine and conceptus development in the cow: II. Blood flow and nutrient flux. *J. Anim. Sci.* **1991**, *69*, 1954–1965. [CrossRef] [PubMed]
25. Ferrell, C.L.; Reynolds, L.P. Uterine and umbilical blood flows and net nutrient uptake by fetuses and uteroplacental tissues of cows gravid with either single or twin fetuses. *J. Anim. Sci.* **1992**, *70*, 426–433. [CrossRef] [PubMed]
26. Schmidt, M.; Sangild, P.T.; Jacobsen, H.; Greve, T. A method to reduce the invasiveness of fetal catheterization in the cow. *Anim. Reprod. Sci.* **2004**, *80*, 193–200. [CrossRef] [PubMed]
27. Taverne, M.A.M.; Bevers, M.M.; Van der Weyden, G.C.; Dieleman, S.J.; Fontijne, P. Concentration of growth hormone, prolactin and cortisol in fetal and maternal blood and amniotic fluid during late pregnancy and parturition in cows with cannulated fetuses. *Anim. Reprod. Sci.* **1988**, *17*, 51–59. [CrossRef]
28. Szenci, O.; Touati, K.; De Sousa, N.M.; Hornick, J.L.; Van Der Weyden, G.C.; Taverne, M.A.M.; Beckers, J.F. Developing a surgical technique for long term catheterisation of bovine foetuses. *Acta Vet. Hung.* **2020**, *68*, 212–220. [CrossRef] [PubMed]
29. Sangild, P.T.; Schmidt, M.; Jacobsen, H.; Fowden, A.L.; Forhead, A.; Avery, B.; Greve, T. Blood chemistry, nutrient metabolism, and organ weights in fetal and newborn calves derived from in vitro-produced bovine embryos. *Biol. Reprod.* **2000**, *62*, 1495–1504. [CrossRef] [PubMed]
30. Wilson, G.D.; Hunter, J.T.; Derrick, G.H.; Aitken, W.M.; Kronfeld, D.S. Fetal and maternal mineral concentrations in dairy cattle during late pregnancy. *J. Dairy Sci.* **1977**, *60*, 935–941. [CrossRef]
31. Adams, R.; Garry, F.; Holland, M.D.; Hay, W.W.; Wagner, A. Cannulation technique for the late gestation bovine fetus. *Theriogenology* **1998**, *49*, 337. [CrossRef]
32. Aoki, M.; Kimura, K.; Hirako, M.; Hanafusa, Y.; Ishizaki, H.; Kariya, Y. Cannulation for a bovine fetus in late gestation under regional anesthesia. *J. Reprod. Dev.* **2002**, *48*, 455–460. [CrossRef]
33. Szenci, O.; Gálfy, P.; Lajcsák, A. Comparison of carbonic anhydrase in neonatal and maternal red blood cells with different levels of acidosis in newborn calves. *Zbl. Vet. Med. A* **1984**, *31*, 437–440. [CrossRef]

34. Vannucchi, C.I.; Silva, L.G.; Lúcio, C.F.; Veiga, G.A.L. Oxidative stress and acid–base balance during the transition period of neonatal Holstein calves submitted to different calving times and obstetric assistance. *J. Dairy Sci.* **2019**, *102*, 1542–1550. [CrossRef] [PubMed]
35. Szenci, O. Role of acid-base disturbances in perinatal mortality of calves (A summary of thesis). *Acta Vet. Hung.* **1985**, *33*, 205–220. [PubMed]
36. Szenci, O.; Besser, T. Changes in blood gas and acid-base parameters of bovine venous blood during storage. *J. Am. Vet. Med. Assoc.* **1990**, *197*, 471–747. [PubMed]
37. R Core Team. R: A Language and Environment for Statistical Computing. R Foundation for Statistical Computing. 2018. Available online: https://www.R-project.org/ (accessed on 1 January 2022).
38. Mülling, M.; Waizenhöfer, H.; Bratig, B. Glucose-, Laktat- und pH_{akt}-Werte bei Kühen und Kalbern wahrend und unmittelbar nach der Geburt. *Berl. Münch. Tierärtzl. Wschr.* **1979**, *92*, 111–117.
39. Szenci, O.; Taverne, M.A.M.; Bakonyi, S.; Erdődi, A. Comparision between pre- and postnatal acid-base status of calves and their perinatal mortality. *Vet. Q.* **1988**, *10*, 140–144. [CrossRef]
40. Bleul, U.; Schwantag, S.; Kähn, W. Blood gas analysis of bovine fetal capillary blood during stage II labor. *Theriogenology* **2008**, *69*, 245–251. [CrossRef]
41. Ammann, H.; Berchtold, M.; Schneider, F. Blutgas- und Säurenbasenverthältnisse bei normalen und asphyktischen Kälbern. *Berl. Münch. Tierärtzl. Wschr.* **1974**, *87*, 66–68.
42. Szenci, O.; Taverne, M.A.M. Perinatal blood gas and acid-base status of Caesarean-derived calves. *J. Vet. Med. A* **1988**, *35*, 572–577.
43. Vannucchi, C.I.; Silva, L.C.G.; Unruh, S.M.; Lucio, C.F.; Veiga, G.A.L. Calving duration and obstetric assistance influence pulmonary function of Holstein calves during immediate fetal-to-neonatal transition. *PLoS ONE* **2018**, *13*, e0204129. [CrossRef]
44. Schuenemann, G.M.; Nieto, I.; Bas, S.; Galvão, K.N.; Workman, J. Assessment of calving progress and reference times for obstetric intervention during dystocia in Holstein dairy cows. *J. Dairy Sci.* **2011**, *94*, 5494–5501. [CrossRef]
45. Kovács, L.; Kézér, F.L.; Szenci, O. Effect of calving process on the outcomes of delivery and postpartum health of dairy cows with unassisted and assisted calvings. *J. Dairy Sci.* **2016**, *99*, 7568–7573. [CrossRef] [PubMed]
46. Villettaz Robichaud, M.; Pearl, D.L.; Godden, S.M.; LeBlanc, S.J.; Haley, D.B. Systematic early obstetrical assistance at calving: I. Effects on dairy calf stillbirth, vigor, and passive immunity transfer. *J. Dairy Sci.* **2017**, *100*, 691–702. [CrossRef] [PubMed]
47. Vannucchi, C.I.; Silva, L.G.; Lucio, C.F.; Veiga, G.A. Influence of the duration of calving and obstetric assistance on the placental retention index in Holstein dairy cows. *Anim. Sci. J.* **2017**, *88*, 451–455. [CrossRef] [PubMed]

Review

Importance of Monitoring Fetal and Neonatal Vitality in Bovine Practices

Ottó Szenci

Department of Obstetrics and Food Animal Medicine Clinic, University of Veterinary Medicine Budapest, Dóra Major, H-2225 Üllő, Hungary; szenci.otto@univet.hu

Simple Summary: The assessment of fetal and neonatal vitality plays an essential role in preventing stillbirth in bovine cattle. Therefore, an accurate diagnosis of fetal vitality prior to obstetric assistance or neonatal vitality after birth is crucial for ensuring prompt and accurate treatment. If severe asphyxia occurs before the availability of obstetric assistance, a cesarean section can be performed as a less life-threatening alternative for the fetus. In cases where severe asphyxia is diagnosed postnatally, available resuscitation methods can be employed in an attempt to protect the newborn's life. However, it must be noted that the potential for such interventions remains limited in dairy and beef practices, underscoring the importance of preventing stillbirths. Consequently, this review focuses on the diagnostic possibilities and limitations related to the evaluation of fetal and neonatal vitality in dairy and beef practices.

Abstract: Prior to initiating any obstetrical intervention for anterior or posterior presentation, it is imperative to emphasize the need for a precise and accurate diagnosis of fetal viability and to select the most appropriate approach for assistance. In uncertain cases, diagnostic tools such as ultrasonography, pulse oximeter, or measurement of acid–base balance or lactate concentration may be employed to confirm the diagnosis. In situations of severe asphyxia, a cesarean section is preferred over traction, even if the duration of asphyxia is less than 60 s, to maximize the likelihood of the survival of the fetus. Postcalving, several vitality scores have been proposed to evaluate the vigor of the newborn calf. Originally, four different clinical signs were recommended for assessing the well-being of newborn calves. Subsequently, five or more different clinical signs were recommended to evaluate vitality. However, despite the efforts for devising a practical tool to assess newborn calf vitality; a user-friendly and highly accurate instrument that can be used on farms remains elusive. Measuring the acid–base balance or lactate concentration may increase the diagnostic accuracy. It is critical to emphasize the importance of reducing the incidence of dystocia to mitigate the occurrence of severe asphyxia. In instances where asphyxia is unavoidable, adequate treatments should be administered to minimize losses.

Keywords: dairy cow; fetal and neonatal vitality; acid–base and lactate measurement; ultrasonography; pulse oximetry

Citation: Szenci, O. Importance of Monitoring Fetal and Neonatal Vitality in Bovine Practices. *Animals* **2023**, *13*, 1081. https://doi.org/10.3390/ani13061081

Academic Editors: Daniel Mota-Rojas, Julio Martínez-Burnes and Agustín Orihuela

Received: 10 December 2022
Revised: 13 March 2023
Accepted: 14 March 2023
Published: 17 March 2023

Copyright: © 2023 by the author. Licensee MDPI, Basel, Switzerland. This article is an open access article distributed under the terms and conditions of the Creative Commons Attribution (CC BY) license (https://creativecommons.org/licenses/by/4.0/).

1. Introduction

The profitability of a dairy farm depends greatly on the rate of calves being born alive and successfully reared to maturity [1,2]. Perinatal mortality, also referred to as stillbirth, pertains to the death of a mature fetal calf with a gestation period longer than 260 days that occurs during calving or within 24 to 48 h of postnatal life [3,4]. The incidence of perinatal mortality ranges from 3 to 10.3% [5], 2.4 to 9.7% [6], or 2 to 15% [7] across most countries. Despite considerable advancements in animal breeding, the prevalence of perinatal mortality remains high, particularly in Holstein-Friesian dairy farms [6,8–10]. In comparison, lower prevalence rates are mainly observed in relatively small farms with fewer than 65 dairy cows [5] or in dairy farms raising other breeds such as Norwegian Red or Swedish Red [6].

Over the past few decades, there has been an upward trend in stillbirth rates, especially in Holstein Friesian (HF) heifers. In Swedish, Dutch, and US HF heifer populations, the prevalence of stillbirth ranges between 10% and 13.2% [11–16], which has recently shown a static or declining trend [17]. However, there have been some encouraging outcomes in American [18] and Canadian dairy farms [19], with lower prevalence rates (<2%) of stillbirth. These findings highlight the importance of investigating the underlying factors contributing to perinatal mortality that have also been recently reviewed [20–22].

The etiology of stillbirth with a noninfectious origin is likely to be multifactorial, with direct and indirect asphyxia being the primary causes of death in most cases. Pathological changes were not detected in 58.3 to 75% of the calves that died during the perinatal period, as reported by several studies [23–25]. Furthermore, according to a recent review by Mee [26], the diagnostic rate of anoxia is highly variable between studies, ranging from approximately 10 to 80% of cases.

Fetal and maternal cannulation prior to calving allows for the monitoring of changes in the acid–base balance [27]. "These results indicate that the acid-base variables may start to move gradually in the direction of expressed respiratory and metabolic acidosis only after the amniotic sac and fetal feet appear in the vulva during the second stage of labor. However, in spontaneous calving, physiological respiratory or respiratory and metabolic acidosis is used to develop; therefore, it is essential to complete obstetrical assistance in time" [27]. The duration of survivable asphyxia depends on glycogen reserves in the heart [28]. In an experiment involving Hereford fetuses, the survival period was examined by the clamping of the umbilical cords. Four out of six fetuses survived after 4 min of anoxia, while none survived after 6 to 8 min of anoxia [29].

In dairy and beef practice, different methods are used to assess fetal and neonatal vitality as a measurement of the acid–base balance of fetuses or newborn calves is not feasible. Although portable acid–base analyzers are already available in the field, their present costs limit their daily use [30–32]. Concurrently, measuring lactate concentration using a handheld meter is a more cost-effective option [33,34].

The present review focuses on the determination of fetal and neonatal vitality of dairy calves to provide guidance for practitioners in choosing a suitable method to decrease fetal and/or neonatal mortality. Given the importance of reducing perinatal mortality rates in bovine cattle, recent scientific literature, including a published book [35] and several reviews [1,36–38], have emphasized the significance of this topic using different perspectives.

2. Determining Fetal Vitality during Stage Two of Calving

2.1. Determining the Clinical Signs of Vitality in Anterior and Posterior Presentation during Stage Two of Calving

The vitality of a bovine fetus during an obstetrical examination can be determined by initiating different reflexes in either the anterior (interdigital, bulbar, and swallowing reflexes) or posterior presentation (interdigital reflex, anal reflex, and pulse of the umbilical cord) [39].

As acidosis deepened (physiological, moderate, and severe acidosis), fetuses with anterior presentation ($n = 180$) failed to exhibit the interdigital reflex (88%, 65%, and 20%, respectively), bulbar reflex (95%, 91%, and 47%, respectively), and swallowing reflex (100%, 98%, and 80%, respectively). In the posterior presentation, a similar correlation between the palpable clinical signs of life (interdigital reflex: 52%, 25%, and 12%; anal reflex: 76%, 62%, and 50%; and the pulse of the umbilical cord: 95%, 87%, and 50%, respectively) and the acid–base balance could not be established due to the limited number of fetuses ($n = 37$). Nonetheless, there were 12 fetuses (5.5%) that were unable to initiate any reflexes but were still viable, as demonstrated by measuring the acid–base balance and providing obstetric assistance [39]. Therefore, before considering fetotomy, fetal heartbeat evaluation through B-mode/Doppler ultrasonography [40] and fetal pulse rate evaluation using a pulse oximeter [41] are crucial.

2.2. Measuring Fetal Acid–Base Values during Stage Two of Calving

In bovine practice, the measurement of the acid–base balance in fetal scalp blood is not as widely performed as in human practice, despite the commercial availability of portable acid–base analyzers. A study conducted by Bleul and Götz [31] compared 271 venous blood samples (obtained using the iSTAT analyzer) to reference methods and observed high correlations between acid–base parameters (r = 0.965–0.986), except for pO_2 values (r = 0.817). Fetal blood samples for acid–base measurement can be collected by puncturing the *v. metacarpalis volaris superficialis* or *v. digitalis dorsalis communis III* before the onset of traction or the *a.* and *v. umbilicalis* before the onset of extraction during a cesarean (C-) section [1]. Capillary blood sampling is also a viable option for diagnosing fetal asphyxia [42].

Before and after birth, calves can be assigned to one of three groups according to their pH values, as suggested by Eigenmann et al. [43] and implemented by us [1,27]:

"Group 1: blood pH > 7.2–physiological acidosis = slight, combined respiratory and metabolic acidosis.

Group 2: blood pH 7.2–7.0–moderate acidosis = mild to expressed combined respiratory and metabolic acidosis.

Group 3: blood pH < 7.0, severe acidosis = severe, combined respiratory and metabolic acidosis" [43].

Immediately prior to parturition or obstetrical intervention, a substantial proportion of fetuses exhibited physiological metabolic acidosis, ranging from 57.6% to 80%. Moderate metabolic acidosis was observed in 20–24.9% of fetuses, while severe metabolic acidosis was present in 0–17.5% of cases (Table 1).

Table 1. Prevalence of physiological, moderately, and severely acidotic fetuses before starting obstetrical assistance.

Type of Obstetrical Assistance	Site of Blood Sampling before Calving	No. of Examined Fetuses (*n*)	pH > 7.2 *n* (%)	pH 7.2–7.0 *n* (%)	pH < 7.0 *n* (%)	References
Spontaneous or traction	*v. jugularis*	19	15 (78.9)	4 (21.1)	0	Eichler-Steinhauff [44]
Spontaneous or traction	*v. metacarpalis volaris superficial*	20	16 (80)	4 (20)	0	Mülling et al. [45]
Spontaneous or traction	*v. metacarpalis volaris superficial*	58	43 (74.1)	14 (24.1)	1 (1.7)	Szenci et al. [46]
Spontaneous, traction or C-section	*v. digitalis dorsalis communis III*	217 *	125 (57.6)	54 (24.9) [a]	38 (17.5) [b]	Held et al. [47]
Spontaneous, traction, or C-section	Fetal capillary blood	38	29 (76.3)	9 (23.7)		Bleul et al. [42]
C-section	*a.* or *v. umbilicalis*	44	28 (63.6)	10 (22.7)	6 (13.6)	Szenci and Taverne [48]

* Fetuses were grouped according to the BE values: physiological acidosis: BE > −6.0 mmol/L; moderate acidosis: BE: −6.0 and −12.9 mmol/L; severe acidosis: BE < −13 mmol/L [41]. [a] Four calves died within 24 h of birth (8%). [b] Fifteen calves died within 24 h of birth (51%).

In contrast, Held et al. [47] evaluated the degree of prenatal acidosis by measuring the blood base excess (BE) value in 217 parturitions. The study revealed that 57.6% of the fetuses had physiologic (BE > −6.0 mmol/L), 24.9% had moderate (BE: −6.0 and −12.9 mmol/L), and 17.5% had severe acidosis (BE ≤ −13 mmol/L). Similar results were obtained using data collected from calves delivered by C-section, with 59.1% of the calves having physiological acidosis and 40.9% having moderate to severe acidosis [48], which indicated that both parameters (pH and/or BE) were suitable for classifying fetal acidosis prior to initiating calving assistance.

It is essential to mention that the detection of vigorous fetal movements during obstetric examinations or extraction of the fetus from the uterus during a C-section is indicative of the presence of severe metabolic acidosis. Therefore, assistance must be promptly completed. This is supported by our previous findings where four out of six calves showed fetal movements during C-section and were severely acidotic immediately after birth [48]. Additionally, the occurrence of fetal movements was confirmed in an experiment in which the umbilical cord was clamped [29]. The first movements occurred, on average, after 53 s postclamping of the umbilical cord (range, 10–105 s). Among the nine fetuses that died during the study, only one fetus did not show any movements after being subjected to $\geq$4 min of anoxia [29].

2.3. Measuring Fetal Oxygen Saturation during Stage Two of Calving

Electronic fetal heart rate (FHR) monitoring using cardiotocography is widely used to assess intrapartum hypoxia during labor in human practice [49]. In cases of nonreassuring cardiotocography, human fetal pulse oximetry accurately excludes moderate to advanced acidosis when the fetal pulse oximetry is <30% for at least 10 min. Additionally, it reduces the frequency of fetal blood analysis [50].

In bovine practice, continuous measurements of fetal oxygen saturation of arterial hemoglobin (FSpO$_2$) via pulse oximetry (oxygen sensor) can also be utilized [1,51]. "An oxygen sensor designed for human babies was placed in the mouth against the mucosa of the hard palate. The accuracy of predicting asphyxia was the highest when the oxygen saturation was below for a period of at least 2 min. A cutoff value of <30% had the highest positive predictive value. Sporadic individual values <30% were not clinically significant as they might just represent physiological events at calving, e.g., the oxygen saturation can decrease temporarily during uterine contractions because of increased intrauterine pressure. At the same time, it was emphasized by Bleul and Kähn [51] that further studies are needed to determine whether an FSpO$_2$ value of <30% over a minimum of 2 min is a valuable predictor of neonatal asphyxia in the cows". Thus, rapid obstetrical assistance such as extraction or C-section would help prevent peripartal death of the calf by reducing the duration of asphyxia.

According to Kanz et al. [41], measuring pulse rate (>120 bpm for at least 2 min) by pulse oximetry during the last 25 min of calving could predict neonatal acidosis more accurately (area under the curve [AUC]: 0.764) than measuring fetal oxygen saturation (FSpO$_2$ < 40% for at least 50% of the measurement: AUC = 0.613). The oximetric sensor was interdigitally fixed using a homemade latex cover. It was emphasized that improving the hardware of the device was necessary to obtain immediate results during the examination. Furthermore, improved fixation of the pulse oximeter to the fetus is essential for reducing the risk of intrapartal detachment.

2.4. Measuring Fetal Lactate Concentration during Stage Two of Calving

Recent advancements in dairy farming have led to the adoption of portable devices for measuring L-lactate concentration. Commercially available portable lactate meters, such as Lactate Scout and Biosen C line [33], as well as Accutrend Plus, i-STAT, Lactate Pro, and Lactate Scout [34], have been tested for their accuracy and found to be highly consistent with the reference methods (r = 0.98–0.99). A study by Karapinar et al. [34] reported that i-STAT had the highest sensitivity (100%) and specificity (98.6%) compared to the other three devices.

In human obstetrics, a fetal scalp blood lactate level of $\leq$5.4 mmol/L measured using the StatstripLactate$^{®}$/StatstripXpress$^{®}$ system is considered the cutoff for labor monitoring [52]. However, there is currently no established cutoff for bovine practice, and further research is needed in this area.

2.5. Determining Fetal Heart Rate (FHR) during Stage Two of Calving

An FHR of 90–120 bpm prior to calving was established through transabdominal Doppler ultrasonographic examinations, which was calculated by visual inspection of paper recordings [53,54]. Subsequently, Breukelman et al. [40] used a computer-assisted analysis after an analog–digital conversion of the FHR, establishing the reference values (108.6 ± 1.9 bpm) for FHR two days before expected calving. Cardiotocographic recordings of 84 fetuses during the dilatation stage of parturition identified normocardia (heart rate: 80–155 bpm) and tachycardia (heart rate: >155 bpm), with a close correlation being found between heart rates and the degree of acidosis. Moderate or severe acidosis (BE < −8.9 mmol/L) was observed in all fetuses with tachycardia [55].

Calf births in poor condition were observed when decelerations (periodic decreases in FHR) occurred during the end of a contraction [56]. "In the acidotic group, the mean baseline FHR increased from 113.5 to 138.6 bpm during the last 55 min before birth, while in the normal group, it changed from 116.9 to 121.3 bpm. Decelerations occurred during uterine contractions in acidotic fetuses and in the majority of normal fetuses (12 of 16) during the expulsive stage of parturition; accelerations (periodic increase in FHR) were hardly recorded" [56]. The mean decrease in FHR during uterine contractions increased significantly toward birth, indicating that intrapartum continuous FHR measurements might provide additional information on the acidotic state of fetuses [57]. However, short-term measurements held no diagnostic value [56]. Therefore, at least 30 min of continuous recording is recommended for dairy cows [40].

Ultrasonic transit-time measurement of blood flow in the umbilical arteries and veins of the bovine fetus during stage II of calving facilitates direct and continuous measurement of umbilical blood flow volume per unit of time. It also allows for the investigation of the relationship between umbilical blood flow, uterine contractions, and fetal heart rate [58]. During uterine contractions, a decrease in blood flow is most likely caused by the compression of the umbilical vessels. Venous blood flow is more severely affected than arterial blood flow during muscular contractions due to the thinner walls and fewer muscle fibers of the umbilical veins than those of the umbilical arteries. In a study by Bleul et al. [58], lower umbilical arterial and venous blood flows in acidotic calves were found to be higher than those in nonacidotic calves during the last 30 min before birth.

3. Diagnosis of Neonatal Vitality after Calving

3.1. Evaluating Clinical Signs of Vitality after Birth

Several parameters have been proposed as alternative vitality classification systems for field use after calving, including respiration and reflexes [59], time from birth until head-righting [29], until sternal recumbency [29,60,61], until the first apparent efforts to stand up [29,62,63], until the calf stands up [61–64]; until the first suckling [62,63,65–67], or a combination of attitude, vital signs, feeding behavior, and locomotion [68]. However, in most of these cases, the blood gases and acid–base parameters of newborn calves were not assessed, which can provide more accurate predictions of calf vitality [36].

In human obstetrical practice, a numerical scoring system has been suggested to assess a newborn baby's clinical condition [69], which is based on heart rate, respiratory effort, muscle tone, reflex irritability, and skin color. Clinical signs are assessed for each variable, and a score of 0, 1, or 2 is assigned. The total score is used to calculate a final score (0–10), which is named Apgar points after the inventor [69].

In bovine practice, Mülling [70] was the first to adapt the human Apgar score (muscle tone and movement, reflex activity, respiration, and mucous membrane color) and proposed a neonatal status diagnosis in calves (Table 2), which was subsequently adopted by others [41,71–75]. Instead of the five clinical signs, four were assessed and scores were assigned ranging from zero to two. The higher the score, the more robust the calf: scores 7–8 indicate a healthy calf, scores 4–6 indicate a calf at risk, and scores ≤ 3 indicate a weak calf that is alive. In addition, healthy calves have better acid–base parameters than at-risk calves [41,71].

Table 2. Apgar score system modified for newborn calves by Mülling [70].

Parameters/Scores	0	1	2
Muscle tone and movement	Absence	Reduced	Spontaneous, active movement
Reflex activity	Being absent	Reduced	Fully available
Respiration	Being absent	Slow, irregular	Rhythmic, normal
Mucous membrane color	Bluish-white	Blue	Pink

Scores 7–8: normal, Scores 4–6: calf at risk, Scores 0–3: live weak.

In contrast to previous recommendations, Mauer-Schweizer et al. [76] suggested the use of heart rate detection instead of assessing mucous membrane color for determining the health status of newborn calves (Table 3). Furthermore, they reported a definitive correlation between modified Apgar scores and acid–base parameters [77,78]. Palmer [79] adopted the same modified Apgar score system suggested by Mauer-Schweizer et al. [76]; however, it included the detection of reflex activity in response to nasal stimulation (no response, grimace, and sneeze/cough) and ear tickling (no response, weak ear flick, and ear flick/head shake).

Table 3. Modification of Mülling's Apgar score system by Maurer-Schweizer et al. [76].

Parameters/Scores	0	1	2
Respiration	Being absent	Slow, irregular,	Rhythmic, normal
Heart rate	Not measurable	<100, >150	100–150
Muscle tone and movement	Absence	Reduced	Active movement
Reflex activity	Being absent	Reduced	Fully available

Scores 7–8: normal, Scores 4–6: calf at risk, Scores 0–3: live weak.

Another modification of the Apgar score system developed by Mülling [70] was proposed by Born [80], who recommended evaluating the effect of splashing cold water on the calf's head, eye and interdigital reflexes, respiration, and mucous membrane color after a C-section (Table 4). This system has been adopted by other researchers [39,81–88].

Table 4. Modification of Mülling's Apgar score system by Born [80].

Parameters/Scores	0	1	2
Effect of splashing cold water on the head	Being absent	Reduced	Spontaneous, active movement
Eye and interdigital reflexes	Absence	One reflex is positive	Both reflexes are positive
Respiration	Being absent	Irregular	Rhythmic
Mucous membrane color	Bluish-white	Blue	Pink

Scores 7–8: normal, Scores 4–6: calf at risk, Scores 0–3: live weak.

Essmeyer [89] suggested "evaluating the reaction to cold water on the head (no reaction; reduced, late reaction; lifting, shaking of the head), interdigital reflex (no reaction; pulling away slowly, weak; pulling back strongly, immediate reaction), mucosal membrane color (white; pale pink, cyanotic; pink) and respiration (absent; irregular frequency and intensity; regular frequency and intensity)" which was subsequently employed by Sorge et al. [90].

However, discrepancies between the measured pH values and modified Apgar scores were discovered by Born [80], who found agreement with Maurer-Schweizer and Walser [77]. Herfen and Bostedt [75] then demonstrated that the modified Mülling Apgar score [70] had only a marginal correlation with blood gas analysis results since 13 of 98 newborn calves had physiological pH values immediately after birth, but their modified Apgar scores were ≤3.

In contrast, Vollhardt [91] was the first to propose the use of five clinical parameters (respiration within 1 min, muscle tone and movement (head raising, extremities), reflex excitability (eyelid and claw reflex), conjunctival color, and suckling reflex) to assess the vitality of newborn calves (Table 5). Subsequently, Schulz and Vollhardt [92] found a strong correlation (r = 0.67) between vitality scores and venous blood pH values, while Gürtler et al. [93] found statistically significant relationships between vitality scores and venous blood pH (r = 0.69) and BE values (r = 0.61), as well as an indirect relationship with lactate levels (r = −0.66), following difficult births.

Table 5. Calf vitality score sheet recommended by Vollhardt [91].

Parameters/Scores	0	1	2
Respiration within 1 min	Absent	Irregular	Spontaneous
Muscle tone and movement (head raising, extremities)	Absent	Reduced	Vigorous
Reflex excitability (Eyelid and claw reflex)	Absent	Present	Very good
Conjunctival color	Bluish-white	White	Pink
Suckling reflex	Absent	Present	Vigorous

Scores 8–10: normal, Scores 5–7: calf at risk; scores < 5: live weak.

Later research conducted by Torres and Gonzales [94] introduced five distinct clinical criteria for evaluating neonatal vitality: responsiveness to exogenic stimuli, time to the raising of the head, sucking reflex, interest in the environment, and time needed to successfully stand up, and found a strong correlation with the acid-base status of newborn calves. A vitality score of 8–10 was considered indicative of a healthy calf, scores 6–7 suggested a calf at risk, and scores ≤ 5 indicated a weak calf that was still alive. Probo et al. [95] recommended the assessment of the following five variables: heart rate and rhythm (absent; irregular rhythm or <100 bpm; ≥100 bpm and regular rhythm), respiratory rate and rhythm (absent; irregular rhythm or <30 rpm; ≥30 rpm and regular rhythm), body tone (atonic, hypertonic, and sternal/active), the color of the mucous membranes of the eyes and mouth (hyperemic/cyanotic; pale; pink), and response to nasal and ear stimulation (absent response, grimace or weak response, and avoidance of stimulation). An index of ≥7 was considered normal. Subsequently, Vannucchi et al. [96] proposed the evaluation of five variables, such as "heart rate (absent; bradycardia/irregular: <120 bpm; normal/regular: 120–220 bpm), respiratory rate and effort (absent; irregular < 35 rpm; regular 35–90 rpm), muscle tone (flaccidity; some flexion; flexion), irritability reflex (absent; some movement; hyperactivity) and mucous color of eyes and gums (cyanotic; pale; normal)". An index of ≥7 was considered normal. Concurrently, Kovács et al. [97] proposed the evaluation of newborn calf vitality through the following parameters: muscle tone (toneless, low, normal); erection of the head (head dropping; head requiring support; erected head); muscle reflexes (limbs extended; reduced number and intensity of reflectory movements; normal reflectory movements); heart rate (absent; bradycardia/irregular < 120 bpm; normal/regular 120–220 bpm), and sucking drive (absent; reduced; intensive). Higher scores indicate greater vigor.

Schuijt and Taverne [98] proposed evaluating newborn calf vitality based on the time from birth to the attainment of sternal recumbency (T-SR). "Calves were vital when they received routine care without medical treatment and survived seven days from birth without any symptoms of illness". Nonvital calves were those who failed to meet the aforementioned criteria. "The mean ± SD T-SR values of the healthy calves were 4.0 ± 2.2 min (born spontaneously), 4.5 ± 3.1 min (C-section), 5.4 ± 3.3 min (usual traction), and 9.0 ± 3.3 min (forced traction), respectively. Calves delivered by forceful extraction had longer T-SR, more severe acidosis, recovered more slowly from acidosis, showed higher mortality, and exhibited trauma more frequently. A moderate correlation was reported between T-SR values and 10-min pH and BE values, while there was a weak correlation between T-SR and pCO_2 values".

It is worth noting that although the interval from birth to sternal recumbency is objective and usually short, it has less practicality during daily practice since, immediately after birth, an assisted calf has to be placed into sternal recumbency to help respiration [99]. Nevertheless, Uystepruyst et al. [99] have also measured the time between birth and sternal recumbency in newborn calves to evaluate the vitality delivered by elective C-section, as suggested by Schuijt and Taverne [98]. Similarly, Barrier et al. [62], Murray et al. [100], and Probo et al. [95] measured the time between birth and sternal recumbency in newborn calves to assess calf vigor.

Mee [101] suggested evaluating newborn calf vitality based on the assessment of the following variables: "the presence of meconium staining, peripheral edema, cyanosis of the mucous membranes, as well as heart and respiration rates, muscle tone, stimulation reflexes, rectal temperature, time to sternal recumbency and attempts to stand and suckle" (Table 6).

Table 6. Calf vitality score sheet recommended by Mee [101].

Criterion	Good Vitality	Poor Vitality
Respiration	50–75 bpm and thoracic breathing	Gasping, primary apnea, irregular, abdominal breathing, bellowing, and secondary apnea
Hair coat appearance	Placental fluid-covered	Meconium-stained
Peripheral edema	None	Capital, lingual, or limb edema
Mucous membranes	Pink and normal capillary refill time	Cyanotic, pale, and slow capillary refill time
Response to reflex stimulation	Vigorous head shake, strong corneal suck, or pedal reflex	Weak or no response
Muscle tone	Active with head-righting within minutes	Inactivity and flaccid musculature
Heart rate	100–150 bpm and regular	>150 bpm followed by bradycardia (<80 bpm) and an irregular, decreasing rate
Rectal temperature	102–103 °F (39–39.5 °C) after calving declining to 101–102 °F (38.5–39 °C) by 1 h and stable	103–104 °F (39.5–40 °C) after calving declining to 101 °F (<38.5 °C) by 1 h and decreasing
Sternal recumbence	Achieved within 5 min	Prolonged lateral recumbence
Attempts to rise	Attempting to stand within 15 min Standing within 1 h	Delayed or no attempts to rise
Suckling	Commences within 2 h	Delayed or no attempts to suckle

Murray [102] worked out a Dairy Calf VIGOR score based on ten physical exam parameters in five categories comprising the acronym VIGOR (Table 7):

"V" (visual appearance): evaluating the presence of meconium staining and the appearance of the tongue and head.

"I" (initiation of movement): detecting time to achieve certain postural behaviors.

"G" (general responsiveness): pricking the nasal mucosa with a straw, pinching the tongue, and touching the eyeball.

"O" (oxygenation): evaluating mucous membrane color and length of protruding tongue.

"R" (rates): measuring heart rate and respiratory rate.

Table 7. Calf VIGOR score sheet recommended by Murray [102].

	Visual Appearance			
Score	**0**	**1**	**2**	**3**
1. Meconium staining	Normal: no staining	Slight: around anal/tail head area	Moderate: extending over body	Severe: completely covered
2. Tongue/Head	Normal (no swelling, tongue not protruding)	Tongue protruding but not swollen	Tongue protruding and swollen	Head and tongue swollen, tongue protruding
	Initiation of Movement			
3. Calf movement Taken within	Standing/walking 0–30 min	Attempts to stand 30 min–1.5 h	Sternal 1.5 h–3 h	On side, no efforts to rise >3 h
	General Responsiveness			
4. Head shake in response to straw in nasal cavity	Shakes head vigorously	Moves head away	Twitches or flinches	Does not respond
5. Tongue pinch	Actively withdraws tongue	Attempts to withdraw	Twitches tongue	Does not respond
6. Eye reflex (in response to touching eyeball)	Actively blinks and closes eye	Slow to blink	Does not respond	-
	Oxygenation			
7. Mucous membrane color	Bright Pink	Light Pink	Brick Red	White/blue
8. Length of tongue *	<50 mm	50–61 mm	>62 mm	-
	Rates			
9. Heart rate **	80–100 bpm	>100 bpm	<80 bpm	-
10. Respiration rate ***	~24–36 rrpm	~24 rrpm	~>36 rrpm	-

* Measure from lips. This measurement was recorded only within 5 min of calving. ** Place hand on the calf's chest. The pulse was recorded for 15 s and then multiplied by 4 to obtain beats per minute (bpm). *** View and/or place hand on the calf's abdomen to count the approximate number of breaths for 15 s and multiply by 4 to get respiration rates per minute (rrpm). A lower score indicates greater vigor.

Villettaz Robichaud et al. [19] recently proposed modifications to the vigor score sheet (Table 7), which included the removal of the calf movement evaluation and the measurement of time to sternal recumbency and tongue length. Conversely, heart rate values were replaced with scores of one and two (Score 1: <80 bpm, Score 2: >100 bpm). "Time to sternal recumbency was treated as a continuous variable and analyzed separately from the vigor score, with shorter time indicating greater calf vigor". However, as mentioned previously, an assisted calf must be placed into sternal recumbency immediately after birth if respiration initiation is delayed [99].

Homerosky et al. [103] examined additional parameters besides the traditional Apgar parameters (heart rate, respiration rate, and mucous membrane color) to identify newborn beef calves with acidosis, which included "meconium staining (visual assessment of hair coat and amniotic fluid: absent/present), tongue withdrawal (visual assessment of rostral aspect of the tongue: normal size within oral cavity/protruding or swollen) and suckle reflex (response to placing two fingers longitudinally in calf's mouth: strong/weak). Nasal prick (reaction to pricking the nasal mucosa with straw: actively shakes head/minimal movement) and corneal reflex (response to touching conjunctiva with forefinger: complete blink/ incomplete blink)" were also evaluated and compared with the venous blood acid–base parameters obtained 10 min after birth. The study found that the traditional Apgar parameters, meconium staining, nasal prick test, and suckle and corneal reflex did not help identify newborn beef calves with acidosis. However, tongue withdrawal, calving ease, and parity may be useful in these assessments.

On dairy farms, where obstetrician assistants are familiar with providing continuous surveillance, a simple vitality score system is necessary for immediate and accurate determination of neonatal vitality without requiring laboratory tests. Thus, immediate treatment can be provided whenever required [104]. The degree of neonatal vitality was evaluated based on muscle tone (active with head-righting within seconds), and in problematic cases, cardiac status was also considered. Newborn calves ($n = 147$) were examined immediately after birth and characterized as follows [104]:

"V-III.: Normal tonicity, head erect, and normal reflectoric movements

V-II.: Low tonicity, abdominal recumbency with head requiring support, and reduced number and intensity of reflectoric movements

V-I.: Toneless, head dropping, limbs extended, and cardiac activity present

V-0.: Toneless, head dropping, limbs extended, and cardiac activity absent [104]"

Significant differences were observed between the vitality scores of newborn calves assessed immediately after birth and their associated acid–base parameters, suggesting that this scoring system can provide valuable insights into the neonatal calf's overall health status without the need for laboratory measurements. This approach allows for timely and appropriate administration of treatment, which is particularly important for Cesarean-derived calves, as these animals may take 2 to 3 min to lift their head and turn from lateral to sternal recumbency. In such cases, potential errors can be excluded when assessing the response of the calf to splashing cold water over its head [78,87] and/or turning them immediately into sternal recumbency [99]. In contrast, Schulz et al. [105] suggested evaluating suckling behavior as a criterion alone or as part of the modified Apgar score.

Despite several attempts to develop practical tools for assessing newborn calf vitality, a highly accurate and easy-to-use method suitable for on-farm use remains elusive [102].

3.2. Measuring Neonatal Acid–Base Values after Birth

Following spontaneous calving or obstetrical assistance with one assistant, a considerable proportion of newborn calves experience physiological metabolic acidosis, ranging from 39.7 to 80% [44–46]. Similarly, after C-section, acidosis is observed in 50% to 63.5% of cases [43,48,106]. The prevalence of moderate metabolic acidosis in newborn calves following spontaneous calving or obstetrical assistance ranges between 20% and 50%, with only a small proportion of cases (0–10.3%) being severe [44–46]. Additionally, after C-section, moderate metabolic acidosis prevalence rates are observed in 23.8% to 34.1% of cases, with severe metabolic acidosis ranging from 12.7% to 22.8% [43,48,106] (refer to Table 8). It is important to note that a C-section is a more cautious procedure than traction, particularly forced traction, which has been confirmed by other studies [43,48,93,107].

At birth, upon initiation of respiration, vasoconstriction ceases, and the accumulated acids enter circulation, resulting in a further decline in pH, and the acid–base variables are observed during the first 10 min postcalving. These metabolic reductions are more pronounced following a traction-assisted delivery [76,108–110] compared to a C-section [48,111]. Furthermore, posterior presentation deliveries exhibit more substantial decreases in metabolic values compared to anterior presentation deliveries [109,112].

In cases of severe respiratory metabolic acidosis, postnatal compensation for metabolic acidosis is relatively slower, taking between 1 and 6 h after birth, while respiratory acidosis may persist for up to 24 and 48 h postpartum [77,93,108,110,113]. Therefore, it is crucial to managing severe metabolic acidosis promptly by administering sodium bicarbonate or carbicarb infusion immediately after delivery [43,114].

Table 8. Prevalence of physiological, moderately, and severely acidotic newborn calves immediately after obstetrical assistance.

Type of Obstetrical Assistance	Site of Blood Sampling before Calving	No. of Examined Calves (n)	pH > 7.2 n (%)	pH 7.2–7.0 n (%)	pH < 7.0 n (%)	Reference
Spontaneous or traction	*v. jugularis*	15	12 (80)	3 (20)	0	Eichler-Steinhauff [44]
Spontaneous or traction	*v. metacarpalis volaris superficial*	20	15 (75)	5 (25)	0	Mülling et al. [45]
Spontaneous or traction	*v. jugularis*	58	23 (39.7)	29 (50) [a]	6 (10.3) [b]	Szenci et al. [46]
Spontaneous or C-section	*v. jugularis*	25	12 (48)	10 (40)	3 (12)	Herfen and Bostedt [74]
Spontaneous, traction, or C-section	*v. jugularis*	98	39 (39.8)	48 (49)	11 (11.2)	Herfen and Bostedt [75]
Spontaneous, traction, or C-section	*v. jugularis*	336	192 (57.1)	119 (35.4)	25 (7.4)	Leister [88]
Spontaneous, traction, or C-section	*v. jugularis*	38	26 (68.4)	12 (31.6)		Bleul et al. [42]
C-section	*v. jugularis*	57	30 (52.6)	14 (24.6)	13 (22.8) [c]	Eigenmann et al. [43]
C-section	*v. jugularis*	44	22 (50)	15 (34.1)	7 (15.9) [d]	Szenci and Taverne [48]
C-section	*v. jugularis*	126	80 (63.5)	30 (23.8)	16 (12.7) [e]	Szenci et al. [106]

[a] One calf died within 48 h of birth (3.4%). [b] Four calves died during calving, while two died within 48 h of birth (100%). [c] Nine calves died within 24 h of birth (69.2%). [d] Two calves died within 24 h of birth (28.6%). [e] Six calves died within 48 h of birth (37.5%).

3.3. Measuring Neonatal Oxygen Saturation after Birth

In human medicine [50], pulse oximetry is widely used to measure the percentage of oxygen saturation of arterial hemoglobin (SpO_2) in a noninvasive and precise manner. Conversely, in bovine practice, the accuracy of a pulse oximeter in newborn calves was first examined by Uystepruyst et al. [115] by comparing SpO_2 and arterial oxyhemoglobin saturation (SaO_2), measured with a blood gas analyzer. Two transmission-type sensors of the pulse oximeter were placed on the proximal region of the tail (one on the dorsal surface and the second on the ventral surface), where the skin was nonpigmented and shaved. The study reported a highly significant correlation (r = 0.87) between $mSpO_2$ (SpO_2 values were recorded for 1 min and averaged) and SaO_2 values [115].

Despite its accuracy and portability, as well as being a noninvasive, easy-to-use, and inexpensive technique suitable for dairy farm use, pulse oximetry cannot provide a precise absolute measurement of SaO_2 in calves after birth due to the overestimation of SpO_2 values when SaO_2 values are lower. Additionally, its accuracy may be affected by several variables, including the pulse oximeter device, type of transducer, site of measurement, tissue perfusion, pigmentation of the site, ambient light (i.e., infrared heat lamps), or animal movement [115]. Despite these limitations, the use of pulse oximetry enables the objective evaluation of pulmonary function effectiveness in newborn calves during their transition to extrauterine life [115].

Kanz et al. [116] examined the accuracy of a pulse oximeter (Radius-7) after calving. The sensor was placed in the interdigital space of the front legs of the calves and fixed using a custom-made latex hoof cover. SpO_2 values were compared with arterial SaO_2 values measured using a portable blood gas analyzer as a reference, while pulse rates were measured and compared to the heart rate belt reference. The Spearman correlation coefficients for SpO_2 and pulse rate were 93.8% and 97.7%, respectively. However, these results could be considered valid only for motionless calves with sternal recumbency. Therefore, purpose-built equipment is required for dairy practice.

3.4. Measuring Neonatal Lactate Concentration after Birth

Burfeind and Heuwieser [33] conducted a study to validate the accuracy of a hand-held meter (Lactate Scout) in measuring L-lactate concentration in newborn calves (age: 17 ± 12 days). They found a high correlation (r = 0.98) between the measurements obtained using the Lactate Scout and those obtained through reference laboratory methods.

In another study, Homerosky et al. [103] found a strong negative correlation (r = −0.86) between L-lactate concentration and blood pH in neonatal calves 10 min after birth using a Lactate-Pro handheld meter. The authors concluded that lactate meters are practical and should be strongly considered for on-farm use, as they provide accurate estimates of blood pH. Similarly, Bleul and Götz [31] found a high correlation (r = 0.984) between blood L-lactate and pH (r > 0.95) measured by i-STAT and the reference laboratory methods in blood samples withdrawn immediately after birth. However, in 12 out of 46 newborn calves, the i-STAT lactate concentrations around the reference limit of 20 mmol/L could not be confirmed by measuring blood pH and BE values. Therefore, it was emphasized that there is a need for cautious interpretation of high L-lactate concentration due to suboptimal agreement between the higher L-lactate and reference laboratory levels. Simultaneously, if these calves were treated, they would not have had a detrimental effect since they also suffered from metabolic acidosis. Sorge et al. [88] reported that 8 out of 281 newborn calves had blood L-lactate concentrations of up to 19.8 mmol/L (Lactate-Pro) within 5 min after birth, despite a maximum Apgar score of eight. Therefore, in doubtful cases, it is suggested to confirm L-lactate concentration with the measurement of a new blood sample [103]. Although hand-held devices allow rapid, reliable, and accurate point-of-care blood analysis in dairy farms due to the high variability of L-lactate values, determination of the cut-off values is also warranted.

4. Future Perspectives

Currently, the main emphasis in obstetrical assistance must be placed on the prevention of asphyxia due to the lack of instruments that can reliably clear respiratory passages, maintain this state, and perform artificial respiration, as well as the insufficient competency-based skill of calving assistants to manage artificial respiration in the field. Although a calf aspirator/resuscitator for the suction of bronchial secretions is already available in bovine practice [1,117], intrauterine hypoxia during obstetrical assistance may develop depending on the degree of metabolic acidosis, organ injuries (hemorrhage), or meconium aspiration, which may increase the prevalence of neonatal mortality [118,119].

Reducing the need for calving assistance is the most critical breeding objective, particularly since calving assistance may shift the fetal acid–base balance toward acidosis [1]. However, in many cases, there are no visible clinical signs of calving onset, making it challenging to recognize. Using different sensors to predict the onset of calving may contribute to reducing stillbirth, delayed calving assistance, and its consequences in the field [2]. In cases of dystocia, the mode (traction or C-section) and the time of calving assistance should be chosen considering profitability factors and in a manner that minimizes the fetal acid–base balance towards acidosis [1,27,120]. Obstetric assistance should be initiated within 70 min after the appearance of the amniotic sac or 65 min after the appearance of hooves in the vulva [121]. Early intervention (within 15 min after the first sight of both front hooves in the vulva) does not necessarily increase the stillbirth rates, although the prevalence of injuries to the soft birth canal and retained fetal membranes can be increased [122]. Similarly, Villettaz Robichaud et al. [19] could not confirm the negative effect of early intervention on the stillbirth rate; nevertheless, sterile obstetrical lubricant was applied to the dam's soft birth canal around the fetus before providing obstetrical assistance. Concurrently, the prevalence of injuries to the soft birth canal and retained fetal membranes has not been reported. Delayed obstetrical assistance can increase the rate of severely acidotic calves (pH < 7.0) (<2 h: 0%, 2–4 h: 19%, 4–7 h: 44%) [39]. Furthermore, while Vannucchi et al. [123] reported normal mean metabolic parameters even when the duration of calving was longer than 4 h, significant respiratory depression (altered lung gas exchange and delayed lung clearance) has also been reported [124].

Prior to conducting any obstetric traction, it is crucial to assess the degree of dilatation of the soft birth canal. If the dilatation is found to be insufficiently dilated, nonsurgical or surgical expansion techniques, such as episiotomy lateralis, must be performed [1]. Obstetric lubricants should also be utilized [18,19] to prevent tractions exceeding 2–3 min [46], as excessive traction can lead to rib and vertebral fractures [125].

A C-section should be performed if prolonged traction is expected to preserve the calf's life and prevent maternal birth canal injuries. Recent research suggests that prior to deciding on the mode of calving assistance, the results of the acid–base balance or L-lactate measurements from blood samples should be considered [103,120]. The routinely applied complex treatment for asphyxiated newborn calves, which involves the initiation of respiration through physical stimulation or respiratory stimulants, providing oxygen/air supplementation, alleviation of pain and inflammation following dystocia, compensation of acidosis through buffer therapy, ensuring thermal support, and administering umbilical treatment, may reduce postnatal calf losses [36,101,117,126]. In addition to adequate treatment, special attention should be paid to the consumption and absorption of sufficient amounts of colostrum in asphyxiated newborn calves, as the lack of colostrum uptake is accompanied by an increased susceptibility to gastrointestinal disorders [19,37,68,100,127–129].

5. Conclusions

Several diagnostic methods are available to evaluate the clinical signs of fetal/neonatal vitality during and after calving, as well as to measure the acid–base balance or L-lactate concentrations. The use of ultrasonography to detect heart rate, or pulse oximetry to continuously measure fetal/neonatal oxygen saturation of arterial hemoglobin and heart rate on dairy or beef farms was observed. The implementation of these methods may contribute to recognizing and eliminating threats to the fetal/neonatal calf's vitality in a timely manner. Hence, it is essential for farm management to select and apply these methods, considering current economic aspects, to prevent damage caused by dystocia, which frequently contributes to fetal/neonatal mortality. Nevertheless, even in today's circumstances, primary emphasis must be placed on prevention, with medical treatments being a secondary consideration.

Funding: This research received no external funding.

Institutional Review Board Statement: Not applicable.

Informed Consent Statement: Not applicable.

Data Availability Statement: Data sharing not applicable.

Conflicts of Interest: The author declares no conflict of interest.

References

1. Szenci, O. Diagnosis of prerinatal well-being of dairy calves. *Lucr. Ştiinţifice—Ser. Zooteh.* **2012**, *57*, 3–11.
2. Szenci, O. Accuracy to predict the onset of calving in dairy farms by using different precision livestock farming devices. *Animals* **2022**, *12*, 2006. [CrossRef] [PubMed]
3. Szenci, O.B.; Kiss, M. Perinatal calf losses in large cattle production units. *Acta Vet. Hung.* **1982**, *30*, 85–95.
4. Mee, J.F. Perinatal calf mortality—Recent findings. *Ir. Vet. J.* **1991**, *44*, 80–85.
5. Bleul, U. Respiratory distress syndrome in calves. *Vet. Clin. Food Anim.* **2009**, *25*, 179–193. [CrossRef]
6. Mee, J.F. Why do so many calves die on modern dairy farms and what can we do about calf welfare in the future? *Animals* **2013**, *3*, 1036–1057. [CrossRef]
7. Cuttance, E.; Laven, R. Perinatal mortality risk factors in dairy calves. *Vet. J.* **2019**, *253*, 105394. [CrossRef]
8. Silva del Río, N.; Stewart, S.; Rapnicki, P.; Chang, Y.M.; Fricke, P.M. An observational analysis of twin births, calf sex ratio, and calf mortality in Holstein dairy cattle. *J. Dairy Sci.* **2007**, *90*, 1255–1264. [CrossRef]
9. Mee, J.F. Bovine perinatology: Current understanding and future developments. In *Animal Reproduction: New Research Developments*; Dahnof, L.T., Ed.; Nova Science Publishers, Inc.: Hauppauge, NY, USA, 2009; pp. 67–106.
10. Hoedemaker, M.; Ruddat, I.; Teltscher, M.; Essmeyer, K.; Kreienbrock, L. Influence of animal, herd, and management factors on perinatal mortality in dairy cattle—A survey in Thuringia, Germany. *Berl. Münch. Tierärztl. Wochenschr.* **2010**, *123*, 130–136.
11. Gustafsson, H.; Kornmatitsuk, B.; Königsson, K.; Kindahl, H. Peripartum and early post partum in the cow—Physiology and pathology. *Le Médecin Vétérinaire Du Québec* **2004**, *34*, 64–65.
12. Meyer, C.L.; Berger, P.J.; Koehler, K.J. Interactions among factors affecting stillbirths in Holstein cattle in the United States. *J. Dairy Sci.* **2000**, *83*, 2657–2663. [CrossRef]

13. Berglund, B.; Steinbock, L.; Elvander, M. Causes of stillbirth and time of death in Swedish Holstein calves examined post mortem. *Acta Vet. Scand.* **2003**, *44*, 111–120. [CrossRef]

14. Meyer, C.L.; Berger, P.J.; Koehler, K.J.; Thompson, J.R.; Sattler, C.G. Phenotypic trends in incidence of stillbirth for Holsteins in the United States. *J. Dairy Sci.* **2001**, *84*, 515–523. [CrossRef]

15. Steinbock, L.; Näsholm, A.; Berglund, B.; Johansson, K.; Philipsson, J. Genetic effects on stillbirth and calving difficulty in Swedish Holsteins at first and second calving. *J. Dairy Sci.* **2003**, *86*, 2228–2235. [CrossRef]

16. Harbers, A.; Segeren, L.; De Jong, G. Genetic parameters for stillbirth in the Netherlands. *Interbull Bull.* **2000**, *25*, 117–122.

17. Mee, J.F. Epidemiology of bovine perinatal mortality. In *Bovine Prenatal, Perinatal and Neonatal Medicine*; Szenci, O., Mee, J.F., Bleul, U., Taverne, M.A.M., Eds.; Hungarian Association for Buiatrics: Budapest, Hungary, 2021; pp. 224–232.

18. Niles, D. The modern dairy maternity ward. *Magy. Állatorvosok Lapja* **2016**, *138* (Suppl. 1), 275–279.

19. Villettaz Robichaud, M.; Pearl, D.L.; Godden, S.M.; LeBlanc, S.J.; Haley, D.B. Systematic early obstetrical assistance at calving: I. Effects on dairy calf stillbirth, vigor, and passive immunity transfer. *J. Dairy Sci.* **2017**, *100*, 691–702. [CrossRef]

20. Mee, J.F. Investigation of bovine abortion and stillbirth/perinatal mortality—Similar diagnostic challenges, different approaches. *Ir. Vet. J.* **2020**, *73*, 20. [CrossRef]

21. Mee, J.F.; Jawor, P.; Stefaniak, T. Role of infection and immunity in bovine perinatal mortality: Part 1. Causes and current diagnostic approaches. *Animals* **2021**, *11*, 1033. [CrossRef]

22. Jawor, P.; Mee, J.F.; Stefaniak, T. Role of infection and immunity in bovine perinatal mortality: Part 2. Fetomaternal response to infection and novel diagnostic perspectives. *Animals* **2021**, *11*, 2102. [CrossRef]

23. Hahnsdorf, A. Die wichtigsten Todesursachen des Kalbes Anhand der Sektionstatistik mit Besonderer Berücksichtigung der Pathologisch-Anatomischen Veränderungen. Inaugural Dissertation, Justus Liebig-Universität Giessen, Veterinärmedizinische Fakultät, Giessen, Germany, 1967.

24. Greene, H.J. Causes of dairy calf mortality. *Ir. J. Agric. Res.* **1978**, *17*, 295–301.

25. Schuijt, G. Aspects of Obstetrical Perinatology in Cattle. Ph.D. Thesis, Faculty of Veterinary Medicine, Utrecht University, Utrecht, The Netherlands, 1992.

26. Mee, J.F. Aetiology and timing of bovine perinatal mortality. In *Bovine Prenatal, Perinatal and Neonatal Medicine*; Szenci, O., Mee, J.F., Bleul, U., Taverne, M.A.M., Eds.; Hungarian Association for Buiatrics: Budapest, Hungary, 2021; pp. 233–245.

27. Szenci, O.; Van Der Weyden, G.C.; Lénárt, L.; Taverne, M.A.M. Changes of acid-base variables in dairy cows with chronically implanted fetal and maternal catheters during late gestation and calving. *Animals* **2022**, *12*, 2448. [CrossRef] [PubMed]

28. Dawes, G.S. *Foetal and Neonatal Physiology*; Year Book Medical Publishers: Chicago, IL, USA, 1968.

29. Dufty, J.H.; Sloss, V. Anoxia in the bovine foetus. *Aust. Vet. J.* **1977**, *53*, 262–267. [CrossRef]

30. Erickson, K.A.; Wilding, P. Evaluation of a novel point-of-care system, the i-STAT portable clinical analyzer. *Clin. Chem.* **1993**, *39*, 283–287. [CrossRef] [PubMed]

31. Bleul, U.; Götz, E. Evaluation of the i-STAT portable point-of-care analyzer for determination of blood gases and acid-base status in newborn calves. *J. Vet. Emerg. Crit. Care* **2014**, *24*, 519–528. [CrossRef]

32. Homerosky, E.R. Assessment and Impacts of Newborn Beef Calf Vigour. Ph.D. Thesis, University of Calgary, Calgary, AB, Canada, 2016.

33. Burfeind, O.; Heuwieser, W. Validation of handheld meters to measures blood L-lactate concentration in dairy cows and calves. *J. Dairy Sci.* **2012**, *95*, 6449–6456. [CrossRef]

34. Karapinar, T.; Kaynar, O.; Hayirli, A.; Kom, M. Evaluation of 4 point-of-care units for the determination of blood L-lactate concentration in cattle. *J. Vet. Intern. Med.* **2013**, *27*, 1596–1603. [CrossRef]

35. Szenci, O.; Mee, J.F.; Bleul, U.; Taverne, M.A.M. (Eds.) *Bovine Prenatal, Perinatal and Neonatal Medicine*; Hungarian Association for Buiatrics: Budapest, Hungary, 2021.

36. Murray, C.F.; Leslie, K.E. Newborn calf vitality: Risk factors, characteristics, assessment, resulting outcomes and strategies for improvement. *Vet. J.* **2013**, *198*, 322–328. [CrossRef]

37. Mota-Rojas, D.; López, A.; Martínez-Burnes, J.; Muns, R.; Villanueva-García, D.; Mora-Medina, P.; González-Lozano, M.; Olmos Hernández, A.; Ramírez-Necoechea, R. Is vitality assessment important in neonatal animals? *CAB Rev.* **2018**, *13*, 1–13. [CrossRef]

38. Verdon, M. A review of factors affecting the welfare of dairy calves in pasture-based production systems. *Anim. Prod. Sci.* **2022**, *62*, 1–20. [CrossRef]

39. Held, T. Klinische und Blutgasanalytische Untersuchungen bei Kalbenden Rindern und Deren FETEN. Inaugural Dissertation, Tierärzliche Hochschule Hannover, Hannover, Germany, 1983.

40. Breukelman, S.; Mulder, E.J.H.; van Oord, R.; Jonker, H.; van der Weijden, G.C.; Taverne, M.A.M. Continuous fetal heart rate monitoring during late gestation in cattle by means of Doppler ultrasonography: Reference values obtained by computer-assisted analysis. *Theriogenology* **2006**, *65*, 486–498. [CrossRef]

41. Kanz, P.; Gusterer, E.; Krieger, S.; Schweinzer, V.; Süss, D.; Drillich, M.; Iwersen, M. Pulsoximetric monitoring of fetal arterial oxygen saturation and fetal pulse at stage II of labor to predict acidosis in newborn Holstein Friesian calves. *Theriogenology* **2020**, *142*, 303–309. [CrossRef]

42. Bleul, U.; Schwantag, S.; Kähn, W. Blood gas analysis of bovine fetal capillary blood during stage II labor. *Theriogenology* **2008**, *69*, 245–251. [CrossRef]

43. Eigenmann, U.J.E.; Grunert, E.; Born, E. Untersuchunger über den Einfluss der Schnittentbindung auf den Säurenbasenhaushalt sowie die Plasmaglukosekonzentration neugeborener Kälbern. *Dtsch. Tierärztl. Wschr.* **1981**, *88*, 433–437.

44. Eichler-Steinhauff, H.J. Untersuchungen über den Aktuellen pH-Werte und die Blutgasverhältnisse im Venösem Blut von Kälbern Während des Geburtsablaufes. Inaugural Dissertation, Fakultät für Veterinärmedizin, Freien Universität, Berlin, Germany, 1977.

45. Mülling, M.; Waizenhöfer, H.; Bratig, B. Glucose-, Laktat- und pH_{akt}-Werte bei Kühen und Kalbern wahrend und unmittelbar nach der Geburt. *Berl. Münch. Tierärtzl. Wschr.* **1979**, *92*, 111–117.

46. Szenci, O.; Taverne, M.A.M.; Bakonyi, S.; Erdődi, A. Comparison between pre- and postnatal acid-base status of calves and their perinatal mortality. *Vet. Q.* **1988**, *10*, 140–144. [CrossRef] [PubMed]

47. Held, T.; Eigenmann, U.J.E.; Grunert, E. Blutgasanalytische Untersuchungsbefunde bei bovinen Feten im Aufweitungsstadium der Geburt. *Mh. Vet. Med.* **1985**, *40*, 405–409.

48. Szenci, O.; Taverne, M.A.M. Perinatal blood gas and acid-base status of Caesarean-derived calves. *J. Vet. Med. A* **1988**, *35*, 572–577. [CrossRef]

49. Amer-Wahlin, I.; Kwee, A. Combined cardiotocographic and ST event analysis: A review. *Best Pract. Res. Clin. Obstet.* **2016**, *30*, 48–61. [CrossRef]

50. Nonnenmacher, A.; Hartmut Hopp, H.; Dudenhausen, J. Predictive value of pulse oximetry for the development of fetal acidosis. *J. Perinat. Med.* **2010**, *38*, 83–86. [CrossRef]

51. Bleul, U.; Kähn, W. Monitoring the bovine fetus during stage II of parturition using pulse oximetry. *Theriogenology* **2008**, *69*, 302–311. [CrossRef]

52. Iorizzo, L.; Carlsson, Y.; Johansson, C.; Berggren, R.; Herbst, A.; Wang, M.; Leiding, M.; Isberg, P.-E.; Kristensen, K.; Wiberg-Itzel, E.; et al. Proposed cutoff for fetal scalp blood lactate in intrapartum fetal surveillance based on neonatal outcomes: A large prospective observational study. *BJOG* **2022**, *129*, 636–646. [CrossRef]

53. Jonker, F.H.; van Oord, H.A.; van der Weijden, G.C.; Taverne, M.A.M. Fetal heart rate and the influence of myometrial activity during the last month of gestation in cows. *Am. J. Vet. Res.* **1993**, *54*, 158–163.

54. Jonker, F.H.; van Oord, H.A.; van Geijn, H.P.; van der Weijden, G.C.; Taverne, M.A.M. Feasibility of continuous recording of fetal heart rate in the near term bovine fetus by means of transabdominal Doppler. *Vet. Q.* **1994**, *16*, 165–168. [CrossRef]

55. Held, T.; Scheidegger, A.; Grunert, E. Kardiotokographische Befunde bei lebensfrischen und asphyktischen Rinderfeten während der Aufweitungsphase der Geburt. *Zbl. Vet. Med.* **1986**, *33*, 431–442. [CrossRef]

56. Jonker, F.H.; Taverne, M.A.M.; van der Weijden, G.C. Cardiotocography in cows: A method for monitoring calves during delivery. *Theriogenology* **1989**, *31*, 425–436. [CrossRef]

57. Jonker, F.H.; van Geijn, H.P.; Chan, W.W.; Rausch, W.D.; van der Weijden, G.C.; Taverne, M.A.M. Characteristics of fetal heart rate changes during the expulsive stage of bovine parturition in relation to fetal outcome. *Am. J. Vet. Res.* **1996**, *57*, 1373–1381. [PubMed]

58. Bleul, U.; Lejeune, B.; Schwantag, S.; Kähn, W. Ultrasonic transit-time measurement of blood flow in the umbilical arteries and veins in the bovine fetus during stage II of labor. *Theriogenology* **2007**, *67*, 1123–1133. [CrossRef]

59. Amman, H.; Berchtold, M.; Schneider, F. Blutgas- und Säurenbasenverthältnisse bei normalen und asphyktischen Kälbern. *Berl. Münch. Tierärtzl. Wschr.* **1974**, *87*, 66–68.

60. Kovalcik, K.; Kovalcikova, M.; Brestensky, V. Comparison of the behaviour of newborn calves housed with the dam and in the calf-house. *Appl. Anim. Ethol.* **1980**, *6*, 377–380. [CrossRef]

61. Owens, J.L.; Edey, T.N. Parturient behaviour and calf survival in a herd selected for twinning. *Appl. Anim. Behav. Sci.* **1985**, *13*, 312–333. [CrossRef]

62. Barrier, A.C.; Ruelle, E.; Haskell, M.J.; Dwyer, C.M. Effect of a difficult calving on the vigour of the calf, the onset of maternal behaviour, and some behavioural indicators of pain in the dam. *Prev. Med.* **2012**, *103*, 248–256. [CrossRef] [PubMed]

63. Campler, M.; Munksgaard, L.; Jensen, M.B. The effect of housing on calving behavior and calf vitality in Holstein and Jersey dairy cows. *J. Dairy Sci.* **2015**, *98*, 1797–1804. [CrossRef]

64. Hartwig, K.H. Untersuchungen über Beginn und Verlauf der Geburt beim Rind unter Besonderer Berücksichtigung der Zeitlichen Abläufe und Deren Beeinflussung Durh ein β_2-Mimetikum. Inaugural Dissertation, Tierärzliche Hochschule Hannover, Hannover, Germany, 1983.

65. Edwards, S.A.; Broom, D.M. The period between birth and first suckling in dairy calves. *Res. Vet. Sci.* **1979**, *26*, 255–256. [CrossRef] [PubMed]

66. Edwards, S.A. Factors affecting the time to first suckling in dairy calves. *Anim. Prod.* **1982**, *34*, 339–346. [CrossRef]

67. Fraser, A.F. A monitored study of major physical activities in the perinatal calf. *Vet. Rec.* **1989**, *125*, 38–40. [CrossRef]

68. Boyd, J.W. Relationships between acid-base balance, serum composition and colostrum absorption in newborn calves. *Br. Vet. J.* **1989**, *145*, 249–256. [CrossRef]

69. Apgar, V. A proposal for a new method of evaluation of the newborn infant. *Anesth. Anal. Curr. Res.* **1953**, *32*, 260–267. [CrossRef]

70. Mülling, M. Asphyxie des neugeborenen Kalbes. *Prakt. Tierarzt* **1976**, *58*, 78–80.

71. Köppe, U. Blutgas- und Säure-Basen-Werte bei Vorzeitig und Termingerecht Entwickelten Kälbern in den Ersten 24 Lebensstunden. Inaugural Dissertation, Tierärzliche Hochschule Hannover, Hannover, Germany, 1980.

72. Eigenmann, U.J.E.; Rüdiger, B.; Schoon, H.A.; Grunert, E. Natriumbikarbonat- und Glukosebehandlung bei der Asphyxie des Kalbes. *Dtsch. Tierärtzl. Wschr.* **1982**, *89*, 228–234.

73. Schäfer, S.; Arbeiter, K. Relations between parturition, vitality and some blood parameters of the neonatal calf. *Wien. Tierärztliche Mon.* **1995**, *82*, 271–275.

74. Herfen, K.; Bostedt, H. Acid-base status in newborn calves during the first days of life considering different states of vitality. *Berl. Münch. Tierärztl. Wochenschr.* **1999**, *112*, 166–171.
75. Herfen, K.; Bostedt, H. Correlation between clinical and laboratory diagnostic evaluation of the vitality of newborn calves under particular consideration of length and type of parturition. *Wien. Tierarztl. Mon.* **1999**, *86*, 255–261.
76. Mauer-Schweizer, H.; Wilhelm, U.; Walser, K. Blutgas- und Säuren-Basen-Verhältnisse bei lebensfrischen Kälbern in den ersten 24 Lebensstunden. *Berl. Münch. Tierärtzl. Wschr.* **1977**, *90*, 192–196.
77. Mauer-Schweizer, H.; Walser, K. Azidose and klinischer Zustand bei asphyktischen Kälbern. *Berl. Münch. Tierärtzl. Wschr.* **1977**, *90*, 364–366.
78. Bodenberger, B. Untersuchungen zum Kohlenhydratstoffwechsel Lebensfrischer und Asphyktischer Neugeborener Kälber (Glukose- und Laktatkonzentrationen im Venösen Blut). Inaugural Dissertation, Ludwig-Maximilians-Universität, Tierärztliche Fakultät, München, Germany, 1979.
79. Palmer, J. Perinatal Care of Cloned Calves. Available online: http://nicuvet.com/nicuvet/Equine-Perinatoloy/Web_slides_meetings/ACVIM%202005/PERINATAL%20CARE%20OF%20CLONED%20CALVES.pdf (accessed on 1 December 2022).
80. Born, E. Untersuchungen über den Einfluss der Schnittenbindung auf die Vitalität Neugeborener Kälber. Inaugural Dissertation, Tierärzliche Hochschule Hannover, Hannover, Germany, 1981.
81. Luetgebrune, K. Untersuchungen über die Kolostrumaufnahme und die Immunglobulinabsorption bei Asphyktischen und Lebensfrischen Kälbern. Inaugural Dissertation, Tierärzliche Hochschule Hannover, Hannover, Germany, 1982.
82. Rüdiger, B. Behandlung Asphyktischer Kälber mit Natriumbikarbonat and Glukose. Inaugural Dissertation, Vetsuisse-Faculty University Zurich, Zurich, Switzerland, 1982.
83. Dening, M. Klinische und labordiagnostische Untersuchungen bei Rinderfeten und Neugeborenen Kälbern von Spontan Abkalbenden Primi- und Pluriparen Muttertieren. Inaugural Dissertation, Tierärzliche Hochschule Hannover, Hannover, Germany, 1986.
84. Zaremba, W.; Grunert, E. Zur Asphyxie des Kalbes. *Prakt. Tierarzt.* **1985**, *67*, 17–23.
85. Schmidt, B.R. Vergleich von Arteriellen und Venösen Blutgas- und Säure-Basen-Werten bei Termingerecht Geborenen, Lebensfrischen and Asphyktischen Kälbern. Inaugural Dissertation, Tierärzliche Hochschule Hannover, Hannover, Germany, 1986.
86. Hoyer, C. Die Bedeutung von Stressfaktoren auf den Gesundheits- und Hormonstatus Termingerecht Geborener Kälber in den Ersten sechs Lebenstagen unter Besonderer Berücksichtigung der Asphyxie. Inaugural Dissertation, Tierärzliche Hochschule Hannover, Hannover, Germany, 1988.
87. Zaremba, W.; Guterbock, W.M.; Holmberg, C.A. Efficacy of a dried colostrum powder in the prevention of disease in neonatal Holstein calves. *J. Dairy Sci.* **1993**, *76*, 831–836. [CrossRef]
88. Leister, T.M. Untersuchungen zur Vitalität Neugeborener Kälber in einer Milchviehanlage in Brandenburg bei Optimiertem Geburtsmanagement. Inaugural Dissertation, Tierärzliche Hochschule Hannover, Hannover, Germany, 2009.
89. Essmeyer, K. Aufklärung der Ursachen einer Erhöhten Häufigkeit von Totgeburten in einem Milchviehbetrieb. 2006. Available online: http://elib.tiho-hannover.de/dissertations/essmeyerk_ss06.pdf (accessed on 14 August 2008).
90. Sorge, U.; Kelton, D.; Staufenbiel, R. Neonatal blood lactate concentration and calf morbidity. *Vet. Rec.* **2009**, *164*, 533–534. [CrossRef]
91. Vollhardt, W. Erfassung, Bewertung und Beeinflussung von Kriterien für die Vitalität Neugeborener Kälber. Veterinärmedizinischen Dissertation, Karl-Marx-Universität, Sektion Tierproduktion, Leipzig, Germany, 1983.
92. Schulz, J.; Vollhardt, W. Vitalitätskriterien für Neugeborene Kälber. *Mh. Vet. Med.* **1983**, *38*, 62–64.
93. Gürtler, H.; Zelfel, C.; Schulz, J.; Beyreiss, K. Säure-Basen-Haushalt und Stoffwechselparameter im Blutplasma von Kühen und deren Kälbern in Abhängigkeit vom Geburtsverlauf. *Mh. Vet. Med.* **1989**, *44*, 442–447.
94. Torres, O.; Gonzales, M. Eine Methode zur klinischen Beurteilung des neugeborenen Kalbes. *Mh. Vet. Med.* **1987**, *42*, 27–28.
95. Probo, M.; Giordano, A.; Moretti, P.; Opsomer, G.; Fiems, L.O.; Veronesi, M.C. Mode of delivery is associated with different hematological profiles in the newborn calf. *Theriogenology* **2012**, *77*, 865–872. [CrossRef]
96. Vannucchi, C.I.; Rodrigues, J.A.; Silva, L.C.G.; Lúcio, C.F.; Veiga, G.A.L. Effect of dystocia and treatment with oxytocin on neonatal calf vitality and acid-base, electrolyte and haematological status. *Vet. J.* **2015**, *203*, 228–232. [CrossRef]
97. Kovács, L.; Kézér, F.L.; Albert, E.; Ruff, F.; Szenci, O. Seasonal and maternal effects on acid-base, L-lactate, electrolyte, and hematological status of 205 dairy calves born to eutocic dams. *J. Dairy Sci.* **2017**, *100*, 7534–7543. [CrossRef] [PubMed]
98. Schuijt, G.; Taverne, M.A.M. The interval between birth and sternal recumbency as an objective measure of the vitality of newborn calves. *Vet. Rec.* **1994**, *135*, 111–115. [CrossRef] [PubMed]
99. Uysterpruyst, C.; Coghe, J.; Dorts, T.; Harmegnies, N.; Delsemme, M.H.; Lekeux, P. Sternal recumbency or suspension by the hind legs immediately after delivery improves respiratory and metabolic adaptation to extra uterine life in newborn calves delivered by caesarean section. *Vet. Res.* **2002**, *33*, 709–724. [CrossRef]
100. Murray, C.F.; Veira, D.M.; Nadalin, A.L.; Haines, D.M.; Jackson, M.L.; Pearl, D.L.; Leslie, K.E. The effect of dystocia on physiological and behavioral characteristics related to vitality and passive transfer of immunoglobulins in newborn Holstein calves. *Can. J. Vet. Res.* **2015**, *79*, 109–119. [PubMed]
101. Mee, J.F. Managing the calf at calving time. In Proceedings of the Forty-First Annual Conference, American Association of Bovine Practitioners, Charlotte, NC, USA, 25–27 September 2008; pp. 46–53.
102. Murray, C. Characteristics, Risk Factors and Management Programs for Vitality of Newborn Dairy Calves. Ph.D. Thesis, University of Guelph, Guelph, ON, Canada, 2014.

103. Homerosky, E.R.; Caulkett, N.A.; Timsit, E.; Pajor, E.A.; Kastelic, J.P.; Windeyer, M.C. Clinical indicators of blood gas disturbances, elevated L-lactate concentration and other abnormal blood parameters in newborn beef calves. *Vet. J.* **2017**, *219*, 49–57. [CrossRef]

104. Szenci, O. Correlations between muscle tone and acid-base balance in newborn calves: Experimental substantiation of a simple new score system proposed for neonatal status diagnosis. *Acta Vet. Hung.* **1982**, *30*, 79–84.

105. Schulz, J.; Plischke, B.; Braun, H. Saug- und Trinkverhalten als Vitalitäts Kriterium bei Neugeborener Kalbern. *Tierarztl. Prax.* **1997**, *25*, 116–122.

106. Szenci, O.; Taverne, M.A.M.; Takács, E. A review of 126 Caesarean sections by blood gas and the acid-base status of newborn calves. *Theriogenology* **1989**, *32*, 667–673. [CrossRef]

107. Eigenmann, U.J.E. Der Einfluss geburtshilflicher Massnahmen auf die Lebensfähigkeit neugeborener Kälber. *Prakt. Tierarzt.* **1981**, *62*, 933–942.

108. Schlerka, G.; Petschenig, W.; Jahn, J. Untersuchungen über die Blutgase, den Säure-Basen-Haushalt, Elektrolytgehalt, einige Enzyme und Inhaltsstoffe im Blut neugeborener Kälber. *Dtsch. Tierärtzl. Wschr.* **1979**, *86*, 85–100.

109. Szenci, O.; Tőrös, I.; Sári, A. Changes of acid-base balance in Holstein-Friesian calves during the first two postnatal days. *Acta Vet. Hung.* **1981**, *29*, 143–151.

110. Szenci, O. Role of acid-base disturbances in perinatal mortality of calves (A summary of thesis). *Acta Vet. Hung.* **1985**, *33*, 205–220.

111. Mauer-Schweizer, H.; Wilhelm, U.; Walser, K. Blutgas- und Säuren-Basen-Verhältnisse bei lebensfrischen Kaiserschnittkälbern in den ersten 24 Lebensstunden. *Berl. Münch. Tierärtzl. Wschr.* **1977**, *90*, 215–218.

112. Szenci, O. Effects of type and intensity of assistance on acid-base balance of newborn calves. *Acta Vet. Hung.* **1983**, *31*, 73–79.

113. Kudlac, E.; Schulz, J.; Vedral, V.; Vollhardt, V. Metabolic profile in newborn calves and immunoglobulin levels in the first days of their life (in Czech). *Vet. Med.* **1983**, *28*, 401–412.

114. Bleul, U.; Bachofner, C.; Stocker, H.; Hässig, M.; Braun, U. Comparison of sodium bicarbonate and carbicarb for the treatment of metabolic acidosis in newborn calves. *Vet. Rec.* **2005**, *156*, 202–206. [CrossRef]

115. Uystepruyst, C.; Coghe, J.; Bureau, F.; Lekeux, P. Evaluation of accuracy of pulse oximetry in newborn calves. *Vet. J.* **2000**, *159*, 71–76. [CrossRef]

116. Kanz, P.; Krieger, S.; Drillich, M.; Iwersen, M. Technical note: Evaluation of a wireless pulse oximeter for measuring arterial oxygen saturation and pulse rate in newborn Holstein Friesian calves. *J. Dairy Sci.* **2018**, *101*, 6437–6442. [CrossRef] [PubMed]

117. Bleul, U. Care of the asphyxic perinate. In *Bovine Prenatal, Perinatal and Neonatal Medicine*; Szenci, O., Mee, J.F., Bleul, U., Taverne, M.A.M., Eds.; Hungarian Association for Buiatrics: Budapest, Hungary, 2021; pp. 199–202.

118. Mota-Rojas, D.; Villanueva-García, D.; Solimano, A.; Muns, R.; Ibarra-Ríos, D.; Mota- Reyes, A. Pathophysiology of perinatal asphyxia in humans and animal models. *Biomedicines* **2022**, *10*, 347. [CrossRef] [PubMed]

119. Mota-Rojas, D.; Villanueva-García, D.; Mota-Reyes, A.; Orihuela, A.; Hernández-Ávalos, I.; Domínguez-Oliva, A.; Casas-Alvarado, A.; Flores-Padilla, K.; Jacome-Romero, J.; Martínez-Burnes, J. Meconium aspiration syndrome in animal models: Inflammatory process, apoptosis, and surfactant inactivation. *Animals* **2022**, *12*, 3310. [CrossRef]

120. Szenci, O. Role of acid-base disturbances in perinatal mortality of calves: Review. *Vet. Bull.* **2003**, *73*, 7R–14R. [CrossRef]

121. Schuenemann, G.M.; Nieto, I.; Bas, S.; Galvão, K.N.; Workman, J. Assessment of calving progress and reference times for obstetric intervention during dystocia in Holstein dairy cows. *J. Dairy Sci.* **2011**, *94*, 5494–5501. [CrossRef]

122. Kovács, L.; Kézér, F.L.; Szenci, O. Effect of calving process on the outcomes of delivery and postpartum health of dairy cows with unassisted and assisted calvings. *J. Dairy Sci.* **2016**, *99*, 7568–7573. [CrossRef]

123. Vannucchi, C.I.; Silva, L.G.; Lúcio, C.F.; Veiga, G.A.L. Oxidative stress and acid–base balance during the transition period of neonatal Holstein calves submitted to different calving times and obstetric assistance. *J. Dairy Sci.* **2019**, *102*, 1542–1550. [CrossRef]

124. Vannucchi, C.I.; Silva, L.C.G.; Unruh, S.M.; Lucio, C.F.; Veiga, G.A.L. Calving duration and obstetric assistance influence pulmonary function of Holstein calves during immediate fetal-to-neonatal transition. *PLoS ONE* **2018**, *13*, e0204129. [CrossRef]

125. Schuijt, G. Iatrogenic fractures of ribs and vertebrae during delivery in perinattaly dying calves: 235 cases (1978–1988). *J. Am. Vet. Med. Assoc.* **1990**, *197*, 1196–1202.

126. Mota-Rojas, D.; Wang, D.; Titto, C.G.; Martínez-Burnes, J.; Villanueva-García, D.; Lezama, K.; Domínguez, A.; Hernández-Avalos, I.; Mora-Medina, P.; Verduzco, A.; et al. Neonatal infrared thermography images in the hypothermic ruminant model: Anatomical-morphological- physiological aspects and mechanisms for thermoregulation. *Front. Vet. Sci.* **2022**, *9*, 963205. [CrossRef]

127. Besser, T.E.; Szenci, O.; Gay, C.C. Decreased colostral immunoglobulin absorption in calves with postnatal respiratory acidosis. *J. Am. Vet. Med. Assoc.* **1990**, *196*, 1239–1243.

128. Vasseur, E.; Rushen, J.; de Passillé, A.M. Does a calf's motivation to ingest colostrum depend on time since birth, calf vigor, or provision of heat? *J. Dairy Sci.* **2009**, *92*, 3915–3921. [CrossRef]

129. Kaske, M.; Bleul, U. Feeding of unweaned dairy calves. In *Bovine Prenatal, Perinatal and Neonatal Medicine*; Szenci, O., Mee, J.F., Bleul, U., Taverne, M.A.M., Eds.; Hungarian Association for Buiatrics: Budapest, Hungary, 2021; pp. 291–308.

Disclaimer/Publisher's Note: The statements, opinions and data contained in all publications are solely those of the individual author(s) and contributor(s) and not of MDPI and/or the editor(s). MDPI and/or the editor(s) disclaim responsibility for any injury to people or property resulting from any ideas, methods, instructions or products referred to in the content.

Review

Parturition in Mammals: Animal Models, Pain and Distress

Julio Martínez-Burnes [1],*, Ramon Muns [2], Hugo Barrios-García [1], Dina Villanueva-García [3], Adriana Domínguez-Oliva [4] and Daniel Mota-Rojas [4],*

[1] Animal Health Group, Facultad de Medicina Veterinaria y Zootecnia, Universidad Autónoma de Tamaulipas, Victoria City 87000, Tamaulipas, Mexico; hbarrios@docentes.uat.edu.mx

[2] Agri-Food and Biosciences Institute, Hillsborough, Co Down BT26 6DR, Northern Ireland, UK; rmunsvila@gmail.com

[3] Division of Neonatology, Hospital Infantil de México Federico Gómez, Mexico City 06720, Mexico; dinavg21@yahoo.com

[4] Neurophysiology, Behavior and Animal Welfare Assessment, DPAA, Universidad Autónoma Metropolitana (UAM), Unidad Xochimilco, Mexico City 04960, Mexico; mvz.freena@gmail.com

* Correspondence: jmburnes@docentes.uat.edu.mx (J.M.-B.); dmota100@yahoo.com.mx (D.M.-R.)

Simple Summary: Labour is considered a painful episode where complex physiological, hormonal, morphological, and behavioural changes are present. During animal parturition, the recognition and treatment of pain is not a regular practice, although there are several consequences derived from pain in the mother and the newborn. This review discusses current knowledge about human labour pain, the relevant rat model's contribution to human labour pain, and model parturition pain mechanisms in small and large animals. Parturition's pain represents a potential welfare concern; therefore, pain indicators and appropriate analgesic therapy are also analyzed in this work including the relevance of analgesics and the welfare implications of pain during this physiological stage.

Abstract: Parturition is a complex physiological process and involves many hormonal, morphological, physiological, and behavioural changes. Labour is a crucial moment for numerous species and is usually the most painful experience in females. Contrary to the extensive research in humans, there are limited pain studies associated with the birth process in domestic animals. Nonetheless, awareness of parturition has increased among the public, owners, and the scientific community during recent years. Dystocia is a significant factor that increases the level of parturition pain. It is considered less common in polytocous species because newborns' number and small size might lead to the belief that the parturition process is less painful than in monotocous animal species and humans. This review aims to provide elements of the current knowledge about human labour pain (monotocous species), the relevant contribution of the rat model to human labour pain, and the current clinical and experimental knowledge of parturition pain mechanisms in domestic animals that support the fact that domestic polytocous species also experience pain. Moreover, both for women and domestic animal species, parturition's pain represents a potential welfare concern, and information on pain indicators and the appropriate analgesic therapy are discussed.

Keywords: parturition; delivery; whelping; farrowing; labour; pain; distress

Citation: Martínez-Burnes, J.; Muns, R.; Barrios-García, H.; Villanueva-García, D.; Domínguez-Oliva, A.; Mota-Rojas, D. Parturition in Mammals: Animal Models, Pain and Distress. *Animals* **2021**, *11*, 2960. https://doi.org/10.3390/ani11102960

Academic Editor: Olli A. T. Peltoniemi

Received: 24 September 2021
Accepted: 12 October 2021
Published: 14 October 2021

Publisher's Note: MDPI stays neutral with regard to jurisdictional claims in published maps and institutional affiliations.

Copyright: © 2021 by the authors. Licensee MDPI, Basel, Switzerland. This article is an open access article distributed under the terms and conditions of the Creative Commons Attribution (CC BY) license (https://creativecommons.org/licenses/by/4.0/).

1. Introduction

Parturition is a complex process that involves changes at the hormonal, physiological, morphological, and behavioural levels [1,2]. One of the characteristics of this event is the manifestation of pain that can vary in severity and duration depending on the species studied. Parturition is a significant event for numerous species and is frequently the most painful experience that females suffer [3]. Unlike other acute pain experiences, labour pain is physiological and unique, not associated with disease or trauma, and its presence does not indicate any pathology, but the progression of labour itself [4], and the most basic experience of life, the birthing of a new individual [5]. However, several factors may hinder the normal process of parturition and modify or increase the degree of pain caused by it.

The International Association for the Study of Pain (IASP) defined pain "as an unpleasant sensory and emotional experience usually associated with tissue damage or described in terms of such damage" [6–8], the definition had remained unchanged since 1979. However, in recent years, advances in our understanding of pain in its broadest sense warranted a re-evaluation of the definition [9,10]. Recently in 2019 [11], the current definition was revised to "an unpleasant sensory and emotional experience associated with or resembling that associated with, actual or potential tissue damage". Also, with the following notes: Pain is always an individual experience modified by various degrees by biological, psychological, and social factors, through their life experiences, individuals learn the concept of pain, although pain usually serves an adaptive role, it may have adverse effects on function and social and psychological well-being, and verbal description is only one of several behaviours to manifest pain; incapacity to communicate does not negate the possibility that a human or a nonhuman animal experiences pain. A relevant update is that the pain definition should be used in humans and non-human animals [12].

A large proportion of women describe parturition as severely painful [13,14]; for that reason, labour pain in women has received great scientific interest. Research in the rat model has contributed to elucidate the subjacent neurophysiological and neuropharmacological mechanisms of pain during labour [15].

In small animals, labour pain is generally accepted as a paradoxical and challenging issue, composed of a complex, multidimensional subjective experience involving sensory and affective (emotional) elements [16]. Also, in pigs, pain is described as a perceptual event created from information assembled by specific sensitive receptors for tissue injury, transformed by ascending and descending spinal and supraspinal mechanisms, and combined into an individual sensorial event including negatively valenced emotions [17].

In cows and sows, it is recognised that pain has a beneficial role because it is intimately related to some of the neurohumoral responses required for inflammation and can alter physiological responses, which also help the individual to handle damage [18,19].

The individual's expression and response to pain have an emotional and social interpretation structured by culture [8,20] and visible painful events [21]. The pain-conscious experience challenges the specific anatomical, physiological, and/or pharmacological interpretation. Moreover, it is a subjective sensation that can be suffered even without visible harmful stimulation and can be altered by beliefs, moods, customs or habits of humans who are responsible for animals, coping ability, and by behavioural experiences, including fear, memory, and stress of the different animal species [16].

Contrary to the extensive research in humans, in domestic animals, there are limited pain studies associated with the birth process, reflecting little interest in the subject [19]. Most studies refer to the cascade of hormonal events, morphological and behavioural changes associated with pregnancy, labour, and lactation. Similar neuroanatomical structures associated with pain perception are shared by humans and vertebrates [22,23], including nociceptors, nociceptive pathways, and processing zones in the central nervous system (CNS). Therefore, it is recognised that pain perception in humans and other mammals is similar. Most mammal species have receptors sensitive to noxious stimuli (nociceptors) and have brains with structures analogous to the human cerebral cortex. They have nervous pathways linking nociceptors to the higher brain, and it has been observed that painkillers modify the response physiologically and behaviourally [19].

Nonetheless, during recent years, awareness of the labour process, its impact on welfare, health, and even economic return in domestic animals has increased among the public, owners, and the scientific community. For example, in swine, the health and welfare status of the sow have been shown to reflect her capacity to generate healthy offspring [3]. In dogs, the owners increasing emotional involvement in the birthing process of their pets, and the economic importance of purebred puppies, has produced raised interest in enhancing whelp survival. [24]. Dystocia is usually considered to be less common in polytocous species such as the pig than in monotocous animal species and humans [25]. The number of newborns and smaller size in the former might assume that the labour process is less painful. In this review, the current knowledge about human labour pain (monotocous species), the contribution of the rat model related to human labour pain, and the current clinical and experimental knowledge of parturition pain mechanisms in pigs, with physiological and anatomical similarities with humans for the study of labour pain, support that domestic polytocous species also experience pain. Both for women and domestic animal species, the pain associated with parturition represents a potential welfare concern. The above-mentioned matters justify the need to review the published information of labour pain and current advances drawing on examples from monotocous and polytocous species, based on the availability of research in humans and on pain study models in some domestic animal species.

2. Physiological Response to Labour: Comparative Studies

The timing of labour in humans continues to be an enigma, and more reliable knowledge of these mechanisms is essential to avoid unfavourable consequences of pregnancy. Parturition, a fundamental and species-specific component in the reproductive cycle, is under intense selection pressure. A broad diversity in gestation length and neonatal maturity at birth occurs in species unrelated to the size at birth. Human neonates are similar to other altricial species in terms of maturity at delivery, although primates are usually considered precocial at birth [26].

In humans, a natural time of parturition begins parallel to foetal organs maturity, generally 37–40 weeks gestation. Prevailing hypotheses of labour induction are mainly associated with foetal-maternal endocrine and immune alterations in utero relating to foetal development [27]. Childbirth is when a foetus and placental tissues are released from the uterus through the vagina. In women, the birth process can be divided into two stages. The first is divided into the latent and the active phase. The latent occurs when the woman undergoes regular and painful contractions until the cervix dilates four centimetres. The active one goes from a dilation of the cervix of four centimetres to full dilation. The second stage, in turn, is subdivided into two phases; the descending in which the baby's head drops towards the pelvic floor, and the expulsion when the dam is actively expelling the child [28].

Labour monitoring helps determine its stage and progress, and multiple modalities are used. Consecutive cervical exams are helpful to delimit cervical dilation, effacement, and foetal position, also called tocography. Cardiotocography is used to monitor the foetal heart frequency and capacity of uterine contractions. Also, it is helpful to assess foetal well-being throughout labour continuously. Physicians determine the patient's stage of parturition and monitor the progression of labour based on information from monitoring and cervical examinations [29,30].

In female Dorper sheep, the use of echocardiography has been implemented for accurate readings from the third month of gestation, in which foetal cardiac monitoring allows the identification of congenital or developmental anomalies, and hypoxemia and acidemia due to placental dysfunctions that compromise the viability of the foetus, something that, in production species, especially reproducers, is essential [29,30]. Similarly, foetal abnormalities in horses are identified using the biophysical profile, considered the gold standard in reproduction to assess foetal well-being by determining the foetal heart rate, the diameter of the aorta and the thickness of the uteroplacental unit [29,30].

Diagnosis of labour initiation is a challenging and essential decision for those who provide maternity care. The initial stage of labour, through active uterine contractions, accomplishes the goal of reducing, dilating, or opening the cervix by at least 10 cm in diameter to permit the foetus to pass from the uterus to the vagina. The duration of both stages of labour are clinically significant and hence need uniform strategies to measure. An extended latent phase of parturition is linked with a higher risk for oxytocin increment of labour, caesarean section, amniotic fluid stained with meconium, an Apgar score less than seven at 5 min, demand for newborn resuscitation and ingress to the neonatal intensive care unit (NICU) [30–33]. There is a significant discrepancy in the definitions of labour initiation in the research reports. Still, there was little consensus between studies referring to the same type of parturition initiation (e.g., active labour phase) and an indication of labour initiation, except that 100% of the definitions of latent labour phase refer to the presence of frequent painful contractions [34].

Based on the essential nature of parturition, there is a lack of understanding of how the way by which a baby is born can have later physiological consequences for newborns. More comprehensive knowledge of how the form, start, sequence, and duration of parturition can influence healthy development and long-term well-being consequences for the infant is crucial because of the global increase in delivery by caesarean section (CS). In 2595 women subjected to CS (50.5% of the total sample), a retrospective study made by Khasawneh et al. [35] found that fetal distress was the main indication of CS (15.5%). From the delivered newborn, 16% required hospitalization due to transient tachypnea and respiratory disease syndrome. The increase in CS is also in domestic animals, particularly in some breeds of dogs, and the Belgian White and Blue cattle [36]. The principal mechanisms associated with explaining why the mode of parturition, spontaneous or induced/increased vaginal birth versus CS, may alter neonate development include (1) physical stress and exposure to stress hormones during parturition, (2) unusual bacterial colonisation of the child's intestine, and (3) epigenetic alteration of gene expression [37].

Additionally, the risk of CS in the newborn includes immune deficits (i.e., asthma) and alterations of the young microbiome [38]. Ensuring normal and safe physiological conditions could evade adverse consequences of the mode of parturition. Assessment, care plan, health training, and improvement of natural processes could establish the framework to increase the capacities of women in preparation for childbirth and maternity [39].

3. Labour Pain (Parturition Pain Stimuli)

Recently, foetal tissue senescence was reported in association with foetal growth causing sterile inflammatory markers that can disseminate as paracrine foetal signals linked with childbirth. Homeostatic irregularities created by endocrine and paracrine constituents generate an inflammatory overload that interferes with the preservation of gestation. These signals transform the inactive myometrium into the active state. Although, characteristics of the foetal or maternal-derived signals and their specific mechanism in initiating delivery remain unclear [40]. More recently, Menon (2019) reported that foetal membrane senescence and the associated inflammation function as a paracrine signalling system during labour [41].

Labour is a dynamic process of delivering a foetus and is characterised by regular, painful uterine contractions that progress in number and severity. Labour pain has visceral and somatic components. The cervix has a primary function in both the first and second stages of delivery. Parturition pain occurs out of a range of physiological factors like uterine contractions and cervical dilation, accompanied by mental and emotional constituents, including fear and anxiety. Other constituents such as maternal age, parity, the physical health of women in labour, and the maternal situation can also influence its severity and duration [34,39]. The behaviours displayed by women in parturition can also induce labour pain and can be associated with the length of the stages and the severity of the pain.

Consequently, it is suggested to recognise this problem simultaneously with physiological signs and clinical tests for better management of women in labour [42]. Based on the several similarities among animals and humans in anatomical and chemical pathways of pain perception, it is commonly admitted that pain perception is comparable in humans and other mammals. Hence, from the dam perspective, parturition in all species is usually admitted as a painful process. Broadly, childbirth associated with difficult parturitions or dystocia may produce unacceptably severe pain levels in the dam. Thus, for example, in sows and cows, it is recognized that pain has a beneficial role because it is closely related to some of the neurohumoral responses required by inflammation, which can alter physiological responses and help the individual to manage the damage [18]. In pigs, as it is in other species, parturition is expected to be painful, and pain probably originates from uterine contractions, piglet ejection, and uterine inflammation due to a piglet litter delivery [18].

Regarding inflammation, during parturition, there are notable alterations in the acute-phase protein concentrations. Haptoglobin (Hp) and Serum Amyloid A levels rise throughout calving, higher in heifers than in pluriparous cows, implying a higher inflammation or trauma around calving, producing higher pain levels [18]. In pigs, C-reactive protein (CRP) and Hp are recognised as the most reliable markers of inflammatory damage, and high levels have been described in sows one week after farrowing [43]. Furthermore, higher Hp values are described in primiparous sows than pluriparous sows [44]. Kostro et al. [45] reported inflammation of the reproductive tract and mammary glands associated with raised levels of CRP in sows with Mastitis, Metritis, and Agalactia (MMA).

Primarily, in the lower uterine segment, the concentrations of interleukin (IL) 1β, 6, and 8 were found to be significantly higher when the cervical dilatation was 4 cm (6.6, 67.7, and 125.8 pg/mg, respectively) and promoted normal and premature labour in humans [46]. Another study compared the presence of cytokines in the uterine segment and the amniotic fluid in patients undergoing CS. The authors found that IL-6 concentrations in the amniotic fluid increased earlier (at 2–3.9 cm) than IL-8 (at 6 cm), and no correlation was found in the uterus for this cytokine [47]. During the labour term and preterm, the increase in other cells, such as adhesion molecules in the lower segment and cervix, have also been reported to trigger the cervical tissue ripening [48]. Myometrial inflammation is also involved in the labour beginning, suppressing progesterone action [49], driving pro-labour gene expression, including prostaglandin biosynthetic enzymes [50] and the oxytocin receptor [51]. Studies emphasize the relationship between labour and inflammation, confirming a severe inflammatory myometrial recruitment of monocytes, neutrophils, and macrophages [52–54]. In more recent publications, the same group demonstrated that inflammatory cytokines, bacterial lipopolysaccharide (LPS), and monocytes could increase myometrial cell contractions, the first by raised prostaglandin synthesis and the second for a direct effect on ROCK [55,56]. Therefore, this is compatible with a common antagonist relationship among progesterone and inflammation, which is the essence of diverse hypotheses for the labour start [49,57]. One of the proposed mechanisms includes the secretion by the mature foetal lung of surfactant protein A (SP-A), which initiates the proinflammatory transcription factor NFκB [58]. The second mechanism includes myometrial stretch, whereby the growing pregnancy rises myometrial wall tension, stimulating the proinflammatory transcription factors NFκB and AP-1, which induce the proinflammatory cytokines expression [59]. Although, some key question remains; does myometrial inflammation occur before the onset of labour, or is it merely a consequence of labour? Interestingly, Singh et al. [60] studied myometrial samples from women at various pregnancy and spontaneous labour stages, evaluating some proinflammatory factors and inflammatory cells, and the authors suggested that myometrial inflammation is a consequence rather than a cause of term labour [60].

4. Origin and Transmission of Parturition Pain Stimuli

Parturition pain is a deeply personal, challenging, sensitive, and meaningful experience, very different from other types of distress. In humans, the main determining and influencing factors in labour pain are cognitive, social, and environmental. Interestingly, if a mother can maintain the feeling that her pain has a goal (i.e., her body working for childbirth), if she translates her pain as productive (i.e., a process for the desired goal), along with secure and supportive delivery conditions, she would be expected to undergo pain as a life-changing and non-threatening event. Transforming the conceptualisation of labour pain to a beneficial and fruitful process could be a start to improving women's experiences of it and diminishing their need for pain interventions [61].

During the stages of the labour process (preparation of the uterus, active labour, or the expulsion phase and delivery or expulsion of the placenta), a series of endocrine events have been triggered that mark the anatomical changes to promote the birth [18]. The preparation of the uterus and cervix for birth is characterised by subclinical myometrial contractions, progressive dilation of the cervix, and the positioning of the foetus for expulsion [62,63]. It is worth noting that uterine contractions are weak but regular [15]. Initial preparation of the birth canal takes place along with the usual prepartum endocrine alterations (relaxin discharge, a decrease of progesterone and increased oestrogen and prostaglandin synthesis) [29,64,65]. In pigs, oestrogens and relaxin cause cervix distension by altering the action of collagen [65,66]. Cervical softening and ripening is also influenced by IL-8, IL-6, and granulocyte colony-stimulating factor for remodelling the cervical tissue and opening the cervical canal [67]. Chemokines and prostaglandins in the amniotic fluid and cervix, and the release of collagenases by the neutrophils, which increase together with macrophages in the lower uterine segment, contribute to the cervical dilatation in the prepartum phase [68]. However, it is important to recognize the differences between species; for example, in equine species, maternal progesterone levels during the late stage of gestation are almost nil, and the production of progesterone and estrogens depends on the foetoplacental unit and steroid precursors formed by the foetus. This event only occurs in mares [69–71].

From a physiological point of view, labour pain in humans is induced by the ischemic myometrium while uterine contraction, cervical maturation and dilation (stretching of the vagina and perineum), and distension and compression of adjacent pelvic structures [69–72]. Myometrial contractions differ between species and also between individuals. Usually, the extent, number, and amplitude of myometrial contractions rise and become regular roughly 12 h before the start of the second stage. Eventually, the foetus responds with movements to prolonged uterine stress created by early-stage myometrial contractions [63]. Labour pain also has a visceral component during the dilation of the cervix via afferent fibres in the central nervous system. In humans, these nociceptive stimuli of the dilation stage are predominantly transmitted between spinal nerves T10 to L1 [72] (Figure 1).

Pain pulses are carried out via the spinal cord mainly by afferent A-delta and C fibres [71,73], so pain can progressively refer to the abdomen, the lumbosacral zone, the iliac crests, the gluteus, and the legs [5]. The second stage, or active labour, usually begins with the rupture of the foetal membranes and ends when the last foetus is expelled. This stage comprises terminal cervical widening, achieved by the propulsive forces of regular uterine contractions through labour [66]. Since labour progresses, distention of the maternal birth canal results in increased oxytocin discharge from the posterior pituitary, increasing myometrial contractions [63]. Oxytocin plays a vital role in the parturition progress coordinating the uterine contractions [66,74] and the foetal expulsion [65,75]. In humans, melatonin also has a supporting role in myometrial contractions. Olcese and Beesley [76] found that melatonin has a synergetic action and enhances oxytocin-induced contractions binding to specific melatonin receptors (MT2R) in the myometrium. This response also follows a circadian pattern where the labor onset is triggered during the night/morning hours. Moreover, physiological administration of the monoamine hormone induces contractibility, together with oxytocin, to induce labor [77]. Domestic animals and humans possess a high number of uterine oxytocin receptors, confirming the high sensitivity to exogenous oxytocin at the active labour stage [78–81].

Additionally, at least for rats and humans, oxytocin receptors rise to 200 times during late pregnancy compared to the non-pregnant state [82]. In humans, during the second stage of labour, pain is believed to be somatic because of the perineal distension and strain on pelvic structures surrounding the vagina and further due to the expansion of the pelvic floor and perineum, with afferent fibres between spinal nerves S2 and S4 [5] (Figure 1). It is described that the perineal stretching before spontaneous vaginal delivery is characterised by severe pain [83]. Therefore, the second stage of parturition is considered the most painful phase of labour [84]. Although practically all parturient women suffer low abdominal pain during contractions, 15% to 74% may additionally experience contraction-related low dorsum pain, sometimes continuous, including between contractions [85]. Labor and Maguire [13] point out that the severity of parturition pain intensifies as cervical dilation increases and is directly associated with strength, duration, and number of myometrial contractions. Stage III generally occurs in parallel to stage II and includes passage of the foetal membranes. In the case of polytocous species (like sows), the foetal membranes along with the foetus are regularly expelled. However, in monotocous species (like women), the complete newborn expulsion resembles the third stage. During this stage, uterine contractions remain, diminishing in amplitude but with the highest and least regular frequency [62]; this phase of labour is believed to be painless [86].

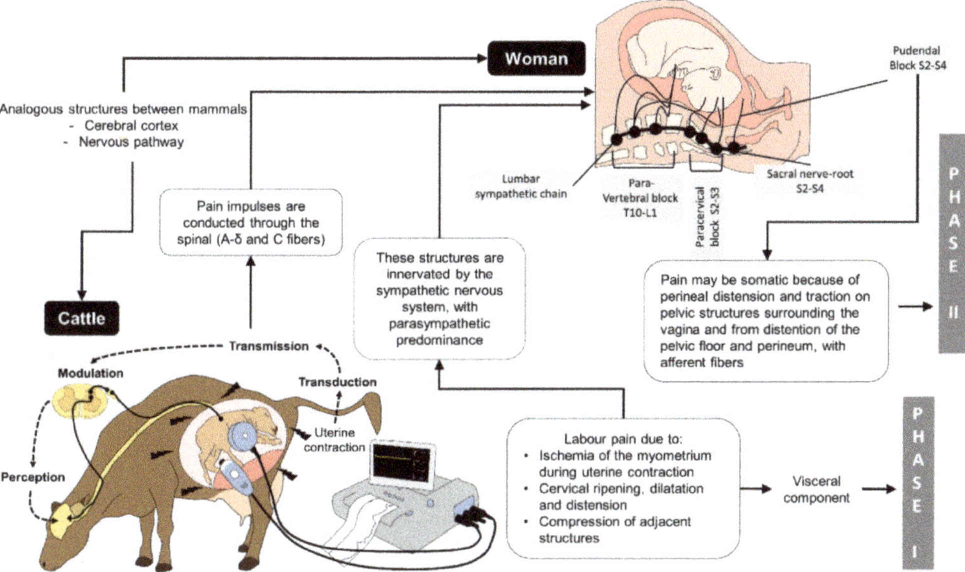

Figure 1. Pathway of pain in labour. In the first stage, there are slight but constant uterine contractions; as the strength of the contractions increases, concomitantly with the distension, effort, and tear of the lower uterine segment and the cervix, it becomes strongest and induces visceral pain with afferent information travelling within the hypogastric and pelvic nerves [15]. In the second labour stage, the expulsion phase is described as the most painful stage due to cervix distension and pressure on the pelvis and perineum, with pudendal nerve innervation. The nociceptive stimuli are processed and transmitted at the dorsal horn of the spinal cord, via the spinothalamic region to the thalamus, brain stem, and cerebellum, where spatial and temporal analysis take place, and also to the hypothalamic and limbic systems [5]. The third stage of labour, the delivery, consists of the expulsion of the placenta and is not painful [86].

Beta-endorphin (β-end) belongs to one of three opioid peptides families, and it is implicated in regulating the body's response to stress, including pain. Endogenous β-end has been determined around parturition and is thought to be linked to reducing pain [87], maintaining passivity [88], and regulating oxytocin release [89]. The naloxone administration diminished the nociceptive threshold of sows, but not wholly, indicating that an endogenous opioid mechanism is likely only partly involved in hypoalgesia around parturition [90].

5. Factors Affecting Labour Pain, Dystocia

As already presented, labour by itself is a cause of pain for any mammal female, even within physiological parameters/conditions [3]. Several factors have been identified that hinder the normal process of parturition and alter the degree of pain caused by it. Dystocia may be defined as parturition difficulty happening from extended natural parturition or extended or severe assisted extraction [19]. Dystocia translates into the inability to expel the foetus(es) through the birth canal, whether due to a physical obstacle or functional defect, associated with intolerable high levels of pain [19,91,92], and in humans, the main cause is due to foetal malposition, as reported by Hautakangas et al. [93] in 5200 women, were 296 required CS. Dystocia also leads to acute foetal distress and mortality and a decrease in the vigour of the newborn. Usually, the causes of dystocia are divided into maternal and foetal [19,62]. The former being the ones that most influence a prolonged birth in women and domestic animal species.

The maternal characteristics of prolonged labour include parity (higher pain rates in primiparous versus pluriparous women), diameter and anatomy of the pelvis, previous obstetric situations, psychological, cultural, and educational aspects. However, the environment (e.g., high temperatures) and management (inappropriate use of ecbolics and excessive obstetric manipulations) can also cause dystocia or alter the birth process [62,94]. Within foetal characteristics are higher body weight and foetal position. Intolerable pain in humans results in a high catecholamine release, which leads to less uterine blood perfusion, reduces uterine contractions' effectiveness, and, consequently, prolongs labour [95]. Also, it induces dystocia, exhaustion, and foetal distress, in addition to postpartum posttraumatic stress disorder [5,96,97]. Sandström et al. [94] highlight the relationship between the duration (prolonged) of the second stage of labour and adverse neonatal outcomes (Table 1).

In cattle, the prevalence of dystocia is calculated between 1.5–22.6% and the leading cause of calf death at birth. It induces mastitis, and also systemic infections such as clinical metritis [98], clinical and sub-clinical endometritis in almost 51.6% of cases [99] due to placental retention in 43.1% and 37.50% of 1300 cows [100]. Barraclough et al. [101] report that one method of preventing complications during parturition is to observe the postures and activity of the animals, which usually increases by up to 80% during stage two of parturition because it is severe pain.

Table 1. Factors associated with dystocia, prolonged labour and pain in women.

Factor That Increases Labour Pain	References
Maternal characteristics: Prolonged labours • Parity. The pattern of parturition pain varies among nulliparous versus pluriparous women, and it is documented that pain rates are higher in the pluriparous contrasted to the pluriparous woman. • Increased "elasticity" and flexibility of the pelvic tissue in pluriparous women may decrease nociceptive stimuli during the dilation phases of labour, but stimuli increase later in childbirth due to the speed and intensity of foetal descent. • The diameter and anatomy of the pelvis, length and mass, myometrium contractile forces, soft tissue stability, ejection efforts, previous obstetric/medical situations, such as hypertension or pregestational/gestational diabetes mellitus. • Psychological aspects. A variety of psychological factors influence pain perception; Anxious, tense women, fear, and loneliness are associated with increased pain during labour. • Cultural aspects and educational level also intervene in the perception of pain. **Foetal characteristics:** • Birth weight (higher foetal body weight increased pain), • Foetal occiput position/degree of flexion, and station at complete cervical dilation.	[102–107]

In female dogs, uterine inertia is the leading cause of dystocia derived from prolonged births [62,108]. Other factors related to dystocia in bitches are older age, smaller litter size, brachycephalic and achondroplastic breeds. A narrow pelvic canal can induce obstruction because of the disability of the foetus to pass normally; the obstruction due to a short pelvic channel is common in brachycephalic and achondroplastic breeds. The higher incidence of dystocia and caesarean section due to primary inertia in these breeds is considered a hereditary condition, and hence the bloodlines in which this problem is shared should be identified and breeders advised avoiding breeding from or combining such lines. Also, to use easy whelpings in their breeding programs to prevent welfare implications in the bitch and her puppies [109]. In bitches, the influence of some genetic factors during parturition has been reported. For example, the expression of oxytocin receptors in the endometrial and myometrial tissue increases during the late phase of the parturition [110], promoting normal delivery without the need for medical intervention. On the other hand, uterine inertia has been associated with an abnormal expression of γ-actin and myosin gene in the smooth muscle of the uterus, leading to dystocia in breeds such as Maltese, German Shepherd, Labrador Retriever, Beagle, and Boxer [111]. Foetal malposture and large foetuses are also considered causes of prolonged labour. It should be noted that obstetric interventions can increase stress and labour pain in addition to inducing more hemodynamic or vascular changes [112,113] (Table 2).

Table 2. Factors associated with dystocia, prolonged farrowing and pain in the bitch.

Factor That Increases Labour Pain	References
Maternal characteristics: Prolonged labours • Primary uterine inertia. Caused by numerous factors including sepsis, disease, age-related abnormalities, or a genetic disability of myometrial contractility. • Age of the bitch. Older age (older than 6 years) predisposes to more single-pup pregnancies, uterine inertia, and prolonged parturition compared with the younger. • Litter size. Pregnancies with few foetuses do not allow the start of labour properly. Low numbers of foetuses (1–2 puppies) represented 21.5% of dystocia. • Breed. Obstruction due to a short pelvic channel is common in brachycephalic and achondroplastic breeds. It is also challenging to give birth to toy dogs, prone to small litters and large foetuses. • Obstructive dystocia. Due to several factors such as uterine torsion, uterine rupture, inguinal hernia, and abnormalities of the soft tissues of the vagina or vulva. • Hypocalcaemia. **Foetal characteristics:** • Foetal Malposture. • Large foetus. Pregnancies with a large foetus, such as, a single puppy, gestational diabetes; or foetal anomalies like ana sarca ("water puppy") or a foetal monstrosity.	[62,108,113]

In felines, the breed has shown a strong correlation with dystocia incidence, with purebred cats such as British shorthairs showing a high incidence rate of 2.5. Likewise, 56% of complications during delivery require caesarean sections and intra- and postsurgical analgesia management in cats [114], in whom the administration of certain types of drugs requires care due to their biotransformation deficiencies.

In pigs, it is recognized that labour is a risky process for both the sow and the piglets and increases in the case of dystocic births due to prolonged parturitions [19]. The delivery in the sow has an average duration of 2.5 to 3 h, and when they last more than 3 h are considered potentially problematic and consequently more painful [90,115]. Maternal characteristics of the sow predisposing prolonged labour and potential pain are considered the breed, short gestation length, larger litter size, sow overweight, and parity; primiparous sows more prone to painful labour than pluriparous, among others. Higher birth weight and posterior presentation at birth are considered as foetal factors associated with dystocia [116–127] (Table 3). The widespread use of pigs in biomedical research is supported based on their physiological and anatomical similarities to humans, also large litters, and their large body size helps multiple samples collection. Pigs are omnivores with comparable digestive systems to humans [128].

Table 3. Factors associated with dystocia, prolonged farrowing and pain in the sow.

Factor That Increases Labour Pain	References
Maternal characteristics: Duration of farrowing (more than 3 h): • Breed. • Short gestation length. • Increased litter size. • Increased number of stillborns. • Sows with overweight. • Constipation. • Lack of exercise. • Parturitions: primiparous sows go through a more painful labour process than pluriparous. **Foetal characteristics:** • Higher birth weight. • Birth order. • Presentation at birth. This is a controversial issue; however, some reports relate the posterior presentation of piglets at birth with the increase of the delivery duration. **Others:** • Misuse in medication control at labour with the use of ecbolics (such as oxytocin). • Excessive obstetric manipulation.	[1,19,116,117,120]

Rats are widely used in biomedical research because of their physiological and anatomical similarity with humans. On the other hand, studies related to parturition, stress, and pain in other mammals such as mares, cows, and mice, have helped to understand neurobiological characteristics of parturition such as the parasympathetic predominance during the expulsion of the foetus. In case of stress and pain during the process can change towards a sympathetic influence with the consequent activation of sympathetic pathways that generate cortisol production, uterine atony, and prolongation of labour and dystocia [129,130]. However, Roussel et al. [131] report in ewes that moderate cortisol levels due to stress in the prenatal period (from 2.5 months of gestation) contribute to offspring with better birth weight, greater activity, and exploratory behaviours. Although pain during delivery has received little attention in cattle, the high prevalence of foetal-pelvic disproportions and the use of traction to remove the calf are events that involve pain and trauma, both for the mother and the calf generating rectovaginal ruptures that require pharmacological intervention for pain treatment [132].

6. Pain Assessment Scales

Pain is a personal subjective emotional state in humans and animals; therefore, it is complex to appreciate what each feel. Although it is feasible to assess pain in humans directly, using a degree scale measured by the subject [133], in non-verbal patients [134] and non-human animals, changes in physiological and behavioural parameters are used to assess the presence and severity of pain [135]. The sympathetic nervous system and the hypothalamic–pituitary–adrenal axis (HPA) mediate most physiological changes associated with painful stimuli. The sympathetic responses can be directly assessed by estimating the circulant catecholamines, adrenalin, and noradrenalin [136], or the resultant autonomic changes [137], body temperature, heart rate variability, and respiratory rate [138]. Glucocorticoids are used as biomarkers of stress, but also HPA changes in response to painful stimuli are commonly evaluated through the measure of glucocorticoid production, such as in rodents [139] (Table 4). Detection and evaluation of pain are crucial to improving welfare in a variety of contexts. Recently, in a review, it was recommended that any pain evaluation system should recognise that the pain event, at least in humans, has three elements: sensory-discriminatory, affective-motivational, and cognitive-evaluative, communicated between them to induce the experience of pain [17]. The experience of pain during labour results from processing multiple physiological, psychosocial, or behavioural factors (Table 5). Individual and subjective interpretation of labour pain by each person [5], and efforts to impartially and accurately estimate it in animals, are also especially challenging. In humans, the self-report is recognised as the "gold standard" for pain assessment, based on either oral or written communication [134]. However, it is described that verbal pain communication has limitations [140]. Several methods are described for assessing labour pain in women, including (a) the McGill Pain Questionnaire [MPQ], (b) The Short-Form MPQ, (c) Visual analogue scales [VAS], however, VAS are considered as the best system to measure pain [141] (Table 5). As mentioned above, while it is feasible a directly evaluate pain in humans using a grade scale scored by the subject [141], informative signs need to be analysed in animals to gather this information [133]. Indirect signs are helpful as indicators for evaluating pain in animals combine alterations in physiological and behavioural parameters [133,135]. Behavioural indicators to assess pain and labour pain across species are presented in Table 5 and Figure 2. Although objective measurements associated with acute pain are used in dogs (heart rate, blood pressure, and plasma cortisol and catecholamine levels), they are not reliable because stress, anxiety, and anaesthetics may influence them; therefore, pain assessment is mostly subjective and based on behavioural signs [133,142]. In the rat model for labour pain, measures include the activity of spinal neurons that receive afferents, using immunodetection of c-Fos protein, automated systems to detect pain behaviour and phase stretching behaviour [15]. In sows, a numerical pain score per animal is calculated by evaluating behavioural, clinical, and physiological patterns in a specific term [19]. However, behaviour is the most frequent indicator recommended for evaluating farrowing pain in sows; although intrauterine pressure may also be included [143] (Table 5).

Table 4. Multi-parametric scales for assessing labour pain (physiological indicators) in human Obstetric and Veterinary Medicine.

Responses	Variables and Scales	References
Sympathetic-Adreno -Medullary System	• Cardiac rate. • Rectal temperature. • Respiratory rate. • Blood pressure.	[135,138]
Hypothalamic Pituitary Adrenocortical System	• Cortisol. • Adrenalin. • Noradrenalin. • β-endorphin (β-end). • Met-encephalin. • Oxytocin. • C-reactive protein (CRP) and cytokines. • Duration of labour. • Number of stillbirths (only for animals).	[135,136,139]

Table 5. Multi-parametric scales for assessing labour pain (behavioural indicators).

Species	Variables and Scales	References
Women	(a) McGill Pain Questionnaire [MPQ], (b) The Short-Form MPQ, (c) Visual analogue scales [VAS]. Verbal rating scales, or simple ordinal scales.	[5,134,141]
Dog	The behavioural representation of pain is species-specific and affected by age, breed, temperament, and additional stressors, including anxiety or fear; hence, it is necessary to identify the normal behaviour of the bitch. The composite measure pain scale (CMPS) has been described for use in dogs with acute pain, based on seven categories: (1) posture, (2) comfort, (3) vocalization, (4) attention to wound, (5) demeanour, (6) mobility, and (7) response to touch.	[16,23,133,142,144]

Table 5. *Cont.*

Species	Variables and Scales	References
Rat	Activity of spinal neurons that receive afferents, using immunodetection of c-Fos protein, an indirect indicator of harmfully activated spinal neurons. Labour induced the expression of neuronal c-Fos in segments T12-S2 of the spinal cord one hour following the delivery of the first pup. Automated systems for the detection of pain behavior. General activities, including food and water consumption, rearing (upright exploration), and chest and head grooming, are evaluated to determine spontaneous behaviour during labour. Phase stretching behaviour during labour included: crushing (an asymmetrical contraction of the lower body and limbs), sidelong contraction (an asymmetric contraction of the lower body and limbs), lengthening (stretching of the abdomen and the four limbs) and phasic humpback posture. Maternal care activities included building nests, licking pups, and eating the placenta.	[15,145,146]
Sow	Numerical pain score per animal is calculated by evaluating behavioural, clinical, and physiological patterns in a specific term. Observation of certain behaviours as potential pain indicators (PPIn) in sows before, during and after farrowing: hind leg forward (in a side-lying posture, the back leg is pulled ahead and/or in towards the body), back arch (in a side-lying position, one or both legs turn rigid and are pushed aside from the body and/or inwards towards the centre, creating a curvature in the back). Putative indicators of pain used to evaluate the pain behaviour in the periparturient sow. Also, it coincides with an activity shift from nest buildings to passivity, an increase of the myometrial electrical activity, and the increment of oxytocin levels before the beginning of the ejection of piglets. Postural changes considered: stand (upright, with all feet on the floor), sit (front legs straight and back end down on the floor), lie lateral (lying on one side with udder exposed), kneel (front knees on the ground, with hinds legs straight), lie ventral (lying with the udder on the floor), tremble (visibly shaking as if shivering when in a lateral lying position), and others; tail-flick (tail is moved rapidly up and down), paw (in a lateral lying position, the front leg scraped in a pawing motion), piglet delivery (piglet completely ejected from the dam).	[3,74,147,148]

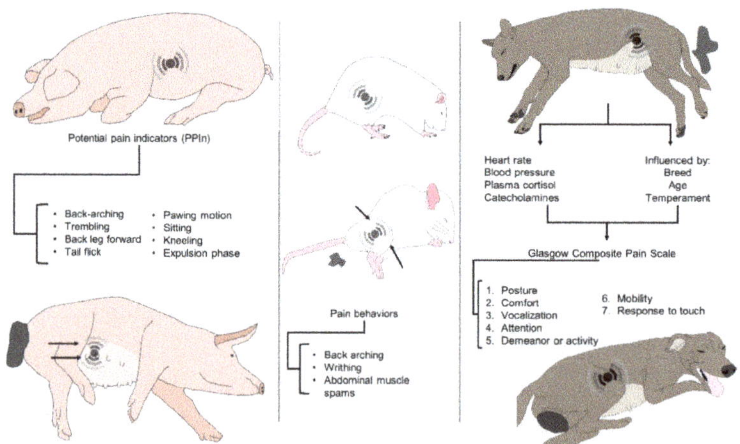

Figure 2. Clinical recognition of pain in sows, rats, and bitches. The recognition of pain during labor in these species can be done by observing altered postures or physiological parameters. In sows, potential pain indicators, such as a back-arching, one back leg forward, sitting, kneeling, trembling, among others, are modified behaviors during parturition. In rats, pain behaviors such as a compact posture, back-arching, writhing, and abdominal muscle contractions are pain signals. Lastly, the Glasgow Composite Scale (CMPS) in dogs uses 7 units in which posture, vocalizations, behavior, mobility, among others, are evaluated together with physiological parameters and plasma cortisol and catecholamine concentrations.

A full evaluation of the pain being experienced is impossible since it is a subjective emotional state and our capacity to identify pain between species is even more challenging [17]. Nonetheless, several indicators of pain have been developed/identified for domestic species. However, they cannot be considered a "Golden Standard" for the presence of pain because detection of specific measures cannot be utilised to confirm or discard pain in non-verbal individuals [149].

Herskin and Di Giminiani [17] listed the seven most common indicators to evaluate pain in pigs (1) Motivational task, (2) Evoked behavioural responses, (3) Vocalisation, (4) Facial expressions, (5) Clinical, (6) Physiology and histopathological measures, (7) Pain scales. Similar indicators are used for most of the domestic species.

Ison et al. [91] evaluated behaviours as potential pain indicators in sows before, during, and post farrowing and two minutes before and after piglet deliveries. Behaviour evaluations included: back leg forward, tremble (shivering), back arch, paw (leg scraped in pawing motion), and tail-flick. All indicators were uncommon or absent pre-farrowing, highest during farrowing, and back leg forward, tremble, and back arches were higher at the immediate post-partum. Significant positive correlations between pain indicators during and post farrowing were found, including the back arch, tail-flick, and paw higher before than after a piglet birth and more frequent earlier in the birth order. However, the back leg forward and tremble did not differ before and after births, and the last increased with birth order. The authors concluded that these behaviours with consistent individual variation might be quantitatively associated with pain [91].

7. Managing Labour Pain

Pain relief in labour promotes maternal comfort and prevents the undesirable consequences of the catecholamine-mediated stress response [150]. Hence, the prevention and relief of pain through an accurate evaluation of this and the appropriate use of analgesics should be a priority for human medicine and animal science and production. The intention of pain relief during labour is to make the parturient relatively free of pain and also be able to participate in the childbirth experience. Preferably, side effects or associated risks for both the dam and newborn should be avoided. As presented in Table 6, there are several methods to relieve labour pain in humans and domestic animal species [21,150–165]. However, adverse effects have been reported, which could make them not ideal [151].

Table 6. Pharmacological analgesia for labour pain in humans and domestic animals.

Species	Route	Drug	Observations	References
Women	Inhalational analgesia (reduce pain perception).	1. Entonox. 2. Isoflurane. 3. Desflurane. 4. Sevoflurane.	Use limited to developed countries.	[13,21,150]
	Parenteral opioids (reduce pain perception).	1. Pethidine. 2. Morphine. 3. Diamorphine. 4. Fentanyl. 5. Remifentanil. 6. Meperidine.	Adverse effects have been observed.	[13,21,150,153–159]
	Regional analgesia (reduce pain transmission).	1. Epidural. 2. Combined spinal-epidural (CSE). Lidocaine. Xilacina. Levobupivacaine. Bupivacaine.	Epidural analgesia is the "gold standard" for labour analgesia.	[13,153,158,160,161]

Table 6. Cont.

Species	Route	Drug	Observations	References
Bitch	Intravenous Intramuscular.	1. Opioids (Methadone, Morphine) 2. AINEs (Meloxicam).	Use limited to caesarean section.	[162,163]
Rat	Intrathecal.	Morphine.	Used in research model for labour pain.	[146]
Sow	Intramuscular, Oral.	Ketoprofen. Meloxicam.		[3,164,165]

First of all, to choose the ideal analgesic to relieve pain during labour should consider: (1) provision of rapid pain relief in both the initial and secondary stage of labour without hazard or side effects to either the dam or the foetus; and (2) preservation of the mother's motion capability and self-sufficiency during labour [69]. There are a wide variety of pharmacological techniques to alleviate pain, including oral pills, inhalation analgesia (i.e., nitrous oxide gas), intravenous and intramuscular opioids (pethidine or diamorphine), or narcotic drugs such as morphine, and various types of local (paracervical or pudendal block) and regional analgesia/anaesthesia (epidural or spinal anaesthetic and combined spinal and epidural (CSE) techniques) [159] (Figure 3). However, the criteria for choosing the analgesic method depend on the patient's medical situation, the course of parturition and especially the in-place support availability [166]. Epidural analgesia (EA) is an effective method for pain management that adapts to the varied pain patterns experienced by female humans during childbirth [167]. Hence, analgesia can be achieved during the first stage of parturition by administering paracervical, paravertebral, or lumbar epidurals with local anaesthetics [13,153]. Although EA is considered the "gold standard", it has been linked with a raised risk of assisted vaginal childbirth, maternal hypotension, motor blockade, maternal fever, urinary retention, oxytocin administration, the more prolonged second stage of parturition, and an increased risk of caesarean delivery due to foetal distress [21]. Therefore, a high number of studies compared the use of EA versus opioids, such as Remifentanil, which is suitable for administration through patient-controlled analgesia (PCA) [154,155]. However, opioids by parenteral administration are not as effective as EA and are not seen as an option when EA is available [157]. Although other drugs have been used, the EA remains the most efficient for releasing labour pain [4]. However, it should be highlighted that this method of analgesia is expensive and not available in many institutions [146]. In recent years, a growing interest has emerged in the combined spinal–epidural method (CSE) for labour analgesia, with the benefit of utilising lower doses of local anaesthetics and rapid analgesia [4,150,161,168]. Advancement in knowledge and developing a novel therapy for labour pain has been partially limited by the absence of an animal model for this kind of pain. This observation supposes that the physiology and pharmacology of labour pain vary from other kinds of pain. As research on various types of pain progresses, it is accepted that analgesia pharmacology significantly diverges [146]. The rat model is the most frequently used in biomedical research. It has been described that intrathecal administration of morphine (0.035–3.5 µg/h) approximately one day before delivery has an antinociceptive effect [146]. Despite the frequent use of conventional analgesic drugs like opioids, NSAIDs, and local anaesthetics to handle pain during a surgical procedure in laboratory animals [169], little is known about their efficacy and adverse effects on labour pain. Pharmacological management of analgesia at delivery (whelping) in bitches is only performed in the case of caesarean section (CS). The CS anaesthetic protocol model should produce good muscle relaxation, analgesia, and narcosis for optimal and secure surgical conditions for the bitch [170,171] and should not harm the vitality and survival of newborns [171,172]. The two classes of analgesics commonly used in veterinary patients are opioids and the nonsteroidal anti-inflammatory analgesics (NSAIDs) [163], despite analgesics and NSAIDs being problematic for CS in pregnant animals and humans [173,174]. The main reason is that opioids produce analgesia; however, they can reach the placenta and induce a critical central nervous system and respiratory depression in newborns. Nonetheless, Mathews [163] observes a consensus on the adequate safety of a 0.1 mg/kg single intravenous dose of meloxicam shortly after the parturition of the puppies. Nonsteroidal anti-inflammatory drugs are used to treat various conditions, such as post-operative pain [175].

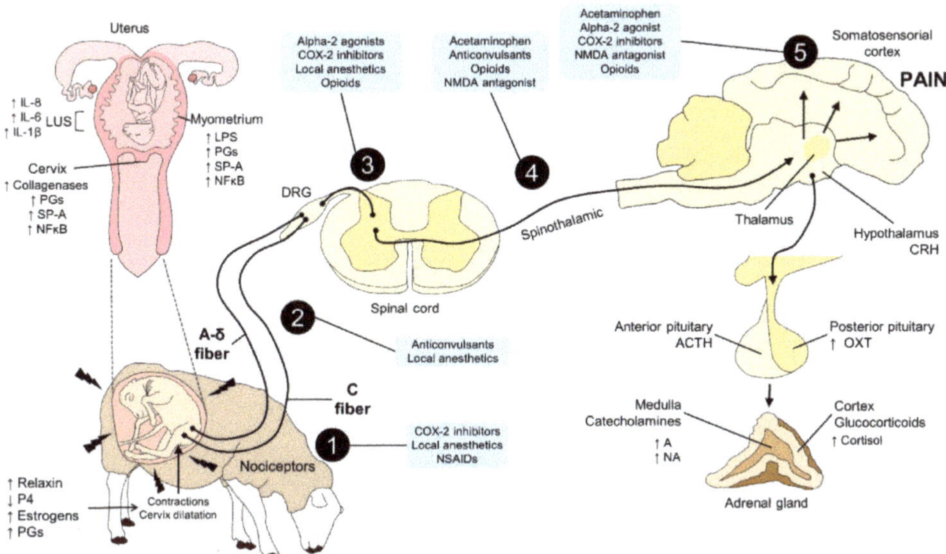

Figure 3. Sites of action of analgesics in the nociceptive pathway of parturition. During labour, hormonal factors such as increases in the concentrations of relaxin, estrogens, and PGs lead to cervical dilation and contractions. In the uterine tissue, LUS, myometrium, and cervix, different pro-inflammatory or degrading substances (such as collagenases) initiate an inflammatory response for the onset of softening and ripening the cervix. All of these factors activate peripheral nociceptors (A-δ and C fibers). These transduce the noxious stimulus into an electrical signal that is transmitted to the DRG of the spinal cord. Once in this structure, nociception modulation occurs, and drugs such as alpha-2 agonists, COX-2 inhibitors, local anesthetics, and opioids act at this site. These spinal neurons project to brain areas such as the thalamus or the somatosensory cortex, where pain perception occurs. Moreover, the activation of thalamic and hypothalamic areas leads to the secretion of other hormones such as ACTH or OXT. The adrenal gland contributes to the increase in circulating levels of A, NA, and cortisol, due to ACTH (cortisol) or by sympathetic influence (A, NA). Local anesthetics, NSAIDs, COX-2 inhibitors, opioids, acetaminophen, NMDA antagonist, and other analgesics act in each of the mentioned stages of the nociceptive pathway, depending on the nature of the drug. 1. transduction; 2: transmission; 3: modulation; 4: projection; 5: perception; A: adrenaline; ACTH: adrenocorticotropic hormone; CRH: corticotropin-releasing hormone; DRG: dorsal root ganglion; IL: interleukin; LPS: lipopolysaccharides; LUS: low uterine segment; NA: noradrenaline; NFκB: nuclear factor κB; NMDA: N-methyl-D-aspartate; NSAIDs: non-steroidal anti-inflammatory drugs; OXT: oxytocin; P4: progesterone; PG: prostaglandins; SP-A: surfactant protein A.

Opioids are not registered for veterinary use in many countries, and therefore availability and accessibility of these drugs should be considered. The FDA in the US has approved some opioids specifically for use in animals, mainly in cats, dogs, horses, and wildlife. Due to the restricted products approved for use in animals, veterinarians use opioids approved for use in humans to control pain in their patients, but they must also follow regulations for extra-label use in animals, along with risk evaluations to assure that the benefits outweigh some risks [176].

Veterinary practitioners frequently use opioids to control pain associated with CS. However, the sensitivity to respiratory and CNS depression of opioids in newborn puppies due to the immaturity of their CNS is higher. Nevertheless, a small sublingual drop of naloxone in depressed puppies after delivery reverses the depressive effects of opioids [163]. While the use of analgesics has been widespread in companion animal medicine, some studies show that cattle often do not receive analgesia during painful procedures or conditions [177]. In pigs, the evidence of the adverse effects, on either the sow or piglets, caused by labour pain has increased the research interest on the use of analgesics around delivery; aiming to enhance/improve health, well-being, and productivity of both the sow and piglets [3,164,178,179]. Most studies are based on the analgesic administration after farrowing. Homedes et al. [180] and Sabaté et al. [178] reported the benefits of NSAIDs administration in the sow within 12 h after farrowing on the reduction of piglet mortality. Mainau et al. [179] reported an increase of the average daily weight gain and immunoglobulin uptake of low-birth-weight piglets (<1180 g) after the administration of meloxicam to the sow intramuscularly 90 min after farrowing.

Similarly, Viitasaari et al. [181] reported higher activity in young sows treated with ketoprofen for 3 days post-partum. In a recent study, 3 mg per kg of body weight or 1ml per 33 kg of body weight of ketoprofen was administered to gilts 90 min after the expulsion of the last piglet. Even though the study failed to prove clear production benefits of ketoprofen, high individual sow variation in piglet mortality, with some with regular performing and most of the piglet mortality happening in a low number of sows indicates a potential for targeted NSAID use [165]. In contrast, Mainau et al. [179] observed an improvement in average daily gain and immunoglobulin transfer in piglets, and reduced time sows spent lying down after oral administration of meloxicam to the sow as soon as possible at the beginning of the farrowing. Authors concluded that the oral meloxicam administration at the beginning of farrowing in pluriparous sows raised the serum IgG concentration of piglets and improved their preweaning growth [179]. The described effects of NSAIDs include reduction of post-partum oedema, pain, and inflammation and anti-endotoxic actions and help reduce preweaning mortality and piglet growth retardation from Post-Partum Dysgalactia Syndrome (PPDS).

8. Consequences of Pain

In female humans, the adverse outcomes of labour pain are assumed to arise primarily from changes in the maternal respiratory pattern and catecholamine-mediated stress response [5]. Brownridge [182] points out that the possible physiological consequences of critical parturition pain may involve increased oxygen consumption and hyperventilation, hypocarbia, and respiratory alkalosis. Furthermore, it may also involve autonomic stimulation and catecholamine release with gastric interference and hyperacidity, lipolysis, raised peripheral vascular resistance, cardiac output, blood pressure, decreased placental perfusion, and reduced uterine activity. These responses are hypothesised to produce maternal metabolic acidaemia, foetal acidosis, and dysfunctional labour. Such effects can be mostly harmless in the course of eutocic delivery [95]. However, pain that an individual is unable to control leads to increased release of catecholamines, reduced uterine contractions and the consequent dystocia, fatigue, foetal suffering, and postpartum posttraumatic stress disorder [5,96]. Several authors have reported that prolonged labour induces distress, fear and exhaustion, and the risk of damage, intrapartum and perinatal mortality. It also leads to increased use of oxytocin, higher frequency of CS and device use in vaginal delivery (i.e., vacuum or forceps), postpartum fever and reduced umbilical pH [95,183,184]. There are also descriptions that stress generated by parturition pain influences the diminished oxytocin level and prolonged labour. Beigi et al. [185] mention that fear of labour pain is one of the most important reasons female humans decide for a CS, even in pluriparous women [95,186]. Usually, a prolonged second phase during labour is associated with greater foetal and maternal complications like the atonic uterus, postpartum bleeding, and perineal trauma, operative vaginal delivery and episiotomy, increased risk of infection, foetal hypoxia, asphyxia, injuries, and perinatal mortality [104,187–189]. Women can experience different types of pain and discomfort following childbirth. It may involve surgical pain after CS, perineal pain after injury, or episiotomy through vaginal delivery; breastfeeding nipple pain and postpartum cramps pains associated with uterine involution. Ultimately, pain can also inhibit normal breastfeeding after delivery, diminishing the mother's ability to care for the newborn and may threaten to establish a good-quality mother-baby relationship/interaction/bond [190]. The use of the rat model in pain investigation has played a pivotal role in recognising the types of nerves, receptors, ion channels, mediators, and biochemical pathways implicated in the origin, transmission, transduction, perception, and management of pain. However, there is no information about the consequences of labour pain in rats. Pregnancies in bitches are unique amongst domestic animal species because parturition is prolonged (up to 24–36 h), and newborns are very susceptible to environmental changes [191]. During parturition, physiological parameters change significantly due to pain, fear, and uterine contractions [192]. A high heart rate in the female dog results from a mixture of stress and myometrial contractions (uterine and abdominal) during labour [191]. As a response to pain, dam hyperventilation has unfavourable foetal outcomes. The consequences include respiratory alkalosis, compensatory progressive metabolic acidosis, becoming critical with the labour progress, and inducing foetal acidosis. Also, there are episodes of hypoventilation, leading to haemoglobin desaturation within contractions; and uterine vasoconstriction [191]. Stress hormones also drive to lipolysis caused by the liberation of free fatty acids (easily transferable through the placenta) and hyperglycaemia, which worsen foetal hypoxia. The above changes increase foetal acidosis, becoming progressively critical as parturition progress [193]. The passage from intrauterine to extrauterine life is usually accompanied by various degrees of hypoxia, which is very well tolerated by newborns [194]. However, if the labour time is prolonged, the survival of the newborn could be compromised. Birth is the most critical phase for neonates of all species and is associated with perinatal mortality and adverse effects on well-being [3,195,196]. In dogs, the stress associated with the birth process is usually an underlying cause of neonatal mortality, reported to be between 5% and 35% [196]. As we have presented, the process of farrowing in pigs, especially when dystocia occurs, has a high impact on sow's well-being. Additionally, the stress of birth might impact the survival of newborns [195–202]. The presence of dystocia increases the risk of various conditions in the dam, including endometritis, vulvar secretion, placental retention, mastitis-metritis-agalactia syndrome, and impaired fertility [13,126,152,203]. On the other hand, piglets are particularly susceptible to intrapartum asphyxia [195,204]. Hypoxia and metabolic acidosis are consequences of asphyxia. They can cause severe effects on piglets' health, vitality, and postnatal performance due to a reduced ability to reach a teat, consequently leading to an inadequate colostrum intake and passive immunity acquisition by the piglet [195]. On the other hand, acidosis can also cause hypothermia and reduce the survival of neonates [197,205]. An early indicator of foetal distress due to intrauterine hypoxia is meconium ejection into the amniotic sac, causing staining on the foetal skin and inhaling and lodging of meconium in severe cases the lungs [25,199,206]. On the other hand, the pain and stress associated with labour inhibit the release of oxytocin, leading to prolonged labour and reduction of colostrum and milk yield, consequently reducing nutritional and immunity supply to piglets [127,207]. Pain and stress of farrowing can also lead to restlessness and even aggressiveness in sows and bitches [208,209]. Insufficient milk production by sows with consequent malnutrition of the piglets can be directly responsible for 6 to 17% of the total pre-weaning mortality in commercial pig farms [210]. Metabolic and endocrine disorders in the sow, bacterial infections such as metritis (e.g., due to improper obstetric management), contribute to the decrease in prolactin secretion, and in primiparous sows, adequate suckling of their piglets is also inhibited [211].

9. Methods to Improve Pain Management during Parturition

Pain management in females during parturition and the controversy over the use of analgesics involves their dose-dependent tocolytic effect at this phase and its impact on uterine activity [212]. It is necessary to consider the pharmacokinetics of the drug and its affinity for the receptor. For example, a drug with a short plasma half-life may be an option to avoid the deleterious effects of pain during labor [213].

Likewise, pharmacokinetics helps to understand the effect that drugs may have on the neuroendocrine control of parturition, which has a fundamental physiological role. An example of this is morphine. This opioid can increase the plasma clearance rate of oxytocin [214], a hormone necessary during parturition. Therefore, drugs that have a higher affinity to its receptor or a prolonged plasma half-life could decrease the elimination time of this hormone.

On the other hand, an increase in uterine activity has been observed after the administration of amines [215], while drugs such as firocoxib do not affect uterine contractility [216]. These observations also suggest the implementation of analgesics with alternative pathways such as the inhibition of Ca2+ channels, such as gabapentinoids [217], NMDA receptor antagonists [213], or cannabinoid receptors [218], whose literature to date has not reported an alteration in uterine activity.

Other approaches that have been proposed to monitor and manage animals during the labour stage, such as motion-based technologies to infer parturition time even in wild species such as the Caribou (*Rangifer tarandus caribou*), are under development and have not yet reported conclusive data [219]. Machine-learning techniques to monitor wildlife and predict parturition have shown high rates of success (76% to 100% in wild ungulates [220]. Likewise, the monitoring of immune function, known as "immunity shift" has been proposed as a technique to evaluate health and immunity function, through rapid blood tests based on nanoparticles, and regularity of the physiological response in companion and farm animals [221].

10. Conclusions

Labour is a crucial moment for numerous species and also considered the most painful episode of females. Pain is fundamentally a psychophysiological phenomenon. However, unfortunately, pain in animals is not regularly recognised and is treated inappropriately. If parturition is a painful process identified in humans, it should also be considered painful in animals. To date, most publications on labour research focus on associated endocrine modifications, and there are fewer studies related to pain in the birth process. More robust comprehension of pain during farrowing in domestic animal species can generate new hypotheses and outcomes concerning its physiology and new pharmacological therapies, mainly concerning the relevance of analgesics and the welfare implications.

Author Contributions: Conceptualization, D.M.-R., R.M. and D.V.-G.; investigation, D.M.-R., R.M., J.M.-B. and H.B.-G.; writing—original draft preparation, D.M.-R., J.M.-B. and D.V.-G.; writing—review and editing, J.M.-B., D.M.-R., D.V.-G. and A.D.-O.; project administration, D.M.-R., J.M.-B. and R.M. All authors have read and agreed to the published version of the manuscript.

Funding: This research received no external funding.

Institutional Review Board Statement: Not applicable.

Conflicts of Interest: The authors declare that the research was conducted in the absence of any commercial or financial relationships that could be construed as a potential conflict of interest.

References

1. Van Rens, B.T.T.M.; Van Der Lende, T. Parturition in gilts: Duration of farrowing, birth intervals and placenta expulsion in relation to maternal, piglet and placental traits. *Theriogenology* **2004**, *62*, 331–352. [CrossRef]
2. Lezama-García, K.; Mariti, C.; Mota-Rojas, D.; Martínez-Burnes, J.; Barrios-García, H.; Gazzano, A. Maternal behaviour in domestic dogs. *Int. J. Vet. Sci. Med.* **2019**, *7*, 20–30. [CrossRef]
3. Ison, S.H.; Jarvis, S.; Hall, S.A.; Ashworth, C.J.; Rutherford, K.M.D. Periparturient behavior and physiology: Further insight into the farrowing process for primiparous and multiparous sows. *Front. Vet. Sci.* **2018**, *5*, 122. [CrossRef]
4. Mędrzycka-Dabrowska, W.; Czyż-Szypenbejl, K.; Pietrzak, J. A review of randomized trials comparisons of epidural with parenteral forms of pain relief during labour and its impact on operative and cesarean delivery rate. *Ginekol. Pol.* **2018**, *89*, 460–467. [CrossRef]

5. Lowe, N.K. The nature of labor pain. *Am. J. Obstet. Gynecol.* **2002**, *186*, S16–S24. [CrossRef]

6. Loeser, J.D.; Treede, R.D. The Kyoto protocol of IASP basic pain terminology. *Pain* **2008**, *137*, 473–477. [CrossRef]

7. IASP. Classification of Chronic Pain. In *Descriptions of Chronic Pain Syndromes and Definitions Of Pain Terms*; Merskey, H., Ed.; IASP Press: Seattle, WA, USA, 1994; p. 222.

8. Treede, R.D.; Rief, W.; Barke, A.; Aziz, Q.; Bennett, M.I.; Benoliel, R.; Cohen, M.; Evers, S.; Finnerup, N.B.; First, M.B.; et al. Chronic pain as a symptom or a disease: The IASP classification of chronic pain for the international classification of diseases (ICD-11). *Pain* **2019**, *160*, 19–27. [CrossRef]

9. Aydede, M. Does the IASP definition of pain need updating? *Pain Rep.* **2019**, *4*, e777. [CrossRef]

10. Hernández-Avalos, I.; Mota-Rojas, D.; Mora-Medina, P.; Martínez-Burnes, J.; Casas Alvarado, A.; Verduzco-Mendoza, A.; Lezama-García, K.; Olmos-Hernandez, A. Review of different methods used for clinical recognition and assessment of pain in dogs and cats. *Int. J. Vet. Sci. Med.* **2019**, *7*, 43–54. [CrossRef]

11. Cohen, M.; Quintner, J.; van Rysewyk, S. Reconsidering the International Association for the Study of Pain definition of pain. *Pain Rep.* **2018**, *3*, e634. [CrossRef]

12. Raja, S.N.; Carr, D.B.; Cohen, M.; Finnerup, N.B.; Flor, H.; Gibson, S.; Keefe, F.J.; Mogil, J.S.; Ringkamp, M.; Sluka, K.A.; et al. The revised International Association for the Study of Pain definition of pain: Concepts, challenges, and compromises. *Pain* **2020**, *161*, 1976–1982. [CrossRef]

13. Labor, S.; Maguire, S. The Pain of Labour. *Rev. Pain* **2008**, *2*, 15–19. [CrossRef]

14. Nanji, J.A.; Carvalho, B. Pain management during labor and vaginal birth. *Best Pract. Res. Clin. Obstet. Gynaecol.* **2020**, *67*, 100–112. [CrossRef]

15. Catheline, G.; Touquet, B.; Besson, J.-M.; Lombard, M.-C. Parturition in the Rat. *Anesthesiology* **2006**, *104*, 1257–1265. [CrossRef]

16. Mathews, K.; Kronen, P.W.; Lascelles, D.; Nolan, A.; Robertson, S.; Steagall, P.V.; Wright, B.; Yamashita, K. guidelines for recognition, assessment and treatment of Pain. *J. Small Anim. Pract.* **2014**, *55*, E10–E68. [CrossRef]

17. Herskin, D.G.P. Pain in pigs: Characterisation, mechanisms an indicators. In *Advances in Pig Welfare*; Spinka, M., Ed.; Woodhead Publishing: Kidlington, UK, 2018; pp. 325–356.

18. Gregory, N.G. (Ed.) *Physiology and Behaviour of Animal Suffering*; Blackwell Publishing: Oxford, UK, 2004; pp. 1–260.

19. Mainau, E.; Manteca, X. Pain and discomfort caused by parturition in cows and sows. *Appl. Anim. Behav. Sci.* **2011**, *135*, 241–251. [CrossRef]

20. da Rocha Pereira, R.; Franco, S.C.; Baldin, N. Pain and the protagonism of women in parturition. *Braz. J. Anesthesiol.* **2011**, *61*, 376–388. [CrossRef]

21. Trehan, G.; Gonzalez, M.N.; Kamel, I. Pain management during labor part 2: Techniques for labor analgesia. *Top. Obs. Gynecol.* **2016**, *36*, 1–10. [CrossRef]

22. Morton, D.B.; Griffiths, P.H. Guidelines on the recognition of pain, distress and discomfort in experimental animals and an hypothesis for assessment. *Vet. Rec.* **1985**, *116*, 431–436. [CrossRef]

23. Buchanan, K.L.; Burt de Perera, T.; Carere, C.; Carter, T.; Hailey, A.; Hubreacht, R.; Jenning, D.J.; Metcalfe, N.B.; Pitcher, T.E.; Péron, F.; et al. Guidelines for the Use of Animals Guidelines for the treatment of animals in behavioural research and teaching. *Anim. Behav.* **2012**, *83*, 301–309. [CrossRef]

24. Veronesi, M.C.; Panzani, S.; Faustini, M.; Rota, A. An Apgar scoring system for routine assessment of newborn puppy viability and short-term survival prognosis. *Theriogenology* **2009**, *72*, 401–407. [CrossRef]

25. Mota-Rojas, D.; Martínez-Burnes, J.; Napolitano, F.; Domínguez-Muñoz, M.; Guerrero-Legarreta, I.; Mora-Medina, P.; Ramírez-Necoechea, R.; Lezama-García, K.; González-Lozano, M. Dystocia: Factors affecting parturition in domestic animals. *CAB Rev. Perspect. Agric. Vet. Sci. Nutr. Nat. Resour.* **2020**, *15*, 1–16. [CrossRef]

26. Menon, R.; Bonney, E.A.; Condon, J.; Mesiano, S.; Taylor, R.N. Novel concepts on pregnancy clocks and alarms: Redundancy and synergy in human parturition. *Hum. Reprod. Update* **2016**, *22*, 535–560. [CrossRef]

27. Hadley, E.E.; Richardson, L.S.; Torloni, M.R.; Menon, R. Gestational tissue inflammatory biomarkers at term labor: A systematic review of literature. *Am. J. Reprod. Immunol.* **2018**, *79*, e12776. [CrossRef]

28. Bernitz, S.; Dalbye, R.; Øian, P.; Zhang, J.; Eggebø, T.M.; Blix, E. Study protocol: The Labor Progression Study, LAPS—Does the use of a dynamic progression guideline in labor reduce the rate of intrapartum cesarean sections in nulliparous women? A multicenter, cluster randomized trial in Norway. *BMC Pregnancy Childbirth* **2017**, *17*, 370. [CrossRef]

29. Liao, J.B.; Buhimschi, C.S.; Norwitz, E.R. Normal labor: Mechanism and duration. *Obstet. Gynecol. Clin. N. Am.* **2005**, *32*, 145–164. [CrossRef]

30. Hutchison, J.; Mahdy, H.; Hutchison, J. *Stages of Labor*; StatPearls Publishing LLC: San Francisco, CA, USA, 2020.

31. Maghoma, J.; Buchmann, E.J. Maternal and fetal risks associated with prolonged latent phase of labour. *J. Obstet. Gynaecol.* **2002**, *22*, 16–19. [CrossRef]

32. Lauzon, L.; Hodnett, E.D. Antenatal education for self-diagnosis of the onset of active labour at term. *Cochrane Database Syst. Rev.* **1998**, *1998*, CD000935. [CrossRef]

33. Chelmow, D.; Kilpatrick, S.J.; Laros, R.K. Maternal and neonatal outcomes after prolonged latent phase. *Obstet. Gynecol.* **1993**, *81*, 486–491.

34. Hanley, G.E.; Munro, S.; Greyson, D.; Gross, M.M.; Hundley, V.; Spiby, H.; Janssen, P.A. Diagnosing onset of labor: A systematic review of definitions in the research literature. *BMC Pregnancy Childbirth* **2016**, *16*, 71. [CrossRef]

35. Khasawneh, W.; Obeidat, N.; Yusef, D.; Alsulaiman, J.W. The impact of cesarean section on neonatal outcomes at a university-based tertiary hospital in Jordan. *BMC Pregnancy Childbirth* **2020**, *20*, 335. [CrossRef] [PubMed]
36. Uystepruyst, C.; Coghe, J.; Dorts, T.; Harmegnies, N.; Delsemme, M.H.; Art, T.; Lekeux, P. Optimal timing of elective caesarean section in Belgian White and Blue breed of cattle: The calf's point of view. *Vet. J.* **2002**, *163*, 267–282. [CrossRef] [PubMed]
37. Tribe, R.M.; Taylor, P.D.; Kelly, N.M.; Rees, D.; Sandall, J.; Kennedy, H.P. Parturition and the perinatal period: Can mode of delivery impact on the future health of the neonate? *J. Physiol.* **2018**, *596*, 5709–5722. [CrossRef]
38. Sandall, J.; Tribe, R.M.; Avery, L.; Mola, G.; Visser, G.H.; Homer, C.S.; Gibbons, D.; Kelly, N.M.; Kennedy, H.P.; Kidanto, H.; et al. Short-term and long-term effects of caesarean section on the health of women and children. *Lancet* **2018**, *392*, 1349–1357. [CrossRef]
39. Renfrew, M.J.; McFadden, A.; Bastos, M.H.; Campbell, J.; Channon, A.A.; Cheung, N.F.; Silva, D.R.A.D.; Downe, S.; Kennedy, H.P.; Malata, A.; et al. Midwifery and quality care: Findings from a new evidence-informed framework for maternal and newborn care. *Lancet* **2014**, *384*, 1129–1145. [CrossRef]
40. Behnia, F.; Taylor, B.D.; Woodson, M.; Kacerovsky, M.; Hawkins, H.; Fortunato, S.J.; Saade, G.R.; Menon, R. Chorioamniotic membrane senescence: A signal for parturition? *Am. J. Obstet. Gynecol.* **2015**, *213*, 359.e1–359.e16. [CrossRef]
41. Menon, R. Initiation of human parturition: Signaling from senescent fetal tissues via extracellular vesicle mediated paracrine mechanism. *Obstet. Gynecol. Sci.* **2019**, *62*, 199. [CrossRef]
42. Hashemi Asl, B.M.; Vatanchi, A.; Golmakani, N.; Najafi, A. Relationship between behavioral indices of pain during labor pain with pain intensity and duration of delivery. *Electron. Physician* **2018**, *10*, 6240–6248. [CrossRef]
43. Kovác, G.; Tóthová, C.; Nagy, O.; Seidel, H. Acute phase proteins during the reproductive cycle of sows. *Acta Vet. Brno.* **2008**, *58*, 459–466. [CrossRef]
44. Verheyen, A.J.M.; Maes, D.G.D.; Mateusen, B.; Deprez, P.; Janssens, G.P.J.; de Lange, L.; Counotte, G. Serum biochemical reference values for gestating and lactating sows. *Vet. J.* **2007**, *174*, 92–98. [CrossRef]
45. Kostro, K.; Wawron, W.; Szczubiał, M.; Luft-Deptula, D.; Gliński, Z. C-reactive protein in monitoring and evaluation of effects of therapy of the MMA syndrome of sows. *Pol. J. Vet. Sci.* **2003**, *6*, 235–238.
46. Winkler, M.; Kemp, B.; Fischer, D.; Maul, H.; Hlubekl, M.; Rath, W. Tissue concentrations of cytokines in the lower uterine segment during preterm parturition. *J. Perinat. Med.* **2001**, *29*, 519–527. [CrossRef] [PubMed]
47. Kemp, B.; Winkler, M.; Maas, A.; Maul, H.; Ruck, P.; Reineke, T.; Rath, W. Cytokine concentrations in the amniotic fluid during parturition at term: Correlation to lower uterine segment values and to labor. *Acta Obstet. Gynecol. Scand.* **2002**, *81*, 938–942. [CrossRef] [PubMed]
48. Winkler, M.; Kemp, B.; Fischer, D.C.; Ruck, P.; Rath, W. Expression of adhesion molecules in the lower uterine segment during term and preterm parturition. *Microsc. Res. Tech.* **2003**, *60*, 430–444. [CrossRef] [PubMed]
49. Condon, J.C.; Hardy, D.B.; Kovaric, K.; Mendelson, C.R. Up-regulation of the progesterone receptor (pr)-c isoform in laboring myometrium by activation of nuclear factor-κb may contribute to the onset of labor through inhibition of PR function. *Mol. Endocrinol.* **2006**, *20*, 764–775. [CrossRef] [PubMed]
50. Lindström, T.M.; Bennett, P.R. The role of nuclear factor kappa B in human labour. *Reproduction* **2005**, *130*, 569–581. [CrossRef] [PubMed]
51. Terzidou, V.; Lee, Y.; Lindström, T.; Johnson, M.; Thornton, S.; Bennett, P.R. Regulation of the human oxytocin receptor by nuclear factor-κB and CCAAT/Enhancer-binding protein-β. *J. Clin. Endocrinol. Metab.* **2006**, *91*, 2317–2326. [CrossRef]
52. Thomson, A.J.; Telfer, J.F.; Young, A.; Campbell, S.; Stewart, C.J.R.; Cameron, I.T.; Greer, I.A.; Norman, J.E. Leukocytes infiltrate the myometrium during human parturition: Further evidence that labour is an inflammatory process. *Hum. Reprod.* **1999**, *14*, 229–236. [CrossRef]
53. Young, A.; Thomson, A.J.; Ledingham, M.; Jordan, F.; Greer, I.A.; Norman, J.E. Immunolocalization of proinflammatory cytokines in myometrium, cervix, and fetal membranes during human parturition at term1. *Biol. Reprod.* **2002**, *66*, 445–449. [CrossRef]
54. Osman, I.; Young, A.; Jordan, F.; Greer, I.A.; Norman, J.E. Leukocyte density and proinflammatory mediator expression in regional human fetal membranes and decidua before and during labot at term. *J. Soc. Gynecol. Investig.* **2006**, *13*, 97–103. [CrossRef]
55. Hutchinson, J.L.; Rajagopal, S.P.; Yuan, M.; Norman, J.E. Lipopolysaccharide promotes contraction of uterine myocytes via activation of Rho/ROCK signaling pathways. *FASEB J.* **2014**, *28*, 94–105. [CrossRef]
56. Rajagopal, S.P.; Hutchinson, J.L.; Dorward, D.A.; Rossi, A.G.; Norman, J.E. Crosstalk between monocytes and myometrial smooth muscle in culture generates synergistic pro-inflammatory cytokine production and enhances myocyte contraction, with effects opposed by progesterone. *Mol. Hum. Reprod.* **2015**, *21*, 672–686. [CrossRef] [PubMed]
57. Hardy, D.B.; Janowski, B.A.; Corey, D.R.; Mendelson, C.R. Progesterone receptor plays a major antiinflammatory role in human myometrial cells by antagonism of nuclear factor-κB activation of cyclooxygenase 2 expression. *Mol. Endocrinol.* **2006**, *20*, 2724–2733. [CrossRef]
58. Condon, J.C.; Jeyasuria, P.; Faust, J.M.; Mendelson, C.R. Surfactant protein secreted by the maturing mouse fetal lung acts as a hormone that signals the initiation of parturition. *Proc. Natl. Acad. Sci. USA* **2004**, *101*, 4978–4983. [CrossRef] [PubMed]
59. Hua, R.; Pease, J.E.; Sooranna, S.R.; Viney, J.M.; Nelson, S.M.; Myatt, L.; Bennett, P.R.; Johnson, M.R. Stretch and inflammatory cytokines drive myometrial chemokine expression via NF-κB activation. *Endocrinology* **2012**, *153*, 481–491. [CrossRef] [PubMed]
60. Singh, N.; Herbert, B.; Sooranna, G.R.; Orsi, N.M.; Edey, L.; Dasgupta, T.; Sooranna, S.R.; Yellon, S.M.; Johnson, M.R. Is myometrial inflammation a cause or a consequence of term human labour? *J. Endocrinol.* **2017**, *235*, 69–83. [CrossRef] [PubMed]

61. Whitburn, L.Y.; Jones, L.E.; Davey, M.A.; McDonald, S. The nature of labour pain: An updated review of the literature. *Women Birth* **2019**, *32*, 28–38. [CrossRef] [PubMed]
62. Runcan, E.E.; Coutinho da Silva, M.A. Whelping and dystocia: Maximizing success of medical management. *Top. Companion Anim. Med.* **2018**, *33*, 12–16. [CrossRef]
63. Taverne, M.; Noakes, D. Parturition and the care of parturient animals, including the newborn. In *Arthur's Veterinary Reproduction and Obstetrics*, 9th ed.; Noakes, D.E., Parkinson, T.J., England, G.C.W., Eds.; Saunders Elsevier: Philadelphia, PA, USA, 2001; pp. 155–187.
64. Challis, J.R.G.; Lye, S.J. Parturition. *Oxf. Rev. Reprod. Biol.* **1986**, *8*, 61–129.
65. Jenkin, G.; Young, I.R. Mechanisms responsible for parturition; the use of experimental models. In Proceedings of the Animal Reproduction Science. *Anim. Reprod. Sci.* **2004**, *82–83*, 567–581. [CrossRef]
66. Taverne, M.A.M. Physiology of parturition. *Anim. Reprod. Sci.* **1992**, *28*, 433–440. [CrossRef]
67. Sennström, M.B.; Ekman, G.; Westergren-Thorsson, G.; Malmström, A.; Byström, B.; Endrésen, U.; Mlambo, N.; Norman, M.; Ståbi, B.; Brauner, A. Human cervical ripening, an inflammatory process mediated by cytokines. *Mol. Hum. Reprod.* **2000**, *6*, 375–381. [CrossRef] [PubMed]
68. Yellon, S.M. Immunobiology of Cervix Ripening. *Front. Immunol.* **2019**, *10*, 3156. [CrossRef]
69. Mousa, O.; Abdelhafez, A.A.; Abdelraheim, A.R.; Yousef, A.M.; Ghaney, A.A.; El Gelany, S. Perceptions and practice of labor pain-relief methods among health professionals conducting delivery in minia maternity units in Egypt. *Obstet. Gynecol. Int.* **2018**, *2018*, 3060953. [CrossRef]
70. Serruya, S.J.; Cecatti, J.G.; di Giacomo do Lago, T. Programa de Humanização no Pré-natal e Nascimento do Ministério da Saúde no Brasil: Resultados iniciais. *Cad. Saude Publica* **2004**, *20*, 1281–1289. [CrossRef] [PubMed]
71. Rowlands, S.; Permezel, M. Physiology of pain in labour. *Baillieres. Clin. Obstet. Gynaecol.* **1998**, *12*, 347–362. [CrossRef]
72. Miquelutti, M.A.; Silveira, C.; Cecatti, J.G. Kinesiologic tape for labor pain control: Randomized controlled trial. *Physiother. Theory Pract.* **2019**, *35*, 614–621. [CrossRef]
73. Stenglin, M.; Foureur, M. Designing out the Fear Cascade to increase the likelihood of normal birth. *Midwifery* **2013**, *29*, 819–825. [CrossRef] [PubMed]
74. Taverne, M.A.M.; Naaktgeboren, C.; Elsaesser, F.; Forsling, M.L.; van der Weyden, G.C.; Ellendorff, F.; Smidt, D. Myometrial electrical activity and plasma concentrations of progesterone, estrogens and oxytocin during late pregnancy and parturition in the miniature pig. *Biol. Reprod.* **1979**, *21*, 1125–1134. [CrossRef]
75. Gilbert, C.L.; Goode, J.A.; McGrath, T.J. Pulsatile secretion of oxytocin during parturition in the pig: Temporal relationship with fetal expulsion. *J. Physiol.* **1994**, *475*, 129–137. [CrossRef]
76. Olcese, J.; Beesley, S. Clinical significance of melatonin receptors in the human myometrium. *Fertil. Steril.* **2014**, *102*, 329–335. [CrossRef] [PubMed]
77. Sharkey, J.T.; Puttaramu, R.; Word, R.A.; Olcese, J. Melatonin synergizes with oxytocin to enhance contractility of human myometrial smooth muscle cells. *J. Clin. Endocrinol. Metab.* **2009**, *94*, 421–427. [CrossRef] [PubMed]
78. Kim, S.H.; Bennett, P.R.; Terzidou, V. Advances in the role of oxytocin receptors in human parturition. *Mol. Cell. Endocrinol.* **2017**, *449*, 56–63. [CrossRef] [PubMed]
79. Gorodeski, G.I.; Sheean, L.A.; Utian, W.H. Myometrial oxytocin receptors levels in the pregnant rat are higher in distal than in proximal portions of the horn and correlate with disparate oxytocin responsive myometrial contractility in these segments. *Endocrinology* **1990**, *127*, 1136–1143. [CrossRef] [PubMed]
80. Fuchs, A.-R. Prostaglandin F2alpha and oxytocin interactions in ovarian and uterine function. *J. Steroid Biochem.* **1987**, *27*, 1073–1080. [CrossRef]
81. Soloff, M.S. Uterine receptor for oxytocin: Effects of estrogen. *Biochem. Biophys. Res. Commun.* **1975**, *65*, 205–212. [CrossRef]
82. Gimpl, G.; Fahrenholz, F. The oxytocin receptor system: Structure, function, and regulation. *Physiol. Rev.* **2001**, *81*, 629–683. [CrossRef]
83. Sanders, J.; Peters, T.J.; Campbell, R. Techniques to reduce perineal pain during spontaneous vaginal delivery and perineal suturing: A UK survey of midwifery practice. *Midwifery* **2005**, *21*, 154–160. [CrossRef]
84. Lowe, N.K.; Roberts, J.E. The convergence between in-labor report and postpartum recall of parturition pain. *Res. Nurs. Health* **1988**, *11*, 11–21. [CrossRef]
85. Melzack, R.; Schaffelberg, D. Low-back pain during labor. *Am. J. Obstet. Gynecol.* **1987**, *156*, 901–905. [CrossRef]
86. Schabel, J.E.; Poppers, P.J. Lumbar Epidural Analgesia for Labor and Vaginal Delivery. *Gynecol. Obstet. Investig.* **1997**, *44*, 73–81. [CrossRef]
87. Jarvis, S.; Lawrence, A..; McLean, K..; Chirnside, J.; Deans, L..; Calvert, S. The effect of environment on plasma cortisol and β-endorphin in the parturient pig and the involvement of endogenous opioids. *Anim. Reprod. Sci.* **1998**, *52*, 139–151. [CrossRef]
88. Jarvis, S.; McLean, K.A.; Calvert, S.K.; Deans, L.A.; Chirnside, J.; Lawrence, A.B. The responsiveness of sows to their piglets in relation to the length of parturition and the involvement of endogenous opioids. *Appl. Anim. Behav. Sci.* **1999**, *63*, 195–207. [CrossRef]
89. Lawrence, A.B.; Petherick, J.C.; McLean, K.; Gilbert, C.L.; Chapman, C.; Russell, J.A. Naloxone prevents interruption of parturition and increases plasma oxytocin following environmental disturbance in parturient sows. *Physiol. Behav.* **1992**, *52*, 917–923. [CrossRef]

90. Jarvis, S.; McLean, K.A.; Chirnside, J.; Deans, L.A.; Calvert, S.K.; Molony, V.; Lawrence, A.B. Opioid-mediated changes in nociceptive threshold during pregnancy and parturition in the sow. *Pain* **1997**, *72*, 153–159. [CrossRef]
91. Ison, S.H.; Clutton, R.E.; Di Giminiani, P.; Rutherford, K.M.D. A review of pain assessment in pigs. *Front. Vet. Sci.* **2016**, *3*, 108. [CrossRef] [PubMed]
92. González-Lozano, M.; Mota-Rojas, D.; Orihuela, A.; Martínez-Burnes, J.; Di Francia, A.; Braghieri, A.; Berdugo-Gutiérrez, J.; Mora-Medina, P.; Ramírez-Necoechea, R.; Napolitano, F. Review: Behavioral, physiological, and reproductive performance of buffalo cows during eutocic and dystocic parturitions. *Appl. Anim. Sci.* **2020**, *36*, 407–422. [CrossRef]
93. Hautakangas, T.; Palomäki, O.; Eidstø, K.; Huhtala, H.; Uotila, J. Impact of obesity and other risk factors on labor dystocia in term primiparous women: A case control study. *BMC Pregnancy Childbirth* **2018**, *18*, 304. [CrossRef]
94. Sandström, A.; Altman, M.; Cnattingius, S.; Johansson, S.; Ahlberg, M.; Stephansson, O. Durations of second stage of labor and pushing, and adverse neonatal outcomes: A population-based cohort study. *J. Perinatol.* **2017**, *37*, 236–242. [CrossRef] [PubMed]
95. Akbarzadeh, M.; Nematollahi, A.; Farahmand, M.; Amooee, S. The effect of two-staged warm compress on the pain duration of first and second labor stages and apgar score in prim gravida women: A Randomized clinical trial. *J. Caring Sci.* **2018**, *7*, 21–26. [CrossRef]
96. Rooks, J.P. Labor pain management other than neuraxial: What do we know and where do we go next? *Birth* **2012**, *39*, 318–322. [CrossRef] [PubMed]
97. Slone, M.E.; James, S.R.; Murray, S.S.; Nelson, K.A.; Aswill, J.W. *Maternal-Child Nursing*, 5th ed.; Elsevier Saunders: St. Louis, MI, USA, 2018; pp. 1–1592.
98. Beagley, J.C.; Whitman, K.J.; Baptiste, K.E.; Scherzer, J. Physiology and treatment of retained fetal membranes in cattle. *J. Vet. Intern. Med.* **2010**, *24*, 261–268. [CrossRef] [PubMed]
99. Mollo, A.; Veronesi, M.C.; Cairoli, F.; Soldano, F. The use of oxytocin for the reduction of cow placental retention, and subsequent endometritis. *Anim. Reprod. Sci.* **1997**, *48*, 47–51. [CrossRef]
100. Zobel, R. Endometritis in simmental cows: Incidence, causes, and therapy options. *Turkish J. Vet. Anim. Sci.* **2013**, *37*, 134–140. [CrossRef]
101. Barraclough, R.A.C.; Shaw, D.J.; Boyce, R.; Haskell, M.J.; Macrae, A.I. The behavior of dairy cattle in late gestation: Effects of parity and dystocia. *J. Dairy Sci.* **2020**, *103*, 14–722. [CrossRef] [PubMed]
102. Cheng, Y.W.; Caughey, A.B. Defining and managing normal and abnormal second stage of labor. *Obstet. Gynecol. Clin. N. Am.* **2017**, *44*, 547–566. [CrossRef]
103. Zaki, M.N.; Hibbard, J.U.; Kominiarek, M.A. Contemporary labor patterns and maternal age. *Obstet. Gynecol.* **2013**, *122*, 1018–1024. [CrossRef]
104. Laughon, S.K.; Berghella, V.; Reddy, U.M.; Sundaram, R.; Lu, Z.; Hoffman, M.K. Neonatal and maternal outcomes with prolonged second stage of labor. *Obstet. Gynecol.* **2014**, *124*, 57–67. [CrossRef]
105. Senécal, J.; Xiong, X.; Fraser, W.D. Effect of fetal position on second-stage duration and labor outcome. *Obstet. Gynecol.* **2005**, *105*, 763–772. [CrossRef]
106. Feinstein, U.; Sheiner, E.; Levy, A.; Hallak, M.; Mazor, M. Risk factors for arrest of descent during the second stage of labor. *Int. J. Gynecol. Obstet.* **2002**, *77*, 7–14. [CrossRef]
107. Sizer, A.R.; Evans, J.; Bailey, S.M.; Wiener, J. A second-stage partogram. *Obstet. Gynecol.* **2000**, *96*, 678–683. [CrossRef]
108. Darvelid, A.W.; Linde-Forsberg, C. Dystocia in the bitch: A retrospective study of 182 cases. *J. Small Anim. Pract.* **1994**, *35*, 402–407. [CrossRef]
109. Linde Forsberg, C.; Persson, G. A survey of dystocia in the Boxer breed. *Acta Vet. Scand.* **2007**, *49*, 8. [CrossRef]
110. Veiga, G.A.L.; Milazzotto, M.P.; Nichi, M.; Lúcio, C.F.; Silva, L.C.G.; Angrimani, D.S.R.; Vannucchi, C.I. Gene expression of estrogen and oxytocin receptors in the uterus of pregnant and parturient bitches. *Braz. J. Med. Biol. Res.* **2015**, *48*, 339–343. [CrossRef]
111. Egloff, S.; Reichler, I.M.; Kowalewski, M.P.; Keller, S.; Goericke-Pesch, S.; Balogh, O. Uterine expression of smooth muscle alpha- and gamma-actin and smooth muscle myosin in bitches diagnosed with uterine inertia and obstructive dystocia. *Theriogenology* **2020**, *156*, 162–170. [CrossRef] [PubMed]
112. Lúcio, C.; Silva, L.; Rodrigues, J.; Veiga, G.; Vannucchi, C. Peripartum haemodynamic status of bitches with normal birth or dystocia. *Reprod. Domest. Anim.* **2009**, *44*, 133–136. [CrossRef]
113. Münnich, A.; Küchenmeister, U. Dystocia in numbers—Evidence-based parameters for intervention in the dog: Causes for dystocia and treatment recommendations. *Reprod. Domest. Anim.* **2009**, *44*, 141–147. [CrossRef] [PubMed]
114. Holst, B.S.; Axnér, E.; Öhlund, M.; Möller, L.; Egenvall, A. Dystocia in the cat evaluated using an insurance database. *J. Feline Med. Surg.* **2017**, *19*, 42–47. [CrossRef] [PubMed]
115. Dalayeun, J.F.; Norès, J.M.; Bergal, S. Physiology of β-endorphins. A close-up view and a review of the literature. *Biomed. Pharmacother.* **1993**, *47*, 311–320. [CrossRef]
116. Farmer, S.; Robert, C. Hormonal, behavioural and performance characteristics of Meishan sows during pregnancy and lactation. *Can. J. Anim. Sci.* **2003**, *83*, 1–12. [CrossRef]
117. van Dijk, A.J.; van Rens, B.T.T.M.; van der Lende, T.; Taverne, M.A.M. Factors affecting duration of the expulsive stage of parturition and piglet birth intervals in sows with uncomplicated, spontaneous farrowings. *Theriogenology* **2005**, *64*, 1573–1590. [CrossRef]

118. Oliviero, C.; Heinonen, M.; Valros, A.; Peltoniemi, O. Environmental and sow-related factors affecting the duration of farrowing. *Anim. Reprod. Sci.* **2010**, *119*, 85–91. [CrossRef] [PubMed]

119. Björkman, S.; Oliviero, C.; Kauffold, J.; Soede, N.M.; Peltoniemi, O.A.T. Prolonged parturition and impaired placenta expulsion increase the risk of postpartum metritis and delay uterine involution in sows. *Theriogenology* **2018**, *106*, 87–92. [CrossRef]

120. Borges, V.F.; Bernardi, M.L.; Bortolozzo, F.P.; Wentz, I. Risk factors for stillbirth and foetal mummification in four Brazilian swine herds. *Prev. Vet. Med.* **2005**, *70*, 165–176. [CrossRef] [PubMed]

121. Fahmy, M.H.; Friend, D.W. Factors influencing, and repeatability of the duration of farrowing in yorkshire sows. *Can. J. Anim. Sci.* **1981**, *61*, 17–22. [CrossRef]

122. Vestergaard, K.H.L. Tethered versus loose sows: Ethological observations and measures of productivity I. Ethological observations during pregnancy and farrowing. *Ann. Rech. Vet.* **1984**, *15*, 245–256. [PubMed]

123. Fraser, D.; Phillips, P.A.; Thompson, B.K. Farrowing behaviour and stillbirth in two environments: An evaluation of the restraint-stillbirth hypothesis. *Appl. Anim. Behav. Sci.* **1997**, *55*, 51–66. [CrossRef]

124. Holm, B.; Bakken, M.; Vangen, O.; Rekaya, R. Genetic analysis of age at first service, return rate, litter size, and weaning-to-first service interval of gilts and sows1. *J. Anim. Sci.* **2005**, *83*, 41–48. [CrossRef]

125. Holm, B.; Bakken, M.; Vangen, O.; Rekaya, R. Genetic analysis of litter size, parturition length, and birth assistance requirements in primiparous sows using a joint linear-threshold animal model1. *J. Anim. Sci.* **2004**, *82*, 2528–2533. [CrossRef]

126. Oliviero, C.; Junnikkala, S.; Peltoniemi, O. The challenge of large litters on the immune system of the sow and the piglets. *Reprod. Domest. Anim.* **2019**, *54*, 12–21. [CrossRef]

127. Pearodwong, P.; Muns, R.; Tummaruk, P. Prevalence of constipation and its influence on post-parturient disorders in tropical sows. *Trop. Anim. Health Prod.* **2016**, *48*, 525–531. [CrossRef] [PubMed]

128. Ison, S.H.; Jarvis, S.; Rutherford, K.M.D. The identification of potential behavioural indicators of pain in periparturient sows. *Res. Vet. Sci.* **2016**, *109*, 114–120. [CrossRef]

129. Nagel, C.; Aurich, C.; Aurich, J. Stress effects on the regulation of parturition in different domestic animal species. *Anim. Reprod. Sci.* **2019**, *207*, 153–161. [CrossRef] [PubMed]

130. Nagel, C.; Erber, R.; Ille, N.; von Lewinski, M.; Aurich, J.; Möstl, E.; Aurich, C. Parturition in horses is dominated by parasympathetic activity of the autonomous nervous system. *Theriogenology* **2014**, *82*, 160–168. [CrossRef] [PubMed]

131. Roussel, S.; Hemsworth, P.H.; Boissy, A.; Duvaux-Ponter, C. Effects of repeated stress during pregnancy in ewes on the behavioural and physiological responses to stressful events and birth weight of their offspring. *Appl. Anim. Behav. Sci.* **2004**, *85*, 259–276. [CrossRef]

132. Ritter, C.; Beaver, A.; von Keyserlingk, M.A.G. The complex relationship between welfare and reproduction in cattle. *Reprod. Domest. Anim.* **2019**, *54*, 29–37. [CrossRef] [PubMed]

133. Landa, L. Pain in domestic animals and how to assess it: A review. *Vet. Med.* **2012**, *57*, 185–192. [CrossRef]

134. Herr, K.; Coyne, P.J.; Key, T.; Manworren, R.; McCaffery, M.; Merkel, S.; Pelosi-Kelly, J.; Wild, L. Pain Assessment in the Nonverbal Patient: Position Statement with Clinical Practice Recommendations. *Pain Manag. Nurs.* **2006**, *7*, 44–52. [CrossRef]

135. Molony, V.; Kent, J.E. Assessment of acute pain in farm animals using behavioral and physiological measurements. *J. Anim. Sci.* **1997**, *75*, 266. [CrossRef]

136. Mellor, D.; Stafford, K.; Todd, S.; Lowe, T.; Gregory, N.; Bruce, R.; Ward, R. A comparison of catecholamine and cortisol responses of young lambs and calves to painful husbandry procedures. *Aust. Vet. J.* **2002**, *80*, 228–233. [CrossRef]

137. Keating, S.C.J.; Thomas, A.A.; Flecknell, P.A.; Leach, M.C. Evaluation of EMLA Cream for Preventing Pain during Tattooing of Rabbits: Changes in Physiological, Behavioural and Facial Expression Responses. *PLoS ONE* **2012**, *7*, e44437. [CrossRef]

138. Flecknell, P.A.; Kirk, A.J.B.; Liles, J.H.; Hayes, P.H.; Dark, J.H. Post-operative analgesia following thoracotomy in the dog: An evaluation of the effects of bupivacaine intercostal nerve block and nalbuphine on respiratory function. *Lab. Anim.* **1991**, *25*, 319–324. [CrossRef]

139. Kalliokoski, O.; Abelson, K.S.P.; Koch, J.; Boschian, A.; Thormose, S.F.; Fauerby, N.; Rasmussen, R.S.; Johansen, F.F.; Hau, J. The effect of voluntarily ingested buprenorphine on rats subjected to surgically induced global cerebral ischaemia. *In Vivo* **2010**, *24*, 641–646.

140. Schiavenato, M.; Craig, K.D. Pain assessment as a social transaction: Beyond the "gold standard". *Clin. J. Pain* **2010**, *26*, 667–676. [CrossRef]

141. Holton, L.; Pawson, P.; Nolan, A.; Reid, J.; Scott, E.M. Development of a behaviour-based scale to measure acute pain in dogs. *Vet. Rec.* **2001**, *148*, 525–531. [CrossRef] [PubMed]

142. Morton, C.M.; Reid, J.; Scott, E.M.; Holton, L.L.; Nolan, A.M. Application of a scaling model to establish and validate an interval level pain scale for assessment of acute pain in dogs. *Am. J. Vet. Res.* **2005**, *66*, 2154–2166. [CrossRef] [PubMed]

143. Gónzalez-Lozano, M.; Trujillo, O.M.E.; Becerril-Herrera, M.; Alonso-Spilsbury, M.; Rosales-Torres, A.M.; Mota-Rojas, D. Uterine activity and fetal electronic monitoring in parturient sows treated with vetrabutin chlorhydrate. *J. Vet. Pharmacol. Therap.* **2009**, *33*, 28–34. [CrossRef] [PubMed]

144. Hansen, B. Analgesia for the critically ill dog or cat: An update. *Vet. Clin. N. Am. Small Anim. Pract.* **2008**, *38*, 1353–1363. [CrossRef] [PubMed]

145. Roughan, J.V.; Wright-Williams, S.L.; Flecknell, P.A. Automated analysis of postoperative behaviour: Assessment of HomeCageScan as a novel method to rapidly identify pain and analgesic effects in mice. *Lab. Anim.* **2009**, *43*, 17–26. [CrossRef] [PubMed]

146. Tong, C.; Conklin, D.R.; Liu, B.; Ririe, D.G.; Eisenach, J.C. Assessment of behavior during labor in rats and effect of intrathecal morphine. *Anesthesiology* **2008**, *108*, 1081–1086. [CrossRef]

147. Castrén, H.; Algers, B.; de Passillé, A.-M.; Rushen, J.; Uvnäs-Moberg, K. Preparturient variation in progesterone, prolactin, oxytocin and somatostatin in relation to nest building in sows. *Appl. Anim. Behav. Sci.* **1993**, *38*, 91–102. [CrossRef]

148. Mainau, E.; Dalmau, A.; Ruiz-de-la-Torre, J.L.; Manteca, X. A behavioural scale to measure ease of farrowing in sows. *Theriogenology* **2010**, *74*, 1279–1287. [CrossRef]

149. Rutherford, K.M.D.; Robson, S.K.; Donald, R.D.; Jarvis, S.; Sandercock, D.A.; Scott, E.M.; Nolan, A.M.; Lawrence, A.B. Pre-natal stress amplifies the immediate behavioural responses to acute pain in piglets. *Biol. Lett.* **2009**, *5*, 452–454. [CrossRef] [PubMed]

150. de Fátima de Assunção Braga, A.; Carvalho, V.H.; da Silva Braga, F.S.; Pereira, R.I.C. Bloqueio combinado raquiperidural para analgesia de parto. Estudo comparativo com bloqueio peridural contínuo. *Braz. J. Anesthesiol.* **2019**, *69*, 7–12. [CrossRef]

151. Leeman, L.; Fontaine, P.; King, V.; Klein, M.C.; Ratcliffe, S. The nature and management of labor pain: Part I. Nonpharmacologic pain relief. *Am. Fam. Physician* **2003**, *68*, 1109–1112.

152. Orihuela, A.; Mota-Rojas, D.; Strappini, A.; Serrapica, F.; Braghieri, A.; Mora-Medina, P.; Napolitano, F. Neurophysiological Mechanisms of Cow–Calf Bonding in Buffalo and Other Farm Animals. *Animals* **2021**, *11*, 1968. [CrossRef]

153. Eisenach, J. Neurophysiology of labour pain. *Eur. Soc. Anaesthesiol.* **2010**, *1*, 1–4.

154. Volmanen, P.; Akural, E.I.; Raudaskoski, T.; Alahuhta, S. Remifentanil in Obstetric Analgesia: A Dose-Finding Study. *Anesth. Analg.* **2002**, *94*, 913–917. [CrossRef]

155. Michelsen, L.G.; Hug, C.C. The pharmacokinetics of remifentanil. *J. Clin. Anesth.* **1996**, *8*, 679–682. [CrossRef]

156. Olofsson, C.; Ekblom, A.; Ekman-Ordeberg, G.; Hjelm, A.; Irestedt, L. Lack of analgesic effect of systemically administered morphine or pethidine on labour pain. *BJOG An. Int. J. Obstet. Gynaecol.* **1996**, *103*, 968–972. [CrossRef] [PubMed]

157. Campbell, D.C. Parenteral opioids for labor analgesia. *Clin. Obstet. Gynecol.* **2003**, *46*, 616–622. [CrossRef] [PubMed]

158. Wilson, M.J.A.; MacArthur, C.; Hewitt, C.A.; Handley, K.; Gao, F.; Beeson, L.; Daniels, J. Intravenous remifentanil patient-controlled analgesia versus intramuscular pethidine for pain relief in labour (RESPITE): An open-label, multicentre, randomised controlled trial. *Lancet* **2018**, *392*, 662–672. [CrossRef]

159. Shaaban, O.M.; Abbas, A.M.; Mohamed, R.A.; Hafiz, H. Lack of pain relief during labor is blamable for the increase in the women demands towards cesarean delivery: A cross-sectional study. *Facts Views Vis. ObGyn* **2017**, *9*, 175–180. [PubMed]

160. Loughnan, B.A.; Carli, F.; Romney, M.; Doré, C.J.; Gordon, H. Randomized controlled comparison of epidural bupivacaine versus pethidine for analgesia in labour. *Br. J. Anaesth.* **2000**, *84*, 715–719. [CrossRef] [PubMed]

161. Sheiner, E.; Shoham-Vardi, I.; Sheiner, E.K.; Press, F.; Hackmon-Ram, R.; Mazor, M.; Katz, M. A comparison between the effectiveness of epidural analgesia and parenteral pethidine during labor. *Arch. Gynecol. Obstet.* **2000**, *263*, 95–98. [CrossRef] [PubMed]

162. Romagnoli, N.; Barbarossa, A.; Cunto, M.; Ballotta, G.; Zambelli, D.; Armorini, S.; Zaghini, A.; Lambertini, C. Evaluation of methadone concentrations in bitches and in umbilical cords after epidural or systemic administration for caesarean section: A randomized trial. *Vet. Anaesth. Analg.* **2019**, *46*, 375–383. [CrossRef]

163. Mathews, K.A. Pain management for the pregnant, lactating, and neonatal to pediatric cat and dog. *Vet. Clin. North Am. Small Anim. Pract.* **2008**, *38*, 1291–1308. [CrossRef] [PubMed]

164. Mainau, E.; Temple, D.; Manteca, X. Experimental study on the effect of oral meloxicam administration in sows on pre-weaning mortality and growth and immunoglobulin G transfer to piglets. *Prev. Vet. Med.* **2016**, *126*, 48–53. [CrossRef]

165. Ison, S.H.; Jarvis, S.; Ashworth, C.J.; Rutherford, K.M.D. The effect of post-farrowing ketoprofen on sow feed intake, nursing behaviour and piglet performance. *Livest. Sci.* **2017**, *202*, 115–123. [CrossRef]

166. Lang, A.J.; Sorrell, J.T.; Rodgers, C.S.; Lebeck, M.M. Anxiety sensitivity as a predictor of labor pain. *Eur. J. Pain* **2006**, *10*, 263–270. [CrossRef]

167. Jones, L.E.; Whitburn, L.Y.; Davey, M.-A.; Small, R. Assessment of pain associated with childbirth: Women's perspectives, preferences and solutions. *Midwifery* **2015**, *31*, 708–712. [CrossRef]

168. Aneiros, F.; Vazquez, M.; Valiño, C.; Taboada, M.; Sabaté, S.; Otero, P.; Costa, J.; Carceller, J.; Vázquez, R.; Díaz-Vieito, M.; et al. Does epidural versus combined spinal-epidural analgesia prolong labor and increase the risk of instrumental and cesarean delivery in nulliparous women? *J. Clin. Anesth.* **2009**, *21*, 94–97. [CrossRef]

169. National Research Council. *Recognition and Alleviation of Pain in Laboratory Animals*; National Academies Press: Washington, DC, USA, 2009; pp. 1–199.

170. De Cramer, K.G.M.; Joubert, K.E.; Nöthling, J.O. Puppy survival and vigor associated with the use of low dose medetomidine premedication, propofol induction and maintenance of anesthesia using sevoflurane gas-inhalation for cesarean section in the bitch. *Theriogenology* **2017**, *96*, 10–15. [CrossRef] [PubMed]

171. Robertson, S. Anaesthetic management for caesarean sections in dogs and cats. *Practice* **2016**, *38*, 327–339. [CrossRef]

172. Gabas, D.T.; Oliva, V.N.L.S.; Matsuba, L.M.; Perri, S.H.V. Estudo clínico e cardiorrespiratório em cadelas gestantes com parto normal ou submetidas à cesariana sob anestesia inalatória com sevofluorano. *Arq. Bras. Med. Vet. Zootec.* **2006**, *58*, 518–524. [CrossRef]

173. Luna, S.P.L.; Basílio, A.C.; Steagall, P.V.M.; Machado, L.P.; Moutinho, F.Q.; Takahira, R.K.; Brandão, C.V.S. Evaluation of adverse effects of long-term oral administration of carprofen, etodolac, flunixin meglumine, ketoprofen, and meloxicam in dogs. *Am. J. Vet. Res.* **2007**, *68*, 258–264. [CrossRef]

174. Mullins, K.B.; Thomason, J.M.; Lunsford, K.V.; Pinchuk, L.M.; Langston, V.C.; Wills, R.W.; McLaughlin, R.M.; Mackin, A.J. Effects of carprofen, meloxicam and deracoxib on platelet function in dogs. *Vet. Anaesth. Analg.* **2012**, *39*, 206–217. [CrossRef]

175. McLean, M.K.; Khan, S.A. Toxicology of frequently encountered nonsteroidal anti-inflammatory drugs in dogs and cats. *Vet. Clin. North Am. Small Anim. Pract.* **2018**, *48*, 969–984. [CrossRef] [PubMed]

176. Cima, G. Synthetic Opioids Put Police Dogs at Risk. 18 January 2018. Available online: https://www.avma.org/javma-news/20 18-02-01/synthetic-opioids-put-police-dogs-risk (accessed on 22 February 2021).

177. Huxley, J.N.; Whay, H.R. Current attitudes of cattle practitioners to pain and the use of analgesics in cattle. *Vet. Rec.* **2006**, *159*, 662–668. [CrossRef]

178. Sabaté, D.; Salichs, M.; Bosch, J.; Ramó, P.H.J. Efficacy of ketoprofen in the reduction of pre-weaning piglet mortality associated with sub-clinical forms of post-partum dysgalactia syndrome in sows. *Pig J.* **2012**, *67*, 1923.

179. Mainau, E.; Ruiz-de-la-Torre, J.L.; Dalmau, A.; Salleras, J.M.; Manteca, X. Effects of meloxicam (Metacam®) on post-farrowing sow behaviour and piglet performance. *Animal* **2012**, *6*, 494–501. [CrossRef] [PubMed]

180. Homedes, J.; Salichs, M.; Sabaté, D.; Sust, M.; Fabre, R. Effect of ketoprofen on pre-weaning piglet mortality on commercial farms. *Vet. J.* **2014**, *201*, 435–437. [CrossRef] [PubMed]

181. Viitasaari, E.; Hänninen, L.; Heinonen, M.; Raekallio, M.; Orro, T.; Peltoniemi, O.; Valros, A. Effects of post-partum administration of ketoprofen on sow health and piglet growth. *Vet. J.* **2013**, *198*, 153–157. [CrossRef]

182. Brownridge, P. The nature and consequences of childbirth pain. *Eur. J. Obstet. Gynecol. Reprod. Biol.* **1995**, *59*, S9–S15. [CrossRef]

183. Frey, H.; Tuuli, M.; Cortez, S.; Odibo, A.; Roehl, K.; Shanks, A.; Macones, G.; Cahill, A. Does delayed pushing in the second stage of labor impact perinatal outcomes? *Am. J. Perinatol.* **2012**, *29*, 807–814. [CrossRef] [PubMed]

184. Akbarzadeh, M.; Moradi, Z.; Hadianfard, M.J.; Zare, N.; Jowkar, A. Comparison of the effect of mono-stage and bi- stage acupressure at sp6 point on the severity of labor pain and the delivery outcome. *Int. J. Community Based Nurs. Midwifery* **2013**, *1*, 165–172.

185. Beigi, N.M.A.; Broumandfar, K.; Bahadoran, P.; Abedi, H.A. Women's experience of pain during childbirth. *Iran. J. Nurs. Midwifery Res.* **2010**, *15*, 77–82.

186. Kolås, T.; Hofoss, D.; Daltveit, A.K.; Nilsen, S.T.; Henriksen, T.; Häger, R.; Ingemarsson, I.; Øian, P. Indications for cesarean deliveries in Norway. *Am. J. Obstet. Gynecol.* **2003**, *188*, 864–870. [CrossRef]

187. Altman, M.; Sandström, A.; Petersson, G.; Frisell, T.; Cnattingius, S.; Stephansson, O. Prolonged second stage of labor is associated with low Apgar score. *Eur. J. Epidemiol.* **2015**, *30*, 1209–1215. [CrossRef]

188. Cheng, Y.W.; Shaffer, B.L.; Nicholson, J.M.; Caughey, A.B. Second Stage of Labor and Epidural Use. *Obstet. Gynecol.* **2014**, *123*, 527–535. [CrossRef]

189. Allen, V.M.; Baskett, T.F.; O'Connell, C.M.; McKeen, D.; Allen, A.C. Maternal and perinatal outcomes with increasing duration of the second stage of labor. *Obstet. Gynecol.* **2009**, *113*, 1248–1258. [CrossRef] [PubMed]

190. Deussen, A.R.; Ashwood, P.; Martis, R. Analgesia for relief of pain due to uterine cramping/involution after birth. *Cochrane Database Syst. Rev.* **2011**, *11*, CD004908. [CrossRef] [PubMed]

191. Kredatusova, G.; Hajurka, J.; Szakallova, I.; Valencakova, A.; Vojtek, B. Physiological events during parturition and possibilities for improving puppy survival: A review. *Vet. Med.* **2011**, *56*, 589–594. [CrossRef]

192. Song, J.; Zhang, S.; Qiao, Y.; Luo, Z.; Zhang, J.; Zeng, Y.; Wang, L. Predicting pregnancy-induced hypertension with dynamic hemodynamics. *Eur. J. Obstet. Gynecol. Reprod. Biol.* **2004**, *117*, 162–168. [CrossRef]

193. Reynolds, F. The effects of maternal labour analgesia on the fetus. *Best Pract. Res. Clin. Obstet. Gynaecol.* **2010**, *24*, 289–302. [CrossRef]

194. Singer, D. Neonatal tolerance to hypoxia: A comparative-physiological approach. *Comp. Biochem. Physiol. Part A Mol. Integr. Physiol.* **1999**, *123*, 221–234. [CrossRef]

195. Muns, R.; Nuntapaitoon, M.; Tummaruk, P. Non-infectious causes of pre-weaning mortality in piglets. *Livest. Sci.* **2016**, *184*, 46–57. [CrossRef]

196. Münnich, A.; Küchenmeister, U. Causes, diagnosis and therapy of common diseases in neonatal puppies in the first days of life: Cornerstones of practical approach. *Reprod. Domest. Anim.* **2014**, *49*, 64–74. [CrossRef]

197. Mota-Rojas, D.; Orozco, H.; Villanueva, D.; Bonilla, H.; Suárez, X.; Hernandez, R.; Roldan, P.; Trujillo, M. Foetal and neonatal energy metabolism in pigs and humans: A review. *Vet. Med.* **2011**, *56*, 215–225. [CrossRef]

198. Mota-Rojas, D.; Villanueva-García, D.; Velázquez-Armenta, E.Y.; Nava-Ocampo, A.A.; Ramírez-Necoechea, R.; Alonso-Spilsbury, M.; Trujillo, M.E. Influence of time at which oxytocin is administered during labor on uterine activity and perinatal death in pigs. *Biol. Res.* **2007**, *40*, 55–63. [CrossRef]

199. Mota-Rojas, D.; Martinez-Burnes, J.; Alonso-Spilsbury, M.L.; Lopez, A.; Ramirez-Necoechea, R.; Trujillo-Ortega, M.E.; Medina-Hernandez, F.J.; Cruz, N.I.; Albores-Torres, V.; Loredo-Osti, J. Meconium staining of the skin and meconium aspiration in porcine intrapartum stillbirths. *Livest. Sci.* **2006**, *102*, 155–162. [CrossRef]

200. Mota-Rojas, D.; Martínez-Burnes, J.; Trujillo, M.E.; López, A.; Rosales, A.M.; Ramírez, R.; Orozco, H.; Merino, A.; Alonso-Spilsbury, M. Uterine and fetal asphyxia monitoring in parturient sows treated with oxytocin. *Anim. Reprod. Sci.* **2005**, *86*, 131–141. [CrossRef] [PubMed]

201. Mota-Rojas, D.; Nava-Ocampo, A.A.; Trujillo, M.E.; Velázquez-Armenta, Y.; Ramírez-Necoechea, R.; Martínez-Burnes, J.; Alonso-Spilsbury, Y.M. Dose minimization study of oxytocin in early labor in sows: Uterine activity and fetal outcome. *Reprod. Toxicol.* **2005**, *20*, 255–259. [CrossRef] [PubMed]
202. Alonso-Spilsbury, M.; Mota-Rojas, D.; Martínez-Burnes, J.; Arch, E.; López Mayagoitia, A.; Ramírez-Necoechea, R.; Olmos, A.; Trujillo, M.E. Use of oxytocin in penned sows and its effect on fetal intra-partum asphyxia. *Anim. Reprod. Sci.* **2004**, *84*, 157–167. [CrossRef]
203. Tummaruk, P. Post-parturient Disorders and Backfat Loss in Tropical Sows in Relation to Backfat Thickness before Farrowing and Postpartum Intravenous Supportive Treatment. *Asian-Australas. J. Anim. Sci.* **2013**, *26*, 171–177. [CrossRef] [PubMed]
204. Muns, R.; Manzanilla, E.G.; Sol, C.; Manteca, X.; Gasa, J. Piglet behavior as a measure of vitality and its influence on piglet survival and growth during lactation. *J. Anim. Sci.* **2013**, *91*, 1838–1843. [CrossRef]
205. Alonso-Spilsbury, M.; Mota-Rojas, D.; Villanueva-García, D.; Martínez-Burnes, J.; Orozco, H.; Ramírez-Necoechea, R.; Mayagoitia, A.L.; Trujillo, M.E. Perinatal asphyxia pathophysiology in pig and human: A review. *Anim. Reprod. Sci.* **2005**, *90*, 1–30. [CrossRef]
206. Martínez-Burnes, J.; Mota-Rojas, D.; Villanueva-García, D.; Ibarra-Rios, D.; Lezama-García, K.; Barrios-García, H.; Lopez, A. Meconium aspiration syndrome in mammals. *CAB Rev. Perspect. Agric. Vet. Sci. Nutr. Nat. Resour.* **2019**, *14*, 1–11. [CrossRef]
207. Edwards, S. Perinatal mortality in the pig: Environmental or physiological solutions? *Livest. Prod. Sci.* **2002**, *78*, 3–12. [CrossRef]
208. Kalantaridou, S.; Makrigiannakis, A.; Zoumakis, E.; Chrousos, G. Stress and the female reproductive system. *J. Reprod. Immunol.* **2004**, *62*, 61–68. [CrossRef]
209. Lezama-García, K.; Orihuela, A.; Olmos-Hernández, A.; Reyes-Long, S.; Mota-Rojas, D. Facial expressions and emotions in domestic animals. *CAB Rev. Perspect. Agric. Vet. Sci. Nutr. Nat. Resour.* **2019**, *14*, 1–12. [CrossRef]
210. van Gelder, K.N.; Bilkei, G. The course of acute-phase proteins and serum cortisol in mastitis metritis agalactia (MMA) of the sow and sow performance. *Tijdschr. Diergeneeskd.* **2005**, *130*, 38–41. [PubMed]
211. Alexopoulos, J.G.; Lines, D.S.; Hallett, S.; Plush, K.J. A review of success factors for piglet fostering in lactation. *Animals* **2018**, *8*, 38. [CrossRef]
212. Kushnir, Y.; Epstein, A. Anesthesia for the pregnant cat and dog. *Isr. J. Vet. Med.* **2012**, *67*, 19–23.
213. Radbrook, C.A.; Clark, L. State of the art analgesia- recent developments in pharmacological approaches to acute pain management in dogs and cats. Part 1. *Vet. J.* **2018**, *238*, 76–82. [CrossRef]
214. Bai, Y.H.; Pak, S.C.; Choi, B.C.; Wilson, L. Tocolytic effect of morphine via increased metabolic clearance of oxytocin in the baboon. *Yonsei Med. J.* **2002**, *43*, 567–572. [CrossRef]
215. Arici, G.; Karsli, B.; Kayacan, N.; Akar, M. The effects of bupivacaine, ropivacaine and mepivacaine on the contractility of rat myometrium. *Int. J. Obstet. Anesth.* **2004**, *13*, 95–98. [CrossRef] [PubMed]
216. Castagnetti, C.; Mariella, J. Anti-inflammatory drugs in equine neonatal medicine. Part I: Nonsteroidal anti-inflammatory drugs. *J. Equine Vet. Sci.* **2015**, *35*, 475–480. [CrossRef]
217. Kukkar, A.; Bali, A.; Singh, N.; Jaggi, A.S. Implications and mechanism of action of gabapentin in neuropathic pain. *Arch. Pharm. Res.* **2013**, *36*, 237–251. [CrossRef]
218. Elikottil, J.; Gupta, P.; Gupta, K. The analgesic potential of cannabinoids. *J. Opioid Manag.* **2009**, *5*, 341–357. [CrossRef]
219. Bonar, M.; Hance Ellington, E.; Lewis, K.P.; Wal, E. Implementing a novel movement-based approach to inferring parturition and neonate caribou calf survival. *PLoS ONE* **2018**, *13*, e192204. [CrossRef] [PubMed]
220. Marchand, P.; Garel, M.; Morellet, N.; Benoit, L.; Chaval, Y.; Itty, C.; Petit, E.; Cargnelutti, B.; Hewison, A.J.M.; Loison, A. A standardised biologging approach to infer parturition: An application in large herbivores across the hider-follower continuum. *Methods Ecol. Evol.* **2021**, *12*, 1017–1030. [CrossRef]
221. Zheng, T.; Moustafa, Y.; Finn, C.; Scott, S.; Haase, C.J.; Carpinelli, N.A.; Osorio, J.S.; McKinstry, K.K.; Strutt, T.M.; Huo, Q. A rapid blood test to monitor immunity shift during pregnancy and potential application for animal health management. *Sensors Int.* **2020**, *1*, 100009. [CrossRef]

Article

Factors Associated with Farrowing Duration in Hyperprolific Sows in a Free Farrowing System under Tropical Conditions

Yosua Kristian Adi [1,2], Rafa Boonprakob [1], Roy N. Kirkwood [3] and Padet Tummaruk [1,*]

1   Centre of Excellence in Swine Reproduction, Department of Obstetrics, Gynaecology and Reproduction, Faculty of Veterinary Science, Chulalongkorn University, Bangkok 10330, Thailand
2   Department of Reproduction and Obstetrics, Faculty of Veterinary Medicine, Universitas Gadjah Mada, Yogyakarta 55281, Indonesia
3   School of Animal and Veterinary Sciences, University of Adelaide, Roseworthy, SA 5371, Australia
*   Correspondence: padet.t@chula.ac.th; Tel.: +66-819271066

Citation: Adi, Y.K.; Boonprakob, R.; Kirkwood, R.N.; Tummaruk, P. Factors Associated with Farrowing Duration in Hyperprolific Sows in a Free Farrowing System under Tropical Conditions. *Animals* **2022**, *12*, 2943. https://doi.org/10.3390/ani12212943

Academic Editors: Daniel Mota-Rojas, Julio Martínez-Burnes and Agustín Orihuela

Received: 24 September 2022
Accepted: 23 October 2022
Published: 26 October 2022

Publisher's Note: MDPI stays neutral with regard to jurisdictional claims in published maps and institutional affiliations.

Copyright: © 2022 by the authors. Licensee MDPI, Basel, Switzerland. This article is an open access article distributed under the terms and conditions of the Creative Commons Attribution (CC BY) license (https:// creativecommons.org/licenses/by/ 4.0/).

Simple Summary: Sows in most breeding herds worldwide have larger litters than several years ago. One of the most important problems when using these genetics is the prolonged duration of farrowing, which can cause postpartum complications in sows and increase the percentage of stillborn piglets per litter. In this retrospective study, we found that the farrowing duration of sows kept in a free farrowing system in a tropical environment was associated with several factors. A high number of piglets born per litter, a high parity number, parturition during working hours, and high temperature and humidity in the 7 days before parturition led to a prolonged farrowing duration. In these sows, farrowing was longer than the acceptable farrowing duration, which may cause a higher number of stillborn piglets. Therefore, management for sows during the perinatal period needs to be considered, especially in hyperprolific and older sows, as well as those that farrow during working hours.

Abstract: The ongoing selection for increased litter size has had significant impacts on sow husbandry practice. The present study investigated factors associated with farrowing duration and the proportion of sows that had prolonged farrowing in modern hyperprolific sows kept in a free farrowing system in a tropical environment. Farrowing data from 2493 Landrace x Yorkshire cross-bred sows in a commercial swine herd in Thailand were included in the study. The time of farrowing, parity number, litter size, and the birth status of each piglet were recorded. Farrowing duration was analysed using multiple analyses of variance. Total number of piglets born per litter (TB), parity, and time onset of farrowing were included in the statistical models. On average, TB, piglets born alive, and farrowing duration were 13.7, 12.1, and 221.0 min, respectively. Of these sows, 26.4% had TB $\geq$ 16 and 21.7% had a prolonged farrowing duration ($\geq$300 min). Farrowing duration was positively correlated with TB (r = 0.141, $p < 0.001$), percentage of stillborn (SB) piglets per litter (r = 0.259, $p < 0.001$), percentage of mummified foetuses (MF) per litter (r = 0.049, $p = 0.015$), piglet birth weight (r = 0.068, $p < 0.001$), and litter birth weight (r = 0.041, $p = 0.043$). The proportion of SB per litter was higher and piglet birth weight lower in litters that had $\geq$16 TB than those with 8–12 TB ($p < 0.05$). The farrowing duration of sows with parity numbers 5–7 (247.7 $\pm$ 5.1 min) and 8–10 (237.1 $\pm$ 5.1 min) was longer than that of sows with parity numbers 1 (188.3 $\pm$ 5.2 min) and 2–4 (214.3 $\pm$ 3.9 min) ($p < 0.05$). Sows that had started farrowing during working hours had longer farrowing durations (229.3 $\pm$ 3.6 min) than those that had started farrowing during non-working hours (217.6 $\pm$ 3.4 min, $p = 0.017$). In multiparous sows, the duration of farrowing was positively correlated with the maximum temperature (r = 0.056, $p = 0.012$) and the maximum temperature–humidity index (r = 0.059, $p = 0.008$) in the 7 days before farrowing. The present data confirm that TB, sow parity, and time of onset of farrowing are significant risk factors for a prolonged farrowing.

Keywords: climate; litter size; parturition; pig; stillbirth

---

1. Introduction

Over the last decade, genetic progress on the litter size of sows had a high impact on sow husbandry practice. As a result, sows in most breeding herds worldwide have larger litters than several years ago [1]. In general, sows with a total number of piglets born per litter (TB) of ≥ 16 can be regarded as hyperprolific sows [2]. In practice, many sows can reach such a high litter size in many parity numbers [3]. Hyperprolific sow genetics have been widely distributed to the swine industry worldwide, including Thailand [1–4]. However, one of the most important problems when using these genetics is the prolonged duration of farrowing [5]. Oliviero et al. [6] reported that the average duration of farrowing in modern swine production can range from 156 to 262 min, and farrowing lasting more than 300 min should be considered prolonged farrowing. Likewise, Ju et al. [7] suggested that a farrowing duration of 240 to 300 min can be considered the ideal cutoff point for Landrace x Yorkshire hybrid sows in a commercial pig farm having 5 to 22 TB. Prolonged farrowing duration can cause postpartum complications in sows and increase the percentage of stillborn piglets per litter (SB) [8]. For instance, the SB in sows with a long duration of farrowing (>4 h) is higher than that in sows with a short duration (<2 h) of farrowing (29.2% versus 7.9%, respectively) [8]. The average farrowing duration of sows without any SB is shorter than that of sows with SB $\geq$ 3 (221.0 versus 373.0 min, respectively) [4]. Therefore, understanding the factors significantly associated with the duration of farrowing in sows under field conditions is important.

Farrowing duration is one aspect that is affected by an increase in litter size in modern sows. The farrowing duration in most of the European domestic pig breeds has increased from about 2 h per 12 TB to 6 h 40 min per 19 TB [5]. However, comprehensive information about the factors related to the duration of farrowing in modern hyperprolific sows in Asia is limited [4]. Factors known to influence farrowing duration are the state of constipation, the body condition of the sow, housing, gestation length, sow age, and genetics [6]. In a tropical environment, hyperprolific sows with an average TB of 17.5 have an average farrowing duration of 330.6 min [4]. Thus, prolonged farrowing duration in hyperprolific sows has become one of the most important issues.

Prolonged farrowing duration can have adverse effects on both sows and piglets [8]. The incidence of intrapartum hypoxia in newborn piglets increases following a prolonged duration of farrowing [5]. Parturition is associated with many physiological processes, including both hormonal and behavioural changes, and is the most painful experience for females [9]. Moreover, the expulsion interval of each piglet is positively associated with SB [4]. As the concentration of colostrum IgG decreases by 50% within 6 h after the birth of the first piglet [10], the access of piglets to good-quality colostrum is reduced in sows with a long farrowing duration. Prolonged parturition may reduce piglet vitality at birth [1,5]. A previous study demonstrated that a higher proportion of piglets that attempted to stand after 5 min (38.5%) died compared to piglets that attempted to stand within 1 min (6.3%) after birth [11]. Moreover, retained placentae and uterine inflammation increase in sows with a long duration of farrowing [12]. Thus, sows with a long farrowing duration may have compromised reproductive performance. Oliviero et al. [13] found that 13% of sows that failed to get pregnant at the first insemination after weaning had a relatively long farrowing duration in their previous farrowing.

In tropical environments, the average environmental temperature generally varies from 20 to 35 °C [14]. On average, the temperature at night is a few degrees lower than that during the day. However, it is still a few degrees above the comfort-zone temperature for a pregnant sow [15]. Although farrowing during the night is favourable because of the lower environmental temperature, this is a nonworking period in most pig farms. During late gestation, sows prefer a temperature of 12.6 to 15.6 °C [15]. Lucy and Safranski [16] found that exposure to heat stress in late gestation in sows resulted in some negative effects, such as reduced piglet birth weight and an increased number of stillborn piglets. Not only high temperature but also high relative humidity and/or a high temperature–humidity index during gestation can reduce TB [17]. Heat stress that comes from the environment not only affects piglet characteristics but also compromises the reproductive performance of sows, e.g., prolonged weaning-to-first-service interval and reduced farrowing rate [18]. Interestingly, the threshold of temperatures leading to a prolonged weaning-to-first-service interval is lower for primiparous than for multiparous sows (17 °C vs. 25 °C), and the threshold temperatures leading to reductions in farrowing rates for gilts, primiparous sows and multiparous sows are 20, 21 and 24 to 25 °C, respectively [18]. However, to our knowledge, the influences of temperature and humidity in the tropical environment on farrowing duration in both sows with normal litter size and hyperprolific sows have never been reported. Therefore, we investigated the factors influencing farrowing duration and the proportion of sows that had a prolonged farrowing duration in modern hyperprolific sows in a free farrowing system under tropical conditions. The influences of the time of the onset of farrowing (working vs. non-working hours) and temperature and relative humidity inside the farrowing house for 7 days before farrowing on farrowing duration and the incidence of sows with prolonged farrowing duration were also investigated.

2. Materials and Methods

2.1. Study Design

The present study included farrowing data from 2493 Landrace x Yorkshire cross-bred sows that farrowed during the period from January to April 2021 in a commercial swine herd in the central region of Thailand. Sows were randomly distributed according to TB into three groups: 8–12 (n = 853), 13–15 (n = 983) and ≥16 piglets/litter (n = 657). Parity number of sows was classified into four groups: 1 (n = 506), 2–4 (n = 919), 5–7 (n = 534) and 8–10 (n = 534). The time when the onset of farrowing occurred was classified into two groups: working hours (0700 h–1700 h) (n = 1176) and non-working hours (1701 h–0659 h) (n = 1317). The average temperature, humidity and temperature–humidity index (THI) during the 7-day period before farrowing were recorded for each individual sow. Farrowing duration was expressed as either a continuous trait (the interval from the first piglet to the last piglet delivered in minutes) or a categorical trait (the proportion of sows that had a farrowing duration of longer than 300 min). Factors including TB, parity number, the time when the onset of farrowing occurred and the average temperature, humidity and THI during the 7-day period before farrowing were analysed to determine their association with the farrowing duration and the proportion of sows that had a prolonged farrowing.

2.2. Data: Inclusion and Exclusion Criteria

The original farrowing data were obtained from 2750 sows. The data included sow identity, breed of sows, parity number, date of farrowing, TB, BA, SB, percentage of mummified foetuses per litter (MF), liveborn piglet birth weight, the time when the onset of farrowing occurred and the end of parturition. The farrowing duration for each individual sow was calculated, defined as the period from the first to the last piglet delivery in minutes. Data were scrutinised for correctness, and data with values too extreme were excluded from the analyses. Errors in the reported farrowing times records were checked by calculating the farrowing duration and constructing the frequency distribution of the farrowing duration. The data of sows with too short farrowing duration (<30 min, n = 15) and farrowing duration too long (>720 min, n = 30) were excluded from the analyses. Old sows (parity numbers ≥ 11, n = 7) and sows that had a TB ≤ 7 (n = 205) were excluded. In total, 9.4% (n = 257) of the raw data were excluded. Thus, the analysed data contained observations on 2493 sows.

2.3. Housing and General Management

The sows and gilts were kept in a group housing system during gestation. The number of sows per group was 40, and the size of the gestating pen was 9 × 11 m. The sows were kept group-housed within 3 days after the last insemination until 108 days of gestation before transfer to the farrowing pen. The average daily minimum to maximum temperatures inside the barn during the experimental period were 26.5 ± 0.9 °C (range

21.8–28.2 °C), and the average daily minimum to maximum humidity levels inside the barn were 71.0 ± 1.0% (range 69.0–73.7%). During the first, middle, and late periods of gestation, the sows were fed approximately 1.8–2.0, 2.0–2.2 and 3.0–3.5 kg feed per sow per day, respectively. Three days before the estimated day of farrowing, the feed was reduced to 2.5–3.0 kg of feed per sow per day. The gestation diet contained 15.0% crude protein, 2700 kcal/kg metabolizable energy, and 0.7% lysine. After farrowing, the sows were fed ad libitum. The lactating sows were fed using an automatic feeding machine that allowed the sows to consume feed freely. The lactation diet contained 16.0% crude protein, 3600 kcal/kg metabolizable energy, and 0.8% lysine. Water was provided ad libitum via a drinking nipple. At 109 ± 2.0 days of gestation, sows and gilts were moved to a free farrowing pen system. Temperature and humidity in the farrowing house were recorded manually by stock persons three times a day at 0600 h, 1300 h and 1600 h. The farrowing process was carefully monitored by stock persons in the barn for 24 h daily. The time onset and the end of farrowing, birth weight of live born piglet and the status of the piglets at birth (i.e., live-born, stillborn, or mummified foetuses) were recoded. During farrowing, the sows and gilts were disturbed as little as possible. Single-dose administration of 20 IU oxytocin via intramuscular route (Phenix Pharmaceuticals N.V. Co., Ltd., Hoogstraten, Belgium) was performed if the sow had a birth interval of >60 min and/or no sign of uterine contraction. In addition, 20 IU oxytocin was routinely administered via intramuscular route to all sows after the 10th piglet was born to initiate placental expulsion and milk let-down. Live-born piglets were weighed individually. Stillborn piglets and mummified foetuses were removed and distinguished based on their skin colour, sucked eyes and skin appearance. Stillborns were dead piglets that had pink skin, non-sucked eyes and wet skin, whereas mummified foetuses were dead piglets with dark skin colour, sucked eyes and dry skin. At the end of parturition, all sows were treated with an antipyretic drug (ketoprofen 3.0 mg/kg intramuscularly using Ketaprofen®, KELA N.V., Hoogstraten, Belgium). The health status of the sows was monitored routinely by the herd veterinarian. Gestating sows were vaccinated against foot and mouth disease (AFTOPOR®, Merial SAS, Lyon, France) and porcine epidemic diarrhoea virus (SUIT-SHOT PT-100®, Choong Ang Vaccine Laboratories Co., Ltd., Deajeon, Korea), porcine circovirus (Circumvent PCV®, Merck Animal Health, Kenilworth, QL, USA) and Aujeszky's disease virus (Porcilis® Ad Begonia, Merck Animal Health, Madison, WI, USA). After farrowing, the sows were vaccinated against classical swine fever (Ceva-Phylaxia Veterinary Biologicals Co., Ltd., Budapest, Hungary) and porcine parvovirus–*Leptospira*–erysipelas (Zoetis ZA, Sandton, South Africa). The piglets were vaccinated against *Mycoplasma hyopneumoniae* (Hyogen®, Ceva Santé Animale S.A, Libourne, France) at 18–22 days of age.

2.4. Statistical Analysis

All statistical analyses were carried out using SAS version 9.4 (SAS Inst., Cary, NC, USA). Descriptive statistics for continuous and categorical variables were analysed using MEANS and FREQ procedures, respectively. Frequency distribution for the duration of farrowing was analysed using the FREQ procedure of SAS. Pearson's correlation was calculated to determine the associations between farrowing duration and other continuous traits including TB, BA, SB, MF, piglet birth weight, litter birth weight, percentage of piglets with birth weight <1.0 kg, temperature, humidity and THI during the period of 7 days before parturition. Farrowing duration was analysed using multiple analyses of variance, applying the general linear model procedure of SAS. The factors included in the statistical models were TB classes (8–12, 13–15 and ≥16), parity number classes (1, 2–4, 5–7 and 8–10), time of the onset of farrowing (working hours and non-working hours) and two-way interactions with $p < 0.10$. Least-square means were obtained from each variable class and compared using the Tukey–Kramer adjustment for multiple comparisons. The proportion of sows with prolonged farrowing (i.e., >300 min) was expressed as a percentage and analysed using logistic regression in the generalised linear mixed model (GLIMMIX) procedure of SAS. The factors included in the statistical models were TB classes (8–12, 13–15 and $\geq$16), parity number classes (1, 2–4, 5–7 and 8–10), onset of farrowing (working hours and non-working hours) and two-way interactions with $p < 0.10$. Least-square means were obtained from the models and compared using the Tukey–Kramer adjustment for multiple comparisons. Additionally, the influences of temperature, humidity and THI during 7 days before farrowing on farrowing duration was analysed using the general linear model procedure of SAS. The factors included in the statistical models were TB classes (8–12, 13–15 and $\geq$16), parity number classes (1, 2–4, 5–7 and 8–10) and two-way interactions with $p < 0.10$. The temperature, humidity and THI values during the 7-day period before parturition were included in the statistical models one at a time, as they were highly correlated. For all analyses, differences at $p < 0.05$ were regarded statistically significant.

3. Results

3.1. Descriptive Data

Table 1 shows the descriptive statistics on sow reproductive performances and farrowing characteristics. On average, sows in the present study had 13.7 TB with farrowing duration of 30.0 to 716.0 min (Table 1). The proportions of sows that had 8–12, 13–15 and $\geq$16 TB were 34.2%, 39.4% and 26.4%, respectively. The proportion of sows with the onset of farrowing during working hours and non-working hours were 47.2% and 52.8%, respectively. Of all sows, 21.7% had a prolonged farrowing duration ($\geq$300 min) (Figure 1). Pearson's correlations between farrowing duration, birth interval and sow reproductive characteristics are shown in Table 2. Farrowing duration was positively correlated with TB ($r = 0.141$, $p < 0.001$), SB ($r = 0.259$, $p < 0.001$), MF ($r = 0.049$, $p = 0.015$), piglet birth weight ($r = 0.068$, $p < 0.001$) and litter birth weight ($r = 0.041$, $p = 0.043$).

Table 1. Descriptive statistics on reproductive performance and farrowing characteristics of 2493 Landrace x Yorkshire cross-bred sows in a commercial swine herd in Thailand.

Variables	Means ± SD	Range
Parity number	4.4 ± 2.9	1–10
Total number of piglets born per litter	13.7 ± 2.8	8–23
Number of piglets born alive per litter	12.1 ± 3.1	0–23
Stillborn piglets per litter (%)	5.9 ± 8.0	0–100
Mummified foetuses per litter (%)	5.8 ± 12.4	0–100
Litter birth weight (kg)	16.2 ± 4.0	1.9–41.4
Piglet birth weight (kg)	1.4 ± 0.2	0.8–3.2
Piglet with birth weight <1.0 kg (%)	11.0 ± 13.4	0–87.5
Duration of farrowing (min)	221.0 ± 119.3	30.0–716.0
Birth interval (min)	16.6 ± 9.4	2.1–71.9

Table 2. Pearson's correlations between farrowing duration (mean ± SD = 221.0 ± 119.3 min), birth interval (mean ± SD = 16.6 ± 9.4 min) and sow reproductive characteristics ($n = 2493$).

Variables	Correlation Coefficient (r)	
	Farrowing Duration	**Birth Interval**
Total number of piglets born per litter	0.141 ***	−0.239 ***
Stillborn piglets per litter (%)	0.259 ***	0.167 ***
Mummified foetuses per litter (%)	0.049 **	NS
Piglet birth weight (kg)	0.068 ***	0.181 ***
Litter birth weight (kg)	0.041 *	−0.177 ***
Piglets with birth weight <1.0 kg	NS	−0.065 ***
Number of piglets born alive per litter	NS	−0.280 ***

NS = not significant ($p > 0.05$); significance is indicated by * $0.01 < p < 0.05$, ** $0.001 < p < 0.01$ and *** $p < 0.001$.

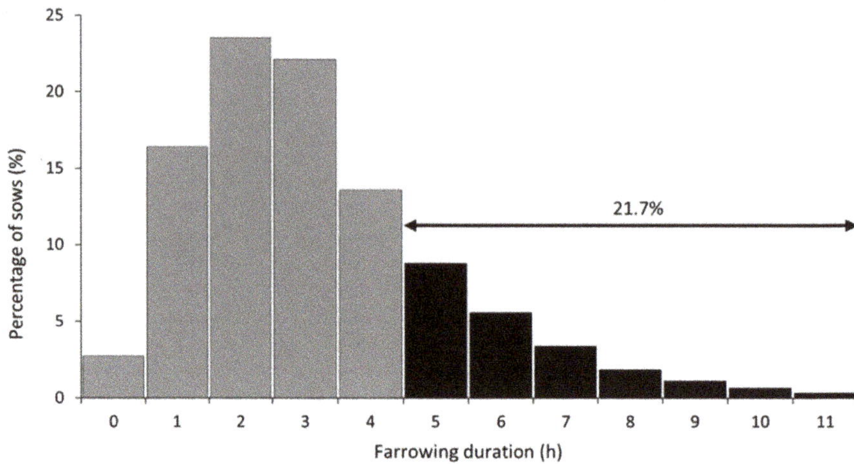

Figure 1. Frequency distribution of sows based on farrowing duration (n = 2493).

3.2. Effect of Litter Size

The reproductive performance and farrowing characteristics of sows that had 8–12, 13–15 and ≥16 TB are presented in Table 3. The proportion of sows with a prolonged farrowing in the litters that had TB ≥ 16 was higher than that in litters with TB 8–12 ($p < 0.001$) and tended to be higher than that in litters with TB 13–15 ($p = 0.071$). Likewise, the farrowing duration of sows that had TB ≥ 16 and 13–15 was longer than that of sows with TB 8–12 (Table 3). However, the birth interval of sows that had TB ≥ 16 was shorter than that of sows that had TB 13–15 and 8–12 (Table 3). In addition, the SB was higher and the piglet birth weight lower in the litters with ≥16 TB compared to the litters with 8–12 TB (Table 3). Farrowing duration and birth intervals among the TB groups by parity classes are presented in Figure 2a,b, respectively. Prolonged farrowing as well as a long birth interval were frequently detected in sow parities 5–7 and 8–10. In addition, farrowing duration and birth interval in sows that had 3 and ≥4 stillborn piglets per litter were longer than in sows that had 0, 1 and 2 stillborn piglets per litter (Figure 3).

Table 3. Reproductive performances and farrowing characteristics of sows by the total number of piglets born per litter.

Variables	Total Number of Piglets Born per Litter		
	8–12	**13–15**	**≥16**
Number of sows	853	983	657
Parity number	4.7 ± 0.1 [a]	4.3 ± 0.1 [b]	4.0 ± 0.1 [c]
Total number of piglets born per litter	10.6 ± 0.0 [a]	14.0 ± 0.0 [b]	17.1 ± 0.0 [c]
Number of piglets born alive per litter	9.4 ± 0.1 [a]	12.4 ± 0.1 [b]	15.1 ± 0.1 [c]
Stillborn piglets per litter (%)	5.5 ± 0.3 [a]	6.0 ± 0.3 [a,b]	6.4 ± 0.3 [b]
Mummified foetuses per litter (%)	6.0 ± 0.4 [a]	5.9 ± 0.4 [a]	5.4 ± 0.5 [a]
Litter birth weight (kg)	13.6 ± 0.1 [a]	16.6 ± 0.1 [b]	19.2 ± 0.1 [c]
Piglet birth weight (kg)	1.43 ± 0.01 [a]	1.34 ± 0.01 [b]	1.27 ± 0.01 [c]
Piglets with birth weight <1.0 kg (%)	7.8 ± 0.5 [a]	10.9 ± 0.4 [b]	15.4 ± 0.5 [c]
Farrowing duration (min)	199.9 ± 4.0 [a]	228.2 ± 3.8 [b]	237.9 ± 4.6 [b]
Proportion of sows farrowed >300 min (%)	17.0 [a]	22.9 [b]	25.9 [b]
Birth interval (min)	19.1 ± 0.3 [a]	16.4 ± 0.3 [b]	13.9 ± 0.4 [c]

[a,b,c] Different superscript letters within a row denote data that differ significantly ($p < 0.05$).

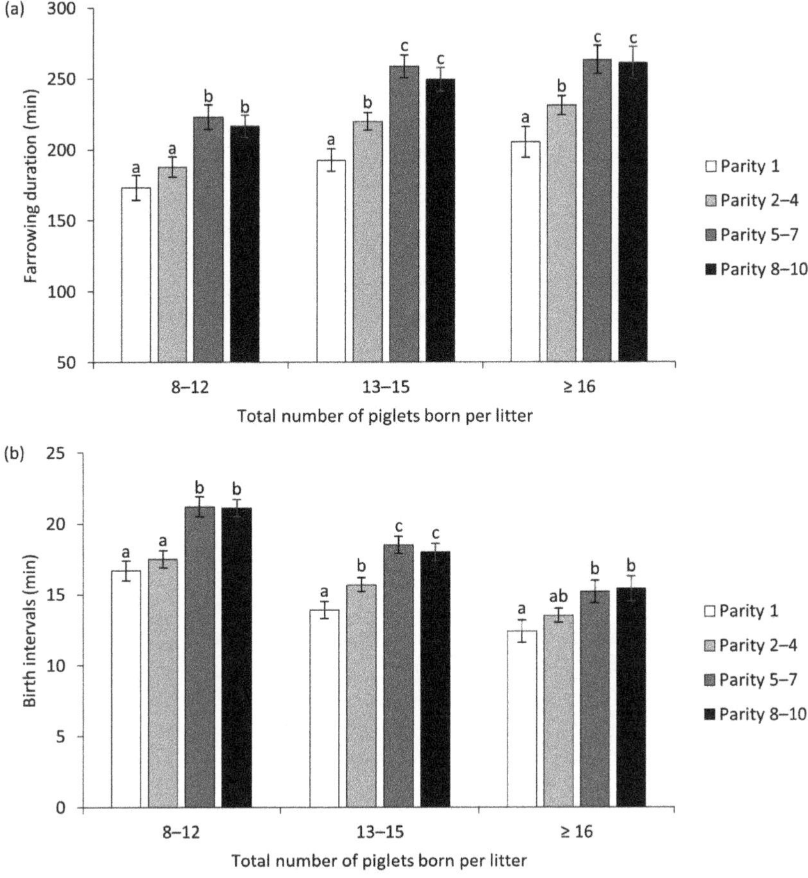

Figure 2. The duration of farrowing (**a**) and birth intervals (**b**) of sows among the total number of piglets born per litters and parities classes. [a,b,c] Different lowercase letters within the class of total number of piglets born per litter denote data that differ significantly ($p < 0.05$). Bars indicate standard error.

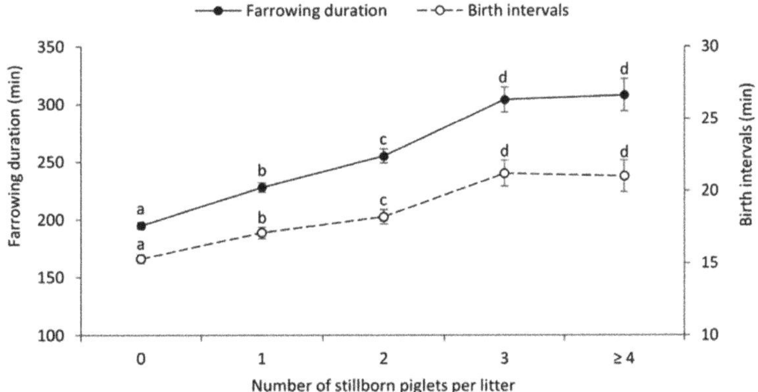

Figure 3. Farrowing duration and birth intervals of sows with different numbers of stillborn piglets per litter. [a,b,c,d] Different superscripts differ significantly ($p < 0.05$).

3.3. Effect of Parity Numbers

The farrowing duration differed among the parity groups ($p < 0.05$). The farrowing duration of sows with parity numbers 5–7 and 8–10 was longer than that of sows with parity numbers 1 and 2–4 (Table 4). Primiparous sows that had TB 8–12 had the shortest farrowing duration (173.2 ± 8.8 min), and sows with parity numbers 5–7 and TB ≥ 16 had the longest farrowing duration (263.2 ± 10.0 min) (Figure 2a). Birth intervals also differed among parities. Birth intervals of sows with parity numbers 5–7 and 8–10 were longer than those of sows with parity numbers 1 and 2–4 (Table 4).

Table 4. Reproductive performance and farrowing characteristics of sows by parity number.

Variables	Parity Number			
	1	2–4	5–7	8–10
Number of sows	506	919	534	534
Total number of piglets born per litter	13.4 ± 0.1 [a]	14.1 ± 0.1 [b]	13.7 ± 0.1 [a]	13.0 ± 0.1 [c]
Number of piglets born alive per litter	12.3 ± 0.1 [a]	12.8 ± 0.1 [b]	11.7 ± 0.1 [c]	10.9 ± 0.1 [d]
Stillborn piglets per litter (%)	3.8 ± 0.3 [a]	4.4 ± 0.3 [a]	7.6 ± 0.3 [b]	8.9 ± 0.3 [c]
Mummified foetuses per litter (%)	4.5 ± 0.5 [a]	4.9 ± 0.4 [a]	7.2 ± 0.5 [b]	7.2 ± 0.5 [b]
Litter birth weight (kg)	15.6 ± 0.2 [a]	17.4 ± 0.1 [b]	16.0 ± 0.2 [a]	15.1 ± 0.2 [c]
Piglet birth weight (kg)	1.28 ± 0.01 [a]	1.37 ± 0.01 [b]	1.36 ± 0.01 [b]	1.38 ± 0.01 [b]
Piglets with birth weight <1.0 kg (%)	12.1 ± 0.6 [a]	10.0 ± 0.4 [b]	12.0 ± 0.6 [a]	10.9 ± 0.6 [a,b]
Farrowing duration (min)	188.3 ± 5.2 [a]	214.3 ± 3.9 [b]	247.7 ± 5.1 [c]	237.1 ± 5.1 [c]
Proportion of sows farrowed >300 min (%)	13.8 [a]	18.3 [a]	29.6 [b]	27.0 [b]
Birth interval (min)	14.5 ± 0.4 [a]	15.5 ± 0.3 [b]	18.6 ± 0.4 [c]	18.7 ± 0.4 [c]

[a,b,c,d] Different superscript letters within a row denote data that differ significantly ($p < 0.05$).

3.4. Onset of Farrowing

Overall, farrowing duration, birth intervals and SB were affected by the onset of farrowing ($p < 0.05$). In general, sows that started farrowing during working hours had a longer farrowing duration (229.3 ± 3.6 min) than sows that started farrowing during non-working hours (217.6 ± 3.4 min, $p = 0.017$). Similarly, sows that started farrowing during working hours also had longer birth intervals (17.0 ± 0.3 min) than sows that started farrowing during non-working hours (16.1 ± 0.3 min, $p = 0.019$). On the other hand, sows that started farrowing during working hours had a lower SB than sows that started to farrow during non-working hours ($5.8 \pm 0.2\%$ vs. $6.7 \pm 0.2\%$, $p = 0.007$). The farrowing duration, birth intervals and SB during working hours and non-working hours in different classes of TB and parity number are presented in Table 5. Interestingly, the difference in the farrowing duration of sows between sows that started farrowing during working hours and non-working hours was significant in only sow parity numbers 5–7 (Table 5). Similarly, the influence of onset of farrowing on the birth interval of piglets was also detected in sow parity numbers 5–7 (Table 5).

3.5. Temperature and Humidity

In the present study, the housing had a good cooling system, resulting in a narrow temperature range (24.2–$27.3\ °C$) and stable relative humidity (69.7%–72.1%) inside the barn. In general, only the average maximum temperature and the maximum temperature–humidity index in the farrowing house for a period of 7 days before farrowing influenced farrowing duration ($p < 0.05$). Farrowing duration in primiparous sows was not correlated with daily mean temperature, daily maximum temperature, relative humidity, temperature–humidity index, or maximum temperature–humidity index ($p > 0.05$) (Table 6). However, in multiparous sows, farrowing duration was positively correlated with maximum temperature ($r = 0.056$, $p = 0.012$) and maximum temperature–humidity index ($r = 0.059$, $p = 0.008$) during the 7-day period before farrowing (Table 6).

Table 5. Farrowing duration, birth intervals and percentage of stillborn piglets during working hours (07.00 h–17.00 h) and non-working hours (17.01 h–06.59 h) by total number of piglets born per litter (TB) and parity number.

Variables	Working Hours	Non-Working Hours	*p* Value
Number of sows	1176	1317	
Farrowing duration (min)			
All sows	229.3 ± 3.6	217.6 ± 3.4	0.017
TB classes			
8–12	206.2 ± 5.9	194.1 ± 5.5	NS
13–15	236.3 ± 5.6	223.9 ± 5.2	NS
≥16	245.5 ± 6.9	234.9 ± 6.7	NS
Parity number classes			
1	194.9 ± 7.8	185.9 ± 7.2	NS
2–4	215.5 ± 5.5	210.4 ± 5.4	NS
5–7	260.7 ± 7.1	235.8 ± 7.3	0.014
8–10	246.2 ± 7.9	238.4 ± 6.9	NS
Birth interval (min)			
All sows	17.0 ± 0.3	16.1 ± 0.3	0.019
TB classes			
8–12	19.6 ± 0.5	18.6 ± 0.4	NS
13–15	17.0 ± 0.4	16.0 ± 0.4	NS
≥16	14.4 ± 0.5	13.8 ± 0.5	NS
Parity number classes			
1	14.7 ± 0.6	13.9 ± 0.6	NS
2–4	15.7 ± 0.4	15.4 ± 0.4	NS
5–7	19.3 ± 0.6	17.3 ± 0.6	0.010
8–10	18.4 ± 0.6	17.9 ± 0.5	NS
Stillborn piglets (%)			
All sows	5.8 ± 0.2	6.7 ± 0.2	0.007
TB classes			
8–12	4.9 ± 0.4	6.1 ± 0.4	0.025
13–15	5.6 ± 0.4	6.9 ± 0.3	0.012
≥16	6.8 ± 0.5	7.0 ± 0.4	NS
Parity number classes			
1	3.3 ± 0.5	4.4 ± 0.5	NS
2–4	4.2 ± 0.4	4.5 ± 0.4	NS
5–7	7.6 ± 0.5	7.6 ± 0.5	NS
8–10	7.9 ± 0.5	10.1 ± 0.5	0.001

NS = not significant ($p > 0.05$).

Table 6. Pearson's correlations between farrowing duration and climatic parameters during the 7-day period before farrowing in sows by parity number.

Climatic Parameters	Mean ± SD	Parity Number of Sows	
		Primiparous	Multiparous (Parities 2–10)
Number of sows		506	1987
Mean temperature (°C)	26.5 ± 0.3	NS	NS
Maximum temperature (°C)	28.4 ± 0.3	NS	0.056 *
Relative humidity (%)	71.0 ± 0.2	NS	NS
Temperature-humidity index	76.4 ± 0.4	NS	NS
Maximum temperature-humidity index	79.2 ± 0.4	NS	0.059 **

NS = not significant ($p > 0.05$); significance levels are indicated by * $0.01 < p < 0.05$, ** $0.001 < p < 0.01$.

4. Discussion

4.1. Effect of Litter Size

In this study, the sows in a commercial swine herd in Thailand currently had 13.7 TB. This indicates that the TB has increased by 38% over the last two decades (TB 9.9) [19]. In another commercial swine herd in Thailand, the average TB was as high as 17.5 [4]. This is mainly due to the import of modern hyperprolific sows from European countries, especially from Denmark. When classifying the litters according to TB, sows that had $\geq$16 TB under the tropical climate accounted for more than a quarter of the sow population. This indicates that sows with $\geq$16 TB are becoming increasingly common in the swine industry. We demonstrated that farrowing duration increased following an increase in TB, while the average birth interval decreased. This can be due to the routine administration of oxytocin in all sows after the birth of the 10th piglet. The administration of oxytocin during parturition can increase the duration and intensity of myometrium contraction, thus decreasing farrowing duration and the average birth interval [20]. However, the proportion of sows with a prolonged farrowing duration (i.e., >300 min) with litters of TB $\geq$ 16 was higher than that with a lower TB. Parturition may last longer in large litters because of the accumulation of expulsion for each piglet [21]. In sows as well as in other species, labour is suspected to be painful due to contraction of the uterus, foetal expulsion, and female reproductive tract inflammation, especially if it lasts for more than 3 h [9]. Therefore, an endogenous opioid-mediated analgesia system exists as a defence mechanism against pain during parturition [21]. However, increasing the release of opioids due to severe pain and stress can interfere with oxytocin [9,21], especially during farrowing in sows with large litters. This is supported by a study in rats, which found an opioid-dependent reduction of oxytocin release during prolonged parturition under stress [22]. There are two known mechanisms for the inhibition of oxytocin release by opioids. First, opioids bind to κ-opioid receptors in the neurohypophysis, which results in the inhibition of neurosecretory terminals [23]; second, opioids bind to μ-opioid receptors in the paraventricular nucleus, resulting in a reduction in the pulse rate of oxytocinergic neurons [24]. Moreover, pain during parturition also activates the autonomic nervous system, which increases catecholamine secretion. High plasma concentration of catecholamine has been considered to affect uterine motility by reducing myometrial contractibility and promoting muscular relaxation. These mechanisms can lead to prolonged farrowing and increased number of nociceptive signals [25]. Decreasing the oxytocin secretion can reduce uterine contraction and results in a prolonged piglet expulsion. In addition, parturition requires energy, and in large litters, the energy demand may be greater. Uterine and mother fatigue due to insufficient energy can cause delivery difficulties or even stop farrowing in sows [21]. Thus, sows with large litters are more susceptible to experiencing severe pain and stress, leading to a decrease in oxytocin release. Moreover, insufficient maternal energy in sows with large litters during parturition may lead to slowing down uterine contractions, further prolonging the farrowing duration. In the present study, sows with prolonged farrowing had more stillborn piglets than sows with shorter farrowing duration. Therefore, various procedures to increase the uterine contractions of sows with large litters need to be comprehensively investigated [26].

4.2. Effect of Parity Number

In the present study, farrowing duration and birth intervals were longer in old sows (parity numbers 5–7 and 8–10) than in young sows (parity numbers 1 and 2–4). This agrees with a previous study that demonstrated longer farrowing durations in sows with higher parity numbers compared to those with lower ones [27]. On the other hand, van Dijk et al. [28] demonstrated that parity number did not affect the duration of the expulsive stage. Additionally, Yang et al. [29] found a shorter farrowing duration in sows with higher parity numbers (6–9) compared to sows with lower parity numbers (i.e., 1 and 2–5). However, in their study, sows with parity numbers 6–9 had a smaller total number of piglets born per litter (12.4 $\pm$ 3.2) than sows with parity numbers 2–5 and 1 (15.5 $\pm$ 3.1 and 16.8 $\pm$ 1.8) [29]. Litter size is positively correlated with farrowing duration, and thus, in sows with large litters, farrowing may last longer. In the present study, farrowing duration and birth intervals were longer in sows with higher parity numbers within the same classes of the total number of piglets born per litter. It is suspected that primiparous sows are more susceptible to experiencing painful parturition than multiparous sows [9], with effects on the duration of labour. However, a previous study demonstrated that sows experienced more pain than gilts due to the uterine activity during farrowing [30]. Ison et al. [30] found that sows showed more frequent back arching as an indicator of pain and had higher salivary cortisol levels than gilts on the day of farrowing. In addition, a study in rats demonstrated that young rats exhibited greater spontaneous contractile activity in myometrium tissue and tended to be stronger than during labour than old rats [31]. Also, integral activity and rate of contraction were greater in young rats than in old rats [31]. Similarly, Mota-Rojas et al. [25] reported that gilts had a better uterine contraction and greater contraction intensity than old sows (i.e., sixth parity). This supports our finding that farrowing duration in sows with lower parity numbers is shorter than in those with higher parity numbers. With less pain during parturition and primed myometrium, the duration of labouring is expected to be shorter in younger animals. These findings indicate that prolonged farrowing duration in hyperprolific sows is a serious issue in multiparous rather than primiparous sows. Pain management as well as improved myometrial activity in multiparous sows should therefore be focused on.

4.3. Onset of Farrowing

In the modern swine industry, farrowing management is important to optimise pig production, with sows starting farrowing during working hours becoming preferable compared to those starting farrowing during non-working hours. In the present study, we found that the farrowing duration and birth interval lasted longer when the sows had started farrowing during working hours compared to sows that had started farrowing during non-working hours. Surprisingly, even though the farrowing duration was longer, the stillbirth rate was lower when sows had started farrowing during working hours. This can be explained by the noise in the farrowing house during working hours, either from the sows themselves (e.g., screaming during feeding times) or the environment (e.g., mechanical ventilation, high-pressure cleaning, feed mixing and manure removal lines) [32]. A previous study found that pigs are sensitive to prolonged or intermittent noise, which can cause increased cortisol levels [33]. A previous study in women found that cortisol and oxytocin levels are reciprocal: when cortisol increases, then oxytocin decreases and vice versa [34]. Therefore, sows that start farrowing during working hours are likely susceptible to noise stress, which can lead to an increase in cortisol levels and a decrease in oxytocin levels, resulting in delayed foetal expulsion. A previous study also confirmed that maternal stress can lead to hyperactivity of the hypothalamic–pituitary–adrenal axis and the sympathetic–adrenal–medullary system, which is related to catecholamine release [25]. As stated before, catecholamine compromises uterine activity during parturition. However, farrowing during working hours is preferable due to the ease of monitoring, and staff can assist immediately if dystocia occurs. In a previous study, the administration of exogenous hormones to control the onset of farrowing was investigated [35,36]. The authors demonstrated that altrenogest supplementation in combination with double administrations of PGF2α successfully synchronised the onset of farrowing in sows, with the proportion of sows that started farrowing during working hours tending to be higher in the treatment group than in the control group. However, the use of pharmacology to control parturition in sows needs to be carefully considered due to its side effects, e.g., reduced colostrum intake and high stillbirth rates [35].

4.4. Temperature and Humidity

Interestingly, the maximum temperature and maximum temperature–humidity index were positively correlated with farrowing duration in multiparous sows (parity number 2–10), but this was not the case in primiparous sows. In contrast, Iida et al. [18] demonstrated that gilts had a lower critical temperature threshold than sows, making them more susceptible to heat stress than sows. In a previous study, sows kept at lower room temperatures (15 °C) had shorter farrowing durations and birth intervals than sows kept at higher room temperatures (20 and 25 °C) [37]. Similarly, Muns et al. [38] demonstrated that second-parity sows kept at an initial room temperature of 20 °C, with a gradual increase to 25 °C, from d 112 to 115 of gestation had a longer farrowing duration compared to sows kept at a room temperature of 20 °C. Failure to perform thermoregulatory behaviour around farrowing may result in heat stress, which has negative effects that lead to prolonged farrowing [38]. Pigs require different housing temperatures for different reproductive stages. Gestating sows need housing temperatures of 15–24 °C, whereas lactating sows require slightly lower housing temperatures: 15–21 °C [39]. However, suckling piglets need much higher housing temperatures: 28–32 °C [39]. Thus, in addition to the mother's requirements, it is important to provide a heater behind the sows during the first few hours after birth, which can reduce hypothermia in newborn piglets. Our findings indicate that heat stress due to rising temperatures and/or humidity in the 7-day period before parturition can cause prolonged farrowing durations in multiparous sows. However, temperature or humidity may not be the only factors influencing the farrowing duration of sows, since they have a relatively low correlation coefficient. Therefore, other factors, such as parity number and maternal stress, may also contribute to the prolonged farrowing duration problem in sows under tropical conditions.

5. Conclusions

The significant risk factors associated with the farrowing duration of Landrace x Yorkshire sows kept in free farrowing pens under tropical conditions included TB, SB, MF, parity number, litter birth weight, piglet birth weight, and time of onset of farrowing. The proportion of sows with a prolonged farrowing in the litters that had TB $\geq$ 16 was higher than that in litters with TB 8–12. Sows that started farrowing during working hours had a longer farrowing duration than sows that started farrowing during non-working hours. The farrowing duration of sows with parity numbers 5–7 and 8–10 was longer than that of sows with parity numbers 1 and 2–4. For multiparous sows, the maximum temperature and the maximum temperature–humidity index also influenced the farrowing duration.

Author Contributions: Conceptualisation, P.T. and Y.K.A.; methodology, P.T.; software, Y.K.A.; validation, P.T., Y.K.A. and R.B.; formal analysis, P.T. and Y.K.A.; investigation, Y.K.A. and R.B.; resources, P.T. and R.B.; data curation, Y.K.A. and R.B.; writing—original draft preparation, Y.K.A.; writing—review and editing, P.T. and R.N.K.; visualisation, Y.K.A. and P.T.; supervision, P.T. and R.N.K.; project administration, P.T.; funding acquisition, P.T. All authors have read and agreed to the published version of the manuscript.

Funding: This research was funded by the Thailand Science Research and Innovation Fund, Chulalongkorn University (FOOD66310009). Yosua Kristian Adi's PhD scholarship is supported by the Second Century Fund (C2F), Chulalongkorn University. Chulalongkorn University supported the One Health Research Cluster.

Institutional Review Board Statement: Ethics review and approval were waived for this study due to the it being a retrospective study conducted based on data obtained from the herd. All animal management followed standard industry animal husbandry techniques. Animals were cared for in compliance with local legal standards. The health and welfare of all animals were monitored by the herd veterinarian according to the standard operating protocols in the farm.

Informed Consent Statement: Not applicable.

Data Availability Statement: The data presented in this study are available on request from the corresponding author. The data are not publicly available due to the original data belonging to the swine breeding herd.

Conflicts of Interest: The authors declare no conflict of interest. The funders had no role in the design of the study, collection, analyses, or interpretation of data, writing of the manuscript, or the decision to publish the results.

References

1. Ward, S.A.; Kirkwood, R.N.; Plush, K.J. Are larger litters a concern for piglet survival or an effectively manageable trait? *Animals* **2020**, *10*, 309. [CrossRef] [PubMed]
2. Oliviero, C. Offspring of hyper prolific sows: Immunity, birthweight and heterogeneous litters. *Mol. Reprod. Dev.* **2022**, 1–5. [CrossRef] [PubMed]
3. Freyer, G. Maximum number of total born piglets in a parity and individual ranges in litter size expressed as specific characteristics of sows. *J. Anim. Sci. Technol.* **2018**, *60*, 13. [CrossRef]
4. Udomchanya, J.; Suwannutsiri, A.; Sripantabut, K.; Pruchayakul, P.; Juthamanee, P.; Nuntapaitoon, M.; Tummaruk, P. Association between the incidence of stillbirths and expulsion interval, piglet birth weight, litter size and carbetocin administration in hyperprolific sows. *Livest. Sci.* **2019**, *227*, 128–134. [CrossRef]
5. Peltoniemi, O.; Oliviero, C.; Yun, J.; Grahofer, A.; Björkman, S. Management practices to optimize the farrowing process in the hyperprolific sow. *J. Anim. Sci.* **2020**, *98*, S96–S106. [CrossRef]
6. Oliviero, C.; Heinonen, M.; Valros, A.; Peltoniemi, O. Environmental and sow-related factors affecting the duration of farrowing. *Anim. Reprod. Sci.* **2010**, *119*, 85–91. [CrossRef] [PubMed]
7. Ju, M.; Wang, X.; Li, X.; Zhang, M.; Shi, L.; Hu, P.; Zhang, B.; Han, X.; Wang, K.; Li, X.; et al. Effects of Litter Size and Parity on Farrowing Duration of Landrace × Yorkshire Sows. *Animals* **2022**, *12*, 94. [CrossRef]
8. Tummaruk, P.; Pearodwong, P. Postparturient disorders and backfat loss in tropical sows associated with parity, farrowing duration and type of antibiotic. *Trop. Anim. Health Prod.* **2015**, *47*, 1457–1464. [CrossRef]
9. Martínez-Burnes, J.; Muns, R.; Barrios-García, H.; Villanueva-García, D.; Domínguez-Oliva, A.; Mota-Rojas, D. Parturition in mammals: Animal models, pain and distress. *Animals* **2021**, *11*, 2960. [CrossRef]
10. Le Dividich, J.; Charneca, R.; Thomas, F. Relationship between birth order, birth weight, colostrum intake, acquisition of passive immunity and pre-weaning mortality of piglets. *Span. J. Agric. Res.* **2017**, *15*, e0603. [CrossRef]
11. Nuntapaitoon, M.; Muns, R.; Tummaruk, P. Newborn traits associated with pre-weaning growth and survival in piglets. *Asian-Australas. J. Anim. Sci.* **2018**, *31*, 237–244. [CrossRef] [PubMed]
12. Björkman, S.; Oliviero, C.; Kauffold, J.; Soede, N.; Peltoniemi, O. Prolonged parturition and impaired placenta expulsion increase the risk of postpartum metritis and delay uterine involution in sows. *Theriogenology* **2018**, *106*, 87–92. [CrossRef] [PubMed]
13. Oliviero, C.; Kothe, S.; Heinonen, M.; Valros, A.; Peltoniemi, O.A.T. Prolonged duration of farrowing is associated with subsequent decreased fertility in sows. *Theriogenology* **2013**, *79*, 1095–1099. [CrossRef]
14. Pidwirny, M. *Understanding Physical Geography*; Our Planet Earth Publishing: Kelowna, BC, Canada, 2021; p. 75.
15. Robbins, L.A.; Green-Miller, A.R.; Lay, D.C., Jr.; Schinckel, A.P.; Johnson, J.S.; Gaskill, B.N. Evaluation of sow thermal preference across three stages of reproduction. *J. Anim. Sci.* **2021**, *99*, skab202. [CrossRef]
16. Lucy, M.C.; Safranski, T.J. Heat stress in pregnant sows: Thermal responses and subsequent performance of sows and their offspring. *Mol. Reprod. Dev.* **2017**, *84*, 946–956. [CrossRef]
17. Tummaruk, P.; Tantasuparuk, W.; Techakumphu, M.; Kunavongkrit, A. Seasonal influences on the litter size at birth of pigs are more pronounced in the gilt than sow litters. *J. Agri. Sci.* **2010**, *148*, 421–432. [CrossRef]
18. Iida, R.; Piñeiro, C.; Koketsu, Y. Timing and temperature thresholds of heat stress effects on fertility performance of different parity sows in Spanish herds. *J. Anim. Sci.* **2021**, *99*, skab173. [CrossRef]
19. Tummaruk, P.; Tantasuparuk, W.; Techakumphu, M.; Kunavongkrit, A. Effect of season and outdoor climate on litter size at birth in purebred Landrace and Yorkshire sows in Thailand. *J. Vet. Med. Sci.* **2004**, *66*, 477–482. [CrossRef] [PubMed]
20. Mota-Rojas, D.; Villanueva-García, D.; Velazquez-Armenta, E.Y.; Nava-Ocampo, A.A.; Ramírez-Necoechea, R.; Alonso-Spilsbury, M.; Trujillo, M.E. Influence of time at which oxytocin is administered during labor on uterine activity and perinatal death in pigs. *Biol. Res.* **2007**, *40*, 55–63. [CrossRef]
21. Rutherford, K.M.D.; Baxter, E.M.; D'Eath, R.B.; Turner, S.P.; Arnott, G.; Roehe, R.; Ask, B.; Sandøe, P.; Moustsen, V.A.; Thorup, F.; et al. The welfare implications of large litter size in the domestic pig I: Biological factors. *Anim. Welf.* **2013**, *22*, 199–218. [CrossRef]
22. Leng, G.; Mansfield, S.; Bicknell, R.J.; Brown, D.; Chapman, C.; Hollingsworth, S.; Ingram, C.D.; Marsh, M.I.; Yates, J.O.; Dyer, R.G. Stress-induced disruption of parturition in the rat may be mediated by endogenous opioids. *J. Endocrinol.* **1987**, *114*, 247–252. [CrossRef] [PubMed]
23. Bicknell, R.J.; Leng, G. Endogenous opiates regulate oxytocin but not vasopressin secretion from the neurohypophysis. *Nature* **1982**, *298*, 161–162. [CrossRef] [PubMed]
24. Morris, M.S.; Domino, E.F.; Domino, S.E. Opioid modulation of oxytocin release. *J. Clin. Pharmacol.* **2010**, *50*, 1112–1117. [CrossRef] [PubMed]
25. Mota-Rojas, D.; Velarde, A.; Marcet-Rius, M.; Orihuela, A.; Bragaglio, A.; Hernández-Ávalos, I.; Casas-Alvarado, A.; Domínguez-Oliva, A.; Whittaker, A.L. Analgesia during Parturition in Domestic Animals: Perspectives and Controversies on Its Use. *Animals* **2022**, *12*, 2686. [CrossRef] [PubMed]
26. Vongsariyavanich, S.; Sundaraketu, P.; Sakulsirajit, R.; Suriyapornchaikul, C.; Therarachatamongkol, S.; Boonraungrod, N.; Pearodwong, P.; Tummaruk, P. Effect of carbetocin administration during the mid-period of parturition on farrowing duration, newborn piglet characteristics, colostrum yield and milk yield in hyperprolific sows. *Theriogenology* **2021**, *172*, 150–159. [CrossRef]
27. Tummaruk, P.; Sang-Gassanee, K. Effect of farrowing duration, parity number and the type of anti-inflammatory drug on postparturient disorders in sows: A clinical study. *Trop. Anim. Health Prod.* **2013**, *45*, 1071–1077. [CrossRef]

28. van Dijk, A.J.; van Rens, B.T.; van der Lende, T.; Taverne, M.A. Factors affecting duration of the expulsive stage of parturition and piglet birth intervals in sows with uncomplicated, spontaneous farrowings. *Theriogenology* **2005**, *64*, 1573–1590. [CrossRef]
29. Yang, K.Y.; Jeon, J.H.; Kwon, K.S.; Choi, H.C.; Kim, J.B.; Lee, J.Y. Effect of different parities on reproductive performance, birth intervals, and tail behavior in sows. *J. Anim. Sci. Technol.* **2019**, *61*, 147–153. [CrossRef]
30. Ison, S.H.; Jarvis, S.; Hall, S.A.; Ashworth, C.J.; Rutherford, K.M.D. Periparturient behavior and physiology: Further insight into the farrowing process for primiparous and multiparous sows. *Front. Vet. Sci.* **2018**, *5*, 122. [CrossRef]
31. Elmes, M.J.; Szyszka, A.; Pauliat, C.; Clifford, B.; Daniel, Z.; Cheng, Z.; Wathes, D.C.; McMullen, S. Maternal age effects on myometrial expression of contractile proteins, uterine gene expression, and contractile activity during labor in the rat. *Physiol. Rep.* **2015**, *3*, e12305. [CrossRef]
32. Sistkova, M.; Dolan, A.; Broucek, J.; Bartos, P. Time of day and season affect the level of noise made by pigs kept on slatted floors. *Arch. Anim. Breed.* **2015**, *58*, 185–191. [CrossRef]
33. Broucek, J. Effects of noise on performance, stress, and behaviour of animals: A review. *Slovak. J. Anim. Sci.* **2014**, *47*, 111–123.
34. Li, Y.; Hassett, A.L.; Seng, J.S. Exploring the mutual regulation between oxytocin and cortisol as a marker of resilience. *Arch. Psychiatr. Nurs.* **2019**, *33*, 164–173. [CrossRef] [PubMed]
35. Taechamaeteekul, P.; Dumniem, N.; Sang-Gassanee, K.; Tummaruk, P. Control of parturition in hyperprolific sows by using altrenogest and double administrations of PGF2α. *Theriogenology* **2022**, *181*, 24–33. [CrossRef] [PubMed]
36. Boonraungrod, N.; Sutthiya, N.; Kumwan, P.; Tossakui, P.; Nuntapaitoon, M.; Muns, R.; Tummaruk, P. Control of parturition in swine using PGF2α in combination with carbetocin. *Livest. Sci.* **2018**, *214*, 1–8. [CrossRef]
37. Malmkvist, J.; Pedersen, L.J.; Kammersgaard, T.S.; Jørgensen, E. Influence of thermal environment on sows around farrowing and during the lactation period. *J. Anim. Sci.* **2012**, *90*, 3186–3199. [CrossRef]
38. Muns, R.; Malmkvist, J.; Larsen, M.L.; Sørensen, D.; Pedersen, L.J. High environmental temperature around farrowing induced heat stress in crated sows. *J. Anim. Sci.* **2016**, *94*, 377–384. [CrossRef]
39. Ooehnalová, S.; Vlnkler, A.; Novák, P.; Drábek, J. The dynamics of changes in selected parameters in relation to different air temperature in the farrowing house for sows. *Czech J. Anim. Sci.* **2008**, *53*, 195–203. [CrossRef]

Article

Pen Versus Crate: A Comparative Study on the Effects of Different Farrowing Systems on Farrowing Performance, Colostrum Yield and Piglet Preweaning Mortality in Sows under Tropical Conditions

Natchanon Dumniem [1], Rafa Boonprakob [1], Thomas D. Parsons [2] and Padet Tummaruk [1,3,*]

[1]   Department of Obstetrics, Gynaecology and Reproduction, Faculty of Veterinary Science, Chulalongkorn University, Bangkok 10330, Thailand
[2]   School of Veterinary Medicine, University of Pennsylvania, Kennett Square, Philadelphia, PA 19348-1692, USA
[3]   Center of Excellence in Swine Reproduction, Chulalongkorn University, Bangkok 10330, Thailand
*   Correspondence: padet.t@chula.ac.th

Citation: Dumniem, N.; Boonprakob, R.; Parsons, T.D.; Tummaruk, P. Pen Versus Crate: A Comparative Study on the Effects of Different Farrowing Systems on Farrowing Performance, Colostrum Yield and Piglet Preweaning Mortality in Sows under Tropical Conditions. *Animals* 2023, 13, 233. https://doi.org/10.3390/ani13020233

Academic Editors: Daniel Mota-Rojas, Julio Martínez-Burnes and Agustín Orihuela

Received: 21 December 2022
Revised: 5 January 2023
Accepted: 6 January 2023
Published: 8 January 2023

Copyright: © 2023 by the authors. Licensee MDPI, Basel, Switzerland. This article is an open access article distributed under the terms and conditions of the Creative Commons Attribution (CC BY) license (https://creativecommons.org/licenses/by/4.0/).

Simple Summary: Loose-housed pens are being implemented as alternative farrowing systems in the swine industry worldwide. This system allows sows to express natural behaviour and reduces stress during the peripartum period. However, most intensive swine farms in Thailand still confine sows in crates during lactation to minimise piglet mortality due to crushing. The present study was performed to compare the reproductive performance of sows kept in the farrowing crate and in the free-farrowing system under tropical conditions. Sows kept in the free-farrowing system produced more colostrum than crated sows. Piglet preweaning mortality rate and the proportion of piglet loss due to crushing in free-farrowing sows were greater than in crated sows. Sow farrowing performance, newborn piglet characteristics and milk production did not differ between the two farrowing systems. Interestingly, in the free-farrowing system, the incidence of crushing in sows with high backfat thickness was significantly higher than in those with moderate and low backfat thickness. These findings imply that free-farrowing pens can be applied in tropical environments without impairing sow farrowing and can enhance sow colostrum production. However, intensive management strategies should focus on adjusting the body conditions of sows prior to farrowing to avoid crushing piglets.

Abstract: The present study was performed to determine the farrowing performance of sows, newborn piglet characteristics, colostrum yield, milk yield and piglet preweaning mortality in a free-farrowing pen compared to a conventional farrowing crate system in a tropical environment. A total of 92 sows and 1344 piglets were included in the study. The sows were allocated by parity into two farrowing systems, either a free-farrowing pen (n = 54 sows and 805 piglets) or a crate (n = 38 sows and 539 piglets). Backfat thickness and loin muscle depth of sows at 109.0 ± 3.0 days of gestation were measured. Reproductive performance data including total number of piglets born (TB), number of piglets born alive (BA), percentage of stillborn piglets (SB) and percentage of mummified foetuses (MF) per litter, farrowing duration, piglet expulsion interval, time from onset of farrowing to the last placental expulsion, piglet preweaning mortality rate, percentage of piglets crushed by sows and number of piglets at weaning were analysed. In addition, piglet colostrum intake, colostrum yield, Brix index and milk yield of sows were evaluated. On average, TB, BA, farrowing duration, colostrum yield and milk yield during 3 to 10 and 10 to 17 days of lactation were 14.7 ± 2.8, 12.8 ± 3.1, 213.2 ± 142.2 min, 5.3 ± 1.4 kg, 8.6 ± 1.5 kg, and 10.4 ± 2.2 kg, respectively. Sows kept in the free-farrowing pen tended to produce more colostrum than crated sows (5.5 ± 0.2 vs. 4.9 ± 0.2 kg, p = 0.080). Piglets born in the free-farrowing pen had a higher colostrum intake than those in the crate system (437.0 ± 6.9 and 411.7 ± 8.3 g, p = 0.019). However, the piglet preweaning mortality rate (26.8 ± 2.9 vs. 17.0 ± 3.8, p = 0.045) and the proportion of piglets crushed by sows (13.1 ± 2.1 vs. 5.8 ± 2.7, p = 0.037) in the free-farrowing pen were higher than those in the crate system. Interestingly, in the free-farrowing pen, piglet preweaning mortality rate in sows with high backfat thickness was higher than that in sows with moderate (37.8 ± 5.1% vs. 21.6 ± 3.6%, $p = 0.011$) and low (21.0 ± 6.2%, $p = 0.038$) backfat thickness. Moreover, the incidence of crushing in sows with high backfat thickness was higher in the free-farrowing pen than in the crate system (17.6 ± 3.6 vs. 4.0 ± 5.7, $p = 0.049$), but this difference was not detected for sows with moderate and low backfat thickness ($p > 0.05$). Milk yield of sows during 3 to 10 days (8.6 ± 0.2 vs. 8.6 ± 2.3, $p > 0.05$) and 10 to 17 days (10.2 ± 0.3 vs. 10.4 ± 0.4, $p > 0.05$) did not differ between the two farrowing systems. In conclusion, piglets born in the free-farrowing pen had a higher colostrum intake than those in the crate system. However, the piglet preweaning mortality rate and the proportion of piglets crushed by sows in the free-farrowing pen were higher than in the crate system. Interestingly, a high proportion of piglet preweaning mortality in the free-farrowing system was detected only in sows with high backfat thickness before farrowing but not in those with low and moderate backfat thickness. Therefore, additional management in sows with high backfat thickness (>24 mm) before farrowing should be considered to avoid the crushing of piglets by sows.

Keywords: animal welfare; backfat thickness; colostrum; farrowing; piglet preweaning mortality

1. Introduction

In recent decades, animal welfare has become an issue of interest in the intensive swine industry. In many European countries, the use of gestation crates has been limited or prohibited in most periods of pregnancy except for the first month of gestation and the week before farrowing [1]. The gestation crate fails to meet all of a sow's biological requirements in part by limiting her ability to perform several natural behaviours, including simply turning around. These altered behavioural responses result from the central nervous system processing of both internal and external stimuli and can frustrate a sow, evoking negative emotional responses and potentially compromising her well-being [2]. The farrowing crate places similar limitations on the periparturient sow. Compromised behavioural responses of sows prior to farrowing is associated with untoward physiological and/or endocrine responses during parturition and lactation periods.

Norway, Finland, Sweden and Switzerland have banned the use of farrowing crates and replaced them with pen-based farrowing systems for lactating sows [1,3]. These so-called free-farrowing systems feature a loose-housed design that allows sows to move freely during the transition and lactation periods and are designed to be an alternative to conventional crated farrowing systems [4]. Sows in free-farrowing pens bedded with straw have lower cortisol responses to the corticotropic-releasing hormone challenge test than crated sows, indicated a lower stress response [5]. Moreover, penned sows tended to have a higher oxytocin pulse than crated sows, which benefits the farrowing process [6]. These findings provide evidence for possible animal welfare benefits to sows housed in free-farrowing systems.

Litter size at farrowing, which has been dramatically increased in modern swine genetics via selective breeding [7], portends several challenges for sows farrowing in either conventional or alternative facilities. This includes prolonged farrowing duration, a variation in newborn piglet birthweight, insufficient colostrum intake and an increased piglet preweaning mortality rate [7]. A previous study over almost three decades revealed the trend of increasing litter size and farrowing duration in modern hyperprolific sows [7]. The process of delivering foetuses causes visceral pain, and its magnitude will be proportional to the number of offspring and the length of parturition [8]. Prolonged farrowing impairs placenta expulsion and increases the risks of postpartum metritis and retained placenta [7]. Continual uterine contraction can cause umbilical cord rupture, meconium staining and peripartum death of piglets [9]. These findings indicate that a long farrowing duration compromises sow welfare as well as health in postpartum and lactation periods. Postpartum complications due to prolonged farrowing duration are exacerbated by barren farrowing environments, for example, confinement or the lack of nest building material [10].

Another consequence of large litter size is that a certain proportion of newborn piglets suffer from intrauterine growth restriction (IUGR). This results in high variation in piglet birthweight within the litter; high competition for colostrum intake, often compromising their ability to achieve sufficient colostrum consumption; and the increased number of low-viability piglets. Large litter size also results in an increased frequency of sows experiencing a negative energy balance during lactation due to their need to produce large volumes of milk [11,12]. Backfat thickness and loin muscle depth are the body-condition parameters associated with sow feed intake, milk yield and lactation performance [11–14]. Sows with high backfat thickness before farrowing have an increased farrowing duration and piglet expulsion interval [15], leading to a high backfat loss during lactation [13]. However, backfat thickness and loin muscle depth at farrowing are positively correlated with both sow milk yield [13] and milk fat content during lactation [11].

Increased litter size also has begotten increased preweaning mortality. In Thailand, the piglet preweaning mortality average is 11.2% and varies from 4.8% to 19.2% among herds [16], with 78.5% of preweaning mortality occurring within the first 72 h postpartum [17]. Even a short period of peri-parturient asphyxia and hypoxia can lead to brain damage, increase the piglet's risk of being crushed by a sow and compromise piglet vitality during early postnatal life [18]. The distinct elevation of piglet preweaning mortality rate has become a both a production and welfare concern within the modern swine industry, especially in free-farrowing systems [19]. In the loose-housed system, the primary cause of piglet mortality based on post-mortem examination is trauma, which is most likely associated with crushing by sows [20].

Taken together, these challenges highlight the opportunities for improvement in managing the modern hyperprolific sow of today. Much less is known about how free-farrowing systems impact these factors and to our knowledge have never been examined in a tropical environment. They are all important issues to be considered in the design and adoption of pen-based farrowing systems. Thus, before the further implementation of loose-housed farrowing pens in the large-scale swine industry under tropical conditions, additional knowledge associated with both sow health and piglet characteristics is required. Hence, the present study determined the farrowing performance of sows including the dynamics of backfat thickness and its role on newborn piglet characteristics, colostrum yield, milk yield and piglet preweaning mortality in a free-farrowing system compared to a crated system in a tropical environment.

2. Materials and Methods

2.1. Animals and Experimental Design

The experiment was conducted in a commercial breeding farm with a herd size of 5000 sows, located in central Thailand, in May to August 2022. A total of 101 crossbred sows (Canadian Landrace × Yorkshire) were randomly allotted to one of two farrowing systems, (i) farrowing crates (n = 45) and (ii) free-farrowing pens (n = 56), from entering the farrowing unit until weaning. Parity number of sows averaged 2.1 ± 0.6 (range 1 to 3). The experiment was carried out from 7 days before parturition until weaning in two consecutive replicates. The average lactation length was 22.1 ± 0.9 days (range 21 to 24 days). For multiparous sows, the type of the farrowing structures that the sows had previously experienced was the conventional crate system. The farrowing processes of the sows were monitored closely from the start to the end by the research team. Data on sow farrowing characteristics, piglet birthweight, body weight at 24 h postpartum and piglet preweaning mortality were collected. Piglet colostrum intake, sow colostrum yield and sow milk yield were determined. Sow colostrum IgG was estimated by using the Brix refractometer [21]. The experiment was reviewed and approved by the Institutional Animal Care and Use Committee (IACUC) in accordance with the university regulations and policies governing the care and use of experimental animals (protocol number 2131053).

2.2. Housing and General Management

Pregnant gilts and sows that were raised in a group-housed system with 280 females per pen, equipped with six electronic sow feeders, were included in the experiment. Gilts and sows were moved to the indoor farrowing house with an evaporative cooling system and temperature control facilities 7 days before the expected parturition date (109 ± 3 days of gestation). After entering the farrowing house, the sows were randomly divided into two groups: farrowing crate and free-farrowing pen. The farrowing pen was designed with an adjustable metal swing hinge and a fully plastic slatted floor, measuring $2.00 \times 2.35 \times 0.90$ m and providing the total area of 4.7 m^2 per pen. In the farrowing crate system, the metal swing hinge was permanently closed, and the sows were kept in individual crates ($1.80 \times 0.60 \times 0.90$ m) with a space allowance of 1.08 m^2 per sow. In the free-farrowing system, to create a loose farrowing environment, the swing hinge was completely opened, providing a space allowance of 3.25 m^2 per sow during the whole experimental period. The space allowance for sows in the free-farrowing system in Thailand was designed following the criteria of the minimum space requirement for sows that are able to turn around in their nest space for piglet inspection and gathering behaviour (i.e., 3.17 m^2) in the farrowing pen [22,23]. In the creep area, a heating lamp, a rubber mattress and a feeding bowl were installed. The schematic diagram of the farrowing pen design is illustrated in Figure 1. This farrowing pen design has been used as an alternative farrowing system in a commercial swine herd in Thailand for over 2 years [24]. Gilts and sows were fed with a commercial lactation diet (907 BTG, Betagro Public Co., Ltd., Lopburi, Thailand) via an automatic feeding pipeline, with an averaged feed allowance of 3.0 to 3.5 kg/sow/day before farrowing, and the feed was provided to ad libitum from parturition date until weaning to meet or exceed their nutritional requirements. The lactation diet contained 13.1% crude protein, 3.68 Mcal/kg metabolisable energy and 0.8% lysine. Drinking water was provided ad libitum via nipples for sows and piglets. After parturition, sows were intramuscularly administered an anti-inflammatory drug (6 mg/kg of ketoprofen, Bezter Ketofen Tec 100®, Siam Bioscience Co., Ltd., Nonthaburi, Thailand) and an antibiotic drug (10 mg/kg of amoxicillin, Vetrimoxin L.A.®, Ceva Santé Animale, Libourne, France). The creep feed was provided for all litters from 3 days postpartum onwards. Piglet general husbandry included iron injection (200 mg/piglet of iron dextran, Bezter Irondex 100®, Thainaoka Pharmaceutical Co., Ltd., Samut Sakhon, Thailand), and teeth clipping was carried out at 1 day of age. Additionally, antiprotozoal drug provision (20 mg/kg of 5% toltrazuril, Better Pharma Co., Ltd., Lopburi, Thailand) and castration were performed at 3 days of age. Cross-fostering was performed to balance the sow functional teats and the number of nursing piglets in the same treatment within 2 days postpartum. Routine sow health care and the vaccination programme were handled by a veterinarian. All gilts and sows were vaccinated against foot and mouth disease virus, porcine circovirus, classical swine fever virus, pseudorabies virus, porcine reproductive and respiratory virus and porcine parvovirus. All sows were kept in a close-housed system equipped with an evaporative cooling system and temperature control facilities (DOL-532, SKOV A/S, Roslev, Denmark) to maintain an optimal temperature inside the barn. The average indoor temperature and humidity during the experimental period were 28.1 ± 1.5 °C (range 25.0 to 32.4 °C) and $74 \pm 5.4\%$ (range 67% to 91%), respectively. The proportion of days when the average temperature inside the barn rose above 25.0 °C during the experimental period was 97.4%. In addition, the average maximum daily temperature inside the barn during the experimental period was 30.7 ± 0.9 °C (range of 28.9 to 32.4 °C).

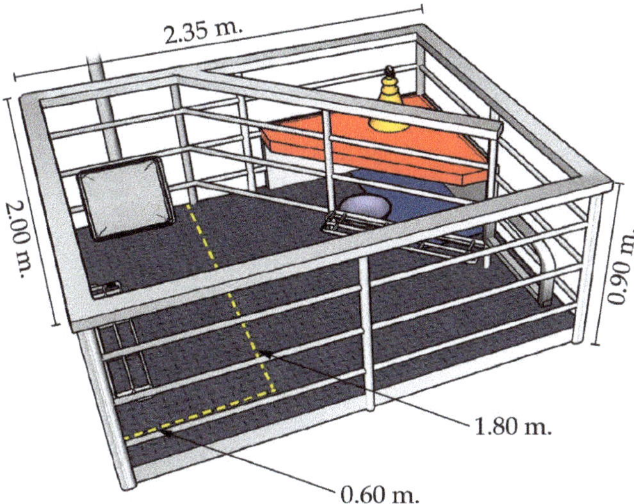

Figure 1. Schematic diagram of the farrowing pen with a lockable swing hinge providing the total area of 4.7 m^2 per pen (2.0 × 2.35 × 0.9 m). In the crate system, the swing hinge was closed (yellow dashed line) from entering the farrowing unit until weaning (sow space allowance = 1.08 m^2). In the free-farrowing system, the hinge was completely opened and locked with one side of the pen. The pen remained opened from entering to the farrowing unit until weaning (sow space allowance = 3.25 m^2). The feeding box was located in the front of the pen. The creeping area was covered with a red plastic roof and consisted of a heating lamp, a rubber mattress and a piglet feeding bowl.

2.3. Farrowing Supervision and Characteristics

The farrowing process was carefully supervised by the research team for 24 h a day. Farrowing induction was not applied in this study. Farrowing assistance was performed only when dystocia was clearly identified. Sow dystocia was defined when an interval of over 45 min elapsed from the birth of the previous piglet and the sow showed intermittent straining, accompanied by the paddling of her legs without any piglet being delivered. Birth assistance included the manual extraction of the piglet and the intramuscular administration of oxytocin (20 IU/sow, Oxytocin Synth, Kela N.V., Hoogstraten, Belgium). The newborn piglet was grabbed immediately after birth and evaluated for the meconium staining score according to Mota-Rojas et al. [9]. Thereafter, the piglet was gently rubbed with a dry towel to remove the remaining amniotic sac, the umbilical cord was cut and tied with a sterilised thread, and the piglet was covered with hygienic powder (Farmasec, Farmapro, Plestan, France). All liveborn, stillborn and mummified foetuses were counted and numbered to determine the birth order. Subsequently, individual liveborn piglets were weighed using a digital scale (SDS® IDS701–CSERIES, SDS Digital Scale Co. Ltd., Yangzhou, China). Intrauterine growth restriction (IUGR) was scored according to Bahnsen et al. [25]. Briefly, the piglets were classified as '0' when their physiological appearance was normal (i.e., normal head shape). The piglets were defined as '1' when they experienced mild IUGR (i.e., steep or dolphin-like forehead, narrow hind part, with a maximum of one secondary parameter) and '2' when they experienced severe IUGR (i.e., steep or dolphin-like forehead, distinctively narrow hind part, with at least one secondary parameter) [25]. The secondary parameters included bulging eyes, wrinkles perpendicular to the mouth, spiky hair and unstable mobility. All neonatal management procedures were done within 3 min after delivery. Farrowing duration was defined as the time interval between the delivery of the first and last piglets. The piglet expulsion interval was defined as the time between the births of two consecutive piglets. The cumulative expulsion interval was defined as the difference in the time between the delivery of the first piglet and the time noted for piglet delivery within the same sow. Time from onset of farrowing to the last placental expulsion was defined as the time between the delivery of the first piglet and the expulsion of the last compartment of the placenta. The coefficient of variance (CV) of piglet birthweight was calculated for each litter.

2.4. Sow Measurement and Data Collection

Sow identities, parity number, insemination date, farrowing date, weaning date, total number of piglets born (TB), number of piglets born alive (BA), number of stillborn piglets, number of mummified foetuses per litter, number of nursed and weaned piglets and weaning-to-service interval were recorded. The percentages of stillborn piglets (SB) and mummified foetuses (MF) per litter were calculated by dividing the number of stillborn piglets or number mummified foetuses per litter with TB and multiplying it by 100. All gilts and sows were evaluated for backfat thickness and loin muscle depth twice at entering the farrowing house and at 21 days of lactation, using a linear array probe and a real-time B mode ultrasonography (HS–2200, Honda Electronics Co., Ltd., Toyohashi, Aichi, Japan). To measure backfat thickness and loin muscle depth, the ultrasound probe was placed approximately 6.5 cm from the dorsal midline at the last rib curve. Lactational backfat loss was calculated by dividing the difference between backfat thickness at entering the farrowing house and at 21 days of lactation with backfat thickness at entering the farrowing house multiplied by 100. Likewise, lactational loin muscle loss was calculated by dividing the difference between loin muscle depth at entering the farrowing house and at 21 days of lactation with loin muscle depth at entering the farrowing house multiplied by 100.

2.5. Piglet Measurement and Preweaning Mortality Data

All live piglets were weighed individually at birth and 24 h after birth. Piglet weight gain at 1 day old was calculated and used to estimate piglet colostrum intake [26]. During lactation, the piglets were weighed at 3, 10, 17 and 21 days of life. Litter weight was calculated by summing all individual piglet body weights. The date and cause of death were recorded for all dead piglets from 1 to 21 days of lactation. Post-mortem examination was not performed because of the farm disease-control policies. However, the cause of death was determined by the observation of the external lesions of the piglet. Piglet preweaning mortality was classified as 'crush' if the piglet presented external traumas or lacerations or fractures of major bones, 'weak' if the piglet was dead with low birthweight and no external lesions were found and 'miscellaneous' if the piglet was dead from other causes not mentioned above. Dead piglets were noted on a daily basis. Piglet preweaning mortality was considered for two periods, including early mortality (the first 3 days of postnatal life) and late mortality (from 4 to 21 days of postnatal life). The piglet preweaning mortality rate of each period was calculated by dividing the total number of dead piglets in the timeframe with BA and multiplying it by 100. Likewise, the proportion of piglets crushed by sows in each period was calculated by dividing the total number of crushed piglets by BA and multiplying it by 100. The piglet preweaning mortality rate during lactation was derived from the piglet preweaning mortality rate of early and late mortality.

2.6. Colostrum and Milk

The colostrum intake of individual piglets was calculated using the equation reported by Thiel et al. [26]: $-106 + 2.26\,WG + 200\,BWB + 0.111\,D - 1414\,WG/D + 0.0182\,WG/BWB$. Sow colostrum yield was calculated by summing the colostrum intake of all piglets within the litter. Milk yield was estimated using the equation reported by Hansen et al. [27]: milk yield day 3 to 10 (g) = 1.93 + 0.07 × (litter size − 9.5) + 0.04 × (litter gain, kg/day − 2.05). Milk yield day 10 to 17 (g) = 2.23 + 0.05 × (litter size − 9.5) + 0.23 × (litter gain, kg/day − 2.05). Furthermore, within 1 h after the onset of parturition, the Brix refractometer (Pocket PAL–1 refractometer, Atago, Tokyo, Japan) was used to estimate the colostrum IgG [21]. The colostrum sample (0.3 mL) was collected manually from the first three pair of teats of the sows and was dropped into the prism chamber of the Brix refractometer using a disposable plastic dropper. The Brix index value was determined immediately after testing.

2.7. Statistical Analysis

All analyses were performed using the statistical analysis system (SAS) software version 9.4 (SAS Institute Inc., Cary, NC, USA). Of the 101 sows, data of sows with litter size less than 8 ($n = 8$) and incomplete farrowing supervision ($n = 1$) were excluded from the analyses. Based on these exclusion criteria, 7 sows in the crate system and 2 sows in the free-farrowing system were excluded, leaving 92 sows and 1344 piglets for data analyses. Descriptive statistics on reproductive data were determined using the MEAN and FREQ procedures of SAS. To differentiate sow lipid deposition, backfat thickness prior to parturition was classified as low (<18 mm), moderate (18 to 24 mm) or high (>24 mm). Continuous data of sows including gestation length, TB, BA, SB, MF, farrowing duration, time from farrowing onset to the last placental expulsion, colostrum yield, Brix index, milk yield from 3 to 10 days and 10 to 17 days of lactation, CV of piglet birthweight within the litter, piglet preweaning mortality rate, proportion of piglets crushed by sow, number of weaned piglets and weaning-to-service interval were analysed by the general linear model (GLM) procedure of SAS. The factors included in the statistical models included farrowing systems (farrowing crate and free-farrowing pen), classes of backfat thickness prior to parturition (low, moderate and high) and their interaction. Least square means of each class of variables were compared using the Tukey–Kramer test. Moreover, sow metabolic parameters, including backfat thickness prior to parturition and at 21 days of lactation, loin muscle depth prior to parturition and at 21 days of lactation and lactational backfat thickness and loin muscle depth loss, were analysed using the GLM procedure of SAS. Piglet characteristics including individual piglet birthweight, piglet expulsion interval, cumulative expulsion interval and colostrum intake were analysed by the general linear mixed model (MIXED) procedure of SAS. The statistical models included the farrowing system (crate and pen), classes of backfat thickness prior to parturition (low, moderate and high) and their interaction as a fixed effect. Sow identities were included in the statistical models to adjust for repeated measurement of the piglet parameters for each sow. Least square means in each class of the variables were compared by using the Tukey–Kramer test. According to Tummaruk and Sang-Gassanee [28], a farrowing duration exceeding 240 min was considered a prolonged farrowing duration. The proportion of sows that had a prolonged farrowing duration (>240 min) for the crate and free-farrowing systems was compared using Chi-square tests. Additionally, the proportions of meconium-stained piglets (score 0 vs. score 1 and 2) and IUGR piglets (score 0 vs. score 1 and 2) were compared between farrowing systems by using Chi-square tests. For all analyses, a p value below 0.05 was considered statistically significant, and a p value between 0.05 and 0.10 indicated a tendency.

3. Results

Across groups, the average TB, BA, SB and MF levels were 14.7 ± 2.8, 12.8 ± 3.1, 9.2% and 3.7%, respectively. Furthermore, farrowing duration, colostrum yield, Brix index and milk yield during 3 to 10 and 10 to 17 days of lactation (means $\pm$ SD) were 213.2 ± 142.2 min, 5.3 ± 1.4 kg, $25.7 \pm 3.4\%$, 8.6 ± 1.5 kg and 10.4 ± 2.2 kg, respectively.

3.1. Sow Characteristics

3.1.1. Gestation Length, Litter Traits and Sow Metabolic Parameters

Gestation length did not differ between sows kept in the farrowing pen compared with those in the farrowing crate (114.4 ± 0.3 vs. 114.8 ± 0.2 days, $p > 0.05$). Litter traits and metabolic parameters of sows in the free-farrowing system compared to those in the crate system are presented in Table 1, and the different classes of backfat thickness prior to parturition are presented in Table 2. Sows with low backfat thickness prior to parturition had a TB 1.7 higher than that of sows with moderate backfat thickness ($p = 0.025$, Table 2).

However, backfat thickness and loin muscle depth prior to parturition were not different between the farrowing systems ($p > 0.05$). Furthermore, sows with low backfat thickness prior to parturition lost less backfat during lactation than those with moderate ($16.5 \pm 3.1\%$ vs. $28.3 \pm 1.9\%$, $p = 0.002$) and high backfat thickness ($33.4 \pm 3.2\%$, $p < 0.001$).

Table 1. Gestation length, litter traits and metabolic parameters in sows kept in the crate system compared to sows kept in the free-farrowing system in a tropical environment (Lsmeans $\pm$ SEM).

Variables	Crate System	Free-Farrowing System	*p* Value
Number of sows	38	54	
Parity number [1]	2.0 ± 0.5	2.2 ± 0.6	
Gestation length (d)	114.4 ± 0.3	114.8 ± 0.2	0.320
Total number of piglets born per litter	14.7 ± 0.5	15.3 ± 0.4	0.365
Number of piglets born alive per litter	12.9 ± 0.6	13.2 ± 0.5	0.699
Stillborn piglets per litter (%)	9.8	8.9	0.757
Mummified foetuses per litter (%)	2.1	4.8	0.191
Backfat thickness prior to parturition (mm)	20.7 ± 0.6	21.2 ± 0.5	0.549
Loin muscle depth prior to parturition (mm)	48.2 ± 0.7	49.4 ± 0.6	0.202
Backfat thickness at 21 days of lactation (mm)	15.1 ± 0.5	15.0 ± 0.4	0.907
Loin muscle depth 21 days of lactation (mm)	41.5 ± 0.8	41.2 ± 0.6	0.822
Lactational backfat loss (%)	25.1	28.3	0.288
Lactational loin muscle loss (%)	14.0	15.8	0.445
Sow loss backfat during lactation >20% (%)	62.9	75.5	0.204
Sow loss loin muscle during lactation >10% (%)	62.9	69.8	0.497
Weaning-to-service interval (days)	4.5	5.1	0.435

[1] Means $\pm$ SD.

Table 2. Gestation length, litter traits and metabolic parameters of sows with low (<18 mm), moderate (18 to 24 mm) and high (>24 mm) backfat thickness prior to parturition (Lsmeans $\pm$ SEM).

Variables	Backfat Thickness Prior to Parturition (mm)		
	Low	Moderate	High
Number of sows	19	51	22
Parity number [1]	2.2 ± 0.5	2.0 ± 0.5	2.2 ± 0.7
Gestation length (d)	114.3 ± 0.4	114.8 ± 0.2	114.6 ± 0.4
Farrowing duration (min)	176.9 ± 33.6	208.9 ± 20.5	258.2 ± 32.4
Total number of piglets born per litter	15.7 ± 0.6 [a]	14.0 ± 0.4 [b]	15.2 ± 0.6 [ab]
Number of born alive piglets per litter	13.6 ± 0.7	12.2 ± 0.4	13.2 ± 0.7
Stillborn piglets per litter (%)	10.0	8.6	9.4
Mummified foetuses per litter (%)	3.1	3.5	3.7
Backfat thickness prior to parturition (mm)	15.1 ± 0.4 [a]	21.2 ± 0.3 [b]	25.7 ± 0.4 [c]
Loin muscle depth prior to parturition (mm)	46.3 ± 0.9 [a]	48.4 ± 0.6 [a]	52.2 ± 0.9 [b]
Backfat thickness at 21 days of lactation (mm)	12.5 ± 0.7 [a]	15.2 ± 0.4 [b]	17.1 ± 0.7 [c]
Loin muscle depth 21 days of lactation (mm)	39.3 ± 1.0 [a]	41.4 ± 0.6 [ab]	43.6 ± 1.1 [b]
Lactational backfat loss (%)	16.5 ± 3.1 [a]	28.3 ± 1.9 [b]	33.4 ± 3.2 [b]
Lactational loin muscle loss (%)	15.0 ± 2.6	14.3 ± 1.6	16.6 ± 2.7
Weaning-to-service interval (days)	4.9 ± 0.3	4.6 ± 0.7	5.0 ± 1.0

[1] Means $\pm$ SD. a, b and c superscripts indicate statistical significance ($p < 0.05$).

3.1.2. Farrowing Performance

Farrowing duration and time from the onset of farrowing to the last placental expulsion of sows did not differ between crate and free-farrowing systems (Table 3). Besides, farrowing duration did not differ among those with low, moderate or high backfat thickness prior to parturition ($p > 0.05$). The proportion of sows that had a prolonged farrowing duration was similar for the two systems ($p > 0.05$). Regarding piglet traits, piglet expulsion interval, cumulative expulsion interval, individual piglet birthweight and proportion of low-body-weight piglets (<1.0 kg), the proportion of meconium-stained piglets and IUGR piglets were not different between the two farrowing systems ($p > 0.05$) (Table 3).

Table 3. Farrowing performance and piglet characteristics in the litters in the crate system compared to the litters in the free-farrowing system in a tropical environment (Lsmeans $\pm$ SEM).

Variables	Crate System	Free-Farrowing System	*p* Value
Number of sows	38	54	
Farrowing duration (min)	229.9 $\pm$ 26.5	199.3 $\pm$ 21.3	0.371
Time from onset of farrowing to the last placental expulsion (min)	471.5 $\pm$ 51.2	384.4 $\pm$ 42.8	0.196
Proportion of sow farrowed longer than 240 min (%)	21.1	22.2	0.894
Coefficient of variance of piglet birthweight (%)	21.4	21.3	0.974
Number of piglets	539	805	
Piglet expulsion interval (min)	12.9 $\pm$ 0.8	12.4 $\pm$ 0.7	0.612
Cumulative expulsion interval (min)	101.8 $\pm$ 4.3	96.7 $\pm$ 3.5	0.361
Individual birthweight (g)	1297 $\pm$ 15	1308 $\pm$ 12	0.570
Proportion of piglets with body weight <1.0 kg (%)	16.2	18.6	0.281
Individual piglet body weight at 1 day old (g)	1388 $\pm$ 17	1420 $\pm$ 14	0.142
Body weight gain during the first 24 h (g)	85.3 $\pm$ 6.2	105.5 $\pm$ 5.1	0.012
Meconium-stained piglets (%)	49.4	46.6	0.347
IUGR piglets (%) [1]	13.6	13.2	0.830
Number of piglets at weaning per litter	11.0 $\pm$ 0.5	10.0 $\pm$ 0.3	0.080
Litter weight at weaning (kg)	55.3 $\pm$ 2.7	49.1 $\pm$ 2.2	0.078

[1] Intrauterine growth restriction.

3.1.3. Colostrum Yield, Milk Yield and Brix Index

Sows kept in the free-farrowing system tended to produce more colostrum than confined sows ($p = 0.080$, Figure 2a). However, the Brix index did not differ between sows kept in the crate and those in the free-farrowing systems (25.7 ± 0.7 vs. 25.6 ± 0.5, $p > 0.05$). Regardless of the farrowing system, sows with high backfat thickness prior to parturition had a higher colostrum yield than those with low backfat thickness (5.7 ± 0.3 vs. 4.8 ± 0.3 kg, $p = 0.065$). Milk yield of sows during 3 to 10 days (8.6 ± 0.2 vs. 8.6 ± 2.3 kg, $p > 0.05$) and 10 to 17 days (10.2 ± 0.3 vs. 10.4 ± 0.4 kg, $p > 0.05$) of lactation did not differ between the two farrowing systems. In addition, in high-backfat sows, milk production during 3 to 10 days (8.3 ± 0.4 vs. 9.6 ± 0.6 kg, $p = 0.059$) and 10 to 17 days (9.6 ± 0.6 vs. 11.7 ± 0.9 kg, $p = 0.050$) of lactation were lower in the free-farrowing system compared to the crate system.

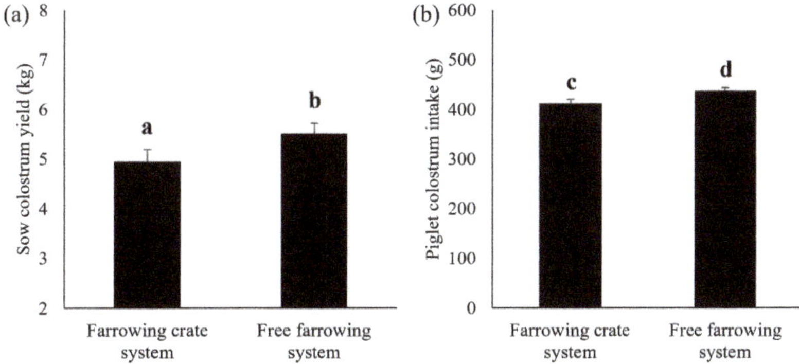

Figure 2. (**a**) Sow colostrum yield and (**b**) piglet colostrum intake for sows kept in the crate system and the free-farrowing system. Data are presented as Lsmeans and SEM. a and b superscripts indicate a tendential difference ($p = 0.080$). c and d superscripts indicate a significant difference ($p < 0.05$).

3.2. Piglet Characteristics

3.2.1. Piglet Measurement and Colostrum Intake

Individual piglet birthweight, CV of the piglet birthweight within the litter and piglet body weight at 1 day did not differ between the two systems (Table 3). However, at 1 day old, the piglets raised in the free-farrowing system had gained more weight than those in the crate system ($p = 0.012$, Table 3). Piglets born in the free-farrowing system ingested more colostrum than those in the crate system ($p = 0.019$, Figure 2b). The number of piglets at weaning tended to be higher in sows kept in the crate system than in those kept in the free-farrowing system ($p = 0.080$, Table 3). Similarly, the litter weight of piglets at weaning for the sows kept in the crate system also tended to be higher than that of piglets from sows kept in the free-farrowing system ($p = 0.078$, Table 3).

3.2.2. Piglet Preweaning Mortality

The total piglet preweaning mortality rate and the mortality rate classified by causes in the free-farrowing system compared with the crate system are presented in Table 4. Interestingly, the piglet preweaning mortality rate ($26.8 \pm 2.9\%$ vs. $17.0 \pm 3.8\%$, $p = 0.045$) and the proportion of piglets crushed by sows ($13.1 \pm 2.1\%$ vs. $5.8 \pm 2.7\%$, $p = 0.037$, Figure 3) were higher in the free-farrowing than in the crate system. The proportion of piglets crushed by sows did not differ between the two farrowing systems during the first 3 days postpartum ($p > 0.05$), but a difference was observed after 4 to 21 days of lactation ($p = 0.008$, Table 4). In the free-farrowing system, the piglet preweaning mortality rate in sows with high backfat thickness was higher than that in sows with moderate ($37.8 \pm 5.1\%$ vs. $21.6 \pm 3.6\%$, $p = 0.011$) and low ($21.0 \pm 6.2\%$, $p = 0.038$) backfat thickness. In addition, piglet preweaning mortality for high-backfat sows in the free-farrowing system was greater than that for crated sows at both 3 days ($17.9 \pm 3.9\%$ vs. $4.8 \pm 5.8\%$, $p = 0.065$) and 21 days of age ($37.8 \pm 5.1\%$ vs. $10.9 \pm 8.0\%$, $p = 0.006$). Similarly, the incidence of crushing in sows with high backfat thickness was higher in the free-farrowing system than in the crate system at both 3 days ($11.1 \pm 3.3\%$ vs. $2.6 \pm 4.9\%$, $p = 0.055$, Figure 4a) and 21 days of age ($17.6 \pm 3.6\%$ vs. $4.0 \pm 5.7\%$, $p = 0.049$, Figure 4b).

Table 4. Causes of piglet mortality during the lactation period in the crate system and the free-farrowing system.

Causes of Piglet Mortality	Crate System	Free-Farrowing System	*p* Value
All piglet preweaning mortality ($n = 260$)			
- Total mortality (%)	17.0 ± 3.8	26.8 ± 2.9	0.045
- 0 to 3 days of age (%)	10.9 ± 2.8	12.6 ± 2.3	0.628
- 4 to 21 days of age (%)	5.3 ± 2.1	14.2 ± 1.6	0.001
Crushing by sow ($n = 104$)			
- Total mortality (%)	5.8 ± 2.7	13.1 ± 2.1	0.037
- 0 to 3 days of age (%)	4.1 ± 2.4	7.4 ± 1.9	0.279
- 4 to 21 days of age (%)	1.4 ± 1.2	5.7 ± 1.0	0.008
Weak ($n = 121$)			
- Total mortality (%)	17.0 ± 3.8	26.8 ± 2.9	0.050
- 0 to 3 days of age (%)	6.1 ± 1.5	4.8 ± 1.2	0.501
- 4 to 21 days of age (%)	3.0 ± 1.3	5.7 ± 1.0	0.116
Miscellaneous causes ($n = 35$)			
- Total mortality (%)	1.4 ± 1.6	3.2 ± 1.2	0.376
- 0 to 3 days of age (%)	0.7 ± 0.5	0.5 ± 0.4	0.743
- 4 to 21 days of age (%)	0.6 ± 1.3	2.7 ± 1.0	0.197

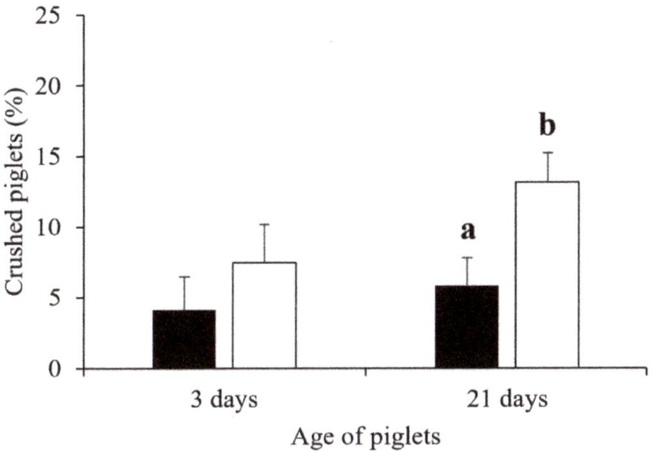

Figure 3. Proportion of piglets dead due to crushing by sows from litters raised either in the crate system (black bar) or the free-farrowing system (white bar) between 0 and 3 days of age and 0 to 21 days of age. Data are presented as Lsmeans and SEM. a and b superscripts indicate a significant difference ($p < 0.05$).

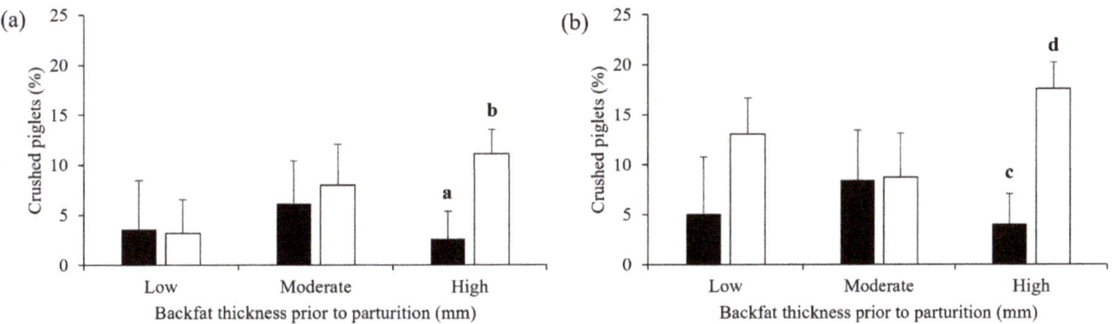

Figure 4. Preweaning mortality of piglets from litters raised either in the crate system (black bar) or the free-farrowing system (white bar) from sows of three different backfat classes before farrowing, low (<18 mm), moderate (18 to 24 mm) and high (>24 mm). (**a**) Between 0 to 3 days of age and (**b**) between 0 to 21 days of age. Data are presented as Lsmeans and SEM. a and b superscripts indicate a tendential difference ($p = 0.055$). c and d superscripts indicate a significant difference ($p < 0.05$).

4. Discussion

In the present study, the farrowing performance of sows, newborn piglet characteristics, colostrum yield, milk yield and piglet preweaning mortality were compared between two different farrowing systems, i.e., crate vs. free-farrowing pen, within the same herd and in the same farrowing house. The sows were under moderate heat stress because the average 24 h indoor temperature and humidity during the experimental period were $28.1 \pm 1.5\ ^\circ\mathrm{C}$ and $74 \pm 5.4\%$, respectively. Moreover, the proportion of days when the average temperature inside the barn rose above $25.0\ ^\circ\mathrm{C}$ was 97.4%. A previous study has demonstrated that the sow thermal preference during the late gestation period was only $14.0\ ^\circ\mathrm{C}$ [29], which is much lower than that observed in the present study. In general, heat stress in sows can occur when the ambient temperatures rises above $25\ ^\circ\mathrm{C}$. This is one of the major problems that decreases daily feed intake and compromises the milk yield of sows under tropical conditions [30]. Furthermore, sow reproductive performance under tropical conditions can be compromised due to the effect of heat stress on the intestinal barrier function, which can limit digestive ability and allow potential pathogens and/or toxins to become systemic [30]. Therefore, all farrowing performance and piglet characteristics demonstrated herein represented those of sows kept in a tropical environment, different from previous studies in temperate areas [6,10,15,31–33]. Moreover, the free-farrowing system has recently been introduced to the Thai swine industry, and scientific data concerning the advantages and disadvantages of this new farrowing system are insufficient. The differences in both farrowing performance and piglet characteristics from birth until weaning between the free-farrowing system and the crate system are presented below.

4.1. Colostrum and Milk Yield

The colostrum yield of sows in the free-farrowing system was higher than that of sows in the crate system. Interestingly, the sows in the free-farrowing system produced 0.5 kg more colostrum than those in the crate system. Oxytocin plays a crucial role as the mediator for mammary myoepithelial cell contraction [34], and an increase in oxytocin around parturition is important for both colostrum production and secretion [34]. Oliviero et al. [6] demonstrated that the levels of oxytocin during farrowing in sows kept in pens were significantly higher than those of sows kept in crate systems. Thus, sows kept in the farrowing pen had a shorter farrowing duration than those in the crate system [6]. Moreover, Yun et al. [33] found that the concentration of prepartum plasma oxytocin of sows in the free-farrowing system with provision of nesting materials was 26.4% higher than that of sows in the crate system. These studies indicate that an increase in oxytocin concentration during pre- and peri-partum periods may attribute to a higher colostrum yield in sows kept in the free-farrowing system compared to the crate system. In the present study, exogenous oxytocin was frequently used in either the crated or the free-farrowing systems. The use of exogenous oxytocin during the peripartum period could be an important factor that influences the colostrum consumption of piglets. Previous studies have demonstrated that exogenous oxytocin administration can increase the number of stillborn and number of live-born piglets with ruptured umbilical cord, meconium staining and neonatal asphyxia [35,36]. These characteristics can influence piglet vitality and hence compromise their colostrum consumption ability. However, in the present study, the proportion of stillborn and meconium-stained piglets did not differ significantly between the crated and the free-farrowing systems.

In the present study, the average milk yield of sows in the free-farrowing system did not differ significantly compared to that of sows kept in the crate system. This indicates that the free-farrowing system has no deleterious effect on sow milk yield. Regardless of the farrowing system, the average milk yield of sows between 3 and 10 days and 10 and 17 days of lactation were 8.6 and 10.4 kg/day, respectively. These values are lower than those reported in an earlier study under tropical conditions, i.e., 10.4 and 12.8 kg/day, respectively [13]. In a previous study in Denmark, the average milk yield of sows at lactation peak was 9.23 kg [27]. In addition, backfat thickness before parturition influences sow milk yield. In the previous study, the milk yield of sows between 3 and 10 days of lactation increased as backfat thickness before parturition increased [13]. However, in the present study, high-backfat sows in the free-farrowing system had a lower milk yield than high-backfat sows in the crate system. The reason could be because sows with high backfat thickness had more mammary parenchymal tissue and more total protein and total DNA than sows with moderate and low backfat thickness [37]. Therefore, increasing parenchymal tissue in late gestation is the major factor that enhances milk production and the growth of suckling piglets [37]. Another reason could be due to a higher piglet mortality rate and a higher proportion of crushed piglets in the high-backfat sows kept in the free-farrowing system, with a consequent reduction in the number of suckling piglets. Therefore, the estimated milk yield was also reduced. Thus, if the number of crushed piglets in the high-backfat sows was reduced, the milk yield of sows in the free-farrowing system might have been increased.

4.2. Piglet Preweaning Mortality

The piglet preweaning mortality rate and the proportion of piglets crushed by sows in the free-farrowing system were 26.8% and 13.1%, respectively. On the other hand, the piglet preweaning mortality rate and the proportion of piglets crushed by sows in the crate system were only 17.0% and 5.8%, respectively. To our knowledge, the present study is the first study demonstrating the piglet preweaning mortality rate and the proportion of piglets crushed by sows in the free-farrowing system under a tropical climate. The average piglet preweaning mortality rate observed in the present study is relatively high but still within the normal range reported earlier in either the crate or the free-farrowing system [16,19,38–42]. In the conventional crate system, the piglet preweaning mortality rate in swine commercial herds in Thailand averages 11.2% and varies among herds from 4.8% to 19.2% [16]. In the free-farrowing system in European countries, the average piglet preweaning mortality rate ranges from 5.1% to 26.0% [19,38–42]. The relatively high piglet preweaning mortality observed in the present study could be related to heat stress in pre- and peri-partum sows because the average temperature inside the farrowing house was, in most cases, above 25.0 °C [43]. A recent study has demonstrated that the risk of piglet mortality in the free-farrowing system was 14% higher than that in the crate system [44]. The present study demonstrated that, within the same herd and the same management, the piglet preweaning mortality rate in the free-farrowing pen was 9.8% higher than that in the crate system. However, a study in Denmark found that the difference in piglet preweaning mortality between free-farrowing and crate systems was only 1.3%, i.e., 13.7% vs. 11.8%, respectively [41]. This indicates that the major disadvantage of the free-farrowing system is the risk of having a high piglet-preweaning mortality. However, the differences among studies indicate that the high piglet-preweaning mortality in the free-farrowing system is a manageable trait and could be overcome by improving various husbandry strategies. For instance, in a previous study, a temporary crate system during some periods of lactation was recommended [42]. However, the total piglet mortality in the temporary confinement system was only slightly decreased compared to that of the complete free-farrowing system, i.e., 25.4% vs. 26.0%, respectively [42]. This indicates that some underlying factors associated with piglet preweaning mortality in the free-farrowing system remain to be further elucidated.

Interestingly, the present study also demonstrated that the proportion of piglets crushed by sows in the free-farrowing pen was 7.3% higher than that in the crate system. This is in agreement with a number of previous studies in temperate areas [31,39,42,45]. For example, in China, the percentage of piglets crushed by sows in the farrowing pen was 14.7% higher than that in the crate system, i.e., 25.5% vs. 10.8%, respectively [45]. However, data regarding the proportion of piglets crushed by sows in the free-farrowing system in the tropics have never been reported. In Finland, the proportion of piglets crushed by sows in the farrowing pen was 14.2% greater than that in the farrowing crate [31]. In Germany, the proportion of piglets crushed by sows in the farrowing pen accounted for up to 70.8% of the piglet preweaning mortality [39]. Similarly, in Switzerland, crushing accounted for 53.4% of the piglet preweaning mortality in the free-farrowing pen [38]. In the present study, crushing by sows accounted for 48.9% of the total piglet preweaning mortality in the free-farrowing pen. On the other hand, Loftus et al. [40] recently demonstrated that the proportion of piglets crushed by sows in the free-farrowing system can be reduced to 3.5%, without a significant difference to the crate system (i.e., 3.3%), by using a large size of the farrowing pen (i.e., 5.6 m^2). In the present study, the total area of the pen was 4.7 m^2, and the space available for a sow in the farrowing pen was only 3.25 m^2, much lower than that recently recommended by the European Food Safety Authority (EFSA), i.e., above 6.6 m^2 for the complete free-farrowing system and 4.3 to 6.3 m^2 for the temporary crating system [46]. Therefore, the incidence of crushing in the free-farrowing system observed in the present study was relatively high (13.1%) compared to that reported in UK (i.e., 3.5%) [40]. This indicates that some husbandry practices, as well as the equipment used in the farrowing house, may help in minimising the proportion of piglets crushed by sows in the free-farrowing system. Additionally, poor maternal behaviours, e.g., rapid postural change and rolling and stepping on the piglets, can also contribute to the high incidence of crushing in the free-farrowing pen [31,39]. Interestingly, the present study found that the proportion of piglets crushed by sows was higher for sows with high backfat thickness (i.e., >24 mm). This is in agreement with Rangstrup-Christensen et al. [47], who observed an increase in piglet preweaning mortality in sows that had a high body-condition score. Most likely, high-backfat sows frequently step on their piglets while lying down [48]. Moreover, these sows frequently change posture and spend more time standing, consequently trapping their piglets [31]. These data indicate that the high prevalence of piglets crushed by sows in the free-farrowing system is usually observed for sows with a relatively high backfat thickness. Thus, additional management strategies to avoid crushing by sows in the free-farrowing system should be focused on sows with high backfat thickness (i.e., >24 mm), trying to minimise the proportion of sows with high body condition before parturition.

4.3. Farrowing Performance and Piglet Characteristics

The farrowing duration of sows in the free-farrowing system did not differ significantly from that of sows in the crate system (199.3 vs. 213.3 min). This is in contrast to a previous study in Finland [15], wherein sows kept in pens had a shorter farrowing duration than those kept in crates, i.e., 212 vs. 301 min, respectively [15]. Furthermore, in a temperate area, Yun et al. [31] revealed that modern hyperprolific sows with an average of 19.3 piglets per litter and kept in a free-farrowing system had a much longer farrowing duration (i.e., 399.4 min) compared to those in the present study. This might be explained by the fact that the total number of piglets born per litter in the Finnish study was four piglets higher than that observed in the present study (15.3 vs. 19.3 piglets/litter) [31]. In the present study, 22.1% and 22.2% of sows in the crate system and the free-farrowing system, respectively, had a prolonged duration of farrowing (i.e., >4 h). Farrowing duration is strongly associated with the concentration of oxytocin, which plays a major role in farrowing progression as it binds to receptors in myometrial cells and stimulates calcium as a second messenger for contraction [8]. During farrowing, the oxytocin concentration was higher in penned sows compared to those kept in crates, i.e., 77.6 vs. 38.1 pg/mL, respectively [6]. Moreover, Blim et al. [49] found that the total calcium concentration in serum at the beginning of the expulsion stage was higher in penned sows compared to crated sows [49]. Thus, if the litter size at birth in the free-farrowing sows under tropical conditions is increased, farrowing duration may also be increased, and the benefits of this farrowing system may be detected.

In the present study, the incidence of stillborn piglets did not differ between the two farrowing systems. This finding agrees with previous studies in temperate environments [15,31]. In additions, the present study is the first that demonstrates the incidence of piglets born with meconium staining (46.6%) and IUGR piglets (13.2%) in the free-farrowing system in a tropical environment. In piglets, meconium staining is associated with umbilical cord rupture and asphyxia [50,51]. Nevertheless, the incidence of either piglets born with meconium staining or IUGR characteristics observed in the present study did not differ between the pen and the crate systems. This indicates that piglets from both systems experienced similar levels of growth retardation and asphyxia. Therefore, the free-farrowing system in tropical environments has no negative impact on newborn piglet characteristics.

4.4. Sow Backfat Thickness and Loin Muscle Depth

Sow backfat thickness and loin muscle depth before parturition and 21 days of lactation and the relative backfat and loin muscle loss during lactation did not significantly differ between sows kept in the farrowing crate and the free-farrowing pen. This finding is in line with the results of a previous study conducted in Europe, which proposed no effects of the farrowing pen on lactational body weight and backfat loss in sows [52]. This indicates a similar backfat and loin muscle loss during lactation in the two different farrowing systems under tropical conditions. Regardless of the farrowing system, lactational backfat loss was higher in sows with high backfat thickness before farrowing compared to those with low and moderate backfat thickness. This is in agreement with a previous study in Thailand [13]. However, the average backfat thickness in the previous study [13] was lower compared to that of the present study. Additionally, our previous study demonstrated that the percentage of sows losing backfat >10% during lactation was higher when backfat was >25.0 mm before farrowing (85.7%) compared to backfat levels of 15.0 to 20.0 mm before farrowing (35.0%) [53]. The difference in sow backfat thickness observed among these studies might be due to the different genetic lines. Therefore, the optimal backfat thickness of sows can vary among herds and genetics lines.

5. Conclusions

Gestation length, stillbirth, farrowing duration, piglet expulsion interval and time from the onset of farrowing to the last placental expulsion in sows kept in farrowing crates did not differ significantly compared to those in sows kept in free-farrowing system. Piglets born in the free-farrowing system had a higher colostrum intake than those in the crate system. However, the piglet preweaning mortality rate and the proportion of piglets crushed by sows in the free-farrowing pen were higher than those in the crate system. Interestingly, a high proportion of piglet preweaning mortality in the free-farrowing pen was detected only in sows with high backfat thickness (>24 mm) before farrowing but not in those with low and moderate backfat thickness. Therefore, special attention, e.g., temporary confinement, can be recommended for sows with high backfat thickness to avoid the crushing of piglets.

Author Contributions: Conceptualisation, N.D. and P.T.; methodology, N.D. and R.B.; software, N.D.; validation, N.D. and P.T.; formal analysis, N.D.; investigation, N.D. and R.B.; data curation, N.D.; writing—original draft preparation, N.D.; writing—review and editing, N.D., T.D.P. and P.T.; supervision, P.T. and T.D.P.; project administration, P.T.; funding acquisition, P.T. All authors have read and agreed to the published version of the manuscript.

Funding: This research project is supported by the Second Century Fund (C2F), Chulalongkorn University and the Thailand Science Research and Innovation Fund, Chulalongkorn University (FOOD66310009).

Institutional Review Board Statement: The present study was performed according to the ethical principles and guidelines for the use of animals for scientific purposes and was approved by the Institutional Animal Care and Use Committee (IACUC) in accordance with the university regulations and policies governing the care and use of experimental animals (protocol number 2131053).

Informed Consent Statement: Not applicable.

Data Availability Statement: The data presented in this study are available on request from the corresponding author.

Acknowledgments: The authors are grateful to the farm staff for helping with animal management.

Conflicts of Interest: The authors declare no conflict of interest.

References

1. Einarsson, S.; Sjunnesson, Y.; Hultén, F.; Eliasson-Selling, L.; Dalin, A.M.; Lundeheim, N.; Magnusson, U. A 25 years experience of group-housed sows–reproduction in animal welfare-friendly systems. *Acta. Vet. Scand.* **2014**, *56*, 37. [CrossRef]
2. Coria-Avila, G.A.; Pfaus, J.G.; Orihuela, A.; Domínguez-Oliva, A.; José-Perez, N.; Hernández, L.A.; Mota-Rojas, D. The neurobiology of behavior and its applicability for animal welfare: A review. *Animals* **2022**, *12*, 928. [CrossRef] [PubMed]
3. Peltoniemi, O.; Oliviero, C.; Yun, J.; Grahofer, A.; Björkman, S. Management practices to optimize the parturition process in the hyperprolific sow. *J. Anim. Sci.* **2020**, *98*, S96–S106. [CrossRef] [PubMed]
4. Peltoniemi, O.; Han, T.; Yun, J. Coping with large litters: Management effects on welfare and nursing capacity of the sow. *J. Anim. Sci. Technol.* **2021**, *63*, 199–210. [CrossRef]

5. Jarvis, S.; D'Eath, R.B.; Robson, S.K.; Lawrence, A.B. The effect of confinement during lactation on the hypothalamic-pituitary-adrenal axis and behaviour of primiparous sows. *Physiol. Behav.* **2006**, *87*, 345–352. [CrossRef] [PubMed]

6. Oliviero, C.; Heinonen, M.; Valros, A.; Hälli, O.; Peltoniemi, O.A. Effect of the environment on the physiology of the sow during late pregnancy, farrowing and early lactation. *Anim. Reprod. Sci.* **2008**, *105*, 365–377. [CrossRef]

7. Oliviero, C.; Junnikkala, S.; Peltoniemi, O. The challenge of large litters on the immune system of the sow and the piglets. *Reprod. Domest. Anim.* **2019**, *54*, 12–21. [CrossRef]

8. Martínez-Burnes, J.; Muns, R.; Barrios-García, H.; Villanueva-García, D.; Domínguez-Oliva, A.; Mota-Rojas, D. Parturition in mammals: Animal models, pain and distress. *Animals* **2021**, *11*, 2960. [CrossRef]

9. Mota-Rojas, D.; Martínez-Burnes, J.; Trujillo-Ortega, M.E.; Alonso-Spilsbury, M.L.; Ramírez-Necoechea, R.; López, A. Effect of oxytocin treatment in sows on umbilical cord morphology, meconium staining, and neonatal mortality of piglets. *Am. J. Vet. Res.* **2002**, *63*, 1571–1574. [CrossRef] [PubMed]

10. Zhang, X.; Li, C.; Hao, Y.; Gu, X. Effects of different farrowing environments on the behavior of sows and piglets. *Animals* **2020**, *10*, 320. [CrossRef] [PubMed]

11. Costermans, N.G.J.; Soede, N.M.; Middelkoop, A.; Laurenssen, B.F.A.; Koopmanschap, R.E.; Zak, L.J.; Knol, E.F.; Keijer, J.; Teerds, K.J.; Kemp, B. Influence of the metabolic state during lactation on milk production in modern sows. *Animal* **2020**, *14*, 2543–2553. [CrossRef] [PubMed]

12. Kirkwood, R.N.; Langendijk, P.; Carr, J. Management strategies for improving survival of piglets from hyperprolific sows. *Thai. J. Vet. Med.* **2021**, *51*, 629–636. [CrossRef]

13. Thongkhuy, S.; Chuaychu, S.H.B.; Burarnrak, P.; Ruangjoy, P.; Juthamanee, P.; Nuntapaitoon, M.; Tummaruk, P. Effect of backfat thickness during late gestation on farrowing duration, piglet birth weight, colostrum yield, milk yield and reproductive performance of sows. *Livest. Sci.* **2020**, *234*, 103983. [CrossRef]

14. Khamtawee, I.; Singdamrong, K.; Tatanan, P.; Chongpaisarn, P.; Dumniem, N.; Pearodwong, P.; Suwimonteerabutr, J.; Nuntapaitoon, M.; Tummaruk, P. Cinnamon oil supplementation of the lactation diet improves feed intake of multiparous sows and reduces pre-weaning piglet mortality in a tropical environment. *Livest. Sci.* **2021**, *251*, 104657. [CrossRef]

15. Oliviero, C.; Heinonen, M.; Valros, A.; Peltoniemi, O. Environmental and sow-related factors affecting the duration of farrowing. *Anim. Reprod. Sci.* **2010**, *119*, 85–91. [CrossRef] [PubMed]

16. Nuntapaitoon, M.; Tummaruk, P. Factors influencing piglet pre-weaning mortality in 47 commercial swine herds in Thailand. *Trop. Anim. Health Prod.* **2018**, *50*, 129–135. [CrossRef]

17. Rootwelt, V.; Reksen, O.; Farstad, W.; Framstad, T. Postpartum deaths: Piglet, placental, and umbilical characteristics. *J. Anim. Sci.* **2013**, *91*, 2647–2656. [CrossRef]

18. Alonso-Spilsbury, M.; Mota-Rojas, D.; Villanueva-García, D.; Martínez-Burnes, J.; Orozco, H.; Ramírez-Necoechea, R.; Mayagoitia, A.L.; Trujillo, M.E. Perinatal asphyxia pathophysiology in pig and human: A review. *Anim. Reprod. Sci.* **2005**, *90*, 1–30. [CrossRef]

19. Olsson, A.C.; Botermans, J.; Englund, J.E. Piglet mortality—A parallel comparison between loose-housed and temporarily confined farrowing sows in the same herd. *Acta Agric. Scand. A Anim. Sci.* **2018**, *68*, 52–62. [CrossRef]

20. Kielland, C.; Wisløff, H.; Valheim, M.; Fauske, A.K.; Reksen, O.; Framstad, T. Preweaning mortality in piglets in loose-housed herds: Etiology and prevalence. *Animal* **2018**, *12*, 1950–1957. [CrossRef]

21. Hasan, S.M.; Junnikkala, S.; Valros, A.; Peltoniemi, O.; Oliviero, C. Validation of Brix refractometer to estimate colostrum immunoglobulin G content and composition in the sow. *Animal* **2016**, *10*, 1728–1733. [CrossRef] [PubMed]

22. Baxter, E.M.; Lawrence, A.B.; Edwards, S.A. Alternative farrowing systems: Design criteria for farrowing systems based on the biological needs of sows and piglets. *Animal* **2011**, *5*, 580–600. [CrossRef] [PubMed]

23. Baxter, E.M.; Moustsen, V.A.; Goumon, S.; Illmann, G.; Edwards, S.A. Transitioning from crates to free farrowing: A road map to navigate key decisions. *Front. Vet. Sci.* **2022**, *14*, 998192. [CrossRef]

24. Adi, Y.K.; Boonprakob, R.; Kirkwood, R.N.; Tummaruk, P. Factors associated with farrowing duration in hyperprolific sows in a free farrowing system under tropical conditions. *Animals* **2022**, *12*, 2943. [CrossRef] [PubMed]

25. Bahnsen, I.; Riddersholm, K.V.; de Knegt, L.V.; Bruun, T.S.; Amdi, C. The effect of different feeding systems on salivary cortisol levels during gestation in sows on herd level. *Animals* **2021**, *11*, 1074. [CrossRef] [PubMed]

26. Theil, P.K.; Flummer, C.; Hurley, W.L.; Kristensen, N.B.; Labouriau, R.L.; Sørensen, M.T. Mechanistic model to predict colostrums intake based on deuterium oxide dilution technique data and impact of gestation and prefarrowing diets on piglet intake and sow yield of colostrum. *J. Anim. Sci.* **2014**, *92*, 5507–5519. [CrossRef]

27. Hansen, A.V.; Strathe, A.B.; Kebreab, E.; France, J.; Theil, P.K. Predicting milk yield and composition in lactating sows: A Bayesian approach. *J. Anim. Sci.* **2012**, *90*, 2285–2298. [CrossRef]

28. Tummaruk, P.; Sang-Gassanee, K. Effect of farrowing duration, parity number and the type of anti-inflammatory drug on postparturient disorders in sows: A clinical study. *Trop. Anim. Health Prod.* **2013**, *45*, 1071–1077. [CrossRef]

29. Robbins, L.A.; Green-Miller, A.R.; Lay, D.C.; Schinckel, A.P.; Johnson, J.S.; Gaskill, B.N. Evaluation of sow thermal preference across three stages of reproduction. *J. Anim. Sci.* **2021**, *99*, skab202. [CrossRef]

30. Tummaruk, P.; De Rensis, F.; Kirkwood, R.N. Managing prolific sows in tropical environments. *Mol. Reprod. Dev.* **2022**. Published electronically December 10. [CrossRef]

31. Yun, J.; Han, T.; Björkman, S.; Nystén, M.; Hasan, S.; Valros, A.; Oliviero, C.; Kim, Y.; Peltoniemi, O. Factors affecting piglet mortality during the first 24 h after the onset of parturition in large litters: Effects of farrowing housing on behaviour of postpartum sows. *Animal* **2019**, *13*, 1045–1053. [CrossRef]
32. Yun, J.; Swan, K.M.; Vienola, K.; Farmer, C.; Oliviero, C.; Peltoniemi, O.; Valros, A. Nest-building in sows: Effects of farrowing housing on hormonal modulation of maternal characteristics. *Appl. Anim. Behav. Sci.* **2013**, *148*, 77–84. [CrossRef]
33. Yun, J.; Swan, K.M.; Farmer, C.; Oliviero, C.; Peltoniemi, O.; Valros, A. Prepartum nest-building has an impact on postpartum nursing performance and maternal behaviour in early lactating sows. *Appl. Anim. Behav. Sci.* **2014**, *160*, 31–37. [CrossRef]
34. Declerck, I.; Sarrazin, S.; Dewulf, J.; Maes, D. Sow and piglet factors determining variation of colostrum intake between and within litters. *Animal* **2017**, *11*, 1336–1343. [CrossRef] [PubMed]
35. Alonso-Spilsbury, M.; Mota-Rojas, D.; Martínez-Burnes, J.; Arch, E.; López Mayagoitia, A.; Ramírez-Necoechea, R.; Olmos, A.; Trujillo, M.E. Use of oxytocin in penned sows and its effect on fetal intra-partum asphyxia. *Anim. Reprod. Sci.* **2004**, *84*, 157–167. [CrossRef]
36. Mota-Rojas, D.; Trujillo, M.E.; Martínez, J.; Rosales, A.M.; Orozco, H.; Ramírez, R.; Sumano, H.; Alonso-Spilsbury, M. Comparative routes of oxytocin administration in crated farrowing sows and its effects on fetal and postnatal asphyxia. *Anim. Reprod. Sci.* **2006**, *92*, 123–143. [CrossRef] [PubMed]
37. Farmer, C.; Martineau, J.P.; Méthot, S.; Bussières, D. Comparative study on the relations between backfat thickness in late-pregnant gilts, mammary development and piglet growth. *Transl. Anim. Sci.* **2017**, *1*, 154–159. [CrossRef]
38. Weber, R.; Keil, N.M.; Fehr, M.; Horat, R. Factors affecting piglet mortality in loose farrowing systems on commercial farms. *Livest. Sci.* **2009**, *124*, 216–222. [CrossRef]
39. Nicolaisen, T.; Lühken, E.; Volkmann, N.; Rohn, K.; Kemper, N.; Fels, M. The effect of sows' and piglets' behaviour on piglet crushing patterns in two different farrowing pen systems. *Animals* **2019**, *9*, 538. [CrossRef] [PubMed]
40. Loftus, L.; Bell, G.; Padmore, E.; Atkinson, S.; Henworth, A.; Hoyle, M. The effect of two different farrowing systems on sow behaviour, and piglet behaviour, mortality and growth. *Appl. Anim. Behav. Sci.* **2020**, *232*, 105102. [CrossRef]
41. Hales, J.; Moustsen, V.A.; Nielsen, M.B.; Hansen, C.F. Higher preweaning mortality in free farrowing pens compared with farrowing crates in three commercial pig farms. *Animal* **2014**, *8*, 113–120. [CrossRef] [PubMed]
42. Hales, J.; Moustsen, V.A.; Nielsen, M.B.; Hansen, C.F. Temporary confinement of loose-housed hyperprolific sows reduces piglet mortality. *J. Anim. Sci.* **2015**, *93*, 4079–4088. [CrossRef] [PubMed]
43. Iida, R.; Piñeiro, C.; Koketsu, Y. Timing and temperature thresholds of heat stress effects on fertility performance of different parity sows in Spanish herds. *J. Anim. Sci.* **2021**, *99*, skab173. [CrossRef] [PubMed]
44. Glencorse, D.; Plush, K.; Hazel, S.; D'Souza, D.; Hebart, M. Impact of non-confinement accommodation on farrowing performance: A systematic review and meta-analysis of farrowing crates versus pens. *Animals* **2019**, *9*, 957. [CrossRef]
45. Gu, Z.; Gao, Y.; Lin, B.; Zhong, Z.; Liu, Z.; Wang, C.; Li, B. Impacts of a freedom farrowing pen design on sow behaviours and performance. *Prev. Vet. Med.* **2011**, *102*, 296–303. [CrossRef]
46. EFSA Panel on Animal Health and Welfare (AHAW); Nielsen, S.S.; Alvarez, J.; Bicout, D.J.; Calistri, P.; Canali, E.; Drewe, J.A.; Garin-Bastuji, B.; Gonzales Rojas, J.L.; Gortázar Schmidt, C.; et al. Welfare of pigs on farm. *EFSA J.* **2022**, *20*, 7421. [CrossRef]
47. Rangstrup-Christensen, L.; Krogh, M.A.; Pedersen, L.J.; Sørensen, J.T. Sow level risk factors for early piglet mortality and crushing in organic outdoor production. *Animal* **2018**, *12*, 810–818. [CrossRef]
48. Bonde, M.; Rousing, T.; Badsberg, J.H.; Sørensen, J.T. Associations between lying-down behaviour problems and body condition, limb disorders and skin lesions of lactating sows housed in farrowing crates in commercial sow herds. *Livest. Prod. Sci.* **2004**, *87*, 179–187. [CrossRef]
49. Blim, S.; Lehn, D.; Scheu, T.; Koch, C.; Thaller, G.; Bostedt, H. Evaluation of the electrolyte status in hyperprolific sows on the farrowing process under different housing conditions. *Theriogenology* **2022**, *193*, 37–46. [CrossRef]
50. Langendijk, P.; Plush, K. Parturition and its relationship with stillbirths and asphyxiated piglets. *Animals* **2019**, *9*, 885. [CrossRef]
51. Castro-Nájera, J.A.; Martínez-Burnes, J.; Mota-Rojas, D.; Cuevas-Reyes, H.; López, A.; Ramírez-Necoechea, R.; Gallegos-Sagredo, R.; Alonso-Spilsbury, M. Morphological changes in the lungs of meconium-stained piglets. *J. Vet. Diagn. Investig.* **2006**, *18*, 622–627. [CrossRef] [PubMed]
52. Kinane, O.; Butler, F.; O'Driscoll, K. Freedom to move: Free lactation pens improve sow welfare. *Animals* **2022**, *12*, 1762. [CrossRef] [PubMed]
53. Tummaruk, P. Post-parturient disorders and backfat loss in tropical sows in relation to backfat thickness before farrowing and postpartum intravenous supportive treatment. *Asian-Australas J. Anim. Sci.* **2013**, *26*, 171–177. [CrossRef] [PubMed]

Disclaimer/Publisher's Note: The statements, opinions and data contained in all publications are solely those of the individual author(s) and contributor(s) and not of MDPI and/or the editor(s). MDPI and/or the editor(s) disclaim responsibility for any injury to people or property resulting from any ideas, methods, instructions or products referred to in the content.

Review

Clinical Experiences and Mechanism of Action with the Use of Oxytocin Injection at Parturition in Domestic Animals: Effect on the Myometrium and Fetuses

Míriam Marcet-Rius [1,*], Cécile Bienboire-Frosini [2], Karina Lezama-García [3], Adriana Domínguez-Oliva [3,*], Adriana Olmos-Hernández [4], Patricia Mora-Medina [5], Ismael Hernández-Ávalos [5], Alejandro Casas-Alvarado [3] and Angelo Gazzano [6]

[1] Animal Behaviour and Welfare Department, Research Institute in Semiochemistry and Applied Ethology (IRSEA), 84400 Apt, France
[2] Department of Molecular Biology and Chemical Communication, Research Institute in Semiochemistry and Applied Ethology (IRSEA), 84400 Apt, France
[3] Neurophysiology, Behavior and Animal Welfare Assessment, DPAA, Universidad Autónoma Metropolitana, Xochimilco Campus, Mexico City 04960, Mexico
[4] Division of Biotechnology—Bioterio and Experimental Surgery, Instituto Nacional de Rehabilitación-Luis Guillermo Ibarra Ibarra (INR-LGII), Tlalpan, Mexico City 14389, Mexico
[5] Facultad de Estudios Superiores Cuautitlán, Universidad Nacional Autónoma de Mexico (UNAM), Cuautitlán 54714, Mexico
[6] Department of Veterinary Sciences, University of Pisa, 56124 Pisa, Italy
* Correspondence: m.marcet@group-irsea.com (M.M.-R.); mvz.freena@gmail.com (A.D.-O.)

Simple Summary: Oxytocin is a hormone that plays an important role in parturition. For this reason, in animal production farms, the exogenous application of this substance is regularly used to help in reducing calving times. However, this hormone can indirectly reduce vitality in the newborn by a direct effect on the intensity of uterine contractions, inducing fetal hypoxia and affecting neonate mortality and thereby a significant economic loss when it is administered in high doses or at times when it is not required. For this reason, the aim is to review the proper administration of oxytocin in domestic animals.

Abstract: Oxytocin is a key hormone for parturition and maternal traits in animals. During the peripartum period, the levels of endogenous oxytocin dictate physiological events such as myometrial contractions, prostaglandin production with the subsequent increase in oxytocin receptors, and the promotion of lactation when administered immediately after birth. While this hormone has some benefits regarding these aspects, the exogenous administration of oxytocin has been shown to have detrimental effects on the fetus, such as asphyxia, meconium staining, ruptured umbilical cords, and more dystocia cases in females. This review aims to analyze the main effects of oxytocin on myometrial activity during parturition, and its potential favorable and negative administration effects reflected in the fetus health of domestic animals. In conclusion, it is convenient to know oxytocin's different effects as well as the adequate doses and the proper moment to administrate it, as it can reduce labor duration, but it can also increase dystocia.

Keywords: oxytocin; domestic animals; parturition; meconium staining; fetal asphyxia

Citation: Marcet-Rius, M.; Bienboire-Frosini, C.; Lezama-García, K.; Domínguez-Oliva, A.; Olmos-Hernández, A.; Mora-Medina, P.; Hernández-Ávalos, I.; Casas-Alvarado, A.; Gazzano, A. Clinical Experiences and Mechanism of Action with the Use of Oxytocin Injection at Parturition in Domestic Animals: Effect on the Myometrium and Fetuses. *Animals* **2023**, *13*, 768. https://doi.org/10.3390/ani13040768

Academic Editor: Robert Dailey

Received: 30 December 2022
Revised: 15 February 2023
Accepted: 17 February 2023
Published: 20 February 2023

Copyright: © 2023 by the authors. Licensee MDPI, Basel, Switzerland. This article is an open access article distributed under the terms and conditions of the Creative Commons Attribution (CC BY) license (https://creativecommons.org/licenses/by/4.0/).

1. Introduction

In all mammal species, birth is characterized by an increase of plasma estradiol, prolactin, and cortisol levels, and by the activation of the oxytocinergic system at the time of delivery [1]. Oxytocin (OXT) is a neuropeptide that presents a dual action neurohormone and neurotransmitter with a chemical structure composed of nine amino acids [2–6]. For this reason, the use of endogenous OXT has become popular in the veterinary field for reproductive management in domestic animals [7].

The synthesis of this hormone has several physiological effects on the central nervous system (CNS). According to Young and Zingg [8], OXT is a behavioral modulator and the abundant presence of oxytogenic neurons in the brain is responsible for the modulation of behavior in both sexes. Additionally, it is pointed out that OXT may be responsible for the stimulation of affiliative relationships in mammals and also in the the generation of bonds of the dam–calf during imprinting [9–11]. During parturition, due to the distribution of OXT receptors, oxytocin also participates in lactation and the development and strengthening of the maternal behavior [10,12–14]. Notably, OXT induces milk ejection and is recognized for its participation in the induction of synchronous and sustained uterine contractions during parturition. For this reason, it is considered the most important hormone during this process [15–18]. Of note, the word "oxytocin" came from Greek words meaning "quick birth" after Dale discovered its uterine-contracting properties in 1906 [19].

On the other hand, it is important to mention that OXT induces contractions in the myometrium in late pregnancy and thereby causes a temporary decrease in blood flow and transient episodes of fetal hypoxia [3]. In addition to this, the number of OXT receptors increases as gestation progresses [20] and it seems important to know that G protein-coupled receptors, such as the oxytocin receptor, could experience desensitization after repeated or prolonged stimulation [21]. More precisely, Robinson et al. [22] demonstrated in vitro that OXT-induced desensitization of myocytes to OXT stimulation occurs within 4.2 h and that this mechanism of desensitization involved OXT receptors.

OXT is present in different systems as it is synthesized in the ovaries, nervous system, testicles, placenta, and even in cardiac tissue [3–6]. OXT is primarily synthesized in magnocellular neurons of the paraventricular and supraoptic nuclei of the hypothalamus. It is stored in vesicles in the posterior pituitary for its further release into peripheral circulation after neurologic stimulation [23–25]. It is also produced by smaller parvocellular neurons of the paraventricular nucleus of the hypothalamus, which directly projects to many regions of the brain, notably the limbic system, with structures such as the amygdala, the hippocampus, the nucleus accumbens, and the medial preoptic area (MPOA) [26].

In dogs, OXT promotes maternal interest in puppies by reducing anxiety and stimulating maternal care [13,27]. Intracerebroventricular infusions of OXT stimulate maternal behavior in sheep [28]. In sows, it has been shown by Gilbert et al. [29] that peripheral OXT is not involved in nest-building behavior because the pre-partum rise in OXT secretion has been linked to the end of sow nest-building behavior, which occurs 1–6 h before birth [30].

It has been seen that in swine production farms, sometimes sows experience prolonged farrowing, postweaning anestrus, and estrus return, which can have a significant impact on the economic profits of the company [31]. Some hormones play an important role in the labor process during farrowing [32] and if the time of farrowing could be reduced by administering them, it would be important for piglet survival, as a delay in farrowing can increase stillbirth numbers [33,34]. In this sense, serum estradiol, as well as the concentration of OXT, prolactin, progesterone, and prostaglandins have been used [3–6,15–17,35,36].

For this reason, the administration of exogenous OXT is one of the most important resources regarding the management of animals during farrowing, and some demographic data mention that 83.1% of pig farms use oxytocin as the first resource in the management of farrowing [37]. This is because its application has been observed to significantly reduce labor time, between 12 and 20 min in most species, and it begins to take effect within the first 10 min after its administration [2,38,39]. This can be recognized as an immediate benefit in relation to costs and care time, because an increase in synchronous contractions facilitates the expulsion of the newborn and even the expulsion of fetal membranes [40].

The benefits of OXT administration could have led to its indiscriminate use due to little control and the apparent absence of adverse effects of this neuropeptide [41–43]. However, due to the greater knowledge of the participation of OXT in other physiological systems, notably in cardiovascular regulation [44], it could make evident the adverse effects that the exogenous administration of this hormone could bring with it [3,16]. In some studies, it has been seen that the increase in uterine contractions caused by exogenous use of OXT can reduce blood flow at the level of the fetal umbilical artery [45]. This can cause hypoxia, which can be associated with states of respiratory stress and lead to suffocation [46], and, due to these events, the fetal vitality in newborns could be decreased [47–51]. In dogs, for example, fetal asphyxia represents 60% of losses of puppies [52]; in swine production, there are reports that have shown that mortality caused by asphyxia is 5–10% when this hormone has been administrated [53], while in dairy cattle there is an incidence of asphyxia between 0.67–9.2% derived from the use of OXT [54].

Therefore, these undesirable effects should be considered prior to OXT administration to reduce the potential dystocia and intrapartum mortality that this would generate. For this reason, the present review aims to analyze the mechanism of action of OXT, its main effects on myometrial activity during parturition, and the potential favorable and negative effects on the fetus of domestic animals.

2. Oxytocin: Pharmacology and Clinical Application at Parturition in Domestic Animals

In small ruminants, the half-life of OXT has been found to range from 1.3 (sheep) to 22 min (goats) [55,56]. In horses, it has been reported that the administration of 25 IU of intravenous OXT allows a half-life of 5.89 min, with a clearance rate of 7.78 min [57]. A comparative study of the pharmacokinetics of OXT suggests that the parameters of bioavailability, clearance time, and half-life are similar in most domestic mammals, which could possibly be taken as a reference [58]. Due to these reported half-life data, it is suggested that OXT in rabbits and goats has extensive penetration into different tissues [59, 60]. It is worth mentioning that a short half-life in some animals leads to questioning if labor can be induced with a single administration of OXT (Table 1) [61,62].

Table 1. Pharmacokinetic parameters of oxytocin in different species.

Dose	Administration Route	Evaluated Parameters	Pharmacokinetics	Species	Reference
		Absorption Kinetics			
25 IU	IV	$T_{1/2}$: 5.89 min., clearance rate of 11.67 L/min., and mean residence time of 7.78 min.	Maximum effective concentration of 0.45 ng/mL and a plasma concentration of 0.25 ng/mL	Mare	Steckler et al. [57]
10 IU	IM	Minimal plasma concentration at 20 min. of administration 4822 pg/h/mL.	Tmax = 1.13 ± 0.91 h, Cmax = 2662 ± 567 pg/mL, AUC = 4822 ± 728 pg/h/mL, $T_{1/2}$ = 1.02 ± 0.33 h.	Rabbit	Zhu and Lal [63]
400 IU	Sublingual	Plasmatic concentration at 20 min. of administration 1234 pg/h/mL.	Tmax = 0.93 ± 0.64 h, Cmax = 1164 ± 1179 pg/mL, AUC = 1234 ± 1001 pg/h/mL, $T_{1/2}$ = 0.90 ± 0.33 h.		
0.33–1.32 IU	Intranasal	$T_{1/2}$ = 12.2 min., Tmax = 12.2 min., Mean plasma concentration = 18.9 pg/mL.	Cmax = 77.3 pg/mL, AUC = 1726 pg/mL/min., mean residence time = 20 min.	Cow	Wagner et al. [64]
0.083, 0.11 and 0.17 IU	IM	$T_{1/2}$ = 1–6 min. Bioavailability: 70–100%, Vss: 40 min.	Clearance rate 7.87 mL.	Sow	Mota-Rojas et al. [65], Hill [66]

Table 1. *Cont.*

Dose	Absorption Kinetics		Pharmacokinetics	Species	Reference
	Administration Route	Evaluated Parameters			
40 IU 20 IU	IM IV Intravulvar	$T_{1/2} = 1.94 \pm 0.21$ min.	Vd = 0.46 ± 0.02 L/kg, $T_{1/2}$ of elimination = 22.3 ± 0.3 min.	Sow	Mota-Rojas et al. [67], Ybarra Navarro [68]

In the case of non-domestic species, the reported values for baboons receiving 500 IU IV OXT were a mean plasma concentration of 10 pg/mL on 10 min., $T_{1/2}$ = 1.1 ± 0.2 min, Vd = 80 mL/Kg, Vss = 200 mL/Kg, fraction of elimination = 34 ± 5%, and mean residence time = 7.7 ± 0.8 min. For rats receiving 200–1000 ng/Kg IV, $T_{1/2}$ = 21.09 min, $T_{1/2}$ elimination = 7.94 min, Vd = 0.13–0.80 L/Kg, Vss = 0.34–0.80 L/Kg, and AUC = 2219–80,087 h/ng/L. AUC: area under the plot of plasma concentration of a drug versus time after dosage; Cmax: peak plasma concentration; IU: international units; IM: intramuscular; IV: intravenous; $T_{1/2}$: half-life; Tmax: time to reach maximum concentration; Vd: volume of distribution; Vss: steady state volume of distribution.

The distribution of OXT and its clinical effect may be related to the route of administration, since it is mentioned that the preferred routes are intravenous (IV) and intramuscular (IM) [69–71]. In a study carried out by Mota-Rojas et al. [67] in 50 sows in which they compared the effect of OXT administered by three different routes: IM, IV, and intravaginal, they observed that in the animals that received OXT intravaginally and intravenously, there was an increase in the number of stillbirths and intrapartum broken cords. However, in the IM administration, umbilical cord rupture could be due to increased uterine contractions. In another study carried out by the same authors, they confirmed that IM administration of OXT significantly increased the intensity of uterine contractions, decreasing their duration and farrowing by up to 40 min, but with adverse effects for the fetus (e.g., meconium-stained piglets when OXT is administered at early phases of farrowing) [48]. This could make evident that the IM and IV routes of administration allow OXT to reach its highest plasmatic concentration and exert its action in the target organ, in this case, the uterus.

In the case of cattle, Wagner et al. [64] studied five bovines (*Bos taurus*) that received a dose of 0.66 IU/kg of OXT intranasally. They found that the concentration of OXT in the blood increased by 63.3 pg/mL at 3.5 min and that it maintained a half-life of 12.1 min. This indicates that the intranasal route could be an alternative for the use of OXT. However, while several behavioral effects have been shown with this route [72,73], as far as we know, to date, there are no reports indicating that it may have any effect on uterine activity.

According to Hill et al. [66], the most used doses of OXT in sows are 5 IU, 10 IU, 20 IU, 25 IU, 30 IU, and 40 IU administered by one of the three common routes of administration, IM, IV and intravulvar. A study was carried out on 200 sows in which they were grouped into four groups according to the dose of OXT administered (1 UI/6 Kg being the high dose, 3.1 UI/9 Kg being the medium dose, 1 UI/12 Kg being the low dose and control group). They observed that there was an increase in the number and intensity of contractions during farrowing in animals treated with the high and medium doses compared to the low dose, but they had greater deleterious effects on the fetuses [74].

In horses, the use of OXT is reported in doses of 3.5 IU up to 75 IU IM for the induction of uterine contractions. Villani and Romano [75], evaluated the effect of three low doses (3.5 IU) of IM OXT in 350 standardbred mares, pregnant at term. They observed that the administration of OXT caused a foaling stimulation effect on the length of gestation: 51.3% of the animals responded to the first dose of OXT, 14.2% to the second dose, and 3.4% to the third dose. This means that the last dose allowed moderate efficacy for labor induction. Opposite results were obtained in another study on 16 mares in which they compared the effect of two doses of oxytocin (1.75 IU and 75 IU) for labor induction. The authors observed that animals receiving the highest dose of OXT had a shorter interval from oxytocin administration to delivery of the fetus compared to animals receiving the higher dose, but resulting in weak foals [76].

Consequently, OXT tends to have a short plasma half-life so that a single dose may not achieve a correct stimulation of myometrial contractions. However, this effect may depend on the route of administration, the IM and IV routes being the most used. The dose may also be responsible for the intensity of the effects at the uterine level and, as studied in human medicine, the mode of administration could also have influence, but it is not well studied in animals.

3. OXT: Its Mechanism of Action and Receptor Signaling in the Myometrium in Domestic Animals

The mechanism of action of exogenous OXT in the myometrium does not differ from its endogenous ligand, promoting uterine contractions due to the arrangement of receptors in this tissue (Figure 1) [77]. In this regard, Taverne et al. [78] evaluated the functionality of uterine OXT receptors in cows around calving, using an intra-arterial dose of 800, 1600, and 3200 mU, making a continuous evaluation of the uterine electromyographic activity. The authors reported that OXT significantly increased myometrial activity and the magnitude of contractions. The production of uterine contractions in domestic animals is induced by the increase in intracellular Ca^{2+} in the myofibrils of the smooth muscle of the myometrium [60]. A close relationship between OXT receptor occupancy and Ca^{2+} entry through voltage-gated or coupled channels has been found, as suggested by in vitro studies on rat myometrial cells [3,79,80]. Furthermore, Riemer and Heymann [81] argue that, along with OXT and prostaglandins, connexin 43 (CX43) also participates in inducing myometrial contractility. Thus, it leads to understanding a possible stimulation of contractions through the activation of other contractile proteins such as CX43, due to the expression of genes encoding for proteins associated with a uterine contraction that cause an increase in said substance, which induces an ionic coupling to allow increased intracellular Ca^{2+} flow into myofibrils and coordinated contractions [82,83]. Immunohistochemical studies performed on neurons of the preoptic nucleus in lactating rats revealed that OXT neurons were positive for CX36, another protein, thus suggesting the involvement of this hormone with OXT [84].

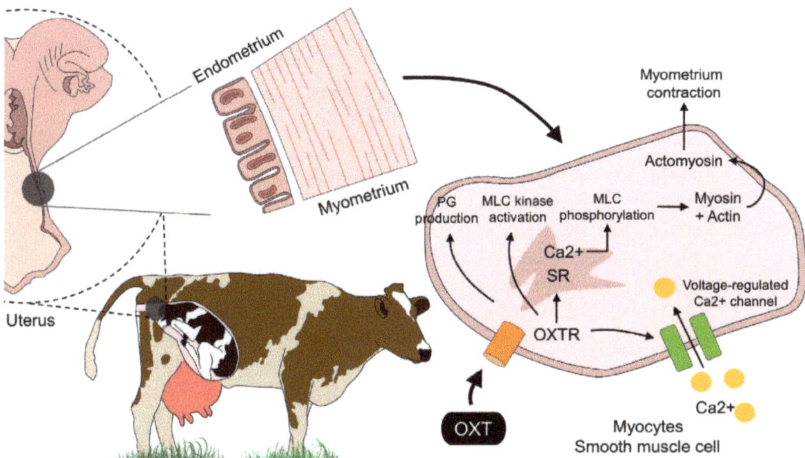

Figure 1. Oxytocin action on the smooth muscle cells in the myometrium. Endogenous and exogenous OXT promotes myometrium contraction by activation of the OXTR and its action on the voltage-regulated Ca^{2+} channels that facilitate Ca influx to the cell. In the SR, the interaction of Ca^{2+} participates in several events that result in muscle contraction, such as PG, production, MLC kinase activation, MLC phosphorylation, and the formation of actomyosin to maintain uterine contractions during calving: myosin light chain; MLC: myosin light-chain; OXT: oxytocin; OXTR: oxytocin receptors; PG: prostaglandin; SR: sarcoplasmic reticulum.

The action of OXT can not only be observed at the uterine level. In fact, due to the universal distribution of its receptor, other physiological effects can be observed [16]. One of the most notorious would be the cardiovascular effect of OXT. Peterson [85] mentions that this neuropeptide induces a paradoxical effect at the vascular level because of its systemic vasodilator effect. In an in vivo study in nephrectomized Wistar rats, it was found that OXT attenuates the serum levels of urea, creatinine, and lactate dehydrogenase, which could be considered a protective effect [86]. However, at the uterine level, it can induce vasoconstriction [49]. Interestingly, given the abundant presence of OXT receptors in neurons, it has been suggested that it may be related to the mechanisms of nociception in females [12,87]. Biurrun Manresa et al. [88] suggest that this is possibly due to oxytocinergic neuron projections at the spinal cord that generate inhibitory modulation of painful stimuli. This makes it evident that OXT shows relevant physiological participation in different systems to consider that this hormone has a limited effect on the myometrium.

In summary, the mechanism of action of exogenous OXT allows an increase in the permeability of Ca^{2+} in the uterine myofibrils, which would stimulate contractions [89]. Despite its obvious mechanism of action, it has been argued that the efficacy of OXT may be affected by conditions such as a lack or decreased number of receptors in the myometrium [60]. In a systematic review, it was reported that the use of OXT reduces up to 3 to 5 min the farrowing interval time in piglets and up to 40 min the labor time of sows in the first stage of farrowing [66]. However, previous reports mention that OXT is not effective at term due to the absence of OXT receptors that develop after a change in the estrogen:progesterone ratio minutes before delivery [90].

Therefore, exogenous OXT promotes uterine contractions by occupying specific receptors and increasing intracellular ion flow in the myometrium. This mechanism of action is primarily responsible for the effect on the uterus. However, due to the dependence on the presence of receptors, the effect of exogenous OXT could be limited in the absence of said receptors. Another aspect relating to the OXT receptors could be of importance: their DNA sequences can be subject to variation, i.e., Single Nucleotide Polymorphisms, that can lead to OXT receptors variants, which in turn, may influence myometrial OXT response in the setting of parturition. For instance, an association between an OXT receptor variant and the development of postpartum hemorrhage in humans has been recently shown [91].

4. Favorable Effects of the Use of Oxytocin during Parturition and Recommendations for Its Use in Veterinary Obstetrics

Exogenous and endogenous OXT also has positive effects both in the dam and the fetus. For example, through placentophagy and consumption of amniotic fluids during parturition, the dam can obtain OXT, and some analgesic effects have also been reported [92]. In bitches, the administration of OXT during partum intervenes in the frequency of myometrial contractions, increasing them. It is usually given when uterine contractions are less than expected and the fetal heart rate is normal. However, the administration of low IV doses is recommended to prevent tetanic effects in the uterus, putting oxygenation of the fetus at risk [93,94]. According to Davidson [95], medical management of dystocia with irregular contractions or primary uterine inertia in bitches could include the administration of 10% calcium gluconate (0.465 mEq/4.5 kg SC) followed by OXT (0.5–1 U.S.P. units/bitch SC).

In the case of intensive pig production farms, the problems of sows at farrowing are resolved indistinctly through the use of oxytocics. However, it is important to point out that the application of exogenous OXT is not recommended in the following cases [96]: when there are regular contractions, when the cervix is not yet fully dilated, when there is a disproportion between the size of the fetus and the pelvis bone, when there is a bad presentation, and when there is any bleeding, vaginal prolapse, or hypocalcemia. In these cases, it can have adverse results if it is used neglectfully and indiscriminately [92]. In a study carried out by González-Lozano et al. [97] in sixty hybrid Yorkshire–Landrace sows (30 with eutocic farrowing and 30 with dystocia farrowing), in which was administrated exogenous OXT, at the dose of 0.083 UI/kg, to 15 eutocic and 15 dystocic sows, intramuscularly after the delivery of the 5th piglet. The results showed that OXT decreases the number of intrapartum deaths by 50%, and the highest viability was observed in the group of eutocic sows treated with OXT. Likewise, Paccamonti [98] points out that the application of OXT in mares for inducing foaling, with a single low dose of 10 IU, IV for a 550 kg mare is preferred because it is sufficient to initiate the cascade of events leading to foaling and provides a more natural course of events than higher doses or repeated administration. Therefore, it can be concluded that both the moment of application, as well as the dose, route, manner, and type of delivery influence whether there are benefits or adverse factors with the exogenous application of this hormone.

On the other hand, it is necessary to point out the importance of uterine contractions due to OXT, since, thanks to them, the exit of the lochia is favored, causing a more rapid uterine involution [99]. In the same way, this hormone contributes to stimulating the descent of milk into the mammary glands together with the stimulation that the offspring make when suckling [100], and also with the establishment of bonding between the dam and newborn [9,13,101–103].

Another important effect of OXT is that it can intervene in helping to establish maternal behaviors and decreasing the incidence of maternal cannibalism. This was confirmed in a study carried out by Kockaya et al. [104] in 30 adult Kangal bitches, 15 with the presentation of cannibalism and 15 that did not present it, in which the levels of OXT in serum were measured and it was found that in bitches with presentation of cannibalism, OXT levels were lower (3.58 ± 0.43 ng/mL) compared to bitches that did not present it (9.68 ± 1.58 ng/mL). In a study conducted by Feldman et al. [105] in humans, it has been shown that plasma and salivary OXT levels are associated with dam–infant bonding. Contrasting to these data, there was a study carried out by Ogi et al. [106] on 25 lactating Labrador Retriever dogs, in which salivary concentration of OXT was unrelated to the amount of maternal care, but had a weak negative correlation with sniffing/poking behavior. Therefore, salivary OXT concentration cannot be considered a strong predictive biomarker of the quantity of maternal care in dogs. Therefore, this study suggests that if what is sought is to improve maternal care in bitches, the exogenous application of OXT is probably not effective. Furthermore, the OXT assay procedures can differ and be themselves a source of discrepancies between the studies. Indeed, the assessment and interpretation of urine and saliva measurements often provide inconsistent findings. To a lesser extent, it can be also true for plasma measurements, as the methodological procedures can deeply impact the outcome of OXT concentrations [107,108].

Thus, it can be deduced that the application of OXT could have beneficial effects, if there are no situations such as those mentioned above in which its application is contraindicated.

5. Effects of the Exogenous Application of Oxytocin on the Fetus and the Umbilical Cord

The induction of myometrial contractions is the main reason behind the use of OXT during the parturition process [109]. Although the expulsion time is reduced, an increase in myometrial contractions due to the administration of OXT is a factor associated with decreased fetal oxygen saturation which can affect the viability and vitality of the fetus [110]. Moreover, its use can have an adverse effect on the fetus due to contractions, resulting in compromised blood flow (Figure 2) [111]. According to Mota-Rojas et al. [96], OXT given to sows at the onset of fetal expulsion significantly increases the rate of fetal distress, anoxia, and intrapartum death in piglets. The rupture of the umbilical cord, as well as the presence of meconium staining, could be other consequences of the excessive or misused exogenous application of OXT [65,112,113]. This effect on the fetus may be related to the type of placentation in different mammalian species [114]. For example, horses, cattle, and pigs present an epitheliochorial-type placenta, and some immunohistochemical studies have shown the presence of OXT receptors, mainly in the endometrium; however, the presence of these receptors in the allantochorionic or cotyledonous region is low, which possibly interferes with its impact on fetal membrane rupture [115,116]. In the case of species thar present hemochorial placenta, such as humans or rats, it has been suggested that OXT may lead to development of oxidative stress effects, due to the interrelationship that exists between the fetal vascular endothelium and the dam. Additionally, uterine contractions could reduce flow due to the rupture of this relationship [117].

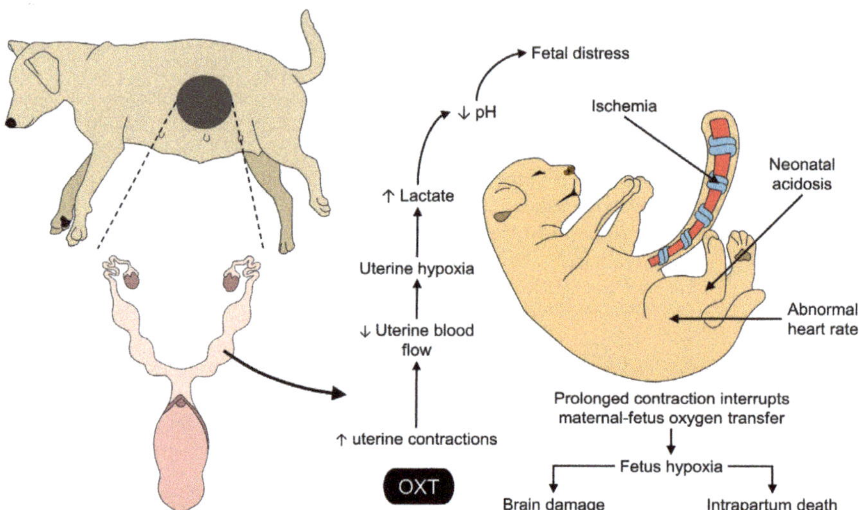

Figure 2. Effect of oxytocin administration on the umbilical cord and fetus health. Although OXT is commonly used during the onset of parturition, adverse effects directly on fetal health have been reported. After the hyperstimulation of the myometrium, prolonged contractions reduce uterine blood flow, resulting in a hypoxia state and accumulation of lactate. These events cause fetal distress with the main consequences including neonatal hypoxia, acidosis, abnormal heart rate, umbilical cord ischemia, and possible brain damage or intrapartum death when administering OXT. OXT: oxytocin.

In Yorkshire x Landrace sows, Alonso-Spilsbury et al. [47] found that the administration of a single dose of OXT immediately after the birth of the first piglet resulted in several health outcomes for the fetus, although the expulsion interval was also reduced from an average of 27.76 ± 0.81 min to a minimum of 22.2 ± 1.80 min. The presentation of stillbirths was higher in the first and third delivery of sows receiving OXT (a range between five and ten stillbirths), and the grade of meconium staining was greater (between 0.27 and 0.30). Additionally, an increased frequency of ruptured umbilical cords was observed in the treated animals compared to the control group (0.42–0.47 vs. 0.07, respectively), and dystocia cases were more prevalent in the sows receiving OXT ($p < 0.01$) [47].

The association between OXT and dystocia is due to the effect that OXT has on the intensity and duration of myometrial contractions [96], reducing uterine blood flow and fetal oxygenation [118]. This has been reported in 300 dairy Holstein calves, resulting in animals with lower mean oxygen partial pressure (39.6 ± 9.3 mmHg) (15 mmHg lower than control animals) and higher sodium concentration (137.7 ± 2.1 mmol/L) than newborns from dams not treated with OXT [118]. Therefore, these effects could affect neonatal vitality and welfare [12].

The direct effect of OXT on the local bloodstream has been suggested as one of the main adverse effects of its exogenous administration. For example, Stenning et al. [49] evaluated the effect of said hormone as an inducer of uterine contractions on the umbilical arterial and venous flow of premature lambs. The authors found that, together with an increase in myometrial contractions and blood pressure (of 5 ± 0.8 mmHg), there was a decrease in arterial and venous umbilical blood flow (by 12.1 ± 2 mmHg and by 20 ± 2 mmHg, respectively). These results show that the increased uterine contraction can diminish the blood flow and result in low systemic oxygenation and low newborn vitality. Likewise, a systematic review by Muro et al. [119] was carried out to evaluate the effect of uterotonics (OXT and carbetocin) on the duration of farrowing, farrowing interval, farrowing assistance, and feasibility of farrowing in sows. These authors observed that OXT reduced the duration of labor by 18% but increased the need for delivery assistance by 137% ($p < 0.01$). These figures were related to the 30% increase in stillborn pigs. For this reason, it seems that OXT would be a useful tool under strict technical guidance. However, there would still exist a risk of some harmful effects on the calf and the dam. The integrity of the umbilical cord and the uterus are other structures that can be affected by OXT administration. For example, in 2099 newborn piglets, the administration of OXT at the onset of fetal expulsion resulted in a higher incidence of stillborns, a greater piglets with ruptured (from 0.55 to 0.73) and hemorrhagic umbilical cords (from 0.21 to 0.36), together with an increase in the severity of meconium staining ($p < 0.001$) [96]. Fetal asphyxia has also been observed in 180 sows treated with OXT, resulting in a high percentage of umbilical cord rupture (76.0%) and an absence of heart rate in 53.5% of newborns [53]. Contrarily, in a murine model of fetal asphyxia (exposition of immature rats to 9% oxygen and 20% CO_2), OXT administered intranasally to the newborn resulted in a neuroprotective role, preventing hippocampal injury [120]. Fetal distress, reported as meconium-stained newborns, was reported in piglets, with 2.5 times the presentation in litters from dams IM-treated with OXT at high doses (0.17 IU/kg), with an opposite effect being observed when low doses of 0.083 IU/kg were administered, also decreasing piglet mortality [65]. A beneficial effect was also reported in foals, in whom a mare receiving low doses of OXT resulted in foaling induction without the birth of immature foals [121]. Therefore, not only OXT but also the dose needs to be considered when deciding to use it during parturition.

The species must be considered in order to evaluate the effect that OXT might have on both the dam and the offspring. However, repeated administration of OXT during parturition causes hyperstimulation, leading to uterine exhaustion and reduced blood flow to the umbilical cord [122], and even uterine rupture when other confounding factors are present, such as an insufficient amniotic fluid [123] and limited cervical dilatation [124]. OXT-induced rupture leads to fetal distress and, consequently, high mortality of neonates and the need for medical intervention [125]. Therefore, the increase in uterine contractions could be associated with an increase in local blood pressure, which would lead to reduced blood flow, fetal stress, and hypoxia, and all these factors could reduce vitality in the newborn, having a high impact on its mortality [121]. In a case of a Great Dane bitch treated with repetitive doses of OXT and manual intervention to help in the dystocia process, a uterine rupture was reported [126]. The indiscriminate use of OXT can cause umbilical cord morphological alterations such as edema, congestion, hemorrhage, and, in extreme cases, umbilical cord rupture. High doses of oxytocin increase fetal heart rate decelerations and the presentation of meconium-stained fetuses. These elements are indicators of fetal hypoxia and can be reflected in intrapartum deaths [64]. Additionally, administering OXT at the beginning of parturition in polytocous species can double the number of meconium-stained fetuses and stillbirths during the first hour after administration [46]. It is relevant to mention the role of progesterone, a hormone that decreases during parturition to promote uterine contractions (known as the removal of the "progesterone block") and fetal expulsion [127]. However, when administered in combination with caffeine at the last week of gestation in sows, the number of live born piglets decreased (11.7 ± 1.03 vs. 14.5 ± 0.73 in the control group) and piglet survival was impaired during the first five days of life and at weaning [128].

In horses, it is recorded that the contractions induced by the administration of OXT help to induce labor or expulsion of the placenta, which avoids the possibility of infections [76,129]. However, Sgorbini et al. [130] evaluated the effect of OXT at a dose of 2.5 IU IV on the incidence of peripartum complications at the behavioral, physical, and blood levels in mares. These authors found that foals born to mares under OXT treatment took longer to stand up and suckle compared to those born with spontaneous partum. In addition to this, they found higher levels of pCO_2 and lactate. This evidence reaffirms that even though these changes are not considered clinically relevant in the dam, they can affect the physical condition of the newborn. In fact, at a high dose (3.5 UI), they can induce the birth of premature foals, thereby altering their vitality [121]. In horses, it is suggested that the presence of complications could possibly be related to pharmacological aspects, either due to the dose used and/or the frequency of administration [131].

Hence, when using OXT in a parturient female, it is essential to consider the benefits but also the possible adverse effects on the newborn, since the use of the hormone can fasten the parturition process but might result in negative outcomes for the fetus.

6. Future Directions

Due to the multiple effects that OXT administration can have on the dam and the fetus, the studies aiming to evaluate this hormone need to focus on different aspects of animal reproduction. One example is the role that OXT might have on behavior in both sexes. It is known that endogenous OXT participates in, and is a key element for, maternal bonding [132]. It has even been pointed out that it can regulate affiliative behavior between congeners. Additionally, alterations in the oxytocinergic system can alter postpartum maternal performance and the administration of exogenous OXT in the early stage has been proposed as a potential treatment to restore the functionality of OXT [133]. In bovines, the administration of OXT can facilitate social interaction, especially during maternal behavior, in addition to generating milk letdown [134,135]. Likewise, neglect and OXT administration to prevent this abnormal maternal behavior is being studied in humans with imaging techniques such as functional magnetic resonance to determine the cerebral regions activated by OXT, such as the inferior frontal junction and putamen [136,137]. In cases of male animals, this could be a field of study when the dam rejects the offspring or shows aggression towards them [25]. The circuit mechanisms by which oxytocin modulates social behavior are receiving increasing attention. Nonetheless, the complexity of the subject requires an entire article that is simultaneously produced by the authors of the present review (11).

The administration route and the pharmacokinetics of OXT depending on this aspect need to be studied to determine if, for example, subcutaneous administration of OXT also increases myometrial contractions and the risk of fetal distress. Currently, intranasal OXT to reach the brain and cerebral structures is being researched in humans [138] and non-human primates [139], but its association with fetal consequences during parturition needs to be determined. As it has been studied in humans, the mode of administration (i.e., bolus or infusion, or a combination of both) and its influence on labor complications [140,141] remains as another field that could be researched in veterinary medicine. In humans, authors such as Grotegut et al. [142] found that infusion of oxytocin was associated with severe postpartum hemorrhage, and Sheehan et al. [143] determined that the combination of bolus + infusion of oxytocin reduced the presentation of major obstetric hemorrhage in women. For this reason, the administration of OXT in different modes and its effect on myometrial activity and obstetric outcome could be assessed.

The implementation of other monitoring techniques, such as tocodynamometry, to evaluate and promptly detect OXT action on the myometrium smooth muscle cells is an alternative to continuing using OXT during parturition in domestic animals and to detect any uterine physiological compromising. Tocodynamometry has been used in the last decade in diverse species to detect primary uterine inertia [33]. Therefore, its implementation could help to identify the presence or absence of uterine myometrial contractions and their relative strength and frequency [95,144], and, if necessary, to determine whether or not the exogenous application of OXT is adequate to promote the presence of contractions. In the same way, it is necessary to study the relationship between the toxicological aspects of OXT with the impact that this hormone can have on the fetus or offspring, since the evidence is scarce in this regard.

Finally, it is important to teach producers and stock people that excessive use of OXT in parturient females must be managed prudently. An alternative could be imparting small courses taught by veterinarians to farmers and people involved with livestock so they can be aware of the advantages and disadvantages of the use of this hormone.

7. Conclusions

There are various functions of OXT around parturition, among which we can mention the establishment of bonding between the dam and offspring; myometrium contractions that, in addition to helping in the expulsion of the fetus, also intervene in the expulsion of lochia and placental residues; establishment of maternal behavior, as well as intervention in lactation by stimulating the descent of milk from the mammary gland. This is why in animal production farms this hormone is used indiscriminately to reduce parturition times, as well as to reduce the incidence of dystocia caused by primary uterine inertia. Before OXT application, it is important to consider the weight of the dam, parity (e.g., primiparous or multiparous), the evolution of parturition, and if the birth canal is free, among other aspects.

There must be adequate management of the dose of OXT supplied, since with an excessive dose, the contractions become powerful and frequent, triggering heart rate deaccelerations and umbilical cord damage, producing partial fetal asphyxia and with each successive contraction, the situation may worsen, causing meconium-stained fetuses and weak, hypoglycemic neonates, until it causes fetal death. For this reason, when this hormone is used, it is convenient to know its effects, because, although in general terms, oxytocics reduce the duration of labor, they have the disadvantage of increasing dystocia. OXT dose-finding studies, as well as studies about administration routes and procedures, could be performed in each species of interest to clarify what doses and procedures are required for adequate effects on uterine contraction without side effects on the fetus.

Author Contributions: All the authors contributed to the conceptualization, writing, reading, and approval of the final manuscript. All authors have read and agreed to the published version of the manuscript.

Funding: This research received no external funding.

Institutional Review Board Statement: Not applicable.

Informed Consent Statement: Not applicable.

Data Availability Statement: Not applicable.

Acknowledgments: The authors want to express their gratitude to Daniel Mota-Rojas, professor in behavior and animal welfare, for the conception and direction of the project that created this manuscript. We are thankful for his continuous support and advice.

Conflicts of Interest: The authors declare no conflict of interest.

References

1. Lévy, F. Neuroendocrine Control of Maternal Behavior in Non-Human and Human Mammals. *Ann. D'endocrinologie* **2016**, *77*, 114–125. [CrossRef] [PubMed]
2. Safdar, A.H.A.; Kia, H.D.; Ramin, F. Physiology of Parturition. *Int. J. Adv. Biol. Biomed. Res.* **2013**, *1*, 214–221. [CrossRef]
3. Gimpl, G.; Fahrenholz, F. The Oxytocin Receptor System: Structure, Function, and Regulation. *Physiol. Rev.* **2001**, *81*, 629–683. [CrossRef] [PubMed]
4. Saller, S.; Kunz, L.; Dissen, G.A.; Stouffer, R.; Ojeda, S.R.; Berg, D.; Berg, U.; Mayerhofer, A. Oxytocin Receptors in the Primate Ovary: Molecular Identity and Link to Apoptosis in Human Granulosa Cells. *Hum. Reprod.* **2010**, *25*, 969–976. [CrossRef]
5. Stadler, B.; Whittaker, M.R.; Exintaris, B.; Middendorff, R. Oxytocin in the Male Reproductive Tract; The Therapeutic Potential of Oxytocin-Agonists and-Antagonists. *Front. Endocrinol.* **2020**, *11*, 565731. [CrossRef]
6. Assinder, S.J.; Carey, M.; Parkinson, T.; Nicholson, H.D. Oxytocin and Vasopressin Expression in the Ovine Testis and Epididymis: Changes with the Onset of Spermatogenesis1. *Biol. Reprod.* **2000**, *63*, 448–456. [CrossRef]

7. Chen, S.; Sato, S. Role of Oxytocin in Improving the Welfare of Farm Animals—A Review. *Asian-Australas. J. Anim. Sci.* **2016**, *30*, 449–454. [CrossRef]

8. Young, L.J.; Zingg, H.H. Oxytocin. In *Hormones, Brain and Behavior*, 3rd ed.; Pfaff, D.W., Joels, M., Eds.; Elsevier Science & Technology: Amsterdam, The Netherlands, 2016; Volume 3, p. 2474.

9. Mota-Rojas, D.; Bienboire-Frosini, C.; Marcet-Rius, M.; Domínguez-Oliva, A.; Mora-Medina, P.; Lezama-García, K.; Orihuela, A. Mother-Young Bond in Non-Human Mammals: Neonatal Communication Pathways and Neurobiological Basis. *Front. Psychol.* **2022**, *13*, 1064444. [CrossRef]

10. Orihuela, A.; Mota-Rojas, D.; Strappini, A.; Serrapica, F.; Braghieri, A.; Mora-Medina, P.; Napolitano, F. Neurophysiological Mechanisms of Cow- Calf Bonding in Buffalo and Other Farm Animals. *Animals* **2021**, *11*, 1968. [CrossRef]

11. Bienboire-Frosini, C.; Marcet-Rius, M.; Orihuela, A.; Domínguez-Oliva, A.; Mora-Medina, P.; Olmos-Hernández, A.; Casas-Alvarado, A.; Mota-Rojas, D. Mother–Young Bonding: Neurobiological Aspects and Maternal Biochemical Signaling in Altricial Domesticated Mammals. *Animals* **2023**, *13*, 532. [CrossRef]

12. Mota-Rojas, D.; Velarde, A.; Marcet-Rius, M.; Orihuela, A.; Bragaglio, A.; Hernández-Ávalos, I.; Casas-Alvarado, A.; Domínguez-Oliva, A.; Whittaker, A.L. Analgesia during Parturition in Domestic Animals: Perspectives and Controversies on Its Use. *Animals* **2022**, *12*, 2686. [CrossRef]

13. Lezama-García, K.; Mariti, C.; Mota-Rojas, D.; Martínez-Burnes, J.; Barrios-García, H.; Gazzano, A. Maternal Behaviour in domestic dogs. *Int. J. Vet. Sci. Med.* **2019**, *7*, 20–30. [CrossRef] [PubMed]

14. Mota-Rojas, D.; Bragaglio, A.; Braghieri, A.; Napolitano, F.; Domínguez-Oliva, A.; Mora-Medina, P.; Álvarez-Macías, A.; De Rosa, G.; Pacelli, C.; José, N.; et al. Dairy Buffalo Behavior: Calving, Imprinting and Allosuckling. *Animals* **2022**, *12*, 2899. [CrossRef] [PubMed]

15. Martínez-Burnes, J.; Muns, R.; Barrios-García, H.; Villanueva-García, D.; Domínguez-Oliva, A.; Mota-Rojas, D. Parturition in Mammals: Animal Models, Pain and Distress. *Animals* **2021**, *11*, 2960. [CrossRef] [PubMed]

16. Phillippe, R.; Francoise, M.; Marie-José, F.-M. Central Effects of Oxytocin. *Am. Physiol. Soc.* **1991**, *71*, 331–355. [CrossRef]

17. Nagel, C.; Aurich, C.; Aurich, J. Stress Effects on the Regulation of Parturition in Different Domestic Animal Species. *Anim. Reprod. Sci.* **2019**, *207*, 153–161. [CrossRef]

18. Lee, H.-J.; Macbeth, A.H.; Pagani, J.; Young, W.S. Oxytocin: The Great Facilitator of Life. *Prog. Neurobiol.* **2009**, *88*, 127–151. [CrossRef]

19. Dale, H.H. On Some Physiological Actions of Ergot. *J. Physiol.* **1906**, *34*, 163–206. [CrossRef]

20. Goodwin, M.; Zograbyan, A. Antagonistas de Receptores de Oxitocina. In *Clínicas De Perinatología*; Strauss, J.M., Ed.; McGrawHill Interamericana: Mexico City, Mexico, 1998; pp. 917–928.

21. Phaneuf, S.; Linares, B.R.; TambyRaja, R.; MacKenzie, I.; Bernal, A.L. Loss of Myometrial Oxytocin Receptors during Oxytocin-Induced and Oxytocin-Augmented Labour. *Reproduction* **2000**, *120*, 91–97. [CrossRef]

22. Robinson, C.; Schumann, R.; Zhang, P.; Young, R.C. Oxytocin-Induced Desensitization of the Oxytocin Receptor. *Am. J. Obstet. Gynecol.* **2003**, *188*, 497–502. [CrossRef]

23. Kustritz, M.V.R. Reproductive Behavior of Small Animals. *Theriogenology* **2005**, *64*, 734–746. [CrossRef]

24. Pageat, P.; Gaultier, E. Current Research in Canine and Feline Pheromones. *Vet. Clin. N. Am. Small Anim. Pract.* **2003**, *33*, 187–211. [CrossRef]

25. Meyer-Lindenberg, A.; Domes, G.; Kirsch, P.; Heinrichs, M. Oxytocin and Vasopressin in the Human Brain: Social Neuropeptides for Translational Medicine. *Nat. Rev. Neurosci.* **2011**, *12*, 524–538. [CrossRef]

26. Numan, M.; Stolzenberg, D.S. Medial Preoptic Area Interactions with Dopamine Neural Systems in the Control of the Onset and Maintenance of Maternal Behavior in Rats. *Front. Neuroendocrinol.* **2009**, *30*, 46–64. [CrossRef]

27. Bridges, R.S. Neuroendocrine Regulation of Maternal Behavior. *Front. Neuroendocrinol.* **2015**, *36*, 178–196. [CrossRef]

28. Kendrick, K.M.; Keverne, E.B.; Baldwin, B.A. Intracerebroventricular Oxytocin Stimulates Maternal Behaviour in the Sheep. *Neuroendocrinology* **1987**, *46*, 56–61. [CrossRef]

29. Gilbert, C.; Burne, T.; Goode, J.; Murfitt, P.; Walton, S. Indomethacin Blocks Pre-Partum Nest Building Behaviour in the Pig (Sus Scrofa): Effects on Plasma Prostaglandin F Metabolite, Oxytocin, Cortisol and Progesterone. *J. Endocrinol.* **2002**, *172*, 507–517. [CrossRef]

30. Castrén, H.; Algers, B.; de Passillé, A.-M.; Rushen, J.; Uvnäs-Moberg, K. Preparturient Variation in Progesterone, Prolactin, Oxytocin and Somatostatin in Relation to Nest Building in Sows. *Appl. Anim. Behav. Sci.* **1993**, *38*, 91–102. [CrossRef]

31. Oliviero, C.; Heinonen, M.; Valros, A.; Peltoniemi, O. Environmental and Sow-Related Factors Affecting the Duration of Farrowing. *Anim. Reprod. Sci.* **2010**, *119*, 85–91. [CrossRef]

32. Frankenfeld, C.L.; Atkinson, C.; Wähälä, K.; Lampe, J.W. Obesity Prevalence in Relation to Gut Microbial Environments Capable of Producing Equol or O-Desmethylangolensin from the Isoflavone Daidzein. *Eur. J. Clin. Nutr.* **2014**, *68*, 526–530. [CrossRef]

33. González-Lozano, M.; Trujillo-Ortega, M.E.; Becerril-Herrera, M.; Alonso-Spilsbury, M.; Rosales-Torres, A.M.; Mota-Rojas, D. Uterine Activity and Fetal Electronic Monitoring in Parturient Sows Treated with Vetrabutin Chlorhydrate. *J. Vet. Pharmacol. Ther.* **2010**, *33*, 28–34. [CrossRef]

34. Herpin, P.; Le Dividich, J.; Hulin, J.C.; Fillaut, M.; De Marco, F.; Bertin, R. Effects of the Level of Asphyxia during Delivery on Viability at Birth and Early Postnatal Vitality of Newborn Pigs. *J. Anim. Sci.* **1996**, *74*, 2067–2075. [CrossRef]

35. Gu, X.; Chen, J.; Li, H.; Song, Z.; Chang, L.; He, X.; Fan, Z. Isomaltooligosaccharide and Bacillus Regulate the Duration of Farrowing and Weaning-Estrous Interval in Sows during the Perinatal Period by Changing the Gut Microbiota of Sows. *Anim. Nutr.* **2021**, *7*, 72–83. [CrossRef]

36. Hydbring, E.; Madej, A.; MacDonald, E.; Drugge-Boholm, G.; Berglund, B.; Olsson, K. Hormonal Changes during Parturition in Heifers and Goats Are Related to the Phases and Severity of Labour. *J. Endocrinol.* **1999**, *160*, 75–85. [CrossRef]

37. Straw, B.E.; Bush, E.J.; Dewey, C.E. Types and Doses of Injectable Medications given to Periparturient Sows. *J. Am. Vet. Med. Assoc.* **2000**, *216*, 510–515. [CrossRef]

38. Wilker, C.E.; Ellington, J.E. Reproductive Toxicology of the Female Companion Animal. In *Small Animal Toxicology*, 2nd ed.; Peterson, M.E., Talcott, P.A., Eds.; Elsevier: Amsterdam, The Netherlands, 2006; pp. 475–499.

39. Plumb, D.C. Oxytocin. In *Plumb´s Veterinary Drugs*; John Wiley & Sons: Tulsa, OK, USA, 2015.

40. Linneen, S.K.; Benz, J.M.; DeRouchey, J.M.; Goodband, R.D.; Tokach, M.D.; Dritz, S.S. A Review of Oxytocin Use for Sows and Gilts. *Kans. Agric. Exp. Stn. Res. Rep.* **2005**, *0*, 1–3. [CrossRef]

41. Assad, N.I.; Pandey, A.K.; Sharma, L.M. Oxytocin, Functions, Uses and Abuses: A Brief Review. *Insight Int. J. Reprod. All Anim.* **2016**, *6*, 1–17. [CrossRef]

42. Ahmad, M. Oxytocin; Effects on Milk Production. *Pure Appl. Biol.* **2021**, *10*, 318–324. [CrossRef]

43. Münnich, A.; Küchenmeister, U. Dystocia in numbers evidence based parameters for intervention in the dog: Causes for dystocia and treatment recommendations. *Reprod. Domest. Anim.* **2009**, *44*, 141–147. [CrossRef]

44. Gutkowska, J.; Jankowski, M. Oxytocin Revisited: Its Role in Cardiovascular Regulation. *J. Neuroendocrinol.* **2012**, *24*, 599–608. [CrossRef]

45. González-Lozano, M.; Trujillo-Ortega, M.E.; Alonso-Spilsbury, M.; Rosales, A.M.; Ramírez-Necoechea, R.; González-Maciel, A.; Martínez-Rodríguez, R.; Becerril-Herrera, M.; Mota-Rojas, D. Vetrabutine Clorhydrate Use in Dystocic Farrowings Minimizes Hemodynamic Sequels in Piglets. *Theriogenology* **2012**, *78*, 455–461. [CrossRef] [PubMed]

46. Mota-Rojas, D.; Villanueva-García, D.; Mota-Reyes, A.; Orihuela, A.; Hernández-Ávalos, I.; Domínguez-Oliva, A.; Casas-Alvarado, A.; Flores-Padilla, K.; Jacome-Romero, J.; Martínez-Burnes, J. Meconium Aspiration Syndrome in Animal Models: Inflammatory Process, Apoptosis, and Surfactant Inactivation. *Animals* **2022**, *12*, 3310. [CrossRef] [PubMed]

47. Alonso-Spilsbury, M.; Mota-Rojas, D.; Martínez-Burnes, J.; Arch, E.; López Mayagoitia, A.; Ramírez-Necoechea, R.; Olmos, A.; Trujillo, M.E. Use of Oxytocin in Penned Sows and Its Effect on Fetal Intra-Partum Asphyxia. *Anim. Reprod. Sci.* **2004**, *84*, 157–167. [CrossRef] [PubMed]

48. Mota-Rojas, D.; Villanueva-García, D.; Velazquez-Armenta, E.Y.; Nava-Ocampo, A.A.; Ramírez-Necoechea, R.; Alonso-Spilsbury, M.; Trujillo, M.E. Influence of Time at Which Oxytocin Is Administered during Labor on Uterine Activity and Perinatal Death in Pigs. *Biol. Res.* **2007**, *40*, 55–63. [CrossRef]

49. Stenning, F.J.; Polglase, G.R.; te Pas, A.B.; Crossley, K.J.; Kluckow, M.; Gill, A.W.; Wallace, E.M.; McGillick, E.V.; Binder, C.; Blank, D.A.; et al. Effect of Maternal Oxytocin on Umbilical Venous and Arterial Blood Flows during Physiological-Based Cord Clamping in Preterm Lambs. *PLoS ONE* **2021**, *16*, e0253306. [CrossRef]

50. Trujillo, O.M.; Mota-Rojas, D.; Juárez, O.; Villanueva-García, D.; Roldan-Santiago, P.; Becerril-Herrera, R.; Hernández-Gónzalez, R.; Mora-Medina, P.; Alonso-Spilsbury, M.; Rosales, A.; et al. Porcine Neonates Failing Vitality Score: Physio-Metabolicprofile and Latency to the First Teat Contact. *Czech J. Anim. Sci.* **2011**, *56*, 499–508. [CrossRef]

51. Mota-Rojas, D.; Villanueva-García, D.; Solimano, A.; Muns, R.; Ibarra-Ríos, D.; Mota-Reyes, A. Pathophysiology of Perinatal Asphyxia in Humans and Animal Models. *Biomedicines* **2022**, *10*, 347. [CrossRef]

52. Mota-Rojas, D.; López, A.; Martínez-Burnes, J.; Muns, R.; Villanueva-García, D.; Mora-Medina, P.; González-Lozano, M.; Olmos-Hernández, A.; Ramírez-Necoechea, R. Is Vitality Assessment Important in Neonatal Animals? *CAB Rev. Perspect. Agric. Vet. Sci. Nutr. Nat. Resour.* **2018**, *2018*, 1–13. [CrossRef]

53. Mota-Rojas, D.; Rosales, A.M.; Trujillo, M.E.; Orozco, H.; Ramírez, R.; Alonso-Spilsbury, M. The Effects of Vetrabutin Chlorhydrate and Oxytocin on Stillbirth Rate and Asphyxia in Swine. *Theriogenology* **2005**, *64*, 1889–1897. [CrossRef]

54. González-Lozano, M.; Mota-Rojas, D.; Orihuela, A.; Martínez-Burnes, J.; Di Francia, A.; Braghieri, A.; Berdugo-Gutiérrez, J.; Mora-Medina, P.; Ramírez-Necoechea, R.; Napolitano, F. Review: Behavioral, Physiological, and Reproductive Performance of Buffalo Cows during Eutocic and Dystocic Parturitions. *Appl. Anim. Sci.* **2020**, *36*, 407–422. [CrossRef]

55. Homeida, A.M.; Cooke, R.G. Biological Half-Life of Oxytocin in the Goat. *Res. Vet. Sci.* **1984**, *37*, 364–365. [CrossRef] [PubMed]

56. Fitzpatrick, R.J. The Estimation of Small Amounts of Oxytocin in Blood. In *Oxytocin*; Caldeyro-Bracia, R., Heller, H., Eds.; Pergamon Press: Oxford, UK, 1961; p. 14.

57. Steckler, D.; Naidoo, V.; Gerber, D.; Kähn, W. Ex Vivo Influence of Carbetocin on Equine Myometrial Muscles and Comparison with Oxytocin. *Theriogenology* **2012**, *78*, 502–509. [CrossRef] [PubMed]

58. Ward, K.W.; Smith, B.R. A Comprehensive Quantitative and Qualitative Evaluation of Extrapolation of Intravenous Pharmacokinetic Parameters from Rat, Dog, and Monkey to Humans. II. Volume of Distribution and Mean Residence Time. *Drug Metab. Dispos.* **2004**, *32*, 612–619. [CrossRef] [PubMed]

59. Fuchs, A.-R.; Dawoodf, Y.M. Oxytocin Release and Uterine Activation during Parturition in Rabbits. *Endocrinology* **1980**, *107*, 1117–1126. [CrossRef] [PubMed]

60. Al-Eknah, M.M.; Homeida, A.M. A Review of Some Aspects of the Pharmacology of Oxytocin in Domestic Animals. *Vet. Res. Commun.* **1991**, *15*, 45–55. [CrossRef] [PubMed]

61. Kowalski, W.B.; Parsons, M.T.; Pak, S.C.; Wilson, L. Morphine Inhibits Nocturnal Oxytocin Secretion and Uterine Contractions in the Pregnant Baboon1. *Biol. Reprod.* **1998**, *58*, 971–976. [CrossRef] [PubMed]

62. Troncy, E.; Morin, V.; Del Castillo, J.R.E.; Authier, S.; Ybarra, N.; Otis, C.; Gauvin, D.; Gutkowska, J. Evidence for Non-Linear Pharmacokinetics of Oxytocin in Anesthetizetized Rat. *J. Pharm. Pharm. Sci.* **2008**, *11*, 12–24. [CrossRef]

63. Zhu, C.; Lal, M. Preclinical Safety and Pharmacokinetics of Heat Stable Oxytocin in Sublingual Fast-Dissolving Tablet Formulation. *Pharmaceutics* **2022**, *14*, 953. [CrossRef]

64. Wagner, B.K.; Relling, A.E.; Kieffer, J.D.; Moraes, L.E.; Parker, A.J. Short communication: Pharmacokinetics of oxytocin administered intranasally to beef cattle. *Domest. Anim. Endocrinol.* **2020**, *71*, 106387. [CrossRef]

65. Mota-Rojas, D.; Nava-Ocampo, A.A.; Trujillo, M.E.; Velázquez-Armenta, Y.; Ramírez-Necoechea, R.; Martínez-Burnes, J.; Alonso-Spilsbury, Y.M. Dose minimization study of oxytocin in early labor in sows: Uterine activity and fetal outcome. *Reprod. Toxicol.* **2005**, *20*, 255–259. [CrossRef]

66. Hill, S.V.; Amezcua, M.D.R.; Ribeiro, E.S.; O'Sullivan, T.L.; Friendship, R.M. Defining the Effect of Oxytocin Use in Farrowing Sows on Stillbirth Rate: A Systematic Review with a Meta-Analysis. *Animals* **2022**, *12*, 1795. [CrossRef] [PubMed]

67. Mota-Rojas, D.; Trujillo, M.E.; Martínez, J.; Rosales, A.M.; Orozco, H.; Ramírez, R.; Sumano, H.; Alonso-Spilsbury, M. Comparative routes of oxytocin administration in crated farrowing sows and its effects on fetal and postnatal asphyxia. *Anim. Reprod. Sci.* **2006**, *92*, 123–143. [CrossRef] [PubMed]

68. Ybarra-Navarro, N.T. Evaluation of Oxytocin Pharmacokinetic: Pharmacodynamic Profile and Establishment of Its Cardiomyogenic Potential in Swine. Ph.D. Thesis, Université of Montréal, Montréal, QC, Canada, 2010.

69. Wiebe, V.J.; Howard, J.P. Pharmacologic Advances in Canine and Feline Reproduction. *Top. Companion Anim. Med.* **2009**, *24*, 71–99. [CrossRef]

70. Brander, G.C.; Pugh, D.M.; Bywater, R.J. *Veterinary Applied Pharmacology and Therapeutics*; Bailliere Tindall Ltd: London, UK, 1982.

71. Bajcsy, Á.; Kindahl, H.; Szenci, O.; van der Weijden, G.; Bartyik, J.; Taverne, M. The Effect of Two Different Routes of Administration of Oxytocin on Peripheral Plasma Prostaglandin F2α Metabolite Levels in Early Post-partum Dairy Cows. *Reprod. Domest. Anim.* **2012**, *47*, 208–212. [CrossRef] [PubMed]

72. Nickles, K.R.; Relling, A.E.; Parker, A.J. Intranasal oxytocin treatment on the day of weaning does not decrease walking behavior or improve plasma metabolites in beef calves placed on pasture. *Transl. Anim. Sci.* **2021**, *5*, txab191. [CrossRef] [PubMed]

73. Rault, J.-L.; Dunshea, F.; Pluske, J. Effects of Oxytocin Administration on the Response of Piglets to Weaning. *Animals* **2015**, *5*, 545–560. [CrossRef] [PubMed]

74. Mota-Rojas, D.; Villanueva, D.; Alonso-Spilsbury, M.; Becerril, H.M.; Ramirez-Necoechea, R.; Gonzalez-Lozano, M.; Trujillo-Ortega, M.E. Effect of Different Doses of Oxytocin at Delivery on Suffering and Survival of Newborn Pigs. *J. Med. Sci.* **2007**, *7*, 170–178. [CrossRef]

75. Villani, M.; Romano, G. Induction of Parturition with Daily Low-dose Oxytocin Injections in Pregnant Mares at Term: Clinical Applications and Limitations. *Reprod. Domest. Anim.* **2008**, *43*, 481–483. [CrossRef] [PubMed]

76. Macpherson, M.L.; Chaffin, M.K.; Carroll, G.L.; Jorgensen, J.; Arrott, C.; Varner, D.D.; Blanchard, T.L. Three methods of oxytocin-induced parturition and their effects of foals. *J. Am. Vet. Med. Assoc.* **1997**, *210*, 799–803. [PubMed]

77. Brunton, P.J. Endogenous opioid signalling in the brain during pregnancy and lactation. *Cell Tissue Res.* **2019**, *375*, 69–83. [CrossRef]

78. Taverne, M.; de Schwartz, N.; Kankofer, M.; Bevers, M.; van Oord, H.; Schams, D.; Gutjahr, S.; van der Weijden, G. Uterine Responses to Exogenous Oxytocin Before and After Pre-partum Luteolysis in the Cow. *Reprod. Domest. Anim.* **2001**, *36*, 267–272. [CrossRef] [PubMed]

79. Sanborn, B.M.; Dodge, K.; Monga, M.; Qian, A.; Wang, W.; Yue, C. Molecular Mechanisms Regulating the Effects of Oxytocin on Myometrial Intracellular Calcium. In *Vasopressin and Oxytocin*; Springer: Boston, MA, USA, 1998; Volume 449, pp. 277–286. [CrossRef]

80. Ku, C.Y.; Qian, A.; Wen, Y.; Anwer, K.; Sanborn, B.M. Oxytocin stimulates myometrial guanosine triphosphatase and phospholipase-C activities via coupling to G alpha q/11. *Endocrinology* **1995**, *136*, 1509–1515. [CrossRef] [PubMed]

81. Riemer, R.K.; Heymann, M.A. Regulation of Uterine Smooth Muscle Function during Gestation. *Pediatr. Res.* **1998**, *44*, 615–627. [CrossRef] [PubMed]

82. Kidder, G.M.; Winterhager, E. Physiological roles of connexins in labour and lactation. *Reproduction* **2015**, *150*, R129–R136. [CrossRef]

83. Shynlova, O.; Tsui, P.; Jaffer, S.; Lye, S.J. Integration of endocrine and mechanical signals in the regulation of myometrial functions during pregnancy and labour. *Eur. J. Obstet. Gynecol. Reprod. Biol.* **2009**, *144*, S2–S10. [CrossRef]

84. Wang, P.; Wang, S.C.; Li, D.; Li, T.; Yang, H.-P.; Wang, L.; Wang, Y.-F.; Parpura, V. Role of Connexin 36 in Autoregulation of Oxytocin Neuronal Activity in Rat Supraoptic Nucleus. *ASN Neuro* **2019**, *11*, 175909141984376. [CrossRef]

85. Petersson, M. Petersson, M. Cardiovascular effects of oxytocin. *Prog. Brain. Res.* **2002**, *131*, 281–288.

86. Tuğtepe, H.; Şener, G.; Bıyıklı, N.K.; Yüksel, M.; Çetinel, Ş.; Gedik, N.; Yeğen, B.Ç. The protective effect of oxytocin on renal ischemia/reperfusion injury in rats. *Regul. Pept.* **2007**, *140*, 101–108. [CrossRef]

87. Lundeberg, T.; Uvnäs-Moberg, K.; Ågren, G.; Bruzelius, G. Anti-nociceptive effects of oxytocin in rats and mice. *Neurosci. Lett.* **1994**, *170*, 153–157. [CrossRef]

88. Manresa, J.A.B.; Schliessbach, J.; Vuilleumier, P.H.; Müller, M.; Musshoff, F.; Stamer, U.; Stüber, F.; Arendt-Nielsen, L.; Curatolo, M. Anti-nociceptive effects of oxytocin receptor modulation in healthy volunteers–A randomized, double-blinded, placebo-controlled study. *Eur. J. Pain* **2021**, *25*, 1723–1738. [CrossRef]

89. Ramagnoli, S. Practical use of hormones in small animal reproduction. *Rev. Bras. Reprod. Anim.* **2017**, *41*, 59–67.

90. Bazer, F.W.; First, N.L. Pregnancy and parturition. *J. Anim. Sci.* **1983**, *57* (Suppl. 2), 425–460. [PubMed]

91. Erickson, E.N.; Krol, K.M.; Perkeybile, A.M.; Connelly, J.J.; Myatt, L. Oxytocin receptor single nucleotide polymorphism predicts atony-related postpartum hemorrhage. *BMC Pregnancy Childbirth* **2022**, *22*, 884. [CrossRef] [PubMed]

92. Mota-Rojas, D.; Martínez-Burnes, J.; Napolitano, F.; Domínguez-Muñoz, M.; Guerrero-Legarreta, I.; Mora-Medina, P.; Ramírez-Necoechea, R.; Lezama-García, K.; González-Lozano, M. Dystocia: Factors affecting parturition in domestic animals. *CABI Rev.* **2020**, *2020*, 1–16. [CrossRef]

93. Feldman, E.C.; Nelson, R.W. *Canine and Feline Endocrinology and Reproduction*; W.B. Saunders: Philadelphia, PA, USA, 2004.

94. Davidson, A.P. Uterine and Fetal Monitoring in the Bitch. *Vet. Clin. N. Am. Small Anim. Pract.* **2001**, *31*, 305–313. [CrossRef] [PubMed]

95. Davidson, A.P. Primary Uterine Inertia in Four Labrador Bitches. *J. Am. Anim. Hosp. Assoc.* **2011**, *47*, 83–88. [CrossRef] [PubMed]

96. Mota-Rojas, D.; Martinez-Burnes, J.; Trujillo-Ortega, M.E.; Alonso-Spilsbury, M.L.; Ramirez-Necoechea, R.; Lopez, A. Effect of oxytocin treatment in sows on umbilical cord morphology, meconium staining, and neonatal mortality of piglets. *Am. J. Vet. Res.* **2002**, *63*, 1571–1574. [CrossRef]

97. González-Lozano, M.; Mota-Rojas, D.; Velázquez-Armenta, E.Y.; Nava-Ocampo, A.A.; Hernández-González, R.; Becerril-Herrera, M.; Trujillo-Ortega, M.E.; Alonso-Spilsbury, M. Obstetric and fetal outcomes in dystocic and eutocic sows to an injection of exogenous oxytocin during farrowing. *Am. Jew. Hist.* **2009**, *50*, 1273–1277.

98. Paccamonti, D. Milk electrolytes and induction of parturition. *Pferdeheilkd. Equine Med.* **2001**, *17*, 616–618. [CrossRef]

99. Olmos-Hernández, A. Caracterización del patrón de comportamiento uterino y fetal a través del monitoreo electrónico en cerdas peri-parturientas en dos sistemas de alojamiento y su efecto sobre la viabilidad y grado de asfixia del neonato. MSc Thesis, Universidad Nacional Autónoma de México, Ciudad de México, Mexico, 2006.

100. Valros, A.; Rundgren, M.; Špinka, M.; Saloniemi, H.; Hultén, F.; Uvnäs-Moberg, K.; Tománek, M.; Krejcı, P.; Algers, B.; Krejcı, P.; et al. Oxytocin, prolactin and somatostatin in lactating sows: Associations with mobilisation of body resources and maternal behaviour. *Livest. Prod. Sci.* **2004**, *85*, 3–13. [CrossRef]

101. Francis, D.D.; Young, L.J.; Meaney, M.J.; Insel, T.R. Naturally Occurring Differences in Maternal Care are Associated with the Expression of Oxytocin and Vasopressin (V1a) Receptors: Gender Differences. *J. Neuroendocrinol.* **2002**, *14*, 349–353. [CrossRef] [PubMed]

102. Kojima, S.; Stewart, R.A.; Demas, G.E.; Alberts, J.R. Maternal Contact Differentially Modulates Central and Peripheral Oxytocin in Rat Pups During a Brief Regime of Mother-Pup Interaction that Induces a Filial Huddling Preference. *J. Neuroendocrinol.* **2012**, *24*, 831–840. [CrossRef] [PubMed]

103. Nowak, R.; Lévy, F.; Chaillou, E.; Cornilleau, F.; Cognié, J.; Marnet, P.-G.; Williams, P.D.; Keller, M. Neonatal Suckling, Oxytocin, and Early Infant Attachment to the Mother. *Front. Endocrinol.* **2021**, *11*, 612651. [CrossRef] [PubMed]

104. Kockaya, M.; Ercan, N.; Demirbas, Y.S.; Pereira, G.D.G. Serum oxytocin and lipid levels of dogs with maternal cannibalism. *J. Vet. Behav.* **2018**, *27*, 23–26. [CrossRef]

105. Feldman, R.; Gordon, I.; Zagoory-Sharon, O. Maternal and paternal plasma, salivary, and urinary oxytocin and parent-infant synchrony: Considering stress and affiliation components of human bonding. *Dev. Sci.* **2011**, *14*, 752–761. [CrossRef] [PubMed]

106. Ogi, A.; Mariti, C.; Pirrone, F.; Baragli, P.; Gazzano, A. The Influence of Oxytocin on Maternal Care in Lactating Dogs. *Animals* **2021**, *11*, 1130. [CrossRef]

107. Bienboire-Frosini, C.; Chabaud, C.; Cozzi, A.; Codecasa, E.; Pageat, P. Validation of a Commercially Available Enzyme ImmunoAssay for the Determination of Oxytocin in Plasma Samples from Seven Domestic Animal Species. *Front. Neurosci.* **2017**, *11*, 524. [CrossRef] [PubMed]

108. MacLean, E.L.; Wilson, S.R.; Martin, W.L.; Davis, J.M.; Nazarloo, H.P.; Carter, C.S. Challenges for measuring oxytocin: The blind men and the elephant? *Psychoneuroendocrinology* **2019**, *107*, 225–231. [CrossRef]

109. Faraz, A.; Waheed, A.; Nazir, M.M.; Hameed, A.; Tauqir, N.A.; Mirza, R.H.; Ishaq, H.M.; Bilal, R.M. Impact of Oxytocin Administration on Milk Quality, Reproductive Performance and Residual Effects in Dairy Animals-A Review. *Punjab Univ. J. Zool.* **2020**, *35*, 61–67. [CrossRef]

110. Simpson, K.R.; James, D.C. Effects of oxytocin-induced uterine hyperstimulation during labor on fetal oxygen status and fetal heart rate patterns. *Am. J. Obstet. Gynecol.* **2008**, *199*, 34.e1–34.e5. [CrossRef]

111. Singhi, S.; Singh, M. Pathogenesis of oxytocin-induced neonatal hyperbilirubinaemia. *Arch. Dis. Child.* **1979**, *54*, 400–402. [CrossRef] [PubMed]

112. Hernández-Ávalos, I.; Mota-Rojas, D.; Mora-Medina, P.; Martínez-Burnes, J.; Casas-Alvarado, A.; Verduzco-Mendoza, A.; Lezama-García, K.; Olmos-Hernández, A. Review of different methods used for clinical recognition and assessment of pain in dogs and cats. *Int. J. Vet. Sci. Med.* **2019**, *7*, 43–54. [CrossRef] [PubMed]

113. Gnanadesikan, G.E.; Hammock, E.A.D.; Tecot, S.R.; Lewis, R.J.; Hart, R.; Carter, C.S.; MacLean, E.L. What are oxytocin assays measuring? Epitope mapping, metabolites, and comparisons of wildtype & knockout mouse urine. *Psychoneuroendocrinology* **2022**, *143*, 105827. [CrossRef]

114. Carter, A.M.; Mess, A.M. The evolution of fetal membranes and placentation in carnivores and ungulates (Ferungulata). *Anim. Reprod.* **2017**, *14*, 124–135. [CrossRef]

115. Rapacz-Leonard, A.; Leonard, M.; Chmielewska-Krzesińska, M.; Siemieniuch, M.; Janowski, T.E. The oxytocin-prostaglandins pathways in the horse (Equus caballus) placenta during pregnancy, physiological parturition, and parturition with fetal membrane retention. *Sci. Rep.* **2020**, *10*, 2089. [CrossRef]

116. Fuchs, A.R.; Helmer, H.; Chang, S.M.; Fields, J.M. Concentration of oxytocin receptors in the palcenta and fetal membranes of cows during pregancy and labour. *J. Reprod. Fert.* **1992**, *96*, 775–783. [CrossRef]

117. Giri, T.; Jiang, J.; Xu, Z.; McCarthy, R.; Halabi, C.M.; Tycksen, E.; Cahill, A.G.; England, S.K.; Palanisamy, A. Labor induction with oxytocin in pregnant rats is not associated with oxidative stress in the fetal brain. *Sci. Rep.* **2022**, *12*, 3143. [CrossRef]

118. Vannucchi, C.I.; Rodrigues, J.A.; Silva, L.C.G.; Lúcio, C.F.; Veiga, G.A.L. Effect of dystocia and treatment with oxytocin on neonatal calf vitality and acid-base, electrolyte and haematological status. *Vet. J.* **2015**, *203*, 228–232. [CrossRef]

119. Muro, B.B.D.; Carnevale, R.F.; Andretta, I.; Leal, D.F.; Monteiro, M.S.; Poor, A.P.; Almond, G.W.; Garbossa, C.A.P. Effects of uterotonics on farrowing traits and piglet vitality: A systematic review and meta-analysis. *Theriogenology* **2021**, *161*, 151–160. [CrossRef]

120. Panaitescu, A.; Isac, S.; Pavel, B.; Illie, A.S.; Ceanga, M.; Totan, A.; Zagrean, L.; Peltecu, G.; Zagrean, A.M. Oxytocin reduces seizure burden and hippocampal injury in a rat model of perinatal asphyxia. *Acta Endocrinol.* **2018**, *14*, 315–319. [CrossRef]

121. Nagel, C.; Aurich, C. Induction of parturition in horses-from physiological pathways to clinical applications. *Domest. Anim. Endocrinol.* **2022**, *78*, 106670. [CrossRef] [PubMed]

122. Hermes, R.; Saragusty, J.; Schaftenaar, W.; Göritz, F.; Schmitt, D.L.; Hildebrandt, T.B. Obstetrics in elephants. *Theriogenology* **2008**, *70*, 131–144. [CrossRef]

123. Thitaram, C.; Pongsopawijit, P.; Thongtip, N.; Angkavanich, T.; Chansittivej, S.; Wongkalasin, W.; Somgird, C.; Suwankong, N.; Prachsilpchai, W.; Suchit, K.; et al. Dystocia following prolonged retention of a dead fetus in an Asian elephant (Elephas maximus). *Theriogenology* **2006**, *66*, 1284–1291. [CrossRef] [PubMed]

124. Schmitt, D. Reproductive System. In *Biology, Medicine, and Surgery of Elephants*; Fowler, M.E., Mikota, S.K., Eds.; Blackwell Publishing Ltd: Oxford, UK, 2006; pp. 347–355.

125. Leung, A.S.; Leung, E.K.; Paul, R.H. Uterine rupture after previous cesarean delivery: Maternal and fetal consequences. *Am. J. Obstet. Gynecol.* **1993**, *169*, 945–950. [CrossRef] [PubMed]

126. Humm, K.R.; Adamantos, S.E.; Benigni, L.; Armitage-Chan, E.A.; Brockman, D.J.; Chan, D.L. Uterine Rupture and Septic Peritonitis Following Dystocia and Assisted Delivery in a Great Dane Bitch. *J. Am. Anim. Hosp. Assoc.* **2010**, *46*, 353–357. [CrossRef] [PubMed]

127. Peltoniemi, O.; Björkman, S.; Oliviero, C. Parturition effects on reproductive health in the gilt and sow. *Reprod. Domest. Anim.* **2016**, *51*, 36–47. [CrossRef]

128. van Wettere, W.H.E.J.; Toplis, P.; Miller, H.M. Effect of oral progesterone and caffeine at the end of gestation on farrowing duration and piglet growth and survival. *Animal* **2018**, *12*, 1638–1641. [CrossRef]

129. Ishii, M.; Kobayashi, S.; Acosta, T.J.; Miki, W.; Matsui, M.; Yamanoi, T.; Miyake, Y.; Miyamoto, A. Effective Oxytocin Treatment on placenta Expulsion after Foaling in Heavy Draft Mares. *J. Vet. Med. Sci.* **2009**, *71*, 293–297. [CrossRef]

130. Sgorbini, M.; Freccero, F.; Castagnetti, C.; Mariella, J.; Lanci, A.; Marmorini, P.; Camillo, F. Peripartum findings and blood gas analysis in newborn foals born after spontaneous or induced parturition. *Theriogenology* **2020**, *158*, 18–23. [CrossRef]

131. Duggan, V.E.; Holyoak, G.R.; MacAllister, C.G.; Confer, A.W. Influence of induction of parturition on the neonatal acute phase response in foals. *Theriogenology* **2007**, *67*, 372–381. [CrossRef]

132. Glasper, E.R.; Kenkel, W.M.; Bick, J.; Rilling, J.K. More than just mothers: The neurobiological and neuroendocrine underpinnings of allomaternal caregiving. *Front. Neuroendocrinol.* **2019**, *53*, 100741. [CrossRef] [PubMed]

133. Rørvang, M.V.; Nielsen, B.L.; Herskin, M.S.; Jensen, M.B. Prepartum Maternal Behavior of Domesticated Cattle: A Comparison with Managed, Feral, and Wild Ungulates. *Front. Vet. Sci.* **2018**, *5*, 45. [CrossRef] [PubMed]

134. Uvnäs-Moberg, K.; Johansson, B.; Lupoli, B.; Svennersten-Sjaunja, K. Oxytocin facilitates behavioural, metabolic and physiological adaptations during lactation. *Appl. Anim. Behav. Sci.* **2001**, *72*, 225–234. [CrossRef] [PubMed]

135. Nevard, R.P.; Pant, S.D.; Broster, J.C.; Norman, S.T.; Stephen, C.P. Maternal Behavior in Beef Cattle: The Physiology, Assessment and Future Directions—A Review. *Vet. Sci.* **2022**, *10*, 10. [CrossRef] [PubMed]

136. Holtfrerich, S.K.C.; Pfister, R.; El Gammal, A.T.; Bellon, E.; Diekhof, E.K. Endogenous testosterone and exogenous oxytocin influence the response to baby schema in the female brain. *Sci. Rep.* **2018**, *8*, 7672. [CrossRef] [PubMed]

137. Witteveen, A.B.; Stramrood, C.A.I.; Henrichs, J.; Flanagan, J.C.; van Pampus, M.G.; Olff, M. The oxytocinergic system in PTSD following traumatic childbirth: Endogenous and exogenous oxytocin in the peripartum period. *Arch. Womens. Ment. Health* **2020**, *23*, 317–329. [CrossRef]

138. Quintana, D.S.; Lischke, A.; Grace, S.; Scheele, D.; Ma, Y.; Becker, B. Advances in the field of intranasal oxytocin research: Lessons learned and future directions for clinical research. *Mol. Psychiatry* **2021**, *26*, 80–91. [CrossRef]

139. Lee, M.R.; Shnitko, T.A.; Blue, S.W.; Kaucher, A.V.; Winchell, A.J.; Erikson, D.W.; Grant, K.A.; Leggio, L. Labeled oxytocin administered via the intranasal route reaches the brain in rhesus macaques. *Nat. Commun.* **2020**, *11*, 2783. [CrossRef]

140. Munn, M. Comparison of two oxytocin regimens to prevent uterine atony at cesarean delivery: A randomized controlled trial. *Obstet. Gynecol.* **2001**, *98*, 386–390. [CrossRef]

141. Carvalho, J.C.A.; Balki, M.; Kingdom, J.; Windrim, R. Oxytocin Requirements at Elective Cesarean Delivery: A Dose-Finding Study. *Obstet. Gynecol.* **2004**, *104*, 1005–1010. [CrossRef]
142. Grotegut, C.A.; Paglia, M.J.; Johnson, L.N.C.; Thames, B.; James, A.H. Oxytocin exposure during labor among women with postpartum hemorrhage secondary to uterine atony. *Am. J. Obstet. Gynecol.* **2011**, *204*, 56.e1–56.e6. [CrossRef] [PubMed]
143. Sheehan, S.R.; Montgomery, A.A.; Carey, M.; McAuliffe, F.M.; Eogan, M.; Gleeson, R.; Geary, M.; Murphy, D.J. Oxytocin bolus versus oxytocin bolus and infusion for control of blood loss at elective caesarean section: Double blind, placebo controlled, randomised trial. *BMJ* **2011**, *343*, d4661. [CrossRef] [PubMed]
144. Davidson, A.P. Obstetrical monitoring in dogs. *Vet. Med.* **2003**, *85*, 508–516.

Disclaimer/Publisher's Note: The statements, opinions and data contained in all publications are solely those of the individual author(s) and contributor(s) and not of MDPI and/or the editor(s). MDPI and/or the editor(s) disclaim responsibility for any injury to people or property resulting from any ideas, methods, instructions or products referred to in the content.

 *animals*

Article

Canine Neonatal Assessment by Vitality Score, Amniotic Fluid, Urine, and Umbilical Cord Blood Analysis of Glucose, Lactate, and Cortisol: Possible Influence of Parturition Type?

Tanja Plavec [1], Tanja Knific [2], Aleksandra Slapšak [1], Sara Raspor [1], Barbara Lukanc [1] and Maja Zakošek Pipan [3,*]

[1] Veterinary Faculty, University of Ljubljana, Small Animal Clinic, Gerbičeva 60, 1000 Ljubljana, Slovenia; tanja.plavec@vf.uni-lj.si (T.P.); aleksandra.slapsak@gmail.com (A.S.); sara.raspor93@gmail.com (S.R.); Barbara.lukanc@vf.uni-lj.si (B.L.)

[2] Veterinary Faculty, Institute of Food Safety, Feed and Environment, University of Ljubljana, 1000 Ljubljana, Slovenia; tanja.knific@vf.uni-lj.si

[3] Veterinary Faculty, University of Ljubljana, Clinic for Reproduction and Large Animals, Gerbičeva 60, 1000 Ljubljana, Slovenia

* Correspondence: maja.zakosekpipan@vf.uni-lj.si; Tel.: +386-1-4779-396

Citation: Plavec, T.; Knific, T.; Slapšak, A.; Raspor, S.; Lukanc, B.; Pipan, M.Z. Canine Neonatal Assessment by Vitality Score, Amniotic Fluid, Urine, and Umbilical Cord Blood Analysis of Glucose, Lactate, and Cortisol: Possible Influence of Parturition Type? *Animals* **2022**, *12*, 1247. https://doi.org/10.3390/ani12101247

Academic Editors: Daniel Mota-Rojas, Julio Martínez-Burnes and Agustín Orihuela

Received: 31 March 2022
Accepted: 10 May 2022
Published: 12 May 2022

Publisher's Note: MDPI stays neutral with regard to jurisdictional claims in published maps and institutional affiliations.

Copyright: © 2022 by the authors. Licensee MDPI, Basel, Switzerland. This article is an open access article distributed under the terms and conditions of the Creative Commons Attribution (CC BY) license (https://creativecommons.org/licenses/by/4.0/).

Simple Summary: Parturition as a stressful event may influence puppies' neonatal morbidity and mortality. The purpose of this study was to investigate the impact of parturition type on stress in newborn puppies, their weight gains, and survival in the first week postpartum. One hundred and twenty-three puppies were divided into three groups: vaginal parturition, emergency, and elective cesarean section. The Apgar score was assessed 5, 15, and 60 min postpartum, and samples of amniotic fluid, umbilical blood, and urine were collected for lactate, glucose, and cortisol concentration measurements. Although emergency cesarean section puppies had the highest cortisol concentration of all groups, their Apgar score at 5 min postpartum was comparable to the vaginal parturition group, which had the highest lactate levels. There were no significant differences between groups regarding relative growth rate. The type of parturition had no impact on puppies' survival in our study, but supportive treatment was provided for non-vital puppies in stress. Non-invasive analysis of amniotic fluid and/or urine could help in the assessment of the neonatal vitality.

Abstract: The objective of this study was to investigate the impact of parturition type on vitality in newborn puppies, their weight gains, and survival in the first week postpartum. One hundred and twenty-three puppies were divided in three groups: vaginal parturition (VP), emergency (EM-CS), and elective cesarean section (EL-CS). Apgar scores were assessed 5, 15, and 60 min postpartum. Lactate and glucose concentrations were measured in amniotic fluid and umbilical blood; cortisol concentrations were measured in amniotic fluid and puppy urine. Puppies' weight gain was tracked daily for 7 days postpartum. Apgar score at 5 and 15 min was significantly better in the VP group. EL-CS puppies had significantly lower umbilical blood and amniotic fluid lactate concentrations compared to the VP group, which also had higher umbilical blood lactate concentration than EM-CS puppies. The cortisol concentration in the amniotic fluid and in urine differed significantly between the groups, with the highest concentration in the EM-CS, followed by the VP group. Glucose concentration in amniotic fluid was higher in the VP group than EM-CS group. The type of parturition had no impact on puppies' weight gain or their survival at birth; however, supportive treatment was provided for non-vital puppies. Non-invasive analysis of puppies' fluids could help in the assessment of the neonatal vitality.

Keywords: neonate; Apgar score; lactate; glucose; cortisol; neonatal fluids

1. Introduction

Parturition is always a challenge for the bitch and her puppies. The neonates are forced to adapt to their new life outside the womb in the first few hours after parturition [1]. Therefore, perinatal disorders and losses in dogs are a common and often unavoidable problem; they can occur in utero, during labor, immediately after, or within the first two to three weeks of life, but most puppy losses occur during whelping and the first 24 h [2]. Many factors may contribute to higher neonatal mortality in dogs [3], ranging from 5% to 35%, depending on complications before, during, and/or after parturition [4–8]. Adequate reproductive management throughout the bitch's reproductive cycle and good veterinary care at the onset of parturition, with intensive neonatal care and monitoring during the first days of life, can significantly reduce neonatal losses [9–11]. At the time of weaning, 91% of 1342 newborn puppies were alive in a study by Münnich and Kuechenmeister, reflecting the aforementioned reproductive management [8].

In human medicine, neonatal assessment is performed using the Apgar score, which immediately assesses the infant's status at birth and targets the need for neonatal resuscitation [12]. In 2009, the Apgar score was modified to adapt it to the needs of canine neonates [13]. Observation of the neonate and measurement of vital signs via the modified Apgar score in conjunction with reflex assessment in the first hour of life has been shown to be satisfactory for assessing neonatal status [12] and short-term survival prognosis [13]. However, in practice, clinical evaluation of neonatal puppies has been largely limited to a subjective examination. Basic diagnostics to objectively evaluate neonatal puppies are readily available to practitioners, but without established parameters, it is difficult to analyze them and determine what is normal versus abnormal and what might indicate future vitality [14]. Further, the type of parturition can also have a major impact on the vitality and survival of puppies after parturition [5,11], with puppies delivered by cesarean section (CS) having better survival than naturally born puppies [5].

In humans, the Apgar score is combined with umbilical blood gas analysis, which provides important information about the condition of the neonate [15]. There are limited data on the blood parameters in neonatal dogs [16,17], but as in humans, measurement of specific biomarkers from umbilical blood or amniotic fluid, e.g., lactate, cortisol, and glucose, could help to discriminate between healthy neonates and those requiring veterinary assistance [10,18]. Since these parameters are good indicators of non-vital neonates in humans, they were chosen in our study to see if they could also be used as indicators of neonatal vitality in puppies.

Clinically, it has been demonstrated that very high cortisol concentrations in puppies in amniotic fluid are a reliable indicator for identifying those neonates that require intensive care in the first 24 h of life [10]. In a recent study, all groups of puppies had cortisol concentrations above the basal concentration at birth. Significantly higher cortisol levels were noticed in fetal dystocia puppies compared to eutocia, maternal dystocia, and CS puppies. The difference between fetal dystocia and CS puppies persisted at 60 min after birth [19].

Increased lactate concentrations in the blood indicate the presence of anaerobic cellular metabolism, which is a sign of tissue hypoperfusion and hypoxia. Lactate concentrations increase before any abnormalities in heart rate, blood pressure, or urine output occur, making blood lactate measurement a superior method for early detection of hypoperfusion [20]. In dogs, there is a study that examined the concentration of lactate in the umbilical vein, which was found to be useful in predicting neonatal mortality within 48 h of birth. The threshold of 5 mmol/L of umbilical vein lactate concentration allowed differentiation between healthy and non-vital puppies. Higher lactate values were associated with non-vital puppies, whereas lower values characterized vital puppies [17]. Mean blood lactate levels of 10.0 mmol/L $\pm$ 4.9 standard deviations (SD) were found in non-surviving puppies [21]. However, this was not confirmed in two other studies where lactate levels did not vary between vital and disturbed puppies [22,23].

Transient hypoglycemia is common in neonates. Insufficient glycogen storage at parturition is a frequent occurrence in puppies with low birthweight and those suffering from respiratory failure or severe respiratory distress during parturition; hence, the measurement of glucose concentration may aid as a prognostic factor [24]. In one study on puppies evaluating glucose concentrations from ear blood, it was found that low blood glucose concentration was found in puppies with higher risk of neonatal mortality [25]. However, Lucio et al. (2021) found that fetal dystocia and CS are hyperglycemic obstetrical conditions for neonatal puppies [19]. There is also a study dealing with amniotic glucose concentration in newborn puppies where it was found that non-surviving puppies had lower median amniotic fluid glucose concentrations compared with neonates that survived. However, two of six amniotic fluid glucose concentrations were below the minimum detectable concentration, so this result should be taken in consideration with caution [23].

The limited data, the different ti"es o' puppy evaluation or collection of samples, and the different vessels used for blood collection make it very difficult to compare the results mentioned above. A thorough evaluation of non-invasive parameters and their correlation with neonatal vitality are still lacking in veterinary neonatology. Therefore, the aim of this study was to assess lactate and glucose concentrations in newborn umbilical blood and fetal amniotic fluid, cortisol concentrations in fetal amniotic fluid, and urine of newborn puppies, and to evaluate the association between these biomarkers with newborn vitality score, parturition type (vaginal parturition (VP) versus elective CS (EL-CS) versus emergency CS (EM-CS)), and puppy survival within the first seven days of life. We also aimed to investigate the association between parturition type and puppy survival and growth. The vitality of the newborn puppies was assessed using the modified Apgar scoring for dogs.

2. Materials and Methods

2.1. Animals

The study was approved by the Animal Welfare Commission of the Veterinary Faculty. The certificate is included as a Non-published Material. All animals were client-owned: the owners volunteered their dogs for the purpose of this research and signed a consent form allowing for the collection of samples from the bitch and the newborns. Samples were collected from March 2017 to May 2018. During this time, 123 puppies born to 20 healthy bitches of 16 different breeds (two Boston terrier bitches whelped twice) were included in our research. The mean age of the bitches was 41.6 months $\pm$ 19.2 SD (range: from 23 to 86 months). All bitches were fed only FDA-approved food for pregnant dogs. Bitches for VP and EM-CS were not fasted before the parturition. The bitches for EL-CS were also allowed to eat a small amount of soft canned food 3 h before the planned surgery. The number of puppies ranged from two to 11 per litter (median 5.5). The mean litter size was 5.59 puppies $\pm$ 2.38 SD. The types of parturition were divided into vaginal parturition (VP) and caesarian section (CS). CS was further divided into emergency CS (EM-CS) and elective CS (EL-CS). Regarding the type of parturition, there were 9 VPs, 8 EL-CSs, and 5 EM-CSs. One bitch gave birth to two alive puppies and one dead puppy via VP and was then presented for the EM-CS (she and the following 4 puppies were accounted to the EM-CS group) (Table 1).

For survival data, each newborn was categorized as stillborn or born alive.

2.2. Cesarean Section

This group consisted of EM-CS and EL-CS. If the medical management of dystocia had failed or was inadvisable, the EM-CS was performed according to indications as previously described [26,27]. The reason for the EM-CS was the primary uterine inertia in one dam, secondary uterine inertia in two, and obstruction in the other two. In one case, the puppy was too big and in the other it was in an incorrect posture. All dams were in good general condition. However, at least one puppy was under stress in all cases, as indicated by its heart rate, measured with ultrasound. The bitch with primary uterine inertia was given oxytocin twice (0.25 IU and 30 min lates 0.5 IU), prior to coming to the clinic for EM-CS.

Table 1. Number and proportion (%) of different breeds included in the study.

Breed Size	Breed	Number of Parturitions (%)	Age of Bitches (Years)	Number of Parity	Number of Female Puppies (%)	Number of Male Puppies (%)	Number of Puppies (%)
	Miniature Schnauzer	2 (9.1)	3–4	1–2	6 (4.9)	4 (3.2)	10 (8.1)
	Yorkshire Terrier	1 (4.5)	2.25	1	1 (0.8)	3 (2.4)	4 (3.3)
	Maltese	1 (4.5)	3.25	2	2 (1.6)	3 (2.4)	5 (4.1)
	Miniature Poodle	1 (4.5)	7	3	1 (0.8)	1 (0.8)	2 (1.6)
	Boston Terrier	4 (18.2)	2–5.5	1–3	11 (8.9)	5 (4.1)	16 (13)
	Jack Russell Terrier–Maltese mix	1 (4.5)	5	1	3 (2.4)	4 (3.3)	7 (5.7)
Total small breeds(<10 kg)		10 (45.4)			24 (19.5)	20 (16.3)	44 (35.8)
	Pembroke Welsh Corgi	2 (9.1)	2.5–2.75	1	3 (2.4)	7 (5.7)	10 (8.1)
	English Bulldog	1 (4.5)	2.5	1	2 (1.6)	2 (1.6)	4 (3.3)
	French Bulldog	2 (9.1)	2–2.5	1	5 (4.1)	8 (6.5)	13 (10.6)
	Beagle	1 (4.5)	4.25	3	3 (2.4)	4 (3.3)	7 (5.7)
	Dachshund	1 (4.5)	3	2	7 (5.7)	2 (1.6)	9 (7.3)
	Whippet	1 (4.5)	2.5	1	4 (3.3)	3 (2.4)	7 (7.3)
Total medium breeds (10–25 kg)		8 (36.4)			24 (19.5)	26 (21.1)	50 (40.6)
	Greater Swiss Mountain Dog	1 (4.5)	3.5	2	1 (0.8)	5 (4.1)	6 (4.9)
	Golden Retriever	1 (4.5)	2.5	1	5 (4.1)	6 (4.9)	11 (8.9)
	Labrador Retriever	1 (4.5)	2.75	2	8 (6.5)	2 (1.6)	10 (8.1)
	German Shepherd	1 (4.5)	3	2	1 (0.8)	1 (0.8)	2 (1.6)
Total large and giant breeds (>25 kg)		4 (18.2)			15 (12.2)	14 (11.4)	29 (23.6)

EL-CS was planned based on progesterone measurements during heat to determine LH peak and ovulation, coupled with ultrasound (US) examinations of the ovaries and uterus. The day of surgical parturition in EL-CS cases was determined using the average 63-day gestation from ovulation guidelines, coupled with a decrease in the dams' rectal temperature and the information collected from the serial blood progesterone concentration monitoring and by fetal ultrasonographic measurements of both the inner chorionic cavity and the biparietal diameter, as reported by Meloni [28]. None of the puppies showed signs of immaturity at birth.

CS was always performed using the same anesthetic protocol. The preoperative health status of the bitch was determined via a clinical examination and complete blood count (ADVIA® 120, Siemens, München, Germany) with blood collected from a peripheral vein. The cephalic vein was used for intravenous cannulation. During the preoxygenation with 100% oxygen for 10 min, the abdomen was shaved and aseptically prepared in a routine manner. Propofol at 4–7 mg/kg of body weight (BW) was given intravenously (Propoven 10 mg/mL, Fresenius Kabi Austria GmbH, Graz, Austria) for induction to anesthesia, followed by orotracheal intubation. General anesthesia was maintained with sevoflurane (Sevorane, AbbVie, Campoverde di Aprilia, Italy) at a concentration of 1.5–3%. Methadone (Synthadon 10 mg/mL, Produlab Pharma B.V., Raamsdonksveer, The Netherlands) at 0.2 mg/kg BW was administered subcutaneously after the skin incision. A balanced isotonic crystalloid solution (Hartmann solution, B. Braun, Melsungen, Germany) at a rate of 10 mL/kg/h intravenously was started 30 min before general anesthesia. CS was performed with a caudal midline abdominal incision, followed by a ventral incision of the uterine body. The most caudal fetus and its fetal membranes were removed first, followed by fetuses and fetal membranes from the left and right uterine horn alternately.

A simple continuous suture pattern using 3/0 or 4/0 Glycomer 631 (Biosyn, Covidien, Dublin, Ireland) was used to close the uterus. To promote uterine contractions and cleaning, oxytocin (Oxytocin Veyx, Veyx-Pharma Gmbh, Schwarzenborn, Germany) at 0.25–1 IU was injected into the uterus at the end of the celiotomy. The abdominal cavity was thoroughly inspected and rinsed with 0.9% sodium chloride (NaCl, B Braun, Melsungen, Germany) and warmed to body temperature before the abdominal wall was sutured routinely in three layers. Lidocaine 10 mg/mL (Lidocaine HCl, University Medical Centre, Ljubljana, Slovenia) at a dosage of 1 mg/kg was administered infiltrative at the incision line.

Four hours after the first dose, methadone (0.2 mg/kg) was administered subcutaneously. When the mothers were awake and demonstrated normal maternal behavior, the bitch and her puppies were discharged from the clinic. Postoperative analgesia was provided with tramadol chloride (Tramal, Stada, Bad Vilbel, Germany) at 3 mg/kg orally every 12 h, as needed, for a maximum of three days postoperatively. When the bitch did not have colostrum at the time of parturition, the puppies were given a drop of 40% glucose under the tongue. If mammary secretions were still not available, newborns were fed with a commercial milk replacer formula (Puppy Protech Colostrum + Milk, Royal Canin, Ljubljana, Slovenia) at suggested dosages every two hours. This was needed only for one litter of EL-CS puppies (for the first day) and for one other puppy that died one day postoperatively.

Puppy Resuscitation

The basic steps for puppies who did not breathe spontaneously and did not vocalize followed the resuscitation guidelines of ABC. First, the airway was cleared by removing the fetal membranes from the face, followed by gentle suctioning with a bulb syringe. The puppies were then gently but briskly dried with a warm towel to promote respiration and prevent chilling. If the neonates were still not breathing, a Jen Chung vacupuncture point in the nasal philtrum with a 26-gauge needle was used. For neonates who were not breathing spontaneously, a constant flow of oxygen was administered through the oxygen mask. Ventilation or endotracheal intubation and cardiac stimulation were not required. The only drug used was naloxone (one drop under the tongue) when the puppies did not respond to environmental stimuli. Puppies were kept warm during resuscitation and in the immediate postpartum period. Water bottles heated to 38 °C were used for this purpose.

2.3. Vaginal Parturition

Vaginal deliveries took place in the home environment of the owner.

2.4. Evaluation of the Puppy, Apgar Score, and Neonatal Reflexes

Each newborn puppy was weighed, assessed using the modified Apgar scoring system [14,17,29], and examined for the presence of congenital malformation (congenital oronasal fistula, cleft palate, atresia ani). When needed, resuscitation was performed by ABC route: rubbing to stimulate breathing, cleaning mucous from the upper airways, oxygen supplementation, heating, glucose supplementation, and, on one occasion, external heart massage. Each puppy was marked immediately after parturition with a colored collar to enable the identification of the puppies and correctly mark the collected samples.

The vitality of the puppies was assessed by observing the color of the mucous membranes, auscultation of the heart to measure the heart rate, and evaluation of the frequency and quality of respiration. The neonatal irritability reflex was assessed with a firm, almost painful stimulus on the back. Active movements of the puppies were observed in the dorsal position to assess muscle tone. Each parameter was scored as 0, 1, or 2 points (Table 2). The Apgar score was interpreted depending on the number of points collected, severe distress from 0 to 3, moderate distress from 4 to 6, and no distress from 7 to 10 points. Apgar score was evaluated 5 (Apgar 5), 15 (Apgar 15), and 60 (Apgar 60) min after parturition.

Table 2. The modified Apgar scoring system used in this study [14,17,29].

Parameter	Points		
	0	**1**	**2**
Mucous membrane color	Pale or cyanotic	Pink	Reddish
Heart rate (bpm [1])	Absent or <120 bpm	120–180 bpm	>180 bpm
Respiratory rate	Absent or <6/min	Weak, irregular, 6–15/min	>15/min, rhythmic
Activity, muscle tone	Flaccid	Some flexions	Active motion
Reflexes irritability	Absent	Weak vocalization and weak reflex	Vigorous vocalization and immediate reflex

[1] bpm: beats/min.

Neonatal reflexes (suckling, rooting, and righting reflex) were also assessed at 5, 15, and 60 min after parturition. The suckling reflex was assessed by inserting a clean, warm tip of the smallest finger into the mouth of the puppies. The righting reflex was assessed by turning the puppies onto their backs and observing how quickly they turned back to the sternal position. The rooting reflex was scored by holding a loosely clenched fist in front of the puppy and observing whether the puppy responded by pressing its snout into the hand. Each reflex was scored from 0 to 2, with 0 representing no reflex and 2 representing a strong reflex. Then the total sum was calculated. The interpretation of the results was as follows: poorly responsive from 0 to 2, moderately responsive from 3 to 4, and adequately responsive from 5 to 6 points [12,14].

2.5. Amniotic Fluid, Umbilical Blood, and Urine Samples

Samples of amniotic fluid were taken immediately after birth from all puppies, but only samples of alive puppies were processed further. The puppy was put in an upward position, the amniotic sac was freed from the puppy's head, and the amniotic fluid samples were collected from the underlying amniotic sac in an aseptic manner using a sterile needle and syringe. Then the amniotic sac was opened completely, and the amniotic fluid was decanted from a syringe in a urine collection tube. The umbilicus was clamped on two sides and a small 25-gauge needle was used to try to obtain umbilical blood within 2 min after birth. At least 25 μL of umbilical blood was taken and was processed immediately. Amniocentesis, sampling of the umbilical blood, and neonatal resuscitation were performed by different individuals.

Amniotic and umbilical blood samples of newborns delivered by spontaneous whelping were collected the same way as in the CS group, if they were born with an intact amniotic sac and umbilical cord. Urine samples were collected within an hour after parturition after manual stimulation of the vulva/prepuce of the neonate with the moistened swab. The first two drops of urine were discarded.

Amniotic fluid analysis included measurement of specific gravity with refractometer (Veterinary refractometer, Eickemeyer, Tuttlingen, Germany), glucose (Wellion Gluco Calea, Med Trust Handelsges.m.b.H., Marz, Austria), lactate (Accutrend Plus, Roche, Basel, Switzerland), and cortisol concentration (MiniVidas analyzer, BioMerieux S.A., Lyon, France) in accordance with the manufacturer's instructions. For cortisol measurement, the amniotic fluid sample was centrifuged at $2000\times g$ for 10 min and frozen within an hour at $-80\ °C$ until analyzed less than 6 months after collection.

Umbilical blood analysis included glucose (measured with Wellion Gluco Callea), and lactate concentration (measured with Accutrend Plus), and the samples were processed within the first 5 min.

After the first suckling, samples of newborn urine were collected. Urine was frozen at $-80\ °C$ within an hour and stored for no longer than 6 months. Cortisol concentrations were then measured using the MiniVidas analyzer.

At the beginning of the study, we performed a partial validation of amniotic glucose and amniotic lactate assay by determining the accuracy and repeatability. The accuracy of the methods for lactate measurements was determined by analysis of a control sample with known low- and high-lactate concentrations, 3.6 and 9.0 mmol/L, respectively. The measured values were 3.32 mmol/L $\pm$ 0.12 SD and 8.49 mmol/L $\pm$ 0.21 SD; the inter-assay coefficients of variation (CV) for the control solution were 3.42% and 2.51%, respectively. By performing 6 measurements of the single amniotic fluid sample, the intra-assay CV for lactate was 3.21% and 2.15% for the two samples with 4.4 (low) and 23.3 mmol/L (high) lactate concentrations, respectively. The procedure of measuring intra-assay CV was repeated several times during the study: If there was enough amniotic fluid, the coefficient of variation stayed in the mentioned frame.

By performing 6 measurements of the single sample, the intra-assay CV for glucose in amniotic fluid was 3.73%. Since there were no puppies with high amniotic glucose concentrations, only one CV was calculated.

At the beginning of the study also a partial validation of amniotic fluid and urine cortisol, determining the accuracy and repeatability was performed. The accuracy of the methods for cortisol measurements was determined by analysis of a control sample with known low- and high-cortisol concentrations, 0.725 and 22.112 ng/mL, respectively. The measured values were 0.729 ng/mL $\pm$ 0.011 SD and 22.009 ng/mL $\pm$ 1.003 SD; inter-assay CVs for the control solution were 1.52% and 2.23%, respectively. By performing 6 measurements of the single sample, the intra-assay CV for cortisol in amniotic fluid was 2.74% and 1.35% for two samples with 8.79 (low) and 16.37 ng/mL (high) cortisol concentrations, respectively. Six measurements of the single urine sample were performed and the intra-assay CV of variation for cortisol in urine was 2.08% and 1.23% for two samples with 11.03 (low) and 18.49 ng/mL (high) cortisol concentrations, respectively.

2.6. Growth Rate of Newborn Puppies

Immediately after birth, the puppies were weighed on a steady scale with an accuracy of $+/-1$ g. After the first day, the puppies were weighed by the owners on the same scale. The growth rate of the puppies was followed until they were eight weeks old. The bodyweight of the puppies was measured each morning in the first week. Later, they were weighted once weekly in a 7-day interval. The day of the parturition was determined as Day 0.

The relative growth rate was calculated as the relative change in the bodyweight compared to the puppy's initial birthweight.

2.7. Statistical Analysis

Data were collected and edited in Microsoft Excel 365. The statistical analyses were performed using R statistical software, version 4.1.2 [30], and $p < 0.05$ was considered significant.

The data are presented with basic descriptive statistics; e.g., median, quartiles, mean $\pm$ standard error of the mean (SEM). The normal distribution of the data was tested by the Shapiro-Wilk test. Differences between blood parameters, amniotic fluid, urine, puppies' vitality estimates, and relative growth rates according to the type of parturition and vitality were tested with a non-parametric Kruskal–Wallis rank-sum test and Wilcoxon rank-sum test, respectively. Due to multiple comparisons, we adjusted the p-values with a Benjamini–Hochberg correction. Correlations were given using the Spearman correlation coefficient, and statistical significance was determined according to the adjusted p-value according to the Holm method. The interpretation of Spearman's correlation coefficient is as follows: 0.00–0.19—very low/weak; 0.20–0.39—low/weak; 0.40–0.59—medium/moderate; 0.60–0.70—high/strong; 0.80–1.0—very high/strong correlation.

3. Results

3.1. Basic Information

The data of different breeds, included in our research, and the number of puppies per breed are presented in Table 1. The number and proportion (%) of puppies according to the type of parturition, sex, and survival of newborns are presented in Table 3. The birthweight data are missing for two puppies that were born at home via VP before the bitch was presented for EM-CS; additional six puppies were lost to follow-up after day 0.

Table 3. Number of puppies according to the type of parturition, sex, survival of newborns and the body weight at birth and on day 7.

Type of Parturition	Number of Puppies			Survival			Body Weight in Grams at Birth (Mean ± SEM)		Body Weight in Grams on Day 7 (Mean ± SEM)	
	Female	Male	Born Alive	Stillborn	Died after Discharge		Female	Male	Female	Male
VP	36	32	63	5	3		279 ± 12 ($n = 33$)	283 ± 17 ($n = 28$)	466 ± 23 ($n = 31$)	464 ± 32 ($n = 27$)
EL-CS	18	20	36	2	4		243 ± 16 ($n = 17$)	234 ± 18 ($n = 19$)	382 ± 32 ($n = 14$)	335 ± 32 ($n = 16$)
EM-CS	9	8	15	2	0		252 ± 48 ($n = 8$)	306 ± 49 ($n = 7$)	439 ± 64 ($n = 5$)	454 ± 47 ($n = 6$)

Legend: EL-CS: elective Cesarean section; EM-CS: emergency Cesarean section, VP: vaginal parturition; SEM: standard error of the mean.

3.2. Survival, Apgar Score, and Neonatal Reflexes

The stillborn puppies ($n = 9$) were excluded from the statistical data analysis. Seven puppies (5.7% of all puppies or 6.1% of born alive) from four bitches died after the discharge from the clinic: They all died within the first week of life (only one died in the first 48 h) and all but one were in severe distress 5 min after parturition. At 15 and 60 min postoperatively, two and five of them, respectively, were in no distress. In one puppy, Apgar 5 was 1, Apgar 15, and Apgar 60 was 3; the puppy was born via VP and died one day postpartum. One puppy showed no distress at 5, 15, and 60 min postoperatively and died on the fourth day postoperatively of an unknown cause. One of the puppies that died on the third day postpartum was diagnosed with neonatal sepsis due to infection with *Staphylococcus pseudintermedius*. Most likely, the puppy was infected through breast milk, although the bitch did not show any clinical signs. Namely, a pure culture of the same bacterium was isolated from its milk [31]. Three Pembroke Welsh Corgi puppies were diagnosed with interstitial pneumonia postmortem, and no infectious agent was found. The others were not examined by autopsy and the cause of death remains unknown.

The results of the Apgar score 5, 15, and 60 min after parturition are presented in Table 4. The data are missing for two puppies that were born at home via VP before the bitch was presented for EM-CS. Data are also missing for 4 puppies at 5 and 15 min and for 3 puppies also at 60 min. These puppies were born with VP before we arrived at the parturition site.

Table 4. Number and proportion of puppies according to Apgar scores 5, 15, and 60 min after parturition and type of parturition.

Type of Parturition	Apgar Score 0–3: Severe Distress (%)	Apgar Score 4–6: Moderate Distress (%)	Apgar Score 7–10: No Distress (%)
	5 min after parturition		
VP	3 (5.3)	0 (0)	54 (94.7)
EM-CS	2 (13.3)	3 (20)	10 (66.7)
EL-CS	26 (72.2)	3 (8.3)	7 (19.4)
	15 min after parturition		
VP	1 (1.8)	1 (1.8)	55 (96.5)
EM-CS	2 (13.3)	2 (13.3)	11 (73.3)
EL-CS	7 (19.4)	9 (25)	20 (55.6)
	60 min after parturition		
VP	1 (1.7)	0 (0)	57 (98.3)
EM-CS	0 (0)	0 (0)	15 (100)
EL-CS	0 (0)	0 (0)	36 (100)

The median Apgar score was 8 for Apgar 5, 9 for Apgar 15, and 10 for Apgar 60.

The most vital puppies were born with VP, followed by puppies born with EM-CS and EL-CS. The Apgar score 5 min after parturition for EL-CS was significantly lower when compared to EM-CS and VP ($p < 0.0001$) (Figure 1A).

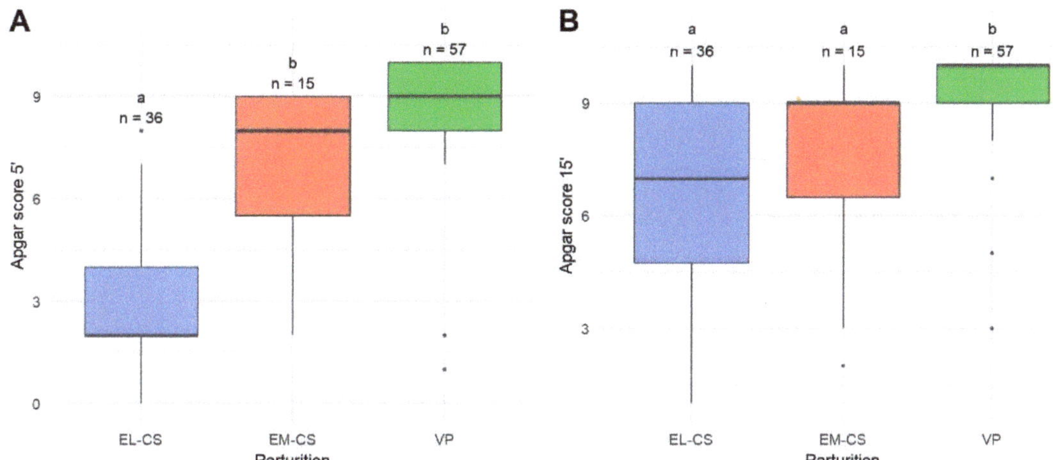

Figure 1. A comparison of Apgar scores 5 min (**A**) and 15 min (**B**) after parturition of puppies born by elective Cesarean section (EL-CS), emergency Cesarean section (EM-CS), and vaginal parturition (VP). Above each boxplot, statistically significant differences ($p < 0.05$) observed between parameters are marked with different lower-case letters, and the number of observations per group is indicated. The box represents the lower and upper quartiles, the line in the box denotes the median values, and whiskers the minimum and maximum values excluding outliers, which are marked with points.

Puppies born with VP still had the highest Apgar scores 15 min after parturition compared with both other groups ($p < 0.0001$), but there was no statistical difference when EL-CS and EM-CS were compared (Figure 1B). There were no statistical differences in Apgar scores between groups 60 min after parturition ($p = 0.2961$). Median Apgar scores of 10 were observed in all three groups.

In the EL-CS group, 26 puppies (72%) were in severe distress 5 min after parturition, whereas 54 (95%) and 10 (67%) of puppies born with either VP or EM-CS showed no distress. Differences in the distribution of Apgar scores were also observed 15 min after parturition, with the distribution of the puppies born with EM-CS and VP similar to 5 min after parturition. However, most puppies born with EL-CS were not in distress and the proportion of severely distressed puppies born with EL CS decreased compared with Apgar scores 5 min after parturition. As a result, the proportion of puppies being moderately distressed 15 min after parturition increased compared with the Apgar score 5 min after parturition.

Puppies in severe distress 5 and 15 min postpartum had statistically significantly worse survival in the first week postpartum, $p = 0.0113$ and $p = 0.0231$, respectively.

Table 5 presents the number of puppies with a certain reflex score according to the type of parturition 5, 15, and 60 min after birth. Stillborn puppies were excluded from the statistical processing of reflex assessment. Data for 6 puppies at 5 and 15 min and data for 5 puppies at 60 min as for Apgar score are missing.

Table 5. Number and proportion (according to the type of parturition) of the poorly, moderately, and adequately responsive puppies at 5, 15, and 60 min after parturition. Suckling, rooting, and righting reflexes were evaluated.

Type of Parturition	Number of Poorly Responsive Puppies (%)	Number of Moderately Responsive Puppies (%)	Number of Adequately Responsive Puppies (%)
	Neonatal reflexes 5 min after parturition		
VP	3 (5.3)	13 (22.8)	41 (71.9)
EM-CS	8 (53.3)	2 (13.3)	5 (33.3)
EL-CS	28 (77.8)	3 (8.3)	5 (13.9)
	Neonatal reflexes 15 min after parturition		
VP	2 (3.5)	3 (5.3)	52 (91.2)
EM-CS	5 (33.3)	2 (13.3)	8 (53.3)
EL-CS	19 (52.8)	12 (33.3)	5 (14.9)
	Neonatal reflexes 60 min after parturition		
VP	1 (2)	1 (2)	56 (96)
EM-CS	0 (0)	1 (7)	14 (93)
EL-CS	0 (0)	3 (8)	33 (92)

3.3. Amniotic Fluid, Umbilical Blood, and Urine Samples

The mean specific gravity of the amniotic fluid was 1.012 kg/L ± 0.004. The mean value for EL-CS was 1.011 kg/L ± 0.005 for EM-CS 1.015 kg/L ± 0.005 and for VP 1.012 ± 0.004. There were no significant differences between the values of the specific gravity of amniotic fluid between the different parturition types.

The mean cortisol concentration of amniotic fluid ($n = 87$) was 8.14 ± 0.48 ng/mL, and the median was 8.70 ng/mL. The lowest concentrations of cortisol were observed in the EL-CS group (3.83 ± 0.44 ng/mL), followed by the group of puppies born via VP (9.24 ± 0.36 ng/mL), and the EM-CS group (14.01 ± 1.10 ng/mL). The difference in cortisol concentration between all three types of parturition was significant ($p < 0.0001$) (Figure 2A).

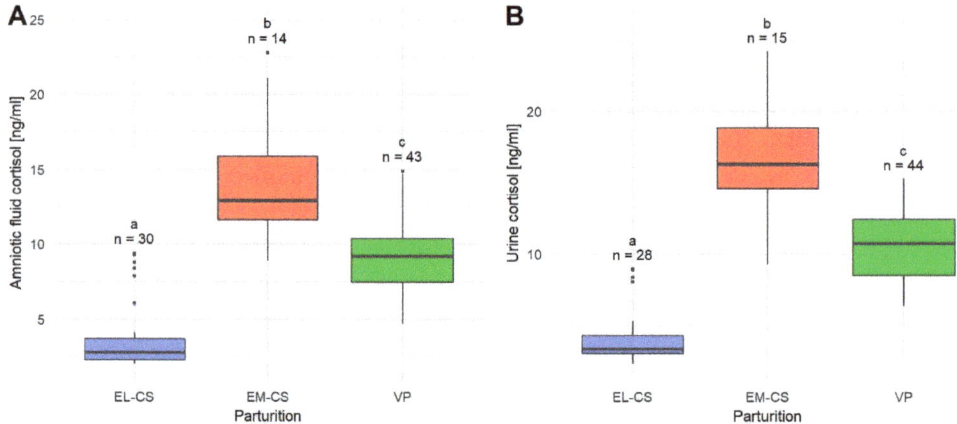

Figure 2. A comparison of amniotic fluid cortisol concentrations (**A**) and urine cortisol concentrations (**B**) in puppies born by elective Cesarean section (EL-CS), emergency Cesarean section (EM-CS), and vaginal parturition (VP). Above each boxplot, significant differences ($p < 0.05$) observed between parameters are marked with different lower-case letters, and the number of observations per group is indicated. The box represents the lower and upper quartiles, the line in the box denotes the median values, and whiskers the minimum and maximum values excluding outliers, which are marked with points.

The mean cortisol concentration in urine ($n = 87$) was 9.64 ± 0.54 ng/mL, and the median was 9.30 ng/mL.

The lowest mean urinary cortisol concentration was observed in puppies born with EL-CS (4.33 ± 0.41 ng/mL), followed by the group of puppies born with VP (10.62 ± 0.37 ng/mL) and EM-CS (16.70 ± 1.05 ng/mL). The difference in urinary cortisol concentration between all three types of parturition was significant ($p < 0.0001$) (Figure 2B).

The mean glucose concentration in the amniotic fluid ($n = 55$) was 3.49 ± 0.19 mmol/L, and the median was 3.70 mmol/L. The lowest mean glucose concentration in the amniotic fluid was observed in puppies born with EM-CS ($n = 11$; 2.73 ± 0.72 mmol/L), followed by puppies born with EL-CS ($n = 6$; 2.80 ± 0.58 mmol/L), and those born with VP ($n = 38$; 3.81 ± 0.15 mmol/L). The difference in the amniotic fluid glucose concentrations between parturition types was not significant ($p = 0.0666$). No statistically significant differences in blood or amnion glucose concentrations ($p = 0.1775$ and $p = 0.2090$, respectively) were found between dead and live pups.

The mean umbilical blood glucose concentration ($n = 59$) was 3.68 ± 0.17 mmol/L, and the median was 4 mmol/L. Puppies born with EL-CS (3.23 ± 0.38 mmol/L) had the lowest mean glucose concentration, followed by puppies born with EM-CS (3.29 ± 0.72 mmol/L) and those born with VP (3.80 ± 0.15 mmol/L). There were no significant differences between the concentrations of umbilical blood glucose between the different types of parturition ($p = 0.1652$).

The mean lactate concentration in the amniotic fluid ($n = 68$) was 11.32 ± 0.54 mmol/L, median 10.63 mmol/L. Puppies born with EL-CS (9.11 ± 0.87 mmol/L) had the lowest mean lactate concentration, followed by puppies born with EM-CS (10.25 ± 1.82 mmol/L) and VP (12.24 ± 0.56 mmol/L). The difference in the amniotic fluid lactate concentration between the EL-CS and VP was significant ($p = 0.0267$) (Figure 3A).

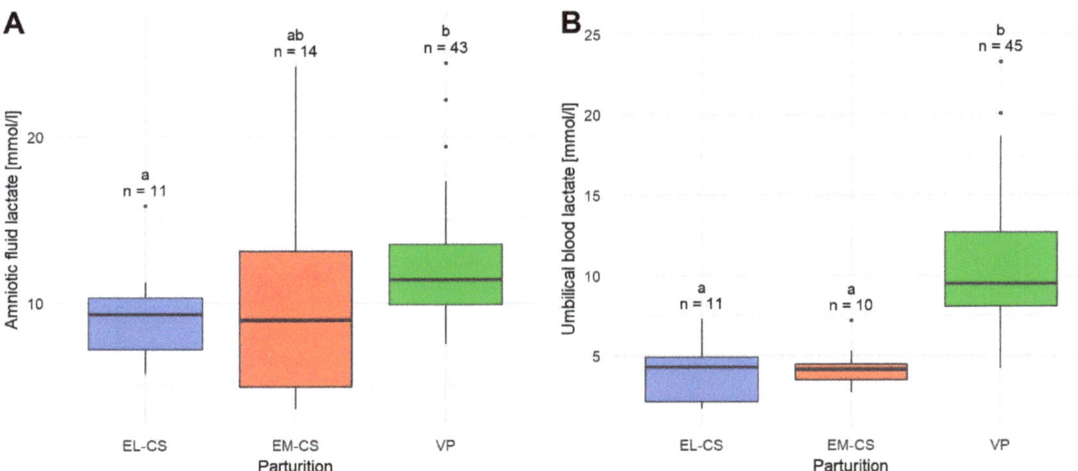

Figure 3. A comparison of amniotic fluid lactate concentrations (**A**) and umbilical blood lactate concentrations (**B**) in puppies born by elective Cesarean section (EL-CS), emergency Cesarean section (EM-CS), and vaginal parturition (VP). Above each boxplot, significant differences ($p < 0.05$) observed between parameters are marked with different lower-case letters, and the number of observations per group is indicated. The box represents the lower and upper quartiles, the line in the box denotes the median values, and whiskers the minimum and maximum values excluding outliers, which are marked with points.

The mean lactate concentration in umbilical blood ($n = 66$) was 8.39 ± 0.55 mmol/L, median 8.15 mmol/L. The lowest mean umbilical blood lactate concentration was observed in puppies born with EL-CS (3.90 ± 0.58 mmol/L), followed by puppies born with EM-CS (4.23 ± 0.41 mmol/L) and VP (10.41 ± 0.57 mmol/L). The difference in umbilical blood lactate concentration between EM-CS and VP as well as between EL-CS and VP was significant ($p < 0.0001$) (Figure 3B).

Puppies born at the end of the parturition did not have significantly different lactate concentrations in either umbilical blood or in amniotic fluid ($p = 0.1775$ and $p = 0.2751$, respectively).

3.4. Growth Rate of Newborn Puppies

The bodyweight of puppies at birth and on day seven by parturition type and separately for female and male puppies are shown in Table 3. In Figure 4, we present the relative growth rate from day one to seven by the type of parturition. The mean relative growth rate on day 1 was −1.95 ± 0.44%, on day 2: 3.52 ± 0.54%, on day 3: 12.55 ± 0.59%, on day 4: 23.76 ± 0.48%, on day 5: 34.49 ± 0.49%, on day 6: 48.26 ± 0.44%, and on day 7: 63.09 ± 0.39%.

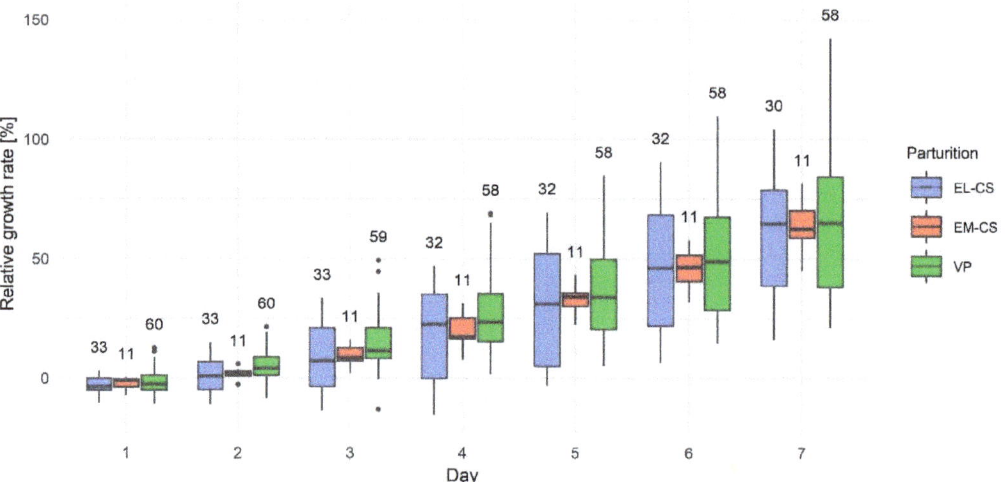

Figure 4. The puppies' relative growth rate in the first week after parturition by elective Cesarean section (EL-CS), emergency Cesarean section (EM-CS), and vaginal parturition (VP). Above each boxplot, the number of observations per group is indicated. The box represents the lower and upper quartiles, the line in the box denotes the median values, and whiskers the minimum and maximum values excluding outliers, which are marked with points.

Puppies born by EL-CS had the lowest relative growth rate, and puppies born by VP had the highest relative growth rate; however, there were no statistically significant differences between the three types of parturition on any given day (p: 0.0878–0.9515) (Figure 4).

The growth rate of the puppies was followed until they were eight weeks old, but a significant difference in relative growth rate between puppies was not observed.

3.5. Correlations

The parameters measured in the blood and in the amniotic fluid were compared with each other and their correlations were determined (Table 6).

Table 6. Spearman's correlation coefficient between measured parameters in the amniotic fluid, umbilical blood, urine, and relative growth rate.

	AM Glucose	AM Lactate	Apgar 5	Apgar 15	AM Cortisol	Cortisol in Urine	UB Glucose	UB Lactate
AM lactate	−0.1520	/						
Apgar 5	**0.4953 ***	−0.2141	/					
Apgar 15	0.3952	−0.1173	**0.7400 *****	/				
AM cortisol	0.2024	0.3513	0.0935	0.0376	/			
Cortisol in urine	0.1744	0.4820	0.1115	0.1087	**0.9199 *****	/		
UB glucose	**0.5307 ***	−0.1361	**0.5536 ****	**0.5854 ****	0.0680	0.0226	/	
UB lactate	0.0055	**0.8741 *****	−0.1602	−0.0013	0.4662	**0.5778 ****	−0.0404	/
Relative growth rate	0.2062	**−0.5378 ***	0.0422	0.0350	−0.2263	−0.3657	0.0901	**−0.5876 ****

Legend: AM—amniotic fluid, UB—umbilical blood, p-value < 0.001 (***), p-value < 0.01 (**), p-value < 0.05 (*), p-value < 0.1. Bold numbers indicate statistically significant results.

Highly statistically significant ($p < 0.0001$) and very strong correlations were found between the cortisol concentrations in urine and in amniotic fluid and also between the lactate concentrations in amniotic fluid and in umbilical blood. There was a strong correlation between Apgar 5 and Apgar 15, which was also highly statistically significant ($p < 0.0001$). However, Apgar 60 did not correlate with any of the observed parameters. The umbilical blood glucose moderately correlated with glucose concentrations in the amniotic fluid ($p = 0.0170$), and with Apgar score 5 min ($p = 0.0093$) and 15 min ($p = 0.0036$) after parturition. Glucose in the amniotic fluid, however, moderately correlated with Apgar score 5 min after parturition ($p = 0.0423$). Lactate in the umbilical blood moderately correlated with cortisol in urine ($p = 0.0045$). The relative growth rate in puppies on day 1 moderately correlated with lactate in the amniotic fluid ($p = 0.0144$) and lactate in the umbilical blood ($p = 0.0035$). However, the relative growth rate from the second and later days did not correlate with any of the other variables.

4. Discussion

The objective of this study was to explore neonatal puppy vitality by investigating common clinical parameters with Apgar score and determining glucose, lactate, and cortisol parameters in fetal and neonatal (urine) fluids while comparing different parturition types. The clinical parameters assessed in this study were selected because they are widely used in veterinary medicine for the subjective assessment of puppies to determine their vitality. These parameters, together with the rapid assessment of the above-mentioned laboratory parameters, could serve as a good predictor for identifying puppies that need special care during the first hours or days after birth in order to improve their survival rates.

In our study, 9 (7.3%) of the puppies were stillborn and 7 (5.7%) died later. This is in accordance with other recent studies [5,6,9,13,17,32] reporting mortality rates of as high as 25% in the clinical setting [5,6]. Better survival rates are reported in controlled pregnancies and parturitions, suggesting that survival can be improved by appropriate reproductive management and support during pregnancy and parturition, and by timely surgical intervention when needed [9,13,17,32]. However, the most important question is how to identify the newborns at risk and help them in the first days of life. Birthweight [6,32–34] and Apgar score [32] are recognized as prognostic factors for early postpartum survival. On the contrary, as reported in [34], the Apgar score seems to be more accurate. Of the seven puppies that died in the first week postpartum in the present study, six were in severe distress 5 min after birth. A statistically lower Apgar score measured at 5 and 15 min after birth was significantly associated with higher mortality. Similarly, an Apgar score between 0–3 measured 5 min after birth was associated with a higher mortality rate than scores between 4–6 and higher in previous studies [13,29,32,35]. In our study, the condition of the puppies improved over the next 10 and 55 min, respectively, indicating the need for an early Apgar score to identify at-risk neonates, as a later measurement may mask the initial stress level, as in our study 99% of puppies showed no distress 60 min after birth. The Apgar score did not correlate with the relative growth rate in our study.

Thirty-one puppies (28.7%) in this study population showed severe distress 5 min after birth, which is significantly more than in some other studies, reporting between 85.3% and 94% normally viable puppies 5 min after birth [10,13], but similar or less than 17% at 5 min [36] or 45.3% of puppies showing severe distress 10 min after birth [17]. In both studies [17,36], puppies born with CS predominantly contributed to Apgar scores 0–3 at 5–10 min after birth, demonstrating the differences in Apgar scores between the different types of parturition. Sixteen of thirty-seven puppies (43.2%) born with CS showed severe distress 5 min after birth [36], but Groppetti et al. (2010) reported as many as 100% and 92% of puppies showing severe distress 10 min postpartum in the EM-CS and EL-CS groups, respectively [17]. Our study confirms significant differences between the groups 5 and 15 min after birth where 72% of the EL-CS puppies, but only 13.3% of the EM-CS and 5.3% of the VP puppies, showed severe distress 5 min postpartum. The results of puppies' responsiveness were very similar to the results of the Apgar score. While anesthesia could be a possible factor for the lower Apgar scores and poor responsiveness in all CS puppies [35,37], the very high proportion of the EL-CS puppies in severe distress in our study compared to two previous studies [10,32] requires further investigation. The reason for the lower Apgar scores could be influenced by the determination of the parturition date, although we used several parameters to determine the best timing for EL-CS. However, their accuracy depends on several factors such as breed, gestational age, and litter size. Therefore, it is challenging to choose the most appropriate parameter for all breeds and each individual animal. Currently, the most accurate method of predicting the parturition date is the prepartum progesterone drop, but the use of ultrasound parameters throughout gestation is still necessary to detect bitches before the onset of parturition [38]. An additional factor was the time that elapsed between neonatal extraction and the start of supportive measures, which was minimally prolonged due to obtaining samples for the study. Although this was minimal, it may have contributed to Apgar scores in highly sensitive, slightly immature EL-CS neonates. Our observation suggests that the appropriate timing of an EL-CS is of paramount importance. To minimize the vulnerability of EL-CS neonates in the immediate postpartum period and the demand for intensive neonatal resuscitation, EL-CS should ideally be performed after the onset of the first stage of parturition.

The expected difficulty in the noninvasive collection of the umbilical blood samples for lactate measurement, especially in small breed dogs [17], was confirmed in our patients. Whereas in a previous study at least 25 µL of umbilical blood was successfully collected from 70 of 94 puppies (74.5%) [17], in our study this was possible in 72 of 108 puppies (66.7%). This is less than mentioned by Groppetti et al. (2010), but in their population, only two out of 21 (9.6%) bitches belonged to small breeds [17] in contrast to 10 out of 22 parturitions (45.5%) in our study. In humans, umbilical blood gas analysis provides important information about the condition of the neonate [15]. In canine neonates, mixed acidosis was confirmed in the jugular venous blood sample 5 min after birth, and metabolic acidosis persisted until 1 h after birth when the second sample was collected [16]. We attempted to obtain 100 µL of blood for umbilical blood cord gas analysis but obtained only 55 complete samples in 108 neonates (51%), similar to Antończyk et al. [22], making this diagnostic tool unsuitable for clinical practice and requiring a learning curve. Further, the umbilical cord has a two-way direction of blood flow and with our technique, we were not able to ascertain that the taken blood sample originated from the neonate. This prompted us to investigate and compare the results of different parameters in fetal fluids other than the umbilical blood (amniotic fluid, urine) and to perform tests requiring less umbilical blood. We collected 87 (80.6%) amniotic fluid samples for the measurement of glucose, lactate, and cortisol concentration and 87 samples for urine cortisol concentration. The reason for the discrepancy in the number of samples collected and tests performed is due to the order and variety of tests that we performed at the beginning of the study.

The highest lactate concentration was found in VP puppies (in blood and amniotic fluid), and lower values were found in EM-CS and EL-CS puppies, which is in agreement with previous studies in dogs [18] and in humans [39]. Most likely, this is a consequence of uterine contractions during labor causing a physiological decrease in placental circulation with hyperlactatemia and acidosis at birth [40]. In another study, an initial lactate level of 6.7 mmol/L was reported in surviving neonates and 10 mmol/L in those who died in the first 24 h after VP [21]. In addition, a cut-off value for umbilical blood lactate of 5 mmol/L has been suggested to distinguish distressed from healthy puppies. Puppies that died within 48 h of birth had a mean lactate value of 12.2 mmol/L $\pm$ 6.7 SD [17]. Antończyk et al. [22] were able to confirm this correlation. In our study, umbilical lactate concentrations were available for three out of four puppies that died 4 days post parturition, and were 15.3, 23.3, and 20.1 mmol/L, respectively. Moreover, the duration of the active second stage of labor correlated significantly with the presence of fetal lactate [40]. In our study, the duration of the parturition was not always recorded, but puppies born at the end of the parturition did not have higher lactate concentrations in either umbilical blood or amniotic fluid. A very strong correlation between lactate concentrations in umbilical blood and amniotic fluid suggests that both fetal fluids can be used for the test, but higher concentrations in amniotic fluid must be considered. Lactate concentration did not correlate with relative growth rate except on day one.

In humans, higher blood glucose concentrations have been found in VP-delivered infants compared to EM-CS infants [41]. Similar to humans and in agreement with previous studies in dogs [12,18], significantly higher amniotic fluid glucose concentrations were found in VP puppies than in EM-CS puppies, but the concentrations reported by Groppetti et al. [18] were lower than in our study. On the contrary, Lucio et al. (2021) reported higher glucose concentrations in blood in fetal dystocia and EM-CS (after fetal or maternal dystocia) puppies when compared to eutocia puppies [19]. While some studies reported higher glucose concentrations in non-surviving compared to surviving neonates [21,22], others found that newborn puppies with glucose concentrations below 2.22 mmol/L usually had low Apgar scores and poor reflexes [12]. Balogh et al. (2018) showed that metabolites of carbohydrate and lipid metabolism in the bitch likely affect fetal concentrations and composition of fetal fluids [42], hence maternal hypoglycemia may affect neonates. Hypoglycemia is a reported syndrome affecting bitches presented for EM-CS and it is a particular risk in smaller dog breeds [42,43], probably due to consumption of glucose, which is reported to be normal or increased in at term dams and at the beginning of the dystocia [42,44]. These small EM-CS bitches often present after many hours in difficult labor as well as after or during episodes of dystocia where serum glucose is depleted. They are also often depressed and inappetent, which prevents them from correcting hypoglycemia. In contrast, one study reported that EM-CS dams had the highest glucose concentrations and the eutocia group had the lowest glucose concentrations, but the weight of the dams was not reported. In this study, a positive and significant correlation was found between maternal cortisol concentrations and glucose [19]. The reason for the lower glucose levels in EL-SC bitches could be that these bitches are often fasted before anesthesia, which would easily lead to lower serum glucose levels in their puppies. The EL-CS dams in our study were not fasted.

In our study, the umbilical blood glucose correlated moderately with amniotic fluid glucose concentrations and with Apgar scores at 5 and 15 min, but not with mortality in the first week. There was also no correlation with the growth rate of the puppies. These results suggest that neonatal hypoglycemia immediately after parturition is a common finding and should not be an immediate clinical concern. In addition, neonates in humans are reported to be much more resistant to hypoglycemia than adults [25]. These lower blood glucose levels early after birth have also been observed in dogs [45], foals [46], and calves [47], leading to the conclusion that they may represent an evolutionary adaptation to early life outside the womb.

In contrast to the previous results, where lower cortisol concentrations in puppies delivered by both types of CS were found [18], our results show the highest cortisol concentrations in puppies delivered by EM-CS and the lowest in those delivered by EL-CS. Again, this only partially supports data from human medicine indicating that vaginal delivery is more stressful for human neonates than CS. In humans, amniotic fluid cortisol concentrations increase abruptly during the last weeks of pregnancy due to activation of the hypothalamic–pituitary–adrenal axis, and the same can be expected in dogs [48]. Therefore, elevated cortisol levels are expected in at-term puppies (VP, EM-CS) compared to EL-CS puppies. Further stress to which the fetuses are subjected during dystocia results in higher cortisol levels in these puppies [19,49], suggesting that the stress of the CS procedure is indeed low. The study from Lucio et al. (2021) even showed that the type of dystocia influences the neonatal cortisol concentrations, with the highest concentrations in fetal dystocia puppies [19].

The correlation of cortisol concentrations in amniotic fluid and fetal urine is very strong, which is consistent with the results of a previous study, where concentrations in amniotic fluid and allantois were also strongly positively correlated. However, higher amniotic fluid but not allantoic cortisol concentrations were found in puppies not surviving 24 h after parturition [10] and lower concentrations were found in neonates with higher Apgar scores [49], which could not be confirmed in our study, because first, only one puppy died in the first 24 h and, second, puppies delivered by EL-CS had the lowest cortisol concentrations and the percentage of lower Apgar scores was the highest in this group. Puppies that died on the first, third, and fourth day (2 puppies) postpartum had amniotic cortisol concentrations of 13.3, 12.1, 2.0, and 9.2 mmol/L, respectively.

In our study, the amniotic fluid mean specific gravity in all parturition types was comparable to previous studies [50,51] which included CS births only. The utility of the amniotic fluid specific gravity as an indicator for neonatal survival was investigated by Fusi et al. who reported no significant differences in specific gravity of amniotic fluid in surviving and non-surviving puppies [51], which we can confirm.

There were no statistically significant differences among the three types of parturition regarding relative growth rate.

Because of various reasons, we had some missing values in the dataset, which is the main limitation of the study. Because of that, we were unable to establish or confirm already published cut-off values of cortisol, lactate, and glucose, which could be predictive of puppies' survival. Further, because of the non-invasive nature of the study, we were possibly evaluating the maternal blood in contrast to the studies where the jugular blood of the neonates was examined. We are also aware that other factors such as breed, size, age of the dam, her vaccination, and health status, as well as low birthweight and litter size [5,6,13,14,32] may influence the survival of the neonates and these factors need to be included in the statistical analysis in the future studies.

5. Conclusions

Research on canine pregnancy and its effects on puppies' vitality at birth, as well as a system for monitoring birth, are necessary to control the vitality of newborn dogs. The development of an easy-to-use monitoring system for at-risk puppies (e.g., orphaned, starved, low birthweight) would be desirable along with a simple Apgar evaluation for dog breeders to reduce the high neonatal mortality rate in kennels. It was shown here that measurement of urinary cortisol and amniotic fluid lactate are rapid tests that could serve as predictors of puppy vitality on the first day of life and help us identify puppies at risk. Our results showed that fetal and neonatal cortisol, lactate, and glucose concentration in puppies differed by the type of parturition. In summary, puppies born by EL-CS had the lowest lactate, glucose, and cortisol concentrations and the lowest Apgar score. However, the highest cortisol concentration was found in the EM-CS group, in which the Apgar score was lower than in the VP group, suggesting that optimal concentration rates for these three parameters should be determined to allow accurate assessment of neonatal vitality in dogs. Further studies are needed to determine whether these parameters can also be used as predictors of neonatal survival during the first weeks of life, when mortality is the highest.

Author Contributions: Conceptualization, T.P. and M.Z.P.; Data curation, A.S., S.R., B.L., T.P. and M.Z.P.; Formal analysis, T.K.; Funding acquisition, T.P. and M.Z.P.; Investigation, T.P., A.S., S.R., M.Z.P. and T.K.; Methodology, T.K. and B.L.; Resources, T.P. and M.Z.P.; Supervision, T.P. and M.Z.P.; Visualization, T.K.; Writing—original draft, T.P.; Writing—review and editing, A.S., S.R., T.K., B.L. and M.Z.P. All authors have read and agreed to the published version of the manuscript.

Funding: The authors acknowledge the financial support from the Slovenian Research Agency (research core funding No. P4-0053 and No. P4-0092).

Institutional Review Board Statement: The study was conducted in accordance with the guidelines of the Declaration of Helsinki and approved by the Institutional Commission for Animal Welfare, Veterinary Faculty, University of Ljubljana on 4 April 2017. They concluded that ethical review and approval were waived for this study, as all interventions on animals were performed for the purpose of treatment and not for experimental reasons.

Data Availability Statement: Data are available from the authors upon request.

Acknowledgments: We would like to thank Alenka Nemec Svete for her help in planning the analytics, colleagues Tina Roškar, Petra Fink and their team at MojVet for their help with collecting patients, Nina Cizej at Felivet for borrowing the iSTAT machine, ARRS Agency for the financing. We thank all the owners with their pets and their willingness to trust us. We would also like to thank Magi Casal for all her help and instructions during the study.

Conflicts of Interest: The authors declare no conflict of interest. The funders had no role in the design of the study; in the collection, analyses, or interpretation of data; in the writing of the manuscript; or in the decision to publish the results.

References

1. Vannucchi, C.I.; Silva, L.C.; Lúcio, C.F.; Regazzi, F.M.; Veiga, G.A.; Angrimani, D.S. Prenatal and neonatal adaptations with a focus on the respiratory system. *Reprod. Domest. Anim.* **2012**, *47*, 177–181. [CrossRef] [PubMed]
2. Gill, M.A. Perinatal and Late Neonatal Mortality in the Dog. Ph.D. Thesis, University of Sydney, Sydney, Australia, March 2001. Available online: https://core.ac.uk/download/pdf/41232423.pdf (accessed on 1 October 2021).
3. Münnich, A. The pathological newborn in small animals: The neonate is not a small adult. *Vet. Res. Commun.* **2008**, *32* (Suppl. S1), S81–S85. [CrossRef]
4. Mosier, J.E. Canine pediatrics: The neonate. In Proceedings of the 48th American Animal Hospital Association Annual Meeting, Montreal, Canada, 11 March 1981; pp. 339–347.
5. Moon, P.F.; Erb, H.N.; Ludders, J.W.; Gleed, R.D.; Pascoe, P.J. Perioperative risk factors for puppies delivered by cesarean section in the United States and Canada. *J. Am. Anim. Hosp. Assoc.* **2000**, *36*, 359–368. [CrossRef] [PubMed]
6. Indrebø, A.; Trangerud, C.; Moe, L. Canine neonatal mortality in four large breeds. *Acta Vet. Scand.* **2007**, *49*, S2. [CrossRef]
7. Tønnessen, R.; Borge, K.S.; Nødtvedt, A.; Indrebø, A. Canine perinatal mortality: A cohort study of 224 breeds. *Theriogenology* **2012**, *77*, 1788–1801. [CrossRef]
8. Münnich, A.; Küchenmeister, U. Causes, diagnosis and therapy of common diseases in neonatal puppies in the first days of life: Cornerstones of practical approach. *Reprod. Domest. Anim.* **2014**, *49* (Suppl. S2), 64–74. [CrossRef]
9. England, G.C.W. Care of the Neonate and Fading Pups. In *Textbook of Veterinary Internal Medicine*, 7th ed.; Ettinger, S.J., Feldman, E.C., Eds.; Saunders: St. Louis, MO, USA, 2010; Volume 2, pp. 1949–1954.
10. Bolis, B.; Prandi, A.; Rota, A.; Faustini, M.; Veronesi, M.C. Cortisol fetal fluid concentrations in term pregnancy of small-sized purebred dogs and its preliminary relation to first 24 hours survival of newborns. *Theriogenology* **2017**, *88*, 264–269. [CrossRef]
11. Roos, J.; Maenhoudt, C.; Zilberstein, L.; Mir, F.; Borges, P.; Furthner, E.; Niewiadomska, Z.; Nudelmann, N.; Fontbonne, A. Neonatal puppy survival after planned caesarean section in the bitch using aglepristone as a primer: A retrospective study on 74 cases. *Reprod. Domest. Anim.* **2018**, *53* (Suppl. S3), 85–95. [CrossRef]
12. Vassalo, F.G.; Simões, C.R.; Sudano, M.J.; Prestes, N.C.; Lopes, M.D.; Chiacchio, S.B.; Lourenço, M.L. Topics in the routine assessment of newborn puppy viability. *Top. Companion. Anim. Med.* **2015**, *30*, 16–21. [CrossRef]
13. Veronesi, M.C.; Panzani, S.; Faustini, M.; Rota, A. An Apgar scoring system for routine assessment of newborn puppy viability and short-term survival prognosis. *Theriogenology* **2009**, *72*, 401–407. [CrossRef]
14. Veronesi, M.C. Assessment of canine neonatal viability-the Apgar score. *Reprod. Domest. Anim.* **2016**, *51* (Suppl. S1), 46–50. [CrossRef] [PubMed]
15. Armstrong, L.; Stenson, B.J. Use of umbilical cord blood gas analysis in the assessment of the newborn. *Arch. Dis. Child. Fetal. Neonatal.* **2007**, *92*, F430–F434. [CrossRef] [PubMed]
16. Lúcio, C.F.; Silva, L.C.; Rodrigues, J.A.; Veiga, G.A.; Vannucchi, C.I. Acid-base changes in canine neonates following normal birth or dystocia. *Reprod. Domest. Anim.* **2009**, *44* (Suppl. S2), 208–210. [CrossRef] [PubMed]

17. Groppetti, D.; Pecile, A.; Del Carro, A.P.; Copley, K.; Minero, M.; Cremonesi, F. Evaluation of newborn canine viability by means of umbilical vein lactate measurement, Apgar score and uterine tocodynamometry. *Theriogenology* **2010**, *74*, 1187–1196. [CrossRef] [PubMed]

18. Groppetti, D.; Martino, P.A.; Ravasio, G.; Bronzo, V.; Pecile, A. Prognostic potential of amniotic fluid analysis at birth on canine neonatal outcomes. *Vet. J.* **2015**, *206*, 423–425. [CrossRef] [PubMed]

19. Lúcio, C.F.; Silva, L.C.G.; Vannucchi, C.I. Perinatal cortisol and blood glucose concentrations in bitches and neonatal puppies: Effects of mode of whelping. *Domest. Anim. Endocrinol.* **2021**, *74*, 106483. [CrossRef]

20. Pang, D.S.; Boysen, S. Lactate in veterinary critical care: Pathophysiology and management. *J. Am. Anim. Hosp. Assoc.* **2007**, *43*, 270–279. [CrossRef]

21. Castagnetti, C.; Cunto, M.; Bini, C.; Mariella, J.; Capolongo, S.; Zambelli, D. Time-dependent changes and prognostic value of lactatemia during the first 24 h of life in brachycephalic newborn dogs. *Theriogenology* **2017**, *94*, 100–104. [CrossRef]

22. Antończyk, A.; Ochota, M.; Niżański, W. Umbilical cord blood gas parameters and Apgar scoring in assessment of new-born dogs delivered by cesarean section. *Animals* **2021**, *11*, 685. [CrossRef]

23. Bolis, B.; Scarpa, P.; Rota, A.; Vitiello, T.; Veronesi, M.C. Association of amniotic uric acid, glucose, lactate and creatinine concentrations and lactate/creatinine ratio with newborn survival in small-sized dogs-prelimi-nary results. *Acta Vet. Hung.* **2018**, *66*, 125–136. [CrossRef]

24. Dysart, K.C. Neonatal hypoglycemia. In *Merck Manual Professional Version*; Merck Sharp & Dohme Corp.: Kenilworth, NJ, USA, 2021; pp. e1–e5. Available online: https://www.merckmanuals.com/professional/pediatrics/metabolic,-electrolyte,-and-toxic-disorders-in-neonates/neonatal-hypoglycemia (accessed on 8 October 2021).

25. Thompson-Branch, A.; Havranek, T. Neonatal Hypoglycemia. *Pediatr. Rev.* **2017**, *38*, 147–157. [CrossRef]

26. Smith, F.O. Challenges in small animal parturition: Timing elective and emergency cesarian sections. *Theriogenology* **2007**, *68*, 348–353. [CrossRef] [PubMed]

27. Traas, A.M. Surgical management of canine and feline dystocia. *Theriogenology* **2008**, *70*, 337–342. [CrossRef] [PubMed]

28. Meloni, T. Some Perinatal Endocrine and Morphological Aspects of Canine Species. Ph.D. Thesis, University of Milan, Milan, Italy, 2015. Available online: https://air.unimi.it/retrieve/handle/2434/265723/370688/phd_unimi_R09461.pdf (accessed on 1 October 2021).

29. Batista, M.; Moreno, C.; Vilar, J.; Golding, M.; Brito, C.; Santana, M.; Alamo, D. Neonatal viability evaluation by Apgar score in puppies delivered by cesarean section in two brachycephalic breeds (English and French bulldog). *Anim. Reprod. Sci.* **2014**, *146*, 218–226. [CrossRef] [PubMed]

30. R Core Team. *R: A Language and Environment for Statistical Computing*; R Foundation for Statistical Computing: Vienna, Austria, 2021; Available online: https://www.R-project.org/ (accessed on 16 January 2022).

31. Zakošek Pipan, M.; Švara, T.; Zdovc, I.; Papić, B.; Avberšek, J.; Kušar, D.; Mrkun, J. Staphylococcus pseudintermedius septicemia in puppies after elective cesarean section: Confirmed transmission via dam's milk. *BMC Vet. Res.* **2019**, *15*, 41. [CrossRef] [PubMed]

32. Fusi, J.; Faustini, M.; Bolis, B.; Veronesi, M.C. Apgar score or birthweight in Chihuahua dogs born by elective Caesarean section: Which is the best predictor of the survival at 24 h after birth? *Acta. Vet. Scand.* **2020**, *62*, 39. [CrossRef]

33. Mila, H.; Grellet, A.; Feugier, A.; Chastant-Maillard, S. Differential impact of birth weight and early growth on neonatal mortality in puppies. *J. Anim. Sci.* **2015**, *93*, 4436–4442. [CrossRef]

34. Mugnier, A.; Mila, H.; Guiraud, F.; Brévaux, J.; Lecarpentier, M.; Martinez, C.; Mariani, C.; Adib-Lesaux, A.; Chastant-Maillard, S.; Saegerman, C.; et al. Birth weight as a risk factor for neonatal mortality: Breed-specific approach to identify at-risk puppies. *Prev. Vet. Med.* **2019**, *171*, 104746. [CrossRef]

35. Doebeli, A.; Michel, E.; Bettschart, R.; Hartnack, S.; Reichler, I.M. Apgar score after induction of anesthesia for canine cesarean section with alfaxalone versus propofol. *Theriogenology* **2013**, *80*, 850–854. [CrossRef]

36. Titkova, R.; Fialkovicova, M.; Karasova, M.; Hajurka, J. Puppy Apgar scores after vaginal delivery and caesarean section. *Vet. Med.* **2017**, *62*, 488–492. [CrossRef]

37. Vilar, J.M.; Batista, M.; Pérez, R.; Zagorskaia, A.; Jouanisson, E.; Díaz-Bertrana, L.; Rosales, S. Comparison of 3 anesthetic protocols for the elective cesarean-section in the dog: Effects on the bitch and the newborn puppies. *Anim. Reprod. Sci.* **2018**, *190*, 53–62. [CrossRef] [PubMed]

38. Siena, G.; Milani, C. Usefulness of maternal and fetal parameters for the prediction of parturition date in dogs. *Animals* **2021**, *11*, 878. [CrossRef] [PubMed]

39. Zaigham, M.; Helfer, S.; Kristensen, K.H.; Isberg, P.E.; Wiberg, N. Maternal arterial blood gas values during delivery: Effect of mode of delivery, maternal characteristics, obstetric interventions and correlation to fetal umbilical cord blood. *Acta Obstet. Gynecol. Scand.* **2020**, *99*, 1674–1681. [CrossRef]

40. Bakker, P.C.; Kurver, P.H.; Kuik, D.J.; Van Geijn, H.P. Elevated uterine activity increases the risk of fetal acidosis at birth. *Am. J. Obstet. Gynecol.* **2007**, *196*, 313.e1-6. [CrossRef] [PubMed]

41. Marom, R.; Dollberg, S.; Mimouni, F.B.; Berger, I.; Mordechayev, N.; Ochshorn, Y.; Mandel, D. Neonatal blood glucose concentra-tions in caesarean and vaginally delivered term infants. *Acta Paediatr.* **2010**, *99*, 1474–1477. [CrossRef] [PubMed]

42. Balogh, O.; Bruckmaier, R.; Keller, S.; Reichler, I.M. Effect of maternal metabolism on fetal supply: Glucose, non-esterified fatty acids and beta-hydroxybutyrate concentrations in canine maternal serum and fetal fluids at term pregnancy. *Anim. Reprod. Sci.* **2018**, *193*, 209–216. [CrossRef]

43. Newfield, A. Reproductive Emergencies. In *Small Animal Emergency and Critical Care for Veterinary Technicians, 4th ed*; Battaglia, A., Steele, A.M., Eds.; Elsevier Health Sciences: St. Louis, MO, USA, 2020; pp. 271–282.

44. Bergström, A.; Fransson, B.; Lagerstedt, A.S.; Olsson, K. Primary uterine inertia in 27 bitches: Aetiology and treatment. *J. Small Anim. Pract.* **2006**, *47*, 456–460. [CrossRef]

45. Mila, H.; Grellet, A.; Delebarre, M.; Mariani, C.; Feugier, A.; Chastant-Maillard, S. Monitoring of the newborn dog and prediction of neonatal mortality. *Prev. Vet. Med.* **2017**, *143*, 11–20. [CrossRef]

46. Aoki, T.; Ishii, M. Hematologic and biochemical profiles in peripartum mares and neonatal foals (heavy draft horse). *J. Eq. Vet. Sci.* **2012**, *32*, 170–176. [CrossRef]

47. Knowles, T.G.; Edwards, J.E.; Bazeley, K.J.; Brown, S.N.; Butterworth, A.; Warriss, P.D. Changes in the blood biochemical and haematological profile of neonatal calves with age. *Vet. Rec.* **2000**, *147*, 593–598. [CrossRef]

48. Mastorakos, G.; Ilias, I. Maternal and fetal hypothalamic-pituitary-adrenal axes during pregnancy and postpartum. *Ann. N. Y. Acad. Sci.* **2003**, *997*, 136–149. [CrossRef] [PubMed]

49. Fusi, J.; Carluccio, A.; Peric, T.; Faustini, M.; Prandi, A.; Veronesi, M.C. Effect of delivery by emergency or elective Cesarean section on nitric oxide metabolites and cortisol amniotic concentrations in at term normal newborn dogs: Preliminary results. *Animals* **2021**, *11*, 713. [CrossRef] [PubMed]

50. Balogh, O.; Roch, M.; Keller, S.; Michel, E.; Reichler, I.M. The use of semi-quantitative tests at Cesarean section delivery for the differentiation of canine fetal fluids from maternal urine on the basis of biochemical characteristics. *Theriogenology* **2017**, *88*, 174–182. [CrossRef] [PubMed]

51. Fusi, J.; Bolis, B.; Probo, M.; Faustini, M.; Carluccio, A.; Veronesi, M.C. Clinical trial on the usefulness of on-site evaluation of canine fetal fluids by reagent test strip in puppies at elective Caesarean section. *Biology* **2021**, *11*, 38. [CrossRef]

Article

Early Blood Analysis and Gas Exchange Monitoring in the Canine Neonate: Effect of Dam's Size and Birth Order

Brenda Reyes-Sotelo [1,2], Asahi Ogi [3], Patricia Mora-Medina [4], Chiara Mariti [3], Adriana Olmos-Hernández [5], Ismael Hernández-Ávalos [4], Adriana Domínguez-Oliva [2,*], Marcelino Evodio Rosas [4], Antonio Verduzco-Mendoza [2,5] and Angelo Gazzano [3,*]

[1] Master in Science Programme "Maestria en Ciencias Agropecuarias", Xochimilco Campus, Universidad Autónoma Metropolitana, Ciudad de México 04960, Mexico; breyess_20@yahoo.com.mx
[2] Neurophysiology, Behavior and Animal Welfare Assessment, Xochimilco Campus, Universidad Autónoma Metropolitana, Ciudad de México 04960, Mexico; cnrverduzco@hotmail.com
[3] Department of Veterinary Sciences, University of Pisa, 56124 Pisa, Italy; asahi.ogi@vet.unipi.it (A.O.); chiara.mariti@unipi.it (C.M.)
[4] Facultad de Estudios Superiores Cuautitlán, Universidad Nacional Autónoma de México (UNAM), Cuautitlán 54714, Mexico; mormed2001@yahoo.com.mx (P.M.-M.); mvziha@hotmail.com (I.H.-Á.); rosasgme@hotmail.com (M.E.R.)
[5] Division of Biotechnology—Bioterio and Experimental Surgery, Instituto Nacional de Rehabilitación—Luis Guillermo Ibarra Ibarra (INR-LGII), Ciudad de México 14389, Mexico; adrianaolmos05@gmail.com
* Correspondence: mvz.freena@gmail.com (A.D.-O.); angelo.gazzano@unipi.it (A.G.)

Simple Summary: The complications that are observed during parturition are events that affect the vitality of the newborn and can also compromise their health by predisposing them to fetal hypoxia, increasing newborn mortality. Blood gas analysis to measure the main biomarkers associated with hypoxia evaluates the physiological and metabolic alterations derived from this state, and these could help identify if said markers respond to maternal or neonatal causes. This study aimed to assess the effect of the dam's size, the birth order, and the presentation of blood gas alterations. Recognizing if these elements are intertwined may enhance newborns' life expectancy by enabling the planning of a perinatal protocol to avoid serious metabolic consequences that are derived from prolonged hypoxia.

Abstract: In canines, size at birth is determined by the dam's weight, which would probably affect the newborn's viability due to litter size and birth order. Fetal hypoxia causes distress and acidemia. Identifying physiological blood alterations in the puppy during the first minute of life through the blood gas exchange of the umbilical cord could determine the puppy's risk of suffering asphyxiation during labor. This study aimed to evaluate the effect of the birth order and dam's size during spontaneous labor and the alterations during the first minute of life. The results indicate that the dam's size and the birth order have considerable physiological and metabolic effects in the puppies, mainly in birth order 1 (BO1) in small-size dogs, while in the medium size, the last puppy presented more alterations, probably because of a prolonged whelping which could have fostered hypoxic processes and death. Likewise, with large-size dogs, intrapartum asphyxiation processes were registered during the first minute of life in any birth order.

Keywords: puppy welfare; hypoxia; size; physiological blood profile; asphyxia; whelping

Citation: Reyes-Sotelo, B.; Ogi, A.; Mora-Medina, P.; Mariti, C.; Olmos-Hernández, A.; Hernández-Ávalos, I.; Domínguez-Oliva, A.; Rosas, M.E.; Verduzco-Mendoza, A.; Gazzano, A. Early Blood Analysis and Gas Exchange Monitoring in the Canine Neonate: Effect of Dam's Size and Birth Order. *Animals* **2022**, *12*, 1508. https://doi.org/10.3390/ani12121508

Academic Editor: Maria Cristina Veronesi

Received: 3 May 2022
Accepted: 6 June 2022
Published: 9 June 2022

Publisher's Note: MDPI stays neutral with regard to jurisdictional claims in published maps and institutional affiliations.

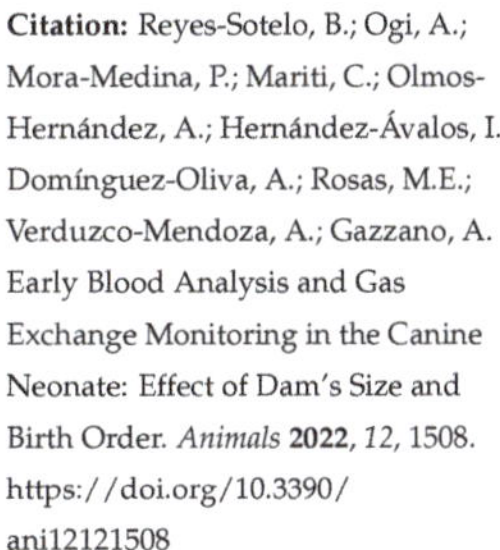

Copyright: © 2022 by the authors. Licensee MDPI, Basel, Switzerland. This article is an open access article distributed under the terms and conditions of the Creative Commons Attribution (CC BY) license (https://creativecommons.org/licenses/by/4.0/).

1. Introduction

Birth represents a great challenge for neonatal survival in a new environment with different conditions. The transition from fetal to neonatal life involves an efficient multi-systemic adaptation where the most critical change is related to the start of breathing [1,2]. Parturition induces significant physiological and adaptive events for mammal species due to the cardiovascular, respiratory, and thermoregulation changes the newborn will confront in the postnatal period [3]. During parturition, transient hypoxia periods are present in the fetus due to the uterine contractions and the mechanical pressure inherent to birth, causing a decrease in blood flow and placental perfusion, compromising gas interchange in the fetus [4]. The state of transitory fetal hypoxia that the pups endure during the perinatal period is part of the natural process of birth [5]. Recognizing a hypoxic process in the newborn is critical for its survival; however, the absence of evident signs in pups, such as hyperventilation, limits its identification [6].

To obtain the physiological stability profile of the newborn, parameter assessment related to the morphological characteristics of the umbilical cord, such as edematous, hemorrhagic, congested or ruptured cord [7] and blood metabolites is suggested. Lactate plays a central role in diagnosing neonatal and fetal distress when the oxygen supply is interrupted, tissular oxygen deprivation develops, and acids start to accumulate, developing a state of acidemia [8].

In the beginning, James and collaborators [9] found that evaluating blood gases in the umbilical cord added fetal hypoxic stress data and provided relevant information during the perinatal period. Moreover, they give a panorama of the acid-base state of the neonate at the moment of birth, when the female blood circulation is interrupted by the occlusion or tear of the umbilical cord [10,11]. Recently, gasometric assessment in the field of neonatal veterinary is an important tool to assess newborn health. The umbilical cord blood analysis allows the evaluation of the concentrations of glucose, lactate, partial pressure of oxygen (pO_2), partial pressure of carbon dioxide (pCO_2), pH, hematocrit, sodium, potassium, and ionized calcium [12]. The blood evaluation of the umbilical cord through gasometry allows the obtention of a physiometabolic profile to guarantee neonatal health [5].

A mature fetus can have abundant reserves of glycogen, making him more tolerant to intermittent periods of hypoxia. However, it is considered that the morphological variability in dogs could affect fetal development [2,13–15]. For newborn puppies, the presence of other factors, such as maternal care [16,17], uterine inertia [18,19], and prolonged labor predisposes them to hypoxia and Type II stillborn (SB) [20–23]. Type II stillborn (SB) is considered to be an effect that is associated with the oxytocinergic system of the dam and its regulation. Hypoxia and SB could have a negative effect on the postnatal adaptation or survival of the neonate [23,24] and it can be reflected in an altered blood physiological profile.

The challenges of the parturition process, along with its risk factors might determine the proportion of the liveborn pups (LBP) in comparison to the stillbirths (SB), as well as the viability of the former [25–27]. Therefore, the morphological variability in canines could be associated with the dam's size and weight [28], and these factors might affect the pups' weight and physiological state at birth.

In pigs, it has been reported that the integration of physiological, neurological, and behavioral factors at birth to recognize hypoxia and its repercussions has been of great importance because of the high incidence of stillborn piglets [29]. According to a study on dogs, this is also related to litter size and birth order, mainly affecting those belonging to the last quarter of the litter [26]. The major risk that has been identified is placental insufficiency and a greater risk of suffering intrauterine hypoxia [27,30,31]. Because of the aforesaid, birth order has been highlighted in other species, such as pigs, as being an indicator of survival [32–34]; however, in canines, research is still limited and has not included other risk factors, such as the dam's size, which is related in this species with the size of the litter. This study aims to evaluate the effect of the dam's size and birth order on the physiological responses of the canine neonate during the first minute of life.

2. Materials and Methods

2.1. Infrastructure, Animals, and Management

2.1.1. Infrastructure

A network of 5 veterinarian clinics located in Mexico City was gathered to recruit pregnant dogs from January to June 2019. A prenatal control was held from the 25th day of pregnancy to the first 24 h post-birth.

2.1.2. Study Population

A total of 58 young multiparous female dogs (2 to 4 births) were recruited. The inclusion criteria were, clinically healthy female dogs, updated vaccination/deworming schedule, fed with commercial formulas, no history of reproductive problems, and radiographic and ultrasound evaluation that ensured eutocic delivery. The exclusion criteria included, primiparous females, a background of dystocia or uterine infections, the presence of type I stillbirths, fetal malformations, the use of delivery inducers or accelerators, females with 8 and 9 body condition scores (obese) according to the WSAVA scale [35], and animals with extremely aggressive behavior. Brachycephalic and gigantic breeds were excluded from the study due to their high incidence of dystocia [22]. The 58 pregnant females were classified in 3 size categories according to their weight: small (<10 kg, n = 18); medium (11–25 kg, n = 20); and large (26–45 kg, n = 20) [15,36,37].

2.1.3. Clinical History

The females' clinical history included the following data: age, weight, feeding, prevention medicine conditions, and a description of the environment in which they inhabited. All the information was recorded in the Qvet® Ed. Professional 2016 database for veterinary clinics.

2.1.4. Pregnancy Diagnosis

The gestation diagnosis was confirmed between the 24 and 28 days post-insemination for each female. The fetal structures and cardiac activity were detected inside the gestational sacs. Using Logiq 400 MD (General Electric®, Yokohama, Japan) ultrasound, probable birth dates were established with a 3.5 MHz convex transductor. The monitoring of fetal maturation and vitality was performed in the 40–50 days of progression when the fetal structure is completely defined, which allows the early identification of pyometra cases, type I stillbirths, and malformations. Afterward, a simple abdomen radiographic study was performed after day 45 of gestation once the bone calcification of the fetus had been reached to discard early fetal-dam dystocia, as well as to provide evidence of cephalopelvic disproportion, which would imply the need for cesarean-section [31], another motive to exclude the dog from the study. On day 60 of pregnancy, the female dogs were assessed by ultrasound to corroborate heartbeat and biparietal fetal diameter. The delivery monitoring was performed through a Sino-Hero® vital signs monitor, model S80Vet (Guangdong, China), to evaluate the dam's physiological parameters. Clinical signs in the intrapartum were observed, such as anorexia, distress, and nesting behavior.

2.2. Puppies and Evaluated Variables during the First Minute of Life

The number of puppies, considering those born alive and dead (stillbirths type II and antepartum deaths—Type I), according to the female dog's size, birth order, and physiological blood profile (of the sample taken from the umbilical cord) was registered for every birth.

However, a total of 310 live-born puppies (LBP) were studied in this stage, distributed in 3 categories according to the dam's size: small n = 75, medium n = 102, large n = 133, considering the following criteria:

(a) Live-born puppies (LBP): neonates who presented breathing and heart frequency during the first minute of life;

(b) Stillborn Puppies (SBP): fetuses classified as dead intrapartum (type II) presented the same appearance as their litter partners, except for the absence of breathing;

(c) Antepartum deaths (type I): those with brown-grayish discoloration due to the initial state of mummification; and the most advanced cases in a clear dehydration state and with hair loss.

The fetuses (type I) and Stillborn (type II) were excluded from the study.

2.2.1. Blood Physiological Profile

Blood Sampling

Veterinarian staff took blood samples from LBP immediately after birth in less than 10 s. Hemostasis was performed in the distal portion of the umbilical cord with a clamp. Subsequently, the index finger and thumb were placed on the base of the abdominal insertion of the umbilical cord and moved up to three centimeters from its distal portion. Once the turgidity and dilatation of the vessel were observed, blood sampling was obtained from the vein. The identification of the umbilical vein was according to its superficial location and appearance, where the vein can be recognized as a thick cyanotic blood vessel, while the artery is deeper with a more intense-bright red color and a smaller caliber. A total of 0.3 mL of venous blood from the umbilical cord was obtained with a tuberculin syringe with a 26G needle that was impregnated with lithium heparin to prevent coagulation and alteration of the sample values. A volume of 150 μL was processed through the analyzer of blood critical variables GEM Premier® 3000 (Instrumentation Laboratory Diagnostics; Lexington, KY, USA/Milano, Italy) to obtain the values of metabolites; glucose (mg/dL); lactate (mg/dL); blood gases pCO_2 (mmHg); pO_2 (mmHg); acid-base balance pH; HCO_3^- (mmol/L) and base excess (BE) (mEq/L); Ca^{2+} (mmol/L); and hematocrit (Htc %). The physiological blood profile was evaluated for all the LBP. The pups did not receive any resuscitation before blood sampling so as not to interfere with gas exchange.

2.2.2. Birth Order

The birth order was registered for every puppy (LBP), obtaining a classification by the dam's size. For the small size, the total number of puppies by litter was 5 (BO1–BO5); for the medium size, 7 puppies (BO1–BO7); and for the large-size, 9 puppies (BO1–BO9).

2.3. Statistical Analysis

Data were organized in means ± SE. The effect of the dam's size category (small, medium, and large) and birth order, as well as the interaction between these factors were obtained using variance analysis (ANOVA) by the GLM procedure (General Linear Model) [38] under the following model:

$\text{Metabolites}_{ijk} = \mu + T_i + BO_i(T_i BO_i) + e_{ijk}$

Metabolites = pH, pCO_2, pO_2, glucose, Ca^{2+}, lactate, hematocrit, HCO_3^-, EB

μ = General mean

T_i = Fixed effect size (small, medium, large)

BO_i = birth order 1, 2, 3, 4 . . . e = error

The multiple comparison of means was performed with the Tukey test.

A Pearson rank test was used to establish the correlation between physiologic blood variables (pH, pCO_2, pO_2, glucose, lactate, HCO_3^-, Ca^{2+}, hematocrit) and the dam's size.

2.4. Ethics Note

The study was held on private property female dogs, and informed consent from every owner was obtained to gather the data. During the study, all the animals were managed according to the guidelines and rules of the Mexican Official Norm NOM-062-ZOO-1999, technical specifications for the production, care and use of laboratory animals, besides those of the competence of the field of applied ethologic studies [39].

The experimental protocol with number CAMCA.32.18 was approved by the Commission of the Master's in Agriculture and Livestock Sciences of the Universidad Autonoma Metropolitana, Xochimilco, Mexico City.

3. Results

The results of this project included the variable of birth order (BO) in female dogs of different sizes and its relation to the litter size. The BO was classified according to the size as follows: for small-size bitches (BO1–BO5), medium-size (BO1–BO7), and large-size (BO1–BO9).

3.1. Physiological Parameters According to Birth Order in Canine Neonates (LBP) from Small-Size Female Dogs

Table 1 shows that, regarding the acid-base and energetic balance, the puppies did not present significant differences in lactate ($p = 0.12$); glucose ($p = 0.20$); pH ($p = 0.53$); HCO_3^- ($p = 0.35$); and BE ($p = 0.16$) levels in the blood, among the groups (BO1, BO2, BO3, BO4, BO5). However, in Table 2, it is observed that blood gas exchange levels are different among groups. The pO_2 levels differed between BO1 and BO2 ($p = 0.04$). Moreover, for the pCO_2 levels, the group BO1 presented the highest concentration, while the group BO3 presented the minimum ($p = 0.02$). In the results of Table 3 for calcium, the puppies born BO1 presented the highest levels and a statistically significant difference ($p = 0.005$) in comparison to BO2, BO3 and BO5, this last group being the one which registered the lowest percentage in its levels ($49.25 \pm 1.77\%$) (Table 3).

3.2. Physiological Parameters According to Birth Order in Live-Born Canine Puppies from Medium-Size Female Dogs

In Table 1, in medium-size female dogs with the largest litter size (BO1–BO7), it is observed that there is no significant difference in birth order in the levels of lactate and glucose ($p = 0.49$). For the case of acid-base balance, in Table 1, it was observed that pH did not show any significant difference among the groups ($p = 0.65$) or pCO_2 ($p = 0.52$) (Table 2). It was only observed that pO_2 values were statistically different between BO1 (10.25 ± 0.75 mmHg) and BO6 (12.98 ± 1.09 mmHg) ($p = 0.02$) with an increase in the concentrations of pCO_2 in BO1 (74.96 ± 2.6 mmHg). On the other hand, only BE showed differences between BO4 (-9.37 ± 0.78 mEq/L) and BO7 (-13.70 ± 1.90 mEq/L) ($p = 0.03$) (Table 3).

3.3. Physiological Parameters According to Birth Order in Live-Born Canine Puppies from Large-Size Female Dogs

In Table 1, no statistical changes were found regarding lactate levels among the canine neonates from large-size female dogs considering birth order (BO1–BO9) ($p = 0.41$). However, the canine neonates presented hypoglycemia immediately after the birth, except for group BO8 (82.16 ± 6.86 mg/dL), considering normal reference ranks, which reflects a drastic decrease in fetal circulation as a consequence of delivery, reducing the hepatic glycogen rapidly, thus reflecting inefficient glucose homeostasis in the newborn [40].

It is important to point out that, for blood pH, no significant difference was found among the groups ($p = 0.33$). In Table 2, the neonates that presented more alterations in pO_2 were the puppies of BO4 (9.30 ± 0.71 mmHg) compared to BO7 (11.75 ± 0.90 mmHg) with a statistically significant difference ($p = 0.03$); likewise, BO4 presented a higher concentration of pCO_2 (78.93 ± 2.48 mmHg), and the lowest was obtained by group BO8 (68.83 ± 4.08 mmHg) ($p = 0.03$).

3.4. Physiological Parameters According to Birth Order in Live-Born Canine Puppies from Female Dogs of Different Sizes

The comparison of puppies with the same birth order was evaluated with the varying size of the dam (small, medium, and large). However, it must be highlighted that it was not possible to compare all the groups due to the number of puppies within the litters.

Table 1 shows that, for lactate metabolite, there were significant differences in BO2, BO3, BO4, and BO5 ($p < 0.05$), while for glucose, only BO3 had statistical differences of $p < 0.05$ among the groups because glucose presented a minimum value of 62.17 ± 4.21 mg/dL in the large size.

Table 1. Mean and standard error of blood physiological profile of live-born canine puppies according to birth order and dam's size.

Blood Traits	Dam's Size	BO 1 (Mean ± SE)	BO 2 (Mean ± SE)	BO 3 (Mean ± SE)	BO 4 (Mean ± SE)	BO 5 (Mean ± SE)	BO 6 (Mean ± SE)	BO 7 (Mean ± SE)	BO 8 (Mean ± SE)	BO 9 (Mean ± SE)	Value of p
Lactate (mg/dL)	Small $n = 75$	8.75 ± 0.44 [a,1]	7.87 ± 0.39 [a,2]	8.06 ± 0.41 [a,2]	7.99 ± 0.50 [a,2]	7.57 ± 0.73 [a,2]	...	...	...	...	0.12
	Medium $n = 102$	9.49 0.38 [a,1]	9.59 ± 0.39 [a,1]	10.04 ± 0.37 [a,1]	9.19 ± 0.41 [a,1,2]	9.90 ± 0.42 [a,1]	8.89 ± 0.55 [a,1]	10.00 ± 1.016 [a,1]	...	...	0.49
	Large $n = 133$	9.77 ± 0.32 [a,1]	10.04 ±0.34 [a,1]	10.44 ± 0.36 [a,1]	10.05 ± 0.36 [a,1]	9.55 ± 0.355 [a,1]	9.50 ± 0.377 [a,1]	10.45 ± 0.45 [a,1]	10.66 ± 0.58 [a]	10.44 ± 0.84 [a]	0.41
	Value of p	$p > 0.05$	$p < 0.05$	$p < 0.05$	$p < 0.05$	$p < 0.05$	$p > 0.05$	$p > 0.05$			
Glucose (mg/dL)	Small $n = 75$	66.21 ± 5.14 [a,1]	75.69 ± 4.64 [a,1]	70.85 ± 4.88 [a,1,2]	74.92 ± 5.90 [a,1]	78.49 ± 8.54 [a,1]	...	...	...	...	0.20
	Medium $n = 102$	70.47 ± 4.44 [a,1]	65.35 ± 4.60 [a,1]	74.18 ± 4.40 [a,1]	71.52 ± 4.89 [a,1]	69.18 ± 5.00 [a,1]	78.84 ± 6.46 [a,1]	54.00 ± 11.88 [a,1]	...	...	0.49
	Large $n = 133$	74.49 ± 3.81 [a,1]	65.59 ± 4.08 [b,1]	62.17 ± 4.21 [b,2]	65.53 ± 4.23 [b,1]	71.77 ± 4.16 [a,b,1]	72.16 ± 4.41 [a,b,1]	74.29 ± 5.33 [a,b,1]	82.16 ± 6.86 [a]	74.00 ± 9.91 [a,b]	0.02
	Value of p	$p > 0.05$	$p > 0.05$	$p < 0.05$	$p > 0.05$	$p > 0.05$	$p > 0.05$	$p > 0.05$			
pH	Small $n = 75$	7.05 ± 0.03 [a,1]	7.08 ± 0.03 [a,1]	7.10 ± 0.03 [a,1]	7.08 ± 0.03 [a,1]	7.05 ± 0.05 [a,1]	...	...	..	...	0.53
	Medium $n = 102$	7.08 ± 0.02 [a,1]	7.068 ± 0.03 [a,1]	7.06 ± 0.029 [a,1]	7.06 ± 0.03 [a,1]	7.08 ± 0.03 [a,1]	7.07 ±0.04 [a,1]	6.97 ± 0.07 [a,1]	...	...	0.65
	Large $n = 133$	7.10± 0.02 [a,1]	7.07 ± 0.02 [a,1]	7.04 ± 0.02 [a,1]	7.04 ± 0.02 [a,1]	7.03 ± 0.02 [a,1]	7.07 ± 0.29 [a,1]	7.05 ± 0.03 [a,1]	7.12 ± 0.04 [a]	7.14 ± 0.06 [a]	0.33
	Value of p	$p > 0.05$	$p > 0.05$	$p > 0.05$	$p > 0.05$	$p > 0.05$	$p > 0.05$	$p > 0.05$			
HCO$_3^-$ (mmol/L)	Small $n = 75$	19.42 ± 0.49 [a,1]	20.36 ± 0.44 [a,1]	19.97 ± 0.47 [a,1]	20.18 ± 0.56 [a,1]	19.61 ± 0.82 [a,1]	...	...	...	...	0.35
	Medium $n = 102$	19.77 ± 0.42 [a,1]	19.71 ± 0.44 [a,1,2]	19.75 ± 0.42 [a,1]	19.81 ± 0.47 [a,1]	19.21 ± 0.48 [a,1]	19.78 ± 0.62 [a,1]	18.75 ± 1.14 [a,1]	...	...	0.61
	Large $n = 133$	19.64 ± 0.36 [a,1]	18.94 ± 0.39 [a,2]	19.20 ± 0.40 [a,1]	18.75 ± 0.40 [a,2]	19.27 ± 0.40 [a,1]	19.12 ± 0.42 [a,1]	18.86 ± 0.51 [a,1]	19.01 ± 0.66 [a]	19.00 ± 0.95 [a]	0.03
	Value of p	$p > 0.05$	$p < 0.05$	$p > 0.05$	$p < 0.05$	$p > 0.05$	$p > 0.05$	$p > 0.05$			
BE (mEq/L)	Small $n = 75$	−10.23 ± 0.82 [a,1]	−9.58 ± 0.74 [a,1]	−9.09 ± 0.78 [a,2]	−9.15 ± 0.94 [a,2]	−9.32 ± 1.37 [a,1]	...	...	...	...	0.16
	Medium $n = 102$	−9.90 ± 0.71 [a,b,1]	−10.91 ± 0.73 [a,b,1]	−10.95 ± 0.70 [a,b,1,2]	−9.37 ± 0.78 [b,2]	−10.51 ± 0.80 [a,b,1]	−10.01 ± 1.03 [a,b,1]	−13.70 ± 1.90 [a,1]	...	...	0.03
	Large $n = 133$	−10.48 ± 0.612 [a,1]	−10.68 ± 0.65 [a,1]	−11.82 ± 0.67 [a,1]	−11.85 ± 0.68 [a,1]	−10.97 ± 0.66 [a,1]	−11.28 ± 0.72 [a,1]	−10.08 ± 0.85 [a,1]	−12.05 ± 1.10 [a]	−11.96 ± 1.58 [a]	0.41
	Value of p	$p > 0.05$	$p > 0.05$	$p < 0.05$	$p < 0.05$	$p > 0.05$	$p > 0.05$	$p > 0.05$			

SE: Standard error. BO: birth order; [a,b] indicate significant difference $p < 0.05$ among rows in the same dam's size, depending on birth order; [1,2] indicate difference among columns in the same birth order, depending on dam's size; Dam's size: according to the category: Small (<10 kg), Medium (11–25.0 kg), Large (26–45 kg); n = number of newborn puppies according to the dam's size; ANOVA, Mixed General Linear Model, Tukey.

Table 2. Mean and standard error of blood gas exchange of live-born canine puppies according to birth order and dam's size.

Blood Traits	Dam's Size	BO 1 (Mean ± SE)	BO 2 (Mean ± SE)	BO 3 (Mean ± SE)	BO 4 (Mean ± SE)	BO 5 (Mean ± SE)	BO 6 (Mean ± SE)	BO 7 (Mean ± SE)	BO 8 (Mean ± SE)	BO 9 (Mean ± SE)	Value of p
pO$_2$ (mmHg)	Small $n = 75$	11.44 ± 0.87 [b,1]	13.32 ± 0.78 [a,1]	12.54 ± 0.82 [a,b,1]	12.86 ± 0.99 [a,b,1]	12.47 ± 1.44 [a,b,1,2]	—	—	—	—	0.04
	Medium $n = 102$	10.25 ± 0.75 [c,1]	10.92 ± 0.78 [a,b,c,2]	10.78 ± 0.76 [a,b,c,1,2]	11.80 ± 0.82 [a,b,c,1]	12.47 ± 0.84 [b,1]	12.98 ± 1.09 [a,1]	9.00 ± 2.01 [a,b,c,1]	—	—	0.02
	Large $n = 133$	10.98 ± 0.64 [a,b,1]	10.34 ± 0.69 [a,b,2]	9.64 ± 0.71 [a,b,2]	9.30 ± 0.71 [b,2]	9.99 ± 0.70 [a,b,2]	9.95 ± 0.74 [a,b,2]	11.75 ± 0.90 [a,1]	11.83 ± 1.62 [a,b]	8.88 ± 1.67 [a,b]	0.0
	Value of p	$p > 0.05$	$p < 0.05$	$p > 0.05$	$p < 0.05$	$p < 0.05$	$p < 0.05$	$p > 0.05$			
pCO$_2$ (mmHg)	Small $n = 75$	69.19 ± 3.02 [a,1]	63.72 ± 2.73 [a,b,2]	62.05 ± 2.87 [b,2]	65.81 ± 3.47 [a,b,2]	63.93 ± 5.02 [a,b,1]	...	...	...	...	0.02
	Medium $n = 102$	74.96 ± 2.61 [a,1]	73.33 ± 2.70 [a,1]	74.42 ± 2.58 [a,1]	70.27 ± 2.87 [a,2]	74.28 ± 2.94 [a,1]	70.31 ± 3.80 [a,1]	77.00 ± 6.98 [a,1]	...	...	0.52
	Large $n = 133$	74.90 ±2.24 [a,b,1]	74.11 ± 2.40 [a,b,1]	74.80 ± 2.47 [a,b,1]	78.93 ±2.48 [a,1]	74.41 ± 2.44 [a,b,1]	75.33 ± 2.59 [a,b,1]	75.13 ± 3.13 [a,b,1]	68.83 ± 4.03 [b]	71.25 ± 5.82 [a,b]	0.03
	Value of p	$p > 0.05$	$p < 0.05$	$p < 0.05$	$p < 0.05$	$p > 0.05$	$p > 0.05$	$p > 0.05$			

SE: Standard error. BO: birth order; [a,b,c] indicate significant difference $p < 0.05$ among rows in the same dam's size, depending on birth order; [1,2] indicate difference among columns in the same birth order, depending on dam's size; Dam's size: according to the category: Small (<10 kg), Medium (11–25.0 kg), Large (26–45 kg); n = number of newborn puppies according to the dam's size; ANOVA, Mixed General Linear Model, Tukey.

Table 3. Mean and standard error of calcium concentrations and hematocrit in live-born canine puppies according to birth order and dam's size.

Blood Traits	Dam's Size	BO 1 (Mean ± SE)	BO 2 (Mean ± SE)	BO 3 (Mean ± SE)	BO 4 (Mean ± SE)	BO 5 (Mean ± SE)	BO 6 (Mean ± SE)	BO 7 (Mean ± SE)	BO 8 (Mean ± SE)	BO 9 (Mean ± SE)	Value of p
Ca^{2+} (mmol/L)	Small $n = 75$	1.72 ± 0.02 [a,1]	1.63 ± 0.02 [b,2]	1.60 ± 0.02 [b,2]	1.65 ± 0.03 [a,b,2]	1.67 ± 0.04 [a,b1]	…	…	…	…	0.005
	Medium $n = 102$	1.73 ± 0.02 [a,1]	1.71 ± 0.02 [a,1]	1.75 ± 0.02 [a,1]	1.69 ±0.02 [a,2]	1.72 ± 0.02 [a,1]	1.71 ± 0.03 [a,1]	1.78 ± 0.06 [a,1]	…	…	0.00
	Large $n = 133$	1.71 ± 0.02 [b,1]	1.76 ± 0.02 [a,b,1]	1.79 ± 0.02 [a,1]	1.77 ±0.02 [a,1]	1.75 ± 0.02 [a,b,1]	1.72 ± 0.02 [b,1]	1.77 ± 0.03 [a,b,1]	1.70 ± 0.03 [b]	1.72 ± 0.05 [a,b]	0.04
	Value of p	$p > 0.05$	$p < 0.05$	$p < 0.05$	$p < 0.05$	$p > 0.05$	$p > 0.05$	$p > 0.05$			
Htc (%)	Small $n = 75$	53.76 ± 1.06 [a,1]	51.24 ± 0.96 [b,2]	51.40 ± 1.01 [b,2]	52.76 ± 1.22 [a,b,2]	49.25 ±1.77 [b,2]	…	…	…	…	0.02
	Medium $n = 102$	53.53 ± 0.92 [a,1]	53.60 ± 0.95 [a,1,2]	54.04 ± 0.91 [a,1]	52.72 ± 1.01 [a,1,2]	55.31 ± 1.04 [a,1]	54.58 ± 1.34 [a,1]	57.10 ± 2.47 [a,1]	…	…	0.39
	Large $n = 133$	53.94 ± 0.79 [a,1]	54.51 ± 0.85 [a,1]	54.87 ± 0.87 [a,1]	55.41 ± 0.88 [a,1]	54.78 ± 0.86 [a,1]	53.22 ± 0.91 [a,1]	53.41 ± 1.10 [a,1]	53.60 ± 1.42 [a]	51.08 ± 2.06 [a]	0.49
	Value of p	$p > 0.05$	$p < 0.05$	$p < 0.05$	$p < 0.05$	$p < 0.05$	$p > 0.05$	$p > 0.05$			

SE: Standard error. BO: birth order; [a,b] indicate significant difference $p < 0.05$ among rows in the same dam's size, depending on birth order; [1,2] indicate difference among columns in the same birth order, depending on dam's size; Dam's size: according to the category: Small (<10 kg), Medium (11–25.00 kg), Large (26–45 kg); n = number of newborn puppies according to the dam's size; ANOVA, Mixed General Linear Model, Tukey.

On the other hand, in Table 1, the blood pH levels differed neither among the dam's size groups, nor the birth order groups. However, in Table 2, it was observed that for the pO_2 levels, the birth orders BO2, BO4, BO5, and BO6 showed a significant difference with the dam's size ($p < 0.05$), as in pCO_2, the difference was shown in groups BO2, BO3, and BO4 ($p < 0.05$). Regarding HCO_3^- concentrations, there was a statistically significant difference in groups BO2 and BO4. Lastly, for EB, the difference ($p < 0.05$) was found in BO3 and BO4 (Table 1).

Moreover, in Table 3, the concentrations of calcium in the blood showed a difference ($p < 0.05$) in BO2, BO3, and BO4, while for hematocrit, again BO2, BO3, BO4, and BO5 were different ($p < 0.05$).

In Table 4, the Pearson rank test shows that, except for glucose ($p = 0.05$), the values of the physiological profile variables (pH, pCO_2, pO_2, Ca^{2+}, Lactate, Htc, HCO_3^- and BE) show a slight strength of association to birth order (BO) ($p < 0.05$) without considering the size of the mother. However, when the variable size of the mother (small, medium, and large) is introduced, as shown in Table 5, it is observed that the values of the physiological profile variables (pH, pCO_2, pO_2, Glucose, Ca^{2+}, Lactate, HCO_3^- and BE), except for Htc ($p = 0.15$), are associated with birth order only in offspring from large-sized bitches ($p < 0.05$).

Table 4. Correlations between birth order (BO) and blood physiological profile in live-born canine puppies.

(y)	(x)	*r* Value	*p* Value
BO	pH	−0.19	$p = 0.00$
	pCO_2 (mmHg)	0.17	$p = 0.00$
	pO_2 (mmHg)	−0.15	$p = 0.00$
	Glucose (mg/dL)	−0.10	$p = 0.05$
	Ca^{2+} (mmol/L)	0.20	$p = 0.00$
	Lactate (mg/dL)	0.24	$p < 0.00$
	Htc (%)	0.18	$p = 0.00$
	HCO_3^- (mmol/L)	−0.21	$p = 0.00$
	BE (mEq/L)	0.24	$p < 0.00$

Dependent variable (y), birth order (BO); independent variable (x), dam size, blood physiological profile. The data are presented as a correlation coefficient (*r* value); $n = 310$; n = number of newborn puppies.

Table 5. Correlations between birth order (BO) and blood physiological profile according to dam's size.

(y)	(x)	Dam's Size		
		Small $n = 75$	**Medium** $n = 102$	**Large** $n = 133$
BO	pH	−0.015 $p = 0.89$	−0.04 $p = 0.65$	−0.25 $p = 0.00$
	pCO_2 (mmHg)	−0.06 $p = 0.59$	−0.02 $p = 0.83$	0.20 $p = 0.02$
	pO_2 (mmHg)	0.03 $p = 0.79$	0.13 $p = 0.17$	−0.20 $p = 0.01$
	Glucose (mg/dL)	0.09 $p = 0.43$	−0.12 $p = 0.20$	−0.17 $p = 0.04$
	Ca^{2+} (mmol/L)	−0.07 $p = 0.53$	0.01 $p = 0.88$	0.18 $p = 0.03$
	Lactate (mg/dL)	−0.056 $p = 0.63$	0.02 $p = 0.82$	0.28 $p = 0.00$
	Htc (%)	−0.14 $p = 0.21$	0.12 $p = 0.22$	0.12 $p = 0.15$
	HCO_3^- (mmol/L)	−0.07 $p = 0.54$	−0.08 $p = 0.42$	−0.21 $p = 0.01$
	BE (mEq/L)	0.13 $p = 0.25$	−0.03 $p = 0.70$	−0.27 $p = 0.00$

Dependent variable (y), birth order (BO); independent variable (x), dam size, blood physiological profile. The data are presented as a correlation coefficient (*r* value). Dam's size: according to the category: Small (<10 kg), Medium (11–25.00 kg), Large (26–45 kg). n = number of newborn puppies according to the dam's size.

Finally, in Table 6, considering the sex of the liveborn puppies' group (LBP), the percentage of males is higher compared to females (53.23% and 46.77%). Within males, the higher percentage was observed in litters from large sized bitches (42.90%) from the total of 310 registered. Nonetheless, the X^2 test did not show statistically significant differences ($p > 0.05$).

Table 6. Frequency and percentage of canine liveborn puppies (LBP) by sex according to the size of the bitch.

Dam's size	Female $n = 145$ (46.77%)	Male $n = 165$ (53.23%)	Total $n = 310$
	LBP	LBP	
Small	38 (46.67%) [a,1]	37 (41.33%) [a,1]	75 (24.20%)
Medium	49 (45.10%) [a,1]	53 (41.18%) [a,1]	102 (32.90%)
Large	58 (39.10%) [a,1]	75 (38.35%) [a,1]	133 (42.90%)

[a,b] indicate significant difference ($p < 0.05$) among columns in the same sex, to the x^2 test; [1,2] indicate significant difference ($p < 0.05$) among rows in the same dam's size, to the x^2 test; LBP: liveborn puppies; dam's size = according to the category: Small (<10 kg), Medium (11–25.0 kg), Large (26–45 kg).

4. Discussion

Birth in all mammalian species is accompanied by a period of obligatory or transitory asphyxia in the newborn. The uterine contractions during normal parturition increase the intrauterine pressure, compromising placental perfusion and oxygenation and gas interchange. Umbilical cord blood gas analysis provides valuable information about the neonate's condition immediately after birth in all fetuses. The results that were obtained in this study permitted the identification or the degree of alteration of the neonatal physiological blood profile according to the birth order of the pups and the dam's size.

4.1. Physiological Parameters According to Birth Order in Canine Puppies from Small-Size Female Dogs

Although the uterine contractions during term labor result in a rise in intrauterine pressure, compromising placental perfusion and oxygenation, affecting gas interchange in all fetuses, the results, in general, suggest that the degree of alteration of the neonatal physiological profile is modified according to birth order. In the group of small-size dams, it was observed that the newborn puppies of BO1 presented more blood physiological alterations than those born in BO2, BO3, BO4, and BO5. This repeated impairment of gaseous exchange leads to a slight but consistent reduction in pH (7.05 ± 0.03), pO2 (11.44 ± 0.87 mmHg), and an increase in pCO2 (69.19 ± 3.02), besides presenting higher concentrations of calcium (1.72 ± 0.02 mmol/L). These variations could indicate the development of a process of intrapartum hypoxia, as referred to by Ferreiro [41] and Mota-Rojas et al. [42,43].

As stated by Uchańska et al. [44], hypoxia can cause almost 60% of neonatal deaths; the interruption of umbilical circulation during prolonged parturition or dystocia is one of the main reasons. This is similar to what was reported in other studies that mention maternal factors such as duration in the expulsion stage or the high stress of the peripartum period [32,45,46], which could result in oxygen reduction, as it has been observed in piglets [42]. However, it is important to point out that the decrease in the interchange of gasses during the delivery of these pups led to an O_2 deficit (hypoxemia) which gave way to a rise in CO_2 in the extracellular liquid (hypercapnia). This could conditionate to a rise in the concentration of bicarbonate (HCO_3) which would eventually culminate in the presence of a respiratory acidosis [45]. However, the pups born in BO1 did not show a rise in HCO3 concentration; therefore, they did not show changes in blood pH, so the blood lactate and glucose values were not observed as altered.

These hemodynamic alterations in the fetus are established to enable breathing, decreasing the flow in the pulmonary vascular resistance [47]. This translates to an increase in the pulmonary vascular flow with a rise in oxygen concentration in blood and then oxygen saturation, producing the elimination of placental circulation [48]. This is why it is possible to consider that, in this study, this period of brief hypoxia results to be normal during the delivery process [49]. Only in the case of persistent hypoxia would it cause a delay in breathing and a possible metabolic acidosis, as it relates to neonatal morbidity and mortality [50]. Nevertheless, the percentage of mortality in pups of the group of small-size dams was not significant. Experimental studies with the asphyxia model confirm that the healthy fetus to be born has an impressive tolerance to hypoxemia. However, it has been reported that puppies from small-size dams present a lesser vulnerability and mortality risk than puppies from large-size dams; nevertheless, if SBP is detected, the LBP requires extra monitoring during the perinatal period [22]. Likewise, some authors have reported an influence of the birth position on the presentation of hypoxia during parturition [51]. This factor was not assessed in the present study but could be a relevant factor for further studies.

4.2. Physiological Parameters According to Birth Order in Canine Puppies from Medium-Size Female Dogs

For this category, group BO7 presented more acid-base, energetic and calcium balance alterations than the puppies from large-size dams. Our study states that this could be explained by the development of an intrapartum hypoxic process in the litters, especially in those born at the end of the delivery. The alteration in gas exchange can generate different degrees of hypoxia, hypercapnia, and acidosis according to the duration and severity of the interruption in the oxygen flux [52]. The delay at the beginning of the pulmonary ventilation at birth provoked a reduction in the blood's oxygen saturation of the pups belonging to the group of medium-size dams; the persistent and higher degree of hypoxia caused a metabolic change to glycolysis with the consequent increased production of lactate, a result that can be observed when compared to the pups of small-size dams. The latter could be due to fatigue in the female dog in the expulsion of more numerous litters, in which a decrease in the number and intensity of uterine contractions could be observed, findings similar to what is reported by Mota-Rojas et al. [42] in litters of piglets. In dogs, the concentration of oxytocin and the expression of oxytocin receptors are key elements for the onset of parturition. Since this hormone maintains the synchronized uterine contractions and dilatation of the cervix to facilitate a eutocic parturition [17], abnormalities in the oxytocinergic system could extend the time of parturition and, therefore, the consequences of delayed birth. For example, Cornelius et al. [53] determined that canine pups with dystocia were 2.35 times more likely to be stillborn due to one of the leading causes of delayed parturition: uterine inertia. Further, the decrease in pH (6.97 ± 0.07) and the significant increase in the concentrations of lactate (10.00 ± 1.01 mg/dL) in this study suggest a state of metabolic acidosis [53], which could be considered to be severe, which is similar to the findings mentioned by Mota-Rojas et al. [54,55] and van Dijk et al. [56]. Moreover, prolonged hypoxia could have contributed to the low glucose levels (54.00 ± 11.88 mg/dL), predisposing toy and small breed puppies to hypoglycemia, as stated by Münnich and Küchenmeister [57]. This differs from our results, and it could be attributed to the fast consumption of the newborn's energy reserves. Similarly, the finding of a high concentration of calcium in the blood (1.78 ± 0.06 mmol/L) suggests that the muscular activity of the puppy was increased at the moment of labor, which fostered the mobilization of the metabolite from the newborn's bones, as pointed out in the studies made by Rydhmer et al. [58] and Mota-Rojas et al. [42].

The analysis indicates that the alterations in the metabolic profile of the late-born puppies (BO7) of medium-size female dogs in our study are a sign of the consumption of muscular glycogen, which could be due to the restriction of oxygen in the uterus, as explained by Mota-Rojas et al. [54], with the possibility that fetal suffering could also be present. This suggests that the long-term births would increase the risk of hypoxia in late-born puppies, with a more frequent Type II SBP and weak puppies that die rapidly after birth, as reported by Indrebφ et al. [59] and Münnich and Küchenmeister [57], and observed in our results with a percentage of 26.42% of intrapartum deaths Type II.

4.3. Physiological Parameters According to Birth Order in Neonate Canine Puppies from Large-Size Female Dogs

In this group, the reduction in base excess and the increase in pCO_2 and lactate during the glucogenesis, and the evidence of hypoxia due to respiratory failure compared to puppies from medium-sized dams (BO6 or later) can be attributed to acidosis during birth, of metabolic and respiratory types. Acidosis can be a consequence of stress even in a normal delivery where tissue hypoxia and placentary insufficiency occur. This was mentioned by Plavec et al. [60], who reported hyperlactatemia of 12.24 ± 0.56 mmol/L and acidosis in pups from vaginal parturition due to the decreased placental circulation caused by uterine contractions. This results in a mixture of metabolic and respiratory acidosis [61] when the passage of oxygen is restricted and the carbohydrate metabolism is affected [62], shifting to anaerobic activity and the consequent accumulation of lactic acid which decreases blood pH [63]. Additionally, a hypothesis to this effect could be that an increase in the size of the products and the number of puppies in the litter could cause a higher number of uterine contractions of higher intensity, triggering a lower blood flow which reduces the oxygen supply to the fetus. This factor negatively impacts the gas interchange: the metabolites and gases of the physiological profile are, in fact, altered, as was observed in the puppies of our study. This evidence could also be promoted by amniotic liquid at the pulmonary level during the first minute of life, as reported by Vannucchi et al. [64]. Additionally, this acidosis could have increased blood calcium due to muscular mobilization [65], mainly in the puppies in the birth order BO1, BO3, BO4, and BO8 ($p = 0.04$). The alteration of the acid-base balance involves the complex interaction of several organ systems, including the brain, lungs, and liver [63]. Therefore, evaluating this parameter could help to reduce the systemic alterations that the newborn can develop.

4.4. Physiological Parameters According to Birth Order in Canine Puppies from Female Dogs of Different Sizes

Based on the results, it was possible to establish that the birth orders BO2, BO4, and BO7 in puppies of small and medium-size dams, and the last third of BO, presented the most maladjustments with an inadequate gas interchange. Generating hypoxia and hypercapnia due to the accumulation of carbon dioxide resulted in biochemical changes inside the neonates' bodies, which would probably generate the death of neuronal cells and brain damage, as pointed out by Pitsawong and Panichkul [66].

According to the distribution of LP and SB type II, it is considered that, despite not observing significant differences in the dam's size, this could be due to the litter's size and the breed, which could be predisposing factors to stillbirths. In our research, the large size registered 56.60% of the highest percentage, with more than a half of the individuals ($n = 53$). Regarding the sex of the puppies, the percentages of small-sized male puppies obtained in the present study were close to that reported in a study analyzing the different types of parturitions (5.35% vs. 3.3%) [60]. Although the present results show a higher percentage of SB type II males than females in medium and large sizes, regardless of the sex, to date, studies report no significant association between sex and stillbirth risk in canine pups [53].

When considering the Pearson rank test and the correlations between large sized dams and the physiological profile variables (pH, pCO_2, pO_2, glucose, Ca^{2+}, lactate, HCO_3^- and BE), Mila et al. [67] also reported a significant effect of breed size on the lactate concentration of large breed puppies (1.4 mmol/L). Nonetheless, the authors concluded that the effect of the physiological moderate metabolic acidosis makes it unclear if this factor could help during the assessment of newborn puppies.

Finally, it is important to underline that alteration in the neonate's respiratory profile could be due to the regulation of breathing, which involves a selective reduction in the consumption of oxygen during the hypoxic process, restricting the function of chemoreceptors and being capable of redistributing the blood flow to the heart, brain, diaphragm, and adrenal glands, but not to the spleen, gastrointestinal tract, skin, and kidneys. Therefore, in severe hypoxia, there would be a decrease in heart rate and an increase in intestinal motility, leading to the expulsion and aspiration of meconium and Meconium Aspiration Syndrome (MAS) and damage to the intestinal mucosa [68,69], as well as failure in other tissues that require great supplies of oxygen [70]. Other factors that compromise the viability of the canine neonate must be considered and not analyzed in this study, such as the fetus's position at birth, the umbilical cord morphology, degree of meconium staining, and premature deliveries, which have already been reported in human medicine [71], would also have an impact in the presence of intrapartum asphyxiation in canine puppies.

5. Conclusions

During the eutocic delivery process, fetuses are exposed to intermittent periods of hypoxia. However, they can activate adaptive mechanisms to protect the organism from hypoxemia. The results obtained in the present study show that the size of the mother and the order of birth are risk factors for newborn puppies since they may present greater physiological blood alterations that affect adaptation to extrauterine life. It was identified that BO1 puppies of small-sized bitches, as well as BO7 puppies (last litter) of medium-sized bitches, presented more significant blood alterations. In small-sized females, this effect can be attributed to uterine morphology and activity, such as intensity, duration, or the number of contractions. Contrarily, in the last pups of medium-sized bitches, it may be due to the hypoxic process of exhaustion in the newborn due to prolonged labor time, causing weak puppies at the first minute of life. Notably, no effects were observed by birth order in puppies from large-sized female dogs, since all presented considerable blood physiological alterations, leading to intrapartum asphyxiation processes and prolonged hypoxia, affecting neonatal survival.

Author Contributions: Investigation, A.G., B.R.-S., P.M.-M., A.O. and C.M.; writing—original draft preparation, B.R.-S., A.O., M.E.R., A.D.-O., A.O.-H., C.M. and A.G.; writing—review and editing, A.O., A.G., P.M.-M., A.D.-O., I.H.-Á., A.V.-M. and B.R.-S.; project administration, P.M.-M., A.O.-H., I.H.-Á.; and A.G. All authors have read and agreed to the published version of the manuscript.

Funding: This research received no external funding.

Institutional Review Board Statement: The study was conducted according to the guidelines of the Declaration of Helsinki, and the experimental protocol was approved by the Committee of the Master's Program in Agricultural Sciences (protocol code CAMCA.32.18) of the Universidad Autónoma Metropolitana-Xochimilco campus, Mexico City.

Informed Consent Statement: Written informed consent was obtained from all animal owners.

Data Availability Statement: Data is contained within the article.

Acknowledgments: The authors want to express their gratitude to Daniel Mota-Rojas Professor of studies in behavior and animal welfare, for the conception and direction of the project that created this manuscript. We are thankful for his continuous support and advice.

Conflicts of Interest: The authors declare that they have no conflict of interest.

Abbreviations

pCO_2	partial carbon dioxide saturation
pO_2	partial oxygen saturation
O_2	oxygen
BE	base excess
HCO_3^-	bicarbonate
Ca^{++}	calcium
Htc	hematocrit

References

1. Grundy, S.A. Clinically relevant physiology of the neonate. *Vet. Clin. N. Am. Small Anim. Pract.* **2006**, *36*, 443–459. [CrossRef] [PubMed]
2. Veronesi, M.C. Assessment of canine neonatal viability—the Apgar score. *Reprod. Domest. Anim.* **2016**, *51*, 46–50. [CrossRef] [PubMed]
3. Morton, S.U.; Brodsky, D. Fetal Physiology and the Transition to Extrauterine Life. *Clin. Perinatol.* **2016**, *43*, 395–407. [CrossRef]
4. Yli, B.M.; Kjellmer, I. Pathophysiology of foetal oxygenation and cell damage during labour. *Best Pract. Res. Clin. Obstet. Gynaecol.* **2016**, *30*, 9–21. [CrossRef] [PubMed]
5. Vassalo, F.G.; Simões, C.R.B.; Sudano, M.J.; Prestes, N.C.; Lopes, M.D.; Chiacchio, S.B.; Lourenço, M.L.G. Topics in the Routine Assessment of Newborn Puppy Viability. *Top. Companion Anim. Med.* **2015**, *30*, 16–21. [CrossRef] [PubMed]
6. Swan, H.G.; Christian, J.J.; Hamilton, C. The process of anoxic death in newborn pups. *Surg. Gynecol. Obs.* **1984**, *99*, 5–8.
7. Mota-Rojas, D.; Martinez-Burnes, J.; Trujillo-Ortega, M.E.; Alonso-Spilsbury, M.L.; Ramirez-Necoechea, R.; Lopez, A. Effect of oxytocin treatment in sows on umbilical cord morphology, meconium staining, and neonatal mortality of piglets. *Am. J. Vet. Res.* **2002**, *63*, 1571–1574. [CrossRef]
8. Blickstein, I.; Green, T. Umbilical cord blood gases. *Clin. Perinatol.* **2007**, *34*, 451–459. [CrossRef]
9. James, L.S.; Weisbrot, I.; Prince, C.; Holaday, D.; Apgar, V. The acid-base status of human infants in relation to birth asphyxia and the onset of respiration. *J. Pediatr.* **1958**, *52*, 379–394. [CrossRef]
10. Armstrong, L.; Stenson, B.J. In the Assessment of the Newborn. *Arch. Dis. Child. Fetal Neonatal Ed.* **2007**, *92*, 430–434. [CrossRef]
11. Rodríguez Fernández, V.; López Ramón y Cajal, C.N.; Marín Ortiz, E.; Couceiro Naveira, E. Intrapartum and perinatal results associated with different degrees of staining of meconium stained amniotic fluid. *Eur. J. Obstet. Gynecol. Reprod. Biol.* **2018**, *228*, 65–70. [CrossRef] [PubMed]
12. Rootwelt, V.; Reksen, O.; Farstad, W.; Framstad, T. Associations between intrapartum death and piglet, placental, and umbilical characteristics. *J. Anim. Sci.* **2012**, *90*, 4289–4296. [CrossRef] [PubMed]
13. Kelley, R. Canine Reproductive Management: Factors Affecting Litter Size. In Proceedings of the Annual Conference of the Society for Theriogenology and American College of Theriogenology, Colorado Springs, CO, USA, 7–11 August 2002; pp. 291–301.
14. Sonntag, Q.; Overall, K.L. Key determinants of dog and cat welfare: Behaviour, breeding and household lifestyle. *Rev. Sci. Tech.* **2014**, *33*, 213–220. [CrossRef] [PubMed]
15. Mugnier, A.; Mila, H.; Guiraud, F.; Brévaux, J.; Lecarpentier, M.; Martinez, C.; Mariani, C.; Adib-Lesaux, A.; Chastant-Maillard, S.; Saegerman, C.; et al. Birth weight as a risk factor for neonatal mortality: Breed-specific approach to identify at-risk puppies. *Prev. Vet. Med.* **2019**, *171*, 104746. [CrossRef] [PubMed]
16. Guardini, G.; Bowen, J.; Mariti, C.; Fatjó, J.; Sighieri, C.; Gazzano, A. Influence of maternal care on behavioural development of domestic dogs (*Canis familiaris*) living in a home environment. *Animals* **2017**, *7*, 93. [CrossRef]
17. Lezama-García, K.; Mariti, C.; Mota-Rojas, D.; Martínez-Burnes, J.; Barrios-García, H.; Gazzano, A. Maternal behaviour in domestic dogs. *Int. J. Vet. Sci. Med.* **2019**, *7*, 20–30. [CrossRef] [PubMed]
18. Bergström, A.; Fransson, B.; Lagerstedt, A.-S.; Kindahl, H.; Olsson, U.; Olsson, K. Hormonal concentrations in bitches with primary uterine inertia. *Theriogenology* **2010**, *73*, 1068–1075. [CrossRef]
19. Darvelid, A.W.; Linde-Forsberg, C. Dystocia in the bitch: A retrospective study of 182 cases. *J. Small Anim. Pract.* **1994**, *35*, 402–407. [CrossRef]
20. Ettinger, S.; Feldman, E.C. *Textbook of Veterinary Internal Medicine*; Elsevier Saunders: Philadelphia, PA, USA, 2005; pp. 1655–1667.
21. Münnich, A.; Küchenmeister, U. Dystocia in Numbers-Evidence-Based Parameters for Intervention in the Dog: Causes for Dystocia and Treatment Recommendations. *Reprod. Domest. Anim.* **2009**, *44*, 141–147. [CrossRef]
22. Tønnessen, R.; Borge, K.S.; Nødtvedt, A.; Indrebø, A. Canine perinatal mortality: A cohort study of 224 breeds. *Theriogenology* **2012**, *77*, 1788–1801. [CrossRef]
23. Reyes-Sotelo, B.; Mota-Rojas, D.; Mora-Medina, P.; Ogi, A.; Mariti, C.; Olmos-Hernández, A.; Martínez-Burnes, J.; Hernández-Ávalos, I.; Sánchez-Millán, J.; Gazzano, A. Blood Biomarker Profile Alterations in Newborn Canines: Effect of the Mother's Weight. *Animals* **2021**, *11*, 2307. [CrossRef] [PubMed]
24. Ogi, A.; Naef, V.; Santorelli, F.M.; Mariti, C.; Gazzano, A. Oxytocin receptor gene polymorphism in lactating dogs. *Animals* **2021**, *11*, 3099. [CrossRef] [PubMed]
25. Le Cozler, Y.; Guyomarc'h, C.; Pichodo, X.; Quinio, P.Y.; Pellois, H. Factors associated with stillborn and mummified piglets in high-prolific sows. *Anim. Res.* **2002**, *51*, 261–268. [CrossRef]
26. Lucia, T.; Corrêa, M.N.; Deschamps, J.C.; Bianchi, I.; Donin, M.A.; Machado, A.C.; Meincke, W.; Matheus, J.E.M. Risk factors for stillbirths in two swine farms in the south of Brazil. *Prev. Vet. Med.* **2002**, *53*, 285–292. [CrossRef]
27. Vanderhaeghe, C.; Dewulf, J.; de Kruif, A.; Maes, D. Non-infectious factors associated with stillbirth in pigs: A review. *Anim. Reprod. Sci.* **2013**, *139*, 76–88. [CrossRef]
28. Wiberg, N.; Källén, K.; Olofsson, P. Physiological development of a mixed metabolic and respiratory umbilical cord blood acidemia with advancing gestational age. *Early Hum. Dev.* **2006**, *82*, 583–589. [CrossRef]
29. Islas-Fabila, P.; Mota-Rojas, D.; Martínez-Burnes, J.; Mora-Medina, P.; González-Lozano, M.; Santiago, R.; González, M.; Vega, X.; Gregorio, H. Physiological and metabolic responses in newborn piglets associated with the birth order. *Anim. Reprod. Sci.* **2018**, *197*, 247–256. [CrossRef]

30. Alonso-Spilsbury, M.; Mota-Rojas, D.; Villanueva-García, D.; Martínez-Burnes, J.; Orozco, H.; Ramírez-Necoechea, R.; Mayagoitia, A.L.; Trujillo, M.E. Perinatal asphyxia pathophysiology in pig and human: A review. *Anim. Reprod. Sci.* **2005**, *90*, 1–30. [CrossRef]

31. Johnson, C.A. Pregnancy management in the bitch. *Theriogenology* **2008**, *70*, 1412–1417. [CrossRef]

32. Baxter, E.M.; Jarvis, S.; D'Eath, R.B.; Ross, D.W.; Robson, S.K.; Farish, M.; Nevison, I.M.; Lawrence, A.B.; Edwards, S.A. Investigating the behavioural and physiological indicators of neonatal survival in pigs. *Theriogenology* **2008**, *69*, 773–783. [CrossRef]

33. Holland, A. Piglet Vitality and Mortality within 48h of Life from Farrowing Sows Treated with Carbetocin, Oxytocin or without Intervention. Available online: https://dspace.library.uu.nl/handle/1874/289549dspace.library.uu.nl.file:///C:/ (accessed on 12 March 2020).

34. Björkman, S.; Oliviero, C.; Rajala-Schultz, P.J.; Soede, N.M.; Peltoniemi, O.A.T. The effect of litter size, parity, farrowing duration on placenta expulsion and retention in sows. *Theriogenology* **2017**, *92*, 36–44. [CrossRef] [PubMed]

35. World Small Animal Veterinary Association. Global Nutritional Assesment Guidelines. Available online: http://wsava.org/wp-content/uploads/2020/01/Global-Nutritional-Assesment-Guidelines-Spanish.pdf (accessed on 12 March 2020).

36. Gandini, G.; Cizinauskas, S.; Lang, J.; Fatzer, R.; Jaggy, A. Fibrocartilaginous embolism in 75 dogs: Clinical findings and factors influencing the recovery rate. *J. Small Anim. Pract.* **2003**, *44*, 76–80. [CrossRef] [PubMed]

37. Royal Canin Breeds. Available online: https://www.royalcanin.com/mx/dogs/breeds/breed-library (accessed on 18 June 2020).

38. Chartier, S.; Faulkner, A. General Linear Models: An Integrated Approach to Statistics. *Tutor. Quant. Methods Psychol.* **2008**, *4*, 65–78. [CrossRef]

39. Sherwin, C.M.; Christiansen, S.B.; Duncan, I.J.; Erhard, H.W.; Lay, D.C.; Mench, J.A.; O'Connor, C.E.; Petherick, J.C. Guidelines for the ethical use of animals in applied ethology studies. *Appl. Anim. Behav. Sci.* **2003**, *81*, 291–305. [CrossRef]

40. Nowak, R.; Poindron, P. From birth to colostrum: Early steps leading to lamb survival. *Reprod. Nutr. Dev.* **2006**, *46*, 431–446. [CrossRef]

41. Ferreiro, D. Neonatal Brain Injury. *N. Engl. J. Med.* **2004**, *351*, 1985–1995. [CrossRef]

42. Mota-Rojas, D.; Fierro, R.; Santiago, P.; Gregorio, H.; González-Lozano, M.; Bonilla, H.; Martínez-Rodríguez, R.; García-Herrera, R.; Mora-Medina, P.; Flores-Peinado, S.; et al. Outcomes of gestation length in relation to farrowing performance in sows and daily weight gain and metabolic profiles in piglets. *Anim. Prod. Sci.* **2015**, *55*, 93–100. [CrossRef]

43. Mota-Rojas, D.; López, A.; Martínez-Burnes, J.; Muns, R.; Villanueva-García, D.; Mora-Medina, P.; González-Lozano, M.; Olmos-Hernández, A.; Ramírez-Necoechea, R. Is vitality assessment important in neonatal animals? *CAB Rev.* **2018**, *13*, 1–13. [CrossRef]

44. Uchańska, O.; Ochota, M.; Eberhardt, M.; Niżański, W. Dead or alive? A review of perinatal factors that determine canine neonata viability. *Animals* **2022**, *12*, 1402. [CrossRef]

45. Tummaruk, P.; Tantasuparuk, W.; Techakumphu, M.; Kunavongkrit, A. Seasonal influences on the litter size at birth of pigs are more pronounced in the gilt than sow litters. *J. Agric. Sci.* **2010**, *148*, 421–432. [CrossRef]

46. Muns, R.; Silva, C.; Manteca, X.; Gasa, J. Effect of cross-fostering and oral supplementation with colostrums on performance of newborn piglets. *J. Anim. Sci.* **2014**, *92*, 1193–1199. [CrossRef] [PubMed]

47. Yates, R.W. Cardiovascular Diseases. In *Rennie & Robertson's Textbook of Neonatology*; Rennie, J.M., Ed.; Elsevier: St. Louis, MO, USA, 2012.

48. Goldsmith, J. Overview and Initial Management of Delivery Room Resuscitation. In *Fanaroff and Martin's Neonatal-Perinatal Medicine Diseases of the Fetus and Infant*; Martin, R., Fanaroff, A., Walsh, M., Eds.; Elsevier: St. Louis, MO, USA, 2015; pp. 460–470.

49. Massip, A. Relationship between pH, plasma, cortisol and glucose concentrations in the calf at birth. *Br. Vet. J.* **1980**, *136*, 597–601. [CrossRef]

50. Andres, R.L.; Saade, G.; Gilstrap, L.C.; Wilkins, I.; Witlin, A.; Zlatnik, F.; Hankins, G.V. Association between umbilical blood gas parameters and neonatal morbidity and death in neonates with pathologic fetal acidemia. *Am. J. Obstet. Gynecol.* **1999**, *181*, 867–871. [CrossRef]

51. Von Dehn, B. Pediatric Clinical Pathology. *Vet. Clin. Small Anim. Pract.* **2014**, *44*, 205–219. [CrossRef]

52. Mota-Rojas, D.; Villanueva-García, D.; Solimano, A.; Muns, R.; Ibarra-Ríos, D.; Mota-Reyes, A. Pathophysiology of Perinatal Asphyxia in Humans and Animal Models. *Biomedicines* **2022**, *10*, 347. [CrossRef]

53. Cornelius, A.J.; Moxon, R.; Russenberger, J.; Havlena, B.; Cheong, S.H. Identifying risk factors for canine dystocia and stillbirths. *Theriogenology* **2019**, *128*, 201–206. [CrossRef]

54. Mota-Rojas, D.; Villanueva-García, D.; Hernández, R.; Roldán-Santiago, P.; Martínez-Rodríguez, R.; Mora-Medina, P.; González-Meneses, B.; Sánchez-Hernández, M. Assessment of the vitality of the newborn: An overview. *Sci. Res. Essays* **2012**, *7*, 712–718. [CrossRef]

55. Mota-Rojas, D.; Martinez-Burnes, J.; Villanueva-Garcia, D.; Santiago, P.; Trujillo-Ortega, M.E.; Gregorio, H.; Bonilla, H.; Lopez-Mayagoitia, A. Animal welfare in the newborn piglet: A review. *Vet. Med.* **2012**, *57*, 338–349. [CrossRef]

56. Van Dijk, A.J.; van Rens, B.T.T.M.; van der Lende, T.; Taverne, M.A.M. Factors affecting duration of the expulsive stage of parturition and piglet birth intervals in sows with uncomplicated, spontaneous farrowings. *Theriogenology* **2005**, *64*, 1573–1590. [CrossRef]

57. Münnich, A.; Küchenmeister, U. Causes, Diagnosis and Therapy of Common Diseases in Neonatal Puppies in the First Days of Life: Cornerstones of Practical Approach. *Reprod. Domest. Anim.* **2014**, *49*, 64–74. [CrossRef]

58. Rydhmer, L.; Lundeheim, N.; Canario, L. Genetic correlations between gestation length, piglet survival and early growth. *Livest. Sci.* **2008**, *115*, 287–293. [CrossRef]

59. Indrebø, A.; Trangerud, C.; Moe, L. Canine neonatal mortality in four large breeds. *Acta Vet. Scand.* **2007**, *49*, S2. [CrossRef]

60. Plavec, T.; Knific, T.; Slapšak, A.; Raspor, S.; Lukanc, B.; Pipan, M.Z. Canine Neonatal Assessment by Vitality Score, Amniotic Fluid, Urine, and Umbilical Cord Blood Analysis of Glucose, Lactate, and Cortisol: Possible Influence of Parturition Type? *Animals* **2022**, *12*, 1247. [CrossRef] [PubMed]

61. Brouillette, R.T.; Waxman, D.H. Evaluation of the newborn's blood gas status. *Clin. Chem.* **1997**, *43*, 215–221. [CrossRef] [PubMed]

62. Connett, R.J.; Honig, C.R.; Gayeski, T.E.; Brooks, G.A. Defining hypoxia: A systems view of VO2, glycolysis, energetics, and intracellular PO2. *J. Appl. Physiol.* **1990**, *68*, 833–842. [CrossRef] [PubMed]

63. Seifter, J.L.; Chang, H.-Y. Disorders of Acid-Base Balance: New Perspectives. *Kidney Dis.* **2016**, *2*, 170–186. [CrossRef]

64. Vannucchi, C.I.; Rodrigues, J.A.; Silva, L.C.G.; Lúcio, C.F.; Veiga, G.A.L. A clinical and hemogasometric survey of neonatal lambs. *Small Rumin. Res.* **2012**, *108*, 107–112. [CrossRef]

65. Manning, M. Electrolyte disorders. *Vet. Clin. North Am. Small Anim. Pract.* **2001**, *17*, 8–13. [CrossRef]

66. Pitsawong, C.; Panichkul, P. Risk factors associated with birth asphyxia in Phramonggkutklao Hospital. *Thai J. Obstet. Gynaecol.* **2012**, *19*, 165–171.

67. Mila, H.; Grellet, A.; Delebarre, M.; Mariani, C.; Feugier, A.; Chastant-Maillard, S. Monitoring of the newborn dog and prediction of neonatal mortality. *Prev. Vet. Med.* **2017**, *143*, 11–20. [CrossRef]

68. Davidson, A. Approaches to Reducing Neonatal Mortality in Dogs. In *Recent Advances in Small Animal Reproduction*; Concannon, P., England, G., Verstegen, J., Linde-Forsberg, C., Eds.; Ivis: Ithaca, NY, USA, 2003; pp. 1226–1303.

69. Martínez-Burnes, J.; Mota-Rojas, D.; Villanueva-García, D.; Ibarra, R.D.; Barrios, G.H.N.; López, A. Meconium aspiration syndrome in mammals. *CAB Rev.* **2019**, *14*, 1–11. [CrossRef]

70. Nayci, A.; Atis, S.; Ersoz, G.; Polat, A. Gut Decontamination Prevents Bronchoscopy-Induced Bacterial Translocation. *Respiration* **2004**, *71*, 66–71. [CrossRef] [PubMed]

71. Aslam, H.M.; Saleem, S.; Afzal, R.; Iqbal, U.; Saleem, S.M.; Shaikh, M.W.A.; Shahid, N. Risk factors of birth asphyxia. *Ital. J. Pediatr.* **2014**, *40*, 94. [CrossRef] [PubMed]

Review

Dead or Alive? A Review of Perinatal Factors That Determine Canine Neonatal Viability

Oliwia Uchańska, Małgorzata Ochota *, Maria Eberhardt and Wojciech Niżański

Department of Reproduction and Clinic of Farm Animals, Faculty of Veterinary Medicine, Wroclaw University of Environmental and Life Sciences, 50-366 Wroclaw, Poland; o.uchanska@gmail.com (O.U.); maria.eberhardt@upwr.edu.pl (M.E.); wojciech.nizanski@upwr.edu.pl (W.N.)
* Correspondence: malgorzata.ochota@upwr.edu.pl

Simple Summary: The article summarizes the current knowledge on factors related to pregnancy, parturition, and newborns that affect the health status of a puppy and determine its chances for survival and development. The detailed information is provided in terms of breed predispositions, objectives of pregnancy monitoring, potential sources of complications, and veterinary advances in care and treatment of perinatal conditions. Successful pregnancy outcomes still pose challenges in veterinary neonatology; thus, publications presenting the current state of knowledge in this field are in demand.

Abstract: The perinatal period has a critical impact on viability of the newborns. The variety of factors that can potentially affect the health of a litter during pregnancy, birth, and the first weeks of life requires proper attention from both the breeder and the veterinarian. The health status of puppies can be influenced by various maternal factors, including breed characteristics, anatomy, quality of nutrition, delivery assistance, neonatal care, and environmental or infectious agents encountered during pregnancy. Regular examinations and pregnancy monitoring are key tools for early detection of signals that can indicate disorders even before clinical signs occur. Early detection significantly increases the chances of puppies' survival and proper development. The purpose of the review was to summarize and discuss the complex interactions between all elements that, throughout pregnancy and the first days of life, have a tangible impact on the subsequent fate of the offspring. Many of these components continue to pose challenges in veterinary neonatology; thus, publications presenting the current state of knowledge in this field are in demand.

Keywords: dog; pregnancy; parturition; puppy; neonatology

Citation: Uchańska, O.; Ochota, M.; Eberhardt, M.; Niżański, W. Dead or Alive? A Review of Perinatal Factors That Determine Canine Neonatal Viability. *Animals* **2022**, *12*, 1402. https://doi.org/10.3390/ani12111402

Academic Editors: Daniel Mota-Rojas, Julio Martínez-Burnes and Agustín Orihuela

Received: 1 May 2022
Accepted: 27 May 2022
Published: 30 May 2022

Publisher's Note: MDPI stays neutral with regard to jurisdictional claims in published maps and institutional affiliations.

Copyright: © 2022 by the authors. Licensee MDPI, Basel, Switzerland. This article is an open access article distributed under the terms and conditions of the Creative Commons Attribution (CC BY) license (https://creativecommons.org/licenses/by/4.0/).

1. Introduction

The main goal for dog breeders and their veterinary associates is the successful rearing of puppies. High mortality rates can affect even litters that receive the best care.

Depending on the cause, it can range from 8 to 20% [1–5]. This paper summarizes the main findings regarding the maternal, pregnancy, and newborn-related components that eventually would determine puppy survival. Analyzing maternal and pregnancy-related factors emphasizes the need for tailored care for each pregnant female. This is particularly true for pedigree bitches, which can be prone to obstetric complications due to the anatomy of their birth canal [6–8], as is observed with brachycephalic bitches [9,10]. The combination of female age and body size also affects the potential number of puppies per litter [11,12]. This is an important risk factor, especially for females older than 6 years, for miniature and giant breed females with singleton pregnancies, and giant breed bitches carrying more than 11 fetuses [13,14]. These cases are exposed to primary and secondary uterine inertia, which, as the data in the literature indicate, would pose a direct threat to puppies' lives [3]. The health status of the dam and the level of care are also of considerable importance.

Pregnancy is a demanding state that can be significantly affected by the development of any metabolic and hormonal disorders that, if diagnosed too late, can cause serious health complications or even death of both the mother and the litter. Although not all complications encountered by a pregnant or delivering bitch will be fatal, many of them may significantly affect the health of the offspring in those first crucial days and weeks of life. This paper aims to provide and discuss up-to-date knowledge concerning the most important pre- and postnatal factors that may affect the health and survival of puppies during the first weeks of life.

2. Maternal and Pregnancy-Related Factors

2.1. Anatomy and Breed Predispositions

There is no doubt that of all the factors that affect puppies' health at birth, maternal factors are the first to play the most important role right from the very beginning of pregnancy. The health status and age of the bitch at the time of conception are most important for the subsequent embryonic and fetal development. Moreover, a full clinical examination should be performed, including a complete blood count and vaccination status, and the general health status should be considered before deciding on breeding a particular animal. During pregnancy, some hematological changes usually occur and should be closely monitored, especially red cell and platelet counts. Gestational anemia and thrombocytopenia are normal findings; however, if excessively low, these may affect the course of pregnancy and lead to clotting problems during the caesarean section.

Undoubtedly, special attention should also be paid to the breed of the dam and the sire breed. Breed identity determines typical anatomical structure and is closely related to the predisposition to perinatal complications [6]. The percentage of these complications increases with the prevalence of certain anatomical characteristics. Among these, the pelvic structure and shape and the skull size are very important for the passage of the fetus through the birth canal during delivery [10]. Serious problems in this regard, due to the disproportionately large size of the fetal head in relation to the size of the birth canal, are encountered in brachycephalic breeds. Bitches of these breeds have a predisposition to perinatal complications almost 11 times higher than in other breeds [10]. This results in an elective cesarean section becoming the norm as natural delivery poses a too high risk for the litter. CT pelvimetry studies have shown that English bulldogs have a significantly smaller pelvis and pelvic canal compared to non-brachycephalic dogs of the same weight. Furthermore, the pelvic conformation is characterized by a significantly shorter pelvis and pelvic canal and a significantly narrower caudal opening of the pelvis [9]. Problems related to pelvic anatomy have also been reported in some medium-sized breeds, such as Scottish terriers and Boston terriers [6,7]. X-ray measurements in bitches diagnosed with birth complications due to fetal-pelvic disproportion showed a smaller pelvis and a dorsal-ventral flattening of the pelvic canal. Whereas, in Boston terrier bitches, the problem was also caused by a combination of the pelvic shape described above and the relatively large head size of mature fetuses. Such characteristics significantly increase predisposition to obstructive dystocia and secondary uterine inertia [6,7,15]. According to a study conducted on a population of ~200,000 bitches registered with Swedish Kennel Club (Stockholm, Sweden) between 1995 and 2002, Scottish terriers were the most susceptible to dystocia resulting in emergency cesarean section [16]. The welfare of breeding dogs has gained increased interest in recent years. Improved understanding that dystocia may pose a greater risk in particular breeds of dogs should contribute to reducing the general popularity of such breeds and help veterinarians and kennel clubs to better focus their resources on strategies to limit breeds with high risk of dystocia [1,17].

In addition to the pelvic bone shape, the structure of the soft tissues in the birth canal can also create a problem. It has been proven that abnormal vaginal anatomy can affect fertility, as it prevents natural mating. The presence of adhesions, septations, or double cervical orifices should be diagnosed before the bitch is intended for breeding [18]. Moreover, if detected, such females should be excluded from breeding, as these defects might be transmitted to the next generation [19]. A study comparing the fertility of bitches undergoing surgical vaginal correction and those without surgery showed that although there was a relatively non-significant effect on reproductive performance, pregnancy rates were substantially lower in the group of bitches with severe abnormalities. Furthermore, the same group of bitches showed a markedly greater predisposition to dystocia and thus cesarean section than the group of bitches with mild abnormalities [8].

2.2. Inbreeding

When analyzing the causes that could affect the litter size, inbreeding should also be considered. According to research, a high inbreeding rate can result in impaired fertility and therefore a reduction in the number of pups in the litters that are born [20]. In pedigree dogs, breeding decisions on mating closely related individuals are motivated by the need to obtain the desired phenotype or behavioral traits, leading to genetic bottlenecks. Targeted selection in pedigree dogs for traits of morphology and performance has greatly decreased fertility rates, increased reproductive problems, and significantly contributed to dystocia due to anatomical disproportions in some breeds [10]. The most notable among these are brachycephalic breeds, in which reproduction without assisted mating and a planned caesarean section is today almost not feasible [21]. Focusing on breeding for desired appearance alone raises a lot of public concern today, which hopefully will lead to better awareness and more responsible dog breeding in future.

The availability of genetic tests dedicated to particular breeds is also worth mentioning. Today, the problem of selecting pairs of animals for breeding rests mainly on the responsible approach of the owner, and any poorly considered actions can lead to the perpetuation of traits detrimental to the health of dogs and the narrowing of their genetic pool [22]. The number of breed-specific conditions increased at more than 100% between year 2013 and 2020 [23]. The range of dedicated tests for their identification is also growing steadily [23]. Among the 10 breeds for which test availability is the greatest, are Labrador Retriever, Beagle, Australian Shepherd, German Shepherd, Standard and Miniature Poodle, Golden Retriever, Collie, Pembroke Welsh Corgi, and Dachshund [23]. Detailed information on the range of genetic tests for detecting inherited diseases can be accessed online at: The WSAVA-PennGen DNA Testing Database, which is part of "A project of the WSAVA Hereditary Disease Committee" [24].

2.3. The Impact of Litter Size

When considering the influence of maternal factors, one cannot forget to mention their effect on the litter size. The litter size directly affects the safety of puppies during birth and depends on the mother's age, breed, and the breeding method [11,12]. The larger the breed, the more pups a litter would potentially contain (up to 11–12 pups), whereas in small and miniature breeds on average, up to 3–5 pups are more common [11,13,14].

Large females have a larger uterine capacity and are able to accommodate more fetuses; hence, their litters are naturally more numerous than those of smaller females. In larger breeds, the small size of the litter is especially important because a small number of pups may not give enough strong signals to induce labor [25]. Furthermore, multiple litters of 12 or more pups can cause primary uterine inertia by overstretching the myometrium, leading to weak uterine contractions, or secondary uterine inertia, where the uterine myometrium is exhausted by prolonged labor [6,11,13].

Regardless of the litter size, the last pup is at the highest risk of stillbirth [6]. Some authors pointed out that high perinatal mortality most often affected the first pregnancy of a given dam; in addition, the risk of early neonatal mortality nearly doubled in litters with a concurrent incidence of stillborn pups during birth [13].

Large litters can also be associated with risks to maternal health. Among other things, it can be one of the factors contributing to the bitch's disease during pregnancy and it can increase the risk of dystocia. Gestational toxemia, for example, occurs most often secondarily to a negative energy balance associated with inadequate nutrition and largely with the presence of a large litter of puppies [26,27].

The importance of maternal age played a role in very young (1 year and under) and older (6 years and older) bitches, that tend to have fewer pups (one or two) than bitches aged 2 to 5 years [13,14]. It is worth noting here that primiparous bitches older than 6 years are predisposed to single-puppy pregnancies, uterine disorders, and prolonged labor [3,13]. Hence, such factors would significantly increase the risk of dystocia, and it is highly inadvisable to breed bitches over 6 years of age [13].

The mating method can also be a limiting factor in terms of the number of offspring. The most commonly used method is natural mating or artificial insemination (AI). AI can be intravaginal, intracervical, or surgical, using fresh, chilled, or frozen semen [28,29]. The highest efficiency in obtaining numerous litters was reported through natural mating rather than with the artificial insemination method [11]. However, natural mating has become less and less popular among dog breeders, mainly due to the need to travel long distances for a chosen male. This makes breeders much more eager to choose the artificial insemination methods and shipped semen [29]. The use of the artificial insemination method also provides highly satisfactory results in terms of litter size, especially when using fresh semen rather than chilled or frozen-thawed [12,30].

2.4. Maternal Health Condition

The course of a pregnancy fundamentally determines the health of a newborn during its first weeks of life. The number of factors that may affect the mother's condition during this period is innumerable. They can be divided into those related to the bitch, such as current health state or ability to maintain hormonal balance, and those that are external: mainly the diet and the quality of pregnancy monitoring. During the pregnancy, several disease entities can also develop such as gestational toxemia, diabetes, and eclampsia. The pregnancy can also be affected by corpus luteum insufficiency, which is a serious disorder at the ovarian level [26,31]. Awareness of the potential risks to the mother and the level of care provided if complications arise are vital and can significantly impact the future of any litter.

2.4.1. Pregnancy Toxemia and Diabetes Mellitus

Canine pregnancy toxemia can pose a serious life threat to both the bitch and her fetuses. The disorder, which is mainly seen in late pregnancy, is characterized by hypoglycemia, ketosis, ketonuria, and liver lipidosis, presenting with weakness, which can later progress to seizures, collapse, and even death [32]. Bitches with multiple litters are more at risk of developing toxemia, especially if they show signs of anorexia during the last two weeks of pregnancy [18]. The diagnosis is based on clinical signs and the presence of hypoglycemia combined with high concentrations of ketone bodies in blood and urine, and treatment involves glucose supplementation [15,32,33].

Symptoms similar to gestational toxemia may also be observed during gestational diabetes, which is often developed without the typical clinical signs evident. A syndrome resembling gestational diabetes in humans may occur in older female dogs during pregnancy. Insulin resistance is a normal feature of pregnancy [33]. Two factors promoting insulin resistance, progesterone and growth hormone, are present in similar concentrations in pregnant and non-pregnant bitches in diestrus [34]. Progesterone stimulates the secretion of growth hormones from the canine mammary gland, and both hormones cause insulin resistance and carbohydrate intolerance in the dog [35]. However, pregnant bitches are more insulin-resistant than non-pregnant diestrus bitches [36]. The diagnosis of gestational diabetes mellitus (GDM) is based on history and clinical findings and is confirmed by documenting the persistent hyperglycemia with glucosuria. In humans, GDM is associated with excessive fetal growth (macrosomia), which contributes to dystocia and birth trauma, and, as a result, increased maternal, fetal, and neonatal morbidity and mortality [37,38]. In humans, the prevalence of congenital malformations is higher than in the general population if GDM occurs early in pregnancy, but not in mid-pregnancy or later [39]. In bitches diagnosed with gestational diabetes mellitus, fetal viability should be assessed regularly with ultrasound; knowing whether the pups are large enough to be at risk of dystocia or are no longer alive would guide treatment decisions, such as the schedule of the cesarean section. Such puppies would be at increased risk of premature delivery if CS was decided too early, while if delayed too long, it would lead to fatal distress and metabolic disorders. Published reports of managing canine GDM are rare [35,36,40]. Similar to humans, in case of GDM neonatal hypoglycemia, hyperbilirubinemia, hypocalcemia, and poor suckling can occur more frequently in puppies, and proper management involves intensive fluid and insulin therapy to correct water, electrolyte, acid-base, and glucose derangements [33,40].

2.4.2. Hypoluteoidism

Hypoluteoidism is described as insufficient production and secretion of progesterone by the corpus luteum [41]. Necessary to maintain pregnancy, proper corpus luteum function and progesterone production, if defective, can determine the fate of any pregnancy at its very early stage. A serum progesterone level of at least 2 ng/mL is considered by many authors to be the key to maintaining a canine pregnancy [34,42]. A decrease in progesterone concentration below 2 ng/mL in a pregnant bitch has been shown to lead to abortion [34,42]. Interestingly, some authors reported that miscarriage can occur even when the progesterone level drops below 10 ng/mL [43]. Since serum progesterone measurements can be performed using a variety of methods and devices on the market, with RIA being considered the reference assay, the exact progesterone values may vary between published studies and reports [44]. In all cases, the interpretation of the results should be approached with caution and the device and method used should always be considered when analyzing the progesterone concentration.

Distinguishing between corpus luteum insufficiency as the primary cause of fetal death, rather than its consequence, is certainly worth attention, especially in bitches who lose litters in the second half of pregnancy despite the exclusion of other causes. The reasons and pathogenesis of corpus luteum insufficiency, despite many studies, are still incompletely investigated and diagnosis often poses a problem because many factors, such as maternal health status, fetal death, infectious agents, trauma, or poor nutrition, may contribute to the primary decrease in progesterone level [34,45,46]. Some studies confirmed the association of corpus luteum insufficiency with the presence of IgE antibodies against endogenous progesterone in bitches [47]. These authors suggested a genetic link to the occurrence of hypoluteoidism in specific breeds, such as German Shepherds.

The diagnosis of primary corpus luteum insufficiency as a cause of pregnancy loss is complex. It requires confirmation of a gradual decrease in progesterone levels while assessing fetal viability. The possibility that the decrease in serum progesterone levels occurred due to fetal distress or abortion must be excluded. The function of the corpus luteum can be influenced by many endo- and exogenous factors that control its lifespan, and among the most important substances with luteolytic effects are PGF2a and antigestagens such as aglepristone [46,48].

If the diagnosis is confirmed, exogenous progestin supplementation therapy is recommended, of which medroxyprogesterone acetate (MPA), alternogest, or progesterone in oil are the most commonly used in dogs [34,49]. The administration of progestins in pregnant bitches may be associated with side effects such as the development of pyometra, septicemia, or placentitis, and the risk of prolonged pregnancy-causing dystocia [34,42,50]. Furthermore, progestogens, which have androgenic effects, may lead to masculinization of female fetuses [42].

Recently, the first documented case of mammary gland fibroadenoma in dogs, so far reported only in cats, has also been described. An Istrian Shorthaired Hound bitch was diagnosed with primary corpus luteum insufficiency in three consecutive pregnancies [51]. Based on the diagnosis established during the first pregnancy, in the second one with repeated suspicion of corpus luteum insufficiency, the bitch was treated with 1.65 mg/kg of progesterone in oil intramuscularly (PROGEST-E®, Fort Dodge Animal Health S.p.A., Bologna, Italy) daily, from day 19 to day 22, and every 48 h from day 23 to day 58. During the third pregnancy, based on previous experience, the bitch also received 0.075 mg/kg alternogest (Regumate®, Intervet Italia S.r.l., Peschiera Borromeo, MI, Italy) from day 8 and then every 24 h until day 52 of pregnancy. The dose was reduced to 0.058 mg/kg PO from day 53 to day 57 [51]. Both times, an increase in the size of the mammary glands was observed. One of the pregnancies required a cesarean section due to fetal macrosomia. Whereas, females born in the other pregnancy, according to the owner's report, showed no signs of estrus cycle during the first 3 years of life; moreover, in one female, clitoral hypertrophy and a blindly ending vagina were diagnosed. Such abnormalities may be related to progestin treatment in the embryogenesis stage, as they affect the development of the genital tract. If progestins are used as a supplemental treatment for hypoluteoidism, administration should not begin before days 30–35 to avoid genital abnormalities [49,52]. Since hypoluteoidism is an ovarian dysfunction and its treatment with progestins might lead to severe unwarranted effects, it is recommended to exclude affected bitches from breeding.

2.4.3. Hypothyroidism

An equally interesting topic is the relation between pregnancy and hypothyroidism. In general, hypothyroidism is a common endocrine disease in dogs. Many previous studies have demonstrated a heritable tendency in dogs of certain breeds, such as Toy Fox Terriers, Giant Schnauzers, Boxers, or Scottish Deerhounds [53]. Bitches with untreated hypothyroidism often exhibit a significant reduction in fertility that prevents natural pregnancy, as well as variable interestrus intervals, abortions, and stillbirths [54]. Studies have shown that they are also at risk of developing dystocia as a result of a prolonged duration of labor contractions and a decrease in their intensity. Puppies have been reported to have a higher incidence of low birth weight, low viability, and overall increased litter mortality [54]. The situation changed when levothyroxine supplementation was started, which was shown to reverse the negative effect of hormone deprivation on neonates' birth weight and high mortality [55]. The question of changes that occur in thyroid hormone concentrations during the course of pregnancy in bitches certainly requires further study. Some data indicated insignificant fluctuations that do not require adaptive hormonal supplementation during pregnancy [56,57]. In contrast, others suggested that significant changes may occur, especially during the second half of pregnancy, when the rate of fetal development increases, and therefore the rate of maternal metabolism should follow to meet the energy demands [42,58]. The same is true in human pregnancy, where adequate supplementation during pregnancy is necessary for its maintenance and must be gradually increased, as well as the fetal development and metabolic rates [59]. In bitches experiencing miscarriage, a significant decrease in thyroid hormones and a correlating decrease in progesterone levels were observed even one week before the first clinical symptoms [42]. The problem of levothyroxine supplementation recommendations and the need to increase the dose with the progression of pregnancy certainly require further research and attention; this is still a matter of debate according to different study results [56–58], because decisions made during treatment can have a significant impact on the outcome of the bitch's pregnancy and therefore the health status of the whole litter.

2.4.4. Maternal Microbiota and Its Effect on Puppy Survival

The development of the newborn microbiota is a gradual process, influenced by many factors. The environment, the health status of the mother and, above all, the intake of colostrum, are essential in the first hours of life [60].

The colostrum in mammals is the first food of crucial importance for the health of the offspring in the early neonatal period. In canine species, only 5–10% of antibodies can cross the placenta, so newborns are almost completely immune-deficient [61]. The timing and amount of colostrum intake immediately after birth are important for their survival and future health. The intestinal barrier is only permeable to IgG from the gastrointestinal tract during the first 12–16 h of life, so it is important to feed newborns with colostrum as soon as possible shortly after birth [61]. Although the exact mechanisms that determine the formation of the colostrum microbiota are not fully understood, studies seem to support the enteromammary hypothesis that the colostrum microbiota is shaped by gut bacteria present in the dam [61].

The relationship between colostrum microflora and intestinal microflora of neonates is confirmed by studies in which the composition of meconium collected from neonates was found to be similar to that of colostrum collected from dams [61]. Adequate microflora composition determines colonization of neonatal intestines, and this directly translates into their chances of survival, adaptation to the ectopic environment, and development of health status [60].

Studies on supplementation of pregnant bitches with prebiotics containing, among others, *E. faecium* and *L. acidophilus* proved their influence on producing better quality colostrum, which improved the immunity of puppies. Furthermore, the maternal microbiota, positively altered by supplementation, was transferred to newborns, making the offspring more resistant to gastroenteritis [62].

The type of parturition can also be a factor that affects the composition of the microflora of the colostrum [61]. Whether it is vaginal delivery, emergency cesarean section, or elective cesarean section, the colostrum samples contained different microorganisms. The colostrum of the bitches with vaginal delivery and elective cesarean section was characterized by a similar composition, unlike the colostrum collected from the bitches undergoing emergency cesarean section [61]. It should be noted that the differences may result from the fact that bitches in the latter group usually experience higher levels of stress as well as some degree of exhaustion.

Recent findings concerning the placental microbiome in healthy pregnant dams also indicate the possibility of early intrauterine bacterial colonization of the fetuses [63]. The bacterial composition of the neonatal meconium resembled that of the maternal vagina in the pups born by vaginal delivery. The ones born via CS had microflora similar to the oral and vaginal microbiota of the mother [63].

It should also be noted that the presence of the gut microbiota can contribute to better weight gain in puppies, which is essential for their future development and health status in adulthood [61,63]. Puppies in which bacteria could not be cultured from meconium had a slower growth rate compared to puppies with some bacteria present in their meconium [63].

It is also extremely important for the health of the offspring to maintain the balanced composition of their intestinal microbiota. The underdeveloped immune system can be easily affected by the gut microflora imbalance, which may lead to the growth and multiplication of undesirable pathogens such as Proteobacteria and Pasteurellaceae, and may progress towards serious disorders, mainly Fading Puppy Syndrome (FPS), which can often cause fatal results [64]. The evaluation and culture of the intestinal microflora composition and the fecal samples of puppies can therefore become a helpful tool in the diagnosis of neonatal problems.

2.4.5. Challenges in Pregnancy Nutrition

Proper nutrition has an important impact on the health of the mother and her offspring during pregnancy and the subsequent lactation. The connection between reproduction and nutrition is undeniable and awareness of key aspects of bitch nutrition during pregnancy, perinatal period, and lactation is a prerequisite for success in healthy breeding. Facing differences in reproductive animals' nutritional needs, Association of American Feed Control Officials (AAFCO) has published separate nutrient profiles with different feeding trial protocols for maintenance, growth, and gestation-lactation periods [65]. Energy requirements gradually increase in the bitches from day 40 of gestation, on average by nearly 10% for each additional week, to about 1.25–1.5 times those of normal maintenance requirements. The amount of food provided should be properly recalculated to prevent excessive weight gain [66]. Many owners believe that large amounts of energy are required from the very beginning of the pregnancy. They start to overfeed pregnant bitches, which can result in increased fat deposition [66]. Unfortunately, obesity is one of the factors that strongly predispose to dystocia, which most likely decreases uterine contractions during labor due to excessive fat deposition in muscle tissue [6,67]. It is also worth noting the practice of dividing the daily dose into several smaller meals. It prevents the stomach from overloading with too much food, especially in late pregnancy, when a large part of the abdominal cavity is filled with the gravid uterus [66].

Just as important as quantity is the quality and type of food provided. A study conducted among a group of breeders in Canada about their feeding practices has shown that a substantial percentage of breeders were unable to determine whether the commercial food they were feeding met the AAFCO recommended standards. In addition, nearly 16.9% of breeders participating in that survey, who fed commercial diets, reported offering diets to pregnant and lactating bitches that are not intended for the gestation-lactation period. Moreover, almost 9% of breeders weaned puppies onto a diet not designed for the growing puppies [68]. An alternative to commercial diets available on the market is the bones and raw food diet (BARF), prepared by owners or breeders at home, sometimes with specific additives [69]. Its nutritional value for pregnant and lactating bitches and growing pups is difficult to assess and very likely BARF diet would not meet the demands of breeding animals [68,69]. This type of diet may also increase the risk of infection with several pathogens and parasites responsible for reproductive disorders, including *Salmonella* spp., *Campylobacter* spp., *Neorickettsia* spp., Shiga toxin and hemorrhagic *Escherichia coli*, *Toxoplasma* sp., and *Neospora* sp. [68].

When selecting foods for pregnant and lactating bitches, attention should also be paid to the content of the particular ingredients and the use of various additives (e.g., mineral supplements, dairy products); breed-related demands should also be considered here, especially regarding calcium content [66,70]. Some breeders believe that calcium addition can ensure adequate lactation and healthy development of puppies [66]. However, a lack of consideration in calcium supplementation may cause more harm than benefit. Calcium homeostasis is tightly regulated by parathyroid hormone (PTH) and calcitonin [71]. During pregnancy, a physiologic decrease in calcium serum levels leads to a series of reactions designed to elevate it again. Under the action of parathyroid hormone (PTH), an increased release of calcium from bones and enhanced absorption from the gastrointestinal tract takes place [70,71]. When a bitch receives extra calcium with food, the serum calcium concentration remains sufficiently high, and there is no need to activate the mechanisms of calcium release from bones and increase the intestinal absorption, which downregulates the PTH activity. Then, with the onset of lactation, a bitch cannot precisely respond to the high calcium demand and the risk of developing eclampsia increases [66]. Eclampsia most often occurs in small and miniature breeds, usually after birth, but sometimes might also be observed during pregnancy [15]. Clinical signs are initially nonspecific, but can progress to muscle tremors, ataxia, elevated rectal temperature above 39.7 °C, tetany, and in severe cases, even death [70]. Treatment is based primarily on intravenous calcium administration.

If possible, the litter should be weaned permanently and hand-reared to prevent the risk of hypocalcemia relapse.

Another nutrient that has also attracted interest is folic acid, recommended for pregnant women. Experiments conducted on mice as experimental animal models also demonstrated that folic acid is necessary for normal embryonic-fetal development [72,73]. Some studies have shown its important role in the prevention of cleft lip and/or palate in brachycephalic puppies [74–76]. However, no similar relationship was proven in a study conducted on bitches of breeds considered as non-predisposed (Labrador, Golden Retriever, and Labrador/Golden crosses) to the cleft palate problem [77]. If considering supplementation in dogs, studies suggest one should start folic acid when a bitch is intended for breeding/enters the estrus phase, since the medullary tube closes during the first part of gestation to reduce the risk of developing cleft lip or palate [66].

The crucial importance of a proper diet for the health of the pregnant bitch and her offspring remains undisputed. In addition to portioning daily food intake and administering it in appropriate amounts to avoid excessive weight gain, the trend of administering certain supplements deserves attention, especially calcium and folic acid supplementation. A rational approach to the use of nutritional supplements and the application of sound feeding practices are key aspects in providing adequate care to the pregnant bitch.

2.4.6. Pregnancy Monitoring

In routine care of pregnant females, periodic examinations are recommended to evaluate the pregnancy development. US pregnancy scanning, if performed by an experienced person, would allow for early detection of abnormalities. Defective trophoblast invasion is believed to be the basis for abnormal blood flow in the uterine artery, which in turn leads to uteroplacental insufficiency [78]. Alarming changes include reduced diameter and abnormalities in the contour of the gestational vesicle, lack of viability, increased placental thickness, increased fluid echogenicity, and increases in RI (resistivity index) and PI (pulsatility index) of uteroplacental arteries of conceptuses [78]. The appearance of changes correlates with the onset of embryo resorption around weeks 2 and 4 of gestation [78]. The umbilical artery and the fetal renal artery were also useful in evaluating the quality of fetal-maternal flow; its abnormal parameters suggest gestational pathology [79]. Differences were also reported in the blood flow of the uterine and umbilical arteries in small, medium, and large breeds of dogs during the second half of pregnancy. Conversely, during this same period, FHR did not vary between small, medium, and large breeds [80]. This indicates that physiological variations should also be considered when a gestational ultrasound is interpreted in different breeds of dogs. Doppler ultrasound also has a prognostic value in assessing the effect of the morphology of the ductus venosus (DV) waveform on canine neonatal mortality. One study showed that litters in which three-phase waves of DV (tDVw) were recorded had a higher (almost 21 times more) chance of neonatal mortality (one or more dead pups per litter) than those with only two-phase waves (dDVw) [81]. DV postnatal patency could also lead to a congenital portosystemic shunt (CPSS). Non-invasive color flow mapping (CFM), which is used routinely to diagnose CPSS in adult dogs, has been shown to be equally useful to confirm or exclude DV closure within the first ten days after delivery in canine neonates. This might become useful for early screening tests evaluating the DV patency, especially in puppies of breeds predisposed to congenital portosystemic shunts [82]. Changes observed during ultrasound examination may indicate the development of pregnancy pathology long before the first visible symptoms appear [83].

In addition to ultrasonography, electrocardiography can sometimes be useful to detect pathologies in specific electrocardiographic parameters, including QRS waveforms, maternal heart rate (MHR), or fetal heart rate (FHR) [84]. The correct MHR should be around 70–120 bpm, while the fetal heart rate is between 180 and 220 bpm [85]. Any factor causing fetal distress is most often manifested by fetal bradycardia and, therefore, changes in FHR. A more complete picture of the disorder can be obtained by analyzing the FHR/MHR ratio, which more accurately reflects fetal health in relation to maternal health than single FHR values [85]. This is especially the case when combined with Doppler examination, which enables one to visualize and compare the blood flow of the uteroplacental and umbilical arteries in normal and abnormal conceptus [78]. Accurate understanding of the changes that occur can contribute to early diagnosis for identification and, if possible, exclusion of the problem to ensure the safety of the developing offspring.

3. Type of Delivery, Perinatal Complications, and Proper Management of the Bitch and Her Offspring

3.1. Normal Parturition

The pregnancy in the bitch is expected to last between 57 and 72 days from the day of mating [40,86]. Estimating the exact date of parturition in dogs is sometimes difficult. The timing of ovulation is usually delayed in relation to both the LH surge and the liberation of the immature ovum, which needs to mature further to be competent for fertilization in the oviduct [87]. It is also necessary to consider the 24–48 h period in which the ovum can be fertilized and the survival of the sperm in the reproductive tract, depending on the date of mating or the type of insemination.

Serum P4 concentration is considered an essential parameter for the detection of the LH peak, ovulation, and parturition date. Serum P4 concentration begins to increase before the onset of the luteal phase, and the first day when P4 is >1.5 ng/mL is considered to indicate the peak of LH [88]. On the day of ovulation, which occurs almost 48–60 h after the peak of LH, the length of pregnancy is reported to be 63 ± 1 days (62–64 days) [40,88]. The accuracy of parturition timing using prebreeding P4 concentration is described to be 67, 90, and 100% within 65 ± 1, ±2, and ±3 days, respectively [88]. Instead, vaginal cytology for continuous monitoring of the reproductive cycle could be helpful in determining the appropriate time to start serum P4 monitoring [86]. Vaginal cytology can also be used to determine the first day of diestrus (D1), after which the pregnancy should last 57 ± 3 days (54–60 days) [86]. The determination of D1 does not present difficulties because the image in this phase of the cycle clearly differs from that in estrus by the decrease in the percentage of cornified cells from 80–100% to at least 20%, and the appearance of parabasal cells and neutrophils. These changes occur in less than 24–48 h [86].

A normal parturition in a dam is divided into three phases. In phase one, the progesterone block is lifted with the onset of uterine contractions and cervical dilation. The duration varies between 6 and 24 h, possibly up to 36 in a primiparous or very nervous bitch [40]. During the second phase of labor, uterine contractions are combined with abdominal wall contractions, and the fetuses are expelled sequentially from the uterus. Although there may be a longer interval between the first and subsequent pups, births usually occur every 15 to 120 min, depending on the size and breed of the mother and the number of puppies. The third phase involves placental expulsion and occurs immediately or within 15 min after the birth of each pup [40]. However, several placentas may be delivered at once [89].

3.2. First Aid Delivery

Labor, for a variety of reasons, does not always proceed as expected and veterinary assistance may be required when complications occur. A fast and accurate diagnosis of dystocia is essential to increase the chances of survival of a newborn. The most important predisposing factors to dystocia are those previously discussed, including maternal age, size, breed, and the number of puppies [6]. Difficult labor can be approached with conservative methods or surgical intervention. Conservative methods involve the use of pharmacological substances and/or manual attempts to correct the fetal position in the birth canal [5]. When trying to pharmacologically resolve difficult labor, oxytocin is administered most frequently [2]. Calcium gluconate and glucose are also recommended [89]. If there is no response to drug therapy or obstetric maneuvers aimed at correcting dystocia, and fetuses are confirmed to be alive, the dam should immediately be submitted for an emergency cesarean section [5]. When considering the potential impact of the cesarean section on the health of puppies, two key things should be remembered. One is related to whether the cesarean section is performed as an emergency or a scheduled procedure, and the other is the type of anesthetic protocol used during the intervention.

3.3. Emergency Cesarean Section

The emergency cesarean section is a procedure associated with a high risk to the life of both the bitch and the puppies. Mortality rates can be as high as 20% of puppies and 1% of bitches, especially in small and predisposed breeds, including predominantly brachycephalic breeds like French Bulldog, Boston Terrier, Chihuahua, or Pugs [1–3].

The most common maternal causes include difficult fetal passage (narrow birth canal, presence of undetected earlier congenital vaginal defects), disturbed labor (inertia, spasms), prolonged pregnancy, uterine disturbances (torsion, malformations), and poor psychogenic status (restlessness, abnormal or aggressive behavior) [1,3]. Fetal factors include puppy size, especially oversized ones, fetal malpresentation, and the total number of the puppies [3]. Single pup pregnancies or abnormally high numbers of pups should always be treated with special care because they are serious predisposing factors to delivery complications.

The longer it takes to deliver a puppy and the later the surgery is started, the less chance of survival a puppy has. Severe fetal distress translates into heart rate per minute; a drop below 180 bpm is an immediate indication for the surgical intervention, whereas below 120 is a very poor prognosis for the puppy's survival [21,90]. The published data directly indicate that puppies born via emergency CS are more than 7.3 times less likely to survive than those born via elective CS [21,91]. The most common cause of the pup death is hypoxia due to prolonged labor and placental separation, often associated with large numbers of fetuses, malpresentation, and wedging in the birth tract [3,92].

3.4. Planned Cesarean Section

One of the ways to prevent puppy loss due to dystocia is to schedule a cesarean section. Specific indications for an elective CS are breed predisposition, maternal age, and litter size. Females with previous birth complications and those of high breeding value should also be considered. It has been proven that in some situations, scheduling a cesarean section is the safest method of pregnancy termination both for the dam and her litter [91]. However, performing a cesarean section without clear indications at the owner's own request is the subject of much controversy and is considered unethical by some members of the veterinary community. In the case of scheduled operation and for the safe surgical outcome, the appropriate anesthetic protocol should be used, which considers both the dam and the fetuses in the uterus.

In each of the types of delivery, the puppy viability assessment should be performed and adequate treatment instituted if abnormalities are detected [5,93]. Timing is very important for puppy survival, further development, and overall health status.

3.5. Neonatal Assessment

Neonatal viability can be evaluated using a modified APGAR score proposed by Veronesi [94]. Five parameters are assessed, including mucus color, heart rate, reflex irritability, motility, and vocalization. Points (0–2) are given for every parameter depending on the condition of the newborn [94]. A heart rate above 220 bpm (beats per minute) gets 2 points, between 220 and 180 gets 1 point, and below 180 gets 0 points. Respiratory effort is assessed when the newborn is crying (>15 respiratory rate (rr)), moderately crying (between 6 and 15 rr). Reflex irritability can be vigorous, a grimace alone can be visible, or there may be no response. When assessing motility, it is possible to observe active motion, some reflections, or flaccid movements. The last parameter, mucus color, can be assessed as dark pink, pale, or cyanotic. The total amount determines the final APGAR score, which identifies the degree of a newborn's distress: 7 to 10 points mean no distress; 4 to 6, moderate distress; and 0 to 3, severe distress.

For brachycephalic breeds born by cesarean section, an exclusive scale has been designed for the evaluation of brachycephalic newborns [95]. The scale was specially modified after the Veronesi et al. scoring system and adapted to their characteristics due to the lower degree of vitality after birth often observed in newborns of these breeds. A heart rate above 180 bpm gets 2 points, between 120 and 180 gets 1 point, and below 120 gets 0 points [95]. Other parameters including respiratory effort, reflex irritability, motility and mucus color are scored the same as in the APGAR scale proposed by Veronesi et al. [94].

Neonatal viability reflexes (NVR) can also be assessed. The purpose of this scale is to evaluate a newborn's postnatal depression based on active searching for the mammary gland and the strength of the sucking, measured as weak, moderate, or normal [5].

These reflexes are essential in the early postnatal period to ensure newborn feeding and survival. The suckling reflex can be assessed by inserting the gloved tip of the smallest digit into the mouth of the neonate and checking the suckling force; it can be described as strong (5 suckles/min), weak (>3 suckles/min) or absent. The rooting reflex can be assessed by approaching the nose of the neonate with the forefinger and thumb shaped into a circle and checking whether the neonate inserts its nose into this circle; immediate, slow or absent muzzle fitting inside the circle can be observed. The righting reflex of the neonate can be evaluated by placing it on its back on a soft surface and verifying whether it is able to return to the sternal recumbence. Those reflexes are also scored from 0 to 2 points, and the interpretation is as follows: 0–2 points: weak viability; 3–4 points: moderate viability; 5–6 means normal viability. Weak reflexes are often the primary sign of hypoxia in puppies [5].

The better the viability of the newborn, the higher score it receives on each scale mentioned above, and therefore the prognosis of its short-term survival would be better. The transient decline in vital functions immediately after birth is often observed even in eutocic pups [5]. Healthy newborns quickly regain their vitality; however, due to the special vulnerability in the first few hours of life, each should receive special attention and care. The critical and weak puppies (0–6 on the APGAR score) need even more attention and care, as they are much more likely to have a fatal outcome within the first 2 h of life [5].

To increase the chances for survival of weak newborns, prompt and efficient resuscitation is crucial. The protocol proposed by Traas et al. adopted the following sequence of actions according to their importance: warmth, airway, breathing, circulation, and drug administration [96]. In each case, the treatment should consist of clearing airways, drying of a puppy, respiratory stimulation (cardiopulmonary resuscitation), oxygenation, ventilation by a mask or endotracheal tube and, when necessary, fluid administration [5,96]. Intravenous administration of drugs such as doxapram, aminophylline, or epinephrine could also be considered [96]. Naloxone can be given to reverse the anesthetic effect of opioids administered to the dam prior to the cesarean section. In distressed newborns at risk of sepsis due to hypoxia-induced bacterial translocation, the administration of safe antibiotics (these include cephalosporins, penicillins, clavulanic acid, macrolides, trimethoprim-sulfonamide, and amikacin) should be considered [18,96].

Blood glucose measurements are also helpful in assessing the risk in frail puppies. Hypoglycemia (<40 mg/dL) within the first 8 h of life was associated with high mortality in newborns during the first 24 h of life. At 24 h of life, the level of 92 mg/dL or below was associated with a higher risk of mortality during the entire neonatal period (1–21 days) [4].

A second useful parameter for predicting neonatal mortality within the first 48 h of life is the lactate level. Its concentration in the umbilical cord blood reflects the presence of acidosis in a neonate, with high levels noted in distressed pups and low levels in healthy, vigorous ones [97]. Both parameters combined with the neonatal APGAR score can be a useful tool for the early identification of weak newborns that require special care, thus reducing neonatal mortality.

Awareness must also be given to the fact that each evaluation should always be objectively performed by experienced practitioners. Low scores do not always determine a negative outcome for a puppy; however, high neonatal mortality is more frequently correlated with it [90]. When performing a viability evaluation, each newborn should also be checked for the presence of birth defects such as cleft lip/palate, presence of a hernia, or atresia ani.

3.6. Postoperative Pain Management

An important aspect that should not be overlooked is proper pain management for the dam after a cesarean section [98]. In dogs, as with all mammalian species, the mother-newborn bond is crucial for the suitable development of maternal behavior that guarantees offspring survival. In dogs, both natural birth and assisted delivery might be a stressful and painful experience, which could affect the way a bitch cares for her litter [99]. A female in pain may present a weaker maternal instinct and refuse to care for or feed her puppies. This, in turn, puts them at risk of developing hypoglycemia and hypothermia, both very dangerous in newborns [100,101]. However, when administering analgesics to the bitch, one must be aware of their ability to pass through the mammary gland into milk. The most commonly used analgesics belong to nonsteroidal anti-inflammatory drugs (NSAID) or opioids [101]. Due to the risk of damaging the immature kidneys and liver of the newborns in the case of using NSAID, or being more sensitive to the sedative and respiratory depressant effects of opioids, especially in pups under three weeks of age, not many drugs are approved for use in pregnant and lactating bitches, significantly affecting their post-surgical recovery and welfare [101]. Studies have shown that cimicoxib and carprofen exhibit low penetration through the mammary gland barrier [101,102]. Treatment of lactating bitches for a short period of time seems to be safe for the offspring; however, carprofen should be used with caution in bitches with diagnosed mastitis, in which case the drug can penetrate the milk in higher amounts [101].

All of the previously mentioned issues are only a fraction of the knowledge necessary to ensure a safe pregnancy outcome and the health of the litter. In the case of emerging complications, the correct diagnosis and prompt treatment are essential. Providing proper care for both the mother and the offspring would significantly reduce the risk of neonatal mortality.

4. Newborn's Related Factors

Birth is the moment of fetal-to-neonatal transition which requires very complex adaptive changes which need to occur in a short period of time. Adaptation failures will result in newborn compromise and homeostasis disturbances, and will determine the success of its rearing. During the transition period, the neonatal condition is self-occurring or caused by factors beyond the newborn's control, but also the health status and behavior of the mother, as well as improper care and environmental conditions, may play a role in the newborn's welfare and development [92,103].

4.1. Fetal Congenital Malformations

An important issue that can affect intrauterine life or the health status of the newborns is the presence of congenital malformations. The canine congenital malformations are structural or functional abnormalities present at birth that may interfere with the viability of newborns, thus contributing to neonatal mortality [89,104]. They can be caused by genetic factors, inherited and/or breed-related, or exposure during pregnancy to teratogenic agents (toxins, chemicals, irradiation, or excessive supply of vitamins A and D) [104]. Studies that examine the frequency of malformations have shown that the most common ones include: cleft palate (2.8%), hydrocephalus (1.5%), anasarca (0.7%), cleft lip (0.6%), polydactyly (0.5%), segmental intestinal aplasia (0.4%) or atresia ani (0.4%), and others [104]. The pups with diagnosed malformations in the above study were 6.7% of 803 animals examined. The mortality rate caused by these abnormalities among the pups during the first two days of life was recorded at 61.4%, while from day 3 to day 30 it was 38.6%. In the vast majority of cases, single malformations were present [104]; much less frequently, simultaneous occurrence of two or more in one fetus was observed [105]. The risk of malformations is much more common in pedigree dogs, and the brachycephalic dogs are among the most predisposed breeds [2,21]. The accumulation of genetic defects over the years due to inbreeding or selective breeding leads to reduced genetic diversity and increased predisposition of these animals to present anatomical anomalies [10]. However, not all anomalies are associated with an increased risk of mortality. For example, puppies born with additional fingers (polydactyly) are unlikely to suffer from other health problems. The vast majority of congenital malformations do, however, have a marked effect on longevity; in the case of anasarca, hydrocephalus, cleft palate, atrophied intestine fragments, or atresia ani, the life span will be substantially affected. Puppies with a cleft palate often develop aspiration pneumonia due to milk entering the airways during sucking [106]. Those with intestinal fragment atrophy are unable to pass meconium and feces, which initially leads to painful constipation and over time death [107]. For some anatomical malformations, surgical correction may be considered. For example cleft palate surgery [106,108] and surgical operation of the posterior gastrointestinal obstruction [107,109] have been reported with variable success. However, in most cases and with significant severity of defects, euthanasia of affected puppies is often chosen to avoid unnecessary suffering.

The diagnosis of malformations is often difficult and many of them remain undetected mainly due to a lack of clinical examination or inaccurate clinical examination or the initial absence of clinical signs. Investigation of malformations in newborns immediately after birth is fundamental because an early diagnosis of these conditions could lead to timely clinical interventions and help to minimize mortality.

4.2. Birth Weight

Birth weight that reflects intrauterine growth is one of the most important determinants of neonate survival [103,110]. Puppies weighing 25% less than the average weight of a newborn in the given breed are suspected to have a significantly higher mortality rate [111]. The risk of death for puppies characterized by low birth weight is 12 times higher compared to other newborns of normal birth weight from the same litter [112]. Low birth weight is usually followed by fetal immaturity that limits the adaptation of the newborn to the postnatal environment. Such puppies in comparison to average neonates of the same litter have a lower ability to maintain proper body temperature due to a higher body surface/weight ratio and are prone to hypothermia [113]. Moreover, puppies with low birth weight have reduced energy supplies, are less vigorous, and as a result, cannot compete with the rest of the puppies in suckling and thus do not take a sufficient amount of colostrum, which makes them prone to hypoglycemia and dehydration [103,111]. As a result, such newborns would be susceptible to developing growth and neurocognitive deficiencies, cerebral palsy, intracranial hemorrhage, sepsis, hyaline membrane disease, apnea, and retrolental fibroplasia [114].

Factors that influence birth weight could be classified as maternal, fetal, and placental [115]. Dam size, weight, body condition, age and breed, environmental conditions, as well as litter size were described as important variables that affect fetal weight [103,116]. Due to the fact that birth weight reflects intrauterine nutrition of the fetus, the condition of the placenta is one of the basic factors affecting the newborn [113]. Numerous studies have been carried out on human and other animal species that provide information on the relation between the birth weight and placenta [117–121]. Several studies were conducted on dogs [113,122]. In small and toy breeds, authors revealed a positive correlation between the birth weight of a puppy and the placental weight and its total area [113]. The total area of the placenta, transfer zone area, and total vascular area were also proven to correlate strongly with the placental weight [113]. The research focused on the histological examination of the placenta revealed that necrosis was a frequent finding in dogs. However, it was shown that only multifocal-confluent necrosis was associated with a higher risk of newborn death [122]. The placental examination should always be considered as it may provide prognostic information about the puppy's future development. However, further investigation on a larger scale on breeds from the remaining groups should still be carried out.

4.3. Noninfectious Conditions in Newborns

All newborns deserve great attention at this critical moment shortly after birth. The birth weight and daily gains should be carefully recorded during the first weeks of life because adequate growth reflects the puppy's vitality, health, and proper development [103,111]. Stasis in weight gain is often the first alarming signal in the case of most neonatal diseases. It indicates that the newborn is, for some reason, weakened and does not take enough food [92].

The body of a newborn dog works very differently from an adult dog's body. It is worth noting that in almost every case, the particular disorders of homeostasis in newborns occur together and their symptoms overlap.

4.3.1. Hypoxia

Hypoxia is the first emerging state responsible for 60% of all neonatal deaths. The oxygen deficits in neonates usually start from the dysfunction of umbilical circulation that can be caused by prolonged and/or complicated parturition, i.e., umbilical cord vessel compression or rupture, as well as too early placental detachment while a puppy is still in the birth canal [92]. In addition, it was reported that puppies born in posterior presentation were more susceptible to respiratory and metabolic acidosis than those born in anterior presentation [123]. Furthermore, administration of oxytocin, anesthetic agents, surgical preparation, and the cesarean section may also contribute to newborn hypoxia [92]. Mildly hypoxic neonates are able to shift circulating blood from intestines, kidneys, spleen, or skin and towards heart, brain, diaphragm, and adrenal glands. Severe oxygen deprivation decreases fetal heartbeat, leading to tissue hypoxia and ischemia and finally multiorgan failure. Affected puppies are more prone to amniotic fluid aspiration, and their mucous membranes become less resistant to the pathogens' penetration [92].

4.3.2. Hypothermia

The next noninfectious condition that may threaten the life of the newborn is hypothermia. Immediately after birth, any neonate is introduced into the adversely cool environment in comparison to the intrauterine conditions [111,124]. The cooling sensation is escalated by the amniotic fluid residuals. Hypothermia could lead to a significant drop in the heart rate (200–250 bpm at 35.6 °C vs. 40–50 bpm at 21.1 °C), respiratory rate decrease, and loss of suckling reflex, which in turn can cause dehydration and gastrointestinal disorders. Such puppies would be much more susceptible to infections (e.g. herpes virus, bacteria, and opportunistic pathogens) [92]. As thermoregulation is not fully developed in newborn puppies and they are not able to shiver until day 7 of life, to compensate for the temperature loss, they can only use the high energy-consuming thermal conduction [111]. Therefore, the body temperature of a puppy strictly depends on the efficiency of the mother's care. If the dam's maternal instincts are inadequate, the proper room temperature and additional heat sources are essential for puppies' survival [111].

4.3.3. Hypoglycemia

Hypoglycemia might be the consequence of the low core temperature or occur separately with the suckling failure due to a variety of reasons. A neonate is born with immature liver and the energy stored as hepatic glycogen is usually enough only for the first day of life [92,125]. Moreover, some factors such as mother's malnutrition or insufficient nutrition during pregnancy could reduce the newborn's glycogen supplies. Within 8–12 h after being born, a puppy is forced to rely on colostrum/milk intake to maintain the proper glucose blood level. Reduced or no food intake would result in rapid glycogen reserves depletion and the development of hypoglycemia with typical symptoms of nervousness, vocalization, irritability, and intense hunger that, if not corrected, would be followed by lethargy, mental dullness, depression or stupor, seizures, tremors, and finally the death of the puppy [92]. It is worth noticing that the severity of hypoglycemia symptoms does not always correspond to relatively low blood glucose levels. Moreover, the clinical condition reflects the puppy's energy reserve. That explains why empirically based treatment with intravenous or oral or glucose in many cases does not lead to a clinical improvement in the neonate [123].

4.3.4. Dehydration

Dehydration represents another homeostatic disorder which is usually a result of a non-properly functioning excretory system, but can also occur with inadequate milk intake. Newborn's kidneys are not fully developed at birth and need a minimum 2–3 weeks to undergo nephrogenesis to become fully functional [111,125]. The early kidney filtration is characterized by a slow clearance of fluids, increased sodium loss, and, most importantly, the inability to conserve water. That is why neonates are extremely susceptible to dehydration [111]. It is important to remember that water turnover rate in pups is double than in adult dogs and they demand an intake of approximately 132–220 mL/kg/day [111]. Usually, a sick puppy, independently of cause, brought to a veterinary clinic presents a set of common symptoms such as low body temperature, malnutrition, and dehydration [111].

4.4. Neonatal Diarrhea

Neonatal diarrhea most commonly results from improper postnatal care and nutrition; however, it can also be caused by infectious agents, especially in the case of high pathogen exposure or adverse environmental conditions. The most common cause of non-infectious diarrhea is overfeeding the puppy or offering unsuitable food [92,126]. Most often it is noted in orphaned puppies and underfed neonates, which require complementary nutrition [92,126]. Thus, it is essential to choose a well-composed milk replacer and carefully calculate the intake for the puppy weight, according to the following rule: 20% of actual body weight per 24 h divided into 6–8 portions [92]. Poorly balanced diet, wrong feeding schedule, or inaccurate amounts would contribute to illnesses and other health problems.

4.5. Passive Immune Transfer and Colostrum Intake

It has to be remembered that the primary reason for the majority of postnatal disorders is the insufficient consumption of colostrum, and thus ineffective passive immunity transfer [14,127,128]. Only 10–20% of the mother's IgG goes through the endotheliochorial placenta in dogs. However, other authors report that puppy serum IgG concentration before colostrum intake contains only 5% of immunoglobulins G compared to adult dogs [129]. The colostrum intake immediately after birth is essential for neonatal survival, development, and further health [130]. Colostrum, apart from IgG, IgA, and IgM, essential for passive immunity of a newborn, also contains nutrients, lysozymes, lactoferrins, white blood cells, cytokines, hormones (cortisol, insulin, thyroxin, somatotropin, growth hormone), specific microbiota, and several growth factors (insulin-like growth factors, epidermal growth factor, nerve growth factor), all needed for proper development [61,129,130]. The most important factor in colostrum intake is the time of ingestion which should not be later than 12–16 h of life [61,130]. This limitation is an effect of neonate digestive wall differentiation. From the moment of birth up to 4 h of life, around 40% of colostrum immunoglobulin (Ig) could be absorbed. This ability decreases gradually up to 12–16 h after parturition, when the junctions between enterocytes become tight and the intestinal wall becomes impermeable for Ig and thus its absorption is not possible any longer [130]. Since the sufficient intake of colostrum controls the risk of newborn puppies' mortality, colostrum deprivation greatly increases the risk of necrotizing enteric disease and septicemia [130].

In the case of the lack of colostrum or its insufficient supply, puppies need to be hand-reared using commercial or homemade milk replacer. However, balancing the formula to cover the needs for nutrients, energy, and volume might be quite challenging, especially for the homemade one. Breeders can also look for a foster dam with a litter of similar age to ensure sufficient immune protection [130]. It was also reported that the oral administration of canine serum immediately after birth resulted in sufficient protective IgG level (2.3 g/L) in puppies' serum [131]. It is recommended that plasma donors live in the same environment. Moreover, promising results were obtained with the use of serum containing specific antibodies against common canine pathogens [131]. Many products are available on the market; unfortunately, no milk substitute can completely replace the colostrum or natural milk. Immunoglobulins present in some of the formulae can, to some extent, replace the effect of colostrum by coating the intestinal epithelium and creating a barrier against pathogens' adhesion and translocation to the bloodstream [131]. Some authors also suggested the establishment of dog colostrum banks, similar to farm animals' husbandry [131].

Toxic Milk Syndrome

Another non-infectious risk factor which could appear in newborns is toxic milk syndrome (TMS), caused by the milk being contaminated with bacterial toxins [132]. The most common cause of TMS is acute mastitis or metritis present in dams [132]. Toxic milk syndrome usually affects puppies from birth to two weeks of age. The pups become weak and cry intensively. In most cases, diarrhea and intestinal gas accumulation occur. Frequently, due to intensive diarrhea, the puppy's anus becomes red and swollen. In the authors' own observations, often the strongest and the most eager to suckle puppies are affected first. If TMS is suspected, suckling should be immediately discontinued, puppies should be fed with milk replacer, and suitable treatment in a dam should be started.

4.6. Maternal Care

Dam's behavior after parturition is strictly related to the offspring's survival, and the lack of effective maternal care exposes pups to environmental factors dangerous to their health [133,134]. The most common signs of maternal neglect are behavioral: the refusal to permit nursing or abandoning the litter; and physiological: the complete lack of lactation or insufficient lactation. For that reason, the most common effect of maternal neglect is initially hypothermia followed by hypoglycemia and dehydration [92]. On the other hand, some bitches may present excessive maternal behavior characterized by very intense licking and cleaning, which could also lead to hypothermia of a newborn [133]. Breeders can be assured that in abandoned puppies, accurate environmental conditions, scrupulous nurture, and suitable diet should be sufficient to raise a healthy litter. It should also be remembered that maternal care is essential not only for puppies' health, but also for their socialization and behavioral development [133].

The most important conclusion regarding postnatal factors interfering with puppy survival is that most of those factors could be avoided with proper and conscious postnatal care. That is why understanding the mechanism of neonatal problems is essential for successful breeding [103].

5. Infectious Factors

A physiologically immature puppy without proper neonatal care will be more prone to suffer from inadequate colostrum intake, and thus will have deficient immunity and increased risk of infection. A bitch may come into contact with different microorganisms at different stages of life. The outcome of pregnancy and the viability of the offspring are most influenced by dams' immunization (vaccinations), mating hygiene, care during pregnancy and perinatal period, as well as the conditions in which the mother and her puppies are housed for the first few weeks of their lives. Infectious agents are considered the second main cause of a high mortality rate after dystocia [92]. Microorganisms can be transmitted from bitch to pups during pregnancy through the placenta, during delivery, and later from vaginal and oronasal discharges, feces, urine, or milk [135].

5.1. Bacterial Agents

Bacterial agents most commonly found in pregnancy loss and increased neonatal death include *Escherichia coli, Staphylococcus* spp. *and Streptococcus* spp., *and Klebsiella pneumoniae* [92, 135–137]. Those detected less frequently include bacteria such as *Brucella canis, Proteus mirabilis, Pseudomonas aeruginosa, Mycoplasma* spp., and *Campylobacter jejuni* [138–142]. Their presence can be associated with abortion, stillbirths, reduced neonatal viability, and even temporary or permanent infertility in adult animals [135]. In the course of most bacterial infections, septicemia is considered the leading cause of death in puppies under 21 days of age [135]. The course of the disease is usually hyperacute and the death occurs shortly thereafter. However, subacute cases have also been observed [4]. The care of a newborn suffering from septicemia is very difficult due to the very vague clinical signs and the rapid progress of the disease, which significantly hinders diagnostic and therapeutic efforts; hence, the prognosis in such cases is very cautious [92]. Among many indicators reviewed for their usefulness in the assessment of mortality risk, the viability assessment carried out using the modified APGAR scale and the colostrum intake measured by blood glucose concentration during the first 24 h of life has proven to be useful for canine neonates [4,94]. It should also be mentioned that ticks can be a source of infection with dangerous bacteria. Cases of fetuses being infected by their mother suffering from tick-borne diseases caused by the *Anaplasma platys* bacterium have been documented [143,144]. The detection of bacterial DNA in newborn pups of seropositive dams in the absence of clinical signs is suggestive of the ability of the bacterium to cross the placental barrier. Moreover, its capability to infect fetuses as early as the first half of pregnancy has also been demonstrated [143]. More research is needed to determine the effects of infection during the fetal period on the life and health of the offspring, as well as to raise awareness among vets and owners to administer appropriate bacterial prophylaxis to their pets.

5.2. Viral Agents

Viruses are a common cause of reproductive failure in companion animals. Their small size facilitates crossing the placental barrier, leading to pregnancy losses either by transplacental transmission itself and direct infection of embryos or fetuses or, less frequently, by severe debilitation of pregnant animals in the absence of congenital infection [145]. Canine parvovirus (CPV-1) and canine herpesvirus (CaHV-1), among others, are mainly responsible for such problems. In the case of herpesvirus infection, litter size disproportion often develops, neonates are born dead or weak, and mortality in the first 8 weeks of life is usually high [146]. In the case of canine parvovirus infection, depending on the stage of pregnancy in which the infection occurred, embryonic death and resorption are possible in the early stages, while stillborn or weak pups are observed in the later stages. Acute infection in puppies in the first days of life is present with vague general signs (no food intake, hypoglycemia, fever) and usually soon leads to the puppy's death [147]. The morbidity of the virus is usually 100%, and mortality among puppies younger than 6 months without medical treatment reaches 91%, which can be reduced with suitable treatment [148]. The effective management of viral infections in breeding dogs is primarily based on extensive preventive measures. Various immunization protocols are available, including vaccination against canine parvovirus [149] and canine herpesvirus [146]. Vaccination is the most effective way to control the spread of both infections in dogs and to prevent the development of clinical forms of diseases. Vaccination eliminates the risk of infection during and after pregnancy, increasing the pup's chances of survival.

5.3. Parasitic Agents

When talking about infectious factors, one cannot forget the problem associated with infections with intracellular parasites such as *Toxoplasma gondii* and *Neospora caninum*. Toxoplasmosis in dogs is much less common than in cats. Most often it leads to immunosuppression and the onset of neurological symptoms: seizures, nerve deficits, ataxia, or paralysis can be evident [150]. Infection-prone pregnant bitches are a serious concern. Parasitemia can cause placentitis, followed by the spread of tachyzoites to the fetus, and can cause miscarriage. Fetal resorptions and sudden infant death have also been reported [151]. *T. gondii* has also been isolated from pups of seropositive dams without clinical signs [150]. Toxoplasmosis is frequently associated with secondary infections in dogs. When combined with viral infections, such as the distemper virus (CDV), it can cause the death of the entire litter due to a complete immune failure in very young animals (up to 30 days of age) [152]. The second intracellular parasite is the protozoan *Neospora caninum* [153]. Dogs can be intermediate or definitive hosts, while infection occurs mainly through contact with contaminated water or food containing cysts. The horizontal transmission of the parasite from mother to fetus through the placenta is also possible [153]. Robbe et al. reported that pregnancy might be a predisposing factor for the *Neospora caninum* infection, which usually results in abortion or the birth of weak puppies that die shortly after [154]. In Australia, a case of a bulldog litter born from a seropositive mother was reported, in which one of the seven pups died from a multisystemic infection caused by *Neospora caninum* [155]. The pup was the smallest of the litter and was reported to have signs of weakness, lack of sucking reflex, and difficulty breathing. When necropsy was performed, diffuse pulmonary edema, inflammatory changes in internal organs, and acute myocarditis were detected. Studies conducted in Italy [154], Iran [156], Brazil [157], and Australia [158] indicated that parasitic diseases in breeding dogs are underestimated and often overlooked despite the significant prevalence of toxoplasmosis and neosporosis in the companion dog population. It should be noted that they are related to the trend of feeding raw animal meat in the BARF diet, which, at least for this reason, should not be recommended for feeding pregnant females, as it can be a potential source of infection with these protozoa [159,160].

6. Conclusions

The neonatal period is a challenging time of adaptation for any puppy to life outside their mother's body. In our review, we have discussed a variety of factors that affect puppy viability in the early stages of life. The order of discussion was guided by the natural course of pregnancy, birth, and the neonatal period. The viability of puppies depends first on the health of the mother and the environment that affects her in different ways, and then on the maturity of the adaptive mechanisms developed by those puppies during embryonic and fetal development.

For the early detection of the first signs of any emerging abnormality, the fundamental factor is conducting examinations and observing the pregnant bitch. Regular ultrasound and electrocardiographic monitoring enable rapid diagnosis of abnormalities, which, in turn, increases the chance of effective implementation of therapeutic measures [5,78–82,84]. This is especially true in the case of potential perinatal complications, because the impact of delivery on pup viability should always be kept in mind. The emergency cesarean section carries far greater risks than natural birth or planned surgery, and delaying the decision can often be fatal to the viability of the pups [5]. Therefore, the safest solution would be a planned cesarean section in the case of bitches known to be prone to perinatal complications, either due to breed predisposition or previous history. After the surgery, it is necessary to perform a quick evaluation of newborns, usually using the modified APGAR score [94] and the NVR [5], which are highly effective in identifying weak pups that require special care.

The cesarean section is also a challenging procedure due to the adequate intra- and postoperative care. For the safety of milk-sucking puppies, medications given to the mother are often highly restricted. However, after all, they are essential for dams' welfare and could significantly reflect the quality of care a female provides for her puppies [100,133,134]. In the authors' opinion, the problem of providing adequate analgesia protection to the bitch after a cesarean section is urgent and certainly requires further research to improve current surgical protocols. However, it should be remembered that once a puppy does not take milk, especially not colostrum, the risk of developing immunity problems would increase. Any factors that weaken a newborn's immunity drastically increase its susceptibility to infections. Thus, it is crucial to provide diligent care and hygiene for the bitch and her offspring. The risk of certain infections can be reduced through regular vaccinations.

In conclusion, all collected information focuses on factors influencing the viability of puppies during their most critical infancy period. The authors hope that this review would be helpful both for scientists and practitioners to grasp the picture of the vastness of the interrelationships during the perinatal period that determine neonatal health and welfare. It should also be mentioned that canine reproduction currently focuses on pedigree dogs only. Hence, inbreeding and focusing solely on exterior characteristics significantly increased susceptibility to perinatal complications [10]. This leads to a situation observed more and more often nowadays, where reproduction in some breeds, mainly brachycephalic, is impossible due to their anatomy and physiology, without medical assistance performed by veterinarians [21]. The ethics of such a practice is difficult to assess due to divergent opinions among scientists, physicians, and breeders involved in small animal reproduction [17,161]. Cooperation between science and veterinary medicine can contribute to a more effective accumulation of knowledge in the field of canine neonatology and thus a further improvement in the quality of services provided.

Author Contributions: O.U.—writing and editing the manuscript; M.O.—conceptualization and corrections of the manuscript; M.E.—writing the manuscript; W.N.—supervising and consulting the manuscript. All authors have read and agreed to the published version of the manuscript.

Funding: The costs of publications were financed under the Leading Research Groups support project from the subsidy increased for the period 2020–2025 in the amount of 2% of the subsidy referred to Art. 387 (3) of the Law of 20 July 2018 on Higher Education and Science, obtained in 2019. The APC is financed by Wroclaw University of Environmental and Life Sciences.

Institutional Review Board Statement: Not applicable.

Data Availability Statement: Not applicable.

Conflicts of Interest: Authors have no conflict of interest to declare.

References

1. O'neill, D.G.; O'sullivan, A.M.; Manson, E.A.; Church, D.B.; Boag, A.K.; Mcgreevy, P.D.; Brodbelt, D.C. Canine Dystocia in 50 UK First-Opinion Emergency Care Veterinary Practices: Prevalence and Risk Factors. *Vet. Rec.* **2017**, *181*, 88. [CrossRef] [PubMed]
2. O'Neill, D.G.; O'Sullivan, A.M.; Manson, E.A.; Church, D.B.; McGreevy, P.D.; Boag, A.K.; Brodbelt, D.C. Canine Dystocia in 50 UK First-Opinion Emergency Care Veterinary Practices: Clinical Management and Outcomes. *Vet. Rec.* **2019**, *184*, 409. [CrossRef] [PubMed]
3. Münnich, A.; Küchenmeister, U. Dystocia in Numbers-Evidence-Based Parameters for Intervention in the Dog: Causes for Dystocia and Treatment Recommendations. *Reprod. Domest. Anim.* **2009**, *44*, 141–147. [CrossRef]
4. Mila, H.; Grellet, A.; Delebarre, M.; Mariani, C.; Feugier, A.; Chastant-Maillard, S. Monitoring of the Newborn Dog and Prediction of Neonatal Mortality. *Prev. Vet. Med.* **2017**, *143*, 11–20. [CrossRef]
5. Vassalo, F.G.; Simões, C.R.B.; Sudano, M.J.; Prestes, N.C.; Lopes, M.D.; Chiacchio, S.B.; Lourenço, M.L.G. Topics in the Routine Assessment of Newborn Puppy Viability. *Top. Companion Anim. Med.* **2015**, *30*, 16–21. [CrossRef]
6. Cornelius, A.J.; Moxon, R.; Russenberger, J.; Havlena, B.; Cheong, S.H. Identifying Risk Factors for Canine Dystocia and Stillbirths. *Theriogenology* **2019**, *128*, 201–206. [CrossRef] [PubMed]
7. Eneroth, A.; Linde-Forsberg, C.; Uhlhorn, M.; Hall, M. Radiographic Pelvimetry for Assessment of Dystocia in Bitches: A Clinical Study in Two Terrier Breeds. *J. Small Anim. Pract.* **1999**, *40*, 257–264. [CrossRef]
8. Moxon, R.; England, G.C.W. Fertility and Whelping Complications in Bitches Following Correction of Vaginal Abnormalities. *Vet. Rec.* **2011**, *168*, 642. [CrossRef]
9. Dobak, T.P.; Voorhout, G.; Vernooij, J.C.M.; Boroffka, S.A.E.B. Computed Tomographic Pelvimetry in English Bulldogs. *Theriogenology* **2018**, *118*, 144–149. [CrossRef]
10. Marelli, S.P.; Beccaglia, M.; Bagnato, A.; Strillacci, M.G. Canine Fertility: The Consequences of Selection for Special Traits. *Reprod. Domest. Anim.* **2020**, *55*, 4–9. [CrossRef]
11. Borge, K.S.; Tønnessen, R.; Nødtvedt, A.; Indrebø, A. Litter Size at Birth in Purebred Dogs—A Retrospective Study of 224 Breeds. *Theriogenology* **2011**, *75*, 911–919. [CrossRef]
12. Hollinshead, F.K.; Hanlon, D.W. Factors Affecting the Reproductive Performance of Bitches: A Prospective Cohort Study Involving 1203 Inseminations with Fresh and Frozen Semen. *Theriogenology* **2017**, *101*, 62–72. [CrossRef] [PubMed]

13. Tønnessen, R.; Borge, K.S.; Nødtvedt, A.; Indrebø, A. Canine Perinatal Mortality: A Cohort Study of 224 Breeds. *Theriogenology* **2012**, *77*, 1788–1801. [CrossRef] [PubMed]
14. Chastant-Maillard, S.; Guillemot, C.; Feugier, A.; Mariani, C.; Grellet, A.; Mila, H. Reproductive Performance and Pre-Weaning Mortality: Preliminary Analysis of 27,221 Purebred Female Dogs and 204,537 Puppies in France. *Reprod. Domest. Anim.* **2017**, *52*, 158–162. [CrossRef] [PubMed]
15. Frehner, B.L.; Reichler, I.M.; Keller, S.; Goericke-Pesch, S.; Balogh, O. Blood Calcium, Glucose and Haematology Profiles of Parturient Bitches Diagnosed with Uterine Inertia or Obstructive Dystocia. *Reprod. Domest. Anim.* **2018**, *53*, 680–687. [CrossRef]
16. Bergström, A.; Nødtvedt, A.; Lagerstedt, A.S.; Egenvall, A. Incidence and Breed Predilection for Dystocia and Risk Factors for Cesarean Section in a Swedish Population of Insured Dogs. *Vet. Surg.* **2006**, *35*, 786–791. [CrossRef]
17. Farstad, W. Ethics in Animal Breeding. *Reprod. Domest. Anim.* **2018**, *53*, 4–13. [CrossRef]
18. Johnston, S.D.; Root Kustritz, M.V.; Olson, P.N.S. *Canine and Feline Theriogenology*; Saunders: Philadelphia, PA, USA, 2001; ISBN 0721656072.
19. Arlt, S.P.; Rohne, J.; Ebert, A.D.; Heuwieser, W. Endoscopic Resection of a Vaginal Septum in a Bitch and Observation of Septa in Two Related Bitches. *N. Z. Vet. J.* **2012**, *60*, 258–260. [CrossRef]
20. Schrack, J.; Dolf, G.; Reichler, I.M.; Schelling, C. Factors Influencing Litter Size and Puppy Losses in the Entlebucher Mountain Dog. *Theriogenology* **2017**, *95*, 163–170. [CrossRef]
21. De Cramer, K.G.M.; Nöthling, J.O. Towards Scheduled Pre-Parturient Caesarean Sections in Bitches. *Reprod. Domest. Anim.* **2020**, *55*, 38–48. [CrossRef]
22. Bovenkerk, B.; Nijland, H.J. The Pedigree Dog Breeding Debate in Ethics and Practice: Beyond Welfare Arguments. *J. Agric. Environ. Ethics* **2017**, *30*, 387–412. [CrossRef]
23. Rokhsar, J.L.; Canino, J.; Raj, K.; Yuhnke, S.; Slutsky, J.; Giger, U. Web Resource on Available DNA Variant Tests for Hereditary Diseases and Genetic Predispositions in Dogs and Cats: An Update. *Hum. Genet.* **2021**, *140*, 1505–1515. [CrossRef] [PubMed]
24. The WSAVA-PennGen DNA Testing Database. Available online: http://research.vet.upenn.edu/WSAVA-LabSearch (accessed on 17 May 2022).
25. Forsberg, C.L.; Persson, G. A Survey of Dystocia in the Boxer Breed. *Acta Vet. Scand.* **2007**, *49*, 8. [CrossRef] [PubMed]
26. Lamm, C.G.; Njaa, B.L. Clinical Approach to Abortion, Stillbirth, and Neonatal Death in Dogs and Cats. *Vet. Clin. North Am.-Small Anim. Pract.* **2012**, *42*, 501–513. [CrossRef] [PubMed]
27. Dejneka, J.G.; Ochota, M.; Bielas, W.; Niżański, W. Dystocia after Unwanted Mating as One of the Risk Factors in Non-Spayed Bitches—A Retrospective Study. *Animals* **2020**, *10*, 1697. [CrossRef]
28. Mason, S.J.; Rous, N.R. Comparison of Endoscopic-Assisted Transcervical and Laparotomy Insemination with Frozen-Thawed Dog Semen: Aretrospective Clinical Study. *Theriogenology* **2014**, *82*, 844–850. [CrossRef]
29. Mason, S.J. Current Review of Artificial Insemination in Dogs. *Vet. Clin. North Am.-Small Anim. Pract.* **2018**, *48*, 567–580. [CrossRef]
30. Nizański, W. Intravaginal Insemination of Bitches with Fresh and Frozen-Thawed Semen with Addition of Prostatic Fluid: Use of an Infusion Pipette and the Osiris Catheter. *Theriogenology* **2006**, *66*, 470–483. [CrossRef]
31. Lamm, C.G.; Makloski, C.L. Current Advances in Gestation and Parturition in Cats and Dogs. *Vet. Clin. North Am.-Small Anim. Pract.* **2012**, *42*, 445–456. [CrossRef]
32. Nak, D.; Nak, Y.; Shahzad, A.H. Pregnancy Toxemia in a Golden Retriever Bitch, a Case Report. *Adv. Anim. Vet. Sci.* **2020**, *8*, 1184–1187. [CrossRef]
33. Balogh, O.; Bruckmaier, R.; Keller, S.; Reichler, I.M. Effect of Maternal Metabolism on Fetal Supply: Glucose, Non-Esterified Fatty Acids and Beta-Hydroxybutyrate Concentrations in Canine Maternal Serum and Fetal Fluids at Term Pregnancy. *Anim. Reprod. Sci.* **2018**, *193*, 209–216. [CrossRef] [PubMed]
34. Johnson, C.A. High-Risk Pregnancy and Hypoluteoidism in the Bitch. *Theriogenology* **2008**, *70*, 1424–1430. [CrossRef] [PubMed]
35. Nelson, R.W.; Reusch, C.E. Animal Models of Disease: Classification and Etiology of Diabetes in Dogs and Cats. *J. Endocrinol.* **2014**, *222*, T1–T9. [CrossRef] [PubMed]
36. Fall, T.; Johansson Kreuger, S.; Juberget, Å.; Bergström, A.; Hedhammar, Å. Gestational Diabetes Mellitus in 13 Dogs. *J. Vet. Intern. Med.* **2008**, *22*, 1296–1300. [CrossRef]
37. Johns, E.C.; Denison, F.C.; Norman, J.E.; Reynolds, R.M. Gestational Diabetes Mellitus: Mechanisms, Treatment, and Complications. *Trends Endocrinol. Metab.* **2018**, *29*, 743–754. [CrossRef]
38. Szmuilowicz, E.D.; Josefson, J.L.; Metzger, B.E. Gestational Diabetes Mellitus. *Endocrinol. Metab. Clin. North Am.* **2019**, *48*, 479–493. [CrossRef]
39. Johnson, C.A. Glucose Homeostasis during Canine Pregnancy: Insulin Resistance, Ketosis, and Hypoglycemia. *Theriogenology* **2008**, *70*, 1418–1423. [CrossRef]
40. Norman, E.J.; Wolsky, K.G.; MacKay, G.A. Pregnancy-Related Diabetes Mellitus in Two Dogs. *N. Z. Vet. J.* **2006**, *54*, 360–364. [CrossRef]
41. Arlt, S.P. The Bitch around Parturition. *Theriogenology* **2020**, *150*, 452–457. [CrossRef] [PubMed]
42. Hinderer, J.; Lüdeke, J.; Riege, L.; Haimerl, P.; Bartel, A.; Kohn, B.; Weber, C.; Müller, E.; Arlt, S.P. Progesterone Concentrations during Canine Pregnancy. *Animals* **2021**, *11*, 3369. [CrossRef]

43. Thuróczy, J.; Müller, L.; Kollár, E.; Balogh, L. Theriogenology Thyroxin and Progesterone Concentrations in Pregnant, Nonpregnant Bitches, and Bitches during Abortion. *Theriogenology* **2016**, *85*, 1186–1191. [CrossRef] [PubMed]

44. Hussein, H.A.; Schuler, G.; Conze, T.; Wehrend, A. Comparison of Three Progesterone Quantification Methods Using Blood Samples Drawn from Bitches during the Periovulatory Phase. *Vet. World* **2022**, *15*, 119–123. [CrossRef] [PubMed]

45. Ilicic, M.; Zakar, T.; Paul, J.W. Epigenetic Regulation of Progesterone Receptors and the Onset of Labour. *Reprod. Fertil. Dev.* **2019**, *31*, 1035–1048. [CrossRef] [PubMed]

46. Papa, P.C.; Kowalewski, M.P. Factors Affecting the Fate of the Canine Corpus Luteum: Potential Contributors to Pregnancy and Non-Pregnancy. *Theriogenology* **2020**, *150*, 339–346. [CrossRef]

47. Krachudel, J.; Bondzio, A.; Einspanier, R.; Einspanier, A.; Gottschalk, J.; Kuechenmeister, U.; Muennich, A. Theriogenology Luteal Insuf Fi Ciency in Bitches as a Consequence of an Autoimmune Response against Progesterone? *Theriogenology* **2013**, *79*, 1278–1283. [CrossRef]

48. Kowalewski, M.P. Luteal Regression vs. Prepartum Luteolysis: Regulatory Mechanisms Governing Canine Corpus Luteum Function. *Reprod. Biol.* **2014**, *14*, 89–102. [CrossRef]

49. Urhausen, C.; Wolf, K.; Einspanier, A.; Oei, C.; Piechotta, M. Serum Progesterone in Pregnant Bitches Supplemented with Progestin Because of Expected or Suspected Luteal Insufficiency. *Reprod. Domest. Anim.* **2012**, *47*, 55–60. [CrossRef]

50. Galac, S.; Kooistra, H.S.; Okkens, A.C.; Go, S. Hypoluteoidism in a Bitch. *Theriogenology* **2005**, *64*, 213–219. [CrossRef]

51. Zedda, M.T.; Bogliolo, L.; Antuofermo, E.; Falchi, L.; Ariu, F.; Burrai, G.P.; Pau, S. Hypoluteoidism in a Dog Associated with Recurrent Mammary Fibroadenoma Stimulated by Progestin Therapy. *Acta Vet. Scand.* **2017**, *59*, 55. [CrossRef]

52. Kustritz, M.V.R. Pregnancy Diagnosis and Abnormalities of Pregnancy in the Dog. *Theriogenology* **2005**, *64*, 755–765. [CrossRef]

53. Graham, P.A.; Refsal, K.R.; Nachreiner, R.F. Etiopathologic Findings of Canine Hypothyroidism. *Vet. Clin. North Am.-Small Anim. Pract.* **2007**, *37*, 617–631. [CrossRef] [PubMed]

54. Panciera, D.L.; Purswell, B.J.; Kolster, K.A. Effect of Short-Term Hypothyroidism on Reproduction in the Bitch. *Theriogenology* **2007**, *68*, 316–321. [CrossRef] [PubMed]

55. Panciera, D.L.; Purswell, B.J.; Kolster, K.A.; Werre, S.R.; Trout, S.W. Reproductive Effects of Prolonged Experimentally Induced Hypothyroidism in Bitches. *J. Vet. Intern. Med.* **2012**, *26*, 326–333. [CrossRef] [PubMed]

56. Cardinali, L.; Troisi, A.; Verstegen, J.P.; Menchetti, L.; Elad Ngonput, A.; Boiti, C.; Canello, S.; Zelli, R.; Polisca, A. Serum Concentration Dynamic of Energy Homeostasis Hormones, Leptin, Insulin, Thyroid Hormones, and Cortisol throughout Canine Pregnancy and Lactation. *Theriogenology* **2017**, *97*, 154–158. [CrossRef]

57. Cecere, J.; Purswell, B.; Panciera, D. Levothyroxine Supplementation in Hypothyroid Bitches during Pregnancy. *Theriogenology* **2020**, *142*, 48–53. [CrossRef]

58. Reimers, T.J.; Mummery, L.K.; McCann, J.P.; Cowan, R.G.; Concannon, P.W. Effects of Reproductive State on Concentrations of Thyroxine, 3,5,3′-Triiodothyronine and Cortisol in Serum of Dogs. *Biol. Reprod.* **1984**, *31*, 148–154. [CrossRef]

59. Sullivan, S.A. Hypothyroidism in Pregnancy. *Clin. Obstet. Gynecol.* **2019**, *62*, 308–319. [CrossRef]

60. Steiner, M.; Mariani, C.; Suchodolski, J.S.; Guard, B.C.; Mila, H.; Steiner, J.M.; Mariani, C.; Suchodolski, J.S.; Chastant-Maillard, S. Characterization of the Fecal Microbiome during Neonatal and Early Pediatric Development in Puppies. *PLoS ONE* **2017**, *12*, e0175718.

61. Kajdič, L.; Plavec, T.; Zdovc, I.; Kalin, A.; Pipan, M.Z. Impact of Type of Parturition on Colostrum Microbiota Composition and Puppy Survival. *Animals* **2021**, *11*, 1897. [CrossRef]

62. Melandri, M.; Aiudi, G.G.; Caira, M.; Alonge, S. A Biotic Support during Pregnancy to Strengthen the Gastrointestinal Performance in Puppies. *Front. Vet. Sci.* **2020**, *7*, 1–6. [CrossRef]

63. Zakošek Pipan, M.; Kajdič, L.; Kalin, A.; Plavec, T.; Zdovc, I. Do Newborn Puppies Have Their Own Microbiota at Birth? Influence of Type of Birth on Newborn Puppy Microbiota. *Theriogenology* **2020**, *152*, 18–28. [CrossRef] [PubMed]

64. Tal, S.; Tikhonov, E.; Aroch, I.; Hefetz, L.; Turjeman, S.; Koren, O.; Kuzi, S. Developmental Intestinal Microbiome Alterations in Canine Fading Puppy Syndrome: A Prospective Observational Study. *Npj Biofilms Microb.* **2021**, *7*, 52. [CrossRef] [PubMed]

65. Dzanis, D.A. The Association of American Feed Control Officials Dog and Cat Food Nutrient Profiles: Substantiation of Nutritional Adequacy of Complete and Balanced Pet Foods in the United States. *J. Nutr.* **1994**, *124*, 2535–2539. [CrossRef]

66. Fontaine, E. Food Intake and Nutrition during Pregnancy, Lactation and Weaning in the Dam and Offspring. *Reprod. Domest. Anim.* **2012**, *47*, 326–330. [CrossRef] [PubMed]

67. Gram, A.; Boos, A.; Kowalewski, M.P. Uterine and Placental Expression of Canine Oxytocin Receptor during Pregnancy and Normal and Induced Parturition. *Reprod. Domest. Anim.* **2014**, *49*, 41–49. [CrossRef] [PubMed]

68. Connolly, K.M.; Heinze, C.R.; Freeman, L.M. Feeding Practices of Dog Breeders in the United States and Canada. *J. Am. Vet. Med. Assoc.* **2014**, *245*, 669–676. [CrossRef] [PubMed]

69. Kölle, P.; Schmidt, M. Raw-Meat-Based Diets (RMBD) as a Feeding Principle for Dogs. *Tierarztl. Prax. Ausgabe K Kleintiere-Heimtiere* **2015**, *43*, 409–419.

70. Schmitt, S.; Dobenecker, B. Calcium and Phosphorus Metabolism in Peripartal Dogs. *J. Anim. Physiol. Anim. Nutr.* **2020**, *104*, 707–714. [CrossRef]

71. Hollinshead, F.K.; Hanlon, D.W.; Gilbert, R.O.; Verstegen, J.P.; Krekeler, N.; Volkmann, D.H. Calcium, Parathyroid Hormone, Oxytocin and PH Profiles in the Whelping Bitch. *Theriogenology* **2010**, *73*, 1276–1283. [CrossRef]

72. Spiegelstein, O.; Lu, X.; Le, X.C.; Troen, A.; Selhub, J.; Melnyk, S.; James, S.J.; Finnell, R.H. Effects of Dietary Folate Intake and Folate Binding Protein-2 (Folbp2) on Urinary Speciation of Sodium Arsenate in Mice. *Environ. Toxicol. Pharmacol.* **2005**, *19*, 1–7. [CrossRef]

73. Kappen, C. Folate Supplementation in Three Genetic Models. *Am. J. Med. Genet. C Semin. Med. Genet.* **2005**, *15*, 24–30. [CrossRef] [PubMed]

74. Elwood, J.M.; Colquhoun, T.A. Observations on the Prevention of Cleft Palate in Dogs by Folic Acid and Potential Relevance to Humans. *N. Z. Vet. J.* **1997**, *45*, 254–256. [CrossRef] [PubMed]

75. Malandain, E. Nutrition and Reroduction in Bitches and Queens. In Proceedings of the 5th Biannual Congress, European Veterinary Society for Small Animal Reproduction (EVSSAR), Budapest, Hungary, 7–9 April 2006; pp. 180–224.

76. Domosławska, A.; Jurczak, A.; Janowski, T. Oral Folic Acid Supplementation Decreases Palate and/or Lip Cleft Occurrence in Pug and Chihuahua Puppies and Elevates Folic Acid Blood Levels in Pregnant Bitches. *Pol. J. Vet. Sci.* **2013**, *16*, 33–37. [CrossRef] [PubMed]

77. Gonzales, K.L.; Famula, T.R.; Feng, L.C.; Power, H.M.N.; Bullis, J.M. Folic Acid Supplementation Does Not Decrease Stillbirths and Congenital Malformations in a Guide Dog Colony. *J. Small Anim. Pract.* **2021**, *62*, 286–292. [CrossRef]

78. de Freitas, L.A.; Mota, G.L.; Silva, H.V.R.; Carvalho, C.F.; Silva, L.D.M. da Can Maternal-Fetal Hemodynamics Influence Prenatal Development in Dogs? *Anim. Reprod. Sci.* **2016**, *172*, 83–93. [CrossRef]

79. Blanco, P.G.; Rodríguez, R.; Rube, A.; Arias, D.O.; Tórtora, M.; Díaz, J.D.; Gobello, C. Doppler Ultrasonographic Assessment of Maternal and Fetal Blood Flow in Abnormal Canine Pregnancy. *Anim. Reprod. Sci.* **2011**, *126*, 130–135. [CrossRef]

80. Blanco, P.G.; Huk, M.; Lapuente, C.; Tórtora, M.; Rodríguez, R.; Arias, D.O.; Gobello, C. Uterine and Umbilical Resistance Index and Fetal Heart Rate in Pregnant Bitches of Different Body Weight. *Anim. Reprod. Sci.* **2020**, *212*, 106225. [CrossRef]

81. Barella, G.; Faverzani, S.; Faustini, M.; Groppetti, D.; Pecile, A. Neonatal Mortality in Dogs: Prognostic Value of Doppler Ductus Venosus Waveform Evaluation-Preliminary Results. *Vet. World* **2016**, *9*, 356–360. [CrossRef]

82. Van den Bossche, L.; van Steenbeek, F.G. Canine Congenital Portosystemic Shunts: Disconnections Dissected. *Vet. J.* **2016**, *211*, 14–20. [CrossRef]

83. Barstow, C.; Wilborn, R.R.; Johnson, A.K. Breeding Soundness Examination of the Bitch. *Vet. Clin. N. Am.-Small Anim. Pract.* **2018**, *48*, 547–566. [CrossRef]

84. Blanco, P.G.; Batista, P.R.; Re, N.E.; Mattioli, G.A.; Arias, D.O.; Gobello, C. Electrocardiographic Changes in Normal and Abnormal Canine Pregnancy. *Reprod. Domest. Anim.* **2012**, *47*, 252–256. [CrossRef] [PubMed]

85. Alonge, S.; Mauri, M.; Faustini, M.; Luvoni, G.C. Feto-Maternal Heart Rate Ratio in Pregnant Bitches: Effect of Gestational Age and Maternal Size. *Reprod. Domest. Anim.* **2016**, *51*, 688–692. [CrossRef] [PubMed]

86. Siena, G.; Milani, C. Usefulness of Maternal and Fetal Parameters for the Prediction of Parturition Date in Dogs. *Animals* **2021**, *11*, 878. [CrossRef]

87. Reynaud, K.; Fontbonne, A.; Saint-Dizier, M.; Thoumire, S.; Marnier, C.; Tahir, M.Z.; Meylheuc, T.; Chastant-Maillard, S. Folliculogenesis, Ovulation and Endocrine Control of Oocytes and Embryos in the Dog. *Reprod. Domest. Anim.* **2012**, *47*, 66–69. [CrossRef] [PubMed]

88. Kutzler, M.A.; Mohammed, H.O.; Lamb, S.V.; Meyers-Wallen, V.N. Accuracy of Canine Parturition Date Prediction from the Initial Rise in Preovulatory Progesterone Concentration. *Theriogenology* **2003**, *60*, 1187–1196. [CrossRef]

89. Smith, F.O. Guide to Emergency Interception During Parturition in the Dog and Cat. *Vet. Clin. North Am.-Small Anim. Pract.* **2012**, *42*, 489–499. [CrossRef]

90. Veronesi, M.C. Assessment of Canine Neonatal Viability—The Apgar Score. *Reprod. Domest. Anim.* **2016**, *51*, 46–50. [CrossRef]

91. De Cramer, K.G.M.; Nöthling, J.O. Curtailing Parturition Observation and Performing Preparturient Cesarean Section in Bitches. *Theriogenology* **2019**, *124*, 57–64. [CrossRef]

92. Münnich, A.; Küchenmeister, U. Causes, Diagnosis and Therapy of Common Diseases in Neonatal Puppies in the First Days of Life: Cornerstones of Practical Approach. *Reprod. Domest. Anim.* **2014**, *49*, 64–74. [CrossRef]

93. Antończyk, A.; Ochota, M.; Niżański, W. Umbilical Cord Blood Gas Parameters and Apgar Scoring in Assessment of New-born Dogs Delivered by Cesarean Section. *Animals* **2021**, *11*, 685. [CrossRef]

94. Veronesi, M.C.; Panzani, S.; Faustini, M.; Rota, A. An Apgar Scoring System for Routine Assessment of Newborn Puppy Viability and Short-Term Survival Prognosis. *Theriogenology* **2009**, *72*, 401–407. [CrossRef] [PubMed]

95. Batista, M.; Moreno, C.; Vilar, J.; Golding, M.; Brito, C.; Santana, M.; Alamo, D. Neonatal Viability Evaluation by Apgar Score in Puppies Delivered by Cesarean Section in Two Brachycephalic Breeds (English and French Bulldog). *Anim. Reprod. Sci.* **2014**, *146*, 218–226. [CrossRef] [PubMed]

96. Traas, A.M. Resuscitation of Canine and Feline Neonates. *Theriogenology* **2008**, *70*, 343–348. [CrossRef] [PubMed]

97. Groppetti, D.; Pecile, A.; Del Carro, A.P.; Copley, K.; Minero, M.; Cremonesi, F. Evaluation of Newborn Canine Viability by Means of Umbilical Vein Lactate Measurement, Apgar Score and Uterine Tocodynamometry. *Theriogenology* **2010**, *74*, 1187–1196. [CrossRef] [PubMed]

98. Martínez-Burnes, J.; Muns, R.; Barrios-García, H.; Villanueva-García, D.; Domínguez-Oliva, A.; Mota-Rojas, D. Parturition in Mammals: Animal Models, Pain and Distress. *Animals* **2021**, *11*, 2960. [CrossRef]

99. Lezama-García, K.; Mariti, C.; Mota-Rojas, D.; Martínez-Burnes, J.; Barrios-García, H.; Gazzano, A. Maternal Behaviour in Domestic Dogs. *Int. J. Vet. Sci. Med.* **2019**, *7*, 20–30. [CrossRef]

100. Mathews, K.A. Analgesia for the Pregnant, Lactating and Neonatal to Pediatric Cat and Dog. *J. Vet. Emerg. Crit. Care* **2005**, *15*, 273–284. [CrossRef]

101. Ferrari, D.; Lundgren, S.; Holmberg, J.; Edner, A.; Ekstrand, C.; Nyman, G.; Bondesson, U.; Hagman, R. Concentration of Carprofen in the Milk of Lactating Bitches after Cesarean Section and during Inflammatory Conditions. *Theriogenology* **2022**, *181*, 59–68. [CrossRef]

102. Schneider, M.; Kuchta, A.; Dron, F.; Woehrlé, F. Disposition of Cimicoxib in Plasma and Milk of Whelping Bitches and in Their Puppies. *BMC Vet. Res.* **2015**, *11*, 4–9. [CrossRef]

103. Kusse, G.; Birhanu, M.; Ayele, T. Review on Dog Neonatal Mortality. *Adv. Life Sci. Technol.* **2018**, *65*, 19–26.

104. Nobre Pacifico Pereira, K.H.; Cruz dos Santos Correia, L.E.; Ritir Oliveira, E.L.; Bernardo, R.B.; Nagib Jorge, M.L.; Mezzena Gobato, M.L.; Ferreira de Souza, F.; Rocha, N.S.; Chiacchio, S.B.; Gomes Lourenço, M.L. Incidence of Congenital Malformations and Impact on the Mortality of Neonatal Canines. *Theriogenology* **2019**, *140*, 52–57. [CrossRef] [PubMed]

105. Kim, M.J.; Oh, H.J.; Kim, G.A.; Jo, Y.K.; Choi, J.; Kim, H.J.; Choi, H.Y.; Kim, H.W.; Choi, M.C.; Lee, B.C. Reduced Birth Weight, Cleft Palate and Preputial Abnormalities in a Cloned Dog. *Acta Vet. Scand.* **2014**, *56*, 18. [CrossRef] [PubMed]

106. Davidson, A.P.; Gregory, C.; Dedrick, P. Successful Management Permitting Delayed Operative Revision of Cleft Palate in a Labrador Retriever. *Vet. Clin. North Am.-Small Anim. Pract.* **2014**, *44*, 325–329. [CrossRef]

107. Mahler, S.; Williams, G. Preservation of the Fistula for Reconstruction of the Anal Canal and the Anus in Atresia Ani and Rectovestibular Fistula in 2 Dogs. *Vet. Surg.* **2005**, *34*, 148–152. [CrossRef] [PubMed]

108. Fiani, N.; Verstraete, F.J.M.; Arzi, B. Reconstruction of Congenital Nose, Cleft Primary Palate, and Lip Disorders. *Vet. Clin. North Am.-Small Anim. Pract.* **2016**, *46*, 663–675. [CrossRef]

109. Ellison, G.W.; Papazoglou, L.G. Long-Term Results of Surgery for Atresia Ani with or without Anogenital Malformations in Puppies and a Kitten: 12 Cases (1983-2010). *J. Am. Vet. Med. Assoc.* **2012**, *240*, 186–192. [CrossRef]

110. Mila, H.; Grellet, A.; Feugier, A.; Chastant-Maillard, S. Differential Impact of Birth Weight and Early Growth on Neonatal Mortality in Puppies. *J. Anim. Sci.* **2015**, *93*, 4436–4442. [CrossRef]

111. Ki, O.; So, O.; Mma, D.; Mh, A.; Eo, A.; Jg, G. A Review of Neonatal Mortality in Dogs. *Int. J. Life Sci.* **2016**, *4*, 451–460.

112. Mugnier, A.; Mila, H.; Guiraud, F.; Brévaux, J.; Lecarpentier, M.; Martinez, C.; Mariani, C.; Adib-Lesaux, A.; Chastant-Maillard, S.; Saegerman, C.; et al. Birth Weight as a Risk Factor for Neonatal Mortality: Breed-Specific Approach to Identify at-Risk Puppies. *Prev. Vet. Med.* **2019**, *171*, 104746. [CrossRef]

113. Tesi, M.; Miragliotta, V.; Scala, L.; Aronica, E.; Lazzarini, G.; Fanelli, D.; Abramo, F.; Rota, A. Relationship between Placental Characteristics and Puppies' Birth Weight in Toy and Small Sized Dog Breeds. *Theriogenology* **2020**, *141*, 1–8. [CrossRef]

114. Groppetti, D.; Pecile, A.; Palestrini, C.; Marelli, S.P.; Boracchi, P. A National Census of Birthweight in Purebred Dogs in Italy. *Animals* **2017**, *7*, 43. [CrossRef] [PubMed]

115. Mugnier, A.; Chastant-Maillard, S.; Mila, H.; Lyazrhi, F.; Guiraud, F.; Adib-Lesaux, A.; Gaillard, V.; Saegerman, C.; Grellet, A. Low and Very Low Birth Weight in Puppies: Definitions, Risk Factors and Survival in a Large-Scale Population. *BMC Vet. Res.* **2020**, *16*, 354. [CrossRef] [PubMed]

116. Reyes-Sotelo, B.; Mota-Rojas, D.; Mora-Medina, P.; Ogi, A.; Mariti, C.; Olmos-Hernández, A.; Martínez-Burnes, J.; Hernández-ávalos, I.; Sánchez-Millán, J.; Gazzano, A. Blood Biomarker Profile Alterations in Newborn Canines: Effect of the Mother's Weight. *Animals* **2021**, *11*, 2307. [CrossRef] [PubMed]

117. Sanin, L.H.; López, S.R.; Olivares, E.T.; Terrazas, M.C.; Silva, M.A.R.; Carrillo, M.L. Relation between Birth Weight and Placenta Weight. *Biol. Neonate* **2001**, *80*, 113–117. [CrossRef] [PubMed]

118. Dombrowski, M.P.; Berry, S.M.; Johnson, M.P.; Saleh, A.A.A.; Sokol, R.J. Birth Weight–Length Ratios, Ponderal Indexes, Placental Weights, and Birth Weight–Placenta Ratios in a Large Population. *Arch. Pediatr. Adolesc. Med.* **1994**, *148*, 508–512. [CrossRef]

119. Katheria, A.C.; Lakshminrusimha, S.; Rabe, H.; McAdams, R.; Mercer, J.S. Placental Transfusion: A Review. *J. Perinatol.* **2017**, *37*, 105–111. [CrossRef]

120. Elliott, C.; Morton, J.; Chopin, J. Factors Affecting Foal Birth Weight in Thoroughbred Horses. *Theriogenology* **2009**, *71*, 683–689. [CrossRef]

121. Dwyer, C.M.; Calvert, S.K.; Farish, M.; Donbavand, J.; Pickup, H.E. Breed, Litter and Parity Effects on Placental Weight and Placentome Number, and Consequences for the Neonatal Behaviour of the Lamb. *Theriogenology* **2005**, *63*, 1092–1110. [CrossRef]

122. Sarli, G.; Castagnetti, C.; Bianco, C.; Ballotta, G.; Tura, G.; Caporaletti, M.; Cunto, M.; Avallone, G.; Benazzi, C.; Ostanello, F.; et al. Canine Placenta Histological Findings and Microvascular Density: The Histological Basis of a Negative Neonatal Outcome? *Animals* **2021**, *11*, 1418. [CrossRef]

123. Lawler, D.F. Neonatal and Pediatric Care of the Puppy and Kitten. *Theriogenology* **2008**, *70*, 384–392. [CrossRef]

124. Reyes-Sotelo, B.; Mota-Rojas, D.; Martínez-Burnes, J.; Olmos-Hernández, A.; Hernández-Ávalos, I.; José, N.; Casas-Alvarado, A.; Gómez, J.; Mora-Medina, P. Thermal Homeostasis in the Newborn Puppy: Behavioral and Physiological Responses. *J. Anim. Behav. Biometeorol.* **2021**, *9*, 1–25. [CrossRef]

125. Von Dehn, B. Pediatric Clinical Pathology. *Vet. Clin. North Am.-Small Anim. Pract.* **2014**, *44*, 205–219. [CrossRef] [PubMed]

126. Fitzgerald, K.T.; Newquist, K.L. Husbandry of the Neonate. In *Small Animal Pediatrics*; Elsevier: Amsterdam, The Netherlands, 2011; pp. 44–52. ISBN 9781416048893.

127. Mila, H.; Feugier, A.; Grellet, A.; Anne, J.; Gonnier, M.; Martin, M.; Rossig, L.; Chastant-Maillard, S. Inadequate Passive Immune Transfer in Puppies: Definition, Risk Factors and Prevention in a Large Multi-Breed Kennel. *Prev. Vet. Med.* **2014**, *116*, 209–213. [CrossRef] [PubMed]

128. Mila, H.; Grellet, A.; Mariani, C.; Feugier, A.; Guard, B.; Suchodolski, J.; Steiner, J.; Chastant-Maillard, S. Natural and Artificial Hyperimmune Solutions: Impact on Health in Puppies. *Reprod. Domest. Anim.* **2017**, *52*, 163–169. [CrossRef] [PubMed]

129. Rossi, L.; Lumbreras, A.E.V.; Vagni, S.; Dell'Anno, M.; Bontempo, V. Nutritional and Functional Properties of Colostrum in Puppies and Kittens. *Animals* **2021**, *11*, 3260. [CrossRef] [PubMed]

130. Chastant-Maillard, S.; Aggouni, C.; Albaret, A.; Fournier, A.; Mila, H. Canine and Feline Colostrum. *Reprod. Domest. Anim.* **2017**, *52*, 148–152. [CrossRef] [PubMed]

131. Chastant, S.; Mila, H. Passive Immune Transfer in Puppies. *Anim. Reprod. Sci.* **2019**, *207*, 162–170. [CrossRef]

132. Kaszak, I.; Ruszczak, A.; Kanafa, S.; Piłaszewicz, O.W.; Sacharczuk, M.; Jurka, P. New Insights of Canine Mastitis—A Review. *Anim. Sci. Pap. Rep.* **2018**, *36*, 33–44.

133. Guardini, G.; Bowen, J.; Raviglione, S.; Farina, R.; Gazzano, A. Maternal Behaviour in Domestic Dogs: A Comparison between Primiparous and Multiparous Dogs. *Dog Behav.* **2015**, *1*, 23–33. [CrossRef]

134. Santos, N.R.; Beck, A.; Fontbonne, A. A Review of Maternal Behaviour in Dogs and Potential Areas for Further Research. *J. Small Anim. Pract.* **2020**, *61*, 85–92. [CrossRef]

135. Meloni, T.; Martino, P.A.; Grieco, V.; Pisu, M.C.; Banco, B.; Rota, A.; Veronesi, M.C. Studio Del Ruolo Delle Infezioni Batteriche Nella Mortalità Neonatale Dei Cani. *Vet. Ital.* **2014**, *50*, 293–299. [CrossRef] [PubMed]

136. Corrò, M.; Skarin, J.; Börjesson, S.; Rota, A. Occurrence and Characterization of Methicillin-Resistant Staphylococcus Pseudintermedius in Successive Parturitions of Bitches and Their Puppies in Two Kennels in Italy. *BMC Vet. Res.* **2018**, *14*, 308. [CrossRef] [PubMed]

137. Guerrero, A.E.; Stornelli, M.C.; Jurado, S.B.; Giacoboni, G.; Sguazza, G.H.; de la Sota, R.L.; Stornelli, M.A. Vaginal Isolation of Beta-Haemolytic Streptococcus from Bitches with and without Neonatal Deaths in the Litters. *Reprod. Domest. Anim.* **2018**, *53*, 609–616. [CrossRef] [PubMed]

138. Fernández, A.G.; Hielpos, M.S.; Ferrero, M.C.; Fossati, C.A.; Baldi, P.C. Proinflammatory Response of Canine Trophoblasts to Brucella Canis Infection. *PLoS ONE* **2017**, *12*, e0186561. [CrossRef]

139. Goericke-Pesch, S.; Fux, V.; Prenger-Berninghoff, E.; Wehrend, A. Bacteriological Findings in the Canine Uterus during Caesarean Section Performed Due to Dystocia and Their Correlation to Puppy Mortality at the Time of Parturition. *Reprod. Domest. Anim.* **2018**, *53*, 889–894. [CrossRef]

140. Montgomery, M.P.; Robertson, S.; Koski, L.; Salehi, E.; Stevenson, L.M.; Silver, R.; Sundararaman, P.; Singh, A.; Joseph, L.A.; Weisner, M.B.; et al. Multidrug-Resistant Campylobacter Jejuni Outbreak Linked to Puppy Exposure—United States, 2016–2018. *MMWR. Morb. Mortal. Wkly. Rep.* **2018**, *67*, 1032–1035. [CrossRef]

141. Lashnits, E.; Grant, S.; Thomas, B.; Qurollo, B.; Breitschwerdt, E.B. Evidence for Vertical Transmission of Mycoplasma Haemocanis, but Not Ehrlichia Ewingii, in a Dog. *J. Vet. Intern. Med.* **2019**, *33*, 1747–1752. [CrossRef]

142. Santos, R.L.; Souza, T.D.; Mol, J.P.S.; Eckstein, C.; Paíxão, T.A. Canine Brucellosis: An Update. *Front. Vet. Sci.* **2021**, *8*, 1–17. [CrossRef]

143. Latrofa, M.S.; Dantas-Torres, F.; De Caprariis, D.; Cantacessi, C.; Capelli, G.; Lia, R.P.; Breitschwerdt, E.B.; Otranto, D. Vertical Transmission of Anaplasma Platys and Leishmania Infantum in Dogs during the First Half of Gestation. *Parasites Vectors* **2016**, *9*, 269. [CrossRef]

144. Matei, I.A.; Stuen, S.; Modrý, D.; Degan, A.; D'Amico, G.; Mihalca, A.D. Neonatal Anaplasma Platys Infection in Puppies: Further Evidence for Possible Vertical Transmission. *Vet. J.* **2017**, *219*, 40–41. [CrossRef]

145. Decaro, N.; Carmichael, L.E.; Buonavoglia, C. Viral Reproductive Pathogens of Dogs and Cats. *Vet. Clin. N. Am.-Small Anim. Pract.* **2012**, *42*, 583–598. [CrossRef] [PubMed]

146. Ström Holst, B.; Hagberg Gustavsson, M.; Grapperon-Mathis, M.; Lilliehöök, I.; Johannisson, A.; Isaksson, M.; Lindhe, A.; Axnér, E. Canine Herpesvirus During Pregnancy and Non-Pregnant Luteal Phase. *Reprod. Domest. Anim.* **2012**, *47*, 362–365. [CrossRef] [PubMed]

147. Mila, H.; Grellet, A.; Feugier, A.; Desario, C.; Decaro, N.; Buonavoglia, C.; Mariani, C.; Chastant-Maillard, S. General and Type 2 Parvovirus-Specific Passive Immune Transfer in Puppies-Evaluation by Early Growth. *Reprod. Domest. Anim.* **2018**, *53*, 96–102. [CrossRef] [PubMed]

148. Muñoz, A.I.; Vallejo-Castillo, L.; Fragozo, A.; Vázquez-Leyva, S.; Pavón, L.; Pérez-Sánchez, G.; Soria-Castro, R.; Mellado-Sánchez, G.; Cobos-Marin, L.; Pérez-Tapia, S.M. Increased Survival in Puppies Affected by Canine Parvovirus Type II Using an Immunomodulator as a Therapeutic Aid. *Sci. Rep.* **2021**, *11*, 19864. [CrossRef]

149. Decaro, N.; Buonavoglia, C.; Barrs, V.R. Canine Parvovirus Vaccination and Immunisation Failures: Are We Far from Disease Eradication? *Vet. Microbiol.* **2020**, *247*, 108760. [CrossRef]

150. Calero-Berna, R.; Gennari, S.M. Clinical Toxoplasmosis in Dogs and Cats: An Update. *Front. Vet. Sci.* **2019**, *6*, 54. [CrossRef]

151. Dubey, J.P.; Murata, F.H.A.; Cerqueira-Cézar, C.K.; Kwok, O.C.H.; Yang, Y.; Su, C. Toxoplasma Gondii Infections in Dogs: 2009-2020. *Vet. Parasitol.* **2020**, *287*, 2009–2020. [CrossRef]

152. Headley, S.A.; Alfieri, A.A.; Fritzen, J.T.T.; Garcia, J.L.; Weissenböck, H.; da Silva, A.P.; Bodnar, L.; Okano, W.; Alfieri, A.F. Concomitant Canine Distemper, Infectious Canine Hepatitis, Canine Parvoviral Enteritis, Canine Infectious Tracheobronchitis, and Toxoplasmosis in a Puppy. *J. Vet. Diagn. Investig.* **2013**, *25*, 129–135. [CrossRef]

153. Anvari, D.; Saberi, R.; Sharif, M.; Sarvi, S.; Hosseini, S.A.; Moosazadeh, M.; Hosseininejad, Z.; Chegeni, T.N.; Daryani, A. Seroprevalence of Neospora Caninum Infection in Dog Population Worldwide: A Systematic Review and Meta-Analysis. *Acta Parasitol.* **2020**, *65*, 273–290. [CrossRef]

154. Robbe, D.; Passarelli, A.; Gloria, A.; Di Cesare, A.; Capelli, G.; Iorio, R.; Traversa, D. Neospora Caninum Seropositivity and Reproductive Risk Factors in Dogs. *Exp. Parasitol.* **2016**, *164*, 31–35. [CrossRef]

155. McAllister, M.M.; Funnell, O.; Donahoe, S.L.; Šlapeta, J. Unusual Presentation of Neosporosis in a Neonatal Puppy from a Litter of Bulldogs. *Aust. Vet. J.* **2016**, *94*, 411–414. [CrossRef] [PubMed]

156. Gharekhani, J.; Yakhchali, M.; Berahmat, R. Neospora Caninum Infection in Iran (2004–2020): A Review. *J. Parasit. Dis.* **2020**, *44*, 671–686. [CrossRef] [PubMed]

157. Cerqueira-Cézar, C.K.; Calero-Bernal, R.; Dubey, J.P.; Gennari, S.M. All about Neosporosis in Brazil Tudo Sobre Neosporose No Brasil. *Electron. Braz. J. Vet. Parasitol.* **2017**, *26*, 253–279. [CrossRef] [PubMed]

158. Sloan, S.; Šlapeta, J.; Jabbar, A.; Hunnam, J.; De Groef, B.; Rawlin, G.; McCowan, C. High Seroprevalance of Neospora Caninum in Dogs in Victoria, Australia, Compared to 20 Years Ago. *Parasites Vectors* **2017**, *10*, 503. [CrossRef]

159. van Bree, F.P.J.; Bokken, G.C.A.M.; Mineur, R.; Franssen, F.; Opsteegh, M.; van der Giessen, J.W.B.; Lipman, L.J.A.; Overgaauw, P.A.M. Zoonotic Bacteria and Parasites Found in Raw Meat-Based Diets for Cats and Dogs. *Vet. Rec.* **2018**, *182*, 50. [CrossRef] [PubMed]

160. Ahmed, F.; Cappai, M.G.; Morrone, S.; Cavallo, L.; Berlinguer, F.; Dessì, G.; Tamponi, C.; Scala, A.; Varcasia, A. Raw Meat Based Diet (RMBD) for Household Pets as Potential Door Opener to Parasitic Load of Domestic and Urban Environment. Revival of Understated Zoonotic Hazards? A Review. *One Health* **2021**, *13*, 100327. [CrossRef]

161. Broeckx, B.J.G. The Dog 2.0: Lessons Learned from the Past. *Theriogenology* **2020**, *150*, 20–26. [CrossRef]

Review

Analgesia during Parturition in Domestic Animals: Perspectives and Controversies on Its Use

Daniel Mota-Rojas [1,*], Antonio Velarde [2], Míriam Marcet-Rius [3], Agustín Orihuela [4], Andrea Bragaglio [5], Ismael Hernández-Ávalos [6], Alejandro Casas-Alvarado [1], Adriana Domínguez-Oliva [1] and Alexandra L. Whittaker [7]

[1] Neurophysiology, Behavior and Animal Welfare Assessment, DPAA, Universidad Autónoma Metropolitana, (UAM), Mexico City 04960, Mexico

[2] Institute of Agrifood Research and Technology (IRTA), Animal Welfare Program, Veinat Sies S-N, 17121 Monells, Spain

[3] Animal Behaviour and Welfare Department, IRSEA (Research Institute in Semiochemistry and Applied Ethology), Quartier Salignan, 84400 Apt, France

[4] Facultad de Ciencias Agropecuarias, Universidad Autónoma del Estado de Morelos, Cuernavaca 62209, Mexico

[5] Department of Veterinary Medicine, University of Bari A. Moro, 70010 Valenzano, Italy

[6] Clinical Pharmacology and Veterinary Anesthesia, Facultad de Estudios Superiores Cuautitlán, Universidad Nacional Autónoma de México (UNAM), Mexico 54714, Mexico

[7] School of Animal and Veterinary Sciences, Roseworthy Campus, University of Adelaide, Roseworthy, SA 5116, Australia

* Correspondence: dmota100@yahoo.com.mx

Citation: Mota-Rojas, D.; Velarde, A.; Marcet-Rius, M.; Orihuela, A.; Bragaglio, A.; Hernández-Ávalos, I.; Casas-Alvarado, A.; Domínguez-Oliva, A.; Whittaker, A.L. Analgesia during Parturition in Domestic Animals: Perspectives and Controversies on Its Use. *Animals* **2022**, *12*, 2686. https://doi.org/10.3390/ani12192686

Academic Editor: Peter White

Received: 14 September 2022
Accepted: 4 October 2022
Published: 6 October 2022

Publisher's Note: MDPI stays neutral with regard to jurisdictional claims in published maps and institutional affiliations.

Copyright: © 2022 by the authors. Licensee MDPI, Basel, Switzerland. This article is an open access article distributed under the terms and conditions of the Creative Commons Attribution (CC BY) license (https://creativecommons.org/licenses/by/4.0/).

Simple Summary: Although the pain associated with parturition performs the primary physiological function of maintaining frequent, strong myometrial contractions, its biological consequences can affect the health of both mother and fetus. Whilst analgesic therapy may be recommended to avoid pain, the evidence indicates that the mechanisms of some analgesic drugs can interfere with the biological process of labor. Opioids and non-steroidal anti-inflammatory drugs have been shown to indirectly inhibit myometrial contractions by decreasing oxytocin secretion, while local analgesics decrease the number of contractions, although its intensity increases, improving maternal performance. This article analyzes the physiological role of pain during labor and the use and efficacy of opioids, non-steroidal anti-inflammatory drugs, and local analgesics in treatments to manage parturition in domestic mammals.

Abstract: This article analyzes the physiological role of pain during parturition in domestic animals, discusses the controversies surrounding the use of opioids, non-steroidal anti-inflammatory drugs (NSAIDs), and local analgesics as treatments during labor, and presents the advantages and disadvantages for mother and offspring. Labor is a potentially stressful and painful event, due to the contractions that promote expulsion of the fetus. During labor, neurotransmitters such as the prostaglandins contribute to the sensitization of oxytocin receptors in the myometrium and the activation of nociceptive fibers, thus supporting the physiological role of pain. Endogenously, the body secretes opioid peptides that modulate harmful stimuli and, at the same time, can inhibit oxytocin's action in the myometrium. Treating pain during the different stages of parturition is an option that can help prevent such consequences as tachycardia, changes in breathing patterns, and respiratory acidosis, all of which can harm the wellbeing of offspring. However, studies have found that some analgesics can promote myometrial contractility, increase expulsion time, affect fetal circulation, and alter mother–offspring recognition due to hypnotic effects. Other data, however, indicate that reducing the number of uterine contractions with analgesics increases their potency, thus improving maternal performance. Managing pain during labor requires understanding the tocolytic properties of analgesics and their advantages in preventing the consequences of pain.

Keywords: pain; labor pain; calving; farrowing; analgesics

---

1. Introduction

It is well established that the avoidance of pain is beneficial for animal welfare. In order to achieve this, adequate, timely management techniques are required to ensure animal wellbeing and mitigate the sensory and physiological consequences of pain [1,2]. In domestic animals, parturition is an event that normally involves pain, but controversy surrounds the use of analgesics during labor because of conflicting reports as to their benefits in controlling pain, their ability to maintain productive functions and adequate maternal performance [3], their tocolytic effects, and the inhibition of uterine contractions that may prolong labor and affect fetal circulation and/or the dam's recognition of her newborn(s) [4–6]. A further challenge is in the practical implementation of analgesic protocols, since animals may show few signs that parturition is imminent and may give birth away from stockpeople who can intervene.

Eutocic parturition is considered a stressful and painful event, mainly because of the frequent, strong uterine contractions that occur during this process [7]. The physiological consequences of pain during the various phases of parturition derive from chemical mediators such as prostaglandins (PG) and cortisol, which sensitize oxytocin receptors in the myometrium to promote contractions [8] but can also trigger autonomic responses such as tachycardia, hypertension, changes in breathing patterns, and a state of acidosis [9].

Pain is also involved in the maintenance of myometrial contractions via positive feedback [10] in association with the release of endogenous opioid peptides (EOP) that decrease oxytocin sensitization and the frequency of uterine contractions, with the consequence that an eutocic delivery could be converted into a dystocic one [11,12]. In contrast, some studies have shown that a decrease in the number of contractions as a result of analgesic administration is not necessarily associated with complications during labor, but may actually increase the strength of contractions and lessen pain-related consequences [13–15]. Given this controversy, in which some authors sustain that the role of pain during parturition is limited [16], while others argue that pain is an essential element of the normal course of labor [10], we set out to analyze the physiological role of pain during parturition and discuss the usefulness of local anesthetics, non-steroidal anti-inflammatory drugs (NSAIDs), and opioids as treatments during labor in domestic animals, as well as the potential advantages and disadvantages of their use for both mother and offspring.

2. Controversy on the Participation of Pain during Parturition

Mainau and Manteca [16] established that parturition is a physiological stage in which the perception of acute pain results from uterine contractions, the expulsion of the fetus, and inflammation of the uterine tract. Uterine contractions are necessary, especially during the expulsion phase, when they increase in strength and duration, increasing pain intensity (Figure 1) [17,18].

Martínez-Burnes et al. [7] described the physiological role of pain that, together with the degree of fetal maturity, transform the uterus from a quiescent organ to an active one with an abundant presence of oxytocin receptors and proteins such as connexin 43 (Cx43), the elements that cause sustained, synchronic contractions of the myometrium [19,20]. Reports on mammals suggest that the increase in binding sites in the uterus occurs 24 h before the onset of parturition, and that their number increases over 300-fold once the phase of parturition begins [21].

Secretion of oxytocin is required for contractions to occur. Experimental models of rodents have shown that pain has a positive effect on oxytocin secretion [22]. In rodents, oxytocin neurons produce a series of synchronic bursts of this hormone that later act on myometrial oxytocin receptors, while oxytocin neurons generate signals at the level of the spinal cord and vagal nerve directed to α2 noradrenergic cells. This creates a self-sustaining positive feedback cycle that culminates with the release of oxytocin and maintains contractions. The cycle is known as the Ferguson reflex [10,23–26]. In addition, increased noradrenaline concentrations in the supraoptic nucleus (SON) and magnocellular region of the paraventricular nucleus (where the α1-adrenoreceptors are located) activate oxytocin neurons during the phases of parturition, promoting oxytocin synthesis and secretion [27].

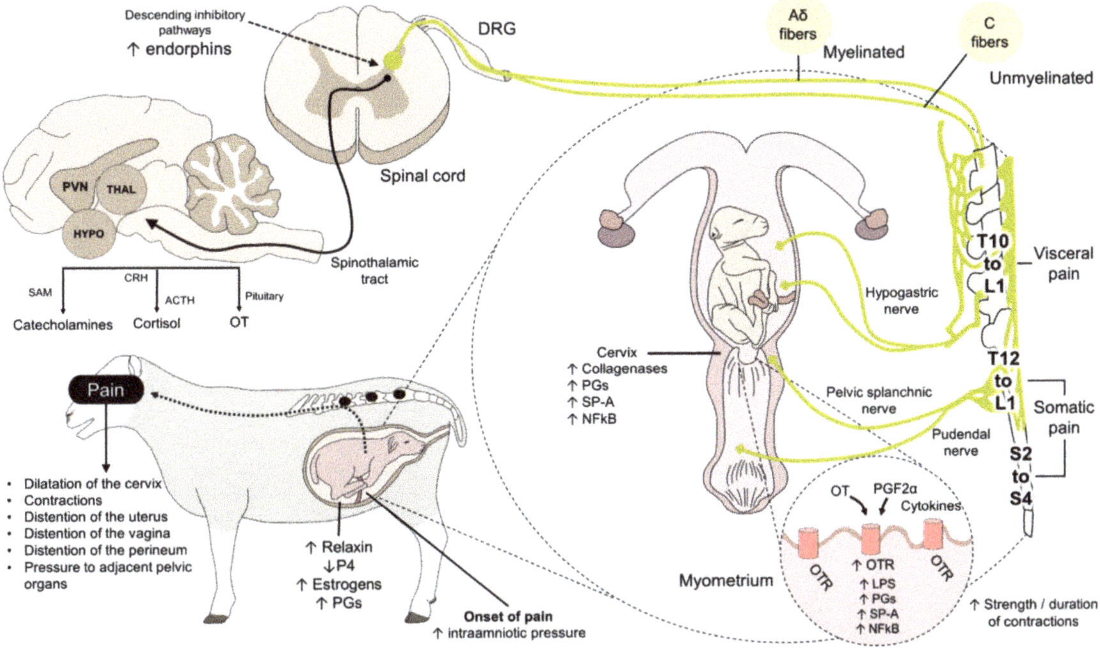

Figure 1. Neurobiology of pain during parturition. With the onset of labor and its clinical signs, such as cervical dilation, myometrial contractions, uterine distention, and increased concentrations of PG, estrogen, collagenase, and other mediators, a nociceptive response is detected by Aδ and C fibers. These neurons transmit sensations of visceral and somatic pain to the dorsal horn of the spinal cord, where interneurons modulate and project to supraspinal centers to trigger the conscious perception of pain. Brain areas such as the HYPO and PVN are responsible for generating physiological responses that trigger secretion of catecholamines, glucocorticoids (cortisol), and OT. OT is the main hormone associated with the onset and maintenance of labor. Its secretion, together with PGF2α and cytokines, activates OTR to increase the strength, frequency, amplitude, and duration of myometrial contractions. ACTH: adrenocorticotropic hormone; CRH: corticotropin-releasing hormone; DRG: dorsal root ganglion; HYPO: hypothalamus; LPS: lipopolysaccharides; NFkB: nuclear factor-kB; OT: oxytocin; OTR: oxytocin receptors; P4: progesterone; PG: prostaglandins; PVN: paraventricular nucleus; SAM: sympathetic-adrenomedullary axis; SP-A: surfactant protein A; TAL: thalamus.

Oxytocin is associated with an analgesic effect, achieved by changing the permeability of the chloride ion in GABAergic neurons [28]. The peripheral, spinal, and supraspinal analgesic action of oxytocin is also linked to hyperpolarization, mediated by a Ca2+-dependent mechanism [29] and by the interaction with transient potential vanilloid 1 receptors (TRPV1). Subcutaneous administration of oxytocin reduces capsaicin-induced nociception in rats (*Rattus norvegicus*) and mice (*Mus musculus*), due to a desensitization effect on the ion channels [30,31]. The function of pain, oxytocin secretion and its potential analgesic role are, however, thought to be limited when an acute process generates an attenuation response by the hypothalamus–pituitary–adrenal axis (HPA). Activation of the HPA releases endogenous opioid peptides (EOP) such as dynorphin and β-endorphins that inhibit oxytocin secretion by occupying κ receptors in the neuronal terminals of the anterior pituitary. As a direct consequence, this reduces secretion of the adrenocorticotropic hormone (ACTH), causing a decrease in the number and intensity of uterine contractions [11,32–35].

Another consequence of labor pain is the activation of the autonomic nervous system (ANS). This leads to an increase in catecholamine secretion. The increase in plasma concentrations has been considered to have a tocolytic effect on uterine motility that reduces myometrial contractibility by occupying β-adrenergic receptors in the sympathetic nervous system (SNS), and promoting muscular relaxation [36]. Studies in human medicine have demonstrated that β-adrenergic stimulation also reduces the synthesis of PGF2α and PGE2 [36], two mediators that participate in the expulsion of the fetus and placenta [37]. The relaxation of the myometrium and uterine contractions prevention result in prolonged farrowing and increased number of nociceptive signals. Contrarily, the antagonism of β-receptors with (carbazolyl), in the Sabuncu et al. [38] analysis of 150 German Landrace sows at 111 days of gestation, showed a reduction in farrowing time from 4.7 ± 0.3 h to 4.0 ± 0.3 h. However, Bostedt and Rudloff [39] found that the action of catecholamines on the myometrium depends on α and β receptors, since their relative abundance can vary in relation to the reproductive cycle stages. It is well known that α receptors predominate during estrus, while β receptors increase in number as gestation advances, so it can be inferred that catecholamines action and the presence of predominant adrenergic receptors can alter myometrial activity.

The physiological consequences described in relation to the decrease in myometrial motility have repercussions not only for the condition of the gestating female, but also for the health and vitality of her offspring. In this regard, Olmos-Hernández et al. [40] evaluated the effect of parity number on the presentation of dystocic farrowing in 120 hybrid Yorkshire × Landrance sows. In that study, gilts presented more contractions than sixth-parity sows, greater contraction intensity (12.30 mmHg), and more intrapartum stillbirths. Those results were attributed to the deleterious effect of prolonged, sustained myometrial activity on fetal circulation, which significantly decreased gas exchange within the placenta and produced neonatal hypoxia in the fetus. Later studies confirmed that maternal stress produces hyperactivity of the HPA and the sympathetic–adrenal–medullary system, related to catecholamine release, elevated fetal blood pressure, and poor placenta-uterine perfusion [41,42].

The relationship between catecholamines and pain and stress responses during lambing was investigated in a study where 2 h of psychosocial stress was imposed through isolation of the gestating ewes. The response was measured in uterine blood flow that utilized an adrenergic block with labetalol to determine the influence and impact of catecholamines on the fetus. The results showed fetal cortisol levels 8.1 ± 2.1% higher than in the dams, elevated noradrenaline concentrations, and a 22% reduction in uterine blood flow that produced prolonged anaerobic metabolism and fetal hypertension. It is important to note that the adrenergic block impeded the reduction in fetal blood flow, indicating the role that hormones and neurotransmitters play in the health outcomes of fetuses [43].

As Mainau et al. [44] discuss, the occurrence of uncontrolled pain during parturition alters the physiological effect of this condition and triggers a cascade of events ending with the secretion of glucocorticoids (cortisol) that, in conjunction with EOP, inhibit oxytocin release. When concentrations of oxytocin are altered during the expulsion phase, direct or indirect consequences for the fetus can include hypoxia, aggression by the dam, and rejection of the offspring, as well as inhibition of colostrum and milk ejection, increasing pre-weaning mortality. Analgesic therapy of dams during labor might be justified to avoid these problems. However, as we discuss in the following sections, the site and mechanism of action, and adverse effects of the analgesics often employed during birth, must be well understood and evaluated before deciding to apply any analgesic protocol (Figure 2).

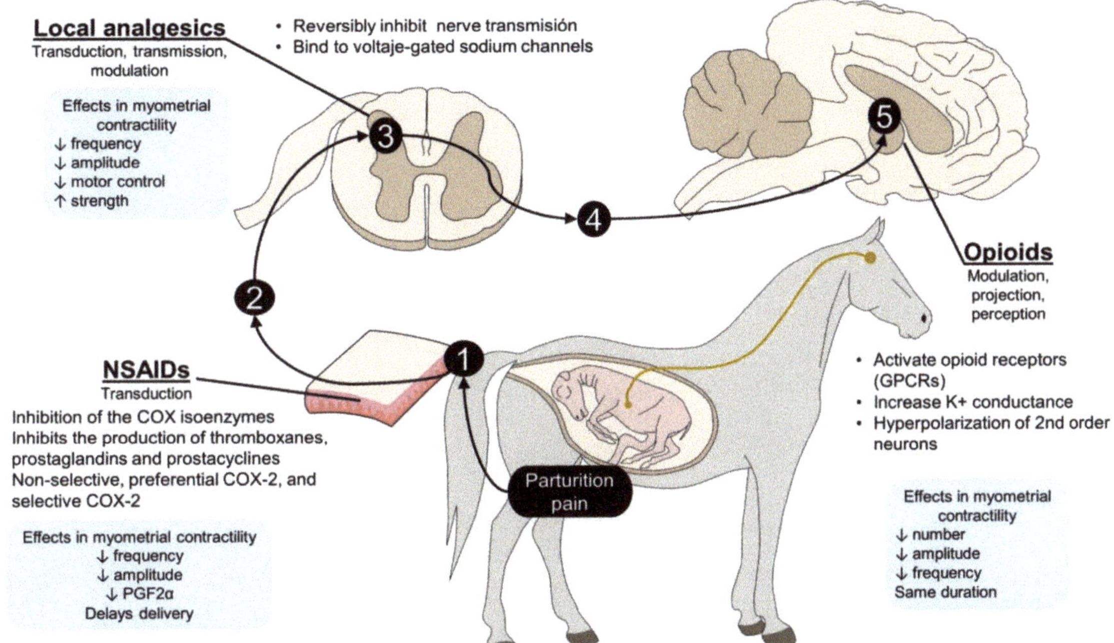

Figure 2. Action site of analgesics commonly used during parturition. Depending on their action mechanism, these drugs intervene at different levels of the nociceptive pathway. The NSAIDs block harmful impulse transduction by inhibiting COX and PG production, while LA can intervene in the first three stages of the nociceptive pathway by blocking voltage-gated sodium channels in the spinal cord. Secondly, opioids are centrally acting analgesics that hyperpolarize neurons to prevent the projection and perception of pain. In general, all three types of drugs can potentially decrease the frequency and amplitude of contractions of the uterus, but this effect depends on the drug and dose administered. 1: transduction; 2: transmission; 3: modulation; 4: projection; 5; perception; COX: enzyme cyclooxygenase. GPCRs: G protein-coupled receptors; PGF2α: prostaglandin 2α.

3. Opioids

These analgesics are generally considered superior for control of moderate-to-severe acute pain through agonism to µ, κ, and λ receptors [45]. Their action inhibits the action of G protein-coupled receptors (GPCRs) of cyclic adenosine monophosphate (cAMP), generating an increase in the conductance to K+ that promotes hyperpolarization of the membrane in second-order neurons (Figure 3) [46,47]. Morphine, fentanyl, methadone, and meperidine are the most commonly applied opioids for pain management [7]. However, their effects on parturition are controversial, and may be species dependent.

The analgesic action of morphine at a dosage of 30 µg/10 µL administered with subcutaneous oxytocin was evaluated in 62 gestating albino Sprague Dawley rats based on observing nocifensive behaviors. In the animals that received morphine, the researchers registered a reduction in the number of stretches, a behavior associated with the nociceptive component of uterine contractions, compared to a control group that received an epidural saline solution (frequency of 8 ± 2 vs. 57 ± 12 stretches). The authors did not report significant alterations in mean labor duration, but the rodents treated with the opioid had lower amounts of spinal c-Fos-positive neurons (80 ± 21 vs. 165 ± 17) [48]. C-Fos expression is considered a neuronal marker of nociceptive activation by the spinal afferent fibers that responds to visceral uterine pain during labor in the first 30 min after stimulus onset [49].

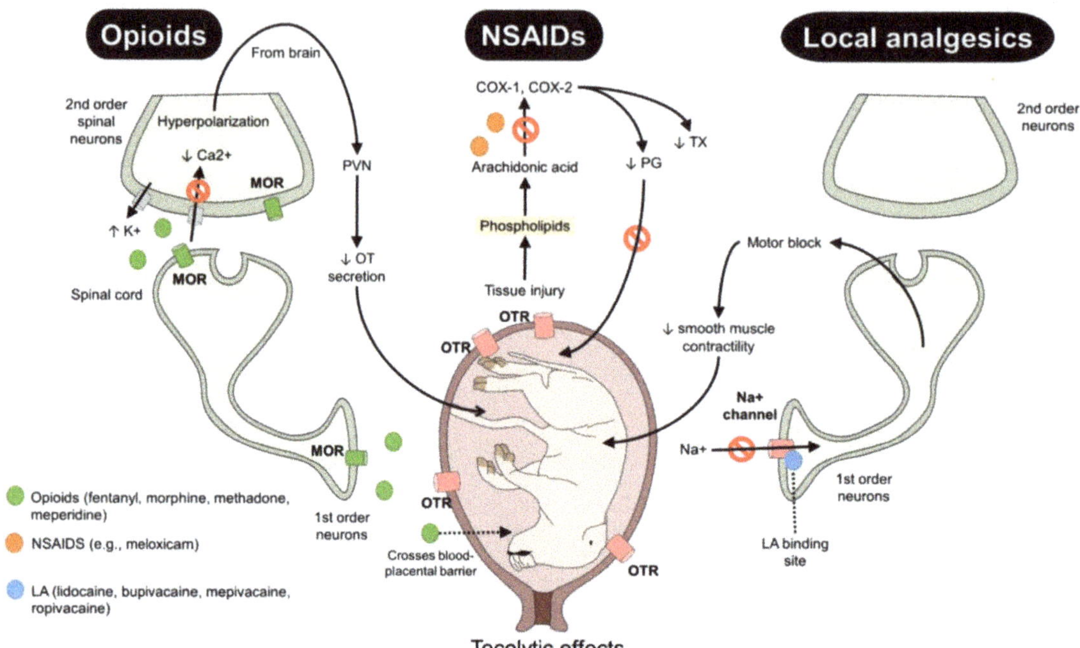

Figure 3. Action mechanisms of opioids, NSAIDs, and local analgesics. During labor, all three types of analgesics exert their action by interacting with various receptors and metabolic pathways. Opioids activate MOR in first- and second-order neurons to prevent Ca2+ influx, increase K conductivity, and hyperpolarize the membrane. NSAIDs block the COX enzyme and production of PGs. Local analgesics block Na+ channels, preventing the transmission of harmful stimuli, but also exert a tocolytic effect by generating a motor blockade that reduces the smooth muscle contractility of the myometrium. Similarly, the decreases in PG production and OT secretion affect the contractility of the uterus. COX: cyclooxygenase; LA: local analgesics; MOR: mu-opioid receptors; NSAIDs: non-steroidal anti-inflammatory drugs; OT: oxytocin; OTR: oxytocin receptors; PVN: paraventricular nucleus.

Tong et al. [50] determined that intrathecal administration of morphine at infusion rates above 0.035 microg/h administered one day before the onset of parturition, eliminated phasic stretching in 25 primiparous gestating Sprague Dawley rats. By blocking the pain caused by the distension and inflammation of the pelvic viscera, morphine reduced the perception of the harmful stimulus, but showed no adverse effects on the duration of labor, on the presentation of normal activities (e.g., eating, drinking, grooming), or on maternal behaviors such as nest-building, pup-licking, or eating the placenta. The authors thus suggested intrathecal application of morphine as a pharmacological option for treating pain during the birth process.

In spite of these in vivo studies finding no difference in labor duration with opioid administration, other in vivo and tissue-based studies have shown that contractility of smooth muscle is modified by opioids. Nacitarhan et al. [4], evaluated the cumulative effect of alfentanil, meperidine, and remifentanil on strips of the longitudinal uterine smooth muscle of gestating rats (18–21 days). After inducing contractions with 1 mU/mL of oxytocin, the authors reported that the opioids and local analgesics (LA) such as bupivacaine and ropivacaine significantly reduced myometrial contractions. Similarly, the tocolytic action of morphine and its impact on labor have also been reported by Javadi-Paydar et al. [51], who studied NMRI (Naval Medical Research Institute) mice at 15 days of gestation. Those authors evaluated the effect of a single intraperitoneal (IP) dose of morphine (10 or 20 mg/kg) vs. a double dose (5 or 10 mg/kg) in lipopolysaccharide (LPS)-induced births. The opioid administered in the double dose reduced the incidence of LPS-induced preterm delivery by $50 \pm 17.7\%$ compared to the control group, a result that meant an interruption or delay of parturition due to diminished oxytocin secretion [6]. It has been suggested that the myometrial effect of exogenous administration of opium-derived analgesics is related to a mechanism similar to that of the EOP during parturition, since the presence of a large amount of opioid receptors in PVN neurons decreases oxytocin secretion with the expected consequences [27]. For example, morphine increases the rate of the metabolic clearance of oxytocin, which has been shown to reduce uterine contractions by up to 17%, 30–45 min after administration [52].

In contrast to the findings in rodents, the use of opioids in humans has been associated with respiratory depression, altered neonatal vitality scores, higher mortality, and problems in mother–offspring interaction due to hypnotic effects [53]. There have also been reports of a tocolytic effect on myometrial activity, suggesting that this could affect maternal performance at birth [54–56]. These findings are corroborated by a study in baboons (*Papio anubis*) in their third trimester of gestation (140–150 days). The animals were treated with one 5 mg dose of morphine intravenously (IV) after receiving a bolus injection of oxytocin (500 mU by intravenous infusion) to induce uterine contractions. Morphine increased the metabolic clearance rate by 40%, and significantly reduced the amplitude and frequency of contractions (27 ± 8 mmHg and 3.2 ± 0.5 contractions/15 min, respectively) but did not affect their mean duration [57].

The delay in fetus expulsion may have physiological consequences for both, the dam and her neonate, manifested in myometrial dysfunction, fetal asphyxia, stillbirths, maternal and fetal morbidity, and fetal cardiorespiratory depression [58]. The occurrence of these adverse effects is attributed, in some cases, to the pharmacokinetics of the drugs. One example of this was observed in a study of 16 gestating ewes that received a 5 µg/kg-dose of fentanyl by the IV route. The measured transplacental rate of the drug was 77% [59], which is greater than that seen with morphine, likely due to the greater liposolubility of fentanyl. For the neonates, the opioids' capacity to cross the blood–placental barrier is considered the main cause of respiratory depression, fetal cardiovascular depression, mood depression, inhibition of the sucking reflex, and delayed colostrum intake [6,60]. Butorphanol and pethidine may be particularly problematic for the neonate since these drugs easily cross the blood–placental barrier due to the umbilical vein/maternal vein ratio [61]. The elimination half-life of butorphanol in milk is 36 h, so its administration to the mothers could be considered a risk factor for the safety of the neonate [62], as was reported in newborn colts, where a sedated state [63] with significant respiratory depression was observed. Furthermore, in the offspring, the half-life and distribution volume of this drug were greater (half-life = 2.1 h, VDss = 3.86 L/kg, respectively) than in adults (half-life = 0.74 h, VDss = 1.13 L/kg) [64].

In an effort to prevent negative effects on myometrial contractility and depressor effects in the neonate, some authors have proposed the use of atypical opioids such as tramadol, which work through inhibiting serotonin and norepinephrine reuptake in sensitive fibers [65], as an option that does not alter uterine activity. A study by Yakovleva et al. [66] evaluated tramadol's effect on uterine contractile activity in chinchillas (*Oryctolagus cuniculus*) at 28 days of gestation. Animals received 1 U of oxytocin by IV to initiate labor, and then 1 mL of tramadol in infusion at a concentration of 5 mg/mL, 10 min later. In that study, tramadol administration did not affect the number of contractions, but increased their amplitude and duration compared to the control group. Their results thus differed from those reported for common opioids such as morphine, fentanyl, and butorphanol [4,50].

In conclusion, even though opioids offer efficacious pain management during parturition, their adverse effects–depressed mental states, reduced oxytocin secretion–hinder mother–offspring interaction in some species and can indirectly affect uterine activity and neonate survival during the first days of life. For these reasons, other therapeutic options are available that do not exert their action at the central level. These include the NSAIDs.

4. Non-Steroidal Anti-Inflammatory Drugs

The NSAIDs are a group of drugs widely used in veterinary medicine to manage acute pain [65,67,68]. They work through inhibiting the enzyme cyclooxygenase (COX), which has three isoforms (COX-1, COX-2, COX-3) that modulate the inflammatory response which follows the production of chemical mediators such as PG, thromboxane A, and the prostacyclin [69,70]. However, the use of NSAIDs during parturition is controversial, and some adverse effects on uterine contractions have been attributed to them (Figure 3) [71].

Yousif and Thulesius [72] evaluated meloxicam, a preferential COX-2 inhibitor, and indomethacin on the in vitro uterine motility of uterine strips from rats. They found that both drugs reduced the frequency (pre-treatment = 4.5 contractions/min vs. post-treatment = 0.3 contractions/min) and amplitude (pre-treatment = 90 mmHg vs. post-treatment = 10 mm Hg) of contractions in pregnant and non-pregnant animals. These results were corroborated by Rac et al. [71], who analyzed meloxicam's capacity to inhibit preterm parturition in nine Polled Dorset ewes at 121 days of gestation. After application of a contraction promotor to induce parturition, meloxicam's action reduced the amplitude and frequency of myometrial contractions (2.25 ± 0.48 contractions per minute and 5.21 ± 0.43 mmHg, respectively), thus inhibiting parturition. The same effect on the amplitude and frequency of contractions has been observed in in vitro studies of the uteri of dairy cattle [73]. The biochemical explanation of this phenomenon lies in the mechanism of action that inhibits the COX-1 and COX-2 enzymes.

Expression of the COX-1 and COX-2 enzymes during gestation increases in uterine tissue, as has been demonstrated in sheep models [5]. Both of these isoforms are fundamental for producing PGs, such as PGF2α and PGE2, which are catalyzed by COX-2 to transform arachidonic acid into PGG2 and final products such as PGF2α, PGE2, and PGH2 by means of peroxides [8,74]. In mice, direct inhibition of COX-2 expression negatively regulates PGF2α synthesis during gestation, triggering the indirect suppression of the expression of c-Fos, which is considered a transcriptional activator of COX-2 [75] that is essential for inducing oxytocin receptors in the myometrium [76]. In a similar study, interference in PG secretion in adult female rats medicated orally with meloxicam at doses of 7.5 and 10 mg/kg from day 20–22 of gestation, prolonged delivery times following a dose-dependent pattern (37–51 h of delay), and produced more stillborn pups (5.3 ± 0.67 fetuses in the 7.5 mg treatment vs. 5.6 ± 0.42 in the dams medicated with 10 mg/kg) [77]. A similar prolongation effect has been seen in dogs when NSAIDs were administered during the final stage of pregnancy [78].

The selectivity of the NSAIDs has been associated with both benefits and complications for dams. In dairy cows, for example, administering NSAIDs of the inhibitor of COX-1 type produced adverse effects such as retained placenta, metritis, and culling, because COX-1 is constitutive and participates in diverse physiological functions. In contrast, preferential inhibitors of COX-2 have been related to benefits for the health and productivity of females [79,80].

In a study of 237 dairy cows, administration of meloxicam (1 mg/kg) before and after calving had no effect on the health of dams, but increased milk production in 6.8 kg/day more than in the control [81]. Similarly, in Holstein cows, those that were treated with meloxicam had increased milk compared to a control group. There were other benefits with the probability of the medicated cows suffering subclinical mastitis being reduced by 0.05 times, whilst the probability of euthanasia or death was 0.46 times lower [3]. Similar results have been suggested for the use of firocoxib, another selective COX-2 inhibitor [82].

The consequences of firocoxib administration for the fetus, compared to meloxicam or flunixin meglumine, was analyzed in the embryos of 30 mares. In that study, Okada et al. [83] evaluated embryonic movement using transrectal ultrasonography every 5 min for 1 h before treatment. Firocoxib did not interfere with embryo mobility at

12 days after pregnancy confirmation, with the embryos maintaining constant movements (measured in movements/hour) before, immediately after, and 24 h after treatment. Contrarily, embryo mobility decreased with the use of flunixin meglumine (pre-treatment = 5.9 ± 0.3; post-treatment = 1.9 ± 0.3) and meloxicam (pre-treatment = 5.8 ± 0.5; post-treatment = 2.3 ± 0.5 movements/h). In contrast, in a study of ewes, meloxicam increased fetal blood flow, albeit with no effect on osmolarity, uterine blood pressure, or maternal–fetal gas exchange [71]. Nevertheless, firocoxib is likely the drug of choice for administration to gestating animals and has been considered the best non-steroidal option for treating pathologies such as endometritis (inflammation of the endometrium) in the early stages of gestation [83].

One drug that has been shown to have a harmful effect on fetal viability is flunixin meglumine. Newby et al. [84] studied 34 Holstein cows treated with flunixin and 38 with a placebo, before and after calving. The results showed that the offspring of the animals treated with the flunixin 24 h before parturition had higher mortality rates and an increased probability of placental retention and fever, coupled with lower milk production and a higher risk of metritis development. These effects were attributed to the drug's property of inhibiting endogenous production of PG by as much as 80% (at a dose of 2.2 mg kg^{-1} for 10 days). Therefore, if the level of PG required to promote contractions of the smooth uterine muscle is reduced, myometrial motility also decreases, and the response of oxytocin is greater, as are the adverse effects [85].

In summary, administration of NSAIDs during parturition highlights the selective influence of isoforms on COX with COX-2 selective agents being preferential. The decrease in PG concentration and the tocolytic effect of these analgesics, which are associated with risks for the offspring of some species, have shown no risk to others. Although they are contraindicated for use in gestating females or during dystocia births in rodents, in some instances [86], these drugs have also been proposed for some species as options that do not affect dams or their neonates either biologically or in terms of productivity.

5. Local Analgesics

Local analgesics (LA) are considered the gold standard for controlling acute pain [87,88], especially during parturition in both, animals and humans [7]. Lidocaine, bupivacaine, mepivacaine, and ropivacaine are the most commonly used [89]. Their mechanism of action is through the reversible blockade of voltage-gated Na2+ channels that impedes the transduction of harmful signals and their ensuing transmission [90]. As occurs with sensitive signals, when LA are administered motor signals are also inhibited, generating a decrease in the contraction of both skeletal and smooth muscle, as numerous studies have shown [13,91,92]. To date, however, it is unclear whether these drugs could affect the performance of mothers during labor. In a study of humans, Qian et al. [14] evaluated 213 pregnant women under different pain control treatments (levobupivacaine, ropivacaine, and controls). They found that levobupivacaine did not impede the normal progress of labor, while both levobupivacaine and ropivacaine efficaciously controlled pain and helped maintain all physiological functions intact.

Studies in non-human animals have demonstrated that, like opioids, LA present tocolytic activity in experimental models. A study by Karsli et al. [93] compared the effect of opioids (meperidine, alfentanil, remifentanil) to those of LA (mepivacaine, ropivacaine, bupivacaine) in cell cultures of gestating rats. In those tests, the amplitude and frequency of the contraction of uterine cells decreased significantly with the use of both groups of drugs [4]. The reduction in the number of contractions of the smooth uterine after administration of LA is dose dependent, and the degree of muscular relaxation that the analgesics cause increases as the dosage is increased. This was pointed out by Arici et al. [94] in their evaluation of the effect of 10 accumulated doses of (−8) to 10 (−4) mol/L of mepivacaine, ropivacaine, and bupivacaine on strips of uterine muscle isolated from female Wistar rats at 18–21 days of gestation. Upon stimulating the uterine tissue with oxytocin, both bupivacaine and ropivacaine reduced uterine contractions in a dose-dependent pattern, but mepivacaine exerted the opposite effect by significantly increasing the number of contractions. These results were similar to those published by Li et al. [92], who used full-thick myometrial strips of gestating and non-gestating rats, exposed to cumulative amounts of levobupivacaine and bupivacaine (10-8 mol/L to 10-4 mol/L). Those authors found that at higher amounts both LA caused inhibition of contractibility, with levobupivacaine increasing the amplitude of contractions and bupivacaine increasing their frequency.

The uterine response to progressively greater doses of procaine, lidocaine, or ropivacaine—from 0.1 mg/mL to 0.5 mg/mL and then to 1 mg/mL—was studied using an experimental pig model. In general, the LA increased intrauterine pressure in the isthmus and corpus that was dose dependent, although the authors also observed a reduction in the frequency and amplitude of contractions [95]. These results support the proposal that the reduction in contractibility does not necessarily have a negative effect on performance during parturition but may help regulating the strength of contractions and make them more effective, with the additional advantage of no nociceptive response.

On the other hand, the muscular relaxation effect of local analgesics leads to a vasodilation effect at the systemic level that reduces blood flow to various systems [96]. This event was observed in a series of studies in pregnant ewes, where epidural administration of LA reduced uterine blood flow by as much as 65 ± 9 mmHg. The consequent reduced fetal blood flow and fetal gas exchange [97] was similar to that observed with opioids, but with the advantage that no mental state or hypnotic effect was induced in the offspring. By the same token, administering epidural analgesia with LA blocks the Ferguson reflex and reduces blood oxytocin level, altering the main beneficial effects of this hormone. These include the promotion of social interaction, formation of the mother–offspring bond, and reduced anxiety, pain, and stress. Based on the fact that LA interrupt the full functioning of the neural pathways, the adverse effects of their use during parturition are attributed to those characteristics [98].

Similarly to opioids and NSAIDs, LA can reduce uterine contractions due to their tocolytic action. That reduction has an inverse relation to the strength of the contractions, making them more effective and opening the possibility of improving the performance of dams during parturition. However, the hypotension they cause is an element that must be considered when deciding whether to use these drugs to avoid risks for the fetus and the mother [99]. A summary of the discussed drugs and their effect or benefits during parturition can be seen in Table 1.

Table 1. Summary of the different analgesic drugs used during parturition in domestic animals.

Drug	Action Mechanism	Effect	Benefit	Species	References
Opioids Morphine Fentanyl Tramadol	Agonism to μ, κ, and δ opioid receptors in the cerebral cortex and spinal cord.	Therapeutic: Systemic analgesia. In myometrium: Decreases the amplitude and frequency of uterine contractions.	Modulates and reduces pain perception.	Bovine, ovine, caprine, equid, and leporids.	[4,50,51,66]
Non-steroidal anti-inflammatory drugs (NSAIDs) Meloxicam Carprofen Firocoxib Flunixin meglumine	COX inhibitor and nociceptive modulator.	Therapeutic: Systemic analgesia. In myometrium: Tocolytic effect. Diminishes the amplitude and frequency of uterine contractions due to a reduction in PG secretion.	Inhibits the production of COX and proinflammatory mediators.	Bovine, equid, and ovine.	[71,72,83]
Local Analgesics Lidocaine Bupivacaine Ropivacaine Levobupivacaine	Sodium channel blockers. Interferes with the transduction and transmission of the nociceptive stimuli.	Therapeutic: Local analgesia. In myometrium: Reduces the amplitude and number of uterine contractions.	Inhibits the transduction and transmission of pain.	Bovine, equid, and pig.	[13,14,93,94]

COX: cyclooxygenases; PG: prostaglandins.

6. Perspectives

The control of pain in animals is fundamental for preserving their welfare [1]. The negative impact of pain perception on states of health, hemodynamic stability, the immune system, and the capacity to response to infections are justifications for preventing this process in animals [100–102]. The pain produced during physiological events such as parturition requires additional studies that deepen and broaden existing perspectives. This is especially so given the controversy around this topic with some authors supporting the proposition that the presence of pain sustains uterine contractions, whereas others emphasize the potentially harmful effects of the processes triggered when perception of painful sensations is prolonged, the latter increasing the possibility of dystocic births with severe consequences for both mother and fetus (Figure 4) [16].

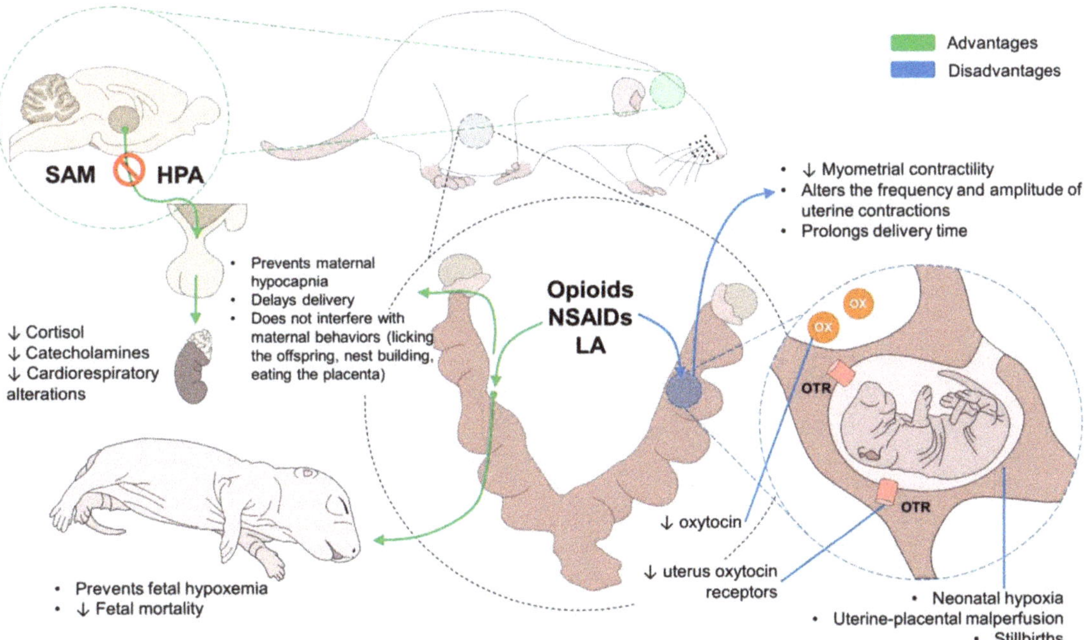

Figure 4. Summary of advantages and disadvantages of analgesia in parturition. Opioids, NSAIDs, and LA can have beneficial or detrimental effects on the mother and the newborn. However, since these actions differ between species and individuals, an appropriate analgesic protocol must consider these factors from a pharmacological point of view. Advantages such as the prevention of hypocapnia and hypoxemia in the dam and the offspring, respectively, are marked in green inside the figure. Disadvantages such as a decrease in OX and OTR are marked with blue. HPA: hypothalamus–pituitary–adrenal axis; LA: local analgesics; NSAIDs: non-steroidal anti-inflammatory drugs; OTR: oxytocin receptors; OX: oxytocin; SAM: sympathetic–adreno–medullary axis.

The analgesics currently utilized during labor have shown an ability to inhibit the activity of the myometrium [4]. This is an obstacle for establishing a balanced analgesic strategy that would prevent pain but also avoid interfering with the normal course of parturition. On the one hand, additional studies are needed to substantiate the idea and importance of pain management and its potential benefits in relation to production. On the other, it is necessary to determine whether the inhibition of uterine contractions has a negative effect on fetal blood flow and causes fetal asphyxia or could improve the effectiveness of contractions and thus enhance performance during parturition [82,103,104]. One option for achieving these objectives (i.e., preventing pain without affecting fetal blood flow or maternal performance) could entail using alternative drugs such as firocoxib, ketamine, cannabinoids, or gabapentinoids, which help inhibit the perception of pain through distinct mechanisms without affecting uterine activity in such a potentially stressful event as parturition [82,103–106].

7. Conclusions

Pain has a dual role during parturition, as the positive feedback to myometrial contractions is essential for expulsion of the fetus, but at the same time, this harmful stimulus may trigger serious physiological consequences. Most of the analgesics mentioned herein have a tocolytic effect, but if clinicians understand the adverse effects that may occur when establishing an analgesic protocol, this effect should not impede managing pain in specific events. Opioids indirectly inhibit uterine contractions. Research has shown that their hypnotic and depressant effects can affect birth performance and offspring vitality. In contrast, the indirect tocolytic effect of NSAIDs may not exceed therapeutic efficacy in maintaining the productive performance of animals after parturition. Finally, the local analgesics considered the gold standard for controlling pain during labor have been shown to decrease the number of contractions but increase their strength, thus facilitating labor. Therefore, the aim of developing a pain management protocol should be to reduce the physiological consequences of these effects. Their administration, however, requires constant monitoring of the course of labor to identify the moment at which the physiological benefit of pain is lost, or when the benefits that analgesia provides during parturition could begin to have detrimental effects on the animals involved.

Author Contributions: All authors contributed to the conceptualization, writing, reading, and approval of the final manuscript. All authors have read and agreed to the published version of the manuscript.

Funding: This research received no external funding.

Institutional Review Board Statement: Not applicable.

Informed Consent Statement: Not applicable.

Data Availability Statement: Not applicable.

Conflicts of Interest: The authors declare that they have no conflict of interest.

References

1. Hernández-Avalos, I.; Mota-Rojas, D.; Mora-Medina, P.; Martínez-Burnes, J.; Casas Alvarado, A.; Verduzco-Mendoza, A.; Lezama-García, K.; Olmos-Hernandez, A. Review of different methods used for clinical recognition and assessment of pain in dogs and cats. *Int. J. Vet. Sci. Med.* **2019**, *7*, 43–54. [CrossRef] [PubMed]
2. Hernández-Avalos, I.; Flores-Gasca, E.; Mota-Rojas, D.; Casas-Alvarado, A.; Miranda-Cortés, A.E.; Domínguez-Oliva, A. Neurobiology of anesthetic-surgical stress and induced behavioral changes in dogs and cats: A review. *Vet. World* **2021**, *14*, 393–404. [CrossRef] [PubMed]
3. Shock, D.A.; Renaund, D.L.; Roche, S.M.; Poliquin, R.; Thomson, R.; Olson, M.E. Correction: Evaluating the impact of meloxicam oral suspension administered at parturition on subsequent production, health, and culling in dairy cows: A randomized clinical field trial. *PLoS ONE* **2018**, *13*, e0210326. [CrossRef] [PubMed]
4. Nacitarhan, C.; Sadan, G.; Kayacan, N.; Ertugrul, F.; Arici, G.; Karsli, B.; Erman, M. The effects of opioids, local anaesthetics and adjuvants on isolated pregnant rat uterine muscles. *Methods Find. Exp. Clin. Pharmacol.* **2007**, *29*, 273. [CrossRef]
5. Rac, V.E.; Scott, C.A.; Small, C.; Lee Adamson, S.; Rurak, D.; Challis, J.R.; Lye, S.J. Dose-dependent effects of meloxicam administration on cyclooxygenase-1 and cyclooxygenase-2 protein expression in intrauterine tissues and fetal tissues of a sheep model of preterm labor. *Reprod. Sci.* **2007**, *14*, 750–764. [CrossRef]
6. Alper, M.H.; Brown, W.U.; Ostheimer, G.W.; Scanlon, J.W. Effects of maternal analgesia and anaesthesia on the newborn. *Clin. Obstet. Gynaecol.* **1975**, *2*, 661–671. [CrossRef]
7. Martínez-Burnes, J.; Muns, R.; Barrios-García, H.; Villanueva-García, D.; Domínguez-Oliva, A.; Mota-Rojas, D. Parturition in mammals: Animal models, pain and distress. *Animals* **2021**, *11*, 2960. [CrossRef] [PubMed]
8. Nagel, C.; Aurich, C.; Aurich, J. Stress effects on the regulation of parturition in different domestic animal species. *Anim. Reprod. Sci.* **2019**, *207*, 153–161. [CrossRef] [PubMed]

9. Heavner, J.E.; Cooper, D.M. Pharmacology of analgesics. In *Anesthesia and Analgesia in Laboratory Animals*, 2nd ed.; Fish, R., Danneman, P., Brown, M., Karas, A., Eds.; Academic Press: San Diego, CA, USA, 2008; pp. 97–123.

10. Welsh, T.; Johnson, M.; Yi, L.; Tan, H.; Rahman, R.; Merlino, A.; Zakar, T.; Mesiano, S. Estrogen receptor (ER) expression and function in the pregnant human myometrium: Estradiol via ERα activates ERK1/2 signaling in term myometrium. *J. Endocrinol.* **2012**, *212*, 227–238. [CrossRef] [PubMed]

11. Katz, N.; Mazer, N.A. The impact of opioids on the endocrine system. *Clin. J. Pain* **2009**, *25*, 170–175. [CrossRef] [PubMed]

12. Vuong, C.; Van Uum, S.H.M.; O'Dell, L.E.; Lutfy, K.; Friedman, T.C. The effects of opioids and opioid analogs on animal and human endocrine systems. *Endocr. Rev.* **2010**, *31*, 98–132. [CrossRef]

13. Wei, J.-S.; Jin, Z.-B.; Yin, Z.-Q.; Xie, Q.-M.; Chen, J.-Q.; Li, Z.-G.; Tang, H.-F. Effects of local anesthetics on contractions of pregnant and non-pregnant rat myometrium in vitro. *Acta Physiol. Hung.* **2014**, *101*, 228–235. [CrossRef] [PubMed]

14. Qian, X.; Wang, Q.; Ou, X.; Li, P.; Zhao, B.; Liu, H. Effects of ropivacaine in patient-controlled epidural analgesia on uterine electromyographic activities during labor. *Biomed. Res. Int.* **2018**, *2018*, 7162865. [CrossRef] [PubMed]

15. Zhao, B.; Qian, X.; Wang, Q.; Ou, X.; Lin, B.; Song, X. The effects of ropivacaine 0.0625% and levobupivacaine 0.0625% on uterine and abdominal muscle electromyographic activity during the second stage of labor. *Minerva Anestesiol.* **2019**, *85*, 854–861. [CrossRef] [PubMed]

16. Mainau, E.; Manteca, X. Pain and discomfort caused by parturition in cows and sows. *Appl. Anim. Behav. Sci.* **2011**, *135*, 241–251. [CrossRef]

17. Alimoradi, Z.; Kazemi, F.; Valiani, M.; Gorji, M. Comparing the effect of auricular acupressure and body acupressure on pain and duration of the first stage of labor: Study protocol for a randomized controlled trial. *Trials* **2019**, *20*, 766. [CrossRef] [PubMed]

18. Hashemi Asl, B.M.; Vatanchi, A.; Golmakani, N.; Najafi, A. Relationship between behavioral indices of pain during labor pain with pain intensity and duration of delivery. *Electron. Physician* **2018**, *10*, 6240–6248. [CrossRef] [PubMed]

19. Kidder, G.M.; Winterhager, E. Physiological roles of connexins in labour and lactation. *Reproduction* **2015**, *150*, R129–R136. [CrossRef] [PubMed]

20. Purohit, G. Parturition in domestic animals: A review. *WebmedCentral Reprod.* **2010**, *1*, WMC00748.

21. Mota-Rojas, D.; Villanueva-García, D.; Velazquez-Armenta, E.Y.; Nava-Ocampo, A.A.; Ramírez-Necoechea, R.; Alonso-Sspilsbury, M.; Trujillo, M.E. Influence of time at which oxytocin is administered during labor on uterine activity and perinatal death in pigs. *Biol. Res.* **2007**, *40*, 55–63. [CrossRef]

22. Boll, S.; Almeida de Minas, A.C.; Raftogianni, A.; Herpertz, S.C.; Grinevich, V. Oxytocin and pain perception: From animal models to human research. *Neuroscience* **2018**, *387*, 149–161. [CrossRef]

23. Meddle, S.; Leng, G.; Selvarajah, J.; Bicknell, R.; Russell, J. Direct pathways to the supraoptic nucleus from the brainstem and the main olfactory bulb are activated at parturition in the rat. *Neuroscience* **2000**, *101*, 1013–1021. [CrossRef]

24. Meddle, S.L.; Bishop, V.R.; Gkoumassi, E.; van Leeuwen, F.W.; Douglas, A.J. Dynamic changes in oxytocin receptor expression and activation at parturition in the rat brain. *Endocrinology* **2007**, *148*, 5095–5104. [CrossRef]

25. Ferguson, J.K.W. A Study of the motility of the intact uterus at term. *Surg. Gynecol. Obstet.* **1941**, *73*, 359–366.

26. Murata, T.; Narita, K.; Honda, K.; Matsukawa, S.; Higuchi, T. Differential regulation of estrogen receptor alpha and beta mRNAs in the rat uterus during pregnancy and labor: Possible involvement of estrogen receptors in oxytocin receptor regulation. *Endocr. J.* **2003**, *50*, 579–587. [CrossRef] [PubMed]

27. Brunton, P.J. Endogenous opioid signalling in the brain during pregnancy and lactation. *Cell Tissue Res.* **2019**, *375*, 69–83. [CrossRef] [PubMed]

28. Mazzuca, M.; Minlebaev, M.; Shakirzyanova, A.; Tyzio, R.; Taccola, G.; Janackova, S.; Gataullina, S.; Ben-Ari, Y.; Giniatullin, R.; Khazipov, R. Newborn analgesia mediated by oxytocin during delivery. *Front. Cell. Neurosci.* **2011**, *5*, 3. [CrossRef] [PubMed]

29. Gonzalez-Hernandez, A.; Charlet, A. Oxytocin, GABA, and TRPV1, the analgesic triad? *Front. Mol. Neurosci.* **2018**, *11*, 398. [CrossRef] [PubMed]

30. Street, L.M.; Harris, L.; Curry, R.S.; Eisenach, J.C. Capsaicin-induced pain and sensitisation in the postpartum period. *Br. J. Anaesth.* **2019**, *122*, 103–110. [CrossRef] [PubMed]

31. Nersesyan, Y.; Demirkhanyan, L.; Cabezas-Bratesco, D.; Oakes, V.; Kusuda, R.; Dawson, T.; Sun, X.; Cao, C.; Cohen, A.M.; Chelluboina, B.; et al. Oxytocin modulates nociception as an agonist of pain-sensing TRPV1. *Cell Rep.* **2017**, *21*, 1681–1691. [CrossRef] [PubMed]

32. Dobrinski, I.; Aurich, J.E.; Grunert, E.; Hoppen, H.O. Endogenous opioid peptides in cattle during pregnancy, birth and the newborn period. *Deutsche Tierarztliche Wochenschrift* **1991**, *98*, 224–226. [CrossRef] [PubMed]

33. Douglas, A.J.; Russell, J.A. Chapter 5 Endogenous opioid regulation of oxytocin and ACTH secretion during pregnancy and parturition. In *Progress in Brain Research*; Russell, J.A., Douglas, R.J., Windle, R.J., Ingram, C.D., Eds.; Elsevier: Amsterdam, The Netherlands, 2001; Volume 133, pp. 67–82.

34. Hapidou, E.G. Perception of pain during pregnancy and labor. In *Pathophysiology of Pain Perception*; Lautenbacher, S., Fillingim, R.B., Eds.; Springer: Boston, MA, USA, 2004; pp. 199–214.

35. Morris, M.S.; Domino, E.F.; Domino, S.E. Review: Opioid modulation of oxytocin release. *J. Clin. Pharmacol.* **2010**, *50*, 1112–1117. [CrossRef]

36. Quaas, L.; Zahradnik, H.P. The effects of α- and β-adrenergic stimulation on contractility and prostaglandin (prostaglandins E2 and F2α and 6-keto-prostaglandin F1α) production of pregnant human myometrial strips. *Am. J. Obstet. Gynecol.* **1985**, *152*, 852–856. [CrossRef]
37. Fuchs, A.-R. Prostaglandin F2alpha and oxytocin interactions in ovarian and uterine function. *J. Steroid Biochem.* **1987**, *27*, 1073–1080. [CrossRef]
38. Sabuncu, A.; Tek, C.; Bademkiran, S.; Kasikci, G.; Horoz Kaya, H.; Senunver, A. Effect of beta adrenergic blocker carazolol on the duration of parturitionin gilts. *Indian Vet. J.* **2008**, *85*, 21–22.
39. Bostedt, H.; Rudloff, P.R. Prophylactic administration of the beta-blocker carazolol to influence the duration of parturition in sows. *Theriogenology* **1983**, *20*, 191–196. [CrossRef]
40. Olmos-Hernández, A.; Trujillo-Ortega, M.E.; Alonso-Spilsbury, M.; Sánchez-Aparicio, P.; Ramírez-Necoechea, R.; Mota-Rojas, D. Foetal monitoring, uterine dynamics and reproductive performance in spontaneous farrowings in sows. *J. Appl. Anim. Res.* **2008**, *33*, 181–185. [CrossRef]
41. Dreiling, M.; Bischoff, S.; Schiffner, R.; Rupprecht, S.; Kiehntopf, M.; Schubert, H.; Witte, O.W.; Nathanielsz, P.W.; Schwab, M.; Rakers, F. Stress-induced decrease of uterine blood flow in sheep is mediated by alpha 1-adrenergic receptors. *Stress* **2016**, *19*, 547–551. [CrossRef] [PubMed]
42. Dreiling, M.; Schiffner, R.; Bischoff, S.; Rupprecht, S.; Kroegel, N.; Schubert, H.; Witte, O.W.; Schwab, M.; Rakers, F. Impact of chronic maternal stress during early gestation on maternal-fetal stress transfer and fetal stress sensitivity in sheep. *Stress* **2018**, *21*, 1–10. [CrossRef]
43. Rakers, F.; Bischoff, S.; Schiffner, R.; Haase, M.; Rupprecht, S.; Kiehntopf, M.; Kühn-Velten, W.N.; Schubert, H.; Witte, O.W.; Nijland, M.J.; et al. Role of catecholamines in maternal-fetal stress transfer in sheep. *Am. J. Obstet. Gynecol.* **2015**, *213*, 684.e1–684.e9. [CrossRef] [PubMed]
44. Mainau, E.; Temple, D.; Llonch, P.; Manteca, X. Pain caused by parturition in sows the farm animal welfare fact sheet farrowing: A painful and stressful process. *Farm Anim. Welf. Fact Sheet* **2018**, *20*, 1–2.
45. Domínguez-Oliva, A.; Casas-Alvarado, A.; Miranda-Cortés, A.E.; Hernández-Avalos, I. Clinical pharmacology of tramadol and tapentadol, and their therapeutic efficacy in different models of acute and chronic pain in dogs and cats. *J. Adv. Vet. Anim. Res.* **2021**, *8*, 404. [CrossRef] [PubMed]
46. De Leon-Casasola, O.A.; Lema, M.J. Postoperative epidural opioid analgesia. *Anesth. Analg.* **1996**, *83*, 867–875. [CrossRef]
47. McKune, C.M.; Murrell, J.C.; Nolan, A.M.; White, K.L.; Wright, B.D. Nociception and pain. In *Veterinary Anesthesia and Analgesia*, 5th ed.; Grimm, K.A., Laomnt, L.A., Tranquilli, W.J., Greene, S.A., Robertson, S.A., Eds.; John Wiley & Sons: Chichester, UK, 2017; pp. 584–623.
48. Catheline, G.; Touquet, B.; Besson, J.-M.; Lombard, M.-C. Parturition in the rat. *Anesthesiology* **2006**, *104*, 1257–1265. [CrossRef] [PubMed]
49. Yang, T.; Zhuang, L.; Rei Fidalgo, A.M.; Petrides, E.; Terrando, N.; Wu, X.; Sanders, R.D.; Robertson, N.J.; Johnson, M.R.; Maze, M.; et al. Xenon and sevoflurane provide analgesia during labor and fetal brain protection in a perinatal rat model of hypoxia-ischemia. *PLoS ONE* **2012**, *7*, e37020. [CrossRef] [PubMed]
50. Tong, C.; Conklin, D.R.; Liu, B.; Ririe, D.G.; Eisenach, J.C. Assessment of behavior during labor in rats and effect of intrathecal morphine. *Anesthesiology* **2008**, *108*, 1081–1086. [CrossRef] [PubMed]
51. Javadi-Paydar, M.; Lesani, A.; Vakilipour, R.; Ghazi, P.; Tavangar, S.M.; Hantoushzadeh, S.; Norouzi, A.; Dehpour, A.R. Evaluation of the tocolytic effect of morphine in a mouse model of lipopolysaccharide-induced preterm delivery: The role of nitric oxide. *Eur. J. Obstet. Gynecol. Reprod. Biol.* **2009**, *147*, 166–172. [CrossRef] [PubMed]
52. Kowalski, W.B.; Parsons, M.T.; Pak, S.C.; Wilson, L. Morphine inhibits nocturnal oxytocin secretion and uterine contractions in the pregnant baboon. *Biol. Reprod.* **1998**, *58*, 971–976. [CrossRef]
53. Bricker, L. Parenteral opioids for labor pain relief: A systematic review. *Am. J. Obstet. Gynecol.* **2002**, *186*, S94–S109. [CrossRef]
54. Choi, S.K.; Kim, Y.H.; Kim, S.M.; Wie, J.H.; Lee, D.-G.; Kwon, J.Y.; Song, J.H.; Lee, S.J.; Park, I.Y. Opioid analgesics are the leading cause of adverse drug reactions in the obstetric population in South Korea. *Medicine* **2019**, *98*, e15756. [CrossRef]
55. Lee, H.-L.; Lu, K.-C.; Foo, J.; Huang, I.-T.; Fan, Y.-C.; Tsai, P.-S.; Huang, C.-J. Different impacts of various tocolytic agents on increased risk of postoperative hemorrhage in preterm labor women undergoing Cesarean delivery. *Medicine* **2020**, *99*, e23651. [CrossRef] [PubMed]
56. Jørgensen, J.S.; Weile, L.K.K.; Lamont, R.F. Preterm labor: Current tocolytic options for the treatment of preterm labor. *Expert Opin. Pharmacother.* **2014**, *15*, 585–588. [CrossRef] [PubMed]
57. Bai, Y.H.; Pak, S.C.; Choi, B.C.; Wilson, L. Tocolytic effect of morphine via increased metabolic clearance of oxytocin in the baboon. *Yonsei Med. J.* **2002**, *43*, 567–572. [CrossRef]
58. Wray, S. Insights from physiology into myometrial function and dysfunction. *Exp. Physiol.* **2015**, *100*, 1468–1476. [CrossRef] [PubMed]
59. Musk, G.C.; Catanchin, C.S.M.; Usuda, H.; Woodward, E.; Kemp, M.W. The uptake of transdermal fentanyl in a pregnant sheep model. *Vet. Anaesth. Analg.* **2017**, *44*, 1382–1390. [CrossRef]
60. Varcoe, T.J.; Darby, J.R.T.; Gatford, K.L.; Holman, S.L.; Cheung, P.; Berry, M.J.; Wiese, M.D.; Morrison, J.L. Considerations in selecting postoperative analgesia for pregnant sheep following fetal instrumentation surgery. *Anim. Front.* **2019**, *9*, 60–67. [CrossRef]

61. Hodgkinson, R.; Huff, R.W.; Hayashi, R.H.; Husain, F.J. Double-blind comparison of maternal analgesia and neonatal neurobehaviour following intravenous butorphanol and meperidine. *J. Int. Med. Res.* **1979**, *7*, 224–230. [CrossRef]
62. Court, M.H.; Dodman, N.H.; Levine, H.D.; Richey, M.T.; Lee, J.W.; Hustead, D.R. Pharmacokinetics and milk residues of butorphanol in dairy cows after single intravenous administration. *J. Vet. Pharmacol. Ther.* **1992**, *15*, 28–35. [CrossRef]
63. Arguedas, M.G.; Hines, M.T.; Papich, M.G.; Farnsworth, K.D.; Sellon, D.C. Pharmacokinetics of butorphan ol and evaluation of physiologic and behavioral effects after intravenous and intramuscular administration to neonatal foals. *J. Vet. Intern. Med.* **2008**, *22*, 1417–1426. [CrossRef]
64. Mattingly, J.E.; Alessio, J.D.; Ramanathan, J. Effects of obstetric analgesics and anesthetics a review. *Paediatr. Drugs* **2003**, *5*, 615–627. [CrossRef]
65. Bradbrook, C.A.; Clark, L. State of the art analgesia- recent developments in pharmacological approaches to acute pain management in dogs and cats. Part 1. *Vet. J.* **2018**, *238*, 76–82. [CrossRef] [PubMed]
66. Yakovleva, A.A.; Nazarova, L.A.; Prokopenko, V.M.; Pavlova, N.G. Effect of tramadol on rabbit uterine contractile activity induced in late pregnancy. *Bull. Exp. Biol. Med.* **2017**, *162*, 349–352. [CrossRef] [PubMed]
67. Brideau, C.; Van Staden, C.; Chan, C.C. In vitro effects of cyclooxygenase inhibitors in whole blood of horses, dogs, and cats. *Am. J. Vet. Res.* **2001**, *62*, 1755–1760. [CrossRef] [PubMed]
68. Moeremans, I.; Devreese, M.; De Baere, S.; Croubels, S.; Hermans, K. Pharmacokinetics and absolute oral bioavailability of meloxicam in guinea pigs (*Cavia porcellus*). *Vet. Anaesth. Analg.* **2019**, *46*, 548–555. [CrossRef] [PubMed]
69. Simmons, D.L.; Botting, R.M.; Hla, T. Cyclooxygenase isozymes: The biology of prostaglandin synthesis and inhibition. *Pharmacol. Rev.* **2004**, *56*, 387–437. [CrossRef] [PubMed]
70. Monteiro-Steagall, B.P.; Steagall, P.V.M.; Lascelles, B.D.X. Systematic review of nonsteroidal anti-inflammatory drug-induced adverse effects in dogs. *J. Vet. Intern. Med.* **2013**, *27*, 1011–1019. [CrossRef] [PubMed]
71. Rac, V.E.; Small, C.; Scott, C.A.; Adamson, S.L.; Rurak, D.; Challis, J.R.; Lye, S.J. Meloxicam effectively inhibits preterm labor uterine contractions in a chronically catheterized pregnant sheep model: Impact on fetal blood flow and fetal-maternal physiologic parameters. *Am. J. Obstet. Gynecol.* **2006**, *195*, 528–534. [CrossRef]
72. Yousif, M.H.; Thulesius, O. Tocolytic effect of the cyclooxygenase-2 inhibitor, meloxicam: Studies on uterine contractions in the rat. *J. Pharm. Pharmacol.* **2011**, *50*, 681–685. [CrossRef]
73. Das, Y.K.; Aksoy, A.; Yavuz, O.; Guvenc, D.; Management, P. Tocolytic effects of meloxicam on isolated cattle myometrium tocolytic effects of meloxicam on isolated cattle myometrium. In Proceedings of the 12th International Congress of the European Association for Veterinary Pharmacology and Toxicology (EAVPT), Holland, The Netherlands, 8–12 July 2012; Volume 2, pp. 8–12.
74. Verma, A.D.; Panigrahi, M.; Baba, N.A.; Sulabh, S.; Sadam, A.; Parida, S.; Narayanan, K.; Sonwane, A.A.; Bhushan, B. Differential expression of ten candidate genes regulating prostaglandin action in reproductive tissues of buffalo during estrous cycle and pregnancy. *Theriogenology* **2018**, *105*, 7–14. [CrossRef]
75. Li, H.; Zhou, J.; Wei, X.; Chen, R.; Geng, J.; Zheng, R.; Chai, J.; Li, F.; Jiang, S. miR-144 and targets, c-fos and cyclooxygenase-2 (COX2), modulate synthesis of PGE2 in the amnion during pregnancy and labor. *Sci. Rep.* **2016**, *6*, 27914. [CrossRef]
76. Thornton, J.M.; Browne, B.; Ramphul, M. Mechanisms and management of normal labour. *Obstet. Gynaecol. Reprod. Med.* **2020**, *30*, 84–90. [CrossRef]
77. Jaffal, S.M.; Salhab, A.S.; Disi, A.M.; Al-Qaadan, F. Effects of meloxicam on implantation and parturition of rat. *Jordan Med. J.* **2006**, *40*, 88–95.
78. Landsbergen, N.; Pellicaan, C.H.; Schaefers-Okkens, A.C. The use of veterinary drugs during pregnancy of the dog. *Tijdschrift voor Diergeneeskunde* **2001**, *126*, 716–722. [PubMed]
79. Trimboli, F.; Ragusa, M.; Piras, C.; Lopreiato, V.; Britti, D. Outcomes from experimental testing of nonsteroidal anti-inflammatory drug (nsaid) administration during the transition period of dairy cows. *Animals* **2020**, *10*, 1832. [CrossRef] [PubMed]
80. Giammarco, M.; Fusaro, I.; Vignola, G.; Manetta, A.C.; Gramenzi, A.; Fustini, M.; Palmonari, A.; Formigoni, A. Effects of a single injection of Flunixin meglumine or Carprofen postpartum on haematological parameters, productive performance and fertility of dairy cattle. *Anim. Prod. Sci.* **2018**, *58*, 322. [CrossRef]
81. Swartz, T.H.; Schramm, H.H.; Bewley, J.M.; Wood, C.M.; Leslie, K.E.; Petersson-Wolfe, C.S. Meloxicam administration either prior to or after parturition: Effects on behavior, health, and production in dairy cows. *J. Dairy Sci.* **2018**, *101*, 10151–10167. [CrossRef]
82. Castagnetti, C.; Mariella, J. Anti-inflammatory drugs in equine neonatal medicine. Part I: Nonsteroidal Anti-inflammatory Drugs. *J. Equine Vet. Sci.* **2015**, *35*, 475–480. [CrossRef]
83. Okada, C.T.C.; Andrade, V.P.; Freitas-Dell'Aqua, C.P.; Nichi, M.; Fernandes, C.B.; Papa, F.O.; Alvarenga, M.A. The effect of flunixin meglumine, firocoxib and meloxicam on the uterine mobility of equine embryos. *Theriogenology* **2019**, *123*, 132–138. [CrossRef]
84. Newby, N.C.; Leslie, K.E.; Dingwell, H.D.P.; Kelton, D.F.; Weary, D.M.; Neuder, L.; Millman, S.T.; Duffield, T.F. The effects of periparturient administration of flunixin meglumine on the health and production of dairy cattle. *J. Dairy Sci.* **2017**, *100*, 582–587. [CrossRef]
85. Thun, R.; Kündig, H.; Zerobin, K.; Kindahl, H.; Gustafsson, B.K.; Ziegler, W. Uterine motility of cattle during late pregnancy, labor and puerperium. III. Use of flunixin meglumine and endocrine changes. *Schweizer Archiv fur Tierheilkunde* **1993**, *135*, 333–344.
86. Narver, H.L. Oxytocin in the treatment of dystocia in mice. *J. Am. Assoc. Lab. Anim. Sci.* **2012**, *51*, 10–17. [PubMed]

87. Steagall, P.V.M. An update on drugs used for lumbosacral epidural anesthesia and analgesia in dogs. *Front. Vet. Sci.* **2017**, *4*, 68. [CrossRef] [PubMed]

88. Lamont, L.A. Multimodal pain management in veterinary medicine: The physiologic basis of pharmacologic therapies. *Vet. Clin. N. Am. Small Anim. Pract.* **2008**, *38*, 1173–1186. [CrossRef] [PubMed]

89. Kushnir, Y.; Epstein, A. Anesthesia for the pregnant cat and dog. *Isr. J. Vet. Med.* **2012**, *67*, 19–23.

90. Rioja-Garcia, E. Local Anesthetics. In *Veterinary Anesthesia and Analgesia*, 5th ed.; Grimm, K.A., Laomnt, L.A., Tranquilli, W.J., Greene, S.A., Robertson, S.A., Eds.; John Wiley & Sons: Chichester, UK, 2017; pp. 332–354.

91. Fanning, R.A.; Campion, D.P.; Collins, C.B.; Keely, S.; Briggs, L.P.; O'Connor, J.J.; Carey, M.F. A Comparison of the inhibitory effects of bupivacaine and levobupivacaine on isolated human pregnant myometrium contractility. *Anesth. Analg.* **2008**, *107*, 1303–1307. [CrossRef]

92. Li, Z.; Zhou, L.; Tang, H. Effects of levobupivacaine and bupivacaine on rat myometrium. *J. Zhejiang Univ. Sci. B* **2006**, *7*, 757–762. [CrossRef]

93. Karsli, B.; Kayacan, N.; Kucukyavuz, Z.; Mimaroglu, C. Effects of local anesthetics on pregnant uterine muscles. *Pol. J. Pharmacol.* **2003**, *55*, 51–56. [PubMed]

94. Arici, G.; Karsli, B.; Kayacan, N.; Akar, M. The effects of bupivacaine, ropivacaine and mepivacaine on the contractility of rat myometrium. *Int. J. Obstet. Anesth.* **2004**, *13*, 95–98. [CrossRef]

95. Weinschenk, F.; Dittrich, R.; Müller, A.; Lotz, L.; Beckmann, M.W.; Weinschenk, S.W. Uterine contractility changes in a perfused swine uterus model induced by local anesthetics procaine, lidocaine, and ropivacaine. *PLoS ONE* **2018**, *13*, e0206053. [CrossRef]

96. Garcia-Pereira, F. Epidural anesthesia and analgesia in small animal practice: An update. *Vet. J.* **2018**, *242*, 24–32. [CrossRef]

97. Greiss, F.C.; Still, J.G.; Anderson, S.G. Effects of local anesthetic agents on the uterine vasculatures and myometrium. *Am. J. Obstet. Gynecol.* **1976**, *124*, 889–899. [CrossRef]

98. Uvnäs-Moberg, K.; Ekström-Bergström, A.; Berg, M.; Buckley, S.; Pajalic, Z.; Hadjigeorgiou, E.; Kotłowska, A.; Lengler, L.; Kielbratowska, B.; Leon-Larios, F.; et al. Maternal plasma levels of oxytocin during physiological childbirth—A systematic review with implications for uterine contractions and central actions of oxytocin. *BMC Pregnancy Childbirth* **2019**, *19*, 285. [CrossRef] [PubMed]

99. Negoita, S.; Gica, N.; Ciobanu, A.M.; Peltecu, G.; Panaitescu, A.M. Anesthesia during pregnancy. *Rom. Med. J.* **2022**, *69*, 88–93. [CrossRef]

100. Bell, A. The neurobiology of acute pain. *Vet. J.* **2018**, *237*, 55–62. [CrossRef] [PubMed]

101. Mansour, C.; Mocci, R.; Santangelo, B.; Sredensek, J.; Chaaya, R.; Allaouchiche, B.; Bonnet-Garin, J.-M.; Boselli, E.; Junot, S. Performance of the Parasympathetic Tone Activity (PTA) index to predict changes in mean arterial pressure in anaesthetized horses with different health conditions. *Res. Vet. Sci.* **2021**, *139*, 43–50. [CrossRef]

102. Hernández-Avalos, I.; Valverde, A.; Antonio Ibancovichi-Camarillo, J.; Sánchez-Aparicio, P.; Recillas-Morales, S.; Rodríguez-Velázquez, D.; Osorio-Avalos, J.; Armando Magdaleno-Torres, L.; Chavez-Monteagudo, J.; Manuel Acevedo-Arcique, C. Clinical use of the parasympathetic tone activity index as a measurement of postoperative analgaesia in dogs undergoing ovariohysterectomy. *J. Vet. Res.* **2021**, *65*, 117–123. [CrossRef]

103. Steagall, P.V.; Benito, J.; Monteiro, B.P.; Doodnaught, G.M.; Beauchamp, G.; Evangelista, M.C. Analgesic effects of gabapentin and buprenorphine in cats undergoing ovariohysterectomy using two pain-scoring systems: A randomized clinical trial. *J. Feline Med. Surg.* **2018**, *20*, 741–748. [CrossRef]

104. Miranda-Cortés, A.E.; Ruiz-García, A.G.; Olivera-Ayub, A.E.; Garza-Malacara, G.; Ruiz-Cervantes, J.G.; Toscano-Zapien, J.A.; Hernández-Avalos, I. Cardiorespiratory effects of epidurally administered ketamine or lidocaine in dogs undergoing ovariohysterectomy surgery: A comparative study. *Iran. J. Vet. Res.* **2020**, *21*, 92–96.

105. Mota-Rojas, D.; Nava-Ocampo, A.A.; Trujillo, M.E.; Velázquez-Armenta, Y.; Ramírez-Necoechea, R.; Martínez-Burnes, J.; Alonso-Spilsbury, Y.M. Dose minimization study of oxytocin in early labor in sows: Uterine activity and fetal outcome. *Reprod. Toxicol.* **2005**, *20*, 255–259. [CrossRef]

106. Mota-Rojas, D.; Villanueva-García, D.; Solimano, A.; Muns, R.; Ibarra-Ríos, D.; Mota-Reyes, A. Pathophysiology of perinatal asphyxia in humans and animal models. *Biomedicines* **2022**, *10*, 347. [CrossRef]

Review

The Role of Oxytocin in Domestic Animal's Maternal Care: Parturition, Bonding, and Lactation

Daniel Mota-Rojas [1,*], Míriam Marcet-Rius [2], Adriana Domínguez-Oliva [1], Julio Martínez-Burnes [3], Karina Lezama-García [1], Ismael Hernández-Ávalos [4], Daniela Rodríguez-González [1] and Cécile Bienboire-Frosini [5]

[1] Neurophysiology, Behavior and Animal Welfare Assessment, DPAA, Xochimilco Campus, Universidad Autónoma Metropolitana, Mexico City 04960, Mexico
[2] Department of Animal Behaviour and Welfare, Research Institute in Semiochemistry and Applied Ethology (IRSEA), 84400 Apt, France
[3] Facultad de Medicina Veterinaria y Zootecnia, Universidad Autónoma de Tamaulipas, Victoria City 87000, Mexico
[4] Facultad de Estudios Superiores Cuautitlán, Universidad Nacional Autónoma de Mexico (UNAM), Cuautitlán 54714, Mexico
[5] Department of Molecular Biology and Chemical Communication, Research Institute in Semiochemistry and Applied Ethology (IRSEA), 84400 Apt, France
* Correspondence: dmota100@yahoo.com.mx

Simple Summary: Oxytocin is one of the most important hormones in birth and lactation. Its importance lies in the fact that it is also involved in various bodily functions around parturition and establishing maternal behavior. We aimed to analyze all the facts involving oxytocin regulation and participation in different peripartum situations in domestic mammals. In addition, the authors examined the impact of altering normal plasma oxytocin values on maternal behavior and care, productive parameters, and imprinting.

Abstract: Oxytocin (OXT) is one of the essential hormones in the birth process; however, estradiol, prolactin, cortisol, relaxin, connexin, and prostaglandin are also present. In addition to parturition, the functions in which OXT is also involved in mammals include the induction of maternal behavior, including imprinting and maternal care, social cognition, and affiliative behavior, which can affect allo-parental care. The present article aimed to analyze the role of OXT and the neurophysiologic regulation of this hormone during parturition, how it can promote or impair maternal behavior and bonding, and its importance in lactation in domestic animals.

Keywords: maternal care; maternal aggression; mother–young bonding; milk ejection

Citation: Mota-Rojas, D.; Marcet-Rius, M.; Domínguez-Oliva, A.; Martínez-Burnes, J.; Lezama-García, K.; Hernández-Ávalos, I.; Rodríguez-González, D.; Bienboire-Frosini, C. The Role of Oxytocin in Domestic Animal's Maternal Care: Parturition, Bonding, and Lactation. *Animals* **2023**, *13*, 1207. https://doi.org/10.3390/ani13071207

Academic Editor: Robert Dailey

Received: 28 February 2023
Revised: 22 March 2023
Accepted: 28 March 2023
Published: 30 March 2023

Copyright: © 2023 by the authors. Licensee MDPI, Basel, Switzerland. This article is an open access article distributed under the terms and conditions of the Creative Commons Attribution (CC BY) license (https://creativecommons.org/licenses/by/4.0/).

1. Introduction

Oxytocin (OXT) is a uterotonic hormone in the endocrine response during parturition [1]. It is synthesized in the hypothalamus, in the paraventricular (mainly from its magnocellular cells) and the supraoptic nucleus [2], and transported through the neuro-hypophysial bundle to the posterior lobe of the pituitary gland, where it is stored in secretory vesicles before being released into the peripheral circulation [3]. Then, it induces myometrium contractions due to its action on myoepithelial cells [4,5]. It regulates, together with other hormones (e.g., estrogen and progesterone (P4)), the reproductive function of domestic animals [6]. In addition, there is a central action of OXT through dendritic and/or axonal releases within the brain. Smaller parvocellular neurons in the paraventricular nucleus also produce OXT and project directly to other regions in the brain, such as the limbic system (notably the amygdala and hippocampus) and the nucleus accumbens [3]. The dual action of this nonapeptide as a peripheral hormone and a central neuromodulator/neurotransmitter [7], as well as the wide distribution of OXT receptors (OXTR) in various brain areas and peripheral organs, indicate that OXT is involved in diverse biological functions and, notably, OXT influences the parturition process, the immediate establishment of maternal recognition, and milk ejection [8–10].

In mammals, several species have been evaluated to study the association between OXT concentration and maternal care [11], social cognition, and affiliative behavior [12,13], including nursing non-filial offspring or allo-parental care [14,15]. According to some authors, the positive relationship with close social partners is associated with alterations in peripheral OXT in dogs (*Canis familiaris*) [16] and chimpanzees (*Pan troglodytes*) [17,18].

The degree of interaction between parturition and maternal attachment to OXT concentrations is suggested to be related to a high density of oxytocinergic neurons in the brain, which would have a beneficial effect on the behavior of the mother in facilitating maternal recognition, protection of the offspring, and even habituation to the environment [19–21]. However, it is unclear whether this behavior is only induced by the action of OXT or is coordinated by the effect of hormonal peaks present before parturition that could sensitize OXTR. In the same way, OXT actively participates in lactation because it facilitates milk ejection and can even help induce the care of offspring from other mothers [22,23].

The present article aimed to analyze the role of OXT and the neurophysiologic regulation of this hormone during parturition, how it can promote or impair maternal behavior and bonding, and its importance in lactation on domestic animals, including pets and livestock.

2. Neurophysiological Regulation of OXT

OXT is a cyclic neuropeptide of nine amino acids whose chemical structure contains a cyclic disulfide essential for its biological effect [24]. It has a structural similarity to vasopressin (AVP), with a minimum difference of two amino acids, which explains why they share physiological effects [25]. One of the main functions of OXT as a neurotransmitter and neuromodulator is during parturition, lactation, and maternal behavior, due to its uterotonic effect [22,26–28] (Figure 1).

From the paraventricular nucleus (PVN), where OXT is synthesized, OXT is transported to the neurohypophysis as a precursor molecule in oxytocin-neurophysin. It is later catalyzed to its active form with the activity of the enzyme monooxygenase peptidyl glycine α-amidating [29].

OXT synthesis does not only occur at the hypothalamic level, since its synthesis has also been recorded, to a lesser extent, in the corpus luteum, ovary [30], and testicles [31]. Other elements, such as estrogens, are also involved in regulating OXT. For example, estrogens downregulated OXT gene expression in the PVN and supraoptic nucleus (SON) [32,33]. This is due to the presence of estrogen receptors (ER) in oxytocinergic neurons of the PVN. In mice, ERbeta is necessary to regulate the expression of OXT and arginine vasopressin (AVP) [34]. This is relevant considering that both hormones participate in social interactions, fear reactions, and stress-related behaviors and depend on their upregulation to activate OXT and AVP systems [35,36]. Neuropeptide synthesis also takes place in cardiovascular tissues, as reported by Jankowski et al. [37] when studying OXT concentrations in the pulmonary artery and vena cava in male rats and the aorta of dogs and sheep. The results indicated concentrations of 2745 ± 180 pg/mg in all tissues studied, confirming their presence by PCR.

Additionally, high levels of OXT mRNA receptors were observed in the vena cava, vein, and pulmonary artery, indicating specific sites of action. These findings suggest the role of OXT in regulating vasomotor tone due to its natriuretic properties. In the heart, Jankowski et al. [38] studied cultures of rat cardiomyocytes with a concentration of OXT 19 times higher than in utero. In an immunostaining test on atrial myocytes and fibroblasts, the intensity was found to parallel atrial natriuretic peptide stores. Its presence in this tissue supports its vasomotor effect and possible regulation of cardiovascular activity [38].

OXT secretion also occurs at the brain level in the dendrites of magnocellular neurons. These neurons express a more significant amount of protein products of the immediate early c-fos gene. This effect was reported by Douglas et al. [39], who used forced swimming in 16- or 17-day-old virgin female rats as a stressor to investigate the influence of the event on OXT levels. The authors found that this factor increased OXT at 5 min post-test. These results suggest that this neuropeptide activates the hypothalamic–pituitary–adrenal (HPA) axis during parturition and potentially stressful events [40].

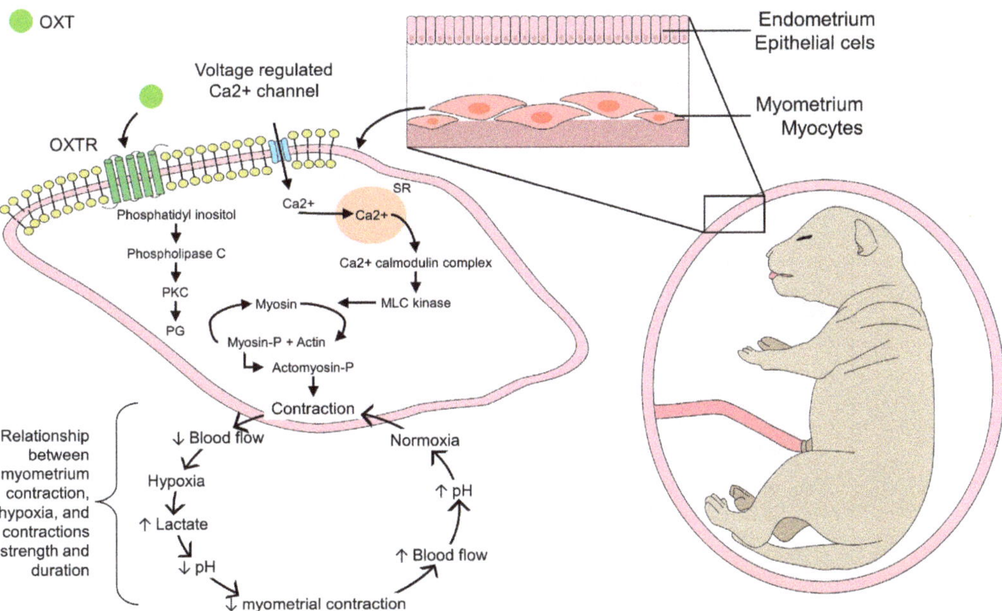

Figure 1. Mechanism of action of oxytocin during myometrial contractions. During parturition, the activation of OXTR at the uterus by OXT starts myometrial contractions. The interaction between OXTR and voltage-regulated Ca^{2+} channels enables the formation of the actomyosin-P complex to produce myocytes' contraction. After contraction starts, a relation between contraction maintenance and uterine blood flow is responsible for maintaining effective contraction movements and, therefore, a normal course during parturition. In the first instance, myometrial contractions cause compression of uterine blood vessels, reducing blood flow. Hypoxia induces an acidemia state, by lactate increase, which decreases Ca^{2+} influx and, consequently, myometrial contractions stop, restoring blood flow and normoxia, in order to start another effective contraction. The ongoing maternal and fetal stimulation that promotes OXT release maintains rhythmic contractions due to initial hypoxia and acidemia of the myocytes, an effect that decreases contractions, and the cycle begins again when reaching normoxia and a regular uterine blood flow. MLC: myosin-light chain kinase; OXTR: oxytocin receptor; PG: prostaglandins; PKC: protein kinase C; SR: sarcoplasmic reticulum.

Concerning this, Ochedalski et al. [41] determined the effect of OXT in 7-day-old ovariectomized rats that received an infusion of intracerebroventricular OXT (100 ng/h), in addition to administration of estradiol (E2) with or without progesterone replacement. These authors found that E2 increased plasma adrenocorticotropic hormone (ACTH) and corticosterone concentrations. This was also associated with increased estrogen and increased corticotropin-releasing factor (CRF) mRNA in the PVN and mRNA of opiomelanocortin (POMC) in the pituitary gland. In another study, the administration of OXT reduced levels of plasmatic ACTH, CRF, and POMC in response to the elevation of E2 due to stress. Therefore, these findings suggest a modulating effect of OXT in response to a stressor due to actions on the HPA axis [42].

The above mentioned studies show the complex functions and interactions of OXT in the neurobiology of the organism. Brunton [28] points out that one of the mechanisms that restrict OXT secretion is regulated through endogenous opioids. These act centrally on the axonal terminals of the posterior pituitary, which has been reported in bovines with opioidergic receptors in neuronal terminals of the neurohypophysis, supraoptic neurons, and paraventricular in the hypothalamus [43]. The presence of these receptors in contiguous regions in oxytocinergic neurons has been demonstrated in other studies, such as that developed by Douglas et al. [44] in pregnant rats from 18 to 21 days. In these animals, a k-receptor agonist and an opioid antagonist, such as naloxone, were administered, reporting that agonism decreased OXT levels while antagonists increased its concentration. These same authors mention that, during the last week of gestation, the sensitivity towards these receptors decreases, probably as a mechanism to allow neurosecretion of OXT during delivery.

In addition, a study on the interaction between OXT and nitric oxide (NO) has shown that the increase in NO can lead to a decrease in c-Fos expression and, consequently, inhibit OXT secretion [45]. Srisawat et al. [46] evaluated the effect of an inhibitor of a NO-synthase in anesthetized virgin rats, identifying that the administration of this compound facilitated the secretion of OXT and that its use, together with nitroprusside (NO donor) in the supraoptic nucleus, inhibited the electrical activity of oxytocinergic neurons.

Therefore, since the chemical structure of OXT is similar to AVP, it is possible to understand its participation in cardiovascular control processes and stress-related events such as parturition, known as a eustress process [1,25]. Likewise, OXT modulation during stress perception indicates the participation of this hormone in the complex physiological process of parturition [8,10].

3. OXT Participation at Parturition

In all mammals, the perinatal period is characterized by various hormonal changes, including increases in plasma E2, prolactin (PRL), and cortisol, as well as activation of the oxytocinergic system at parturition [47,48]. This activation increases OXT during parturition in mammals [49], especially in dogs, where OXT plasma concentration is tightly implicated in uterine contractions. This can be useful to detect dystocic parturition opportunely [50]. Unlike in cattle, plasma cortisol levels in dogs are highly variable during the peripartum, higher than during prepartum luteolysis [51]. Olcese and Beesley [52] found that melatonin has a synergetic action and enhances oxytocin-induced contractions binding to specific melatonin receptors (MT2R) in the myometrium. This change involves physiological, neurological, morphological, hormonal, and behavioral levels [53].

There are important interrelationships between OXT and prostaglandin (PGs). OXT has been shown to stimulate PG release in many animals, primarily in the uterine epithelium/decidua. The effects of OXT are mediated by the presence of the tissue-specific OXT receptor, causing myometrium contraction and synthesis of PG in the decidua [49]. However, uterine contractions involve several hormones, including OXT and prostaglandin E2 (PGE2), connexin, progesterone (P4), and estrogen [54,55]. Connexin 43 (Cx43) is a gap junction protein distributed in the myocardium and uterus [56]. According to Chan et al. [57], OXT can stimulate the synthesis and release of PGE2 and PGF2α, increasing the susceptibility of the uterus to OXT and strengthening subsequent uterine contractions. However, it is widely recognized during labor that OXT is the primary hormone that promotes the synchronization of uterine contractions throughout labor and the subsequent dilation of the cervix [8,27,58], although the control of contractions in the myometrium is innervated by the hypogastric nerve, which provides sympathetic innervation that coordinates uterine contractions during the first phase of labor through agonism of alpha-adrenergic receptors [2].

The OXTR increase during pregnancy in the endometrium and myometrium has been reported in various species, e.g., in humans [59], rats [60], sheep [61], bovines [62,63], bitches [64], and pigs [65]. OXT is responsible for inducing the most violent and constant contractions, so that the fetus is expelled from the uterus to the outside [66].

In a study by Fuchs et al. [62], the endometrial OXT and AVP receptor concentrations were inversely correlated with plasma P4 concentrations ($p = 0.005$) with no correlation to plasma E2. In contrast, the myometrial receptor concentrations showed no correlation with plasma P4 but an inverse correlation with plasma E2 ($p = 0.004$). Similarly, in a study by Ou et al. [60], the authors found that mRNA expression of OXTR in the myometrium of rats during pregnancy and parturition is regulated by coordinated interactions between mechanical and endocrine signals. In another study by Meier et al. [61] in ewes, the authors reported that oxytocin-induced 13,14,dihydro-15-keto PGF2 alpha release during early gestation is minimal, despite the presence of endometrial OXT receptors. In mid-gestation, oxytocin-stimulated 13,14,dihydro-15-keto PGF2 alpha release is increased, with a concomitant increase in uterine OXT receptor concentrations [61].

The clearest example of the participation of different hormones during parturition is the relationship between estrogen and OXT. Masoudi et al. [67] evaluated the effect of E2 and OXT treatment on cervical dilation in three sheep breeds. In animals that were administered estrogens, no effect was observed on the dilation of the cervix. In contrast, animals supplied with estrogens and OXT allowed complete cervix dilation. What was shown by these authors is likely to be that the increase in estrogens would lead to higher concentrations of OXT. However, a study by Veiga et al. [64] evaluated the expression of estrogen mRNA and OXT mRNA genes in bitches during pregnancy or parturition. They observed that the mRNA expression for estrogens did not differ.

Nevertheless, mRNA expression for OXT increased during the last week of gestation. These results demonstrated that the estrogen peak during labor would not be related to the activity of OXT on the myometrium. Fuchs et al. [63] maintain that, at least in sheep and cattle, the increase in estrogens can increase the production of OXT in the magnocellular nuclei of the hypothalamus and facilitate the release of OXT from the posterior pituitary lobe. This might suggest that estrogen secretion may help coordinate the rate and release of OXT during labor.

Some stimuli are involved in the release of OXT; the first is the Ferguson reflex, which consists of the pressure of a puppy's head on the cervix during parturition [68], cervicovaginal stimulation during labor and PRL secretion, along with the presence of E2 and progesterone, releasing OXT by producing reactions in the maternal brain [69]. The second stimulus is breastfeeding stimulation of the mammary glands (MG) by the pups [68] (Figure 2).

It is also important to mention that levels of OXT could influence the interval of expulsion or the total parturition time, as reported in sows and bitches. In sows, low levels of OXT depend on environmental factors influencing the farrowing phase, such as being confined in crates (38.1 ± 24.6 pg/mL) rather than pens (77.6 ± 47.6 ng/mL), having a direct effect on farrowing duration ($p < 0.001$) [70]. This was also reported by Yun et al. [71] when providing nesting material to sows, promoting high non-esterified fatty acid concentrations correlated with OXT levels ($r_s = 0.28$) and piglet colostrum intake, potentially improving their immune system and survival. For example, in sows, low levels of OXT can increase the duration of farrowing. According to Alonso-Spilsbury et al. [72], OXT is at lower levels during prolonged farrowing, causing increased piglet mortality in intensive production systems [73]. Similarly, van Dijk et al. [74] observed that the mortality percentages increased statistically significantly in prolonged intervals of piglet expulsion, and with stillbirths or posteriorly presented fetuses, the mortality percentages also increased statistically significantly. This prolonged farrowing can cause higher stillbirth frequency, thereby increasing economic losses in intensive pig farming [2]. In the case of pregnant bitches, low levels of neuropeptide could cause uterine inertia, which is approximately 75%, the most common maternal cause of dystocia in the pregnant bitch. This problem could be due to the absence of myometrial contractions, some obstruction [75,76], or myometrial exhaustion caused by an obstruction after several pups have been expelled [77]. From all the above, it can be understood that, although various hormones are involved in the birth process, OXT is one of the most important in all the reviewed species.

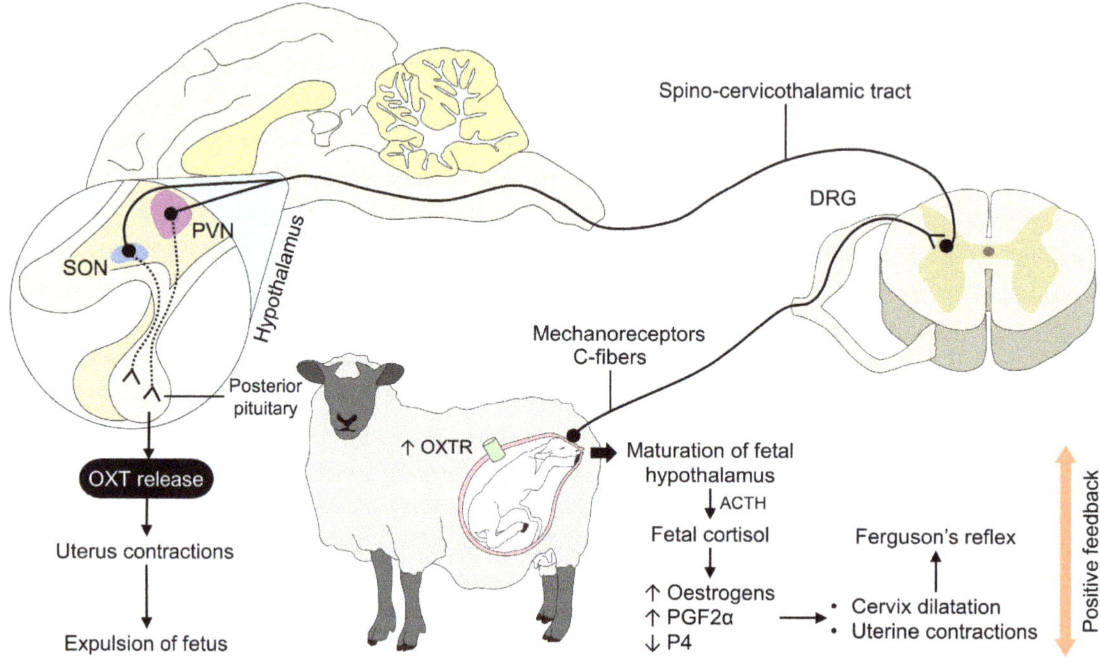

Figure 2. The role of oxytocin during parturition. In the final phase of pregnancy, secretion of ACTH due to the maturation of the hypothalamus leads to the release of fetal cortisol. These hormones also respond to the increase in E2 and PGF2α, and the decrease in P4. These hormones promote cervix dilation and the mechanoreceptors' stimulation in the uterus. The sensory signals are processed in the PVN of the maternal hypothalamus after its transmission through fibers in the spinal cord. The interaction between the hypothalamus and the pituitary causes OXT release by the posterior pituitary into the bloodstream to initiate the physiological process of fetal expulsion. DRG: dorsal root ganglion; OXT: oxytocin; OXTR: oxytocin receptors; P4: progesterone; PGF2α: prostaglandin F2α; PVN: paraventricular nucleus; SON: supraoptic nucleus.

4. The Role of OXT in the Establishment of Maternal Behavior Regulation

Maternal care is a series of species-specific behavioral patterns to ensure the offspring's survival [78]. They include interactions such as nursing, attention, and protection of the neonate [20]. The hormonal cascade and the activation of the MPOA can vary among mammalian species [48]; however, sex steroids synthesized by the ovaries, OXT, AVP, and PRL are present in the maternal brain. High levels of P4 and low E2, PRL, and OXT are present during gestation. Contrarily, in cows, P4 diminishes a couple of days after parturition, and E2, PRL, and OXT increase at the onset of parturition and the first post-partum days [79] (Figure 3).

OXT is involved in the different phases of parturition and is considered an essential hormone to facilitate maternal bonds [80]. In several studies, disrupted brain OXT signaling is linked to poor mothering [81]. OXT evaluation can be performed in domestic animals through plasma, saliva, and milk [82]. In sows, behaviors such as nest building, pawing, and gathering straw are used to appraise maternal aptitude [83]. Hall et al. [84] determined in Landrace x White Large sows that the mean concentration of OXT in milk was higher than in saliva samples (38.32 vs. 29.60 pg/mL). However, both sampling methods registered the highest concentrations during the first four to five post-farrowing days, a trait that can be associated with the specialized maternal care that newborn piglets require during the first days of life [85]. This finding was similar to a study reporting that salivary OXT concentration was significantly higher during the first day after farrowing (1654 pg/mL) than after nine days (1142 pg/mL) [86]. From an endocrine perspective, the first post-partum days are usually accompanied by higher OXT signaling due to the positive feedback loop initiated by suckling [87]. The positive OXT feedback from the newborns' suckling also stimulates OXT release in several mammals [88,89]. Milking behavior predominates during the first seven days post-parturition, as reported in sows and ewes [90,91].

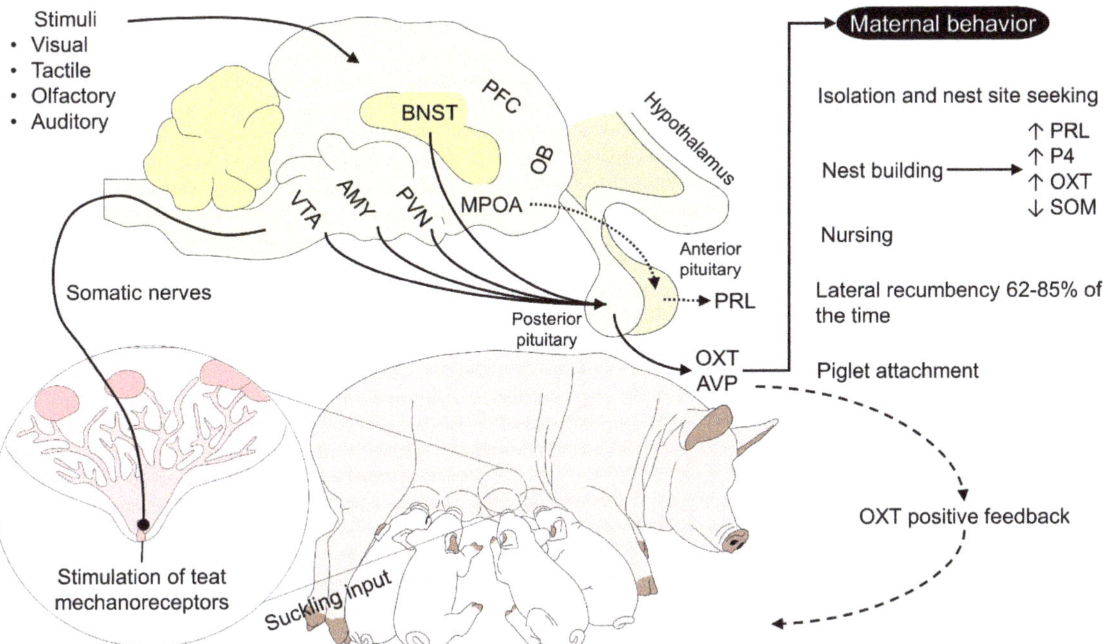

Figure 3. The oxytocinergic positive feedback loop of maternal behavior in sows. The visual, tactile, olfactory, and auditory stimuli that are constantly perceived by the sow participate in maternal responsiveness and attachment due to the activation of vital cerebral regions sensitive to OXT. For example, stimulation and activation of AMY, BNST, PVN, and VTA promote OXT release in the posterior pituitary. After OXT is released, several peripartum maternal behaviors in mammals are influenced by this hormone. In the case of sows, increases in circulating OXT levels before the onset of farrowing motivate nest site seeking and nesting. Particularly, nest building has also been associated with high concentrations of OXT, P4, PRL, and low levels of SOM in aforementioned cerebral regions. After farrowing, the development of maternal attachment and nursing of the piglets requires so-called OXT positive feedback, where piglets stimulate teat mechanoreceptors through suckling, releasing OXT in the maternal brain and promoting maternal behaviors. On the other hand, MPOA axons project to the anterior pituitary to release PRL and participate in milk secretion. AMY: amygdala; AVP: arginine vasopressin; BNST: bed nucleus of the stria terminalis; MPOA: medial preoptic area; OB: olfactory bulb; OXT: oxytocin; P4: progesterone; PFC: prefrontal cortex; PRL: prolactin; PVN: paraventricular nucleus; SOM: somatostatin.

Although OXT is considered essential for maternal care, which includes nest-building, reluctance to leave the nest, genital and overall licking of the newborn, nursing, and direct contact with the litter [92] in mammalian species, the OXT action can pass from crucial to nonessential, depending on the neurodevelopmental stage at birth or the amount of help the dams can receive from other members of the herd [87,93], and can help not only with the establishment but also with the maintenance of the maternal behavior [94]. Ungulates build highly selective maternal care for their offspring. They do not usually receive or allow allomaternal care (known as the nursing of non-filial offspring [95]), although, recently, Orihuela et al. [96] reported that communal rearing was observed in zebu cattle. In this species, maternal bonding and OXT are suggested to have an essential role [97].

For example, licking, nursing, and low-pitched bleats are rapidly observed (during the first minutes post-parturition) in ewes and goats. These behaviors are part of the maternal responsiveness to the hormonal cascade started by vaginocervical stimulation, OXT secretion, and activation of cerebral structures such as the main olfactory system (MOB), MPOA, and PVN [98]. Likewise, vaginocervical stimulation in ewes increases OXT in limbic structures that participate in maternal receptivity. However, in non-gestant animals, intraventricular administration of OXT is indispensable to nursing and accepting suckling [99]. When considering levels of OXT in the cerebrospinal fluid (CSF) of ewes, these were 15% significantly higher than plasma levels during lambing, while plasma concentrations only increased within 15 min post-lambing [100]. Some authors have hypothesized that differences could be found even within species destined for different zootechnical purposes (e.g., beef or dairy cattle). Geburt et al. [80] assessed salivary OXT concentrations in 20 Simmental beef-type and 20 German Black Pied dairy-type cows to correlate its level to maternal behaviors, such as the intention to defend the calf. The results showed that dairy cattle had higher OXT concentrations (88.6 ± 9.2 vs. 62.8 ± 9.2 pg/mL) and higher dam–calf interactions (13.8 ± 1.4) but lower defensive behaviors towards the newborn. Additionally, the differences between both types of cattle were not statistically significant at three days post-calving, so the authors concluded that behavior and OXT concentration need further assessment to consider them objective biomarkers for maternal behavior.

Apart from OXT, PRL and somatostatin have also been studied as markers of maternal behavior in sows and ewes. In sows, blood OXT increases ($p = 0.002$), PRL decreases ($p = 0.005$), and no change in somatostatin ($p = 0.65$) were reported during nursing [101]. Furthermore, the relation between OXT and other hormones is not limited to maternal traits since the levels of OXT were correlated to the piglet's weight gain, showing that this hormone influences the lactation process and piglet growth [101]. As for PRL and cortisol, environmental enrichment in sows before and after farrowing increased OXT and PRL concentrations (44.3 ± 5.67 ng/mL and 19.6 ± 3.61 ng/mL, respectively) while decreasing cortisol levels. These results were correlated to nest construction behavior, in which prepartum pigs in the enrichment group had higher frequency and periods of nesting ($p < 0.01$), as well as longer times in lateral recumbency after farrowing (a behavior that promotes suckling from the piglets) [102]. In ewes, along with OXT, arginine-AVP levels were evaluated, finding that plasma concentrations increased during lambing and 15 min post-lambing [100].

In contrast to livestock, in non-domestic species such as prairie voles, increased plasma OXT levels were obtained in reproductively naïve male individuals after 10 min of pup exposure, activating oxytocinergic neurons in the paraventricular nucleus (PVN) of the hypothalamus [103]. A high quantity of OXT receptors in the nucleus accumbens of virgin female prairie voles has also been associated with females showing spontaneous maternal behaviors such as lick and pup grooming ($p < 0.05$), indicating that OXT concentration has a secondary role in developing maternal traits in these species [104]. Its role during reproductive events has been reported in grey seals, where significantly lower concentrations of OXT (4.3 ± 0.5 pg/mL) were found in non-breeding females than in those during early and late lactation (8.2 ± 0.8 pg/m and 6.9 ± 0.7 pg/mL, respectively) [105]. Therefore, as mentioned in a heuristic model by Taylor and Grieb [93], the importance of OXT to mothering is higher in sheep/goats, cows, rats, mice, prairie voles, and marmosets.

Since OXT is essential to establish positive interactions with newborns, alterations or decreasing levels of this hormone have been associated with maternal aggression and cases of cannibalism. In a study made in Kangal dogs, cannibalism has been associated with low concentrations of plasma OXT (3.58 ± 0.43 pg/mL), as well as cholesterol (125.50 ± 8.6 mg/dL), high-density lipoproteins (52.00 ± 9.34 mg/dL) and low-density lipoproteins in serum (30.45 $\pm$ 3.56 mg/dL) [106]. Negative traits in ewes, such as butting and moving them away from the lamb, were recorded in animals without vaginocervical stimuli that did not receive P4 or estrogen [107]. Contrarily, McLean, et al. [108] reported that sex steroids do not relate to aggressive bouts in parturient gilts. Conversely, in rodents, OXT knockout mice have shown reduced aggression, and administration of OXT increased the number of attacks on pups at the beginning of the lactation. At the same time, high AVP levels and blockade of AVP receptors reduced maternal aggression [109].

According to the species in question, the results can be variable regarding the function of the OXT. Although OXT and other biomarkers such as PRL, AVP, or cortisone are used to study maternal behavior, some limitations exist [78], as Ogi et al. [7] reported in 25 Labrador Retriever dogs. They found no association between salivary OXT levels and maternal behaviors, except for sniffing or poking the newborn. At the same time, other studies relate OXT to interest in the newborn and anxiety reduction [8]. In the present authors' opinion, the conclusions of these studies can be applied to all species: maternal behavior (and aggression) is highly influenced by the simultaneous activation of the endocrine and sensory systems in triggering care for the newborn and developing the mother–newborn bonding.

Bonding is a process by which the dam and the offspring establish a social preference [110,111]. It has been initially studied in birds and later in sheep, buffalo, deer, and other non-domestic species such as insects [112]. In mammals, this preference constitutes mutual recognition to guarantee the young's development through adequate maternal care [20,95]. Since the mother is the main caregiver in most species, a significant part of knowledge about the neurobiological and neuroendocrine control of bonding actions is based on the maternal response [113–116]. Through this bonding, the neonate perceives and acquires a multisensory image of its mother, building a selective and preferential behavior towards her [117,118].

Maternal recognition occurs in precocial and altricial animals; however, the level of autonomy at birth and their early learning ability determines the type of bond in domestic animals, wildlife, and companion animals [119–121]. Three processes are required to establish the female bond: an increased acceptance of the individual, a reduced fear or low rejection reaction, and a motivation to care for the neonate [79]. To accept the offspring immediately after parturition, sensorial stimuli, such as visual, olfactory, gustatory, and auditory cues, activate brain structures such as the locus coeruleus, some areas of the limbic system, olfactory bulb, auditory cortex, and visual cortex [22,122], and release neurotransmitters to modify the learning process [120,121,123].

OXT participates during dam–young bonding and simultaneously acts peripherally and centrally with other hormones from pregnancy to postpartum [124,125]. For this reason, OXT is recognized as the primary neurochemical substance associated with affiliative and learning behaviors [119]. Gamma-aminobutyric acid, glutamate, and acetylcholine are other neurotransmitters which have been studied during this period [111].

The so-called sensitive period, which can last a few hours or days depending on the species, is the interval in which both the female and the offspring are more receptive to selective recognition [123,126]. In ewes, this period implies that early dam-calf bonding depends on maternal and neonatal behaviors, social interaction, and neuroendocrine modifications, mainly led by OXT [122,127–129].

For mammals, olfactory recognition is one of the leading systems to establish the maternal bond, due to the interaction between the central olfactory system and OXT release by the anterior pituitary [130]. OXT and other nonapeptides, such as AVP, promote maternal care by acting in cerebral structures at the bed nucleus of the stria terminalis and medial amygdala [79], emphasizing that only the central olfactory system participates during odor discrimination of offspring [130]. This has been studied in ewes, in which the selective bond through olfactory cues occurs during the first two to four hours after lambing [131], and suppression of olfactory neurogenesis impairs maternal acceptance [132]. Keller et al. [133] reported that visual/auditory recognition is highly affected by the maternal experience.

In contrast, olfactory stimuli occur indistinctly between 30 min to 4 h post-lambing in primiparous and multiparous females, and the release of OXT from the PVN triggers maternal attachment [98].

In this sense, Kojima et al. [134] observed in laboratory rats that OXT brain levels in the litter can be modified to establish recognition of their mother. At the same time, the contact translates into mutual recognition [135]. When the lambs do not have an adequate amount of fetal fluids, the ewes do not need to clean them, and this leads to a decrease in the release of P4, E2, and OXT, because there is not so much physical communication, which stimulates lactation [136]. Therefore, maternal attachment is a process in which OXT has an essential role. However, it also depends on the endocrine control of other elements, such as OXT (opioids, estrogens, among others) and multisensorial cues [137]. As stated by Lévy [48], maternal behavior in non-human and human animals is mediated by hormonal mechanisms, their interaction with crucial cerebral regions such as the MPOA, and the interaction with the newborn.

5. Role of OXT in Lactation

A neuro-hormonal reflex produces milk ejection: this comprises an ascending neuronal pathway from the nipples to the hypothalamus and a descending vascular link that transports pituitary hormones, especially OXT, to the mammary gland (MG) [138]. Milk ejection occurs after stimulating and activating pressure-sensitive receptors in the inguinal canal at a neurophysiological and neuroendocrine level. These have projections towards the dorsal roots of the spinal cord, which in response promote, via the release of OXT into the bloodstream from the posterior pituitary, the contraction of myoepithelial cells at the alveolar and ductal level, increasing intra-alveolar pressure and minimizing resistance to milk flow, descending to the MG cistern and, consequently, ejection [23,139,140] (Figure 4).

OXT plays an essential role in milk ejection, and mice without OXT could not breast-feed their pups [141,142]. Likewise, it has been reported that, during lactation, the activation of the OXT receptor in the MG is observed [143,144]. Intracellular calcium stores are released to generate auto excitation during the next release of OXT and the mobilization of this hormone from dense core vesicles, favoring its more efficient release during the rest of lactation [143,145].

Although OXT is the main hormone associated with milk ejection, hormones such as PRL, cortisol, and the growth hormone, among others, also participate in this process [146,147]. In mammals such as sheep, mice, and cattle, blood OXT concentrations have been associated with a decrease in stress hormones due to the decrease of the action of the HPA axis, reducing the concentration of plasma cortisol plasma [140,148,149]. Wagner et al. [150] evaluated blood cortisol concentration in lactating and non-lactating female *Bos taurus* with or without OXT administration. In non-lactating animals, cortisol concentrations were higher ($p = 0.02$), but the authors concluded that OXT administration alone could not reduce cortisol levels. In contrast, Cook [151] indicated that, in lactating and non-lactating sheep, the variation in blood cortisol could be caused by environmental stress, with or without handling. This factor is also associated with PRL and OXT in the posterior pituitary and the PVN, where there was a lower blood cortisol concentration in non-lactating animals before the same stimuli for both treatments.

For species whose zootechnical purpose is milk production, milk ejection and flow begin with a pronounced increase in plasmatic OXT before milking, highlighting the large reservoir available for milk storage in the udder cistern [152]. Studies in mares have reported that milk ejection occurs 30.8 ± 0.97 s after the suckling foal applies stimuli to the udder, with a peak plasma OXT of 15.8 pmol/L during the 30 to 45 s after feeding initiation of milk letdown [152]. Moreover, it has been described that OXT participates during lactation but also influences maternal behavior during this stage. These effects were described by Valros et al. [101] in 21 Yorkshire lactating sows on the 13th day postpartum during three nursings. The authors investigated the relationship between maternal characteristics and OXT, PRL, and somatostatin levels, finding that udder massage, sow–piglet nasal contacts, and non-stressful nursing periods increase OXT and favor positive behaviors of the mother towards her piglets. Therefore, OXT is a hormone associated with the efficiency of dairy farms, particularly in cattle, the utilization of individual female reserves and increased offspring growth.

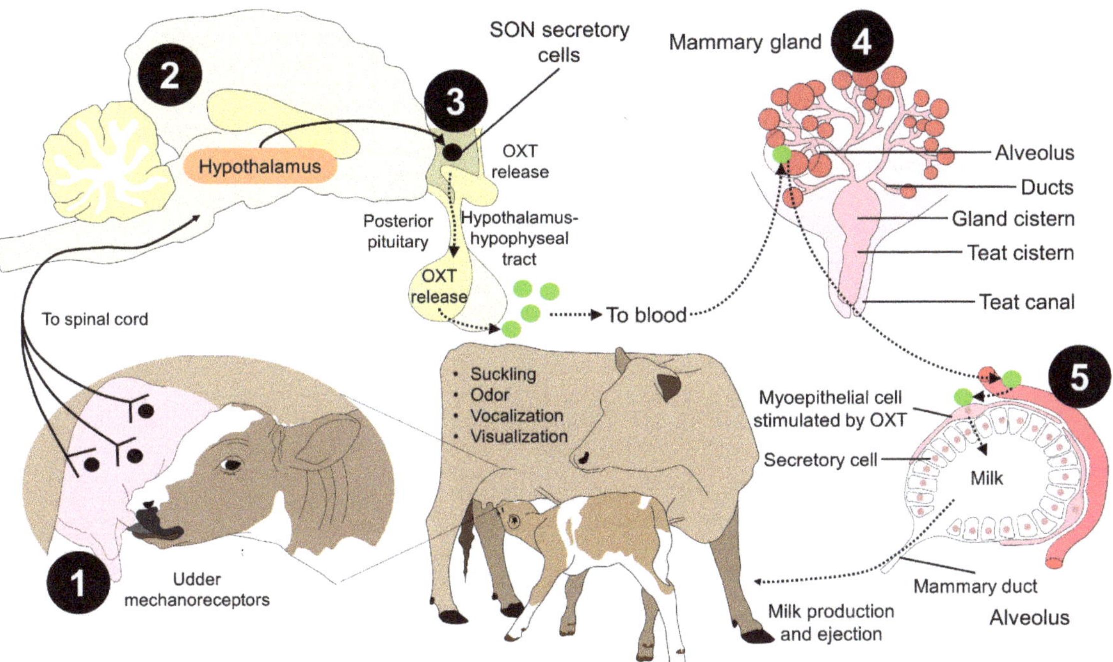

Figure 4. Neurophysiology of milk ejection and oxytocin influence. Milk yield and ejection can be divided into five stages. 1. Environmental stimulus, such as suckling from the calf, activates udder mechanoreceptors in the teat. 2. Through nervous tracts, the sensory signal reaches the hypothalamus. 3. The SON secretory cells are activated to release OXT. 4. When OXT reaches the bloodstream, it acts on the myoepithelial cells of the alveolus, causing the contraction and secretion of milk by the secretory cells (5). This physiological path and the aforementioned stimulus from the offspring are one of the reasons why manual or mechanical stimulation of the udder is important in dairy systems and can depend on OXT administration, regardless of its controversial effects. OXT: oxytocin; SON: supraoptic nucleus.

An essential element during lactation is reducing stressors to achieve adequate milk ejection [8,68,70]. These characteristics depend on the species. In the case of water buffalo, it has been described that avoiding screams, strange sounds, and changes in the milking process and place prevent the activation of the hypothalamic–pituitary axis and the sympathetic–adreno–medullary axis [153,154]. Since both endocrine axes cause the release of catecholamines and glucocorticoids (e.g., adrenaline and cortisol, respectively), this inhibits the circulation and action of OXT in the myoepithelium of MG, where the portion of milk in the cistern is 5%. Moreover, the alveolar portion represents 95%, making OXT even more relevant for evacuating the alveolar portion [23,155,156]. Routines such as maintaining contact with the calf before and during milking facilitate continuous milk ejection [157].

In addition to providing a comfortable environment, an adequate lactation diet during gestation and after parturition can also activate OXT neurons, as reported in Holstein cows consuming 0.67 kg of dry matter fiber per day [158]. At the same time, parity is associated with higher OXT peak values in multiparous females [159–161]. Likewise, allowing the adequate stimulation of MG by the young or establishing other physical tools in dairy farms, such as mechanical brushes to improve well-being, increased affiliative social behaviors and milk yield (1.52 kg) in 72 Holstein (32%) and Swedish Red (68%) cows [158]. The use of this technology in conjunction with vibratory stimulation on MG reported higher plasmatic OXT levels ($p < 0.05$) when comparing Holstein Friesian cows without/with brushing and without/with stimulation by vibration (with 4.96 ± 0.6 pg/mL vs. 25.56 ± 7.3 pg/mL, respectively). However, the amount of residual milk was not affected by treatment and stayed within a normal range (14.9–15.8%) [162]. Exposure to auditory stimuli such as classical, lullaby, or meditation music has also been shown to have physiological benefits in dairy cattle, improving heart and respiratory rates [163].

On the other hand, exogenous administration of OXT in animal production is a common therapeutic practice to treat uterine inertia, placental retention and incomplete abortion, prevent bleeding after parturition, facilitate milk ejection and reduce milking times [5,164–166]. However, there is controversy surrounding its use, since it directly influences the productive parameters and the total kg per lactation curve and is an invasive methodology that might compromise animal welfare [167]. The type and frequency of milking and breed also modify these values, as well as the gender of the offspring, as reported in female lambs that increased milk ejection, compared to newborn males [168].

Some of the reported adverse effects are changes in the physicochemical analysis of the milk and impacts on the reproductive health of females. For example, the effect of long-term administration of OXT on pregnancy rates of 23 Nili Ravi buffaloes was studied by Murtaza et al. [169]. The animals received two doses of OXT twice daily (10 and 30 IU) for 154 days post-calving. In general, in animals receiving the hormone, pregnancy rates decreased ($p < 0.05$), more artificial insemination (AI) per conception was required, and the presentation of fetal losses increased in comparison to animals not receiving OXT [169]. A similar finding was reported in 430 cows (Holstein Friesian and Swedish-Red) by Gümen et al. [170]. After AI and administration of OXT intramuscularly (50 IU), conception rates were lower ($p = 0.02$) than those registered in control groups (between 29.3–35.5% vs. 44.2–57.1%, respectively). In the case of ewes, after cervical insemination in 300 animals, a practice where OXT is used to dilate the cervix, lambing rates decreased from 42–69% to 10–52%. However, in these animals, litter size was not affected by OXT treatment [171]. Contrarily, the pregnancy rate in lactating dairy cows increased after OXT administration. Nonetheless, since the sample size was small (17 animals) [172], it is important to mention that the reported differences in these studies can be attributed to different dosages, routes of administration, period of treatment, time to pregnancy diagnosis, and even species.

Concerning the productive affectations in Sahiwal cattle, when groups that did not receive exogenous OXT (MG) and those with 20 IU were compared, a reduction in the percentages of fat (4.49 ± 0.11 G1 vs. 4.49 ± 0.11 G2), lactose (4.84 ± 0.04 G1 vs. 4.37 ± 0.22 G2), protein (3.75 ± 0.08 G1 vs. 3.49 ± 0.15), total solids (13.84 ± 0.20 G1 vs. 13.84 ± 0.20 G2) and non-fat solids (9.35 ± 0.11 G1 vs. 8.73 ± 0.34 G2) was found in the OXT group. In addition, it was found that the application of OXT considerably influenced the mineral profile in milk, causing negative variations in the manufacture of dairy products and by-products [165]. Morgan et al. [173] reported in 30 primiparous ewes that administration of OXT at three doses (1, 5, and 10 IU) did not affect milk yield. However, differences in fat percentages were found, having the lowest values at milk-out (0 h) (7.01%) in animals receiving 1 IU, rather than higher doses of OXT (7.81%). In contrast, sows treated with 75 IU of intramuscular OXT improved milk composition with higher numbers of solids ($p < 0.05$), protein ($p < 0.01$), energy ($p < 0.05$), and even higher IgA concentrations [174].

According to these results, the effect of OXT on milk composition, milk yield, and reproductive health of females depends on the lactation phase, parity, species, pregnancy periods, dosages, and administration routes. Therefore, these elements must be considered when assessing the advantages and disadvantages of OXT.

Furthermore, if circulating OXT levels are low because the mother is in threatening or challenging environments, insufficient milk ejection results, causing inconsistencies and resulting in incomplete emptying of alveolar and cisternal fractions of MG, harming the productive levels, the health status of the MG (e.g., with the onset of mastitis), and the survival of the offspring [147,151]. For this reason, auditory, visual, and tactile tools, and even the administration of exogenous OXT, have been implemented to avoid said deficiencies. However, continuous analysis of these practices is necessary to know their scope and the positive and negative impact they could have on the lactation process and thus encourage its continued use or eradication.

6. Future Directions

The possible perspectives of the study concerning OXT could be aimed at assessing the synthesis sites of OXT in tissues other than the myometrium. For example, this hormone can be synthesized in vascular tissue, where it may have a local effect, allowing regulation of the vasomotor response. However, it is still unclear if this could affect the myometrium or if it can alter the behavioral response in animals.

The investigations of OXT in different anatomical regions and compounds are another research field. Recent studies in rabbits have shown that OXT antagonist administration does not impair maternal behavior [175]; however, studies by Keverne and Kendrick [99] have reported that OXT's effect on maternal neuroendocrine regulation is affected by the administration of opioid receptor blockers, limiting OXT release, a relevant issue when using analgesic treatment or other therapeutic protocols during parturition [27]. Moreover, nursing and maternal traits are highly motivated by parity, species, experience, and intrapartum events such as dystocia [87]. Therefore, further research must address the hormonal response that elicits maternal behavior as a complex phenomenon, not solely dependent on OXT or other neuropeptides.

Regarding the exogenous management of OXT, although it is used to reduce expulsion times, improve milk yield and ejection, and reduce newborn mortality [47–49], it is a controversial practice because of the potential adverse effects on dairy animals [5], such as, e.g., intervals between calving and retracements in ovulation [169], and productive effects (impact on milk composition) [23]. On the other hand, it is also necessary to investigate positive stimuli that have been promising, such as the use of brushes, massage [176], music [163], synthetic analogues of maternal appeasing pheromones [177], and the acquisition of a pre-calving milking routine [153,154] to improve dairy cow performance and/or milk parameters. A better understanding of the properties of oxytocinergic cells and the physiology of OXT secretion could also help improve milking machines, for instance, by better synchronizing them with the natural bursts of OXT secretion from the PVN OXT neurons to promote an appropriate hormone release and an efficient/complete milk release, especially in small ruminants [178].

The influence of OXT in species where allomaternal care is present could help to understand the importance of this hormone in both domestic animal behavior and physiology.

7. Conclusions

Parturition in mammals is an event coordinated mainly by OXT and its positive feedback loop to maintain constant and synchronous contractions. Although OXT is essential for the onset of parturition, other hormones such as E2, P4, and PG influence the onset of parturition and myometrial contractions.

Immediately after parturition, maternal attachment and behaviors aimed at caring, nursing, and protecting the newborn are highly influenced by increases in serum, milk, and salivary OXT. In several domestic species, low levels of OXT are associated with impaired maternal fitness and delayed or deficient dam–young bonding. The neuroendocrinal management of maternal traits is a relevant topic because it is not only associated with behaviors towards the newborn but also with altruistic care for non-filial offspring, known as allo-nursing, and because neonatal survival is strictly dependent on maternal behavior.

Likewise, the role of OXT in lactation and milk ejection involves the recognition of sensorial stimulus that promotes OXT release and its action on myoepithelial cells located in the alveolus. Due to its importance, exogenous OXT is common in dairy production systems. However, before continuing with its administration, the potential adverse effects must be considered. Therefore, OXT is one of the main hormones influencing physiological and behavioral traits in parturient domestic animals.

Author Contributions: All the authors contributed to the conceptualization, writing, reading, and approval of the final manuscript. All authors have read and agreed to the published version of the manuscript.

Funding: This research received no external funding.

Institutional Review Board Statement: Not applicable.

Informed Consent Statement: Not applicable.

Data Availability Statement: Not applicable.

Conflicts of Interest: The authors declare no conflict of interest.

References

1.	Gram, A.; Boos, A.; Kowalewski, M.P. Uterine and placental expression of canine oxytocin receptor during pregnancy and normal and induced parturition. *Reprod. Domest. Anim.* **2014**, *49*, 41–49. [CrossRef]
2.	Wang, Q.-H.; Zhang, S.; Qin, L.-M.; Zhang, W.-J.; Liu, F.-H.; Xu, J.-Q.; Ma, Y.-F.; Teng, K.-D. Yimu san improves obstetric ability of pregnant mice by increasing serum oxytocin levels and connexin 43 expression in uterine smooth muscle. *J. Zhejiang Univ. B* **2017**, *18*, 986–993. [CrossRef]
3.	Tsingotjidou, A.S. Oxytocin: A multi-functional biomolecule with potential actions in dysfunctional conditions; from animal studies and beyond. *Biomolecules* **2022**, *12*, 1603. [CrossRef] [PubMed]
4.	Soloff, M.S.; Chakraborty, J.; Sadhukhan, P.; Senitzer, D.; Wieder, M.; Fernstrom, M.A.; Sweet, P. Purification and characterization of mammary myoepithelial and secretory cells from the lactating rat. *Endocrinology* **1980**, *106*, 887–897. [CrossRef]
5.	Faraz, A.; Waheed, A.; Nazir, M.M.; Hameed, A.; Tauqir, N.A.; Mirza, R.H.; Ishaq, H.M.; Bilal, R.M. Impact of oxytocin administration on milk quality, reproductive performance and residual effects in dairy animals—A Review. *Punjab Univ. J. Zool.* **2020**, *35*, 61–67. [CrossRef]
6.	Thornton, J.M.; Browne, B.; Ramphul, M. Mechanisms and management of normal labour. *Obstet. Gynaecol. Reprod. Med.* **2020**, *30*, 84–90. [CrossRef]
7.	Ogi, A.; Mariti, C.; Pirrone, F.; Baragli, P.; Gazzano, A. The influence of oxytocin on maternal care in lactating dogs. *Animals* **2021**, *11*, 1130. [CrossRef] [PubMed]
8.	Lezama-García, K.; Mariti, C.; Mota-Rojas, D.; Martínez-Burnes, J.; Barrios-García, H.; Gazzano, A. Maternal behaviour in domestic dogs. *Int. J. Vet. Sci. Med.* **2019**, *7*, 20–30. [CrossRef]
9.	Coria-Avila, G.A.; Pfaus, J.G.; Orihuela, A.; Domínguez-Oliva, A.; José-Pérez, N.; Hernández, L.A.; Mota-Rojas, D. The neurobiology of behavior and its applicability for animal welfare: A review. *Animals* **2022**, *12*, 928. [CrossRef]
10.	Jurek, B.; Neumann, I.D. The oxytocin receptor: From intracellular signaling to behavior. *Physiol. Rev.* **2018**, *98*, 1805–1908. [CrossRef]
11.	Enriquez, M.F.; Pérez-Torres, L.; Orihuela, A.; Rubio, I.; Corro, M.; Galina, C.S. Relationship between protective maternal behavior and some reproductive variables in Zebu-Type cows (*Bos Indicus*). *Front. Neuroendocrinol.* **2021**, *9*, 2124. [CrossRef]
12.	Ross, H.E.; Young, L.J. Oxytocin and the neural mechanisms regulating social cognition and affiliative behavior. *Front. Neuroendocrinol.* **2009**, *30*, 534–547. [CrossRef] [PubMed]
13.	Insel, T.R. The challenge of translation in social neuroscience: A review of oxytocin, vasopressin, and affiliative behavior. *Neuron* **2010**, *65*, 768. [CrossRef] [PubMed]
14.	Bales, K.L.; Pfeifer, L.A.; Carter, C.S. Sex differences and developmental effects of manipulations of oxytocin on alloparenting and anxiety in prairie voles. *Dev. Psychobiol.* **2004**, *44*, 123–131. [CrossRef]
15.	Madden, J.R.; Clutton-Brock, T.H. Experimental peripheral administration of oxytocin elevates a suite of cooperative behaviours in a wild social mammal. *Proc. R. Soc. B Biol. Sci.* **2011**, *278*, 1189–1194. [CrossRef]
16.	Odendaal, J.S.; Meintjes, R. Neurophysiological correlates of affiliative behaviour between humans and dogs. *Vet. J.* **2003**, *165*, 296–301. [CrossRef] [PubMed]
17.	Crockford, C.; Wittig, R.M.; Langergraber, K.; Ziegler, T.E.; Zuberbühler, K.; Deschner, T. Urinary oxytocin and social bonding in related and unrelated wild chimpanzees. *Proc. R. Soc. B Biol. Sci.* **2013**, *280*, 20122765. [CrossRef]
18.	Wittig, R.M.; Crockford, C.; Deschner, T.; Langergraber, K.E.; Ziegler, T.E.; Zuberbühler, K. Food sharing is linked to urinary oxytocin levels and bonding in related and unrelated wild chimpanzees. *Proc. R. Soc. B Biol. Sci.* **2014**, *281*, 20133096. [CrossRef]

19. Mora-Medina, P.; Orihuela-Trujillo, A.; Arch-Tirado, E.; Roldan-Santiago, P.; Terrazas, A.; Mota-Rojas, D. Sensory factors involved in mother-young bonding in sheep: A review. *Vet. Med.* **2016**, *61*, 595–611. [CrossRef]
20. Bienboire-Frosini, C.; Marcet-Rius, M.; Orihuela, A.; Domínguez-Oliva, A.; Mora-Medina, P.; Olmos-Hernández, A.; Casas-Alvarado, A.; Mota-Rojas, D. Mother–young bonding: Neurobiological aspects and maternal biochemical signaling in altricial domesticated mammals. *Animals* **2023**, *13*, 532. [CrossRef]
21. Kent, J.P. The cow–calf relationship: From maternal responsiveness to the maternal bond and the possibilities for fostering. *J. Dairy Res.* **2020**, *87*, 101–107. [CrossRef] [PubMed]
22. Mota-Rojas, D.; Bragaglio, A.; Braghieri, A.; Napolitano, F.; Domínguez-Oliva, A.; Mora-Medina, P.; Álvarez-Macías, A.; De Rosa, G.; Pacelli, C.; José, N.; et al. Dairy buffalo behavior: Calving, imprinting and allosuckling. *Animals* **2022**, *12*, 2899. [CrossRef] [PubMed]
23. Napolitano, F.; Braghieri, A.; Bragaglio, A.; Rodríguez-González, D.; Mora-Medina, P.; Ghezzi, M.D.; Álvarez-Macías, A.; Lendez, P.A.; Sabia, E.; Domínguez-Oliva, A.; et al. Neurophysiology of milk ejection and prestimulation in dairy buffaloes. *Animals* **2022**, *12*, 2649. [CrossRef] [PubMed]
24. Ybarra-Navarro, N.T. Evaluation of Oxytocin Pharmacokinetic: Pharmacodynamic Profile and Establishment of Its Cardiomyogenic Potential in Swine. Ph.D. Thesis, Faculté de Médecine Vétérinaire, Université de Montréal, Saint-Hyacinthe, QC, Canada, 2010.
25. Lippert, T.H.; Mueck, A.O.; Seeger, H.; Pfaff, A. Effects of oxytocin outside pregnancy. *Horm. Res. Paediatr.* **2003**, *60*, 262–271. [CrossRef]
26. Andriolo, A.; Schmidek, W. Suckling behaviour in water buffalo (*Bubalus bubalis*). *Rev. Etol.* **2001**, *3*, 129–136.
27. Mota-Rojas, D.; Velarde, A.; Marcet-Rius, M.; Orihuela, A.; Bragaglio, A.; Hernández-Ávalos, I.; Casas-Alvarado, A.; Domínguez-Oliva, A.; Whittaker, A.L. Analgesia during parturition in domestic animals: Perspectives and controversies on its use. *Animals* **2022**, *12*, 2686. [CrossRef]
28. Brunton, P.J. Endogenous opioid signalling in the brain during pregnancy and lactation. *Cell Tissue Res.* **2019**, *375*, 69–83. [CrossRef]
29. Sheldrick, E.L.; Flint, A.P.F. Post-Translational processing of oxytocin-neurophysin prohormone in the ovine corpus luteum: Activity of peptidyl glycine α-amidating mono-oxygenase and concentrations of its cofactor, ascorbic acid. *J. Endocrinol.* **1989**, *122*, 313–322. [CrossRef]
30. Schams, D.; Kruip, T.A.M.; Koll, R. Oxytocin determination in steroid producing tissues and in vitro production in ovarian follicles. *Eur. J. Endocrinol.* **1985**, *109*, 530–536. [CrossRef]
31. Knickerbocker, J.J.; Sawyer, H.R.; Amann, R.P.; Tekpetey, F.R.; Niswender, G.D. Evidence for the presence of oxytocin in the ovine epididymis. *Biol. Reprod.* **1988**, *39*, 391–397. [CrossRef]
32. Breton, C.; Zingg, H.H. Expression and region-specific regulation of the oxytocin receptor gene in rat brain. *Endocrinology* **1997**, *138*, 1857–1862. [CrossRef] [PubMed]
33. Shughrue, P.J.; Dellovade, T.L.; Merchenthaler, I. Chapter 2 Estrogen modulates oxytocin gene expression in regions of the rat supraoptic and paraventricular nuclei that contain estrogen receptor-β. *Prog. Brain Res.* **2002**, *139*, 15–29. [CrossRef] [PubMed]
34. Patisaul, H.B.; Scordalakes, E.M.; Young, L.J.; Rissman, E.F. Oxytocin, but not oxytocin receptor, is regulated by oestrogen receptor β in the female mouse hypothalamus. *J. Neuroendocrinol.* **2003**, *15*, 787–793. [CrossRef] [PubMed]
35. Murakami, G.; Hunter, R.G.; Fontaine, C.; Ribeiro, A.; Pfaff, D. Relationships among estrogen receptor, oxytocin and vasopressin gene expression and social interaction in male mice. *Eur. J. Neurosci.* **2011**, *34*, 469–477. [CrossRef]
36. Acevedo-Rodriguez, A.; Mani, S.K.; Handa, R.J. Oxytocin and estrogen receptor β in the brain: An overview. *Front. Endocrinol.* **2015**, *6*, 160. [CrossRef] [PubMed]
37. Jankowski, M.; Wang, D.; Hajjar, F.; Mukaddam-Daher, S.; McCann, S.M.; Gutkowska, J. Oxytocin and its receptors are synthesized in the rat vasculature. *Proc. Natl. Acad. Sci. USA* **2000**, *97*, 6207–6211. [CrossRef] [PubMed]
38. Jankowski, M.; Hajjar, F.; Al Kawas, S.; Mukaddam-Daher, S.; Hoffman, G.; Mccann, S.M.; Gutkowska, J. Rat Heart: A site of oxytocin production and action. *Proc. Natl. Acad. Sci. USA* **1998**, *95*, 14558–14563. [CrossRef]
39. Douglas, A.J.; Johnstone, H.; Brunton, P.; Russell, J. Sex-Steroid induction of endogenous opioid inhibition on oxytocin secretory responses to stress. *J. Neuroendocrinol.* **2001**, *12*, 343–350. [CrossRef]
40. Brunton, P.J.; Russell, J.A.; Hirst, J.J. Allopregnanolone in the brain: Protecting pregnancy and birth outcomes. *Prog. Neurobiol.* **2014**, *113*, 106–136. [CrossRef]
41. Ochedalski, T.; Subburaju, S.; Wynn, P.C.; Aguilera, G. Interaction between oestrogen and oxytocin on hypothalamic-pituitary-adrenal axis activity. *J. Neuroendocrinol.* **2007**, *19*, 189–197. [CrossRef]
42. Neumann, I.D.; Krömer, S.A.; Toschi, N.; Ebner, K. Brain oxytocin inhibits the (re)activity of the hypothalamo–pituitary–adrenal axis in male rats: Involvement of hypothalamic and limbic brain regions. *Regul. Pept.* **2000**, *96*, 31–38. [CrossRef] [PubMed]
43. Tančin, V.; Bruckmaier, R.M. Factors affecting milk ejection and removal during milking and suckling of dairy cows. *Vet. Med.* **2001**, *46*, 108–118. [CrossRef]
44. Douglas, A.J.; Dye, S.; Leng, G.; Russell, J.A.; Bicknell, R.J. Endogenous opioid regulation of oxytocin secretion through pregnancy in the rat. *J. Neuroendocrinol.* **1993**, *5*, 307–314. [CrossRef] [PubMed]
45. Stack, E.C.; Balakrishnan, R.; Numan, M.J.; Numan, M. A functional neuroanatomical investigation of the role of the medial preoptic area in neural circuits regulating maternal behavior. *Behav. Brain Res.* **2002**, *131*, 17–36. [CrossRef] [PubMed]

46. Srisawat, R.; Ludwig, M.; Bull, P.M.; Douglas, A.J.; Russell, J.A.; Leng, G. Nitric oxide and the oxytocin system in pregnancy. *J. Neurosci.* **2000**, *20*, 6721–6727. [CrossRef]

47. Purohit, G. Parturition in domestic animals: A review. *WebmedCentral* **2010**, *1*, WMC00748. [CrossRef]

48. Lévy, F. Neuroendocrine control of maternal behavior in non-human and human mammals. *Ann. Endocrinol.* **2016**, *77*, 114–125. [CrossRef]

49. Blanks, A.M.; Thornton, S. The role of oxytocin in parturition. *BJOG Int. J. Obstet. Gynaecol.* **2003**, *110*, 46–51. [CrossRef]

50. Fusi, J.; Veronesi, M. Canine parturition: What is known about the hormonal setting? *Domest. Anim. Endocrinol.* **2022**, *78*, 106687. [CrossRef]

51. Kowalewski, M.P.; Tavares Pereira, M.; Kazemian, A. Canine conceptus-maternal communication during maintenance and termination of pregnancy, including the role of species-specific decidualization. *Theriogenology* **2020**, *150*, 329–338. [CrossRef]

52. Olcese, J.; Beesley, S. Clinical significance of melatonin receptors in the human myometrium. *Fertil. Steril.* **2014**, *102*, 329–335. [CrossRef] [PubMed]

53. Martínez-Burnes, J.; Muns, R.; Barrios-García, H.; Villanueva-García, D.; Domínguez-Oliva, A.; Mota-Rojas, D. Parturition in mammals: Animal models, pain and distress. *Animals* **2021**, *11*, 2960. [CrossRef] [PubMed]

54. Vallet, J.L.; Miles, J.R.; Brown-Brandl, T.M.; Nienaber, J.A. Proportion of the litter farrowed, litter size, and progesterone and estradiol effects on piglet birth intervals and stillbirths. *Anim. Reprod. Sci.* **2010**, *119*, 68–75. [CrossRef] [PubMed]

55. Oliviero, C. Sucessful Farrowing in Sows. Ph.D. Thesis, University of Helsinki, Helsinki, Finland, 2010.

56. Contreras, J.E.; Sáez, J.C.; Bukauskas, F.F.; Bennett, M.V.L. Gating and regulation of connexin 43 (Cx43) hemichannels. *Proc. Natl. Acad. Sci. USA* **2003**, *100*, 11388–11393. [CrossRef]

57. Chan, L.Y.-S.; Fu, L.; Leung, T.N.; Wong, S.F.; Lau, T.K. Obstetric outcomes after cervical ripening by multiple doses of vaginal prostaglandin E2. *Acta Obstet. Gynecol. Scand.* **2004**, *2*, 70–74.

58. Papatsonis, D.; Flenady, V.; Cole, S.; Liley, H. Oxytocin receptor antagonists for inhibiting preterm labour. In *Cochrane Database of Systematic Reviews*; Papatsonis, D., Ed.; John Wiley & Sons, Ltd.: Chichester, UK, 2005.

59. Kimura, T.; Tanizawa, O.; Mori, K.; Brownstein, M.J.; Okayama, H. Structure and expression of a human oxytocin receptor. *Nature* **1992**, *356*, 526–529. [CrossRef]

60. Ou, C.-W.; Chen, Z.-Q.; Qi, S.; Lye, S.J. Increased expression of the rat myometrial oxytocin receptor messenger ribonucleic acid during labor requires both mechanical and hormonal signals. *Biol. Reprod.* **1998**, *59*, 1055–1061. [CrossRef]

61. Meier, S.; Lau, T.M.; Jenkin, G.; Fairclough, R.J. Oxytocin-induced prostaglandin $F_{2\alpha}$ release and endometrial oxytocin receptor concentrations throughout pregnancy in ewes. *J. Reprod. Fertil.* **1995**, *103*, 233–238.

62. Fuchs, A.R.; Behrens, O.; Helmer, H.; Liu, C.-H.; Barros, C.M.; Fields, M.J. Oxytocin and vasopressin receptors in bovine endometrium and myometrium during the estrous cycle and early pregnancy. *Endocrinology* **1990**, *127*, 629–636. [CrossRef]

63. Fuchs, A.-R.; Ivell, R.; Ganz, N.; Fields, M.J.; Gimenez, T. Secretion of oxytocin in pregnant and parturient cows: Corpus luteum may contribute to plasma oxytocin at term. *Biol. Reprod.* **2001**, *65*, 1135–1141. [CrossRef]

64. Veiga, G.A.L.; Milazzotto, M.P.; Nichi, M.; Lúcio, C.F.; Silva, L.C.G.; Vannucchi, C.I. Correlation of oxytocin (OTR) and estrogen receptor (ER) MRNA in the canine placenta with the detected circulating levels of oxytocin and estrogen during pregnancy and parturition. *Am. J. Anim. Vet. Sci.* **2016**, *11*, 11–17. [CrossRef]

65. Lundin-Schiller, S.; Kreider, D.L.; Rorie, R.W.; Hardesty, D.; Mitchell, M.D.; Koike, T.I. Characterization of porcine endometrial, myometrial, and mammary oxytocin binding sites during gestation and labor. *Biol. Reprod.* **1996**, *55*, 575–581. [CrossRef] [PubMed]

66. Norwitz, E.R.; Robinson, J.N.; Challis, J.R. The control of labor. *N. Engl. J. Med.* **1999**, *341*, 660–666. [PubMed]

67. Masoudi, R. Effect of estradiol and oxytocin on ovine cervical relaxation. *Afr. J. Biotechnol.* **2012**, *11*, 2803–2806. [CrossRef]

68. Kustritz, M.V.R. Reproductive behavior of small animals. *Theriogenology* **2005**, *64*, 734–746. [CrossRef]

69. Whitman, D.C.; Albers, H.E. Role of oxytocin in the hypothalamic regulation of sexual receptivity in hamsters. *Brain Res.* **1995**, *680*, 73–79. [CrossRef] [PubMed]

70. Oliviero, C.; Heinonen, M.; Valros, A.; Hälli, O.; Peltoniemi, O.A.T. Effect of the environment on the physiology of the sow during late pregnancy, farrowing and early lactation. *Anim. Reprod. Sci.* **2008**, *105*, 365–377. [CrossRef]

71. Yun, J.; Swan, K.M.; Vienola, K.; Kim, Y.Y.; Oliviero, C.; Peltoniemi, O.A.T.; Valros, A. Farrowing environment has an impact on sow metabolic status and piglet colostrum intake in early lactation. *Livest. Sci.* **2014**, *163*, 120–125. [CrossRef]

72. Alonso-Spilsbury, M.; Mota-Rojas, D.; Martínez-Burnes, J.; Arch, E.; López Mayagoitia, A.; Ramírez-Necoechea, R.; Olmos, A.; Trujillo, M.E. Use of Oxytocin in penned sows and its effect on fetal intra-partum asphyxia. *Anim. Reprod. Sci.* **2004**, *84*, 157–167. [CrossRef]

73. Oliviero, C.; Heinonen, M.; Valros, A.; Peltoniemi, O. Environmental and sow-related factors affecting the duration of farrowing. *Anim. Reprod. Sci.* **2010**, *119*, 85–91. [CrossRef]

74. van Dijk, A.J.; van Rens, B.T.T.M.; van der Lende, T.; Taverne, M.A.M. Factors affecting duration of the expulsive stage of parturition and piglet birth intervals in sows with uncomplicated, spontaneous farrowings. *Theriogenology* **2005**, *64*, 1573–1590. [CrossRef] [PubMed]

75. Darvelid, A.W.; Linde-Forsberg, C. Dystocia in the bitch: A retrospective study of 182 cases. *J. Small Anim. Pract.* **1994**, *35*, 402–407. [CrossRef]

76. Davidson, A.P. Primary uterine inertia in four labrador bitches. *J. Am. Anim. Hosp. Assoc.* **2011**, *47*, 83–88. [CrossRef] [PubMed]

77. Jungmann, C.; Houghton, C.G.; Nielsen, F.G.; Packeiser, E.-M.; Körber, H.; Reichler, I.M.; Balogh, O.; Goericke-Pesch, S. Involvement of Oxytocin and Progesterone Receptor Expression in the Etiology of Canine Uterine Inertia. *Int. J. Mol. Sci.* **2022**, *23*, 13601. [CrossRef]

78. Nevard, R.P.; Pant, S.D.; Broster, J.C.; Norman, S.T.; Stephen, C.P. Maternal behavior in beef cattle: The physiology, assessment and future directions—A review. *Vet. Sci.* **2022**, *10*, 10. [CrossRef]

79. Coria-Avila, G.A.; Herrera-Covarrubias, D.; García, L.I.; Toledo, R.; Hernández, M.E.; Paredes-Ramos, P.; Corona-Morales, A.A.; Manzo, J. Neurobiology of maternal behavior in nonhuman mammals: Acceptance, recognition, motivation, and rejection. *Animals* **2022**, *12*, 3589. [CrossRef]

80. Geburt, K.; Friedrich, M.; Piechotta, M.; Gauly, M.; König von Borstel, U. Validity of physiological biomarkers for maternal behavior in cows—A comparison of beef and dairy cattle. *Physiol. Behav.* **2015**, *139*, 361–368. [CrossRef]

81. Sanson, A.; Bosch, O.J. Dysfunctions of brain oxytocin signaling: Implications for poor mothering. *Neuropharmacology* **2022**, *211*, 109049. [CrossRef]

82. Orihuela, A.; Mota-Rojas, D.; Strappini, A.; Serrapica, F.; Braghieri, A.; Mora-Medina, P.; Napolitano, F. Neurophysiological mechanisms of cow–calf bonding in buffalo and other farm animals. *Animals* **2021**, *11*, 1968. [CrossRef]

83. Algers, B.; Uvnäs-Moberg, K. Maternal behavior in pigs. *Horm. Behav.* **2007**, *52*, 78–85. [CrossRef]

84. Hall, S.A.; Farish, M.; Coe, J.; Baker, E.; Camerlink, I.; Lawrence, A.B.; Baxter, E.M. Minimally invasive biomarkers to detect maternal physiological status in sow saliva and milk. *Animal* **2021**, *15*, 100369. [CrossRef] [PubMed]

85. Rosvold, E.M.; Newberry, R.C.; Andersen, I.L. Early Mother-young interactions in domestic sows–nest-building material increases maternal investment. *Appl. Anim. Behav. Sci.* **2019**, *219*, 104837. [CrossRef]

86. López-Arjona, M.; Mainau, E.; Navarro, E.; Contreras-Aguilar, M.D.; Escribano, D.; Mateo, S.V.; Manteca, X.; Cerón, J.J.; Martínez-Subiela, S. Oxytocin in bovine saliva: Validation of two assays and changes in parturition and at weaning. *BMC Vet. Res.* **2021**, *17*, 140. [CrossRef]

87. Lévy, F. The onset of maternal behavior in sheep and goats: Endocrine, sensory, neural, and experiential mechanisms. In *Patterns of Parental Behavior. Advances in Neurobiology*; González-Mariscal, G., Ed.; Springer: Cham, Switzerland, 2022; pp. 79–117.

88. Lupoli, B.; Johansson, B.; Uvnäs-Moberg, K.; Svennersten-Sjaunja, K. Effect of suckling on the release of oxytocin, prolactin, cortisol, gastrin, cholecystokinin, somatostatin and insulin in dairy cows and their calves. *J. Dairy Res.* **2001**, *68*, 175–187. [CrossRef]

89. Silveira, P.A.; Spoon, R.A.; Ryan, D.P.; Williams, G.L. Evidence for maternal behavior as a requisite link in suckling-mediated anovulation in cows. *Biol. Reprod.* **1993**, *49*, 1338–1346. [CrossRef]

90. Yun, J.; Swan, K.-M.; Farmer, C.; Oliviero, C.; Peltoniemi, O.; Valros, A. Prepartum nest-building has an impact on postpartum nursing performance and maternal behaviour in early lactating sows. *Appl. Anim. Behav. Sci.* **2014**, *160*, 31–37. [CrossRef]

91. Wang, H.; Han, C.; Li, M.; Li, F.; Yang, Y.; Wang, Z.; Lv, S. Effects of parity, litter size and lamb sex on maternal behavior of small Tail Han sheep and their neuroendocrine mechanisms. *Small Rumin. Res.* **2021**, *202*, 106451. [CrossRef]

92. Santos, N.R.; Beck, A.; Fontbonne, A. A review of maternal behaviour in dogs and potential areas for further research. *J. Small Anim. Pract.* **2020**, *61*, 85–92. [CrossRef]

93. Taylor, J.H.; Grieb, Z.A. Species differences in the effect of oxytocin on maternal behavior: A model incorporating the potential for allomaternal contributions. *Front. Neuroendocrinol.* **2022**, *65*, 100996. [CrossRef]

94. Bridges, R.S. Neuroendocrine regulation of maternal behavior. *Front. Neuroendocrinol.* **2015**, *36*, 178–196. [CrossRef]

95. Mota-Rojas, D.; Marcet-Rius, M.; Freitas-de-Melo, A.; Muns, R.; Mora-Medina, P.; Domínguez-Oliva, A.; Orihuela, A. Allonursing in wild and farm animals: Biological and physiological foundations and explanatory hypotheses. *Animals* **2021**, *11*, 3092. [CrossRef] [PubMed]

96. Orihuela, A.; Pérez-Torres, L.I.; Ungerfeld, R. Evidence of cooperative calves' care and providers' characteristics in zebu cattle (*Bos indicus*) raised under extensive conditions. *Trop. Anim. Health Prod.* **2021**, *53*, 143. [CrossRef] [PubMed]

97. Williams, G.L.; Gazal, O.S.; Leshin, L.S.; Stanko, R.L.; Anderson, L.L. Physiological regulation of maternal behavior in heifers: Roles of genital stimulation, intracerebral oxytocin release, and ovarian steroids. *Biol. Reprod.* **2001**, *65*, 295–300. [CrossRef] [PubMed]

98. Poindron, P.; Lévy, F.; Keller, M. Maternal responsiveness and maternal selectivity in domestic sheep and goats: The two facets of maternal attachment. *Dev. Psychobiol.* **2007**, *49*, 54–70. [CrossRef] [PubMed]

99. Keverne, B.; Kendrick, K.M. Maternal behaviour in sheep and its neuroendocrine regulation. *Acta Paediatr.* **1994**, *83*, 47–56. [CrossRef] [PubMed]

100. Kendrick, K.M.; Keverne, E.B.; Hinton, M.R.; Goode, J.A. Cerebrospinal fluid and plasma concentrations of oxytocin and vasopressin during parturition and vaginocervical stimulation in the sheep. *Brain Res. Bull.* **1991**, *26*, 803–807. [CrossRef] [PubMed]

101. Valros, A.; Rundgren, M.; Špinka, M.; Saloniemi, H.; Hultén, F.; Uvnäs-Moberg, K.; Tománek, M.; Krejcí, P.; Algers, B. Oxytocin, prolactin and somatostatin in lactating sows: Associations with mobilisation of body resources and maternal behaviour. *Livest. Prod. Sci.* **2004**, *85*, 3–13. [CrossRef]

102. Wang, C.; Han, Q.; Liu, R.; Ji, W.; Bi, Y.; Wen, P.; Yi, R.; Zhao, P.; Bao, J.; Liu, H. Equipping farrowing pens with straw improves maternal behavior and physiology of min-pig hybrid sows. *Animals* **2020**, *10*, 105. [CrossRef]

103. Kenkel, W.M.; Paredes, J.; Yee, J.R.; Pournajafi-Nazarloo, H.; Bales, K.L.; Carter, C.S. Neuroendocrine and behavioural responses to exposure to an infant in male prairie Voles. *J. Neuroendocrinol.* **2012**, *24*, 874–886. [CrossRef]

104. Olazábal, D.E.; Young, L.J. Oxytocin receptors in the nucleus accumbens facilitate "spontaneous" maternal behavior in adult female prairie voles. *Neuroscience* **2006**, *141*, 559–568. [CrossRef]

105. Robinson, K.J.; Twiss, S.D.; Hazon, N.; Pomeroy, P.P. Maternal oxytocin is linked to close mother-infant proximity in grey seals (*Halichoerus grypus*). *PLoS ONE* **2015**, *10*, e0144577. [CrossRef] [PubMed]

106. Kockaya, M.; Ercan, N.; Demirbas, Y.S.; Da Graça Pereira, G. Serum oxytocin and lipid levels of dogs with maternal cannibalism. *J. Vet. Behav.* **2018**, *27*, 23–26. [CrossRef]

107. Kendrick, K.M.; Keverne, E.B. Importance of progesterone and estrogen priming for the induction of maternal behavior by vaginocervical stimulation in sheep: Effects of maternal experience. *Physiol. Behav.* **1991**, *49*, 745–750. [CrossRef]

108. McLean, K.; Lawrence, A.; Petherick, J.; Deans, L.; Chirnside, J.; Vaughan, A.; Nielsen, B.; Webb, R. Investigation of the relationship between farrowing environment, sex steroid concentrations and maternal aggression in gilts. *Anim. Reprod. Sci.* **1998**, *50*, 95–109. [CrossRef] [PubMed]

109. Bosch, O.J.; Neumann, I.D. Both oxytocin and vasopressin are mediators of maternal care and aggression in rodents: From central release to sites of action. *Horm. Behav.* **2012**, *61*, 293–303. [CrossRef]

110. Hess, E.H. HESS: The natural history of imprinting. *Ann. N. Y. Acad. Sci.* **1972**, *193*, 124–136. [CrossRef]

111. Ohki-Hamazaki, H. Neurobiology of imprinting. *Brain Nerve* **2012**, *64*, 657–664. [CrossRef]

112. Hess, E.H. Imprinting. *Science* **1959**, *130*, 133–141. [CrossRef]

113. Bridges, R.S. Long-term alterations in neural and endocrine processes induced by motherhood in mammals. *Horm. Behav.* **2016**, *77*, 193–203. [CrossRef]

114. Kim, P. Human maternal brain plasticity: Adaptation to parenting. *New Dir. Child Adolesc. Dev.* **2016**, *2016*, 47–58. [CrossRef]

115. Kim, P.; Strathearn, L.; Swain, J.E. The maternal brain and its plasticity in humans. *Horm. Behav.* **2016**, *77*, 113–123. [CrossRef] [PubMed]

116. Pawluski, J.L.; Lambert, K.G.; Kinsley, C.H. Neuroplasticity in the maternal hippocampus: Relation to cognition and effects of repeated stress. *Horm. Behav.* **2016**, *77*, 86–97. [CrossRef] [PubMed]

117. Nowak, R. Suckling, milk, and the development of preferences toward maternal cues by neonates: From early learning to filial attachment? In *Advances in the Study of Behavior*; Elsevier: Amsterdam, The Netherlands, 2006; Volume 36, pp. 1–58.

118. Mota-Rojas, D.; Bienboire-Frosini, C.; Marcet-Rius, M.; Domínguez-Oliva, A.; Mora-Medina, P.; Lezama-García, K.; Orihuela, A. Mother-young bond in non-human mammals: Neonatal communication pathways and neurobiological basis. *Front. Psychol.* **2022**, *13*, 1064444. [CrossRef]

119. Mota-Rojas, D.; Napolitano, F.; Orihuela, A.; Serrapica, F.; Olmos-Hernández, A.; Martínez-Burnes, J.; De Rosa, G. Behavior and welfare of dairy buffaloes: Calving, milking, and weaning. In *Biotechnological Applications in Buffalo Research*; Chauhan, M., Selokar, N., Eds.; Springer: Singapore, 2022; pp. 97–119.

120. Mora-Medina, P.; Napolitano, F.; Mota-Rojas, D.; Berdugo-Gutiérrez, J.; Ruiz-Buitrago, J.; Guerrero-Legarreta, I. Imprinting, sucking and allosucking behaviors in buffalo calves. *J. Buffalo Sci.* **2018**, *7*, 49–57. [CrossRef]

121. Orihuela, A.; Mota-Rojas, D.; Urgerfeld, R.; De Rosa, G.; Mora-Medina, P.; Braghieri, A.; Lezama, K.; Domínguez, A.; Napolitano, F. Imprinting in the buffalo and other farm animals: Neurophysiological mechanisms. In *El Búfalo de Agua en Latinoamérica. Hallazgos Recientes*; Napolitano, F., Mota-Rojas, D., Guerrero-Legarreta, I., Orihuela, A., Eds.; BM Editores: Ciudad de México, México, 2020; pp. 923–958. Available online: https://www.lifescienceglobal.com/journals/journal-of-buffalo-science/97-abstract/jbs/4550-el-bufalo-de-agua-en-latinoamerica-hallazgos-recientes (accessed on 5 June 2020).

122. Bolhuis, J.J. Early learning and the development of filial preferences in the chick. *Behav. Brain Res.* **1999**, *98*, 245–252. [CrossRef] [PubMed]

123. Glatzle, M.; Hoops, M.; Kauffold, J.; Seeger, J.; Fietz, S.A. Development of deep and upper neuronal layers in the domestic cat, sheep and pig neocortex. *Anat. Histol. Embryol.* **2017**, *46*, 397–404. [CrossRef]

124. Olazábal, D.E. Role of oxytocin in parental behaviour. *J. Neuroendocrinol.* **2018**, *30*, e12594. [CrossRef]

125. Yoshihara, C.; Numan, M.; Kuroda, K.O. Oxytocin and parental behaviors. In *Current Topics in Behavioral Neurosciences*; Springer: Berlin/Heidelberg, Germany, 2017; Volume 35, pp. 119–153.

126. Muir, G.D. Early ontogeny of locomotor behaviour: A comparison between altricial and precocial animals. *Brain Res. Bull.* **2000**, *53*, 719–726. [CrossRef]

127. Yamaguchi, S.; Aoki, N.; Kitajima, T.; Iikubo, E.; Katagiri, S.; Matsushima, T.; Homma, K.J. Thyroid hormone determines the start of the sensitive period of imprinting and primes later learning. *Nat. Commun.* **2012**, *3*, 1081. [CrossRef]

128. Lemche, E. Research evidence from studies on filial imprinting, attachment, and early life stress: A new route for scientific integration. *Acta Ethologica* **2020**, *23*, 127–133. [CrossRef]

129. Dwyer, C.M. Maternal behaviour and lamb survival: From neuroendocrinology to practical application. *Animal* **2014**, *8*, 102–112. [CrossRef] [PubMed]

130. Lévy, F.; Keller, M. Olfactory mediation of maternal behavior in selected mammalian species. *Behav. Brain Res.* **2009**, *200*, 336–345. [CrossRef] [PubMed]

131. Pissonnier, D.; Thiery, J.C.; Fabre-Nys, C.; Poindron, P.; Keverne, E.B. The Importance of olfactory bulb noradrenalin for maternal recognition in sheep. *Physiol. Behav.* **1985**, *35*, 361–363. [CrossRef]

132. Corona, R.; Meurisse, M.; Cornilleau, F.; Moussu, C.; Keller, M.; Lévy, F. Disruption of adult olfactory neurogenesis induces deficits in maternal behavior in sheep. *Behav. Brain Res.* **2018**, *347*, 124–131. [CrossRef] [PubMed]

133. Keller, M.; Meurisse, M.; Poindron, P.; Nowak, R.; Ferreira, G.; Shayit, M.; Lévy, F. Maternal experience influences the establishment of visual/auditory, but not olfactory recognition of the newborn lamb by ewes at parturition. *Dev. Psychobiol.* **2003**, *43*, 167–176. [CrossRef]

134. Kojima, S.; Stewart, R.A.; Demas, G.E.; Alberts, J.R. Maternal contact differentially modulates central and peripheral oxytocin in rat pups during a brief regime of mother-pup interaction that induces a filial huddling preference. *J. Neuroendocrinol.* **2012**, *24*, 831–840. [CrossRef]

135. Castro-Sierra, E.; de León, F.C.P.; Domínguez, L.F.G.; Rivera, A.P. Neurotransmisores del sistema límbico. hipocampo, gaba y memoria. primera parte. *Salud Ment.* **2007**, *30*, 7–15.

136. Nowak, R.; Porter, R.H.; Lévy, F.; Orgeur, P.; Schaal, B. Role of mother-young interactions in the survival of offspring in domestic mammals. *Rev. Reprod.* **2000**, *5*, 153–163. [CrossRef]

137. Nowak, R.; Keller, M.; Lévy, F. Mother-Young Relationships in Sheep: A model for a multidisciplinary approach of the study of attachment in mammals. *J. Neuroendocrinol.* **2011**, *23*, 1042–1053. [CrossRef]

138. Tsingotjidou, A.S.; Papadopoulos, G.C. The milk-ejection reflex in the sheep: An anatomical study on the afferent pathway. *Anat. Histol. Embryol.* **2008**, *37*, 245–250. [CrossRef]

139. Dûidić, A. Studies on Milk Ejection and Milk Removal during Machine Milking in Different Species. Ph.D. Thesis, Technische Universität München, München, Germany, 2004; pp. 1–23.

140. Ralph, C.R.; Tilbrook, A.J. The hypothalamo-pituitary-adrenal (HPA) axis in sheep is attenuated during lactation in response to psychosocial and predator stress. *Domest. Anim. Endocrinol.* **2016**, *55*, 66–73. [CrossRef]

141. Antonijevic, I.A.; Douglas, A.J.; Dye, S.; Bicknell, R.J.; Leng, G.; Russell, J.A. Oxytocin antagonists delay the initiation of parturition and prolong its active phase in rats. *J. Endocrinol.* **1995**, *145*, 97–103. [CrossRef]

142. Nishimori, K.; Young, L.J.; Guo, Q.; Wang, Z.; Insel, T.R.; Matzuk, M.M. Oxytocin is required for nursing but is not essential for parturition or reproductive behavior. *Proc. Natl. Acad. Sci. USA* **1996**, *93*, 11699–11704. [CrossRef]

143. Ludwig, M.; Sabatier, N.; Bull, P.M.; Landgraf, R.; Dayanithi, G.; Leng, G. Intracellular calcium stores regulate activity-dependent neuropeptide release from dendrites. *Nature* **2002**, *418*, 85–89. [CrossRef] [PubMed]

144. Brown, C.H. Magnocellular neurons and posterior pituitary function. In *Comprehensive Physiology*; Wiley: Hoboken, NJ, USA, 2016; Volume 6, pp. 1701–1741.

145. Meddle, S.L.; Bishop, V.R.; Gkoumassi, E.; van Leeuwen, F.W.; Douglas, A.J. Dynamic changes in oxytocin receptor expression and activation at parturition in the rat brain. *Endocrinology* **2007**, *148*, 5095–5104. [CrossRef] [PubMed]

146. Bruckmaier, R.M.; Wellnitz, O. Induction of milk ejection and milk removal in different production systems. *J. Anim. Sci.* **2008**, *86*, 15–20. [CrossRef]

147. Boselli, C.; Campagna, M.C.; Amatiste, S.; Rosati, R.; Borghese, A. Pre-Stimulation effects on teat anatomy and milk flow curves in mediterranean italian buffalo cows. *J. Anim. Vet. Adv.* **2014**, *13*, 912–916. [CrossRef]

148. Slattery, D.A.; Neumann, I.D. No stress please! Mechanisms of stress hyporesponsiveness of the maternal brain. *J. Physiol.* **2008**, *586*, 377–385. [CrossRef]

149. Tilbrook, A.J.; Turner, A.I.; Ibbott, M.D.; Clarke, I.J. Activation of the hypothalamo-pituitary-adrenal axis by isolation and restraint stress during lactation in ewes: Effect of the presence of the lamb and suckling. *Endocrinology* **2006**, *147*, 3501–3509. [CrossRef]

150. Wagner, B.K.; Relling, A.E.; Kieffer, J.D.; Parker, A.J. Brief communication: Plasma cortisol concentration is affected by lactation, but not intra-nasal oxytocin treatment, in beef cows. *PLoS ONE* **2021**, *16*, e0249323. [CrossRef]

151. Cook, C.J. Oxytocin and prolactin suppress cortisol responses to acute stress in both lactating and non-lactating sheep. *J. Dairy Res.* **1997**, *64*, 327–339. [CrossRef] [PubMed]

152. Ellendorff, F.; Schams, D. Characteristics of milk ejection, associated intramammary pressure changes and oxytocin release in the mare. *J. Endocrinol.* **1988**, *119*, 219–227. [CrossRef] [PubMed]

153. Polikarpus, A.; Napolitano, F.; Grasso, F.; Di Palo, R.; Zicarelli, F.; Arney, D.; De Rosa, G. Effect of pre-partum habituation to milking routine on behaviour and lactation performance of buffalo heifers. *Appl. Anim. Behav. Sci.* **2014**, *161*, 1–6. [CrossRef]

154. Polikarpus, A.; Grasso, F.; Pacelli, C.; Napolitano, F.; De Rosa, G. Milking behaviour of buffalo cows: Entrance order and side preference in the milking parlour. *J. Dairy Res.* **2014**, *81*, 24–29. [CrossRef]

155. Espinosa, Y.; Ponce, P.; Capdevila, J. Efecto de la estimulación con bucerro, oxitocina y manual sobre los indicarores de ordeño en búfalas. *Rev. Salud Anim.* **2011**, *33*, 90–96.

156. Boselli, C.; De Marchi, M.; Costa, A.; Borghese, A. Study of milkability and its relation with milk yield and somatic cell in mediterranean italian water buffalo. *Front. Vet. Sci.* **2020**, *7*, 432. [CrossRef]

157. Singh, M.; Aggarwal, A. Effect of oxytocin administration on certain minerals in the milk of buffaloes (*Bubalus bubalis*). *Asian-Australas. J. Anim. Sci.* **2001**, *14*, 1523–1526. [CrossRef]

158. Keeling, L.J.; De Oliveira, D.; Rustas, B.-O.; Keeling, L.J.; De Oliveira, D.; Rustas, B.-O. Use of mechanical rotating brushes in dairy cows—A potential proxy for performance and welfare. *Precis. Dairy Farming* **2016**, *9*, 343–347.

159. Armstrong, W.E.; Hatton, G.I. The puzzle of pulsatile oxytocin secretion during lactation: Some new pieces. *Am. J. Physiol. Integr. Comp. Physiol.* **2006**, *291*, R26–R28. [CrossRef]

160. Takayanagi, Y.; Onaka, T. Roles of oxytocin in stress responses, allostasis and resilience. *Int. J. Mol. Sci.* **2021**, *23*, 150. [CrossRef]

161. Yukinaga, H.; Hagihara, M.; Tsujimoto, K.; Chiang, H.-L.; Kato, S.; Kobayashi, K.; Miyamichi, K. Recording and manipulation of the maternal oxytocin neural activities in mice. *Curr. Biol.* **2022**, *32*, 3821–3829. [CrossRef] [PubMed]
162. Mačuhová, J.; Tančin, V.; Bruckmaier, R.M. Oxytocin release, milk ejection and milk removal in a multi-box automatic milking system. *Livest. Prod. Sci.* **2003**, *81*, 139–147. [CrossRef]
163. Ciborowska, P.; Michalczuk, M.; Bień, D. The effect of music on livestock: Cattle, poultry and pigs. *Animals* **2021**, *11*, 3572. [CrossRef]
164. Ballou, L.U.; Bleck, J.L.; Bleck, G.T.; Bremel, R.D. The effects of daily oxytocin injections before and after milking on milk production, milk plasmin, and milk composition. *J. Dairy Sci.* **1993**, *76*, 1544–1549. [CrossRef] [PubMed]
165. Hameed, A.; Anjum, F.M.; ur Rehman, Z.; Akhtar, S.; Faraz, A.; Hussain, M.; Ismail, A. Compositional and mineral profile of sahiwal cow milk at various lactation stages as influenced by oxytocin administration. *Pak. J. Zool.* **2021**, *54*, 1–7. [CrossRef]
166. Assad, N.I.; Pandey, A.K.; Sharma, L.M. Oxytocin, Functions, Uses and Abuses: A Brief Review. *Theriogenology Insight-Int. J. Reprod. All Anim.* **2016**, *6*, 1–17. [CrossRef]
167. Borghese, A.; Rasmussen, M.; Thomas, C.S. Milking management of dairy buffalo. *Ital. J. Anim. Sci.* **2007**, *6*, 39–50. [CrossRef]
168. Abecia, J.A.; Palacios, C. Ewes giving birth to female lambs produce more milk than ewes giving birth to male lambs. *Ital. J. Anim. Sci.* **2018**, *17*, 736–739. [CrossRef]
169. Murtaza, S.; Sattar, A.; Ahmad, N.; Ijaz, M.; Omer, T.; Akhtar, M.; Shahzad, M. Long term administration of exogenous oxytocin: Effects on pregnancy rate, and embryonic and fetal losses in Nili-Ravi Buffaloes. *J. Anim. Plant Sci.* **2020**, *30*, 40–49. [CrossRef]
170. Gümen, A.; Keskin, A.; Yilmazbas-Mecitoglu, G.; Karakaya, E.; Cevik, S.; Balci, F. Effects of GnRH, PGF2α and oxytocin treatments on conception rate at the time of artificial insemination in lactating dairy cows. *Czech J. Anim. Sci.* **2011**, *56*, 279–283. [CrossRef]
171. King, M.E.; McKelvey, W.A.C.; Dingwall, W.S.; Matthews, K.P.; Gebbie, F.E.; Mylne, M.J.A.; Stewart, E.; Robinson, J.J. Lambing rates and litter sizes following intrauterine or cervical insemination of frozen/thawed semen with or without oxytocin administration. *Theriogenology* **2004**, *62*, 1236–1244. [CrossRef] [PubMed]
172. Yildiz, A. Effect of oxytocin on conception rate in cows. *Fırat Univ. J. Health Sci.* **2005**, *19*, 75–78.
173. Morgan, J.; Fogarty, N.M.; Nicol, H. Oxytocin administration and its effect on ewe milk composition. *Asian-Australas. J. Anim. Sci.* **2000**, *13*, 206–208.
174. Farmer, C.; Lessard, M.; Knight, C.H.; Quesnel, H. Oxytocin injections in the postpartal period affect mammary tight junctions in sows1. *J. Anim. Sci.* **2017**, *95*, 3532–3539. [CrossRef] [PubMed]
175. Jiménez, A.; Jiménez, P.; Inoue, K.; Young, L.J.; González-Mariscal, G. Oxytocin antagonist does not disrupt rabbit maternal behavior despite binding to brain oxytocin receptors. *J. Neuroendocrinol.* **2023**, *early view*, e13236. [CrossRef]
176. Monks, D.T.; Palanisamy, A. Oxytocin: At birth and beyond. a systematic review of the long-term effects of peripartum oxytocin. *Anaesthesia* **2021**, *76*, 1526–1537. [CrossRef] [PubMed]
177. Osella, M.C.; Cozzi, A.; Spegis, C.; Turille, G.; Barmaz, A.; Lecuelle, C.L.; Teruel, E.; Bienboire-Frosini, C.; Chabaud, C.; Bougrat, L.; et al. The effects of a synthetic analogue of the Bovine Appeasing Pheromone on milk yield and composition in Valdostana dairy cows during the move from winter housing to confined lowland pastures. *J. Dairy Res.* **2018**, *85*, 174–177. [CrossRef]
178. Marnet, P.G.; McKusick, B.C. Regulation of milk ejection and milkability in small ruminants. *Livest. Prod. Sci.* **2001**, *70*, 125–133. [CrossRef]

Disclaimer/Publisher's Note: The statements, opinions and data contained in all publications are solely those of the individual author(s) and contributor(s) and not of MDPI and/or the editor(s). MDPI and/or the editor(s) disclaim responsibility for any injury to people or property resulting from any ideas, methods, instructions or products referred to in the content.

Review

Vitality in Newborn Farm Animals: Adverse Factors, Physiological Responses, Pharmacological Therapies, and Physical Methods to Increase Neonate Vigor

Cécile Bienboire-Frosini [1], Ramon Muns [2], Míriam Marcet-Rius [3], Angelo Gazzano [4], Dina Villanueva-García [5], Julio Martínez-Burnes [6], Adriana Domínguez-Oliva [2,7], Karina Lezama-García [7], Alejandro Casas-Alvarado [7] and Daniel Mota-Rojas [7,*]

[1] Department of Molecular Biology and Chemical Communication, Research Institute in Semiochemistry and Applied Ethology (IRSEA), 84400 Apt, France; c.frosini@group-irsea.com
[2] Agri-Food and Biosciences Institute, Hillsborough BT 26 6DR, Northern Ireland, UK; rmunsvila@gmail.com (R.M.)
[3] Animal Behaviour and Welfare Department, Research Institute in Semiochemistry and Applied Ethology (IRSEA), 84400 Apt, France; m.marcet@group-irsea.com
[4] Department of Veterinary Sciences, University of Pisa, 56124 Pisa, Italy; angelo.gazzano@unipi.it
[5] Division of Neonatology, Hospital Infantil de México Federico Gómez, Mexico City 06720, Mexico; dinavg21@yahoo.com
[6] Facultad de Medicina Veterinaria y Zootecnia, Universidad Autónoma de Tamaulipas, Victoria City 87000, Mexico; jmburnes@docentes.uat.edu.mx
[7] Neurophysiology, Behavior and Animal Welfare Assessment, DPAA, Universidad Autónoma Metropolitana, Xochimilco Campus, Mexico City 04960, Mexico
[*] Correspondence: dmota100@yahoo.com.mx

Citation: Bienboire-Frosini, C.; Muns, R.; Marcet-Rius, M.; Gazzano, A.; Villanueva-García, D.; Martínez-Burnes, J.; Domínguez-Oliva, A.; Lezama-García, K.; Casas-Alvarado, A.; Mota-Rojas, D. Vitality in Newborn Farm Animals: Adverse Factors, Physiological Responses, Pharmacological Therapies, and Physical Methods to Increase Neonate Vigor. *Animals* **2023**, *13*, 1542. https://doi.org/10.3390/ani13091542

Academic Editor: Maria Cristina Veronesi

Received: 30 March 2023
Revised: 18 April 2023
Accepted: 29 April 2023
Published: 4 May 2023

Copyright: © 2023 by the authors. Licensee MDPI, Basel, Switzerland. This article is an open access article distributed under the terms and conditions of the Creative Commons Attribution (CC BY) license (https://creativecommons.org/licenses/by/4.0/).

Simple Summary: Vitality is a characteristic that in newborn animals demonstrates their vigor and their general state of health (heart rate, respiratory rate, skin color, time that they take to stand up) during the first hours of life. It can be measured by numerical scores based on some scales made for babies and then adapted for various animals. Vitality can be affected by several factors. The objective of this review is to analyze pharmacological and physical therapies used to increase vitality in newborn farm animals, as well as understand the factors affecting their vitality, such as hypoxia, depletion of glycogen, birth weight, dystocia, neurodevelopment, hypothermia, and finally, the physiological mechanism to achieve thermostability. It is essential to evaluate vitality in newborns because it can contribute to implementing interventions to reduce newborn mortality.

Abstract: Vitality is the vigor newborn animals exhibit during the first hours of life. It can be assessed by a numerical score, in which variables, such as heart rate, respiratory rate, mucous membranes' coloration, time the offspring took to stand up, and meconium staining, are monitored. Vitality can be affected by several factors, and therapies are used to increase it. This manuscript aims to review and analyze pharmacological and physical therapies used to increase vitality in newborn farm animals, as well as to understand the factors affecting this vitality, such as hypoxia, depletion of glycogen, birth weight, dystocia, neurodevelopment, hypothermia, and finally, the physiological mechanism to achieve thermostability. It has been concluded that assessing vitality immediately after birth is essential to determine the newborn's health and identify those that need medical intervention to minimize the deleterious effect of intrapartum asphyxia. Vitality assessment should be conducted by trained personnel and adequate equipment. Evaluating vitality could reduce long-term neonatal morbidity and mortality in domestic animals, even if it is sometimes difficult with the current organization of some farms. This review highlights the importance of increasing the number of stock people during the expected days of parturitions to reduce long-term neonatal morbidity and mortality, and thus, improve the farm's performance.

Keywords: domestic animals; vitality; MAS; shivering; BAT; meconium staining

1. Introduction

"Vitality" describes the liveliness or vigor that neonate animals exhibit during the first hours post-parturition [1]. It is a trait that is influenced by several factors, such as intrauterine growth, the physiological immaturity of the newborn, and hypoxia due to dystocia. Moreover, vitality in early postnatal life is affected by the behaviors, maternal factors (e.g., maternal health and nutritional status, maternal care), and environmental conditions during birth [1,2]. Fetuses could be exposed in utero to various deleterious factors that disrupt the blood and oxygen flow through the umbilical cord, often culminating in fetal asphyxia, hypoxia, and metabolic acidosis. These situations are the leading causes of intrapartum and neonatal mortality [3–7]. Prolonged or intermittent asphyxia in utero or during birth physiologically weakens the fetus or newborn and makes them less adaptable to extrauterine life, decreasing their viability and vitality [8,9].

In 1952, Virginia Apgar developed the Apgar scoring system, which provides a method to document the newborn's condition at specific intervals after birth in humans. This is a valuable objective indicator to determine the effectiveness of resuscitative efforts in newborns. Epidemiologists quickly adopted the Apgar score as an outcome in perinatal research because it is straightforward, easily understood, and almost universally recorded in birth-related data sources. The score comprises 5 categories (skin color, heart rate, reflex irritability, activity/flexion, and respiratory effort) that are each scored from 0 to 2, resulting in an overall range of 0 to 10, with 10 indicating that the highest score was given for each clinical indicator of neonatal well-being. In modern practice, the Apgar score is calculated 1 and 5 min after birth, and again at 10 min if the 5 min Apgar is low [10]. Adapted from the Apgar score, neonatal vitality in farm animals is usually assessed through a numerical score, with heart rate, respiratory rate, mucous membranes coloration, the time it took for the offspring to stand up, and meconium staining signs as measurements [11], and different options and measurements have been used to assess neonatal vitality. The total time of parturition and expulsion of the newborn is considered the most important data [12].

The Apgar score has been adapted by diverse authors for pigs [13,14] and ruminants [15,16]. Unlike the human Apgar score, bovine tests combine a variety of indicators divided into three groups: (1) Clinical signs observed on physical examination; (2) Behavior of calves at birth; (3) Blood tests (to evaluate glucose and cortisol levels) [15,17,18].

Proper vitality assessment is essential for newborn survival in humans and domestic animals. The evaluation of the physiological status of the offspring requires trained personnel, adequate equipment, and timely intervention to correct cardiorespiratory depression or resuscitate the newborn when necessary [13,14]. In many farm animal species, it is essential to have trained assistants on farms or in veterinary clinics and adequate physical resources to assess the vitality of the newborn. Early recognition of dystocia and fetal distress is essential for successfully managing labor and neonatal health [19].

Currently, several methods and therapies are applied to farm animals to assess and improve the vitality of newborns. Vitality at birth is often directly related to neonatal viability [2]. Vitality assessment is critical to identifying at-risk newborn animals and implementing interventions to prevent neonatal mortality. When considering pharmacological therapies, drugs such as caffeine [20,21] and naloxone have cardiorespiratory effects, benefiting physiological function. For energy supplementation, glucose administration is a common practice to supply newborns with the energy necessary to thermoregulate and move to the udder immediately after birth [22]. Oxygen administration to reduce or compensate for hypoxia states is another option for treatment. However, oxygen doses and their effect might differ between species, and one must consider individual cases according to the species and the newborn health at birth. Moreover, physical methods to promote vitality and reduce mortality rates, such as natural and artificial colostrum supplementation, provision of external heat sources, and controlled microclimates, are being studied to determine their effect on neonatal survival [23,24]. Nonetheless, because maternal and fetal factors alter neonatal vitality and the ability of the animal to adapt to the extrauterine environment [20], it is essential to create a comprehensive intervention protocol for susceptible animals.

Therefore, newborns are susceptible to hypoglycemia, hypothermia, dehydration, metabolic acidosis, less vitality, and death. The aim of this manuscript is to (1) understand the factors affecting vitality in farms animals (i.e., hypothermia, hypoxia, birth weight, glycogen depletion, neurodevelopment, and dystocia) and (2) review and analyze pharmacological and physical methods used to increase vitality in newborn farm animals.

2. Factors Affecting the Vitality of the Newborn

Significant impact on newborn vitality can have various factors, such as environmental (extrinsic), the environment in which the newborn is born and raised. Factors, such as adverse weather conditions for outdoor species (sheep, goat) or temperature, humidity, ventilation, and cleanliness for indoor species, can affect the health and vitality of the newborn. Management practices used on the farm can also affect the vitality of newborns. Proper nutrition, sanitation, and veterinary care can help ensure that newborns are healthy and vigorous. Other factors (intrinsic) are attributable to the fetus/newborn, such as hypothermia, hypoxia, birth weight, and glycogen depletion, or to neurodevelopment, related to other phenotypes, such as litter size, birth weight [25–28], and variation in birthweight. Others are attributable to the mother, i.e., dystocia, maternal ability, the absence of milk, the parity number, gestational age/length of gestation, the number of teats to nurse the newborn [25], and clinical factors (infectious, hypertensive, and metabolic disorders) and the general health status (notably nutritional) [29]. In the case of this review, we will only focus on the intrinsic factors, which will be discussed below.

2.1. Hypothermia

In all endothermal animals, modulation of thermoregulation is closely related to the stability of various cardiovascular, respiratory, renal, endocrine, nervous, muscular, and cellular functions [30], as well as environmental conditions that are not in the scope of this review. One of the leading causes of neonatal mortality in farm animals (i.e., lambs, calves, foals, piglets) and other mammals is hypothermia caused by a significant loss of heat or inhibition of thermoregulation, resulting from starvation when the offspring is unable to suckle [31]. Another cause of hypothermia is physiological immaturity coupled with the fact that when being expulsed from the uterus, the newborn's temperature can drop up to 3.5 °C [32,33]. This physiological change can trigger hypoglycemia, hypoalbuminemia, growth retardation, or changes in the acid–base ratio, leading to multiorgan failure [34]. For example, pre-weaning mortality in piglets in EU countries remains around 15%, with some as high as 25% [35], with many due to "crushing" by the sow, an outcome often associated with postnatal hypoglycemia and hypothermia [36,37]. In all endothermal animals, modulation of thermoregulation is closely related to the stability of various cardiovascular, respiratory, renal, endocrine, nervous, muscular, and cellular functions [30]. One of the leading causes of neonatal mortality in farm animals (i.e., lambs, calves, foals, piglets) and other mammals is hypothermia, produced by a significant loss of heat or the inhibition of thermoregulation and heat production, resulting from starvation when the offspring is unable to suckle [31].

However, how does a neonate fall into hypothermia? The sudden drop in body temperature at birth reduces vigor and affects the newborn's feeding ability. Consequently, colostrum intake (the only source of immunoglobulins, nutrients, and energy) to fuel for thermogenesis is diminished [38–42]. In addition, factors associated with the activation of the hypothalamic–pituitary–adrenal (HPA) axis [43] increase the concentrations of catecholamines and circulating cortisol, both in the fetus and in the neonates [44], generating changes in blood flow that can compromise the ability to thermoregulate [45].

Newborns use various mechanisms to minimize or compensate for hypothermia: shivering, vasomotor control, and the presence of brown adipose tissue (BAT) as an energy source (discussed in Section 3). However, these mechanisms can require much energy and thus perpetuate the problem. Mechanisms include the presence of thermogenic cells, such as BAT; the presence of hair; thickness of the dermis; body-to-mass ratio; behaviors at birth; locomotor abilities; and organ development [46,47]. In farm animals, the characteristics mentioned above can differ markedly between altricial and precocial species [22]. In many aspects, thermoregulation in altricial species is often further complicated due to the absence or shortage of hair, high body surface-to-mass ratio, and scant adipose tissue [48]. In contrast, in both species, the fact that newborns are born wet with amniotic fluid makes them susceptible to evaporative heat losses [49].

In piglets, shivering is a compensatory mechanism often observed, but with high energy demand. As a result, piglets have increased myofibril mass, blood supply, and triad proliferation (fat intake, maturation of adipocytes from multi to unilocular, improved thermal homeostasis) to improve shivering efficiently [39,50]. The case of newborn piglets differs from most mammals; they have almost no BAT, and their thermoregulation highly depends on shivering to produce heat. In addition, the sow does not lick the piglets to remove the blood and amniotic liquids and placenta fragments, meaning that the piglets are wet during some hours and have more risk of hypothermia, a particularity of the species. That is why, in newborn piglets, various molecular, ultrastructural, biochemical, and physiological adjustments are involved in the maturation of the energy metabolism of the musculoskeletal system at birth [39]. Nonetheless, hypothermia can also affect precocial species, such as ruminants. Despite having a better degree of neurodevelopment at birth, hypothermia in ruminants is a factor occasionally associated with mortality in newborns due to the drastic temperature change when exposed to the extrauterine environment in the first hours after birth [51].

2.2. Hypoxia

All factors that reduce the oxygen supply passing from the mother to the fetus promote hypoxia, i.e., constriction of uterine blood vessels, placentitis, and umbilical cord occlusion, among others [3]. Perinatal decrease in oxygenation through the umbilical cord leads to fetus asphyxia due to inadequate respiratory gas exchange, triggering metabolic acidosis [52]. The severity of perinatal hypoxia can be assessed at birth using various indicators, such as blood pH, CO_2, and blood oxygen concentration, as well as various cardiopulmonary and neurological function tests [53,54]. Pre- and intrapartum uterine contractions can cause varying degrees of hypoxia. When suffering from perinatal stress, the body releases catecholamines, significantly impacting the activated metabolism, glycolysis, and hyperglycemia [36,54]. By increasing the metabolism, oxygen requirements will be higher [55]. Uncontrolled hypoxia can lead to asphyxia, and when this occurs intrapartum, the fetal lungs decrease the amount of surfactant, inducing respiratory acidosis [7]. In other words, when a fetus suffers from hypoxia (caused by umbilical cord or fetal head compressions [3]), there are some changes in fetal circulation, which increase intestinal peristalsis, causing relaxation of the tone of the anal sphincter, allowing meconium to pass into the amniotic sac [53]. In an attempt to survive, the hypoxic fetus initiates forceful respiratory movements with the glottis open, inhaling meconium-contaminated amniotic fluid, which is directed deep into the lungs. In turn, inhaled meconium produces obstruction and inflammation of the airways, translating into inadequate oxygenation and causing alveolar surfactant degradation [54], which is associated with asphyxia and pulmonary hypertension [55]. In a study by Bochenek et al. [56], piglet alterations during the hypoxic period demonstrated clinical signs consistent with encephalopathy. The clinical signs derived from peripartum hypoxia and pulmonary anomalies constitute Meconium Aspiration Syndrome (MAS). MAS defines the respiratory distress in newborns born during labor complicated by meconium-stained amniotic fluid [48]. According to Martínez-Burnes et al. [57], functional and structural repercussions of MAS include airway obstruction, atelectasis, chemical pneumonitis, hypoxemia, acidosis, pulmonary hypertension, and, in some cases, death (Figure 1) [58–62].

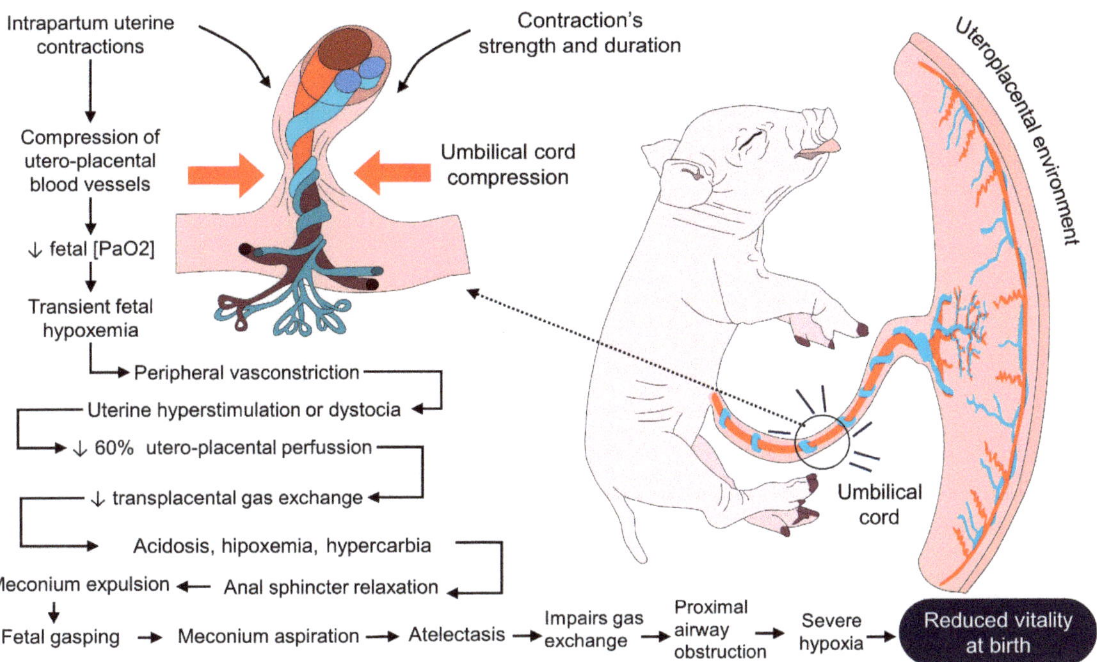

Figure 1. Hypoxia is a factor that reduces vitality. During intrapartum uterine contractions, utero-placental blood vessels and the umbilical cord are compressed. Due to this compression, partial fetal pressure of arterial oxygen (PaO_2) is reduced, causing transient hypoxemia in the fetus. When uterine contractions increase or uterine hyperstimulation occurs, around 60% of the uteroplacental blood circulation is reduced, affecting transplacental gas exchange and fetal oxygen supplementation. Along with hypoxemia, other physiological events, such as acidosis, hypercarbia, and meconium aspiration, can also be present, reducing a newborn's vitality.

On the other hand, anemia in newborn piglets is common and is more frequently observed in large litters [63,64]. This is important because hemoglobin serves as an important non-bicarbonate, essential for oxygen transport [65]. Then, the hypoxic events that the newborn faces can cause a decrease in its vigor since the lack of oxygen can lead to the presentation of MAS. Therefore, one of the first therapeutic lines could be to try to improve the availability of oxygen to reduce the impact of this phenomenon.

2.3. Birth Weight

Birth weight in any species can significantly predict newborn survival [66]. There is controversy on this issue, and some authors point out that low birth weight can be compensated in the weaning period [67], while others point out that low birth weight can cause growth difficulties [68].

Two biological factors could influence neonatal viability in farm animals: birth weight and nutritional status in the perinatal period. Diverse studies suggest that there is a negative association between low birth weight and survival during this period [39,69–71], and it has been demonstrated that a positive correlation exists between the ability to raise thermostability and a newborn's weight. Nonetheless, negative aspects can also be found in newborns with normal weight born from nulliparous females with narrow birth canals (Figure 2) [72–75].

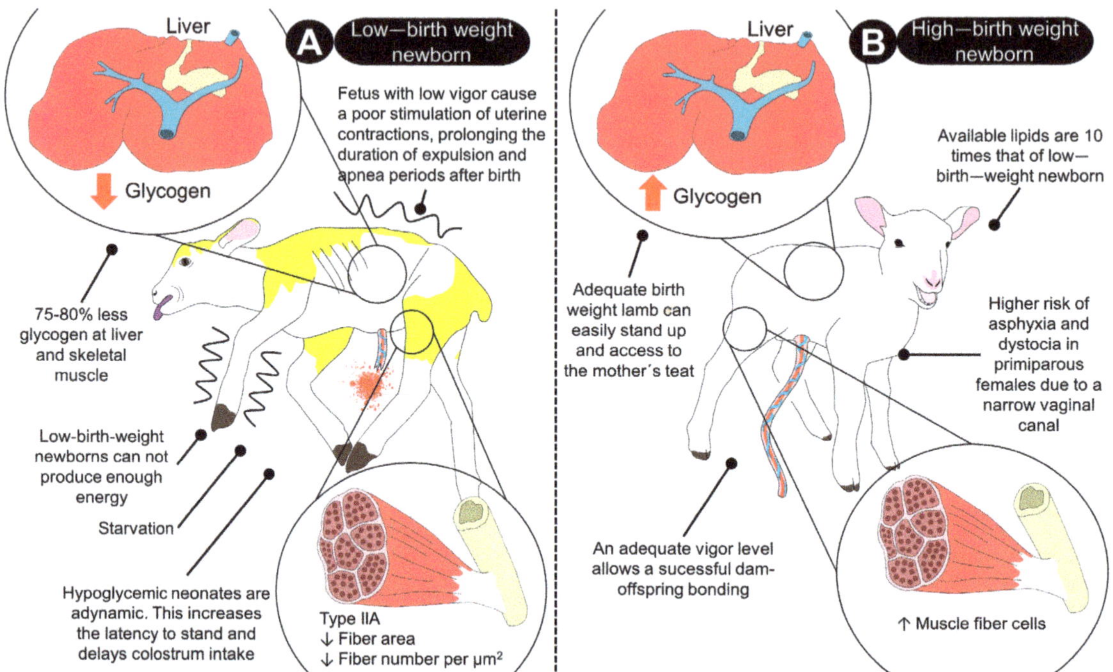

Figure 2. Importance of birth weight in the percentage of newborn survival. (**A**) In newborns with low birth weight, there are many limitations to being able to produce energy from their glycogen reserves in the liver and skeletal muscle because these are reduced between 75 and 80%. Cori's cycle cannot be activated correctly because muscle fibers are reduced per μm^2, and the lamb cannot get up and look for the mother's teat, so it may starve to death. Low birth weight is closely associated with fetal and neonatal morbidity, inhibited growth and cognitive development, and chronic diseases. During oxygen deprivation in the event of fetal hypoxia–ischemia, compensatory mechanisms are responsible for redistributing cardiac output, centralizing blood flow to vital organs, and reducing oxygen consumption. Additionally, an increase in peristalsis with expulsion and staining of meconium on the skin can be present, as illustrated in the lamb on the left side. (**B**) In contrast, high-birth-weight newborns have adequate vigor, can stand up immediately to consume colostrum, and have enough energy reserves. These reserves help to achieve thermoneutrality, together with the higher amount of muscle fibers, and enough hepatic glycogen reserves. In addition, the lipid reserves in lambs born with normal weights is $10\times$ the amount of available lipids in those that have low birth weight. Although normal-weight newborns have increased chances to survive, when the female has a narrow birth canal, these animals can experience dystocia and a prolonged expulsion period that might predispose them to meconium aspiration. Type IIA: muscular fibers type IIA.

In the same way, there have been studies carried out in sheep [76,77], pigs [36,78], and cows [79,80] that have shown how factors, such as the dam's weight and body condition, parity, newborn's weight at birth, breed and age of the mother [47,81], number of siblings in the litter, and maternal and offspring behavior, can influence the survival of the newborn [82]. It is often assumed that low birth weight in newborns is usually associated with prematurity, placental insufficiency (i.e., newborns, piglets, lambs), or maternal undernutrition (i.e., lambs) [83]. In piglets, one of the possible reasons for the lower viability and low birth weight can be immaturity [84].

According to Schmidek et al. [80], reduced birth weight during normal calving may also be associated with poor vitality. They suggest that up to 74% of calves with reduced vigor die before weaning. The availability of liver glycogen and skeletal muscle is decreased by 70–85% in newborns with low birth weight. In the same way, the amount of available lipids in low-birth-weight animals (i.e., pigs, lamb) is only about 10% of the amount available for those born with adequate weight, and these factors can affect the ability of the newborn to generate the energy necessary to stand up or access the mother's teat and suck colostrum [83].

2.4. Glycogen Depletion

In piglets, the risk of hypoglycemia is associated with sibling competition to get better or functional teats, resulting in high colostrum intake variability among piglets [55,84]. The colostrum requirement for newborn piglets is 180–250 g [85], but on some occasions, these requirements are not covered, especially in sows rearing large litters. Animals not reaching their colostrum intake requirements have to rely on using glycogen, which can be inappropriate for weak newborns [86]. Glycogen reserves are stored in the liver and the muscle and are usable by the newborn from 12 to 24 h and 24 to 36 h after birth, respectively [55]. In any case, the liver's glycogen reserves are depleted in most species from 12 h after birth. Then, the newborn depends on the energy supply obtained through colostrum or gluconeogenesis. However, it is important to mention that gluconeogenesis occurs in the newborn's liver simultaneously with the appearance of phosphoenolpyruvate carboxykinase after the rise of plasma glucagon and the fall of plasma insulin [87]. This synthesis of energy requires that activation of the Cori cycle take place, through which the production of lactate in muscle cells arises, its transport to the liver, its conversion into glucose by gluconeogenesis, and its return to the muscle to be converted back into lactate ensuring the functioning of the muscles in periods of great activity, such as shivering. The Cori cycle involves glucose consumption in muscle under anaerobic conditions, producing lactate from pyruvate and nicotinamide adenine dinucleotide (NADH) synthesized during glycolysis, with a considerable expenditure of energy at the liver [88]. The problem with activating this cycle is that lactic acidosis occurs in the muscles, which can decrease the blood's buffer system efficiency, leading to physical fatigue caused by oxygen debt [83]. Moreover, it is a cycle that costs 6 adenosine triphosphate (ATP) in the liver, and it is a cycle that cannot continue indefinitely [6,89].

By depleting glycogen reserves and thus generating hypoglycemia, the newborn is led to a decrease in heat production that causes hypothermia and compromises brain function, which can lead to a coma or even seizures and death [83]. The fact that the animal at birth has limited energy reserves could directly impact its vitality. Therefore, therapeutic strategies should be focused on promoting the vigor of the newborn.

2.5. Neurodevelopment

According to Mellor [90], newborn mammals can be classified according to their neurological maturity in (1) neurologically mature newborns (offspring of guinea pigs, primates, and ungulates, such as deer, cattle, horses, pigs, goats, and sheep); (2) neurologically moderately immature newborns (offspring of cats, dogs, ferrets, bears, rabbits, hamsters, mice, and rats); (3) neurologically exceptionally immature newborns of marsupials (wallabies, opossums, and kangaroos). Most farm animals are neurologically mature at birth, except rabbits, which are moderately mature at birth. In the neurologically mature group, the mother–young bond quickly establishes itself; in the moderately immature group, the female makes nests and burrows to protect her young; and in the exceptionally immature group, immediately after birth, the young hide in the pouch of the mother, where they remain safe and nursing [91,92].

Poor neurodevelopment in the newborn could possibly affect vitality because it could have a limited physiological response. In this sense, it has been reported that limited neurodevelopment in ruminants can decrease their thermoregulatory capacity and, therefore, the response to stimuli at extreme temperatures [93]. Thus, with limited responsiveness, it may fail to respond favorably to events, such as a drop in body temperature, by decreasing heat loss through changes in the superficial microvascular response due to hypothalamic activity, and the autonomic nervous system, which could affect vitality in the newborn [94]. Regarding their sensory capabilities, mature neurological newborns have the proprioception, musculoskeletal system, and vestibular function sufficiently developed to stand up immediately after birth; in the same way, a developed sense of smell and vision allow them to track and focus on objects [95,96]. Neurological moderately immature newborns have developed taste, smell, touch, proprioception, nociception, and thermal sensitivity; however, visual and hearing capacities are absent and not functional until cortical–subcortical connectivity is established at about 10–17 days after birth [97,98].

Intrauterine life gradually prepares the fetus for birth, and it develops the sense of smell and taste to recognize its mother's scent and easily find the udder to consume colostrum. In the same way, the fetus develops essential neuroplastic responses, presenting some outlines of learning and memory that can be useful in extrauterine life [90]. Given the evidence, it is clear that the degree of neurodevelopment directly correlates with the degree of vitality in the newborn animal. Therefore, it is necessary to resort to therapeutic strategies to promote vitality in the different species.

2.6. Dystocia

Previously, we have mentioned that some maternal causes could interfere with the fetus's and newborn's vitality (dystocia, maternal ability, the absence of milk, parity number, or clinical factors). We will focus on dystocia because these maternal factors directly influence the peripartum period and can cause a severe circulatory imbalance, producing low oxygenation in the fetus and newborn [5].

Dystocia compromises fetal circulation, causing intrauterine fetal hypoxia that triggers a redistribution of blood from less sensitive organs, such as the intestine, to vital structures, such as the brain and the heart. Reduced intestinal perfusion promotes peristalsis and reduces the muscle tone of the anal sphincter, resulting in the passage of meconium into the amniotic sac [7], thus causing the MAS mentioned above and its pathophysiology previously described in the hypoxia section [99,100].

In cows, the degree of calving stress depends on several factors, such as calf size, presentation, the dam's pelvic dimensions (correlated with age), the force of contractions, insufficient cervix dilation, and uterine torsion. These factors, alone or in combination, can trigger labor dystocia [101]. In the presence of dystocia, the cow may experience reduced milk production and/or uterine infections, leading to increased production costs and decreased fertility [79]. According to González-Lozano et al. [102,103], in swine production, dystocia increases the risk of piglets with prolonged latency to first udder contact, generating weakness and thereby decreasing their vitality scales.

Commonly, exogenous oxytocin is applied on the farm and domestic animals to reduce long or dystocia births. It has been seen that this could cause an increase in the number of stillbirths if it is not appropriately applied [62,104]. This is because myometrial contractions can cause a decrease in oxygen supply to the fetus, especially in polytocous species [12]; hence, induction protocols should be assessed to better suit newborns' developmental needs [50]. Under all the above, it can be concluded that dystocia, the age of the dam, and birth weight have the most significant impact on newborn vitality and, in the long term, on neonatal survival.

3. Physiological Mechanisms of the Newborn to Achieve Thermostability and Improve Vitality

One of the most adaptive changes the fetus makes to compensate for thermal variations in the perinatal period include the reorganization of the muscular circulation [29] of the hindlimbs and its return to the brain and heart and reduced movements of the limbs or body. After the onset of respiration, the increase in peripheral blood flow in response to the elevation of blood PO_2 releases lactic acid produced in the muscle by anaerobic metabolism. Therefore, transient lactacidemia after the onset of respiration can be expected in normal parturition [105]. Among the relevant hormones in the secretion of pulmonary fluid, we can mention catecholamines, thyroid hormones, and cortisol elevation [6].

3.1. Brown Adipose Tissue

Non-shivering thermogenesis is the production of heat without shivering. This occurs when the neonate's brown adipose tissue (BAT) is activated and produces heat by burning stored fat (unlike white adipose tissue, primarily used for energy storage). BAT gets its name from its dark color, which is caused by numerous mitochondria. These mitochondria contain a protein called thermogenin, also known as uncoupling protein 1 (UCP1), which uncouples the electron transport chain from ATP synthesis, allowing the energy emitted by burning fat to be released as heat instead of being used to produce ATP [106]. The type of species highly influences the degree of thermoregulatory neurodevelopment. Ruminants, classified as precocial, can maintain a constant body temperature during the early postnatal period, even in cold environments [107]. In these species, non-shivering thermogenesis is the most used mechanism by neonates. For example, in lambs (*Ovis aries*), about half of the cold-induced metabolic peak comes from non-shivering thermogenesis. Therefore, metabolic-active BAT during the early postnatal period is essential [108].

Thermogenesis by BAT is crucial in the neonate of most species and represents the first energy resource available during the postnatal period. In response to cold, sympathetic stimulation dramatically increases lipolysis and blood flow through the fat depots to provide direct heat [109].

The quantity of body fat in the newborn varies markedly between species. Precocial animals, such as lambs and calves, are born with well-developed BAT reserves that quickly atrophy and are replaced by white adipocytes shortly after birth [109,110]. In contrast, altricial species are born with zero or minimal amounts of BAT, whose recruitment increases when exposed to low temperatures during the first weeks of birth [109]. BAT replacement or atrophy after birth depends on the species: it disappears in rabbits, sheep/cattle, and goats within a month, from 2 to 3 days, and 2 to 6 days after birth, respectively [109]. BAT is high in rabbits, but there are small amounts in sheep, cattle, and horses, while it is almost non-existent in newborn pigs [105]. Nonetheless, adipose tissue represents only 2% of the body weight of livestock species, and is distributed in the prescapular, inguinal, and prerenal regions (Figure 3) [110–112]. However, in species with limited energy reserves, at birth (such as piglets), colostrum intake and other mechanisms to preserve heat, such as shivering, vasomotor control, and postural behavioral changes, are critical to prevent hypothermia [113]. That is why, in newborn piglets, various molecular, ultrastructural, biochemical, and physiological adjustments are involved in the maturation of the energy metabolism of the musculoskeletal system at birth [39]. Consequently, BAT thermogenesis is essential in the neonate of most species and represents the predominant resort to use during the postnatal period. Nevertheless, colostrum consumption is crucial in species with limited energy reserves to counteract hypothermia.

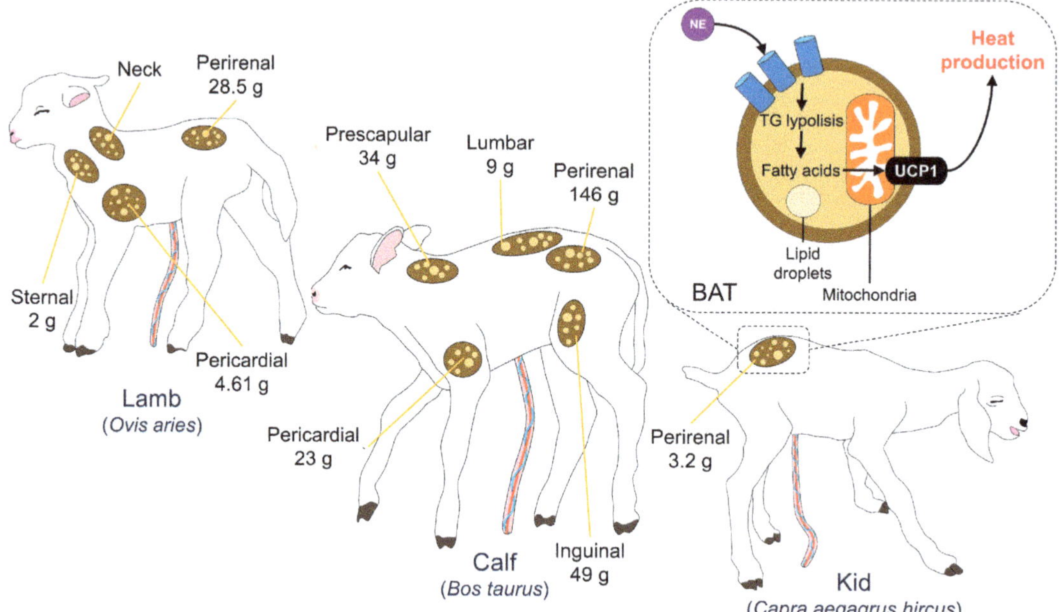

Figure 3. Amount and distribution of brown adipose tissue (BAT) in lambs, calves, and kid goats. According to the species, different levels of BAT are available at birth. For example, concerning the animal's weight at birth, newborn lambs have less pericardial BAT (4.61 g) than calves after birth (23 g). Similarly, perirenal adipose tissue is higher in calves (146 g vs. 28.5 g) and the only reported source in kids, which have 3.2 g of BAT. These values influence the ability of newborns to thermoregulate and prevent critical heat losses. NE: norepinephrine; TG: triglycerides; UCP1: uncoupling protein 1.

3.2. Shivering

Shivering is a thermogenic mechanism of repetitive and rapid skeletal muscle contractions when the body is exposed to cold weather or a fever. One of the main differences between immature and mature species is their morphological characteristics, such as the percentage of body fat, the surface–volume ratio, and the adenosine triphosphate (ATP) necessary to maintain contractions; this could influence the intensity of the shivering [114]. The body core temperature of mammals is regulated by the central nervous system, in which the hypothalamus's preoptic area (POA) plays a pivotal role [115]. Shivering is regulated by structures that connect the POA with the parabrachial nucleus, the dorsomedial hypothalamus, the raphe pallidus, and spinal cord motor neurons. Although it employs thermosensitive neurons in the POA and spinal cord, these can also be activated in metabolic thermogenesis through BAT [116].

The shivering process uses the oxidation of carbohydrates, lipids, and proteins from muscles' reserves and circulating blood [114]. It can increase oxygen consumption 20 times, increasing the aerobic activity of muscle fibers and leading to the oxidation of fatty acids [117].

Although shivering is the most efficient mechanism to produce heat and achieve thermal balance in precocial species, such as ruminants, during exposure to cold [118], it cannot be used as a primary thermogenesis method due to the immaturity of muscle tissue in ruminants [107], especially in lambs of the Merino breed [119]. Thermogenesis through shivering is an efficient method for compensating for hypothermia in the newborn, which helps to complement the metabolic response; however, it can quickly deplete energy reserves.

3.3. Vasomotor Control

Studies in a wide range of species show that the set-point of the major hormonal systems that mediate the stress response (including the autonomic nervous system and hypothalamic–pituitary–adrenal (HPA) axis) can be altered during early life [120]. Vasomotor changes depend on the activation of the sympathetic system [121]. In the case of cold stimuli, the participation of the HPA axis, the secretion of catecholamines (epinephrine and norepinephrine), and their action on the receptors located in the blood vessels generate a vasoconstriction effect [122,123]. The objective of vasoconstriction is to redirect the blood flow of the extremities or peripheral structures toward internal organs and vital centers, limiting heat loss through the skin [124–126].

3.4. Postural and Behavioral Changes

Many postural changes in newborns aim to avoid heat loss and/or provide additional warmth. For instance, it is observed in pigs that, to reduce heat loss at birth, they adopt positions such as snuggling with littermates [127], or they lay in the sternal position to reduce the contact surface with the ground and prevent heat loss [128]. Due to the direct effect of behavioral changes on the newborn's thermoregulatory capacity, they also impact the animal's vitality. For example, in ruminants, it is reported that the average time to stand up is 20–22 min, and to suckle later is 50 min. If this time is longer, the animal may have weakness due to insufficient energy consumption. This could be related to the lack of availability of energy resources, which would perpetuate the difficulty in standing up and suckling [129]. In a study of rabbits (*Oryctolagus cuniculus*) exposed to thermal challenge with cold, it was observed that they had a greater incidence of presenting the behavior of huddling, and lower levels of triglycerides and BAT was also found in them compared to control animals [130]. Although this demonstrates that changes in behavior and posture are compensation mechanisms against a factor that affects vitality, they could also be indirect indicators of energy resource levels. For this reason, once these signs of weakness have been recognized, it is necessary to adopt therapeutic resources to promote vitality in the newborn.

In altricial species, behaviors such as crowding, seeking warmer sites, and calling to the mother are more commonly observed [131]. In the case of the rabbits (*Oryctolagus cuniculus*), they frequently adopt behaviors such as snuggling, rooting, and climbing, in addition to maintaining close contact with the rest of the litter to achieve a better position within the nest; moreover, this ensures a source of heat and food [132]. Through these physiological and behavioral strategies, newborns can increase their vitality and thus achieve survival.

4. Therapies and Methods Applied in Neonates to Promote Vitality

4.1. Pharmacologic Therapies

4.1.1. Energetic Supplements: Dextrose and Colostrum

The vitality and thermoregulatory capacity of the newborn is mainly determined by its energetic reserves, which are needed for thermogenesis [51,128]. For this reason, one suggested therapy is using energy supplements that provide calories to the newborn to achieve thermal stability [22]. Low blood glucose availability is recognized as one of the risk factors for newborn mortality. For example, McCauley [133] suggests that piglets with low birth weight can have low vitality and greater susceptibility to hypothermia because these animals have fewer energy resources to compensate for their heat loss, which has also been verified in ruminants [134].

In ruminants, the intravenous (IV) administration of 5% dextrose solutions at a rate of 500 mg/kg^{-1} has been suggested as a therapy that reestablishes blood glucose levels when they are lower than 60 mg/dL [135]. In this sense, Eales et al. [136] reported that intraperitoneal administration of 10 mL/kg of 20% glucose and rewarming with air at 40 °C improved survival in lambs. This suggests that providing energy resources to the neonate could increase its vitality and thermoregulatory capacity. However, controversies have arisen with pharmacological treatments. Oral administration of 40% dextrose in lambs increased the chance of their survival 3 h after birth, but was not superior to the implementation of a physical warming technique [137]. There is controversy regarding whether it is more effective to treat hypothermia or hypoglycemia in a newborn with low vitality.

Precisely, Engelsman et al. [138] evaluated the effect on the rectal temperature (RT) and glucose levels of administering energy supplements (glucose or colostrum) in three sessions. They evaluated 88 piglets receiving subcutaneous glucose (50 mg/kg) or 20 mL of colostrum at 35 °C, or a combination of both. They found no difference in RT among the three treatments, but the combined use of glucose and colostrum resulted in higher blood glucose levels. This suggests that high-quality energy resources, like colostrum, are related to vitality. Previously, Dividich and Noblet [139,140] mentioned that promoting colostrum consumption is an alternative to synthetic pharmacological therapies because it increases energy reserves by at least 30% in piglets.

The increased availability of energy resources is not the only benefit of this treatment. According to Silva et al. [141], the thermoregulatory response of newborn Holstein calves is altered when fed different volumes of colostrum (10%, 15%, and 20% of their weight). They found that the animals that consumed 15% and 20% colostrum had a higher pre-scapular temperature than those that consumed 10%; in addition, the total leukocyte counts increased. Both benefits could be associated with increased vitality because the thermoregu-latory and immunological capacity would improve newborn survival. Colostrum contains different cell growth factors that promote differentiation, such as insulin-like growth factors (TGF1 and IGF-2), transforming growth factor b (TGF-b1 and TGF-b2), growth hormone, epidermal growth factor, and insulin [142,143]. Similarly, Muns et al. [144,145] observed an increase in IgG levels at days 4 and 5 of life in piglets orally supplemented with colostrum after birth.

Due to the presence of these elements, colostrum could reduce the incidence of diseases in newborns. Therefore, energy supplementation is a viable therapeutic alternative to increase vigor in weak animals with signs of hypothermia.

4.1.2. Caffeine

Caffeine is a methylxanthine antagonist of adenosine receptors in the Central Nervous System. Caffeine acts as a stimulant on physiological variables, such as the heart and respiratory rates (Figure 4) [20,21,146,147]. This drug has been used intravenously, orally, or subcutaneously at 20 mg/kg to promote vitality, increase oxygenation, and reduce metabolic changes due to perinatal asphyxia in farm animals [148].

Some studies have suggested that caffeine administration between 20 and 30 mg/kg 8 h after the birth of piglets with high weight and diagnosed with asphyxia presented a positive response, increasing body weight by 19% at weaning [149]. However, it may have the opposite effect when administered to animals with low weight and lower vitality, as a lower colostrum intake is usually accompanied by lower weight gain [150]. These authors suggested that caffeine as the only therapeutic intervention would have little effect, as was observed with the additional use of glucose. In addition, Robertson et al. [151] evaluated the effect of caffeine administration at 20 mg/kg to improve Merino lambs' vitality. They observed that the lamb mortality was similar in both the control and the treated group, suggesting a nonsignificant effect on vitality.

A possible method to improve caffeine's effect would be combining energy supple-ments, as suggested by Jarrat et al. [152], who evaluated the effect of caffeine and glucose supplementation in 398 piglets at birth. Oral administration was randomly proportioned with 30 mg of glucose, 30 mg of caffeine, and a combination. This treatment did not benefit animals with ideal or high weight, but improved growth between days 1 and 3 by 0.9 kg in animals with low birth weight. These results suggest that the caffeine administration increased the energy capacity so that the animal could remain stable during the critical period. However, the timing and concentration of caffeine might determine its effect. For example, a study evaluated the optimal concentration and duration of caffeine in Merino ewes at 120 and 140 days of gestation. In animals receiving 10 or 20 mg/kg of caffeine in the feed at 120 days of gestation and 20 mg/kg after 140 days of gestation, it was observed that the high administration of caffeine from day 120 allowed the lambs to present a higher rectal temperature, greater suckling attempts, and a longer sucking time [153]. This suggests that caffeine administration during late pregnancy would be a more viable option to guarantee its effectiveness in the newborn.

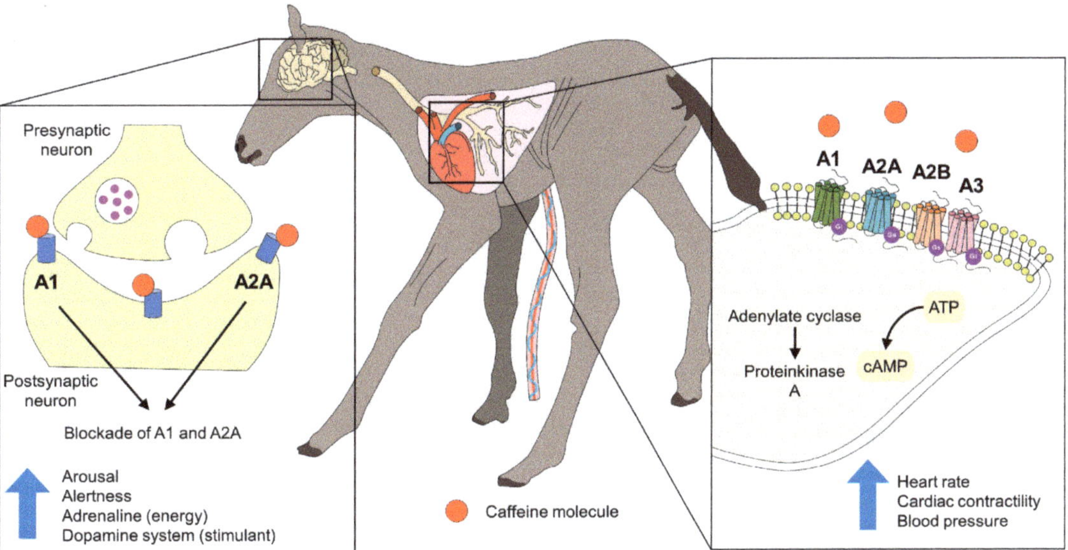

Figure 4. Cardiorespiratory and neural effects of caffeine. The mechanism of action of caffeine, binding to adenosine receptors (A1, A2A, A2B, and A3) in the cell membrane, causes cardiac responses that can improve a newborn's vitality by increasing heart rate, cardiac contractility, and blood pressure. Moreover, it acts on postsynaptic neurons and in the blockade of A1 and A2A. Binding these receptors increases arousal and alertness by accumulating adrenaline and activating the dopaminergic system. ATP: adenosine triphosphate; cAMP: cyclic adenosine monophosphate.

Contrarily, in premature lambs, it was observed that caffeine administration did not affect the carotid flow, heart rate, and oxygen saturation [154]. This is similar to what was reported by Menozzi et al. [155], who carried out a pharmacokinetic study of the administration of oral caffeine in sows and found that 24 h after its administration, plasmatic concentrations of 13.77 ± 0.97 mg/mL were shown.

The literature shows that caffeine administration can be used before birth to improve the vitality and respiratory capacity of newborns. However, more than administration of caffeine alone is needed to achieve this effect, and the combined use of energy supplements would help to achieve better results.

4.1.3. Naloxone

Naloxone is an opioid receptor antagonist that reduces the physiological effect of endogenous opioids, such as dynorphin, enkephalin, and endorphins, at the brain level [156]. The physiological effect of these substances is the decrease in the respiratory rate and tidal volume in the body because the expression of opioid receptors in the respiratory center of the cerebral cortex, thalamus, and baroreceptors in the carotid bodies has become evident [157]. The antagonism of these receptors likely decreases this effect and increases the ventilatory capacity. The therapeutic effect of naloxone has been previously described in newborn animals. For example, Hazinski et al. [158] evaluated the effect of naloxone administration at 4 mg/kg in 19 rabbit pups. They observed that the minute volume increased by 50 mL and the tidal volume increased by 8% in animals administered naloxone compared to the baseline event. In addition, blood gas analysis showed a 5 mmHg decrease in CO_2 levels and a 2 mmHg increase in pO_2. These authors concluded that endogenous opioids participate in respiratory control; therefore, their antagonism can positively intervene in respiratory activity. Interestingly, in a study carried out in paralyzed and vagotomized piglets, the administration of naloxone caused an increase in phrenic neural activity, which increased the respiratory output by $122 \pm 36\%$ and also the output per minute by $54 \pm 12\%$ [159]. This could demonstrate that naloxone has a beneficial effect on the respiratory capacity of the newborn with signs of low vitality or asphyxia. However, the results are limited in whether this could be related to improving thermoregulatory capacity and whether these effects can be prolonged.

4.1.4. Oxygen Therapy

Oxygen is essential for aerobic respiration, yet is potentially toxic, causing what is described as oxidative stress when there is an imbalance between the reactive oxygen species produced by aerobic respiration and the body's ability to detoxify these products. The capability to efficiently deliver sufficient oxygen to the tissues for optimal cellular function while minimizing oxidant-induced tissue damage has been achieved through complex physiological processes developed through evolution [160]. Oxygen therapy has been suggested as an alternative therapy to improve the vitality of newborn animals due to the characteristic of perinatal asphyxia caused by meconium aspiration [61]. For this reason, it has been suggested that the administration of 100% oxygen could counteract the effects of perinatal asphyxia and consequently improve vitality in the newborn [161]. In fact, due to asphyxia, cerebral hypoxia can also become evident due to the decrease in blood oxygen and local hemodynamic changes that would cause the demand for this available resource to be lower [162].

Intranasal oxygen support has been suggested as a form of treatment for perinatal asphyxia. For example, Bleul et al. [163] evaluated the effect of intranasal oxygen administration on blood gas variables in 20 newborn calves with respiratory distress syndrome. They found that this treatment significantly increased the partial pressure of oxygen (PaO_2) by 20.3 ± 8.8 mmHg and SaO_2 by 10% within the first 12 h after birth, which increased the survival rate (treatment group = 9/10 vs. control group = 4/10). This suggests that this simple treatment would be an option to improve blood variables and vitality in the newborn. In fact, in humans, it has been found that the implementation of this therapy, apart from improving the fraction of inspired oxygen, decreases acid–base changes and oxidative stress during neonatal resuscitation due to respiratory stress, so it could be considered a method to counteract the acidosis caused by this event [164].

The possible explanation for this process could be that during birth, there is decreased secretion of the surfactant substance, producing pulmonary atelectasis, which would hinder gas exchange and increase hypoxemia in the individual [165]. Therefore, for improving gas exchange in the newborn, it is necessary to consider that there are also processes of pulmonary hypertension and an increased airway [166]. Thus, a ventilation maneuver would increase the possibility of gas exchange due to the gas pressure that would overcome the increase in airway pressure.

Therefore, oxygen therapy helps increase the newborn's limited energy resources due to asphyxia during this process. In addition, it increases the ventilatory reserve that can be diminished by the aspirated meconium. However, it is necessary to understand that there may be an increase in airway pressure, which would hinder gas exchange, so ventilation maneuvers would help to increase this process.

4.2. Physical Methods

4.2.1. Colostrum: Natural and Artificial Supplementation

Providing good-quality colostrum to newborns is a practice that has improved in the last 20 years, reducing the incidence of insufficient colostrum intake in several species. For example, supplementation has reduced insufficient colostrum intake in dairy calves to 19% in dairy farms [167]. Passive transfer of immunoglobulins (Ig) through the colostrum and the methods to ensure an adequate immune transfer are relevant because they are related to the mortality rate and disease susceptibility [168,169].

Large animals (e.g., ruminants, foals, and piglets) are born agammaglobulinemic due to the placenta structure that cannot pass maternal Ig to the fetus. Therefore, passive immune transfer depends on the intake of Ig through colostrum [170–172]. Colostrum deficiency is associated with increased morbidity and mortality in newborns. In the case of dairy calves, if they consume less than 10 g/L of colostrum, they could have an increased susceptibility to infectious diseases [171].

It is important to ensure colostrum intake from the mother immediately after birth—or provide colostrum from another dam going into parturition or even frozen colostrum—because, for example, in foals, their colostrum absorption efficiency is reduced quickly, reaching only 1/5 of its efficiency at 3 h after birth [23]. In the case of Jersey calves, high-quality colostrum fed immediately after birth and after 12 h resulted in high concentrations of IgG (45.66 mg/mL) [169]. During the first hours of life, it is crucial to consume adequate amounts of colostrum, as reported in piglets, in which the serum IgG concentration affects the survival of animals. In contrast, newborns with less than 1000 mg/dL serum IgG had only a 67% chance of survival during the first 72 h after birth [173]. In the Bragg et al. [157] study, factors associated with the likelihood of having low IgG concentrations at birth in dairy calves were colostrum intake using artificial bottle/tube feeding systems and calving assistance. Therefore, because newborns are not usually considered immunocompetent until reaching weaning ages, at birth, they are exposed to increased mortality rates due to their lack of immunocompetence [174].

To enhance immunocompetence in newborns, an adequate immune transfer and amount of Ig are necessary to improve their vitality. Under this connection, supplying colostrum in the newborn could improve the immune capacity to face infectious events or those that require high energy consumption. There are different methods to provide colostrum, whether from the biological mother, frozen colostrum, other species, and commercial supplements. In Canary goat kids, the administration of refrigerated and frozen colostrum (5% of the kid's body weight) resulted in higher IgG concentrations (25.47 ± 19.89 mg/mL and 15.84 ± 5.91 mg/mL) at 24 h and 36 h post parturition, when compared to commercial sheep colostrum. Additionally, birth weight was related to IgG values, where kids weighing less than 2.5 kg had lower IgG concentrations [170]. In the same species, serum IgG concentrations and daily weight gain did not significantly differ between kid goats receiving natural colostrum and a cow supplement before ingesting the mother's colostrum. However, there was a mortality rate of 4% for the individuals in the supplement group, suggesting that supplementing the mother's colostrum does not benefit the newborn [175].

Contrarily, providing 10 g of IgG in 70 g of colostrum powder in 1-day-old Holstein calves resulted in less diarrhea (6.1% vs. 9.7%) and required fewer antimicrobial treatments than control animals. Although the mortality in the supplemented group was 7.7%, and 26.1% in the control animals, the authors reported no significant influence of colostrum supplementation on mortality and the incidence of respiratory diseases [176]. A study regarding colostrum in standard, thoroughbred, Arabian, and warm-blood foals reported no differences between normal foaling and dystocia. However, the serum IgG concentration at 24 and 48 h was lower in dystocia cases [177]. Therefore, knowing the parturition process in domestic animals is important when considering colostrum supplementation therapies. Supplementation in foals is recommended when the newborn is born from a mare with poor-quality colostrum (less than 20% BRIX). Although the amount of supplemented colostrum might vary, around 500 and 1000 mL is recommended [178]. One common practice is to provide colostrum replacers to newborns. However, in Holstein Friesian calves, neonates fed a replacer had lower body mass and lower blood neutrophils and monocytes (0.08 and 0.06 $\times$ 10^9/L, respectively), as well as lower total serum proteins (44.34 g/L) than calves consuming maternal bovine colostrum [179].

Another factor to consider when deciding to provide colostrum is parity. In the case of sows, it is an important factor that alters colostrum quality and IgG concentrations, which can be 5% higher in multiparous animals [173]. Hyper prolific sows with litters of up to 20 piglets prolong the farrowing and increase the competition for colostrum intake, 2 factors that affect the piglets' vitality and immunocompetence [180]. Similar information has been reported in Holstein cows, where colostrum from multiparous cattle had higher Ig concentrations (around 26%) and total proteins [172]. Maciag et al. [174] evaluated the effect of piglets consuming colostrum from gilts and sows and those bottle-fed with a commercial milk formula during the first 24 h after farrowing. Piglets' serum Ig concentration and lymphocyte proliferation were assessed at 24 h and at 20 days after birth. The authors found that piglets suckling natural colostrum had higher Ig concentrations (particularly from sows with approximately 103.27 $\pm$ 12.8 mg/mL of Ig concentrations) and greater ability to produce B and T cells. At the same time, animals fed the milk replacement presented diarrhea and had lower body weights and a higher mortality rate (8/14 piglets, 57%) [174]. Piglets allowed to suckle naturally, but supplemented with artificial colostrum twice a day resulted in body weight increases after week 1, daily weight, and higher IgG counts [181]. This suggests that letting the newborn suckle ad libitum from their mother and supplementing with artificial formulas provides performance benefits for the animals and, probably, improves their vitality. As mentioned by Uddin et al. [182], in a study with 140 piglets from 10 sows, animals with Apgar vitality scores close to 2—in a scale of 0 to 2 (average of 1.42 $\pm$ 0.07)—consumed higher amounts of colostrum ($p < 0.05$), were born earlier (order 1 to 5), and maintained higher body weight at birth and until weaning ages. It is important to provide these elements to piglets, since factors such as birth order and low colostrum intake are associated with asphyxiation during farrowing, a relevant issue that must be considered when trying to reduce piglet mortality in farms.

Besides the nutritional components of colostrum, in the case of foals, the known neonatal maladjustment syndrome is a disorder that can alter the suckling ability of the foal to consume colostrum. Authors such as Aleman et al. [183] reported that an alternative to address this issue and prevent the consequences of the disorder, namely, the "Madigan squeeze method" applied as pressure for 20 min, helps the newborn to regain alertness and be able to bond with the mare and stand up quickly to consume colostrum.

4.2.2. Temperature Drops Immediately after Birth—Sources of External Heat and Drying

Consuming colostrum immediately after birth is also linked to newborns' ability to maintain body temperature because it serves as a fuel for heat production and thermoregulation [77]. Assessing the rectal temperature (RT) in newborn animals is the gold standard to evaluate their thermal state and how the environment influences their body temperature. Piccione et al. [184] studied the maturation and temperature pattern during the first month of life in foals and lambs. The authors found that animals could maintain a constant high temperature within 10 days post-parturition. In lambs, RT increased from 38.1 °C after birth to 39.4 °C on the 24th day of life. In contrast, foals went from 37.4 °C to 38.3–38.7 °C at 25 days after foaling. In Scottish blackface and Suffolk lambs, RT was lower in animals with low birth weight and those Suffolk animals, showing a difference even between breeds. Animals with a longer latency to suckle (114 min) and ingest colostrum also had lower RT, which was maintained for the first 3 days after lambing [77]. Aleksiev et al. [185] also reported differences between breeds in Pleven black head, Bulgarian, and Bulgarian X East Friesian cross lambs. In all animals, during the first 12 h, there was a fall in RT; however, Bulgarian lambs recorded the lowest temperatures, which was significant at 1 h post-lambing (39.8 °C vs. 40–40.1 °C in the other breeds). In that study, twin lambs also had lower temperature values than single animals. In contrast, in 13 Maltese kid goats and 13 Comisana lambs, Giannetto et al. [186] reported that RT was higher in kids and twins than in singletons and could be due to the physiological maturation at birth. In both species, the RT decreased during the first 33 h post-birth.

Studies by Santiago et al. [11] in 1260 Yorkshire-Landrace x large white piglets determined that a low surface temperature (assessed with an infrared thermographic camera) at birth is associated with low vitality scores. In this study, animals with low vitality scores were related to lower birth weight and lower temperatures (around 1 to 2 °C less than high-vitality piglets at birth and drying). At 24 h after birth, animals born from sows with parities 6 to 7 had higher vitality scores and temperatures (1.5 °C). In this sense, thermal imaging has been proposed as a non-invasive method to assess the thermal state of newborn animals [187]. Labeur et al. [187] studied the influence of shearing on the amount of brown fat tissue deposits in newborn lambs. Through infrared thermography in the dorsum of lambs subjected to cold challenge, the authors reported that shorn lambs had a higher core temperature after 30 min of a cold challenge than control animals. In a preliminary study conducted on piglets by the authors, infrared thermography showed that low surface temperatures at farrowing were associated with low birth weights, birth order, and mortality. In Figure 5, piglets from the same litter show different hindlimb temperatures. Moreover, Figure 6 shows the hypothermia effect in newborn water buffaloes.

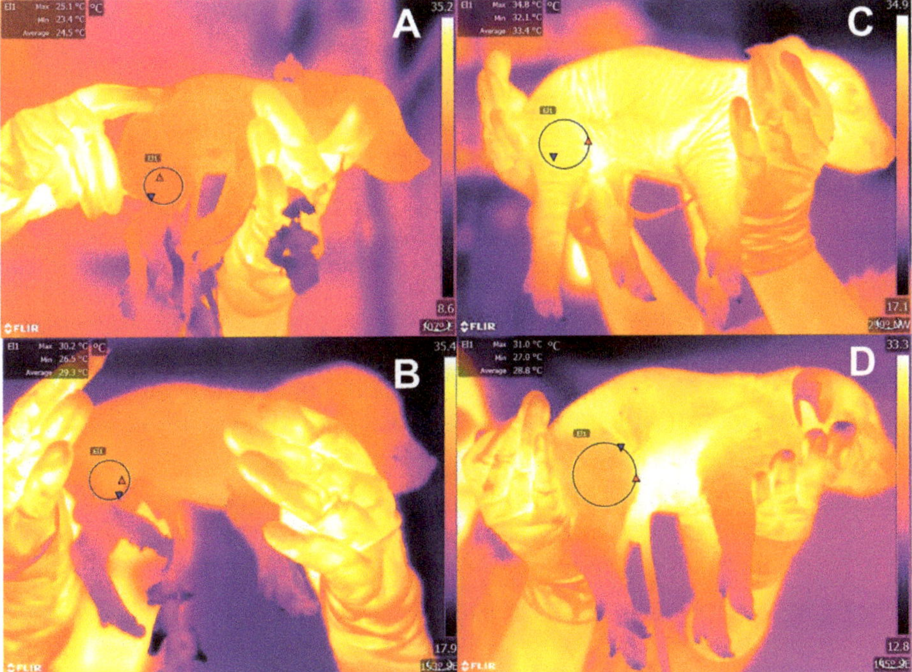

Figure 5. Thermal response of newborn piglets, evaluated by infrared thermography. Through infrared thermography and the delimitation of the region of interest in the pelvic limb (El1), factors such as weight and birth order can be assessed and associated with low birth temperatures, as shown in piglets from the same litter. (**A**) A piglet with a low birth weight (0.60 kg) and birth order eight had a minimum temperature of 23.4 °C. (**B**) A newborn weighing 0.62 kg and birth order five had a minimum temperature of 26.5 °C. In contrast, when comparing piglets from birth order one (**C**) and three (**D**) (weighing 1.58 and 1.63 kg, respectively), the minimum temperature of piglet (**C**) (32.1 °C) is 8.7 °C above piglet (**A**) and 5.6 °C above piglet (**B**). Minimum temperature of piglet (**D**) (27 °C) is maintained higher than (**A**,**B**). Red triangles mark the maximum temperature and blue triangles the minimum values in each thermal window.

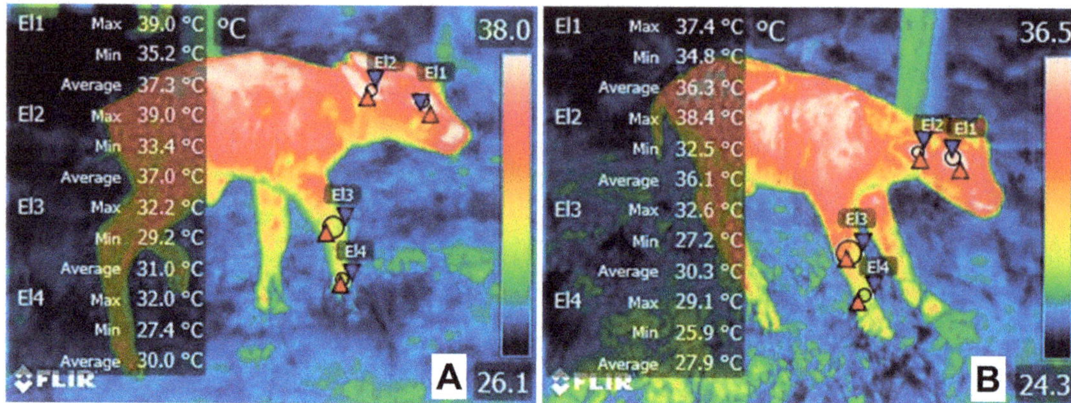

Figure 6. Comparison between central and peripheral thermal windows in newborn calves of water buffaloes with different thermal states. (**A**) Normothermic newborn. Central thermal windows, such as periocular (El1) and auricular (El2), have an average temperature of 37.3 °C and 37.0 °C, respectively. Peripheral regions considered in the elbow (El3) and metacarpal region (El4) show average values of 29.2 °C and 30.0 °C. (**B**) Hypothermic newborn. In contrast to buffalo neonates with thermostability, this animal shows lower temperatures in both central and peripheral temperature. The temperature was lower in the periocular window (El1) by 1 °C and in the auricular region (El2) by 0.9 °C. In peripheral windows at the elbow (El3) and metacarpal region (El4), the temperature dropped by 0.7 °C and 2 °C, respectively. The significantly marked difference between the central and peripheral windows shows that an animal with hypothermia at birth generates microcirculatory changes to preserve body heat in the brain. This leads to the activation of the Autonomic Nervous System, its sympathetic branch, and the neurosecretion of catecholamines that cause peripheral vasoconstriction to prevent further heat loss. This would explain the response observed in the radiometric images of newborn buffaloes with hypothermia. Red triangles mark the maximum temperature and blue triangles the minimum values in each thermal window.

In contrast, the evidence has led to questioning the benefit of using oxygen as a treatment for birth. Vande Pol et al. [188] evaluated the effect of drying and oxygen supply in 485 newborn piglets on rectal temperature during the first 24 h after birth. They observed that drying presented a greater benefit on temperature than using oxygen and drying together. Additionally, oxygen and drying did not affect the vitality or temperature of the piglets. This is possibly due to the obstruction by meconium aspiration, which intervenes in the process of oxygen exchange in the body [61]. An alternative to resolve this phenomenon has been suggested by allowing ventilatory assistance, as mentioned by Donnelly et al. [189], who have evaluated the effect of nasal oxygen insufflation and continuous positive airway pressure on pharmacologically induced respiratory suppression in 10 Holstein calves. They found that only insufflation with oxygen significantly increased pO_2 and carbon dioxide (CO_2), whereas oxygen administration with positive pressure reduced the presence of CO_2, although there was no increase in pO_2. This could indicate a possible improvement in the gas exchange phenomenon.

These results show that newborn mammals cannot completely thermoregulate during the first hours of life. Therefore, providing a suitable environment for the newborn at birth and during the first hours after parturition is essential to improve their vitality. For example, in piglets, although the farrowing houses are set at the recommended temperature for the sows (15–19 °C), this represents a challenge to the piglets that require an environmental temperature between 32 and 35 °C [190]. Moreover, although high temperatures are beneficial to newborn piglets, for sows, floor heating in periparturient sows with temperatures around 33 °C to 34 °C might cause a heat stress response [191]

A study analyzing the benefits of the creep area concerning mortality showed that enriching the area (e.g., insulated bedding and an additional wall to increase heat retention inside the creep) did not influence the mortality rate of animals (between 12.9 and 15.2% for all treatment groups) [192]. Milan et al. [190] evaluated the heat requirements for piglets when using heat lamps during the farrowing process by applying machine learning systems and the factors that can alter the quality of the supplemental heat. Increased air temperature (15–19 °C) and animal weight (between 1 and 2 kg) were two variables that altered the piglet heat requirement. Contrarily, heat lamps are only recommended for lambs while the animals are drying and should not be used after this [193].

Infrared thermography has helped to demonstrate that, apart from placing heat lamps, it is important to consider the lens, the shape, and the height of the lamps. For example, height affects the heated area and can increase/decrease hotspots, altering the net usable area for piglets to acquire their heat needs [194]. Automatic thermal control of heat lamps inside pig farms is being developed. These systems can be applied in newborns and piglets at different growth stages according to the already established temperatures to ensure thermal comfort. In another study, Pedersen et al. [195] applied seven thermal aids to evaluate their effects on reducing hypothermia in piglets on slatted or solid floors, and those with floor heating, a radiant heater from above, and provision of straw. When comparing the RT of the treatment piglets with control animals, all methods reduced the temperature loss and maintained piglets' RT above 35 °C most of the time when providing straw and the radiant heater. In this way, using external heat sources depends on the species, individual elements, and the method characteristics to ensure a benefit for the offspring.

Another factor directly associated with the previous techniques is the thermal challenge due to the exposure to the extrauterine environment and the evaporative heat loss that neonates might present because they are born wet by the amniotic fluid and fetal membranes. After birth, farm animals like piglets are susceptible to chilling, hindering their vitality scores and other measures, such as low birth weight, environmental temperature, housing facilities, or delay in colostrum intake [70]. In the same species, decreases of 3.7 °C in RT during the first 30 min have been reported [188]. It is important to denote that sows do not lick the piglets to remove amniotic fluid; therefore, they remain wet for some hours and have more risk of hypothermia. Physiologically, maternal behaviors, such as licking the newborn immediately after birth, stimulate the offspring to stand, ingest colostrum, and dry their coat, reducing heat loss by evaporation [196]. Drying the newborn is a practice that is recommended in some cases.

Andersen et al. [197] studied the effect of drying newborn piglets or drying and placing animals under a heat lamp. Both treatments reduced the mortality (approximately 6% in both groups vs. 12% in control animals). Additionally, drying and heat lamp treatment resulted in fewer crushed piglets by their mothers (13.6% vs. 47.9% in control groups). This article shows that providing this type of perinatal care reduces the causes of mortality in newborns. Similarly, using a desiccant, a warming box (35 °C), and combining both methods and a control group have been studied [188]. The authors found that animals in the desiccant and the warming box had similar temperatures. However, combining both methods resulted in the highest RT between 10 and 120 min after farrowing (from 37.6 to 38.6 °C, respectively, vs. 36.7 to 37.7 °C in the control group), having the greatest effect on low-birth-weight piglets. Those findings mean that combining both methods reduces piglet temperature decline.

Other studies have shown that drying piglets with a desiccant increased RT up to 2.4 °C from 25 to 180 min after birth, reducing the decline in RT in piglets with low and high weights [198]. Likewise, in another study, Vasdal et al. [199] found that piglets dried and placed at the udder reduced postnatal mortality (by 10%). Therefore, combined techniques, such as colostrum supplementation, providing an external heat source, and a physical approach, are recommended to increase vitality in newborn mammals.

5. Perspectives

Study perspectives about the therapy to promote vitality in the newborn are focused on pharmacological therapies, such as caffeine and naloxone. Due to their direct cardiorespiratory effect, they could help to increase the availability of oxygen or blood resources [147]. However, the evidence does not indicate whether these effects may have a direct or positive relationship with increased vitality or decreased perinatal mortality. Likewise, it would be necessary to explore other drugs, such as sildenafil or doxapram, which have been suggested to have a similar mechanism of action to caffeine and naloxone, but with the advantage of having a cardiovascular effect, which would counteract hypertension, acidemia, and hypoxia during the postpartum period [200,201].

6. Conclusions

Proper assessment of vitality is essential for neonatal survival in farm animals. Assessing vitality immediately after birth is essential to determine the offspring's health state and discover those requiring medical intervention to minimize the deleterious effect of intrapartum asphyxia.

Farms or veterinary clinics must train personnel and have adequate physical resources to assess the health and vitality of the newborn because a correct understanding of neonatal physiology is essential for interpreting vitality scores and detecting sick or weak newborns in need of corrective intervention. Vitality scores should reduce long-term neonatal morbidity and mortality in domestic animals. That is why Apgar scores in animals should be improved by expanding the use of neonatal reflexes, clinical examination, and blood gases. Biochemical and metabolic alterations explain the origin of cardiovascular and neurological abnormalities in newborns that survive intrapartum asphyxia. In addition, veterinarians must consider the type of delivery, the mother's age, birth order, and birth weight when assessing newborn vitality.

It is also essential to use technologies applied to vitality assessment to provide important clinical data on fetal or newborn distress. It could be helpful in all farm and domestic species because it is not an invasive technique, and any person could apply it with basic training.

Author Contributions: All authors contributed to the conceptualization, writing, and reading of the manuscript. All authors have read and agreed to the published version of the manuscript.

Funding: This research received no external funding.

Institutional Review Board Statement: Not applicable.

Data Availability Statement: Data sharing not applicable.

Conflicts of Interest: The authors declare no conflict of interest.

References

1. Farmer, C. *The Suckling and Weaned Piglet*; Wageningen Academic Publishers: Amsterdam, The Netherlands, 2020; 312p.
2. Mota-Rojas, D.; López, A.; Martínez-Burnes, J.; Muns, R.; Villanueva-García, D.; Mora-Medina, P.; González-Lozano, M.; Olmos-Hernández, A.; Ramírez-Necoechea, R. Is vitality assessment important in neonatal animals? *CAB Rev. Perspect. Agric. Vet. Sci. Nutr. Nat. Resour.* **2018**, *13*, 1–13. [CrossRef]
3. Mota-Rojas, D.; Martinez-Burnes, J.; Trujillo-Ortega, M.E.; Alonso-Spilsbury, M.L.; Ramirez-Necoechea, R.; Lopez, A. Effect of oxytocin treatment in sows on umbilical cord morphology, meconium staining, and neonatal mortality of piglets. *Am. J. Vet. Res.* **2002**, *63*, 1571–1574. [CrossRef] [PubMed]
4. Mota-Rojas, D.; Fierro, R.; Santiago, R.; Gonzalez-Lozano, M.; Martínez-Rodríguez, R.; García-Herrera, R.; Mora-Medina, P.; Flores-Peinado, S.; Sánchez, M. Outcomes of gestation length in relation to farrowing performance in sows and daily weight gain and metabolic profiles in piglets. *Anim. Prod. Sci.* **2015**, *55*, 93–100. [CrossRef]
5. Mota-Rojas, D.; López, A.; Muns, R.; Mainau, E.; Martínez, J. Piglet welfare. In *Bienestar Animal. Una Visión Global en Iberoamérica*; Mota-Rojas, D., Velarde, A., Huerta, S., Cajiao, M., Eds.; Elsevier: Barcelona, España, 2016; pp. 51–62.
6. Hillman, N.H.; Kallapur, S.G.; Jobe, A.H. Physiology of transition from intrauterine to extrauterine life. *Clin. Perinatol.* **2012**, *39*, 769–783. [CrossRef] [PubMed]

7. Alonso-Spilsbury, M.; Mota-Rojas, D.; Villanueva-García, D.; Martínez-Burnes, J.; Gregorio, O.; Ramírez-Necoechea, R.; Mayagoitia, A.L.; Trujillo, M.E. Perinatal asphyxia pathophysiology in pig and human: A review. *Anim. Reprod. Sci.* **2005**, *90*, 1–30. [CrossRef]

8. Cavaliere, T.A. From fetus to neonate: A sensational journey. *Newborn Infant Nurs. Rev.* **2016**, *16*, 43–47. [CrossRef]

9. Reyes-Sotelo, B.; Mota-Rojas, D.; Mora-Medina, P.; Ogi, A.; Mariti, C.; Olmos-Hernández, A.; Martínez-Burnes, J.; Hernández-Ávalos, I.; Sánchez-Millán, J.; Gazzano, A. Blood biomarker profile alterations in newborn canines: Effect of the mother's weight. *Animals* **2021**, *11*, 2307. [CrossRef]

10. Bovbjerg, M.L.; Dissanayake, M.V.; Cheyney, M.; Brown, J.; Snowden, J.M. Utility of the 5-minute apgar score as a research endpoint. *Am. J. Epidemiol.* **2019**, *188*, 1695–1704. [CrossRef]

11. Santiago, R.; Martínez-Burnes, J.; Mayagoitia, A.L.; Ramírez-Necoechea, R.; Mota-Rojas, D. Relationship of vitality and weight with the temperature of newborn piglets born to sows of different parity. *Livest. Sci.* **2019**, *220*, 26–31. [CrossRef]

12. Groppetti, D.; Pecile, A.; Del Carro, A.P.; Copley, K.; Minero, M.; Cremonesi, F. Evaluation of newborn canine viability by means of umbilical vein lactate measurement, apgar score and uterine tocodynamometry. *Theriogenology* **2010**, *74*, 1187–1196. [CrossRef]

13. Mota-Rojas, D.; Villanueva-García, D.; Hernández-González, R.; Martínez-Rodríguez, R.; Mora-Medina, P.; Sánchez-Hernández, M.; Trujillo, O.M. Assessment of the vitality of the newborn: An overview. *Sci. Res. Essays* **2012**, *7*, 869. [CrossRef]

14. Mota-Rojas, D.; Martinez-Burnes, J.; Villanueva-Garcia, D.; Santiago, P.; Trujillo-Ortega, M.; Gregorio, O.; Lopez-Mayagoitia, A. Animal welfare in the newborn piglet: A review. *Vet. Med.* **2012**, *57*, 338–349. [CrossRef]

15. Murray, C.F. Characteristics, Risk Factors and Management Programs for Vitality of Newborn Dairy Calves. Ph.D. Thesis, University of Guelp, Guelph, ON, Canada, 2014.

16. Flora, T.; Smallman, M.; Kutzler, M. Developing a modified Apgar scoring system for newborn lambs. *Theriogenology* **2020**, *157*, 321–326. [CrossRef]

17. Murray, C.F.; Leslie, K.E. Newborn calf vitality: Risk factors, characteristics, assessment, resulting outcomes and strategies for improvement. *Vet. J.* **2013**, *198*, 322–328. [CrossRef]

18. Barrier, A.C.; Ruelle, E.; Haskell, M.J.; Dwyer, C.M. Effect of a difficult calving on the vigour of the calf, the onset of maternal behaviour, and some behavioural indicators of pain in the dam. *Prev. Vet. Med.* **2012**, *103*, 248–256. [CrossRef]

19. Moon, P.F.; Massat, B.J.; Pascoe, P.J. Neonatal critical care. *Vet. Clin. N. Am. Small Anim. Pract.* **2001**, *31*, 343–367. [CrossRef]

20. Villanueva-García, D.; Mota-Rojas, D.; Miranda-Cortés, A.E.; Mora-Medina, P.; Hernández-Avalos, I.; Casas-Alvarado, A.; Olmos-Hernández, A.; Martínez-Burnes, J. Neurobehavioral and neuroprotector effects of caffeine in animal models. *J. Anim. Behav. Biometeorol.* **2020**, *8*, 298–307. [CrossRef]

21. Villanueva-García, D.; Mota-Rojas, D.; Miranda-Cortés, A.; Ibarra-Ríos, D.; Casas-Alvarado, A.; Mora-Medina, P.; Martínez-Burnes, J.; Olmos-Hernández, A.; Hernández-Avalos, I. Caffeine: Cardiorespiratory effects and tissue protection in animal models. *Exp. Anim.* **2021**, *70*, 431–439. [CrossRef]

22. Lezama-García, K.; Mota-Rojas, D.; Martínez-Burnes, J.; Villanueva-García, D.; Domínguez-Oliva, A.; Gómez-Prado, J.; Mora-Medina, P.; Casas-Alvarado, A.; Olmos-Hernández, A.; Soto, P.; et al. Strategies for hypothermia compensation in altricial and precocial newborn mammals and their monitoring by infrared thermography. *Vet. Sci.* **2022**, *9*, 246. [CrossRef]

23. Watt, B.; Wright, B. The Importance of Colostrum to Foals. Colostrum and Passive Transfer Assessment. Available online: https://www.equineguelph.ca/pdf/facts/ImportanceofColostrumtoFoalsApr_08.pdf (accessed on 13 March 2023).

24. Ramirez, B.C.; Hayes, M.D.; Condotta, I.C.F.S.; Leonard, S.M. Impact of housing environment and management on pre-/post-weaning piglet productivity. *J. Anim. Sci.* **2022**, *100*, skac142. [CrossRef]

25. Knol, E.F.; van der Spek, D.; Zak, L.J. Genetic aspects of piglet survival and related traits: A review. *J. Anim. Sci.* **2022**, *100*, skac190. [CrossRef]

26. Nel, C.L.; Cloete, S.W.P.; Kruger, A.C.M.; Dzama, K. Long term genetic selection for reproductive success affects neonatal lamb vitality across cold stress conditions. *J. Therm. Biol.* **2021**, *98*, 102908. [CrossRef]

27. Canario, L.; Bidanel, J.-P.; Rydhmer, L. Genetic trends in maternal and neonatal behaviors and their association with perinatal survival in French Large White swine. *Front. Genet.* **2014**, *5*, 410. [CrossRef]

28. Kapell, D.N.R.G.; Ashworth, C.J.; Knap, P.W.; Roehe, R. Genetic parameters for piglet survival, litter size and birth weight or its variation within litter in sire and dam lines using Bayesian analysis. *Livest. Sci.* **2011**, *135*, 215–224. [CrossRef]

29. Swanson, J.R.; Sinkin, R.A. Transition from fetus to newborn. *Pediatr. Clin. N. Am.* **2015**, *62*, 329–343. [CrossRef]

30. Nakamura, K.; Morrison, S.F. Central efferent pathways for cold-defensive and febrile shivering. *J. Physiol.* **2011**, *589*, 3641–3658. [CrossRef]

31. Mellor, D.J.; Stafford, K.J. Animal welfare implications of neonatal mortality and morbidity in farm animals. *Vet. J.* **2004**, *168*, 118–133. [CrossRef]

32. Nakamura, K. Central circuitries for body temperature regulation and fever. *Am. J. Physiol. Integr. Comp. Physiol.* **2011**, *301*, R1207–R1228. [CrossRef]

33. Vande Pol, K.D.; Tolosa, A.F.; Shull, C.M.; Brown, C.B.; Alencar, S.A.S.; Ellis, M. Effect of drying and warming piglets at birth on preweaning mortality. *Transl. Anim. Sci.* **2021**, *5*, txab016. [CrossRef]

34. Lawler, D. Neonatal and pediatric care of the puppy and kitten. *Theriogenology* **2008**, *70*, 384–392. [CrossRef]

35. Schild, S.-L.A.; Foldager, L.; Rangstrup-Christensen, L.; Pedersen, L.J. Characteristics of piglets born by two highly prolific sow hybrids. *Front. Vet. Sci.* **2020**, *7*, 355. [CrossRef] [PubMed]

36. Tuchscherer, M.; Puppe, B.; Tuchscherer, A.; Tiemann, U. Early identification of neonates at risk: Traits of newborn piglets with respect to survival. *Theriogenology* **2000**, *54*, 371–388. [CrossRef] [PubMed]

37. Pedersen, L.J.; Berg, P.; Jørgensen, G.; Andersen, I.L. Neonatal piglet traits of importance for survival in crates and indoor pens. *J. Anim. Sci.* **2011**, *89*, 1207–1218. [CrossRef] [PubMed]

38. Farmer, C.; Edwards, S.A. Review: Improving the performance of neonatal piglets. *Animal* **2022**, *16*, 100350. [CrossRef] [PubMed]

39. Herpin, P.; Damon, M.; Le Dividich, J. Development of thermoregulation and neonatal survival in pigs. *Livest. Prod. Sci.* **2002**, *78*, 25–45. [CrossRef]

40. Villanueva-García, D.; Mota-Rojas, D.; Martínez-Burnes, J.; Olmos-Hernández, A.; Mora-Medina, P.; Salmerón, C.; Gómez, J.; Boscato, L.; Gutiérrez-Pérez, O.; Cruz, V.; et al. Hypothermia in newly born piglets: Mechanisms of thermoregulation and pathophysiology of death. *J. Anim. Behav. Biometeorol.* **2021**, *9*, 2101. [CrossRef]

41. Ahmed, B.M.S.; Younas, U.; Asar, T.O.; Dikmen, S.; Hansen, P.J.; Dahl, G.E. Cows exposed to heat stress during fetal life exhibit improved thermal tolerance. *J. Anim. Sci.* **2017**, *95*, 3497–3503. [CrossRef] [PubMed]

42. Monteiro, A.; Tao, S.; Thompson, I.; Dahl, G. In utero heat stress decreases calf survival and performance through the first lactation. *J. Dairy Sci.* **2016**, *99*, 8443–8450. [CrossRef] [PubMed]

43. Roland, L.; Drillich, M.; Klein-Jöbstl, D.; Iwersen, M. Invited review: Influence of climatic conditions on the development, performance, and health of calves. *J. Dairy Sci.* **2016**, *99*, 2438–2452. [CrossRef] [PubMed]

44. Travain, T.; Colombo, E.S.; Heinzl, E.; Bellucci, D.; Prato Previde, E.; Valsecchi, P. Hot dogs: Thermography in the assessment of stress in dogs (*Canis familiaris*)—A pilot study. *J. Vet. Behav.* **2015**, *10*, 17–23. [CrossRef]

45. Bouwknecht, J.A.; Olivier, B.; Paylor, R.E. The stress-induced hyperthermia paradigm as a physiological animal model for anxiety: A review of pharmacological and genetic studies in the mouse. *Neurosci. Biobehav. Rev.* **2007**, *31*, 41–59. [CrossRef] [PubMed]

46. Pacheco-Cobos, L.; Rosetti, M.; Distel, H.; Hudson, R. To stay or not to stay: The contribution of tactile and thermal cues to coming to rest in newborn rabbits. *J. Comp. Physiol. A* **2003**, *189*, 383–389. [CrossRef] [PubMed]

47. Lezama-García, K.; Martínez-Burnes, J.; Pérez-Jiménez, J.C.; Domínguez-Oliva, A.; Mora-Medina, P.; Olmos-Hernández, A.; Hernández-Ávalos, I.; Mota-Rojas, D. Relation between the dam's weight on superficial temperature of her puppies at different stages of the post-partum. *Vet. Sci.* **2022**, *9*, 673. [CrossRef] [PubMed]

48. Noblet, J.; Le Dividich, J. Energy metabolism of the newborn pig during the first 24 h of life. *Neonatology* **1981**, *40*, 175–182. [CrossRef] [PubMed]

49. Fyda, T.J.; Spencer, C.; Jastroch, M.; Gaudry, M. Disruption of thermogenic *UCP1* predated the divergence of pigs and peccaries. *J. Exp. Biol.* **2020**, *223*, jeb223974. [CrossRef] [PubMed]

50. Tucker, B.S.; Craig, J.R.; Morrison, R.S.; Smits, R.J.; Kirkwood, R.N. Piglet viability: A review of identification and pre-weaning management strategies. *Animals* **2021**, *11*, 2902. [CrossRef]

51. Mota-Rojas, D.; Wang, D.-H.; Gonçalves-Titto, C.; Martinez-Burnes, J.; Villanueva-García, D.; Lezama-García, K.; Domínguez-Oliva, A.; Avalos, I.H.; Mora-Medina, P.; Verduzco-Mendoza, A.; et al. Neonatal infrared thermography images in the hypothermic ruminant model: Anatomical-morphological-physiological aspects and mechanisms for thermoregulation. *Front. Vet. Sci.* **2022**, *9*, 963205. [CrossRef]

52. Kredatusova, G.; Hajurka, J.; Szakallova, I.; Valencakova, A.; Vojtek, B. Physiological events during parturition and possibilities for improving puppy survival: A review. *Vet. Med.* **2011**, *56*, 589–594. [CrossRef]

53. Randall, G.C.B. Changes in fetal and maternal blood at the end of pregnancy and during parturition in the pig. *Res. Vet. Sci.* **1982**, *32*, 278–282. [CrossRef]

54. Mota-Rojas, D.; Villanueva, D.; Suárez, X.; Hernandez, R.; Santiago, R.; Trujillo, M. Foetal and neonatal energy metabolism in pigs and humans: A review. *Vet. Med.* **2011**, *56*, 215–225. [CrossRef]

55. Diehl, B.; Oster, M.; Vernunft, A.; Wimmers, K.; Bostedt, H. Intrinsic challenges of neonatal adaptation in swine. *Arch. Anim. Breed.* **2022**, *65*, 427–438. [CrossRef] [PubMed]

56. Bochenek, L.d.M.S.; Parisotto, E.B.; Salomão, E.d.A.; Maldonado, M.J.M.; Silva, I.S. Characterization of oxidative stress in animal model of neonatal hypoxia. *Acta Cir. Bras.* **2021**, *36*, e361108. [CrossRef] [PubMed]

57. Martínez-Burnes, J.; Mota- Rojas, D.; Villanueva- García, D.; Ibarra- Ríos, D.; Lezama- Garcia, K.; Barrios- Garcia, H.; López-Mayagoitia, A. Meconium aspiration syndrome in mammals. *CAB Rev.* **2019**, *14*, 1–11. [CrossRef]

58. Xodo, S.; Londero, A.P. Is It Time to Redefine Fetal Decelerations in Cardiotocography? *J. Pers. Med.* **2022**, *12*, 1552. [CrossRef] [PubMed]

59. Lear, C.A.; Wassink, G.; Westgate, J.A.; Nijhuis, J.G.; Ugwumadu, A.; Galinsky, R.; Bennet, L.; Gunn, A.J. The peripheral chemoreflex: Indefatigable guardian of fetal physiological adaptation to labour. *J. Physiol.* **2018**, *596*, 5611–5623. [CrossRef] [PubMed]

60. Turner, J.M.; Mitchell, M.D.; Kumar, S.S. The physiology of intrapartum fetal compromise at term. *Am. J. Obstet. Gynecol.* **2020**, *222*, 17–26. [CrossRef] [PubMed]

61. Mota-Rojas, D.; Villanueva-García, D.; Mota-Reyes, A.; Orihuela, A.; Hernández-Ávalos, I.; Domínguez-Oliva, A.; Casas-Alvarado, A.; Flores-Padilla, K.; Jacome-Romero, J.; Martínez-Burnes, J. Meconium Aspiration Syndrome in Animal Models: Inflammatory Process, Apoptosis, and Surfactant Inactivation. *Animals* **2022**, *12*, 3310. [CrossRef]

62. Marcet-Rius, M.; Bienboire-Frosini, C.; Lezama-García, K.; Domínguez-Oliva, A.; Olmos-Hernández, A.; Mora-Medina, P.; Hernández-Ávalos, I.; Casas-Alvarado, A.; Gazzano, A. Clinical Experiences and Mechanism of Action with the Use of Oxytocin Injection at Parturition in Domestic Animals: Effect on the Myometrium and Fetuses. *Animals* **2023**, *13*, 768. [CrossRef]

63. Steinhardt, M.; Bünger, U.; Furcht, G. Zum Eisenbedarf des Schweines in den ersten 2 Lebensmonaten. *Arch. Exp. Vet. Med.* **1984**, *29*, 104–111.

64. Szudzik, M.; Starzyński, R.; Jończy, A.; Mazgaj, R.; Lenartowicz, M.; Lipiński, P. Iron supplementation in suckling piglets: An ostensibly easy therapy of neonatal iron deficiency anemia. *Pharmaceuticals* **2018**, *11*, 128. [CrossRef]

65. Bünger, B.; Bünger, U.; Lemke, E. Verhaltensbiologische Vitalitätseinschätzung von Ferkeln mit hoch- und mittelgradiger konnataler Eisenmangelanämie. *Monatsh. Vet.* **1988**, *43*, 583–587.

66. Gondret, F.; Lefaucheur, L.; Louveau, I.; Lebret, B.; Pichodo, X.; Le Cozler, Y. Influence of piglet birth weight on postnatal growth performance, tissue lipogenic capacity and muscle histological traits at market weight. *Livest. Prod. Sci.* **2005**, *93*, 137–146. [CrossRef]

67. Beaulieu, A.D.; Aalhus, J.L.; Williams, N.H.; Patience, J.F. Impact of piglet birth weight, birth order, and litter size on subsequent growth performance, carcass quality, muscle composition, and eating quality of pork. *J. Anim. Sci.* **2010**, *88*, 2767–2778. [CrossRef] [PubMed]

68. Baschat, A.; Galan, H. Fetal growth restriction. In *Gabbe's Obstretrics: Normal and Problem Pregnancies*; Landon, M., Galan, H., Jauniaux, E., Driscoll, D., Berghella, V., Grobman, W., Kilpatrock, S., Cahil, A., Eds.; Elsevier: Philadelphi, PA, USA, 2020; pp. 555–585.

69. Mota-Rojas, D.; Alonso-Spilsbury, M.L.; Ramírez-Necoechea, R.; Moles y Cervantes, P.; González-Lozano, M. Involved factors in the immune response of neonatal pigs. In *Animal Perinatology: Clinical and Experimental Approaches*; Mota-Rojas, D., Nava-Ocampo, A.A., Villanueva-García, D., Alonso-Spilbury, M.L., Eds.; BM Editores: Mexico City, Mexico, 2008; pp. 443–454.

70. Muns, R.; Nuntapaitoon, M.; Tummaruk, P. Non-infectious causes of pre-weaning mortality in piglets. *Livest. Sci.* **2016**, *184*, 46–57. [CrossRef]

71. England, D.C. Husbandry components in prenatal and perinatal development in swine. *J. Anim. Sci.* **1974**, *38*, 1045–1049. [CrossRef]

72. Vermorel, M.; Dardillat, C.; Vernet, J.; Saido; Dardillat, C.; Demigne, C. Energy metabolism and thermoregulation in the newborn calf. *Ann. Rech. Vet.* **1983**, *14*, 382–389.

73. Mota-Rojas, D.; Rosales, A.M.; Trujillo, M.E.; Gregorio, O.; Ramírez, R.; Alonso-Spilbury, M. The effects of vetrabutin chlorhydrate and oxytocin on stillbirth rate and asphyxia in swine. *Theriogenology* **2005**, *64*, 1889–1897. [CrossRef]

74. Şirin, E.; Aksoy, Y.; Şen, U.; Ulutaş, Z.; Kuran, M. Effect of lamb birth weight on fiber number and type of semitendinosus muscle. *Anadolu Tarım Bilim. Derg.* **2011**, *26*, 63–67.

75. Greenwood, P.L.; Hunt, A.S.; Hermanson, J.W.; Bell, A.W. Effects of birth weight and postnatal nutrition on neonatal sheep: II. Skeletal muscle growth and development. *J. Anim. Sci.* **2000**, *78*, 50. [CrossRef]

76. Gama, L.T.; Dickerson, G.E.; Young, L.D.; Leymaster, K.A. Effects of breed, heterosis, age of dam, litter size, and birth weight on lamb mortality1. *J. Anim. Sci.* **1991**, *69*, 2727–2743. [CrossRef]

77. Dwyer, C.M.; Morgan, C.A. Maintenance of body temperature in the neonatal lamb: Effects of breed, birth weight, and litter size1. *J. Anim. Sci.* **2006**, *84*, 1093–1101. [CrossRef] [PubMed]

78. Kirkden, R.D.; Broom, D.M.; Andersen, I.L. INVITED REVIEW: Piglet mortality: Management solutions. *J. Anim. Sci.* **2013**, *91*, 3361–3389. [CrossRef] [PubMed]

79. Johanson, J.M.; Berger, P.J. Birth weight as a predictor of calving ease and perinatal mortality in holstein cattle. *J. Dairy Sci.* **2003**, *86*, 3745–3755. [CrossRef] [PubMed]

80. Schmidek, A.; da Costa, M.J.R.P.; Mercadante, M.E.Z.; de Toledo, L.M.; Cyrillo, J.N.d.S.G.; Branco, R.H. Genetic and non-genetic effects on calf vigor at birth and preweaning mortality in Nellore calves. *Rev. Bras. Zootec.* **2013**, *42*, 421–427. [CrossRef]

81. Lezama-García, K.; Martínez-Burnes, J.; Marcet-Rius, M.; Gazzano, A.; Olmos-Hernández, A.; Mora-Medina, P.; Domínguez-Oliva, A.; Pereira, A.M.F.; Hernández-Ávalos, I.; Baqueiro-Espinosa, U.; et al. Is the weight of the newborn puppy related to its thermal balance? *Animals* **2022**, *12*, 3536. [CrossRef] [PubMed]

82. Darwish, R.A.; Abou-Ismail, U.A.; El-Kholya, S.Z. Differences in post-parturient behaviour, lamb performance and survival rate between purebred Egyptian Rahmani and its crossbred Finnish ewes. *Small Rumin. Res.* **2010**, *89*, 57–61. [CrossRef]

83. Mellor, D.J.; Cockburn, F. A comparison of energy metabolism in the new-born infant, piglet and lamb. *Q. J. Exp. Physiol.* **1986**, *71*, 361–379. [CrossRef]

84. Devillers, N.; Farmer, C.; Le Dividich, J.; Prunier, A. Variability of colostrum yield and colostrum intake in pigs. *Animal* **2007**, *1*, 1033–1041. [CrossRef]

85. Quesnel, H.; Farmer, C.; Devillers, N. Colostrum intake: Influence on piglet performance and factors of variation. *Livest. Sci.* **2012**, *146*, 105–114. [CrossRef]

86. Szymeczko, R.; Kapelański, W.; Piotrowska, A.; Dybała, J.; Bogusławska-Tryk, M.; Burlikowska, K.; Hertig, I.; Sassek, M.; Pruszyńska-Oszmałek, E.; Maćkowiak, P. Changes in the content of major proteins and selected hormones in the blood serum of piglets during the early postnatal period. *Folia Biol.* **2008**, *57*, 97–103. [CrossRef]

87. Girard, J. Metabolic adaptations to change of nutrition at birth. *Neonatology* **1990**, *58*, 3–15. [CrossRef] [PubMed]

88. Parada, P.R. Cori Cycle. Available online: https://www.lifeder.com/ciclo-de-cori/ (accessed on 13 March 2023).

89. Ophardt, E.C. Cori Cycle. Available online: http://www.elmhurst.edu/~chm/vchembook/615coricycle.html (accessed on 13 March 2023).
90. Mellor, D. Preparing for Life After Birth: Introducing the concepts of intrauterine and extrauterine sensory entrainment in mammalian young. *Animals* **2019**, *9*, 826. [CrossRef] [PubMed]
91. Dwyer, C.M.; Lawrence, A.B. A review of the behavioural and physiological adaptations of hill and lowland breeds of sheep that favour lamb survival. *Appl. Anim. Behav. Sci.* **2005**, *92*, 235–260. [CrossRef]
92. Mellor, D.; Lentle, R. Survival implications of the development of behavioural responsiveness and awareness in different groups of mammalian young. *N. Z. Vet. J.* **2015**, *63*, 131–140. [CrossRef]
93. Mota-Rojas, D.; Martínez-Burnes, J.; Casas-Alvarado, A.; Gómez-Prado, J.; Hernández-Avalos, I.; Domínguez-Oliva, A.; Lezama-García, K.; Jacome-Romero, J.; Rodríguez-González, D.; Pereira, A.M. Clinical usefulness of infrared thermography to detect sick animals: Frequent and current. *CABI Rev.* **2022**, *17*, 1–17. [CrossRef]
94. Gómez-Prado, J.; Pereira, A.M.F.; Wang, D.; Villanueva-García, D.; Domínguez-Oliva, A.; Mora-Medina, P.; Hernández-Avalos, I.; Martínez-Burnes, J.; Casas-Alvarado, A.; Olmos-Hernández, A.; et al. Thermoregulation mechanisms and perspectives for validating thermal windows in pigs with hypothermia and hyperthermia: An overview. *Front. Vet. Sci.* **2022**, *9*, 1023294. [CrossRef]
95. Larson, M.A.; Stein, B.E. The use of tactile and olfactory cues in neonatal orientation and localization of the nipple. *Dev. Psychobiol.* **1984**, *17*, 423–436. [CrossRef]
96. André, V.; Henry, S.; Lemasson, A.; Hausberger, M.; Durier, V. The human newborn's umwelt: Unexplored pathways and perspectives. *Psychon. Bull. Rev.* **2018**, *25*, 350–369. [CrossRef]
97. Fitzgerald, M. The development of nociceptive circuits. *Nat. Rev. Neurosci.* **2005**, *6*, 507–520. [CrossRef]
98. Serra, J.; Nowak, R. Olfactory preference for own mother and litter in 1-day-old rabbits and its impairment by thermotaxis. *Dev. Psychobiol.* **2008**, *50*, 542–553. [CrossRef]
99. Lopez, A.; Bildfell, R. Pulmonary inflammation associated with aspirated meconium and epithelial cells in calves. *Vet. Pathol.* **1992**, *29*, 104–111. [CrossRef] [PubMed]
100. Haakonsen Lindenskov, P.H.; Castellheim, A.; Saugstad, O.D.; Mollnes, T.E. Meconium aspiration syndrome: Possible pathophysiological mechanisms and future potential therapies. *Neonatology* **2015**, *107*, 225–230. [CrossRef] [PubMed]
101. Mee, J.F. Newborn dairy calf management. *Vet. Clin. N. Am. Food Anim. Pract.* **2008**, *24*, 1–17. [CrossRef] [PubMed]
102. González-Lozano, M.; Trujillo-Ortega, M.E.; Becerrill- Herrera, M.; Alonso-Spilsbury, M.L.; Ramírez-Necoechea, R.; Hernández-González, R.; Mota-Rojas, D. Effects of oxytocin on critical blood variables from dystocic sows. *Vet. México* **2009**, *40*, 231–245.
103. González-Lozano, M.; Mota-Rojas, D.; Velázquez-Armenta, E.Y.; Nava-Ocampo, A.A.; Hernández-González, R.; Becerril-Herrera, M.; Trujillo-Ortega, M.E.; Alonso-Spilsbury, M. Obstetric and fetal outcomes in dystocic and eutocic sows to an injection of exogenous oxytocin during farrowing. *Can. Vet. J. = Rev. Vet. Can.* **2009**, *50*, 1273–1277.
104. Muro, B.B.D.; Carnevale, R.F.; Andretta, I.; Leal, D.F.; Monteiro, M.S.; Poor, A.P.; Almond, G.W.; Garbossa, C.A.P. Effects of uterotonics on farrowing traits and piglet vitality: A systematic review and meta-analysis. *Theriogenology* **2021**, *161*, 151–160. [CrossRef]
105. Randall, G.C.B. Perinatal adaptation in animals. *Anim. Reprod. Sci.* **1992**, *28*, 309–318. [CrossRef]
106. Petrovic, N.; Walden, T.B.; Shabalina, I.G.; Timmons, J.A.; Cannon, B.; Nedergaard, J. Chronic peroxisome proliferator-activated receptor γ (pparγ) activation of epididymally derived white adipocyte cultures reveals a population of thermogenically competent, ucp1-containing adipocytes molecularly distinct from classic brown adipocytes. *J. Biol. Chem.* **2010**, *285*, 7153–7164. [CrossRef]
107. Rowan, T.G. Thermoregulation in neonatal ruminants. *BSAP Occas. Publ.* **1992**, *15*, 13–24. [CrossRef]
108. Smith, S.; Carstens, G. Ontogeny and metabolism of brown adipose tissue in livestock species. In *Biology of Growing Animals*; Burrin, D., Mersmann, H., Eds.; Elsevier: Philadephia, PA, USA, 2005; pp. 303–322.
109. Oelkrug, R.; Polymeropoulos, E.T.; Jastroch, M. Brown adipose tissue: Physiological function and evolutionary significance. *J. Comp. Physiol. B* **2015**, *185*, 587–606. [CrossRef]
110. Cannon, B.; Nedergaard, J. Brown adipose tissue: Function and physiological significance. *Physiol. Rev.* **2004**, *84*, 277–359. [CrossRef] [PubMed]
111. Ojha, S.; Robinson, L.; Yazdani, M.; Symonds, M.E.; Budge, H. Brown adipose tissue genes in pericardial adipose tissue of newborn sheep are downregulated by maternal nutrient restriction in late gestation. *Pediatr. Res.* **2013**, *74*, 246–251. [CrossRef] [PubMed]
112. Symonds, M.E.; Bryant, M.J.; Clarke, L.; Darby, C.J.; Lomax, M.A. Effect of maternal cold exposure on brown adipose tissue and thermogenesis in the neonatal lamb. *J. Physiol.* **1992**, *455*, 487–502. [CrossRef] [PubMed]
113. Lezama-García, K.; Mota-Rojas, D.; Pereira, A.M.F.; Martínez-Burnes, J.; Ghezzi, M.; Domínguez, A.; Gómez, J.; de Mira Geraldo, A.; Lendez, P.; Hernández-Ávalos, I.; et al. Transient Receptor Potential (TRP) and thermoregulation in animals: Structural biology and neurophysiological aspects. *Animals* **2022**, *12*, 106. [CrossRef] [PubMed]
114. Haman, F. Shivering in the cold: From mechanisms of fuel selection to survival. *J. Appl. Physiol.* **2006**, *100*, 1702–1708. [CrossRef]
115. Nakamura, K. Afferent pathways for autonomic and shivering thermoeffectors. *Handb. Clin. Neurol.* **2018**, *156*, 263–279. [CrossRef]
116. Kerman, I.A.; Enquist, L.W.; Watson, S.J.; Yates, B.J. Brainstem substrates of sympatho-motor circuitry identified using trans-synaptic tracing with pseudorabies virus recombinants. *J. Neurosci.* **2003**, *23*, 4657–4666. [CrossRef]

117. Hohtola, E. Shivering thermogenesis in birds and mammals. In *Life in the Cold: Evolution, Mechanisms, Adaptation, and Application*; Barnes, M., Carey, H.V., Eds.; ARCUS: Fairbanks, AK, USA, 2004; pp. 241–252.

118. Carstens, G.E. Cold thermoregulation in the newborn calf. *Vet. Clin. N. Am. Food Anim. Pract.* **1994**, *10*, 69–106. [CrossRef]

119. Alexander, G.; Williams, D. Shivering and non-shivering thermogenesis during summit metabolism in young lambs. *J. Physiol.* **1968**, *198*, 251–276. [CrossRef]

120. Phillips, D.I.W.; Jones, A. Fetal programming of autonomic and HPA function: Do people who were small babies have enhanced stress responses? *J. Physiol.* **2006**, *572*, 45–50. [CrossRef]

121. Krogstad, A.-L.; Elam, M.; Karlsson, T.; Wallin, B.G. Arteriovenous anastomoses and the thermoregulatory shift between cutaneous vasoconstrictor and vasodilator reflexes. *J. Auton. Nerv. Syst.* **1995**, *53*, 215–222. [CrossRef]

122. Cook, N.; Schaefer, A.; Warren, L.; Burwash, L.; Anderson, M.; Baron, V. Adrenocortical and metabolic responses to ACTH injection in horses: An assessment by salivary cortisol and infrared thermography of the eye. *Can. J. Anim. Sci.* **2001**, *81*, 621.

123. Ulrich-Lai, Y.M.; Herman, J.P. Neural regulation of endocrine and autonomic stress responses. *Nat. Rev. Neurosci.* **2009**, *10*, 397–409. [CrossRef] [PubMed]

124. Stephens, D.P.; Aoki, K.; Kosiba, W.A.; Johnson, J.M. Nonnoradrenergic mechanism of reflex cutaneous vasoconstriction in men. *Am. J. Physiol. Circ. Physiol.* **2001**, *280*, H1496–H1504. [CrossRef] [PubMed]

125. Stephens, D.P.; Saad, A.R.; Bennett, L.A.T.; Kosiba, W.A.; Johnson, J.M. Neuropeptide Y antagonism reduces reflex cutaneous vasoconstriction in humans. *Am. J. Physiol. Circ. Physiol.* **2004**, *287*, H1404–H1409. [CrossRef]

126. Mai, T.C.; Braun, A.; Bach, V.; Pelletier, A.; Seze, R. Low-level radiofrequency exposure induces vasoconstriction in rats. *Bioelectromagnetics* **2021**, *42*, 455–463. [CrossRef]

127. Hrupka, B.J.; Leibbrandt, V.D.; Crenshaw, T.D.; Benevenga, N.J. Effect of sensory stimuli on huddling behavior of pigs. *J. Anim. Sci.* **2000**, *78*, 592–596. [CrossRef]

128. Mota-Rojas, D.; Velarde, A.; Maris-Huertas, S.; Cajiao, M.N. *Animal Welfare, a Global Vision in Ibero-America*, 3rd ed.; Elsevier: Barcelona, Spain, 2016; p. 516.

129. Regueiro, M.; López-Mazz, C.; Jorge-Smeding, E.; Baldi, F.; Banchero, G. Duration of phase II of labour negatively affects maternal behaviour and lamb viability in wool-type primiparous ewes under extensive rearing. *Appl. Anim. Behav. Sci.* **2021**, *234*, 105207. [CrossRef]

130. García-Torres, E.; Hudson, R.; Castelán, F.; Martínez-Gómez, M.; Bautista, A. Differential metabolism of brown adipose tissue in newborn rabbits in relation to position in the litter huddle. *J. Therm. Biol.* **2015**, *51*, 33–41. [CrossRef]

131. Ferner, K.; Schultz, J.A.; Zeller, U. Comparative anatomy of neonates of the three major mammalian groups (monotremes, marsupials, placentals) and implications for the ancestral mammalian neonate morphotype. *J. Anat.* **2017**, *231*, 798–822. [CrossRef]

132. Bautista, A.; García-Torres, E.; Martínez-Gómez, M.; Hudson, R. Do newborn domestic rabbits *Oryctolagus cuniculus* compete for thermally advantageous positions in the litter huddle? *Behav. Ecol. Sociobiol.* **2008**, *62*, 331–339. [CrossRef]

133. McCauley, S.R. Glucose Metabolism in Low Birth Weight Neonatal Pigs. Doctoral Dissertation, Virginia Polytethic Institute and State University, Blacksburg, VA, USA, 2019.

134. Napolitano, F.; Bragaglio, A.; Braghieri, A.; El-Aziz, A.H.A.; Titto, C.G.; Villanueva-García, D.; Mora-Medina, P.; Pereira, A.M.F.; Hernández-Avalos, I.; José-Pérez, N.; et al. The effect of birth weight and time of day on the thermal response of newborn water buffalo calves. *Front. Vet. Sci.* **2023**, *10*, 209. [CrossRef] [PubMed]

135. Wolfe, B.A.; Lamberski, N. Approaches to management and care of the neonatal nondomestic ruminant. *Vet. Clin. N. Am. Exot. Anim. Pract.* **2012**, *15*, 265–277. [CrossRef]

136. Eales, F.; Small, J.; Gilmour, J. Resuscitation of hypothermic lambs. *Vet. Rec.* **1982**, *110*, 121–123. [CrossRef] [PubMed]

137. Hegarty, J.E.; Harding, J.E.; Oliver, M.H.; Gamble, G.; Dickson, J.L.; Chase, G.; Jaquiery, A.L. Oral dextrose gel to improve survival in less vigorous newborn triplet lambs: A randomised controlled trial. *N. Z. J. Agric. Res.* **2017**, *60*, 54–69. [CrossRef]

138. Engelsmann, M.N.; Hansen, C.F.; Nielsen, M.N.; Kristensen, A.R.; Amdi, C. Glucose injections at birth, warmth and placing at a nurse sow improve the growth of IUGR piglets. *Animals* **2019**, *9*, 519. [CrossRef] [PubMed]

139. Le Dividich, J.; Noblet, J. Thermoregulation and energy metabolism in the neonatal pig. *Ann. Rech. Vet.* **1983**, *14*, 375–381. [PubMed]

140. Le Dividich, J.; Noblet, J. Colostrum intake and thermoregulation in the neonatal pig in relation to environmental temperature. *Neonatology* **1981**, *40*, 167–174. [CrossRef]

141. Silva, F.L.M.; Miqueo, E.; da Silva, M.D.; Torrezan, T.M.; Rocha, N.B.; Salles, M.S.V.; Bittar, C.M.M. Thermoregulatory responses and performance of dairy calves fed different amounts of colostrum. *Animals* **2021**, *11*, 703. [CrossRef]

142. Silva, F.L.M.; Bittar, C.M.M. Thermogenesis and some rearing strategies of dairy calves at low temperature—A review. *J. Appl. Anim. Res.* **2019**, *47*, 115–122. [CrossRef]

143. Pakkanen, R.; Aalto, J. Growth factors and antimicrobial factors of bovine colostrum. *Int. Dairy J.* **1997**, *7*, 285–297. [CrossRef]

144. Muns, R.; Silva, C.; Manteca, X.; Gasa, J. Effect of cross-fostering and oral supplementation with colostrums on performance of newborn piglets. *J. Anim. Sci.* **2014**, *92*, 1193–1199. [CrossRef] [PubMed]

145. Muns, R.; Nuntapaitoon, M.; Tummaruk, P. Effect of oral supplementation with different energy boosters in newborn piglets on pre-weaning mortality, growth and serological levels of IGF-I and IgG. *J. Anim. Sci.* **2017**, *95*, 353. [CrossRef]

146. de Mejia, E.G.; Ramirez-Mares, M.V. Impact of caffeine and coffee on our health. *Trends Endocrinol. Metab.* **2014**, *25*, 489–492. [CrossRef] [PubMed]

147. Kumar; Lipshultz Caffeine and Clinical Outcomes in Premature Neonates. *Children* **2019**, *6*, 118. [CrossRef] [PubMed]

148. Swinbourne, A.M.; Kind, K.L.; Flinn, T.; Kleemann, D.O.; van Wettere, W.H.E.J. Caffeine: A potential strategy to improve survival of neonatal pigs and sheep. *Anim. Reprod. Sci.* **2021**, *226*, 106700. [CrossRef]

149. Orozco, G.; Mota-Rojas, D.; Trujillo-Ortega, M.E.; Becerril-Herrera, M.; Hernández-González, R.; Villanueva-García, D. Effects of administration of caffeine on metabolic variables in neonatal pigs with peripartum asphyxia. *Am. J. Vet. Res.* **2010**, *71*, 1214–1219. [CrossRef]

150. Nowland, T.L.; Kind, K.; Hebart, M.L.; van Wettere, W.H.E.J. Caffeine supplementation at birth, but not 8 to 12 h post-birth, increased 24 h pre-weaning mortality in piglets. *Animal* **2020**, *14*, 1529–1535. [CrossRef]

151. Robertson, S.M.; Edwards, S.H.; Doran, G.S.; Friend, M.A. Maternal caffeine administration to ewes does not affect perinatal lamb survival. *Anim. Reprod. Sci.* **2021**, *231*, 106799. [CrossRef]

152. Jarratt, L.; James, S.E.; Kirkwood, R.N.; Nowland, T.L. Effects of caffeine and glucose supplementation at birth on piglet pre-weaning growth, thermoregulation, and survival. *Animals* **2023**, *13*, 435. [CrossRef]

153. Murdock, N.J.; Weaver, A.C.; Kelly, J.M.; Kleemann, D.O.; van Wettere, W.H.E.J.; Swinbourne, A.M. Supplementing pregnant Merino ewes with caffeine to improve neonatal lamb thermoregulation and viability. *Anim. Reprod. Sci.* **2021**, *226*, 106715. [CrossRef] [PubMed]

154. Binder-Heschl, C.; Crossley, K.; te Pas, A.; Polglase, G.; Blank, D.; Zahra, V.; Moxham, A.; Rodgers, K.; Hooper, S. Haemodynamic effects of prenatal caffeine on the cardiovascular transition in ventilated preterm lambs. *PLoS ONE* **2018**, *13*, e0200572. [CrossRef] [PubMed]

155. Menozzi, A.; Mazzoni, C.; Serventi, P.; Zanardelli, P.; Bertini, S. Pharmacokinetics of oral caffeine in sows: A pilot study. *Large Anim. Rev.* **2015**, *21*, 207–210.

156. Mainau, E.; Manteca, X. Pain and discomfort caused by parturition in cows and sows. *Appl. Anim. Behav. Sci.* **2011**, *135*, 241–251. [CrossRef]

157. Dhaliwal, A.; Gupta, M. *Physiology, Opioid Receptor*; StatPearls: Treasure Island, FL, USA, 2022.

158. Hazinski, T.A.; Grunstein, M.M.; Schlueter, M.A.; Tooley, W.H. Effect of naloxone on ventilation in newborn rabbits. *J. Appl. Physiol. Respir. Environ. Exerc. Physiol.* **1981**, *50*, 713–717. [CrossRef]

159. Long, W.A.; Lawson, E.E. Developmental aspects of the effect of naloxone on control of breathing in piglets. *Respir. Physiol.* **1983**, *51*, 119–129. [CrossRef]

160. Hsia, C.C.W.; Schmitz, A.; Lambertz, M.; Perry, S.F.; Maina, J.N. Evolution of Air Breathing: Oxygen Homeostasis and the Transitions From Water to Land and Sky. In *Comprehensive Physiology*; Wiley: New York, NY, USA, 2013; pp. 849–915.

161. Soraci, A.; Decundo, J.; Dieguez, S.; Martinez, G.; Romanelli, A.; Perez Gaudio, D.; Fernandez Paggi, M.; Amanto, F. Practical oxygen therapy for newborn piglets. *N. Z. Vet. J.* **2020**, *68*, 331–339. [CrossRef]

162. Darby, J.R.T.; Berry, M.J.; Quinn, M.; Holman, S.L.; Bradshaw, E.L.; Jesse, S.M.; Haller, C.; Seed, M.; Morrison, J.L. Haemodynamics and cerebral oxygenation of neonatal piglets in the immediate ex utero period supported by mechanical ventilation or ex utero oxygenator. *J. Physiol.* **2021**, *599*, 2751–2761. [CrossRef]

163. Bleul, U.T.; Bircher, B.M.; Kähn, W.K. Effect of intranasal oxygen administration on blood gas variables and outcome in neonatal calves with respiratory distress syndrome: 20 cases (2004–2006). *J. Am. Vet. Med. Assoc.* **2008**, *233*, 289–293. [CrossRef]

164. Kim, E.; Nguyen, M. Oxygen therapy for neonatal resuscitation in the delivery room. *Neoreviews* **2019**, *20*, e500–e512. [CrossRef]

165. Silva, L.C.G.; Angrimani, D.S.R.; Regazzi, F.M.; Lúcio, C.F.; Veiga, G.A.L.; Fernandes, C.B.; Vannucchi, C.I. Pulmonary changes and redox status after fractionalized dose of prophylactic surfactant treatment in preterm neonatal lambs. *J. Appl. Anim. Res.* **2020**, *48*, 220–227. [CrossRef]

166. Rawat, M.; Chandrasekharan, P.; Gugino, S.F.; Koenigsknecht, C.; Nielsen, L.; Wedgwood, S.; Mathew, B.; Nair, J.; Steinhorn, R.; Lakshminrusimha, S. Optimal oxygen targets in term lambs with meconium aspiration syndrome and pulmonary hypertension. *Am. J. Respir. Cell Mol. Biol.* **2020**, *63*, 510–518. [CrossRef] [PubMed]

167. Chigerwe, M.; Hagey, J.V.; Aly, S.S. Determination of neonatal serum immunoglobulin G concentrations associated with mortality during the first 4 months of life in dairy heifer calves. *J. Dairy Res.* **2015**, *82*, 400–406. [CrossRef] [PubMed]

168. Conneely, M.; Berry, D.P.; Sayers, R.; Murphy, J.P.; Lorenz, I.; Doherty, M.L.; Kennedy, E. Factors associated with the concentration of immunoglobulin G in the colostrum of dairy cows. *Animal* **2013**, *7*, 1824–1832. [CrossRef]

169. Jaster, E.H. Evaluation of quality, quantity, and timing of colostrum feeding on immunoglobulin G1 absorption in Jersey calves. *J. Dairy Sci.* **2005**, *88*, 296–302. [CrossRef]

170. Argüello, A.; Castro, N.; Zamorano, M.J.; Castroalonso, A.; Capote, J. Passive transfer of immunity in kid goats fed refrigerated and frozen goat colostrum and commercial sheep colostrum. *Small Rumin. Res.* **2004**, *54*, 237–241. [CrossRef]

171. Bragg, R.; Macrae, A.; Lycett, S.; Burrough, E.; Russell, G.; Corbishley, A. Prevalence and risk factors associated with failure of transfer of passive immunity in spring born beef suckler calves in Great Britain. *Prev. Vet. Med.* **2020**, *181*, 105059. [CrossRef]

172. Turini, L.; Conte, G.; Bonelli, F.; Sgorbini, M.; Madrigali, A.; Mele, M. The relationship between colostrum quality, passive transfer of immunity and birth and weaning weight in neonatal calves. *Livest. Sci.* **2020**, *238*, 104033. [CrossRef]

173. Cabrera, R.A.; Lin, X.; Campbell, J.M.; Moeser, A.J.; Odle, J. Influence of birth order, birth weight, colostrum and serum immunoglobulin G on neonatal piglet survival. *J. Anim. Sci. Biotechnol.* **2012**, *3*, 42. [CrossRef]

174. Maciag, S.S.; Bellaver, F.V.; Bombassaro, G.; Haach, V.; Morés, M.A.Z.; Baron, L.F.; Coldebella, A.; Bastos, A.P. On the influence of the source of porcine colostrum in the development of early immune ontogeny in piglets. *Sci. Rep.* **2022**, *12*, 15630. [CrossRef]

175. Mellado, M.; Pittroff, W.; García, J.E.; Mellado, J. Serum IgG, blood profiles, growth and survival in goat kids supplemented with artificial colostrum on the first day of life. *Trop. Anim. Health Prod.* **2008**, *40*, 141–145. [CrossRef] [PubMed]

176. Berge, A.C.B.; Besser, T.E.; Moore, D.A.; Sischo, W.M. Evaluation of the effects of oral colostrum supplementation during the first fourteen days on the health and performance of preweaned calves. *J. Dairy Sci.* **2009**, *92*, 286–295. [CrossRef] [PubMed]

177. Aoki, T.; Chiba, A.; Itoh, M.; Nambo, Y.; Yamagishi, N.; Shibano, K.; Cheong, S.H. Colostral and foal serum immunoglobulin G levels and associations with perinatal abnormalities in heavy draft horses in Japan. *J. Equine Sci.* **2020**, *31*, 29–34. [CrossRef]

178. Lester, G.D. Colostrum: Assessment of Quality and Artificial Supplementation. In *Equine Reproduction*; McKinnon, A.O., Squires, E.L., Vaala, W.E., Varner, D.D., Eds.; Wiley Blackwell: Oxford, UK, 2011; pp. 342–345.

179. Grigaleviciute, R.; Planciuniene, R.; Prikockyte, I.; Radzeviciute-Valciuke, E.; Baleviciute, A.; Zelvys, A.; Zinkeviciene, A.; Zigmantaite, V.; Kucinskas, A.; Matusevicius, P.; et al. The Influence of feeding with colostrum and colostrum replacer on major blood biomarkers and growth performance in dairy calves. *Vet. Sci.* **2023**, *10*, 128. [CrossRef] [PubMed]

180. Oliviero, C.; Junnikkala, S.; Peltoniemi, O. The challenge of large litters on the immune system of the sow and the piglets. *Reprod. Domest. Anim.* **2019**, *54*, 12–21. [CrossRef] [PubMed]

181. Sampath, V.; Han, K.; Kim, I.H. Impact of artificial colostrum supplement on the growth performance and blood profile in piglets. *Anim. Nutr. Feed Technol.* **2022**, *22*, 665–672. [CrossRef]

182. Uddin, M.K.; Hasan, S.; Peltoniemi, O.; Oliviero, C. The effect of piglet vitality, birth order, and blood lactate on the piglet growth performances and preweaning survival. *Porc. Health Manag.* **2022**, *8*, 52. [CrossRef]

183. Aleman, M.; Weich, K.; Madigan, J. Survey of Veterinarians Using a Novel Physical Compression Squeeze Procedure in the Management of Neonatal Maladjustment Syndrome in Foals. *Animals* **2017**, *7*, 69. [CrossRef]

184. Piccione, G.; Caola, G.; Refinetti, R. Maturation of the daily body temperature rhythm in sheep and horse. *J. Therm. Biol.* **2002**, *27*, 333–336. [CrossRef]

185. Aleksiev, Y.; Gudev, D.; Dimov, G. Thermal status in three breeds of newborn lambs during the first 24 h of postnatal life. *Bulg. J. Agrisultural Sci.* **2007**, *13*, 563–573.

186. Giannetto, C.; Arfuso, F.; Fazio, F.; Giudice, E.; Panzera, M.; Piccione, G. Rhythmic function of body temperature, breathing and heart rates in newborn goats and sheep during the first hours of life. *J. Vet. Behav.* **2017**, *18*, 29–36. [CrossRef]

187. Labeur, L.; Villiers, G.; Small, A.H.; Hinch, G.N.; Schmoelzl, S. Infrared thermal imaging as a method to evaluate heat loss in newborn lambs. *Res. Vet. Sci.* **2017**, *115*, 517–522. [CrossRef]

188. Vande Pol, K.D.; Tolosa, A.F.; Bautista, R.O.; Willard, N.C.; Gates, R.S.; Shull, C.M.; Brown, C.B.; Alencar, S.A.S.; Lents, C.A.; Ellis, M. Effects of drying and providing supplemental oxygen to piglets at birth on rectal temperature over the first 24 h after birth. *Transl. Anim. Sci.* **2021**, *5*, txab095. [CrossRef] [PubMed]

189. Donnelly, C.G.; Quinn, C.T.; Nielsen, S.G.; Raidal, S.L. Respiratory support for pharmacologically induced hypoxia in neonatal calves. *Vet. Med. Int.* **2016**, *2016*, 2129362. [CrossRef] [PubMed]

190. Milan, H.F.M.; Campos Maia, A.S.; Gebremedhin, K.G. Prediction of optimum supplemental heat for piglets. *Trans. ASABE* **2019**, *62*, 321–342. [CrossRef]

191. Malmkvist, J.; Damgaard, B.M.; Pedersen, L.J.; Jørgensen, E.; Thodberg, K.; Chaloupková, H.; Bruckmaier, R.M. Effects of thermal environment on hypothalamic-pituitary-adrenal axis hormones, oxytocin, and behavioral activity in periparturient sows. *J. Anim. Sci.* **2009**, *87*, 2796–2805. [CrossRef] [PubMed]

192. Vasdal, G.; Glærum, M.; Melišová, M.; Bøe, K.E.; Broom, D.M.; Andersen, I.L. Increasing the piglets' use of the creep area—A battle against biology? *Appl. Anim. Behav. Sci.* **2010**, *125*, 96–102. [CrossRef]

193. Menzies, P.I. Lambing Management and Neonatal Care. In *Current Therapy in Large Animal Theriogenology*; Elsevier: Amsterdam, The Netherlands, 2007; pp. 680–695.

194. Davis, J.D.; Xin, H.; MacDonald, R.D. MacDonald infrared thermographic evaluation of commercially available incandescent heat lamps. *Appl. Eng. Agric.* **2008**, *24*, 685–693. [CrossRef]

195. Pedersen, L.J.; Larsen, M.L.V.; Malmkvist, J. The ability of different thermal aids to reduce hypothermia in neonatal piglets1. *J. Anim. Sci.* **2016**, *94*, 2151–2159. [CrossRef]

196. Nowak, R.; Poindron, P. From birth to colostrum: Early steps leading to lamb survival. *Reprod. Nutr. Dev.* **2006**, *46*, 431–446. [CrossRef]

197. Andersen, I.L.; Haukvik, I.A.; Bøe, K.E. Drying and warming immediately after birth may reduce piglet mortality in loose-housed sows. *Animal* **2009**, *3*, 592–597. [CrossRef] [PubMed]

198. Cooper, N.; Vande Pol, K.D.; Ellis, M.; Xiong, Y.; Gates, R. 7 Effect of piglet birth weight and drying on post-natal changes in rectal temperature. *J. Anim. Sci.* **2019**, *97*, 4. [CrossRef]

199. Vasdal, G.; Østensen, I.; Melišová, M.; Bozděchová, B.; Illmann, G.; Andersen, I.L. Management routines at the time of farrowing—Effects on teat success and postnatal piglet mortality from loose housed sows. *Livest. Sci.* **2011**, *2–3*, 225–231. [CrossRef]

200. Cote, A.; Blanchard, P.W.; Meehan, B. Metabolic and cardiorespiratory effects of doxapram and theophylline in sleeping newborn piglets. *J. Appl. Physiol.* **1992**, *72*, 410–415. [CrossRef] [PubMed]
201. Silvera, F.E.; Blasina, M.F.; Vaamonde, L.; Tellechea, S.; Godoy, C.; Zabala, S.; Mañana, G.; Martell, M.; Olivera, W. Sildenafil prevents the increase of extravascular lung water and pulmonary hypertension after meconium aspiration in newborn piglets. *Braz. J. Med. Biol. Res.* **2011**, *44*, 778–785. [CrossRef]

Disclaimer/Publisher's Note: The statements, opinions and data contained in all publications are solely those of the individual author(s) and contributor(s) and not of MDPI and/or the editor(s). MDPI and/or the editor(s) disclaim responsibility for any injury to people or property resulting from any ideas, methods, instructions or products referred to in the content.

Review

Meconium Aspiration Syndrome in Animal Models: Inflammatory Process, Apoptosis, and Surfactant Inactivation

Daniel Mota-Rojas [1,*], Dina Villanueva-García [2], Andrea Mota-Reyes [3], Agustín Orihuela [4], Ismael Hernández-Ávalos [5], Adriana Domínguez-Oliva [1], Alejandro Casas-Alvarado [1], Karla Flores-Padilla [1], Joseline Jacome-Romero [1] and Julio Martínez-Burnes [6,*]

[1] Neurophysiology, Behavior and Animal Welfare Assessment, Universidad Autónoma Metropolitana (UAM), Mexico City 04960, Mexico
[2] Division of Neonatology, National Institute of Health Hospital Infantil de México Federico Gómez, Mexico City 06720, Mexico
[3] School of Medicine and Health Sciences, TecSalud, Instituto Tecnológico y de Estudios Superiores de Monterrey (ITESM), Monterrey 64849, Mexico
[4] Facultad de Ciencias Agropecuarias, Universidad Autónoma del Estado de Morelos, Cuernavaca 62209, Mexico
[5] Clinical Pharmacology and Veterinary Anesthesia, Facultad de Estudios Superiores Cuautitlán, Universidad Nacional Autónoma de México (UNAM), Mexico City 54714, Mexico
[6] Animal Health Group, Facultad de Medicina Veterinaria y Zootecnia, Universidad Autónoma de Tamaulipas, Victoria City 87000, Mexico
* Correspondence: dmota100@yahoo.com.mx (D.M.-R.); jmburnes@docentes.uat.edu.mx (J.M.-B.)

Citation: Mota-Rojas, D.; Villanueva-García, D.; Mota-Reyes, A.; Orihuela, A.; Hernández-Ávalos, I.; Domínguez-Oliva, A.; Casas-Alvarado, A.; Flores-Padilla, K.; Jacome-Romero, J.; Martínez-Burnes, J. Meconium Aspiration Syndrome in Animal Models: Inflammatory Process, Apoptosis, and Surfactant Inactivation. *Animals* **2022**, *12*, 3310. https://doi.org/10.3390/ani12233310

Academic Editor: Maria Cristina Veronesi

Received: 12 October 2022
Accepted: 24 November 2022
Published: 27 November 2022

Publisher's Note: MDPI stays neutral with regard to jurisdictional claims in published maps and institutional affiliations.

Copyright: © 2022 by the authors. Licensee MDPI, Basel, Switzerland. This article is an open access article distributed under the terms and conditions of the Creative Commons Attribution (CC BY) license (https://creativecommons.org/licenses/by/4.0/).

Simple Summary: *Meconium aspiration syndrome* is a pathology that causes hypoxia, acidosis, and neonatal mortality. The mechanisms behind this rely on physiopathology and the interaction of meconium with pulmonary alveolar cells. The inflammatory response, inactivation of surfactant, and processes such as apoptosis or necrosis of alveolar macrophages, epithelial and endothelial cells also participate in this syndrome. In this review, the physiopathology of meconium aspiration syndrome will be discussed in veterinary medicine to understand the inflammatory response and the cellular and biochemical changes at the alveolar level that cause the main outcomes of this pathology.

Abstract: Meconium Aspiration Syndrome is a condition that causes respiratory distress in newborns due to occlusion and airway inflammation, and surfactant inactivation by meconium. This condition has been described in animal species such as canids, sheep, cattle, horses, pigs, and marine mammals. In its pathogenesis, the pulmonary epithelium activates a limited inflammatory response initiated by cytokines causing leukocyte chemotaxis, inhibition of phagocytosis, and pathogen destruction. Likewise, cytokines release participates in the apoptosis processes of pneumocytes due to the interaction of angiotensin with cytokines and the caspase pathway. Due to these reactions, the prevalent signs are lung injury, hypoxia, acidosis, and pneumonia with susceptibility to infection. Given the importance of the pathophysiological mechanism of meconium aspiration syndrome, this review aims to discuss the relevance of the syndrome in veterinary medicine. The inflammatory processes caused by meconium aspiration in animal models will be analyzed, and the cellular apoptosis and biochemical processes of pulmonary surfactant inactivation will be discussed.

Keywords: meconium; pulmonary inflammatory response; apoptosis; animal models; surfactant inactivation

1. Introduction

Meconium aspiration syndrome (MAS) is a condition that causes hypoxemia, acidosis, and respiratory distress in the newborn, increasing neonatal mortality [1,2]. Meconium contains various substances such as mucopolysaccharides, bile acids, cytokines, cholesterol, and cells that, upon entering the airways, interact with the alveolar epithelium, activate the immune system through chemotactic signaling of neutrophils and their extravasation through the alveolar-capillary membrane, additionally stimulating the production of reactive oxygen species (ROS) and apoptosis [3–6]. This inflammatory process may be accompanied by pulmonary edema, events that consequently cause neonatal hypoxia and acidosis [7,8].

Given the importance of this disease in different species of veterinary interest, various studies have been carried out to describe the pathophysiological mechanisms of MAS [9–11]. An example of MAS pathophysiology is the review by Swarman et al. [12], which describes how meconium aspiration provokes airway obstruction and fetal hypoxia. The same authors and coinciding with others mention that meconium microparticles can interact with the alveolar epithelium inducing a local inflammatory response due to the release of cytokines, such as interleukins (IL) IL-1, IL-6, IL-8, IL-10, and tumor necrosis factor (TNF)-α [13,14]. These interleukins and the additional meconium components interact with angiotensin metabolites, causing alveolar cell destruction and inactivation of the pulmonary surfactant [15,16].

The mechanisms described for this pathology have formed the fundamental basis for strategies to address the MAS [17]. Therefore, this article aims to describe and argue the importance of MAS in veterinary medicine. The lung inflammatory process caused by meconium aspiration in animal models will be analyzed, and the cellular and biochemical mechanism of pulmonary surfactant inactivation will be discussed.

2. MAS in Veterinary Medicine

MAS is a condition most frequently described in human newborns, with an incidence from 0.4 to 22% [2,18–20]. Of these cases, 5 to 28% are associated with infant mortality when adequate obstetric intervention is not received [21,22]. In animals, this pathology has been reported in canids, cattle, sheep, pigs, and marine mammals [22–24]. For example, in newborn piglets, 6% of neonatal mortality is associated with this pathology [25,26], while in dolphins, it is 8% [23], 4% in foals [27], and 1–3% in puppies [28].

The frequency in which this pathology may be present during parturition makes it necessary to understand its pathophysiology to assess low vitality or respiratory distress in animals [29]. Even though some clinicians use the degree of meconium staining of the skin to determine severity, it is a poor predictor [30,31]. Therefore, to achieve a definitive diagnosis or early recognition of MAS, complete knowledge is required regarding the pathophysiological mechanisms that trigger the pulmonary and systemic effects of meconium aspiration.

Animal Experimental Models in MAS

Researchers have tried to reproduce MAS experimentally in laboratory animals to study the nature and progression of pulmonary lesions under controlled conditions. Different animal species have been used as MAS models, including pigs [32–37], dogs [28,38], cats [39], guinea pigs [40], rabbits [41–44], rats [45–47], lambs [24,48] and baboons [49]. Pathophysiology and lesions of MAS in humans have been successfully reproduced in experimental animal models and mainly focused on the study of medical interventions and therapies to reduce the impact on newborns' vitality and mortality [31].

3. Meconium Composition

Meconium is a sterile liquid substance or the first excretion accumulated in the fetal intestine during gestation. It has a viscous consistency and green-like color, and it may be expelled from the fetal intestine in response to intrauterine hypoxia [2] (Figures 1 and 2). It is mainly composed of gastrointestinal secretions, biliary salts, pancreatic juice, desquamated epithelial cells, lanugo, amniotic fluid, and blood, in addition to proinflammatory components such as IL-1, IL-6, IL-8, and TNF, as well as proteolytic enzymes such as phospholipase A2, free fatty acids, bilirubin, hemoglobin, and cholesterol. In this sense, Righetti et al. [50], using a proton nuclear magnetic resonance analysis of the components of meconium, found that it is composed of cholesterol and fatty acids, β-glucose, α-glucose, lactate, and β-hydroxybutyrate. These findings allow us to observe the structural and biochemical components of meconium.

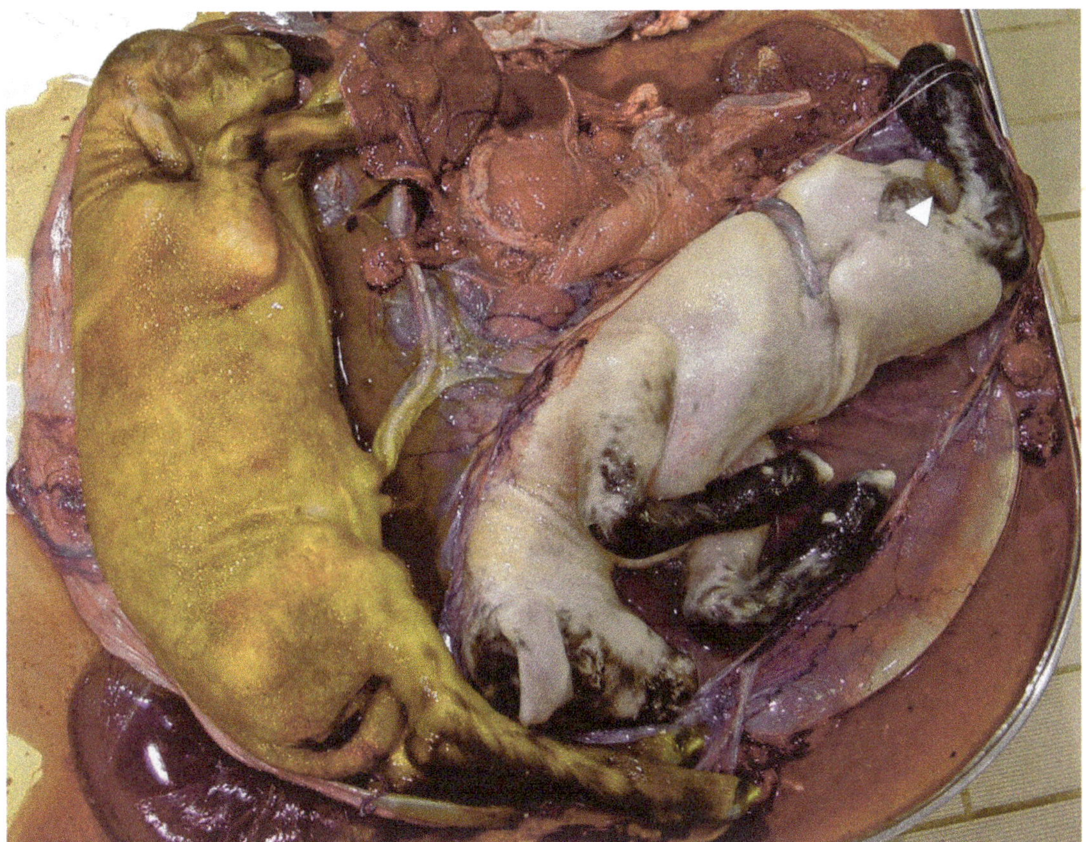

Figure 1. Postmortem inspection of the uterus of a ewe with twin gestation and fetuses that suffered hypoxia. Diffuse severe greenish meconium staining throughout the fetal skin of the left fetus). Large meconium particles expelled from the intestine are on the skin of the second fetus (arrowhead). Courtesy Dr. Carolyn Legge, Atlantic Veterinary College.

Consequently, the components present in the meconium are the main ones responsible for initiating the pulmonary inflammatory response and tissue damage. This event may differ between species, causing the disease mechanisms to be variable, thus leading to a decision of which therapy would be more effective in treating the species.

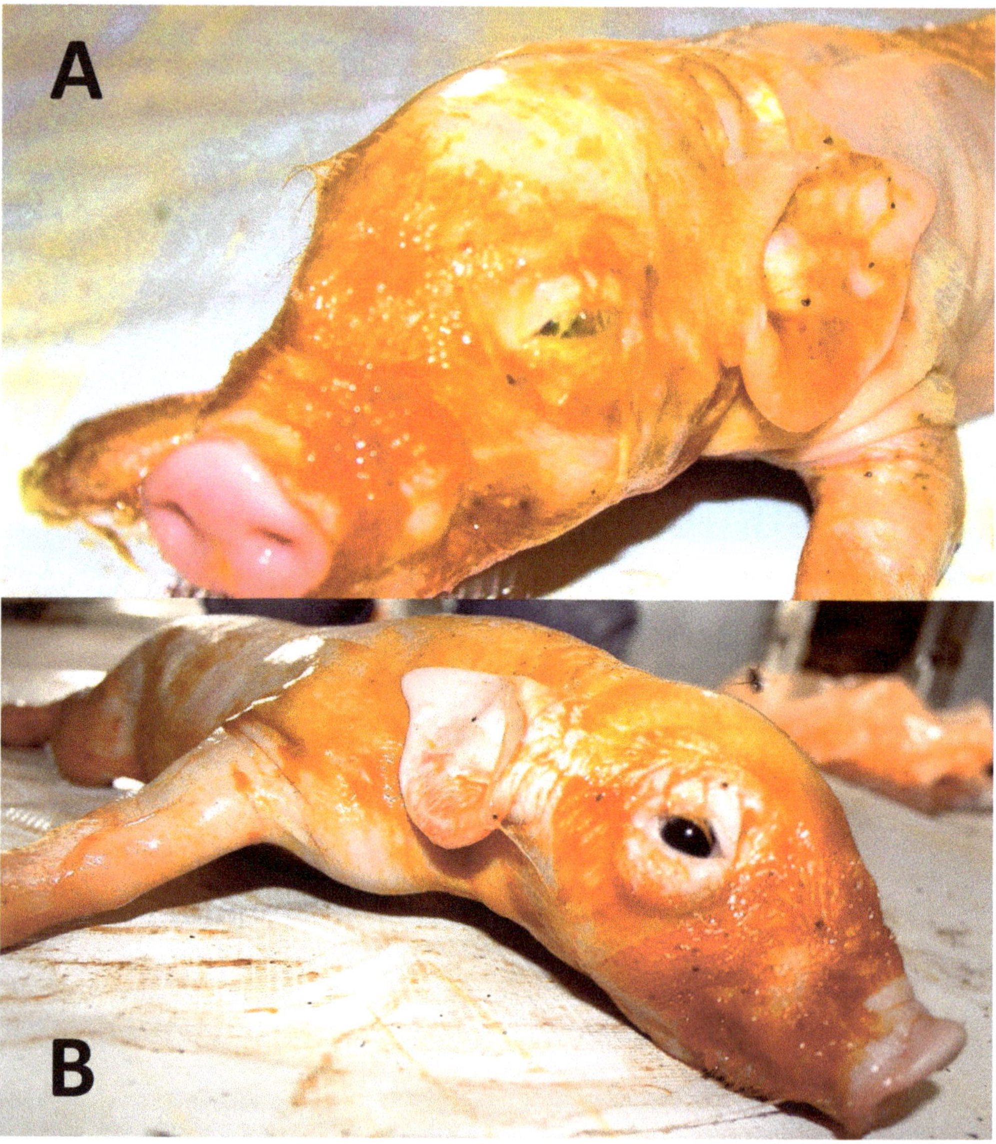

Figure 2. Meconium-stained piglets after dystocia in a primiparous sow. The expulsion interval between neonates was greater than 3 h. Both neonates survived an episode of severe hypoxia. (**A**). The newly born pig was lethargic, adynamic, hypoglycemic, uncoordinated and its vitality score dropped. The newborn was unable to suckle/eat feed as it lost the sucking reflex. It died within the next 24 h. (**B**). Neonate stained with meconium in more than 60% of the body surface. It shows vigor, as it makes attempts to get up despite being weak. After colostrum intake, the piglet survived.

4. MAS and the Pulmonary Inflammatory Process

The systemic inflammatory response induced by meconium inhalation before or during birth is a complex process involving diverse mechanisms of airway damage, such as mechanical airway blockage, epithelial injury, pulmonary hypertension, and surfactant inactivation [14,36]. These lesions involve leukocyte infiltration, activation of alveolar and peritoneal macrophages, myeloperoxidase (MPO) and proinflammatory cytokines release [37,44], cell death by apoptosis, as well as complement and phospholipase A2 activation [5].

The inflammatory process begins as a local lesion in the pulmonary tissue, with an incidence of 13% intraamniotic inflammation and 23% funisitis in the fetus [51]. It can result in a systemic response; however, the importance of the local reaction at the pulmonary level is due to the alterations in the normal physiology of the animal. For example, in the neonatal rat model, airway hyperreactivity leads to neonatal hypoxia, while inflammation of the lung parenchyma can cause atelectasis [2,47].

In the bovine model, a retrospective study analyzing the cause of death in 2-week-old calves showed that 42.5% of the animals had meconium, squamous cells, or keratin in the lung. Histological analysis revealed that the lungs with meconium aspiration developed mild diffuse alveolitis with neutrophils and macrophage infiltration, similar to those reported in humans, by the interaction of cytokines in the pulmonary epithelium [22].

4.1. Local Inflammation

Lung damage in patients with MAS is attributed to different mechanisms, such as airway obstruction, chemical pneumonitis, and inflammation due to immune system activation [5,41,52]. Some studies still discuss whether the inflammation is the cause of death of the animal or if it is due to an additional effect of neonatal hypoxia [5,8].

Once the meconium invades the respiratory tract, two innate immune systems are activated: the Toll-like receptor (TLR) and the complement systems [12,41]. Plasma membrane TLRs are highly expressed in lung cells. They identify meconium components and endogenous ligands (e.g., alarmins) released when the tissular damage causes ischemia, as well as events of alveolitis or pneumonitis [8]. According to Anand et al. [53], MAS pathogenesis is associated with the expression of TLR1, TLR4, TLR9, and TLR7 receptors and the initiation of the inflammatory cascade. Upon recognizing foreign or harmful agents, proinflammatory substances are activated. For example, TRL3 generates a MyD88-dependent signal that activates the NF-kB and promotes cytokine release [54–56].

On the other hand, the activation of the complement system by either of its three pathways (lectin, classical, or alternative) begins after the macrophage infiltration into the pulmonary tissue. The binding of Cq1 activates the classical pathway to antigen-antibody complexes. In contrast, the lectin is activated by mannose-binding lectin or ficolins binding to sugary residues of the antigen to generate a protein complex with the enzymatic activity of the C3 convertase [57]. In vitro studies report that pig meconium causes activation of the alternative pathway [35], which is associated with the activation of some immune mechanisms and the release of IL, as noted by Castellheim et al. [37]. In his experimental study on the lungs of newborn piglets with MAS induction, the levels of a terminal complex sC5b-9 increased significantly compared to the controls ($p < 0.0005$). Likewise, the levels of this complement were significantly correlated with IL-6 ($r = 0.64$, $p < 0.005$) and IL-8 concentrations ($r = 0.32$, $p = 0.03$). This finding demonstrates that the activation of the complement system triggers the production of inflammatory mediators such as cytokines, chemokines, eicosanoids derived from arachidonic acid, and ROS due to the interaction between endothelial cells and leukocytes [58]. The terminal sC5b-9 complex releases more C5a, a potent anaphylatoxin that stimulates vacuole degranulation, releasing intracellular and chemotactic mediators [36].

Therefore, the activation of the inflammatory cascade is a process dependent on cell-to-cell communication of the immune system to stimulate the repair of damaged tissue.

4.2. Cytokines

Meconium contains different cytokines, such as IL-1, IL-6, IL-8, and TNF-α. However, some authors mention that meconium-stained amniotic fluid contains higher concentrations of these substances, therefore, contained in the amniotic fluid [6,59,60]. These findings support the hypothesis that MAS induces the production of cytokines and the chemotaxis of polymorphonuclear cells and macrophages (Figure 3). In this sense, Lindenskov et al. [13] conducted a study on hypoxic piglets after the pulmonary instillation of meconium and receiving 1.4 mL/kg of 30% albumin. They observed that the levels of IL-8 correlated with the deterioration of pulmonary function measured as the oxygenation index (r = 0.71, $p < 0.0001$), pulmonary compliance (r = 0.66, $p < 0.0001$), and ventilation index (r = 0.71, $p < 0.0001$). These results demonstrate that cytokine induction is closely related to the degree of cell and lung injury.

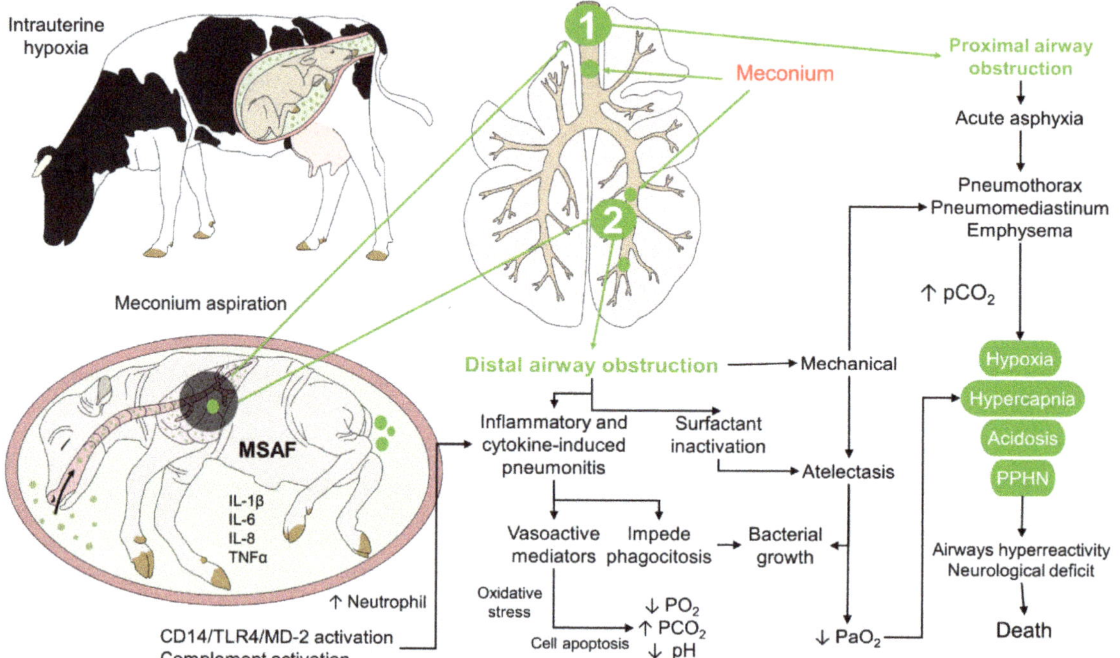

Figure 3. Pathophysiology of meconium aspiration syndrome. In newborns, intrauterine hypoxia is primarily responsible for MSAF and its aspiration. These components generate two primary responses: proximal airway and distal airway obstruction. In the first case, acute asphyxia can lead to pneumothorax, and emphysema, causing hypoxia, hypercapnia, acidosis, and PPHN. Distal airway obstruction induces air trapping and atelectasis. This same effect can be seen after the alveoli's inflammatory response and surfactant inactivation. In MSAF, proinflammatory cytokines such as IL-1β, IL-6, IL-8, and TNF-α are recognized. The main consequence of prolonged hypoxia is airway hyperreactivity and neurological deficit that can cause neonatal mortality. CD14: a cluster of differentiation 14; IL: interleukin; MD-2: protein MD-2; MSAF: meconium-stained amniotic fluid; PaO₂: partial pressure of oxygen; PCO₂: partial pressure of carbon dioxide; PPHN: persistent pulmonary hypertension in the neonate; TLR4: Toll-like receptor 4; TNF-α: tumor necrosis factor-alpha.

Neutrophils or macrophages use cytokines as chemical signals to induce cell growth, differentiation, chemotaxis, and cytotoxicity. They respond to ligands with TRLs and chemokines (e.g., IL-8) [14,44,52]. Thus, cytokine signaling indirectly initiates the inflammatory cascade and indicates the type or degree of cell injury.

In this regard, the instillation of meconium in rabbit pups induced a 52% greater number of apoptotic cells than in milk or saline solution ($p < 0.05$). Likewise, the IL-6 and TNF-α were 1.01 ± 0.32 pg/mL and 0.60 ± 0.34 pg/mL higher in the meconium animals [44]. The above reaffirms that cytokine levels could be considered early indicators to recognize MAS or fetal hypoxia. Since some authors have experimentally proven that the presence of albumin with meconium instillation reduces IL-8 expression and inhibits chemotaxis of polymorphonuclear cells, this could be a future treatment for MAS [13].

In an experimental model of MAS in newborn piglets made by Haakonsen Lindenskov et al. [14], meconium activated the complement system in its C5a' fraction, activating an inflammasome related to the caspase pathway that, in turn, induced a cellular response through the secretion of cytokines such as TNF-α, IL-1β, and IL-6, released by endothelial cells, T lymphocytes, local alveolar, and peritoneal macrophages. These cells favored phagocytosis and sent activation signals to lymphocytes and monocytes to initiate the local inflammation in the pulmonary parenchyma [61–63].

The presence of cytokines in MAS is an essential mechanism in the progression from mild lung disease to severe pneumonia and the development of respiratory distress, cell apoptosis, and severe pulmonary dysfunction.

4.3. Phagocytosis

At the pulmonary level, this mechanism begins with the secretion of cytokines that allow the attraction of other inflammatory cells, such as neutrophils and macrophages, to eliminate pathogens at a lesion site [64]. Craig et al. [65] suggest that meconium could inhibit the phagocytic capacity of alveolar macrophages. This finding was also reported in a trial conducted on rat alveolar macrophages NR8383 exposed to sterile human and equine meconium for a short time. In the latter, meconium induced a decrease in phagocytosis of macrophages to fluorescent latex beads but also reduced the presence of the respiratory burst in response to phorbol myristate acetate. These results suggest that contact with meconium alters the alveolar leukocytes' function and reduces their local defense function [66].

From an immunological point of view, the release of TNF-α by alveolar and peritoneal macrophages can lead to the production of ROS species such as superoxide anion, hydroxyl radicals (OH), as well as non-free radicals such as hydrogen peroxide (H_2O_2) and hypochlorous acid (HOCl) that aim to make the phagocytosed meconium by these cells to undergo oxidative digestion that, in combination with the enzymatic digestion of lysosomes, will cause the foreign antigens of the meconium to degrade [37,67]. The aforementioned is of great importance since the oxidative digestion of the meconium components comes from the granulocytes' respiratory burst. Therefore, macrophages' normal function is to work as a protective barrier for the organism. However, some studies have shown that meconium at low doses (0.2 mg/mL) tends to inhibit neutrophils and their activity, but at higher doses (1 and 2 mg/mL), it progressively stimulates the production of oxygen radicals in these cells [68]. In such a way, meconium could exert a regulatory function of the immune activity depending on its exposure time and the amount it was exposed to [69].

4.4. Apoptosis of Alveolar Cells

The damage induced by meconium aspiration was corroborated in experimental studies in 7 days-old Fisher rats intratracheally instilled with homologous meconium. Meconium-induced exudative alveolitis, deciliation of pseudostratified epithelial cells, recruitment of neutrophils and pulmonary alveolar macrophages to the bronchoalveolar space, intravascular sequestration of neutrophils, platelet aggregation, interstitial edema, escape of red cells and fibrin into the alveolar space [46,47]. Those findings corroborate that MAS may induce cellular damage and even apoptosis of pulmonary parenchyma cells (Figure 4) [70,71].

In the piglet model inoculated with human meconium, high concentrations of the enzyme phospholipase A2 have been identified, considered a biomarker of cellular injury and trigger of local inflammation [72,73]. Therefore, it is clear that meconium aspiration induces apoptosis of alveolar cells, which is suggested to be the starting point for hypoxia, atelectasis, and hemodynamic changes at the pulmonary level [2].

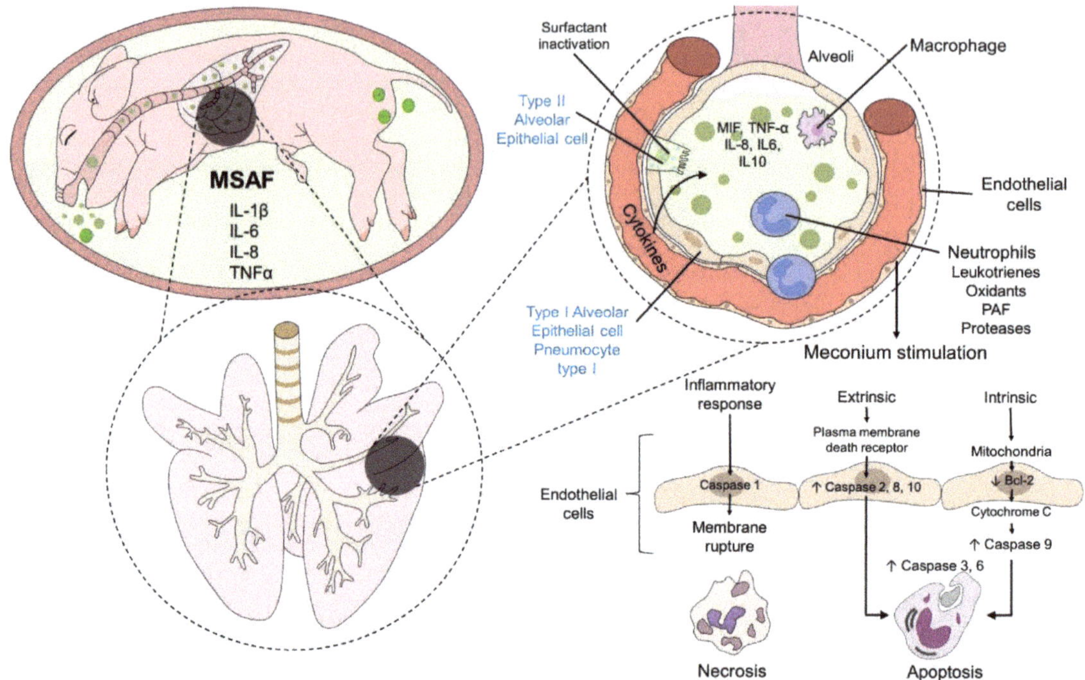

Figure 4. Apoptosis process after meconium aspiration. Derived from the inflammatory process in the lungs and, specifically, in the alveoli, the presence of cytokines, neutrophils, and macrophages evoke the endothelial cells of the alveoli to induce meconium stimulation which can result in three different pathways. The inflammatory response is associated with necrosis of the cells by the intervention of caspase 1. The extrinsic pathway involves the plasma membrane death receptor and the presence of caspase 2, 8, and 10, while caspase 9, together with cytochrome C and Bcl-2, participate in the intrinsic pathway to cause apoptosis of the endothelial cells. Bcl2: b cell lymphoma-2; IL: interleukin; MIF: macrophage migration inhibitory factor; TNF-α: tumor necrosis factor-alpha.

Although this cellular damage is part of the physiopathology of MAS, the precise mechanism of apoptosis induction is still unclear; a hypothesis suggests that the release of proinflammatory cytokines induces the angiotensin II receptors' expression, which is associated with cellular apoptosis [52]. In this context, Zagariya et al. [74] corroborated that after meconium aspiration in rabbit lungs and under an in situ end labeling (ISEL-DNA) assay, 70% of apoptotic cells were of the alveolar epithelium. Additionally, 20% of cells were in an apoptosis state with a significant increase of angiotensinogen ARNm, and caspase- 3 expression. Thus, there could be a link between angiotensin levels and apoptosis induction by cytokine activation. It has been reported that levels less than 5 pg/mL of IL-10 can relate to an apoptosis media of 0.26% of neutrophils in newborns [75]. Thus, the above reaffirms the theory that cytokines may induce cellular death by interacting with angiotensinogen.

The explanation for this theory is that cytokines, such as TNF-α, induce the angiotensinogen gene expression leading to the conversion of angiotensin I to angiotensin II through the angiotensin-converting enzyme (ACE), which later binds to the receptors AT-1, linked to caspase pathways activation [15,16,76]. Under this reasoning, it has been postulated that the inhibition or blockade of ACE could inhibit cell apoptosis, as supported by a study by Zagariya et al. [77]. The authors performed the instillation of 10% sterile meconium in 2-week-old rabbit pups and received a dose of captopril at 500 mg/L in the drinking water prior to the instillation of meconium. They observed that captopril significantly reduced the neutrophils and macrophages expressing IL- 13 and IL- 8, IL-4, and TNF-α levels after the meconium inoculation. This could be explained by inhibiting the angiotensin I transformation, which could hinder the pulmonary lesion. Additionally, these authors maintain that captopril reduces leukocyte recruitment capacity due to pulmonary hyperreactivity, which has been refuted in studies using a mouse model [78].

Hence, apoptotic induction of alveolar cells is a crucial mechanism of the pulmonary lesion in MAS physiopathology related to cytokine release and its interaction with angiotensin II. Understanding this interaction could lead to a way to treat and inhibit lung injury and even promote the repair of the pulmonary parenchyma.

5. Role of the Surfactant

The lung parenchyma comprises various cells that allow oxygen uptake from breathing and the subsequent elimination of waste in the form of CO_2 and water vapor. Therefore, it is essential to note that the blood-air barrier is responsible for this process through pneumocytes type 1, 2, and 3. The first two pneumocytes are involved in the barrier formation and can allow the trespass of oxygen from the alveolar space into the bloodstream [3].

5.1. Biochemistry and Surfactant Function

The pulmonary surfactant is produced by the pneumocytes type 2 at the alveolar level and by Club cells, formerly known as Clara cells, in bronchioles [79]. This substance is a phospholipid complex (80–85%), neutral lipids (2–5%), and specific proteins (10%). It is worth mentioning that phosphatidylcholine is the most abundant phospholipid of the tensioactive agents representing up to 70% of it. It presents two forms, the union of two palmitic acids to the glycerol-phosphorylcholine or dipalmitoyl-phosphatidylcholine (DPCC). In addition to phosphatidylglycerol (12%), phosphatidylethanolamine (5%), phosphatidylinositol (4%), phosphatidylserine (2%), and sphingomyelin (1.5%) [80,81]. Surfactant is stored in lamellar bodies and secreted by exocytosis to the alveolar lumen during the respiratory dynamic, where it exerts its biological function in the alveolar epithelium [82].

The lung surfactant poses two specific biological functions. The first is reducing the surface tension in the air-liquid interface preventing the alveolar collapse during the gaseous interchange [83]. That is, it stabilizes the interface of water and the alveolus because it prevents the surface tension from approaching zero at the end of expiration, seeking to prevent alveolar collapse [84,85]. In this regard, Schenck and Fiegel [86] characterized the tensiometric behavior of a lung surfactant extract of a calf on the air-liquid interface in viscoelastic gels as similar to the mucus of the pulmonary parenchyma inflammation. The authors observed that the viscoelastic gel property inhibited $0.24 \pm 0.42\%$ because of the capacity of lung surfactant to reduce the surface tension.

The second function of lung surfactant is immunity due to the presence of hydrophobic surfactant proteins (SPs), in which surfactant-A (SP-A) and protein surfactant-D (SP-D) are considered calcium-dependent lectins, participating in diverse innate immunity mechanisms in the lung, and in the homeostasis of alveolar surfactant in mammals [85,87]. Han and Mallampalli [82] describe that mentioned proteins allow the elimination of pathogens such as viruses or bacteria. This is due to the presence of domains of C-terminal lectins that bind to pathogen microorganisms' oligosaccharides. It is described that these collectins facilitate their phagocytosis by monocytes and macrophages as part of the function of the innate immune system [88–90].

The presence of collectins in pulmonary surfactant has been the subject of study in experimental models in mice deficient in SP-A, intratracheally inoculated with pathogens such as *Pneumocystis carinii* and group B *Streptococcus*, where the absence of these proteins was associated with an inefficient clearance of these agents [91,92]. Interestingly, other findings were observed in an experiment using mice deficient in SP-A, SP-D, and macrophage TLR receptors, which were associated with a lower presence of connexin-43, IL-1, IL-6, and receptor of oxytocin that caused a 12 h increase in labor time [93]. This means that pulmonary collectins acting via TLR2 serve a modulatory role in the timing of labor, which could play a role in modulating other physiological responses, such as labor.

From all the above, it is evident that pulmonary surfactant has a dual function of maintaining oxygen exchange by reducing surface tension and establishing defense mechanisms against pathogenic microorganisms.

5.2. Surfactant and Lung Maturation

Flageole et al. [94] used 17 fetuses from 9 pregnant ewes at term. Fetuses underwent tracheal tamponade at 93 days by tracheal ligation and were unplugged at 110 days. Morphometric analysis revealed that the tamponade induced pulmonary hyperplasia, in addition to a decrease in the number of type II pneumocytes compared to the control group (tamponade = 4.7 ± 0.1 versus control = 55.9 ± 4, $p = 0.0003$). Although the ovine model studies a complete obstruction, the authors denote the need to study the effect of partial obstruction. Therefore, considering the obstructive properties of aspirated meconium, it would be interesting to evaluate the possible intervention in lung maturation in MAS.

5.2.1. Steroid-Associated Regulatory Mechanisms of Lung Maturation

Mammals at the end of the pregnancy experience an increase in the concentration of fetal adrenocorticosteroids, stimulating the Prostaglandin F2α release, and a decrease in the concentration of progesterone by converting it to estradiol, thus promoting luteolysis, and finally, parturition, in addition to lung maturation and surfactant production [95]. A study by Jobe et al. [96] compared the effect of betamethasone acetate (Beta-Ac) and betamethasone phosphate (Beta-PO4) together with Beta-Ac (a combination clinically used in human fetuses) on fetal lung maturation in merino ewes 48 h before premature delivery at different doses, in which it was observed that fetal lung maturation improved even with a single dose of Beta-Ac. In the same ewes, the treatment influenced surface-active proteins, such as SP-A, SP-B, and SP-C, by significantly increasing their synthesis ($p < 0.05$). Results similar to those obtained by Ballard et al. [97] through the application of betamethasone in sheep in a maximum of 4 doses in 104 days of gestation with parturition at 125 days, showing an increase of 80% of body weight from the second to fourth dose in comparison with their control group. Additionally, there was an ~11% increase in SP-A and ~3% SP-B tissue concentrations.

Similarly, it has been recorded that the fetal lung expresses various estrogen receptors (ER), which participate in its maturation by modulating alveologenesis through ERa and ERb, as well as the expression of vascular endothelial growth factor and SP-B and SP-C in conjunction with progesterone [98]. Likewise, Pepe et al. [99] mention that in baboons, the expression of these proteins increases from day 120–140 of gestation, the same period in which lung maturation depends on estrogen and its control by the placental-fetal pituitary-adrenocortical axis.

5.2.2. Biochemical Signaling of Maturation

At the end of gestation, fetuses undergo various changes, such as lung maturation and surfactant synthesis, as adaptive mechanisms to develop and survive in an extrauterine environment [95], in which various biochemical processes mediated by glucocorticoids are involved, sex steroids, insulin, prolactin, catecholamines, fibroblast-pneumocyte factor, and epidermal growth factor [100]. Lung development proceeds in five phases, (1) embryonic, (2) pseudo-glandular, (3) canalicular, (4) saccular, and (5) alveolar. Particularly in dogs, the canalicular phase is observed between days 48 and 57, which is characterized by the development of type I and II pneumocytes, followed by the saccular phase from day 60 until the onset of labor, where it is believed that, like humans, synthesis of pulmonary surfactant takes place [101].

5.3. Biochemical Signaling of Surfactant Inactivation

The hypothesis that supports the leading cause of mortality from MAS in the newborn is related to the inflammatory process since the presence of meconium causes a severe local immune response [14]. However, some authors consider an additional mechanism: the inactivation of the pulmonary surfactant. This event could be an essential factor that intervenes in this syndrome's pathological process, as Sun et al. [42] studied. Various amounts of meconium, the water-methanol soluble fraction, or the chloroform soluble fraction were added to standard suspensions of porcine surfactant (Curosurf®). In a pulsating-bubble or Wilhelmy-balance system, meconium and its subfractions inhibited surfactant activity, but the chloroform soluble fraction had the highest specific inhibitory activity. Additionally, using an animal model with rabbits that received sterile human meconium observed that the compliance of the lung-thorax was reduced between 27–38%, and histological analysis showed interalveolar accumulation of fine meconium particles but no plugging in larger airways. These findings demonstrated that the inactivation of the surfactant participates significantly in the mechanisms of disease production, remembering that it tends to maintain the elasticity capacity of the parenchyma, but if it is absent, it could be affected.

Different authors recognize that surfactant deficiency in humans and animals can lead to the onset of respiratory distress or severe respiratory disease due to the loss of the ability to lower surface tension and protect against pathogens [80,102]. Given this evidence, it is necessary to mention how the inactivation of the surfactant by meconium occurs, for which two mechanisms are proposed. The first sustains that meconium components can directly alter the structure of the surfactant, according to Pallem et al. [103], while the second, cited by Kopincova and Calkovska [3], states that the presence of meconium alters the structure of DPCC, which undergoes fragmentation of its bilayer and alters the function of liposomes. Similarly, Casals and Cañadas [104] mention that cholesterol and meconium bile acids together inactivate pulmonary surfactant since bile acid micelles are capable of solubilizing cholesterol, facilitating its transfer to the surfactant complexes, such that their membranes become fluidized, altering their structure and function in interfacial absorption [105]. However, it is also anticipated that bile acids can penetrate and disintegrate lamellar structures and the surfactant monolayer, further affecting its function (Figure 5) [106].

Lugones et al. [107] conducted an in vitro study on the exposure of a natural surfactant to serum proteins, meconium, and cholesterol. They observed at the captive bubble surfactometer test under conditions mimicking respiratory dynamics that surfactant is strongly inhibited by serum exposure, but prior exposure to hyaluronic acid protected them from inhibition. This confirms that components such as serum, cholesterol, and triglycerides can directly inactivate the surfactant and induce changes in its dynamics [108]. For this reason, and supported by animal models in lambs and guinea pigs, it has been suggested that surfactant replacement therapy should be accompanied by a drug that allows its mechanism of action to be carried out successfully [109,110].

Likewise, there is a theory that some compounds derived from the inflammatory process, such as Reactive Oxygen Species (ROS) and soluble Phospholipase A2 (sPLA2), indirectly influence the inactivation of pulmonary surfactants. The latter directly inactivate the surfactant by hydrolyzing its phospholipids, generating proinflammatory eicosanoids and lysophospholipids that induce pneumocyte damage and a secondary inflammatory response culminating in the production of bioactive cytokines and ROS by pneumocytes, macrophages, and neutrophils. Those, in turn, modify the phospholipids and surfactant substances in addition to the destruction of pulmonary surfactant complexes, an effect intensified by the simultaneous production of C-reactive protein, which binds to the membranes of the surfactant, modifying its biophysical characteristics, increasing fluidity, as is the case with cholesterol [111].

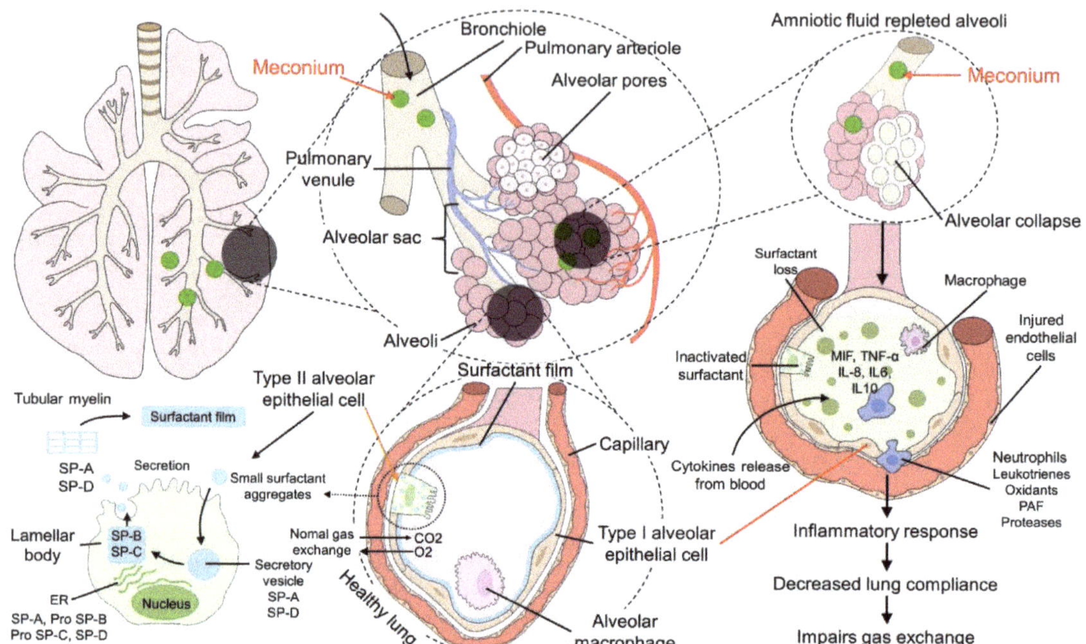

Figure 5. Surfactant inactivation in Meconium Aspiration Syndrome. In a healthy lung, the surfactant film is produced by the type II alveolar epithelial cells (pneumocyte type II) in their lamellar bodies, and this film prevents alveolar collapse. However, when meconium-stained amniotic fluid covers the alveoli, type II alveolar cells are not able to produce surfactant, and the presence of proinflammatory cytokines (e.g., MIF, TNF, IL) promotes the inflammatory response in the lung, as well as a decreased lung compliance, atelectasis, and impairment of the gas exchange, resulting in neonatal hypoxia and acidosis. IL: interleukin; MIF: macrophage inhibitor factor; SP: surfactant protein; TNF: tumor necrosis factor.

Scharama et al. [112] evaluated in vitro the activity of sPLA2 derived from human meconium on the hydrolysis and inactivation of the DPCC of the porcine natural surfactant Curosurf® through the concentrations of lysophosphatidylcholine. After incubation of meconium with the surfactant, the latter hydrolyzed 0.58% of DPCC, but when the temperature increased, this percentage increased to 6.22%, suggesting that the reaction remains stable in the heat. In addition, the surface tension was measured using a pulsating bubble surfactometer, showing an increase when an exclusive PLA2 eluant was applied compared to the eluant composed of other phospholipases, going from 1.7 ± 1.6 mN/m to 19.0 ± 3.58 mN/m ($p = 0.0001$) and 5.9 ± 3.9 mN/m, $p = 0.07$, respectively. In addition, Rodríguez-Capote et al. [113] evaluated the effect of ROS on pulmonary surfactant components by the oxidizing action of hypochlorous acid or Fenton reaction on bovine lipid extract surfactant (BLES) and a synthetic PL and SP-B and SP-C surfactant. A 20% decrease in phosphatidylcholine (PC) was observed after 24 h of exposure to both oxidizing agents, as well as an increase in carbonyls, derived products of protein oxidation six times higher in BLES treated with Fenton's reaction compared to hypochlorous acid ($p = 0.008$), of which a significant increase in carbonyls from SP-B was shown, being 2.6 times higher by the Fenton's reaction.

Therefore, the indirect role of meconium compounds on the inactivation of pulmonary surfactant is corroborated through the oxidation of both PL and surfactant proteins (SP-B and SP-C) mediated by elements resulting from the inflammatory process in the lung after aspiration.

6. Future Directions

The research tendencies regarding MAS have made it clear that it has areas of study that have not been explored. For example, there is still controversy about whether the increase in cytokines produced by meconium or those present in it may be associated with increased cell apoptosis [16] or even ponder the interaction of these cytokines with angiotensin, which could be considered as a direct biomarker of this injury [76]. Similarly, could the inhibition of alveolar macrophage phagocytosis and its respiratory burst response by meconium exposure influence the increase in cell damage? Because this action is known to help remove meconium microparticles and pathogens; however, decreased phagocytosis and its response could promote the apoptosis of these alveolar cells and increase the susceptibility to secondary lung infections in neonates [66].

The treatment of MAS is controversial; since its use does not clearly influence decreasing neonatal mortality incidence, monotherapies such as corticosteroids, antibiotics, and phosphodiesterase inhibitors have presented variable results over time in reducing clinical signs of MAS [7,17,114]. Thus, it might be necessary to explore the combined use or multimodal therapy of the different drugs, as has been observed with the combined use of surfactant with glucocorticoids or the use with antibiotics that can help reduce the inactivation of the surfactant by increasing resistance to it [109,115]. Likewise, explore the use of vasodilators not only to reduce the incidence of pulmonary hypertension but also to consider that they may avoid hypoxia and reduce acidosis in the newborn [116,117].

Finally, an area of opportunity that has not been fully explored is combining treatment with ventilatory assistance with other options. In this regard, it has been observed that the use of standard high-frequency jet ventilation or in combination with low-rate intermittent mandatory ventilation can decrease the level of bronchopneumonia and edema in the lung affected by the MAS in the canine model [118]. Moreover, the combination with surfactant can reduce the presence of lung lesions in the pig model [32,102,119]. It is suggested that using some central nervous system stimulants, such as caffeine, could be beneficial in increasing tidal volume in newborns with signs of hypoxia [120]. In the same way, explore the possibility of the combined use of the treatments previously described with the use of non-invasive ventilatory assistance that could promote recovery or even reduce lung injury. Data has yet to be presented to date, perhaps due to the lack of complete understanding of the pathophysiological mechanisms of MAS.

The contribution of experimental animal models of MAS in advancing knowledge of pathophysiology, the inflammatory process, the inactivation of surfactants, and the different therapeutic options are evident. It is important to note that many experimental studies on the pathophysiology of MAS have used adult laboratory animals without considering that MAS is a disease of the newborn and without considering possible differences in the inflammatory response between fetuses, neonates, and adult animals. In addition, to evaluate the importance of using heterologous instead of homologous meconium in experimental models. Therefore, neonatal animal models should be considered to evaluate the damage and inflammatory response, understand the pathogenesis and prevent MAS.

7. Conclusions

MAS, in veterinary medicine, is a pathological process affecting the newborn's survival; therefore, early recognition could be a tool to reduce mortality in the newborn.

The main event in the MAS is the inflammatory process at the pulmonary level, which is related to the hyperreactivity of the alveolar epithelium due to the presence of meconium. Other meconium components also participate, such as proinflammatory cytokines that generate leukocyte chemotaxis and lead to epithelial destruction by inducing apoptosis. The inhibition of the phagocytic function of alveolar macrophages results from an oxidative process by reactive oxygen radicals. It is also associated with local immunosuppression, which may also be related to meconium components that inactivate and prevent the surface function of pulmonary surfactant and an immunological response by degrading its components.

Recognition of the pathophysiology of this process allows the possibility to design and plan strategies for its prevention and treatment to reduce the adverse effects on lung activity derived from the MAS and its influence on newborn survival.

Author Contributions: D.M.-R., D.V.-G., A.M.-R., A.O., I.H.-Á., A.D.-O., A.C.-A., K.F.-P., J.J.-R., and J.M.-B. contributed to the conceptualization, writing, reading, and approval of the final manuscript. All authors have read and agreed to the published version of the manuscript.

Funding: This research received no external funding.

Institutional Review Board Statement: Not applicable.

Informed Consent Statement: Not applicable.

Data Availability Statement: Not applicable.

Conflicts of Interest: The authors declare no conflict of interest.

References

1. Chiruvolu, A.; Miklis, K.K.; Chen, E.; Petrey, B.; Desai, S. Delivery room management of meconium-stained newborns and respiratory support. *Pediatrics* **2018**, *142*, e20181485. [CrossRef] [PubMed]
2. Martínez-Burnes, J.; Mota-Rojas, D.; Villanueva- García, D.; Ibarra- Ríos, D.; Lezama- García, K.; Barrios- García, H.; López-Mayagoitia, A. Meconium aspiration syndrome in mammals. *CAB Rev.* **2019**, *14*, 1–11. [CrossRef]
3. Kopincova, J.; Calkovska, A. Meconium-induced inflammation and surfactant inactivation: Specifics of molecular mechanisms. *Pediatr. Res.* **2016**, *79*, 514–521. [CrossRef]
4. Yokoi, K.; Iwata, O.; Kobayashi, S.; Muramatsu, K.; Goto, H. Influence of foetal inflammation on the development of meconium aspiration syndrome in term neonates with meconium-stained amniotic fluid. *PeerJ* **2019**, *7*, e7049. [CrossRef]
5. van Ierland, Y.; de Beaufort, A.J. Why does meconium cause meconium aspiration syndrome? Current concepts of MAS pathophysiology. *Early Hum. Dev.* **2009**, *85*, 617–620. [CrossRef] [PubMed]
6. de Beaufort, A.J.; Bakker, A.C.; van Tol, M.J.D.; Poorthuis, B.J.; Schrama, A.J.; Berger, H.M. Meconium is a source of pro-inflammatory substances and can induce cytokine production in cultured A549 epithelial cells. *Pediatr. Res.* **2003**, *54*, 491–495. [CrossRef] [PubMed]
7. Mokra, D.; Mokry, J.; Tonhajzerova, I. Anti-inflammatory treatment of meconium aspiration syndrome: Benefits and risks. *Respir. Physiol. Neurobiol.* **2013**, *187*, 52–57. [CrossRef]
8. Monfredini, C.; Cavallin, F.; Villani, P.E.; Paterlini, G.; Allais, B.; Trevisanuto, D. Meconium aspiration syndrome: A narrative review. *Children* **2021**, *8*, 230. [CrossRef]
9. Mota-Rojas, D.; Martínez-Burnes, J.; Trujillo, M.E.; López, A.; Rosales, A.M.; Ramírez, R.; Orozco, H.; Merino, A.; Alonso-Spilsbury, M. Uterine and fetal asphyxia monitoring in parturient sows treated with oxytocin. *Anim. Reprod. Sci.* **2005**, *86*, 131–141. [CrossRef]
10. Mota-Rojas, D.; Rosales, A.M.; Trujillo, M.E.; Orozco, H.; Ramírez, R.; Alonso-Spilsbury, M. The effects of vetrabutin chlorhydrate and oxytocin on stillbirth rate and asphyxia in swine. *Theriogenology* **2005**, *64*, 1889–1897. [CrossRef]
11. Mota-Rojas, D.; Nava-Ocampo, A.A.; Trujillo, M.E.; Velázquez-Armenta, Y.; Ramírez-Necoechea, R.; Martínez-Burnes, J.; Alonso-Spilsbury, M. Dose minimization study of oxytocin in early labor in sows: Uterine activity and fetal outcome. *Reprod. Toxicol.* **2005**, *20*, 255–259. [CrossRef] [PubMed]
12. Swarnam, K.; Soraisham, A.S.; Sivanandan, S. Advances in the management of meconium aspiration syndrome. *Int. J. Pediatr.* **2012**, *2012*, 359571. [CrossRef] [PubMed]
13. Lindenskov, P.H.H.; Castellheim, A.; Aamodt, G.; Saugstad, O.D. Meconium induced IL-8 production and intratracheal albumin alleviated lung injury in newborn pigs. *Pediatr. Res.* **2005**, *57*, 371–377. [CrossRef] [PubMed]
14. Haakonsen Lindenskov, P.H.; Castellheim, A.; Saugstad, O.D.; Mollnes, T.E. Meconium aspiration syndrome: Possible pathophysiological mechanisms and future potential therapies. *Neonatology* **2015**, *107*, 225–230. [CrossRef]
15. Wang, R.; Zagariya, A.; Ang, E.; Ibarra-Sunga, O.; Uhal, B.D. Fas-induced apoptosis of alveolar epithelial cells requires ANG II generation and receptor interaction. *Am. J. Physiol. Cell. Mol. Physiol.* **1999**, *277*, L1245–L1250. [CrossRef]
16. Filippatos, G.; Tilak, M.; Pinillos, H.; Uhal, B. Regulation of apoptosis by angiotensin II in the heart and lungs (Review). *Int. J. Mol. Med.* **2001**, *7*, 273–280. [CrossRef]
17. Asad, A.; Bhat, R. Pharmacotherapy for meconium aspiration. *J. Perinatol.* **2008**, *28*, S72–S78. [CrossRef]
18. Mota-Rojas, D.; López, A.; Martínez-Burnes, J.; Muns, R.; Villanueva-García, D.; Mora-Medina, P.; González-Lozano, M.; Olmos-Hernández, A.; Ramírez-Necoechea, R. Is vitality assessment important in neonatal animals? *CAB Rev. Perspect. Agric. Vet. Sci. Nutr. Nat. Resour.* **2018**, *13*, 1–13. [CrossRef]
19. Cleary, G.M.; Wiswell, T.E. Meconium-stained amniotic fluid and the meconium aspiration syndrome. *Pediatr. Clin. N. Am.* **1998**, *45*, 511–529. [CrossRef]

20. Fischer, C.; Rybakowski, C.; Ferdynus, C.; Sagot, P.; Gouyon, J.B. A population-based study of meconium aspiration syndrome in neonates born between 37 and 43 weeks of gestation. *Int. J. Pediatr.* **2012**, *2012*, 321545. [CrossRef]

21. Parker, T.A.; Kinsella, J.P. Respiratory Failure in the Term Newborn. In *Avery's Diseases of the Newborn*; Gleason, C.A., Devaskar, S.U., Eds.; Elsevier: Amsterdam, The Netherlands, 2012; pp. 647–657.

22. Lopez, A.; Bildfell, R. Pulmonary inflammation associated with aspirated meconium and epithelial cells in calves. *Vet. Pathol.* **1992**, *29*, 104–111. [CrossRef] [PubMed]

23. Tanaka, M.; Izawa, T.; Kuwamura, M.; Ozaki, M.; Nakao, T.; Ito, S.; Yamate, J. A case of meconium aspiration syndrome in a bottlenose dolphin (*Tursiops truncatus*) calf. *J. Vet. Med. Sci.* **2014**, *76*, 81–84. [CrossRef] [PubMed]

24. Rawat, M.; Chandrasekharan, P.; Gugino, S.F.; Koenigsknecht, C.; Nielsen, L.; Wedgwood, S.; Mathew, B.; Nair, J.; Steinhorn, R.; Lakshminrusimha, S. Optimal oxygen targets in term lambs with meconium aspiration syndrome and pulmonary hypertension. *Am. J. Respir. Cell Mol. Biol.* **2020**, *63*, 510–518. [CrossRef]

25. Randall, G. Observations on parturition in the sow. I. Factors associated with the delivery of the piglets and their subsequent behaviour. *Vet. Rec.* **1972**, *90*, 178–182. [CrossRef] [PubMed]

26. Randall, G. Observations on parturition in the sow. II. Factors influencing stillbirth and perinatal mortality. *Vet. Rec.* **1972**, *90*, 183–186. [CrossRef] [PubMed]

27. Galvin, N.; Collins, D. Perinatal asphyxia syndrome in the foal: Review and a case report. *Ir. Vet. J.* **2004**, *57*, 707. [CrossRef] [PubMed]

28. Kirimi, E.; Tuncer, O.; Kösem, M.; Ceylan, E.; Tas, A.; Tasal, I.; Balahoroğlu, R.; Caksen, H. The effects prednisolone and serum malondialdehyde levels in puppies with experimentally induced meconium aspiration syndrome. *J. Int. Med. Res.* **2003**, *31*, 113–122. [CrossRef] [PubMed]

29. Magdesian, K.G. Neonatology. In *Equine Emergencies*; Orsini, J.A., Divers, T.J., Eds.; Elsevier: Amsterdam, The Netherlands, 2014; pp. 528–564.

30. Mota-Rojas, D.; Martinez-Burnes, J.; Alonso-Spilsbury, M.L.; Lopez, A.; Ramirez-Necoechea, R.; Trujillo-Ortega, M.E.; Medina-Hernandez, F.J.; de la Cruz, N.I.; Albores-Torres, V.; Loredo-Osti, J. Meconium staining of the skin and meconium aspiration in porcine intrapartum stillbirths. *Livest. Sci.* **2006**, *102*, 155–162. [CrossRef]

31. Mota-Rojas, D.; Villanueva-García, D.; Solimano, A.; Muns, R.; Ibarra-Ríos, D.; Mota-Reyes, A. Pathophysiology of perinatal asphyxia in humans and animal models. *Biomedicines* **2022**, *10*, 347. [CrossRef]

32. Wiswell, T.E.; Foster, N.H.; Slayter, M.V.; Hachey, W.E. Management of a piglet model of the meconium aspiration syndrome with high. frequency or conventional ventilation. *Am. J. Dis. Child.* **1992**, *146*, 1287–1293. [CrossRef]

33. Wiswell, T.E.; Peabody, S.S.; Davis, J.M.; Slayter, M.V.; Bent, R.C.; Merritt, T.A. Surfactant therapy and high-frequency jet ventilation in the management of a piglet model of the meconium aspiration syndrome. *Pediatr. Res.* **1994**, *36*, 494–500. [CrossRef] [PubMed]

34. Soukka, H.; Rautanen, M.; Halkola, L.; Kero, P.; Kääpä, P. Meconium aspiration induces ARDS-like pulmonary response in lungs of ten-week-old pigs. *Pediatr. Pulmonol.* **1997**, *23*, 205–211. [CrossRef]

35. Lindenskov, P.H.H.; Castellheim, A.; Aamodt, G.; Saugstad, O.D.; Mollnes, T.E. Complement activation reflects severity of meconium aspiration syndrome in newborn pigs. *Pediatr. Res.* **2004**, *56*, 810–817. [CrossRef] [PubMed]

36. Thomas, A.M.; Schjalm, C.; Nilsson, P.H.; Lindenskov, P.H.H.; Rørtveit, R.; Solberg, R.; Saugstad, O.D.; Berglund, M.M.; Strömberg, P.; Lau, C.; et al. Combined inhibition of C5 and CD14 attenuates systemic inflammation in a piglet model of meconium aspiration syndrome. *Neonatology* **2018**, *113*, 322–330. [CrossRef] [PubMed]

37. Castellheim, A.; Lindenskov, P.H.H.; Pharo, A.; Aamodt, G.; Saugstad, O.D.; Mollnes, T.E. Meconium aspiration syndrome induces complement-associated systemic inflammatory response in newborn piglets. *Scand. J. Immunol.* **2005**, *61*, 217–225. [CrossRef] [PubMed]

38. Gooding, C.A.; Gregory, G.A.; Taber, P.; Wright, R.R. An experimental model for the study of meconium aspiration of the newborn. *Radiology* **1971**, *100*, 137–140. [CrossRef]

39. Mammel, M.C.; Gordon, M.J.; Connett, J.E.; Boros, S.J. Comparison of high-frequency jet ventilation and conventional mechanical ventilation in a meconium aspiration model. *J. Pediatr.* **1983**, *103*, 630–634. [CrossRef]

40. Jovanovic, R.; Nguyen, H.T. Experimental meconium aspiration in guinea pigs. *Obstet. Gynecol.* **1989**, *73*, 652–656. [CrossRef]

41. Tyler, D.C.; Murphy, J.; Cheney, F.W. Mechanical and chemical damage to lung tissue caused by meconium aspiration. *Pediatrics* **1978**, *62*, 454–459. [CrossRef]

42. Sun, B.; Curstedt, T.; Robertson, B. Surfactant inhibition in experimental meconium aspiration. *Acta Paediatr.* **1993**, *82*, 182–189. [CrossRef]

43. Al-Mateen, K.B.; Dailey, K.; Grimes, M.M.; Gutcher, G.R. Improved oxygenation with exogenous surfactant administration in experimental meconium aspiration syndrome. *Pediatr. Pulmonol.* **1994**, *17*, 75–80. [CrossRef] [PubMed]

44. Zagariya, A.; Sierzputovska, M.; Navale, S.; Vidyasagar, D. Role of Meconium and hypoxia in meconium aspiration-induced lung injury in neonatal rabbits. *Mediators Inflamm.* **2010**, *2010*, 204831. [CrossRef]

45. Martínez-Burnes, J.; López, A.; Horney, B.; MacKenzie, A.; Brimacombe, M. Cytologic and biochemical changes associated with inoculation of amniotic fluid and meconium into lungs of neonatal rats. *Am. J. Vet. Res.* **2001**, *62*, 1636–1641. [CrossRef]

46. Martínez-Burnes, J.; López, A.; Wright, G.M.; Ireland, W.P.; Wadowska, D.W.; Dobbin, G. V Microscopic changes induced by the intratracheal inoculation of amniotic fluid and meconium in the lung of neonatal rats. *Histol. Histopathol.* **2002**, *17*, 1067–1076. [CrossRef]

47. Martínez-Burnes, J.; Lopez, A.; Wright, G.M.; Ireland, W.P.; Wadowska, D.W.; Dobbin, G. V Ultrastructural changes in the lungs of neonatal rats intratracheally inoculated with meconium. *Histol. Histopathol.* **2003**, *18*, 1081–1094. [CrossRef] [PubMed]

48. Foust, R.; Tran, N.N.; Cox, C.; Miller, T.F.; Greenspan, J.S.; Wolfson, M.R.; Shaffer, T.H. Liquid assisted ventilation: An alternative ventilatory strategy for acute meconium aspiration injury. *Pediatr. Pulmonol.* **1996**, *21*, 316–322. [CrossRef]

49. Block, M.F.; Kallenberger, D.A.; Kern, J.D.; Nepveux, R.D. In utero meconium aspiration by the baboon fetus. *Obstet. Gynecol.* **1981**, *57*, 37–40. [CrossRef]

50. Righetti, C.; Peroni, D.G.; Pietrobelli, A.; Zancanaro, C. Proton nuclear magnetic resonance analysis of meconium composition in newborns. *J. Pediatr. Gastroenterol. Nutr.* **2003**, *36*, 498–501. [CrossRef]

51. Lee, J.; Romero, R.; Lee, K.A.; Kim, E.N.; Korzeniewski, S.J.; Chaemsaithong, P.; Yoon, B.H. Meconium aspiration syndrome: A role for fetal systemic inflammation. *Am. J. Obstet. Gynecol.* **2016**, *214*, 366.e1–366.e9. [CrossRef]

52. Vidyasagar, D.; Zagariya, A. Studies of meconium-induced lung injury: Inflammatory cytokine expression and apoptosis. *J. Perinatol.* **2008**, *28*, S102–S107. [CrossRef]

53. Anand, V.; Basu, S.; Yadav, S.S.; Narayan, G.; Bhatia, B.D.; Kumar, A. Activation of Toll-like receptors in meconium aspiration syndrome. *J. Perinatol.* **2018**, *38*, 137–141. [CrossRef] [PubMed]

54. Seong, S.-Y.; Matzinger, P. Hydrophobicity: An ancient damage-associated molecular pattern that initiates innate immune responses. *Nat. Rev. Immunol.* **2004**, *4*, 469–478. [CrossRef] [PubMed]

55. Takeda, K.; Akira, S. Regulation of innate immune responses by Toll-like receptors. *Jpn. J. Infect. Dis.* **2001**, *54*, 209–219. [PubMed]

56. Janeway, C.A.; Medzhitov, R. Innate immune recognition. *Annu. Rev. Immunol.* **2002**, *20*, 197–216. [CrossRef] [PubMed]

57. Nozal, P.; López-Trascasa, M. Autoanticuerpos frente a proteínas de la vía alternativa del complemento en enfermedad renal. *Nefrología* **2016**, *36*, 489–495. [CrossRef]

58. Mollnes, T.E.; Castellheim, A.; Lindenskov, P.H.H.; Salvesen, B.; Saugstad, O.D. The role of complement in meconium aspiration syndrome. *J. Perinatol.* **2008**, *28*, S116–S119. [CrossRef] [PubMed]

59. Yamada, T.; Minakami, H.; Matsubara, S.; Yatsuda, T.; Kohmura, Y.; Sato, I. Meconium-stained amniotic fluid exhibits chemotactic activity for polymorphonuclear leukocytes in vitro. *J. Reprod. Immunol.* **2000**, *46*, 21–30. [CrossRef] [PubMed]

60. Yamada, T.; Matsubara, S.; Minakami, H.; Kohmura, Y.; Hiratsuka, M.; Sato, I. Chemotactic activity for polymorphonuclear leukocytes: Meconium versus meconium-stained amniotic fluid. *Am. J. Reprod. Immunol.* **2000**, *44*, 275–278. [CrossRef]

61. Samstad, E.O.; Niyonzima, N.; Nymo, S.; Aune, M.H.; Ryan, L.; Bakke, S.S.; Lappegård, K.T.; Brekke, O.-L.; Lambris, J.D.; Damås, J.K.; et al. Cholesterol crystals induce complement-dependent inflammasome activation and cytokine release. *J. Immunol.* **2014**, *192*, 2837–2845. [CrossRef]

62. Horvath, G.L.; Schrum, J.E.; De Nardo, C.M.; Latz, E. Intracellular sensing of microbes and danger signals by the inflammasomes. *Immunol. Rev.* **2011**, *243*, 119–135. [CrossRef]

63. Klebanoff, S.J. Myeloperoxidase: Friend and foe. *J. Leukoc. Biol.* **2005**, *77*, 598–625. [CrossRef]

64. Hutton, E.K.; Thorpe, J. Consequences of meconium stained amniotic fluid: What does the evidence tell us? *Early Hum. Dev.* **2014**, *90*, 333–339. [CrossRef] [PubMed]

65. Craig, S.; Lopez, A.; Hoskin, D.; Markham, F. Meconium inhibits phagocytosis and stimulates respiratory burst in alveolar macrophages. *Pediatr. Res.* **2005**, *57*, 813–818. [CrossRef]

66. Craig, J.S. The Effect of Meconium and Amniotic Fluid on Alveolar Macrophage Function. Master's Thesis, University of Prince Edward Island, Charlottetown, PE, Canada, 2003.

67. Kojima, T.; Hattori, K.; Fujiwara, T.; Sasai-Takedatsu, M.; Kobayashi, Y. Meconium-induced lung injury mediated by activation of alveolar macrophages. *Life Sci.* **1994**, *54*, 1559–1562. [CrossRef] [PubMed]

68. Soukka, H.R.; Ahotupa, M.; Ruutu, M.; Kääpä, P.O. Meconium stimulates neutrophil oxidative burst. *Am. J. Perinatol.* **2002**, *19*, 279–284. [CrossRef] [PubMed]

69. Bergamini, C.; Gambetti, S.; Dondi, A.; Cervellati, C. Oxygen, reactive oxygen species and tissue damage. *Curr. Pharm. Des.* **2004**, *10*, 1611–1626. [CrossRef]

70. Rahman, S.; Unsworth, J.; Vause, S. Meconium in labour. *Obstet. Gynaecol. Reprod. Med.* **2013**, *23*, 247–252. [CrossRef]

71. Bhutani, V.K. Developing a systems approach to prevent meconium aspiration syndrome: Lessons learned from multinational studies. *J. Perinatol.* **2008**, *28*, S30–S35. [CrossRef] [PubMed]

72. Kääpä, P.; Soukka, H. Phospholipase A2 in meconium-induced lung injury. *J. Perinatol.* **2008**, *28*, S120–S122. [CrossRef]

73. Korhonen, K.; Soukka, H.; Halkola, L.; Peuravuori, H.; Aho, H.; Pulkki, K.; Kero, P.; Kääpä, P.O. Meconium induces only localized inflammatory lung injury in piglets. *Pediatr. Res.* **2003**, *54*, 192–197. [CrossRef]

74. Zagariya, A.; Bhat, R.; Chari, G.; Uhal, B.; Navale, S.; Vidyasagar, D. Apoptosis of airway epithelial cells in response to meconium. *Life Sci.* **2005**, *76*, 1849–1858. [CrossRef] [PubMed]

75. Oei, J.; Lui, K.; Wang, H.; Henry, R. Decreased neutrophil apoptosis in tracheal fluids of preterm infants at risk of chronic lung disease. *Arch. Dis. Child.—Fetal Neonatal Ed.* **2003**, *88*, 245F–249F. [CrossRef]

76. Wang, R.; Alam, G.; Zagariya, A.; Gidea, C.; Pinillos, H.; Lalude, O.; Choudhary, G.; Oezatalay, D.; Uhal, B.D. Apoptosis of lung epithelial cells in response to TNF-alpha requires angiotensin II generation de novo. *J. Cell. Physiol.* **2000**, *185*, 253–259. [CrossRef] [PubMed]

77. Zagariya, A.; Navale, S.; Zagariya, O.; McClain, K.; Vidyasagar, D. IL13-induced lung fibrosis in meconium aspiration. *J. Biomed. Sci. Eng.* **2011**, *04*, 609–619. [CrossRef]

78. Ota, C.; Gopallawa, I.; Evanov, V.; Gewolb, I.H.; Uhal, B.D. Protection of meconium-induced lung epithelial injury by protease inhibitors. *J. Lung Pulm. Respir. Res.* **2017**, *4*, 00145. [CrossRef]

79. Lopez, A.; Martinson, S.A. Respiratory System, Thoracic Cavities, Mediastinum, and Pleurae. In *Pathologic Basis of Veterinary Diease*; Zachary, J.F., Ed.; Elsevier: Amsterdam, The Netherlands, 2022; pp. 547–642.

80. Christmann, U.; Buechner-Maxwell, V.A.; Witonsky, S.G.; Hite, R.D. Role of lung surfactant in respiratory disease: Current knowledge in large animal medicine. *J. Vet. Intern. Med.* **2009**, *23*, 227–242. [CrossRef]

81. Goerke, J. Pulmonary surfactant: Functions and molecular composition. *Biochim. Biophys. Acta—Mol. Basis Dis.* **1998**, *1408*, 79–89. [CrossRef]

82. Han, S.; Mallampalli, R.K. The role of surfactant in lung disease and host defense against pulmonary infections. *Ann. Am. Thorac. Soc.* **2015**, *12*, 765–774. [CrossRef]

83. Nogee, L.M.; Trapnell, B.C. Lung diseases associated with disruption of pulmonary surfactant homeostasis. In *Kendig's Disorders of the Respiratory Tract in Children*; Wilmott, R.W., Deterding, R., Li, A., Ratjen, F., Sly, P., Zar, H.J., Bush, A., Eds.; Elsevier: Philadelphia, PA, USA, 2019; pp. 836–849.

84. Freddi, N.A.; Proença Filho, J.O.; Fiori, H.H. Terapia com surfactante pulmonar exógeno em pediatria. *J. Pediatr.* **2003**, *79*, S205–S212. [CrossRef]

85. Kingma, P.; Jobe, A.H. The Surfactant System. In *Kendig's Disorders of the Respiratory Tract in Children*; Wilmott, R.W., Deterding, R., Li, A., Ratjen, F., Sly, P., Zar, H.J., Bush, A., Eds.; Elsevier: Philadelphia, PA, USA, 2019; pp. 57–62.e2.

86. Schenck, D.M.; Fiegel, J. Tensiometric and phase domain behavior of lung surfactant on mucus-like viscoelastic hydrogels. *ACS Appl. Mater. Interfaces* **2016**, *8*, 5917–5928. [CrossRef]

87. Bernhard, W. Lung surfactant: Function and composition in the context of development and respiratory physiology. *Ann. Anat.—Anat. Anzeiger* **2016**, *208*, 146–150. [CrossRef] [PubMed]

88. Crouch, E.; Hartshorn, K.; Ofek, I. Collectins and pulmonary innate immunity. *Immunol. Rev.* **2000**, *173*, 52–65. [CrossRef] [PubMed]

89. Wright, J.R. Immunomodulatory functions of surfactant. *Physiol. Rev.* **1997**, *77*, 931–962. [CrossRef] [PubMed]

90. Watson, A.; Madsen, J.; Clark, H.W. SP-A and SP-D: Dual functioning immune molecules with antiviral and immunomodulatory properties. *Front. Immunol.* **2021**, *11*, 622598. [CrossRef] [PubMed]

91. Linke, M.J.; Harris, C.E.; Korfhagen, T.R.; McCormack, F.X.; Ashbaugh, A.D.; Steele, P.; Whitsett, J.A.; Walzer, P.D. Immunosuppressed surfactant protein A-deficient mice have increased susceptibility to Pneumocystis carinii infection. *J. Infect. Dis.* **2001**, *183*, 943–952. [CrossRef]

92. LeVine, M.A.; Bruno, M.D.; Huelsman, K.M.; Ross, G.F.; Whitsett, J.A.; Korfhagen, T.R. Surfactent protein A- deficient mice are susceptible to gruop B streptococcal infection. *J. Immunol.* **1997**, *158*, 4336–4340.

93. Montalbano, A.P.; Hawgood, S.; Mendelson, C.R. Mice deficient in surfactant protein A (SP-A) and SP-D or in TLR2 manifest delayed parturition and decreased expression of inflammatory and contractile genes. *Endocrinology* **2013**, *154*, 483–498. [CrossRef]

94. Flageole, H.; Evrard, V.A.; Piedboeuf, B.; Laberge, J.-M.; Lerut, T.E.; Deprest, J.A. The plug-unplug sequence: An important step to achieve type II pneumocyte maturation in the fetal lamb model. *J. Pediatr. Surg.* **1998**, *33*, 299–303. [CrossRef]

95. Bonte, T.; Del Carro, A.; Paquette, J.; Charlot Valdieu, A.; Buff, S.; Rosset, E. Foetal pulmonary maturity in dogs: Estimated from bubble tests in amniotic fluid obtained via amniocentesis. *Reprod. Domest. Anim.* **2017**, *52*, 1025–1029. [CrossRef]

96. Jobe, A.H.; Nitsos, I.; Pillow, J.J.; Polglase, G.R.; Kallapur, S.G.; Newnham, J.P. Betamethasone dose and formulation for induced lung maturation in fetal sheep. *Am. J. Obstet. Gynecol.* **2009**, *201*, 611.e1–611.e7. [CrossRef]

97. Ballard, P.L.; Ning, Y.; Polk, D.; Ikegami, M.; Jobe, A.H. Glucocorticoid regulation of surfactant components in immature lambs. *Am. J. Physiol. Cell. Mol. Physiol.* **1997**, *273*, 1048–1057. [CrossRef] [PubMed]

98. Seaborn, T.; Simard, M.; Provost, P.R.; Piedboeuf, B.; Tremblay, Y. Sex hormone metabolism in lung development and maturation. *Trends Endocrinol. Metab.* **2010**, *21*, 729–738. [CrossRef] [PubMed]

99. Pepe, G.J.; Ballard, P.L.; Albrecht, E.D. Fetal lung maturation in estrogen-deprived baboons. *J. Clin. Endocrinol. Metab.* **2003**, *88*, 471–477. [CrossRef] [PubMed]

100. Laube, M.; Thome, U.H. Y It Matters—Sex differences in fetal lung development. *Biomolecules* **2022**, *12*, 437. [CrossRef] [PubMed]

101. Sipriani, T.; Grandi, F.; da Silva, L.; Maiorka, P.; Vannucchi, C. Pulmonary maturation in canine foetuses from early pregnancy to parturition. *Reprod. Domest. Anim.* **2009**, *44*, 137–140. [CrossRef] [PubMed]

102. Calkovska, A.; Mokra, D. Evaluation of surfactant function: In vitro and in vivo techniques in experimental respirology. *Acta Med. Martiniana* **2011**, *11*, 8–14. [CrossRef]

103. Pallem, V.; Kaviratna, A.S.; Chimote, G.; Banerjee, R. Effect of meconium on surface properties of surfactant monolayers and liposomes. *Colloids Surf. A Physicochem. Eng. Asp.* **2010**, *370*, 6–14. [CrossRef]

104. Casals, C.; Cañadas, O. Role of lipid ordered/disordered phase coexistence in pulmonary surfactant function. *Biochim. Biophys. Acta—Biomembr.* **2012**, *1818*, 2550–2562. [CrossRef]

105. Lopez-Rodriguez, E.; Echaide, M.; Taeusch, H.W.; Perez-Gil, J. Inhibition of pulmonary surfactant by meconium: Biophysical properties and molecular mechanism. *Biophys. J.* **2010**, *98*, 90a. [CrossRef]

106. Mokra, D.; Calkovska, A. How to overcome surfactant dysfunction in meconium aspiration syndrome? *Respir. Physiol. Neurobiol.* **2013**, *187*, 58–63. [CrossRef]

107. Lugones, Y.; Blanco, O.; López-Rodríguez, E.; Echaide, M.; Cruz, A.; Pérez-Gil, J. Inhibition and counterinhibition of Surfacen, a clinical lung surfactant of natural origin. *PLoS ONE* **2018**, *13*, e0204050. [CrossRef] [PubMed]

108. Lopez-Rodriguez, E.; Echaide, M.; Cruz, A.; Taeusch, H.W.; Perez-Gil, J. Meconium impairs pulmonary surfactant by a combined action of cholesterol and bile acids. *Biophys. J.* **2011**, *100*, 646–655. [CrossRef] [PubMed]

109. Bovino, F.; Denadai, D.S.; Ávila, L.G.; Deschk, M.; Santos, P.S.P.; Feitosa, F.L.F.; Peiró, J.R.; Mendes, L.C.N. Influence of dexamethasone and surfactant in the vitality and lung function of preterm lambs born by cesarean section. *Arq. Bras. Med. Veterinária Zootec.* **2019**, *71*, 44–52. [CrossRef]

110. Illia, R.; Solana, C.; Oliveri, P.; Toblli, J.; Imaz, M.U.; Häbich, D. Evidence of fetal pulmonary aspiration of intra-amniotic administered surfactant in animal experiment. *J. Perinat. Med.* **2004**, *32*, 354–358. [CrossRef]

111. Echaide, M.; Autilio, C.; Arroyo, R.; Perez-Gil, J. Restoring pulmonary surfactant membranes and films at the respiratory surface. *Biochim. Biophys. Acta—Biomembr.* **2017**, *1859*, 1725–1739. [CrossRef]

112. Schrama, A.; Beaufort, A.; Sukul, Y.; Jansen, S.; Poorthuis, B.; Berger, H. Phospholipase A2 is present in meconium and inhibits the activity of pulmonary surfactant: An in vitro study. *Acta Paediatr.* **2001**, *90*, 412–416. [CrossRef]

113. Rodríguez-Capote, K.; Manzanares, D.; Haines, T.; Possmayer, F. Reactive oxygen species inactivation of surfactant involves structural and functional alterations to surfactant proteins SP-B and SP-C. *Biophys. J.* **2006**, *90*, 2808–2821. [CrossRef]

114. Mokra, D.; Mokry, J. Glucocorticoids in the treatment of neonatal meconium aspiration syndrome. *Eur. J. Pediatr.* **2011**, *170*, 1495–1505. [CrossRef]

115. Stichtenoth, G.; Jung, P.; Walter, G.; Johansson, J.; Robertson, B.; Curstedt, T.; Herting, E. Polymyxin Bpulmonary surfactant mixtures have increased resistance to inactivation by meconium and reduce growth of gram-negative bacteria in vitro. *Pediatr. Res.* **2006**, *59*, 407–411. [CrossRef]

116. Teixeira-Mendonça, C.; Henriques-Coelho, T. Pathophysiology of pulmonary hypertension in newborns: Therapeutic indications. *Rev. Port. Cardiol.* **2013**, *32*, 1005–1012. [CrossRef]

117. Shekerdemian, L.S.; Ravn, H.B.; Penny, D.J. Intravenous sildenafil lowers pulmonary vascular resistance in a model of neonatal pulmonary hypertension. *Am. J. Respir. Crit. Care Med.* **2002**, *165*, 1098–1102. [CrossRef] [PubMed]

118. Keszler, M.; Klappenbach, S.; Reardon, E. Lung pathology after high frequency jet ventilation combined with low rate intermitten mandatory ventilation in a canine model of meconium aspiration. *Pediatr. Pulmonol.* **1988**, *4*, 144–149. [CrossRef] [PubMed]

119. Calkovska, A.; Sun, B.; Curstedt, T.; Renheim, G.; Robertson, B. Comined effects of high- frequency ventilation and surfactant tratment in experimental meconium aspiration syndrome. *Anaesthesiol. Scand.* **1999**, *43*, 135–145. [CrossRef] [PubMed]

120. Schmidt, B.; Roberts, R.S.; Davis, P.; Doyle, L.W.; Barrington, K.J.; Ohlsson, A.; Solimano, A.; Tin, W. Caffeine therapy for apnea of prematurity. *N. Engl. J. Med.* **2006**, *354*, 2112–2121. [CrossRef] [PubMed]

 *animals*

Article

Effects of Caffeine and Glucose Supplementation at Birth on Piglet Pre-Weaning Growth, Thermoregulation, and Survival

Lillie Jarratt [1], Sarah E. James [2], Roy N. Kirkwood [1] and Tanya L. Nowland [1,2,*]

[1]   School of Animal and Veterinary Sciences, The University of Adelaide, Roseworthy, SA 5371, Australia
[2]   Aquatic and Livestock Sciences, South Australian Research and Development Institute, The University of Adelaide, Roseworthy, SA 5371, Australia
*   Correspondence: tanya.nowland@sa.gov.au

Simple Summary: Genetic selection for larger litters has been a predominant driver for production efficiency improvements in the pork industry. However, the exacerbation of pre-weaning mortalities occurred simultaneously due to uterine crowding and extended farrowing durations, producing less viable piglets. Supplementing newborn piglets with caffeine has previously been shown to have potential for improving the survivability of lower viable piglets by providing neuroprotection. However, caffeine increases energy utilisation so is potentially counterproductive. To counter a possible energy limitation, our study aimed to investigate whether the addition of glucose to a caffeine supplement would improve piglet vigour and encourage quicker milk acquisition for growth and thermoregulation. We found that caffeine and glucose administered together improved early life growth of low birth weight piglets.

Abstract: Piglet pre-weaning mortality of approximately 15% represents a major economic and welfare concern to the pork industry. Supplementing neonatal piglets with glucose and/or caffeine has the potential to counteract hypoxic stress experienced during parturition and provide an energy substrate, which may improve survival to weaning. This study investigated the effects of caffeine and glucose supplementation at birth, in combination or separately, on piglet growth, thermoregulatory ability, and pre-weaning survival. At birth, 398 piglets were assigned to one of four oral treatments: saline, glucose (300 mg), caffeine (30 mg), or caffeine and glucose combined (30 mg caffeine and 300 mg glucose), dissolved in 6 mL saline. Piglets were tagged at birth, and time taken to reach the udder was recorded. Rectal temperatures were recorded at 4 h and 24 h post-partum, and body weights recorded at birth and 1, 3, and 18 days of age. Colostrum intake was estimated using birth and day 1 weights, and all pre-weaning mortalities were recorded. Treatments did not affect rectal temperature, colostrum intake, or pre-weaning mortality ($p > 0.05$). Low birth weight piglets (<0.9 kg) treated with caffeine and glucose had increased growth between 1 and 3 days of age ($p < 0.05$) compared to low birth weight piglets of other treatment groups. Caffeine supplementation alone reduced overall pre-weaning growth in low birth weight piglets compared to all other treatments ($p = 0.05$). Oral caffeine and glucose had no significant effect on piglet performance except in low birthweight piglets, where it improved growth in the first 3 days of life. Caffeine and glucose supplementation in combination may be beneficial for low birth weight piglets.

Keywords: piglets; neonate; colostrum intake; caffeine; glucose; pre-weaning survival

Citation: Jarratt, L.; James, S.E.; Kirkwood, R.N.; Nowland, T.L. Effects of Caffeine and Glucose Supplementation at Birth on Piglet Pre-Weaning Growth, Thermoregulation, and Survival. *Animals* **2023**, *13*, 435. https://doi.org/10.3390/ani13030435

Academic Editors: Daniel Mota-Rojas, Julio Martínez-Burnes and Agustín Orihuela

Received: 9 December 2022
Revised: 12 January 2023
Accepted: 25 January 2023
Published: 27 January 2023

Copyright: © 2023 by the authors. Licensee MDPI, Basel, Switzerland. This article is an open access article distributed under the terms and conditions of the Creative Commons Attribution (CC BY) license (https://creativecommons.org/licenses/by/4.0/).

1. Introduction

Selection for larger litter sizes, targeted to improve farming efficiency, has exacerbated the issue of pre-weaning mortality in the pig industry [1]. Pre-weaning mortality presents both economic and welfare concerns, accounting for 11–20% of piglet deaths [2]. Of these pre-weaning mortalities, 70–90% occur within the first three days postpartum [3]. The main causes are crushing by sows (up to 58%) and low piglet viability (2–30%) [4]. Larger litter sizes lead to an increased in litter weight variation, the proportion of less viable piglets as birth weights decrease, and the number of intrauterine growth restricted piglets. In addition to this, increases in parturition duration result in a higher risk of piglet hypoxia, causing foetal acidaemia and decreased piglet vitality in the first 72 h from birth [5]. Hypoxic pigs often have a lower chance of survival because, in part, they take longer to reach the udder and are less competitive once there. This negatively impacts colostrum intake and reduces immunity due to a lower acquisition of energy and immunoglobulins, decreasing the chance of survival, particularly in low birth weight piglets [6]. The addressing of viability issues of newborn piglets during their most vulnerable time is the appropriate time to target improvements in pre-weaning mortality.

Caffeine, a methylxanthine, has been shown to improve neurological impairment and provide neuroprotection to neonates via its ability to increase pulmonary carbon dioxide sensitivity, increase metabolic rate (increase heart rate), and increase the contractility of the diaphragm, as well as decrease muscle fatigue and diuresis [7]. Caffeine supplemented to rat and mice neonates have been shown to improve myelination defects and reduce brain injuries by increasing oxygen delivery to the brain, which is essential for preventing post-natal parturition-induced hypoxic deaths [8,9]. Studies investigating the efficacy of caffeine supplementation to both low and higher birth weight piglets documented improved metabolic variables of higher birth weight neonates, such as triglyceride, lactate, and blood glucose concentrations, as well as the increased growth of heavier weight piglets, suggesting that caffeine influenced energy utilisation [6,10]. Interestingly, Nowland et al. [10] administered caffeine at birth and 24 h and observed an increase in mortalities of low birth weight piglets, suggesting it to be a result of these piglets having insufficient body energy stores to withstand the increased energy utilisation caused by caffeine supplementation, which, consequently, instead induced a hypoglycaemic state. If this suggestion is accepted, it is plausible that the concurrent addition of glucose to counter the risk of hypoglycaemia may be a potential solution to prevent this issue. Glucose injections have previously been shown to increase blood glucose concentrations in the first 14 h of life and have been associated with an increased piglet body weight during their first 21 d [11].

Therefore, the present study aimed to determine whether supplementing caffeine and/or glucose to piglets at birth is beneficial for piglet viability, growth, thermoregulatory ability, and survival. It was hypothesised that the combined supplementation of glucose and caffeine to piglets at birth will decrease piglet pre-weaning mortality and improve their growth and thermoregulatory ability.

2. Materials and Methods

2.1. Animal Management

This research was conducted at the University of Adelaide's Roseworthy Piggery, South Australia, in accordance with the Australian Code for the care and use of animals for scientific purposes (8th edition, 2013), and was approved by the University of Adelaide ethics committee (Application ID 35136). The study was conducted in two replicates in August and September 2021. In total, 38 multiparous (parity 1–7, average 3.15 ± 1.63) Large White x Landrace sows and their litters were used (total born; n = 12.92 ± 0.53, born alive; n = 11.92 ± 0.49, stillborn; n = 1.00 ± 0.21). Sows were group housed during gestation and were moved into conventional farrowing crates 4.6 ± 0.05 days prior to their expected due date. Farrowing crates were 1.7 × 2.4 m with slatted flooring within temperature-controlled rooms that were maintained at 22 °C prior to farrowing and throughout lactation. Each farrowing crate had a trough feeder and nipple drinkers for the sow and a heated creep area for piglets on one side. Sows were fed a commercial lactation diet formulated to provide 14 MJ DE/kg, 17.0% crude protein, 0.81% total lysine, and 5% crude fibre, with ad libitum access to water. Sows were fed 2.5 kg/day until farrowing and then gradually increased to reach a maximum of 7–8 kg/d by day 7 of lactation, split between a morning and afternoon feed.

Farrowings were induced on day 113 of gestation with split-dose cloprostenol (Juramate®, Jurox Pty, Ltd., Rutherford, NSW, Australia; 250 µg/mL) vulval injections (0.5 mL at 07:00 and 15:00) to induce farrowing during staffed hours. Sows were monitored throughout farrowing, and manual interventions performed if birth intervals exceeded 2 h between each of the first three piglets born or after one hour from the fourth piglet onwards. No trial sows required extra assistance, such as oxytocin administration, above this protocol. Minimal intervention was provided to piglets post-partum. Cross-fostering was performed, based on litter size and teat capacity (12 ± 2), within 48 h of birth. The average litter size post foster was 11.99 ± 0.05, and piglets were weaned at 21 ± 0.5 days of age.

2.2. Experimental Design

At birth, a total of 398 piglets were tagged for individual identification, weighed, their sex recorded, and given one of the following oral treatments:

> 6 mL saline (SAL; n = 101);
> 6 mL saline with 30 mg caffeine (CAFF, n = 98);
> 6 mL saline with 300 mg glucose (GLUC, n = 101);
> 6 mL saline with 30 mg caffeine and 300 mg glucose (GLUC/CAFF, n = 98).

Treatment allocation was predetermined prenatally for each litter and spread evenly across birth order to achieve an even balance across the four treatments. A minimum of four piglets from each litter were treated to ensure each treatment was represented. Treatments were pre-warmed to 36–40 °C to minimise the impact on piglet body temperature and were orally administered to piglets by syringe. Once treated, all piglets were ear tagged for individual identification and then placed back into the farrowing crate behind the sow to minimise any potential influence on time taken to reach the udder.

2.3. Piglet Measurements

Data recorded for each litter were total number of piglets born, born alive, or stillborn and farrowing duration. Piglet rectal temperatures were recorded at 4 and 24 h after birth. All piglets were weighed at birth and again at 1, 3, and 18 days of age. Colostrum intake was estimated from birth and 24 h bodyweights using the equation of Devillers et al. [12]. Video footage was recorded for a subset of 24 litters and later analysed to measure the time taken for piglets to reach the udder following birth, indicated by nose contact with a teat. All piglet fosters and mortalities, including date and reason, were recorded from farrowing to weaning.

2.4. Statistical Analysis

Data analysis was performed using the IBM SPSS statistical package, version 25. Data were assessed for normality and outliers, with transformations of the data implemented where necessary (logarithmic transformations of time to udder and colostrum intake data, and square root transformation for 24 h rectal temperature data). Results were considered statistically significant if $p < 0.05$. A linear mixed model was used to assess and compare treatment effects on piglet growth (kg), body weight (kg), temperature (°C), and colostrum intake (g), along with the behavioural measure of time to udder (min). The fixed effects were treatment (SAL, CAFF, GLUC, or CAFF-GLUC), replicate (1 or 2), room (1, 4, and 5), sex, sow parity group (Group 1: parity 1–3 or Group 2: parity ≥ 4), birth order group, and birth weight category (light: <0.9 kg, medium: 1.0–1.4 kg or heavy: >1.5 kg). Birth sow was fitted as a random term. All two-way interactions were assessed and removed from the model if not significant. A non-parametric Kruskal–Wallis test was used to investigate piglet mortality data. Data are presented as estimated marginal mean ± standard error of the mean (SEM).

3. Results

There were no significant differences among treatments for the time taken for piglets to reach the udder following birth, rectal temperatures at 4 and 24 h, or colostrum intake

($p > 0.05$; Table 1). Additionally, no significant differences between treatments for weight at birth or 1 and 18 days of age were observed ($p > 0.05$; Table 2). However, at 3 days of age, piglets in the saline treatment group were significantly heavier than all other treatments ($p < 0.05$; Table 2).

Table 1. Rectal temperatures at 4 and 24 h (°C), time to udder (min), and colostrum intakes (g) for piglets allocated to oral treatments at birth groups; saline (SAL), caffeine (CAFF), glucose (GLUC) and caffeine plus glucose (CAFF-GLUC).

Variables	n	SAL	CAFF	GLUC	CAFF-GLUC	p-Value
4 h temperature (°C)	218	36.33 ± 0.25	36.61 ± 0.26	36.49 ± 0.27	36.61 ± 0.25	0.705
24 h temperature (°C) #	355	6.16 ± 0.10	6.15 ± 0.01	6.15 ± 0.01	6.14 ± 0.01	0.566
Time to udder (min) *	263	1.24 ± 0.07	1.16 ± 0.80	1.21 ± 0.08	1.21 ± 0.08	0.560
Colostrum intake (g) *	340	2.44 ± 0.30	2.48 ± 0.04	2.47 ± 0.03	2.48 ± 0.30	0.677

* Log transformed data; # sqrt transformed data. n represents the number of animals sampled for each measurement. Data shown as estimated marginal mean ± SEM.

Table 2. Body weights (kg) at birth, 1, 3 and 18 days of age, for piglets allocated to oral treatments at birth groups; saline (SAL), caffeine (CAFF), glucose (GLUC) and caffeine plus glucose (CAFF-GLUC).

Variables	n	SAL	CAFF	GLUC	CAFF-GLUC	p-Value
Day 0 weight (kg)	392	1.42 ± 0.02	1.40 ± 0.02	1.38 ± 0.02	1.42 ± 0.02	0.221
Day 1 weight (kg)	352	1.56 ± 0.02	1.51 ± 0.03	1.50 ± 0.02	1.53 ± 0.02	0.195
Day 3 weight (kg)	334	1.93 ± 0.03 [a]	1.81 ± 0.03 [b]	1.81 ± 0.03 [b]	1.86 ± 0.03 [b]	0.003
Day 18 weight (kg)	324	5.88 ± 0.15	5.60 ± 0.15	5.83 ± 0.15	5.67 ± 0.15	0.131

Values with different superscripts differ significantly ($p < 0.05$). n represents the number of animals sampled at each weighing time point. Data shown as estimated marginal mean ± SEM.

When evaluating piglet growth and performance, interactions between treatment and birth weight category were observed. Piglets born light grew significantly less between 1 and 3 d of age when supplemented with caffeine or glucose at birth ($p = 0.004$; Figure 1). However, treatment had no effect on medium weight pigs, while those born heavy grew significantly faster if they were supplemented with glucose or saline (Figure 1). When evaluating the interaction of treatment with birth weight category for growth between 1 and 18 days of age, piglets born light and treated with caffeine at birth grew significantly less than all other treatment groups, while medium and heavy piglets were not significantly affected by any of the birth treatments (Figure 2).

Overall, treatment did not negatively or positively influence piglet pre-weaning mortality (Table 3). The majority of mortalities occurred in the first 3 days (8.1%), with the least number of mortalities occurring between day 3 and 18 (3.2%; Table 3).

Table 3. Mortality at day 1, 3, and 18 of piglets allocated to oral treatments at birth; saline (SAL), caffeine (CAFF), glucose (GLUC) and caffeine plus glucose (CAFF-GLUC).

Mortality	n	SAL	CAFF	GLUC	CAFF-GLUC	Total	p-Value
Day 0 to 1	28	8/101 (7.9%)	6/98 (6.1%)	5/100 (5%)	9/98 (9.2%)	28/397 (7.1%)	0.667
Day 1 to 3	30	7/93 (7.5%)	6/92 (6.5%)	11/95 (11.6%)	6/89 (6.7%)	30/369 (8.1%)	0.551
Day 3 to 18	11	3/86 (3.5%)	6/86 (7%)	1/84 (1.2%)	1/83 (1.2%)	11/339 (3.2%)	0.109
Total	69	18/280 (6.4%)	18/276 (6.5%)	17/279 (6.1%)	16/270 (5.9%)	18.4%	0.983

n represents the number of animals at each time point.

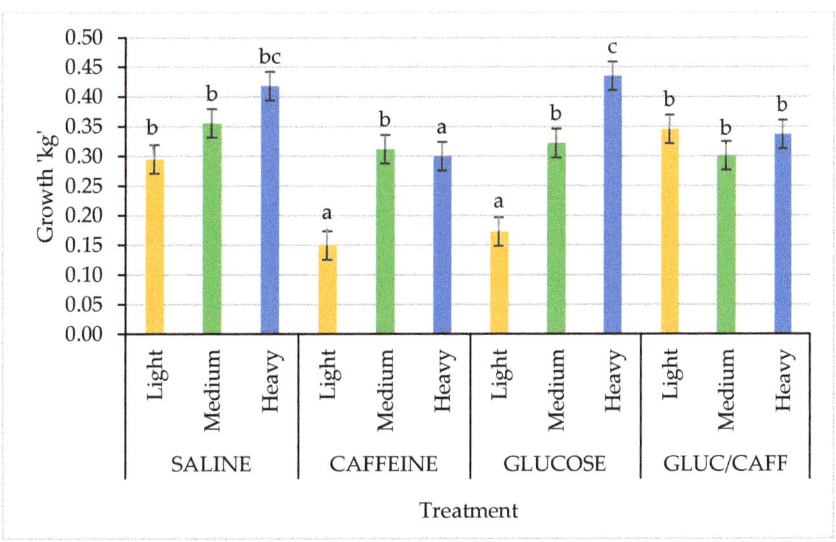

Figure 1. Comparison of growth for birth weight categories (light (yellow): <0.9 kg, medium (green): 1–1.4 kg, heavy (blue): >1.5 kg) from day one to day three of age between the four treatments: saline (SAL), caffeine (CAFF), glucose (GLUC), and caffeine plus glucose (CAFF-GLUC). Means with differing letters (abc) are significantly different ($p < 0.05$). Data shown as mean $\pm$ SEM (error bars).

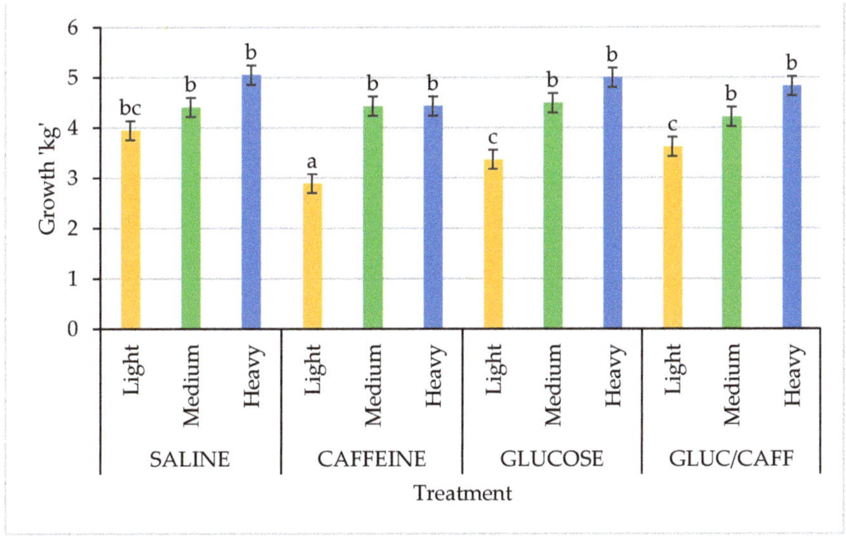

Figure 2. Comparison of growth for birth weight categories (light (yellow): <0.9 kg, medium (green): 1–1.4 kg, heavy (blue): >1.5 kg) from day 1 to day 18 of age between the four treatments: saline (SAL), caffeine (CAFF), glucose (GLUC), and caffeine plus glucose (CAFF-GLUC). Means with differing letters (abc) are significantly different ($p < 0.05$). Data shown as mean $\pm$ SEM (error bars).

4. Discussion

This study investigated the effects of providing glucose and/or caffeine to piglets at birth on factors important for piglet survival. At birth, piglets have relatively low body energy reserves for thermoregulation. Thereafter, litter competition at the udder and additional stressors such as increased litter size and lower colostrum intakes limit the available energy supply and so can adversely affect their survival [6]. In addition, Herpin et al. [13] showed that hypoxic piglets, such as those associated with a prolonged farrowing [14], have an impaired glucose uptake and utilisation in peripheral tissues due to the activation of their sympathetic nervous system. The risk of piglet mortality is highest during the first 3 days of life, especially for low birthweight piglets who have relatively low energy reserves [3,15]. Clearly, effective interventions are needed to support an adequate colostrum intake and pre-weaning survival of these neonatal low birthweight piglets. The approach taken in the present study was a bolus of oral glucose and/or caffeine, since we reasoned that low birthweight piglets have a greater surface area to volume ratio and so a greater propensity to lose heat, which would require energy to correct. In the absence of an adequate colostrum intake, this energy must be supplied by other ways. Studies have also demonstrated that compared to other neonates, piglets are at an increased risk of hypothermia as they are born without brown adipose tissue stores [16]. As these energy stores serve as insulators and are readily utilised to produce heat and raise core body temperature, small piglets are at an added disadvantage [17]. Hence, the addition of oral glucose was hoped to aid the piglet's thermoregulatory ability. However, we detected no treatment effect on piglet rectal temperatures at 4 and 24 h of age, possibly because a single caffeine and/or glucose bolus would have only a relatively short-term effect on blood glucose content and piglet thermoregulatory ability.

In addition to a potentially impaired energy status, smaller piglets are physically less competitive at the udder. With this in mind, we reasoned that a metabolic stimulus, such as provided by caffeine, would provide a short-term energy boost with consequent improved access to colostrum and ongoing competitiveness. However, a previous study noted that when administered in isolation, caffeine compromised piglet survival [10], presumably via an increased metabolic activity and energy expenditure resulting in hypoglycaemia. To counter this possibility, we included glucose with the caffeine, but this was without effect, again, possibly as a consequence of the short term physiological effect of the glucose. We also noted that the caffeine-treated piglets had impaired weight gains, especially in low birthweight piglets. Presumably, this reflects impaired colostrum and subsequent milk intakes. Our ultimate goal was to increase piglet colostrum consumption and success in accessing ongoing milk intakes. Clearly, this was not evident from our data.

Another possible advantage of caffeine is its influence as an adenosine receptor antagonist, potentially providing extra support to those animals that had suffered hypoxic distress during parturition [8,18]. Studies in humans have demonstrated that caffeine administration to premature neonates improved cardiorespiratory effects and reduced the occurrence of apnea [19]. It is important to note that within these studies the participants were very likely given colostrum or formula during and after caffeine administration, and so negating the negative effects we have observed for weight gain of piglets within the current study and pre-weaning survival in a previous study [10].

To our knowledge, this is the first study to investigate the effect of glucose and caffeine in combination, aiming to supply low viable piglets with a readily available energy source while using the action of caffeine to increase their metabolism and energy utilisation. Interestingly, Orozco-Gregorio et al. [6] showed that caffeine has the ability to increase circulating glucose concentration by increasing gluconeogenesis. We hypothesised that the previously noted hypoglycaemic state induced by caffeine would be compensated for by the addition of a glucose supplement when given in combination. In partial support of this, we did note an increased weight gain in low birthweight piglets in response to the combined caffeine/glucose supplement.

Engelsmann et al. [11] and others [17] noted that the ability to maintain body temperature is a key aspect to piglet survival and that warming of piglets has key benefits to their survival. This suggests that the warming of solutions prior to administration could also have a benefit to piglets at birth and aid in thermoregulation. Indeed, it has been demonstrated that an intraperitoneal injection of warm saline markedly increased colostrum intake and subsequent survival of low birthweight piglets [20]. However, it was not possible in that study to separate the effects of warming per se and an increased colostrum intake on piglet survival. Furthermore, the influence of saline on electrolyte balance cannot be ignored; however, although not relevant to the present study, it reinforces the concept that pre-weaning piglet survival is multifactorial, with utilisation of energy and factors leading to low birthweight and/or viability in the external environment being key factors. Future research could examine combining glucose with colostrum to provide for both relatively immediate and ongoing energy support for caffeine-treated piglets on their growth and survival.

5. Conclusions

The improvement of pre-weaning piglet survival will require a focus on the low birth weight/lower viability piglets. Our data suggests that the combination of caffeine and glucose has the potential to improve neonatal performance but that both a rapid and a more prolonged energy support, in addition to improvements in piglet warming, is likely needed for greater effect.

Author Contributions: Conceptualization, R.N.K. and T.L.N.; methodology, R.N.K. and T.L.N.; formal analysis, T.L.N.; data curation, L.J., S.E.J. and T.L.N.; writing—original draft, L.J.; writing—review and editing, L.J., S.E.J., R.N.K. and T.L.N.; supervision, R.N.K. and T.L.N.; project administration, S.E.J., R.N.K. and T.L.N.; funding acquisition, R.N.K. and T.L.N. All authors have read and agreed to the published version of the manuscript.

Funding: This research was supported in part by the University of Adelaide and the South Australian Research and Development Institute.

Institutional Review Board Statement: The animal study protocol was approved by the Institutional Review Board (or Ethics Committee) of the University of Adelaide Animal Ethics Committee (Application ID 35136, approved 03/06/2021).

Data Availability Statement: The data supporting this trial will be shared on reasonable request to the corresponding author.

Acknowledgments: The authors would like to acknowledge the staff and students at the Roseworthy Piggery for their invaluable technical support during the trial and the South Australian Research and Development Institute and the University of Adelaide for financial support.

Conflicts of Interest: The authors declare no conflict of interest.

References

1. Campos, P.H.R.F.; Silva, B.; Donzele, J.; Oliveira, R.; Knol, E. Effects of sow nutrition during gestation on within-litter birth weight variation: A review. *Animal* **2012**, *6*, 797–806. [CrossRef] [PubMed]
2. Swinbourne, A.M.; Kind, K.L.; Flinn, T.; Kleemann, D.O.; van Wettere, W.H. Caffeine: A potential strategy to improve survival of neonatal pigs and sheep. *Anim. Reprod. Sci.* **2021**, *226*, 106700. [CrossRef] [PubMed]
3. Condous, P.; Plush, K.; Tilbrook, A.; Van Wettere, W. Reducing sow confinement during farrowing and in early lactation increases piglet mortality. *J. Anim. Sci.* **2016**, *94*, 3022–3029. [CrossRef] [PubMed]
4. KilBride, A.; Mendl, M.; Statham, P.; Held, S.; Harris, M.; Cooper, S.; Green, L.E. A cohort study of preweaning piglet mortality and farrowing accommodation on 112 commercial pig farms in England. *Prev. Vet. Med.* **2012**, *104*, 281–291. [CrossRef] [PubMed]
5. Randall, G. The relationship of arterial blood pH and pCO2 to the viability of the newborn piglet. *Can. J. Comp. Med.* **1971**, *35*, 141. [PubMed]
6. Orozco-Gregorio, H.; Mota-Rojas, D.; Bonilla-Jaime, H.; Trujillo-Ortega, M.E.; Becerril-Herrera, M.; Hernández-González, R.; Villanueva-García, D. Effects of administration of caffeine on metabolic variables in neonatal pigs with peripartum asphyxia. *Am. J. Vet. Res.* **2010**, *71*, 1214–1219. [CrossRef] [PubMed]
7. Lodha, A.; Seshia, M.; McMillan, D.D.; Barrington, K.; Yang, J.; Lee, S.K.; Shah, P.S.; Canadian Neonatal Network. Association of early caffeine administration and neonatal outcomes in very preterm neonates. *JAMA Pediatr.* **2015**, *169*, 33–38. [CrossRef] [PubMed]
8. Back, S.A.; Craig, A.; Ling, N.L.; Ren, J.; Akundi, R.S.; Riberio, I.; Rivkees, S.A. Protective effects of caffeine on chronic hypoxia-induced perinatal white matter injury. *Ann. Neurol.* **2006**, *60*, 696–705. [CrossRef] [PubMed]

9. Alexander, M.; Smith, A.L.; Rosenkrantz, T.S.; Fitch, R.H. Therapeutic effect of caffeine treatment immediately following neonatal hypoxic-ischemic injury on spatial memory in male rats. *Brain Sci.* **2013**, *3*, 177–190. [CrossRef] [PubMed]

10. Nowland, T.L.; Kind, K.; Hebart, M.; van Wettere, W. Caffeine supplementation at birth, but not 8 to 12 h post-birth, increased 24 h pre-weaning mortality in piglets. *Animal* **2010**, *14*, 1529–1535. [CrossRef] [PubMed]

11. Engelsmann, M.N.; Hansen, C.F.; Nielsen, M.N.; Kristensen, A.R.; Amdi, C. Glucose injections at birth, warmth and placing at a nurse sow improve the growth of IUGR piglets. *Animals* **2019**, *9*, 519. [CrossRef] [PubMed]

12. Devillers, N.; Farmer, C.; Le Dividich, J.; and Prunier, A. Variability of colostrum yield and colostrum intake in pigs. *Animal* **2007**, *1*, 1033–1041. [CrossRef] [PubMed]

13. Herpin, P.; Le Dividich, J.; Hulin, J.C.; Filaut, M.; De Marco, F.; Bertin, R. Effects of the level of asphyxia during delivery on viability at birth and early postnatal vitality of newborn pigs. *J. Anim. Sci.* **1996**, *74*, 2067–2075. [CrossRef] [PubMed]

14. Alonso-Spilsbury, M.; Ramirez-Necoechea, R.; Gonzlez-Lozano, M.; Mota-Rojas, D.; Trujillo-Ortega, M.E. Piglet survival in early lactation: A review. *J. Anim. Vet. Adv.* **2007**, *6*, 76–86.

15. Hole, C.V.; Ayuso, M.; Aerts, P.; Prims, S.; Van Cruchten, S.; Van Ginneken, C. Glucose and glycogen levels in piglets that differ in birth weight and vitality. *Heliyon* **2019**, *5*, e02510.

16. Hou, L.; Hu, C.Y.; Wang, C. Pig has no brown adipose tissue. *FASEB J.* **2018**, *31*, S1. [CrossRef]

17. Lezama-García, K.; Mota-Rojas, D.; Martínez-Burnes, J.; Villanueva-García, D.; Domínguez-Oliva, A.; Gómez-Prado, J.; Mora-Medina, P.; Casas-Alvarado, A.; Olmos-Hernández, A.; Soto, P.; et al. Strategies for hypothermia compensation in altricial and precocial newborn mammals and their monitoring by infrared thermography. *Vet. Sci.* **2022**, *9*, 246. [CrossRef] [PubMed]

18. Villanueva-García, D.; Mota-Rojas, D.; Martínez-Burnes, J.; Olmos-Hernández, A.; Mora-Medina, P.; Salmerón, C.; Gómez, J.; Boscato, L.; Gutiérrez-Pérez, O.; Cruz, V.; et al. Hypothermia in newly born piglets: Mechanisms of thermoregulation and pathophysiology of death. *J. Anim. Behav. Biometeorol.* **2021**, *9*, 2101. [CrossRef]

19. Villanueva-García, D.; Mota-Rojas, D.; Miranda-Cortés, A.; Ibarra-Ríos, D.; Casas-Alvarado, A.; Mora-Medina, P.; Martínez-Burnes, J.; Olmos-Hernández, A.; Hernández-Avalos, I. Caffeine: Cardiorespiratory effects and tissue protection in animal models. *Exp. Anim.* **2021**, *70*, 431–439. [CrossRef] [PubMed]

20. Tucker, B.S.; Petrovski, K.R.; Kirkwood, R.N. Neonatal piglet temperature changes: Effect of intraperitoneal warm saline injection. *Animals* **2022**, *12*, 1312. [CrossRef] [PubMed]

Disclaimer/Publisher's Note: The statements, opinions and data contained in all publications are solely those of the individual author(s) and contributor(s) and not of MDPI and/or the editor(s). MDPI and/or the editor(s) disclaim responsibility for any injury to people or property resulting from any ideas, methods, instructions or products referred to in the content.

Review

Cardiorespiratory and Neuroprotective Effects of Caffeine in Neonate Animal Models

Daniel Mota-Rojas [1,*], Dina Villanueva-García [2], Ismael Hernández-Ávalos [3], Alejandro Casas-Alvarado [1], Adriana Domínguez-Oliva [1], Karina Lezama-García [1], Agatha Miranda-Cortés [3] and Julio Martínez-Burnes [4]

[1] Neurophysiology, Behavior and Animal Welfare Assessment, Universidad Autónoma Metropolitana (UAM), Mexico City 04960, Mexico
[2] Division of Neonatology, National Institute of Health, Hospital Infantil de México Federico Gómez, Mexico City 06720, Mexico; dinavg21@yahoo.com
[3] Clinical Pharmacology and Veterinary Anesthesia, Facultad de Estudios Superiores Cuautitlán, Universidad Nacional Autónoma de México (UNAM), Cuautitlán 54714, Mexico; mvziha@hotmail.com (I.H.-Á.)
[4] Facultad de Medicina Veterinaria y Zootecnia, Universidad Autónoma de Tamaulipas, Victoria City 87000, Mexico
* Correspondence: dmota100@yahoo.com.mx

Simple Summary: Caffeine is a stimulant used in humans and animals to improve newborns' respiratory and neurological responses. The use of caffeine after birth could increase neonate survival. However, due to the immature systems of animals at birth, caffeine use can have different results. This review aims to understand caffeine's effects on respiratory and neurological systems in neonate animal models (rat and mouse pups, goat kids, lambs, and piglets).

Abstract: Caffeine is widely used to improve neonatal health in animals with low vitality. Due to its pharmacokinetics and pharmacodynamics, caffeine stimulates the cardiorespiratory system by antagonism of adenosine receptors and alteration in Ca^{+2} ion channel activity. Moreover, the availability of intracellular Ca^{+2} also has positive inotropic effects by increasing heart contractibility and by having a possible positive effect on neonate vitality. Nonetheless, since neonatal enzymatic and tissular systems are immature at birth, there is a controversy about whether caffeine is an effective therapy for newborns. This review aims to analyze the basic concepts of caffeine in neonatal animal models (rat and mouse pups, goat kids, lambs, and piglets), and it will discuss the neuroprotective effect and its physiological actions in reducing apnea in newborns.

Keywords: positive inotropic effect; hypoxia; newborn; methylxanthine

Citation: Mota-Rojas, D.; Villanueva-García, D.; Hernández-Ávalos, I.; Casas-Alvarado, A.; Domínguez-Oliva, A.; Lezama-García, K.; Miranda-Cortés, A.; Martínez-Burnes, J. Cardiorespiratory and Neuroprotective Effects of Caffeine in Neonate Animal Models. *Animals* **2023**, *13*, 1769. https://doi.org/10.3390/ani13111769

Academic Editor: J. Alberto Montoya-Alonso

Received: 30 March 2023
Revised: 20 May 2023
Accepted: 22 May 2023
Published: 26 May 2023

Copyright: © 2023 by the authors. Licensee MDPI, Basel, Switzerland. This article is an open access article distributed under the terms and conditions of the Creative Commons Attribution (CC BY) license (https://creativecommons.org/licenses/by/4.0/).

1. Introduction

Caffeine or 1,3,7 tri-methylxanthine was first introduced to manage prematurity apnea (AOP) at the McGill University Hospitals in the mid-1970s to produce pharmacologic respirogenesis and reduce the need for intubation and mechanical ventilation in preterm neonates with recurrent AOP. Caffeine has a similar structure to adenosine [1]. Caffeine crosses all biological membranes and distributes into all body fluids [2]. When administered orally, it reaches up to 90% absorption [1].

It is a stimulant of the cerebral cortex and thus modifies the activity of the Central Nervous System (CNS) [3]. Adenosine receptors are present in different tissues, such as the CNS, lung, heart, and skeletal muscle, where caffeine has a stimulating effect. For example, caffeine improves cognitive ability due to the facilitation of synaptic capacity [4]. Caffeine enhances lung performance and capacity in the cardiorespiratory system by increasing both the air volume in each respiratory cycle and the cardiac contractility (positive inotropic effects) [5,6].

The physiological effects of caffeine have also been studied in neonates, where it has been suggested for the treatment of apnea, in addition to having a positive inotropic stimulus in animals with low vitality [7,8]. Studies regarding the effects of caffeine on the neurodevelopment of premature infants have been related to explain their poorly developed respiratory systems [9]. However, there is a need for more clarity about whether these benefits can persist due to the immaturity in the enzymatic and tissular systems that modify the activity of caffeine, possibly reducing its therapeutic efficacy. This review aims to analyze caffeine's pharmacokinetic and pharmacodynamic characteristics in neonatal animal models (rat and mouse pups, goat kids, lambs, and piglets). It will discuss the neuroprotective effect and its physiological actions in reducing apnea in newborns.

2. Pharmacokinetic Characteristics of Caffeine in Neonate Animals

Caffeine is one of the most recognized plant-derived alkaloids for its stimulating effect on the CNS [10]. This methylxanthine is used in neonatology due to its pharmacokinetic characteristics in newborns since its oral bioavailability reaches 99% after 45 min of administration [11]. Given its absorption level, neonates present plasmatic concentrations between 5–20 mg/mL at a dose of 5–10 mg/kg/day [12]. From a comparative perspective, Menozzi et al. [13] studied in sows that the plasma concentration of caffeine was 13.77 ± 0.97 µg/mL with a maximum concentration (Cmax) of 20.02 ± 1.51 µg/mL at 9.51 h after 25 mg/kg orally. This would indicate that caffeine has a lower concentration in neonates due to its larger volume of distribution (Vd), where factors such as postnatal age and birth weight can influence its absorption [14].

Caffeine can readily diffuse into cellular tissues because it has a high Vd in infants (0.8–0.9 L/kg) [14]. According to Noh et al. [15], studies in rats report similar parameters with a Vd of 0.5 ± 0.1 L/kg. These data suggest that caffeine can have a high availability in the tissues and permeation in critical regions. These characteristics also allow caffeine to cross the blood-brain and placental barriers [16]. In this regard, a study on pregnant rats reported that caffeine crosses the placental barrier by passive diffusion, allowing a wide distribution in fetal tissues [17].

This methylxanthine is mainly biotransformed in the liver by the CYP1A2 enzyme, where three active metabolites, N-3 paraxanthine, N-7 theophylline, and N-1 theobromine have been identified in adult humans [18]. However, in infants, the main metabolite produced by demethylation is N-7 theophylline due to the immaturity of the main metabolic pathway [19]. Bienvenu et al. [20] evaluated the hepatic capacity to metabolize caffeine of neonate rats at different ages. The authors found that 25% of the caffeine metabolites at day 1 was N-1 theobromine, while this was 40% for older ages (seven days of age). This suggests that metabolism differs due to the possible immaturity of the enzymatic systems in the neonate. It is important that the hepatic metabolism does not accumulate this substance in excess, especially in the cases of dose dependent toxicities and hypersensitivity reactions [18].

The immaturity of the enzymatic systems in the newborn has a greater impact on the clearance rate of caffeine since it is slower at birth and increases with age as glomerular filtration increases [21]. According to the data in rats, the elimination rate of caffeine is 0.0109 L, with a half-life of 4.19 h in adult animals [22]. Although it is still not clear if this speed and elimination time may be higher in the newborn, it is known that 75% of caffeine is eliminated unchanged in the urine at this age [23].

Therefore, the evidence shows that caffeine in newborns has different pharmacokinetics due to circulatory immaturity. Moreover, the immaturity of the enzymatic systems and the limited renal perfusion can affect this drug's elimination rate. Nonetheless, its physicochemical characteristics promote its distribution and permeation to most tissues, making it a pharmacological alternative for newborns.

3. Caffeine Pharmacodynamics in Neonates

Caffeine has a homologous molecular conformation to adenosine, making it a neuromodulator dependent on adenosine triphosphate (ATP). Its main mechanism is the non-selective antagonism of adenosine (A1, A2a, A2b, and A3) receptors, which are found predominantly in the CNS [24]. Immunohistochemical studies support that A1 and A2a receptors are found in high concentrations in the brain, with A1 receptors being ubiquitous in the region of the hippocampus and neocortex. In contrast, the A2a receptors are identified mainly in the striatum [25–27]. Adenosine is a neurotransmitter with diverse physiological functions, including the control of arousal, sleep, and cerebrovascular homeostasis. It has four known receptors: A1R, A2aR, A2bR, and A3R. Adenosine binding to its receptors leads to the inhibition of inspiratory neurons, resulting in central respiratory depression. Caffeine can non-specifically block these receptors, thereby indirectly stimulating the respiratory center, increasing sensitivity to carbon dioxide, enhancing diaphragm contractility, and improving the respiratory rate and tidal volume [28].

Antagonism of adenosine receptors decreases the activity of the phosphodiesterase enzyme, increasing adenosine $3',5'$-monophosphate cyclase (cAMP) [29]. This substance is an important mediator of second messenger signaling and modulates neurotransmitters. Therefore, it is related to the release of neurotransmitters, such as gamma-aminobutyric acid (GABA), norepinephrine, dopamine, serotonin, acetylcholine, and glutamate. This property influences its mechanism of action on brain adenosine receptors that modulate central noradrenergic, dopaminergic, serotonergic, cholinergic, GABAergic, and glutaminergic systems, which affects neuronal functioning. This facilitates neuronal transmission and higher cognitive performance due to greater chemical synapsing [30].

The decrease in cAMP alters neuronal modulation due to the adenosine inhibition signals on the neurotransmitters [31,32]. However, due to the location of A2a and A2b receptors in the lungs and heart, there is a controversy over whether this would be the primary mechanism of caffeine [33]. It has been described that activating the inhibitory G protein facilitates the mobilization of intracellular Ca^{+2} from the endoplasmic reticulum through the activation of ryanodine channels, which can be considered as a second mechanism of action [34]. Previous studies have described that caffeine reduces the activation threshold of ryanodine channels, facilitating the mobilization of intracellular Ca^{+2} to facilitate sympathetic neuron activity [35,36]. According to Kong et al. [37], caffeine markedly reduced the activation threshold of Ca^{+2} channels at the luminal level in cardiac myocytes. However, it did not affect the action threshold on the activation at the cytosolic level. This suggests that ryanodine channels modify the activation threshold as well as the cellular activity whose mechanism of action was initially suggested.

The modification in the flow of Ca^{+2} affects skeletal muscle, since the increase of this substrate at the intracellular level in the myocyte could improve physical performance [38,39]. This was investigated by Sarbjit-Singh et al. [40], who evaluated the modulating effect of ryanodine on the activation of Na^+ current in skeletal muscle fibers of the murine model. These authors mention that the activation and inactivation of the current-voltage by adding 0.5 and 2 mM of caffeine generated negative changes in the voltage dependence of the Nav 1.4 voltage-gated sodium channel and generated the gradual inactivation of ryanodine receptors that allowed the increase of Ca^{+2} ions at the cytosolic level. This could reaffirm that the negative modulation of the ion channels would positively modify the activity.

In summary, caffeine's mechanism of action is in the antagonism of adenosine, altering the level of intracellular metabolites such as cAMP, which would affect the regulation of neurotransmitters in the neuronal synapse. The increased availability of neurotransmitters could facilitate neural activity. On the other hand, caffeine also alters the activity of Ca^{+2} ion channels, increasing the cytosolic levels of this ion and stimulating muscle activity (an ergonomic effect).

4. Stimulating Effect on the Respiratory Tract of the Newborn

Caffeine is the most commonly used drug in the neonatal intensive care unit (NICU) after antibiotics [8,41]. In recent years, an increasing number of high-quality clinical studies have demonstrated the protective effects of caffeine on the respiratory and nervous systems of premature infants. In this sense, it is argued that four biochemical mechanisms stimulate the respiratory function: (a) the mobilization of intracellular Ca^{+2}, (b) the inhibition of phosphodiesterases, (c) the modulation of $GABA_A$ receptors, and (d) the antagonism of A3 receptors [42]. These mechanisms modify the pulmonary response to hypoxia or the sensitization of the chemoreceptors to O_2 and CO_2 molecules [43]. It is also suggested that the stimulant effects of caffeine increase the sensitivity to CO_2 in the respiratory centers, modifying the respiratory pattern [44].

Due to the chemoreceptor's sensitization, it can reduce the periods of apnea in the newborn, thus improving its ventilatory dynamics. For example, in a pilot study in newborn baboons, Yoder et al. [45] found an association with an enhanced pulmonary mechanical function during the first 24 h of life when administering caffeine. Nevertheless, Crossley et al. [46] evaluated the effects of this drug on kidney and lung function in lambs at 126 days of gestation. These authors reported two main findings; that the administration of caffeine at 40 mg/kg had little effect on lung function —with no differences in PaO_2 and hemoglobin oxygen saturation— and that the dose increased $PaCO_2$ and pulmonary vascular resistance.

The existing controversy questions the mechanism of action of caffeine when administered to newborns. In this sense, caffeine may improve pulmonary vagal afferents due to increased neural activity. A study of neonatal rabbits anesthetized with a barbiturate and receiving caffeine at 10 mg/kg reported an increased respiratory rate, minute volume, and respiratory flow, but caffeine did not improve vagal activity. These changes persisted in animals undergoing vagotomy [47].

The results shown above conclude that the improvement in neuronal synapsing caused by caffeine does not influence lung dynamics. It could possibly improve lung compliance, which would have a greater impact on the newborn's tidal volume and lung capacity. It should be added that, according to what was indicated by Julien et al. [48], exposure to intermittent hypoxic states in rat pups given a 20 mg/kg dose of caffeine may result in an increase in minute volume, which was negatively correlated with apnea frequency in these animals ($r^2 = 0.52$, $p < 0.01$). These results suggest caffeine affects apnea by increasing the central normoxic respiratory drive rather than a hypoxic response. On the other hand, it has been reported that 15 mg/kg of caffeine administered to rat pups had a 22% increase in the response to ventilation and a 15% increase in tidal volume, which led to the suggestion that caffeine possibly modifies the A1 adenosine receptor density [49].

Thus, caffeine modifies and optimizes respiratory control in the newborn due to an improvement in the disposition of adenosine receptors, as reported by Montandon et al. [50]. These authors studied the effect of caffeine on adenosinergic modulation through an antagonist of the A1 and A2 adenosine receptors in rats. The administration of caffeine and the A1 antagonist increased ventilation by 27%, reducing spontaneous apnea frequency with the administration of the A1 antagonist. In contrast, the A2 antagonist did not affect the ventilatory response. These findings support the theory that caffeine modifies the density of A1 receptors, improving ventilatory dynamics. The same authors reported that, in the carotid bodies of newborn rats receiving caffeine, it was possible to increase the mRNA expression of A2 and dopamine 2 receptors, improving the sensitization of respiratory chemoreceptors to hypoxia events [51]. Given the evidence, caffeine improves ventilatory dynamics by sensitizing chemoreceptors that can increase the ventilated tidal volume.

These studies suggest that caffeine increases the air volume that enters the neonate's respiratory tract. From a clinical perspective, this strategy could facilitate the extubation of patients [52]. In addition, some authors have shown that caffeine also participates in the prevention and treatment of neonatal apnea episodes [53,54] (Figure 1). However, whether this benefit may be exclusive to caffeine or any of its metabolites has been questioned.

Skouroliakou et al. [55] observed that standardized doses of caffeine and theophylline, in neonates with younger than 33 weeks of gestation, reduce apnea events significantly. In contrast, caffeine alone only controls apnea in newborns at risk. From these results, it can be inferred that caffeine alone helps to counteract apnea, but the derivatives of its metabolism contribute to the control of this event. Likewise, a meta-analysis focused on the effectiveness of caffeine (compared with amiodarone) as a treatment for apnea and was able to find that it would have similar effectiveness in infants but with the advantage of presenting fewer side effects [56]. Teng et al. [57] reported that hyperoxia exposure increased the expression of Bip, PERK, IRE1, sXBP1, cATF6, and CHOP during the cystic and alveolar stages of lung development, leading to lung injury due to oxidative stress and endoplasmic reticulum (ER) stress. Caffeine is proposed to reverse this oxidative damage, reduce apoptosis, and promote angiogenesis and alveolar development.

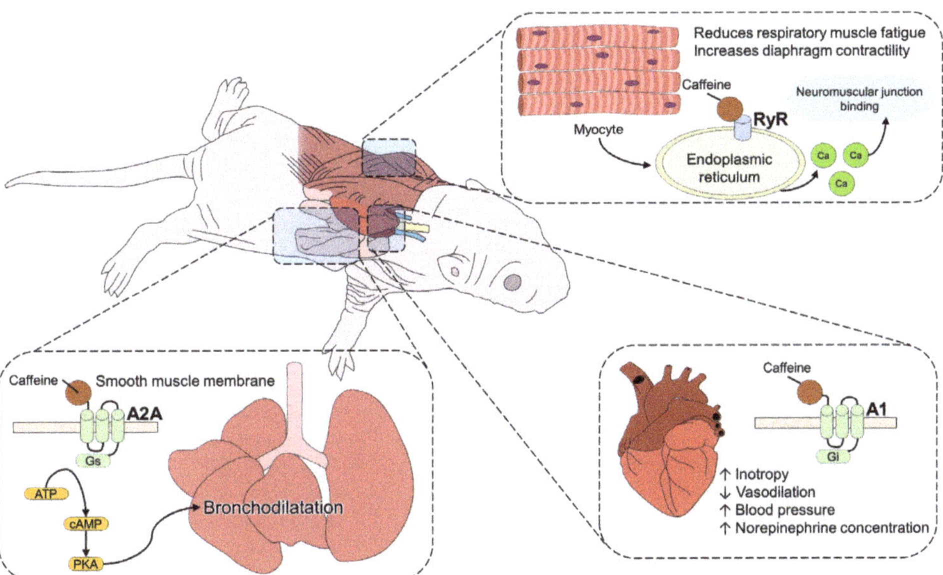

Figure 1. Cardiorespiratory effects of caffeine. Caffeine administration can act at three different levels to improve cardiorespiratory functions. At the skeletal muscle level, it directly induces the activation of RyR located in the muscle cell. This increases Ca^{2+} influx, which facilitates muscle contraction in the neuromuscular junction, reducing respiratory muscle fatigue and increasing diaphragm contractility and minute volume. In the lungs, caffeine blockade of A2A receptors causes bronchodilatation. In the heart, its inotropic effects cause tachycardia and blood pressure to rise. These effects of caffeine are beneficial in cases of neonatal apnea and hypoxia A2A: adenosine receptor 2A; AMP: adenosine monophosphate; cAMP: cyclic adenosine monophosphate; RyR: ryanodine receptors.

Bronchopulmonary dysplasia (BPD) is a chronic respiratory complication that affects a newborn's early life. Three major postnatal pathological factors for BPD are the high-concentration of oxygen inhalation, the inflammatory response, and the mechanical ventilation. Although there have been attempts to adopt therapeutic protocols for this disease in clinical trials, there has been no significant decline in the incidence of BPD and some consequences have been reported with current therapies [58,59]. Lung injury caused by a high concentration of oxygen and mechanical ventilation results in the destruction of the alveolar structure, increased vascular permeability and an inflammatory response as some of the complications [60]. Caffeine has been recognized as a treatment for primary apnea in premature infants, and as a drug to prevent BPD in premature infants. When using caffeine in a placebo-controlled trial with preterm infants, the drug reduced the risk of BPD and inhibited the inflammatory response induced by hyperoxia exposure in infant rats [61]. The potential mechanism behind this reaction is the caffeine's property to prevent lung tissular injury, reduce barotrauma, and improve ventilation and lung compliance. Contrarily, Dayanim et al. [62] showed that caffeine treatment exacerbated hyperoxic lung injury in neonatal rats, an important risk factor of BPD, which is another disorder that increases alveolar cell apoptosis. Caffeine may improve the prognosis of BPD by antagonizing the effect of prostaglandins. Nonetheless, to date, no therapeutic efficacy has been reported in animals (e.g., sheep fetuses) [63], and the timing, dosage, and side effects of caffeine use needs to be further examined.

Therefore, caffeine primarily induces sensitization of the chemoreceptors promoting an increase in the tidal volume of each respiratory cycle. Furthermore, this effect is related to the increased A1 and A2 receptors in the carotid bodies. Thus, these mechanisms can decrease the frequency of apnea and reduce ventilatory support.

5. Positive Inotropic Effect of Caffeine

Methylxanthines have a positive inotropic effect on the heart. Caffeine has also been reported to increase catecholamines and renin, both by peripheral and central effects. Some of the physiological responses are tachycardia, palpitations, rapid hypertension and a small decrease in heart rate in adults [64]. There is controversy on the arrhythmogenic potential of caffeine ingestion; however, the results are inconclusive and this has been proved only in animals and in humans with preexisting premature ventricular beats [6]. In general, methylxanthines have a positive inotropic effect on the heart. Based on a meta-analysis, caffeine at a dose of 400 mg affects the cardiac conduction system, increasing its frequency [6,10,65]. Moreover, this effect could be attributed to the alteration in the flow of Ca^{+2} due to the modification of the activity in the adenosine receptors [66]. This effect was evaluated by Rasmussen et al. [67] in cell cultures of ventricular myocytes of chick embryo. In this study, caffeine caused a 5–12% increase in the contraction amplitude and a 10 mV decrease in the membrane diastolic voltage. These events occurred with the increased Ca^{+2} release from the sarcoplasmic reticulum. Therefore, caffeine has a positive effect on cardiac contractility due to the activity of ryanodine channels.

In neonates, this mechanism could be altered by the immature cardiac tissue that has less Ca^{+2} dependence. In this regard, Miller et al. [68] observed that both mature and immature rabbit myocytes presented a strong rapid contractility response that did not depend on extracellular Ca^{+2} but on the reserves in the sarcoplasmic reticulum, which decreased as these reserves were depleted. The observations made by these authors complement the idea that caffeine induces changes in intracellular Ca^{+2}, stimulating cardiac contractility in newborns [69].

This inotropic effect of caffeine may have a clinical application in animals with low vitality, as indicated by Villanueva-García et al. [70]. These authors suggest that caffeine can stimulate cardiac contractility in newborn animals and thus increase vitality, which could guarantee survival. In addition, this was the main objective in a study by Robertson et al. [71], where they evaluated caffeine's effect on Merino lambs' survival rate. In this study, caffeine treatment reduced daily and first-week mortality compared to a control treatment. The authors attributed this increase in vitality to the stimulation of cardiac contractility and the reduction in hypoxia events upon drug administration. Therefore, caffeine enhances heart contractions by increasing the availability of intracellular Ca^{+2}. This way, it could increase vitality and survival in weak animals at birth. Although caffeine can be considered to have an advantage in inducing a positive inotropic effect, it is discussed that it may present disadvantages, such as the induction of arrhythmias or the increase in blood pressure that can affect microcirculation in peripheral tissues, as at the renal level, however this can be induced dose-dependently [72,73]. For this reason, it is necessary to consider the therapeutic dose, which may be different between adult animals and neonates, as shown in Table 1.

Table 1. Comparative of the doses reported in animal models.

Species	Route Administrated	Dose	Reference
Adults			
Wistar rats	Caffeine (oral/single, bolus)	0.5, 15, or 45 mg/kg	[74]
Sprague Dawley rats	Instant coffee extract (oral/ single dose)	250 or 500 mg/ kg	[75]
Mongrel dogs	Caffeine (IV)	1, 3 and 5 mg/kg	[76]
Arabian horses	Caffeine (IV)	5 mg/kg	[77]
Neonates			
Lactating dairy cows	Caffeine (IV)	2 mg/kg	[78]
Rabbit New Zealand White	Caffeine (PO)	loading dose 20 mg/kg 10 mg/kg daily	[79]
Rabbits New Zealand White	Caffeine (IP)	10 mg/ kg daily	[47]
Wistar rats	Caffeine (IV)	10 mg/kg	[80]

IP: intraperitoneal; IV: intravenous; PO: oral.

6. Caffeine in Neuroprotection

Using caffeine in mouse newborns has been a valuable strategy for reducing neonatal hypoxic-ischemic brain injury [81,82]. It has been used as a standard in all intensive care units (methyl theobromine) [83], replacing other treatments used in cases of apnea, such as theophylline and aminophylline [12]. In a murine model with germinal matrix-intraventricular hemorrhage (GM-IVH), a disorder associated with comorbidities such as cerebral palsy, sensory and motor impairment, learning disabilities, or neuropsychiatric disorders, Alves-Martinez et al. [84] analyzed two doses of caffeine (10 and 20 mg/kg) to treat this disorder. According to the results, both doses reduced hemorrhage burden. The drug showed a general neuroprotective effect in their model while diminishing brain atrophy and ventricle enlargement.

In addition, the therapeutic cardiorespiratory effects of caffeine in the newborn might promote vitality. It has been reported that due to the abundant distribution of adenosine receptors, the neuroprotective effect of caffeine in newborns could be observed since apnea induces hypoxia and ischemia events in the brain, leading to neurotoxicity and degeneration of the white matter [85]. In infants, Schmidt et al. [86] reported that caffeine administration in newborns significantly decreased cerebral palsy, bronchopulmonary dysplasia, patent ductus arteriosus requiring medical and/or surgical treatment, and severe retinopathy of prematurity. However, just as protective and positive effects have been observed in human newborns, animal studies have shown conflicting results regarding caffeine's role in neurodevelopment [87].

Cardiorespiratory effects have been reported at doses of 5–20 mg/kg, and neuroprotective effects have also been observed in newborns at this dose. For example, Winerdal et al. [81] conducted a randomized study in WT C57/bl6 rats in which a single dose of caffeine was administered at 5 mg/kg. Caffeine significantly reduced the presence of CD69+ and CD8 in the brain 24 h after treatment. In addition, there was a 44% decrease in the atrophy or damage to brain functions in the treated rats compared to the control group (treated with phosphate buffered saline). The authors concluded that its administration decreased brain atrophy and improved motor function in the open field test. Similarly, in a systematic review conducted by Bruschettini et al. [87], it was found that caffeine at doses of 5–20 mg/kg had a positive effect on the general functionality of the animals since they were able to observe a better performance in the maze tests carried out in the rats and mice studied.

Yang et al. [88] also reported neuroprotective properties in a neuronal proteomic analysis in newborn rats with induced hypoxia receiving caffeine. These authors found in the immunohistochemical analysis that the levels of myelin basic protein, proteolipid protein, myelin-associated glycoprotein precursor, and sirtiun 2 were reduced significantly with caffeine treatment in hypoxia-ischemic animals. Additionally, caffeine was found to enhance the expression of synaptophysin and postsynaptic density protein. These results demonstrate that caffeine has a neuroprotective effect by reducing the inflammatory process in the CNS. These data were reaffirmed in a subsequent study by these same authors. They evaluated Sprague-Dawley rats that were induced with cerebral hypoxia-ischemic ligation of the common carotid artery and were subsequently treated with caffeine. They observed that caffeine inhibited the activation of the NLRP3 inflammasome, negatively regulated the expression of the CD86 protein and iNOS, and inhibited the transcription of TNF-α and IL-1β, which would support the idea that caffeine reduces the inflammatory process and thus has a positive effect on the cognitive performance in the newborn [89].

In this context, Sun et al. [90] evaluated the effect of brain activity and tissue neuroprotection in newborn rats from mothers treated with caffeine. They found that caffeine reduced brain injury by 1.6 $\pm$ 4.5% and likewise increased the duration and amplitude of activity on electroencephalography. These data again corroborate caffeine's neuroprotective effect by reducing the inflammatory response at the brain level, and consequently, hypoxia. If the pharmacodynamic effects of caffeine are considered, it is possible to understand that adenosine receptor agonism helps in the expression of neurotransmitters and thus improves neuronal activity [70,82] (Figure 2).

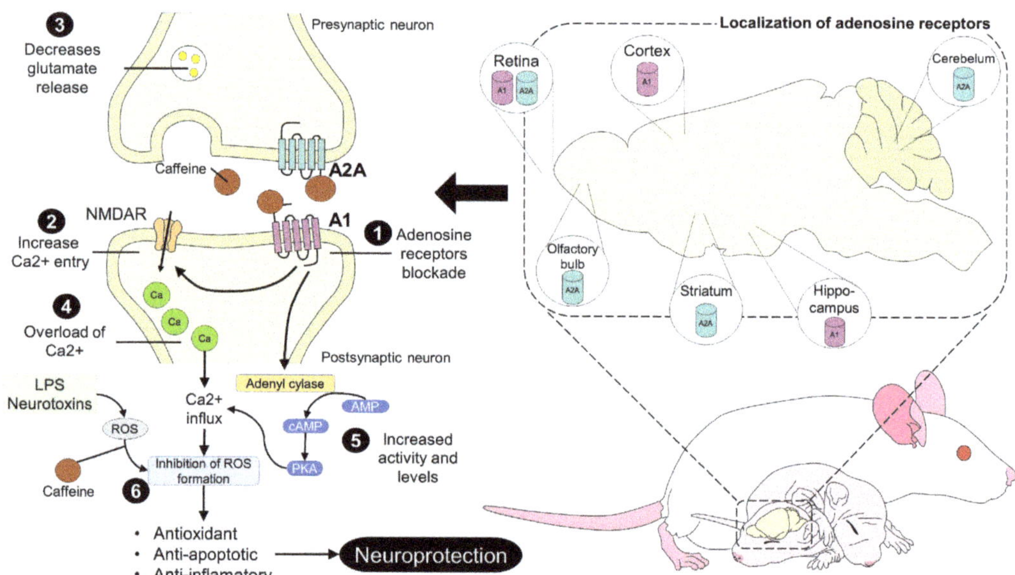

Figure 2. Neuroprotective properties of caffeine in the neonate's brain. Caffeine interacts with adenosine receptors (A2A and A1) located in cerebral structures such as the striatum, cerebellum, and olfactory bulb, among others, by binding to these receptors. In the presynaptic and postsynaptic neurons, the blockade of A2A and A1 receptors, respectively, causes a series of changes to enhance the neuroprotective properties of caffeine. Binding to the receptors causes an increase in Ca^{2+} entry through NMDAR, and increases the activity of adenyl cyclase, cAMP, and PKA while decreasing glutamate release. This causes a Ca^{2+} overload and upregulation of factors that inhibit ROS formation, giving the antioxidant, antiapoptotic, and anti-inflammatory effects of caffeine. AMP: adenosine monophosphate; cAMP: cyclic adenosine monophosphate; LPS: lipopolysaccharides; PKA: protein kinase A; ROS: reactive oxygen species.

Based on all of the above, it can be concluded that caffeine could be a therapy to mitigate the effects of neonatal hypoxic-ischemic brain injury. It may help decrease the burden of morbidities in preterm neonates [87].

7. Future Directions

Besides caffeine use in neonates, other factors, such as the ideal timing of administration and the interaction of caffeine with other drugs used in neonatology are relevant fields of research. Some studies have shown that pre-weaning mortality in piglets increases if caffeine at 30 mg is given orally at birth ($p < 0.05$), contrarily to doses at eight and 12 h post farrowing [91]. The use of caffeine with other drugs, their interaction, and the possible outcome in neonates are other topics that need to be considered. Caffeine and glucose supplementation to piglets at birth is a method used for reducing neonatal mortality and providing energy resources. However, this was only reported in low-birth-weight piglets administered with 30 mg of caffeine and 300 mg of glucose. Growth was improved in the first three days of life without affecting mortality, temperature, or colostrum intake [92]. Through Physiologic Based Pharmacokinetics (PBPK), and its combination with pharmacodynamics, the possible effects of caffeine could be determined to establish the desired and effective drug profile in preterm neonates (e.g., post-anesthesia/post-surgical apnea control, weaning from mechanical ventilation and extubating) or individualized medicine in newborns [2].

The maternal supplementation of caffeine is another approach suggested to improve newborn health. In Merino ewes, caffeine at 20 mg/kg resulted in lambs with higher rectal temperatures ($p = 0.021$), greater immunoglobulin concentrations ($p = 0.041$), and more suckling attempts than control animals and those receiving only 10 mg/kg [93]. A similar result was reported by Dearlove et al. [94] in piglets from sows receiving 2g of caffeine three days before farrowing. In this study, treated sows gave birth to fewer stillborn ($p = 0.05$), but no effect was reported on the viability score. In contrast, ewes receiving 10 mg/kg the day before lambing and those receiving 20 mg/kg in a four-week protocol did not improve lamb mortality and weight gain, concluding that caffeine is not an effective treatment to enhance perinatal survival in the species [95]. Caffeine is known to cross the placental barrier; however, little is known about the long-term impact of gestational caffeine exposure (GCE) on neurodevelopment. In the mouse brain, an alteration of neuron and neural circuits after GCE has been reported. Similarly, in children nine to ten years old, an Adolescent Brain and Cognitive Development^{sm} (ABCD ®) study performed by Christensen et al. [96] registered that GCE alters the developmental trajectory of white matter and neurocognition into adolescence. Nonetheless, further research towards this topic is necessary to validate its effects on neonatal neurocognition. The ambivalence in the results is why caffeine supplementation, not only in neonates, but also in the mother, is a topic that deserves future research.

Other topics that deserve to be fully studied are the exact mechanisms of action of hypothermia, genetics, circadian circle, and caffeine treatment. For hypothermia, it is thought that caffeine promotes energy preservation and reduces cytotoxic edema, free radicals, inflammation, and apoptotic cell death due to its anti-inflammatory properties. However, since factors like behavioral and neuroprotective outcomes might differ according to sex, this comparison needs to be considered to translate from animal to human [97]. On the other hand, in preterm infants with AOP, phenotypes and the expression of certain receptors have been associated with caffeine treatment and efficacy. Guo et al. [98] studied the circadian clock and its relation with the aryl hydrocarbon receptor (AHR) signaling pathways in preterm babies. From 104 individuals, the results showed that AHR genetic variations (rs1476080 and rs2066853), but not AHRR or ARNT genes, influence caffeine therapy's efficacy. Regarding circadian rhythms in premature infants and AOP management, preterm infants experience ultradian or irregular rhythms during early postnatal life [99]. In kid goats, Piccione et al. [100] has reported that the maturity of the circadian rhythm does not mature until the end of the second year of life of the animals. Moreover, 40-day-old foals have blood pressure immaturity [101], a factor that need to be considered when trying to administer caffeine to neonates. Also, caffeine has been shown to alter circadian rhythms in humans and animals [99]. Novel treatment using caffeine could open a field where caffeine treatment could be coordinated with circadian rhythms to improve disease management and care for premature infants.

8. Conclusions

Caffeine and related methylxanthines, in humans and animals, cross all biological membranes and distribute in all body fluids, without accumulating in tissues and organs. Caffeine is essential in stimulating the CNS, lung, heart, and skeletal muscle, improving cognitive abilities, lung capacity, and cardiac contractility. The benefits of caffeine administration to neonates include the reduced risk of hypoxic ischemia, cerebral palsy, bronchopulmonary dysplasia, patent ductus arteriosus, and retinopathy. Moreover, it decreases inflammatory processes and could positively affect the newborn's cognitive development. However, it is important to clarify that its dose, route, time of administration, and species must be considered to define whether its application may be favorable.

Since caffeine has a high availability in tissues and permeation in critical regions, it can cross the blood-brain and placental barriers, still allowing this substance to be absorbed by the fetus. However, as it is metabolized in the liver and fetuses and newborns do not have these processes fully developed, only a quarter of caffeine is metabolized in the newborn due to the immaturity of the enzymatic systems.

Although it is a substance that has been widely used in human and non-human newborns, there is still much research to be done to define its benefits and adverse effects in different species of domestic animals.

Author Contributions: All authors contributed to the conceptualization, writing, reading, and approval of the final manuscript. All authors have read and agreed to the published version of the manuscript.

Funding: This research received no external funding.

Institutional Review Board Statement: Not applicable.

Informed Consent Statement: Not applicable.

Data Availability Statement: Not applicable.

Conflicts of Interest: The authors declare no conflict of interest.

References

1. Fredholm, B.B.; Abbracchio, M.P.; Burnstock, G.; Daly, J.W.; Harden, T.K.; Jacobson, K.A.; Leff, P.; Williams, M. VI. Nomenclature and classification of purinoceptors. *Pharmacol. Rev.* **1994**, *46*, 143–156. [PubMed]
2. Aranda, J.V.; Beharry, K.D. Pharmacokinetics, pharmacodynamics and metabolism of caffeine in newborns. *Semin. Fetal Neonatal Med.* **2020**, *25*, 101183. [CrossRef] [PubMed]
3. Donovan, J.L.; DeVane, C.L. A primer on caffeine pharmacology and its drug interactions in clinical psychopharmacology. *Psychopharmacol. Bull.* **2001**, *35*, 30–48. [PubMed]
4. Kolahdouzan, M.; Hamadeh, M.J. The neuroprotective effects of caffeine in neurodegenerative diseases. *CNS Neurosci. Ther.* **2017**, *23*, 272–290. [CrossRef]
5. Herlenius, E.; Adén, U.; Tang, L.Q.; Lagercrantz, H. Perinatal respiratory control and its modulation by adenosine and caffeine in the rat. *Pediatr. Res.* **2002**, *51*, 4–12. [CrossRef]
6. Pelchovitz, D.J.; Goldberger, J.J. Caffeine and cardiac arrhythmias: A review of the evidence. *Am. J. Med.* **2011**, *124*, 284–289. [CrossRef]
7. Hsieh, E.M.; Hornik, C.P.; Clark, R.H.; Laughon, M.M.; Benjamin, D.K.; Smith, P.B. Medication use in the neonatal intensive care unit. *Am. J. Perinatol.* **2014**, *31*, 811–821. [CrossRef]
8. Schmidt, B.; Roberts, R.S.; Davis, P.; Doyle, L.W.; Barrington, K.J.; Ohlsson, A.; Solimano, A.; Tin, W. Caffeine therapy for apnea of prematurity. *N. Engl. J. Med.* **2006**, *354*, 2112–2121. [CrossRef]
9. Fredholm, B.B. Adenosine, adenosine receptors and the actions of caffeine. *Pharmacol. Toxicol.* **1995**, *76*, 93–101. [CrossRef]
10. Beyer, L.A.; Hixon, M.L. Review of animal studies on the cardiovascular effects of caffeine. *Food Chem. Toxicol.* **2018**, *118*, 566–571. [CrossRef]
11. Blanchard, J.; Sawers, S.J.A. The absolute bioavailability of caffeine in man. *Eur. J. Clin. Pharmacol.* **1983**, *24*, 93–98. [CrossRef]
12. Natarajan, G.; Botica, M.-L.; Thomas, R.; Aranda, J.V. Therapeutic Drug Monitoring for Caffeine in Preterm Neonates: An Unnecessary Exercise? *Pediatrics* **2007**, *119*, 936–940. [CrossRef]

13. Menozzi, A.; Mazzoni, C.; Serventi, P.; Zanardelli, P.; Bertini, S. Pharmacokinetics of oral caffeine in sows: A pilot study. *Large Anim. Rev.* **2015**, *21*, 207–210.

14. Lee, T.C.; Charles, B.; Steer, P.; Flenady, V.; Shearman, A. Population pharmacokinetics of intravenous caffeine in neonates with apnea of prematurity. *Clin. Pharmacol. Ther.* **1997**, *61*, 628–640. [CrossRef]

15. Noh, K.; Oh, D.G.; Nepal, M.R.; Jeong, K.S.; Choi, Y.; Kang, M.J.; Kang, W.; Jeong, H.G.; Jeong, T.C. Pharmacokinetic interaction of chrysin with caffeine in rats. *Biomol. Ther.* **2016**, *24*, 446–452. [CrossRef]

16. Jiritano, L.; Bortolotti, A.; Gaspari, F.; Bonati, M. Caffeine disposition after oral administration to pregnant rats. *Xenobiotica* **1985**, *15*, 1045–1051. [CrossRef]

17. Kimmel, C.A.; Kimmel, G.L.; White, C.G.; Grafton, T.F.; Young, J.F.; Nelson, C.J. Blood flow changes and conceptual development in pregnant rats in response to caffeine. *Toxicol. Sci.* **1984**, *4*, 240–247. [CrossRef]

18. Thorn, C.F.; Whirl-Carrillo, M.; Leeder, J.S.; Klein, T.E.; Altman, R.B. PharmGKB summary. *Pharmacogenet. Genom.* **2012**, *22*, 466–470. [CrossRef]

19. Abdel-Hady, H. Caffeine therapy in preterm infants. *World J. Clin. Pediatr.* **2015**, *4*, 81. [CrossRef]

20. Bienvenu, T.; Pons, G.; Rey, E.; Thiroux, G.; Olive, G. Caffeine metabolism in liver slices during postnatal development in rats. *Drug Metab. Dispos.* **1993**, *21*, 178–180.

21. Le Guennec, J.C.; Billon, B.; Paré, C. Maturational changes of caffeine concentrations and disposition in infancy during maintenance therapy for apnea of prematurity: Influence of gestational age, hepatic disease, and breast-feeding. *Pediatrics* **1985**, *76*, 834–840. [CrossRef] [PubMed]

22. Alshabi, A.M.; Alkahtani, S.A.; Shaikh, I.A.; Habeeb, M.S. Caffeine modulates pharmacokinetic and pharmacodynamic profiles of pioglitazone in diabetic rats. *Saudi Med. J.* **2021**, *42*, 151–160. [CrossRef] [PubMed]

23. Kumar, H.S.B.; Lipshultz, E.S. Caffeine and clinical outcomes in premature neonates. *Children* **2019**, *6*, 118. [CrossRef] [PubMed]

24. Pedata, F.; Dettori, I.; Coppi, E.; Melani, A.; Fusco, I.; Corradetti, R.; Pugliese, A.M. Purinergic signaling in brain ischemia. *Neuropharmacology* **2016**, *104*, 105–130. [CrossRef] [PubMed]

25. Fredholm, B.B.; IJzerman, A.P.; Jacobson, K.A.; Klotz, K.N.; Linden, J. International Union of Pharmacology. XXV. Nomenclature and classification of adenosine receptors. *Pharmacol. Rev.* **2001**, *53*, 527–552.

26. Rivkees, S.A. The ontogeny of cardiac and neural A1 adenosine receptor expression in rats. *Brain Res. Dev. Brain Res.* **1995**, *89*, 202–213. [CrossRef]

27. Wojcik, W.J.; Neff, N.H. Differential location of adenosine A1 and A2 receptors in striatum. *Neurosci. Lett.* **1983**, *41*, 55–60. [CrossRef]

28. Atik, A.; Harding, R.; De Matteo, R.; Kondos-Devcic, D.; Cheong, J.; Doyle, L.W.; Tolcos, M. Caffeine for apnea of prematurity: Effects on the developing brain. *Neurotoxicology* **2017**, *58*, 94–102. [CrossRef]

29. Pardo Lozano, R.; Alvarez García, Y.; Barral Tafalla, D.; Farré Albaladejo, M. Cafeína: Un nutriente, un fármaco, o una droga de abuso. *Adicciones* **2007**, *19*, 225. [CrossRef]

30. Fiani, B.; Zhu, L.; Musch, B.L.; Briceno, S.; Andel, R.; Sadeq, N.; Ansari, A.Z. The Neurophysiology of caffeine as a Central Nervous System stimulant and the resultant effects on cognitive function. *Cureus* **2021**, *13*, e15032. [CrossRef]

31. Owolabi, J.O.; Olatunji, S.Y.; Olanrewaju, A.J. Caffeine and cannabis effects on vital neurotransmitters and enzymes in the brain tissue of juvenile experimental rats. *Ann. Neurosci.* **2017**, *24*, 65–73. [CrossRef]

32. Persad, L.A.B. Energy drinks and the neurophysiological impact of caffeine. *Front. Neurosci.* **2011**, *5*, 116. [CrossRef]

33. Kamp, T.J.; Hell, J.W. Regulation of cardiac L-type calcium channels by protein kinase A and protein kinase C. *Circ. Res.* **2000**, *87*, 1095–1102. [CrossRef]

34. Dias, R.B.; Rombo, D.M.; Ribeiro, J.A.; Henley, J.M.; Sebastião, A.M. Adenosine: Setting the stage for plasticity. *Trends Neurosci.* **2013**, *36*, 248–257. [CrossRef]

35. Friel, D.D.; Tsien, R.W. A caffeine- and ryanodine-sensitive Ca^{2+} store in bullfrog sympathetic neurons modulates effects of Ca^{2+} entry on $[Ca^{2+}]i$. *J. Physiol.* **1992**, *450*, 217–246. [CrossRef]

36. Aronson, R.S.; Cranefield, P.F.; Wit, A.L. The effects of caffeine and ryanodine on the electrical activity of the canine coronary sinus. *J. Physiol.* **1985**, *368*, 593–610. [CrossRef]

37. Kong, H.; Jones, P.P.; Koop, A.; Zhang, L.; Duff, H.J.; Chen, S.R.W. Caffeine induces Ca^{2+} release by reducing the threshold for luminal Ca^{2+} activation of the ryanodine receptor. *Biochem. J.* **2008**, *414*, 441–452. [CrossRef]

38. Tarnopolsky, M.A. Effect of caffeine on the neuromuscular system—Potential as an ergogenic aid. *Appl. Physiol. Nutr. Metab.* **2008**, *33*, 1284–1289. [CrossRef]

39. Tarnopolsky, M.A. Caffeine and creatine use in sport. *Ann. Nutr. Metab.* **2011**, *57*, 1–8. [CrossRef]

40. Sarbjit-Singh, S.S.; Matthews, H.R.; Huang, C.L.-H. Ryanodine receptor modulation by caffeine challenge modifies Na^+ current properties in intact murine skeletal muscle fibers. *Sci. Rep.* **2020**, *10*, 2199. [CrossRef]

41. Stark, A.; Smith, P.B.; Hornik, C.P.; Zimmerman, K.O.; Hornik, C.D.; Pradeep, S.; Clark, R.H.; Benjamin, D.K.; Laughon, M.; Greenberg, R.G. Medication Use in the Neonatal Intensive Care Unit and Changes from 2010 to 2018. *J. Pediatr.* **2022**, *240*, 66–71.e4. [CrossRef] [PubMed]

42. Oñatibia-Astibia, A.; Martínez-Pinilla, E.; Franco, R. The potential of methylxanthine-based therapies in pediatric respiratory tract diseases. *Respir. Med.* **2016**, *112*, 1–9. [CrossRef] [PubMed]

43. Henderson-Smart, D.J.; Steer, P.A. Caffeine versus theophylline for apnea in preterm infants. *Cochrane Database Syst. Rev.* **2010**, *20*, CD000273. [CrossRef]

44. Sawynok, J.; Yaksh, T.L. Caffeine as an analgesic adjuvant: A review of pharmacology and mechanisms of action. *Pharmacol. Rev.* **1993**, *45*, 43–85. [PubMed]

45. Yoder, B.; Thomson, M.; Coalson, J. Lung function in immature baboons with respiratory distress syndrome receiving early caffeine therapy: A pilot study. *Acta Paediatr.* **2007**, *94*, 92–98. [CrossRef]

46. Crossley, K.J.; Allison, B.J.; Polglase, G.R.; Morley, C.J.; Harding, R.; Davis, P.G.; Moss, T.J.M.; Hooper, S.B. Effects of caffeine on renal and pulmonary function in preterm newborn lambs. *Pediatr. Res.* **2012**, *72*, 19–25. [CrossRef]

47. Trippenbach, T.; Zinman, R.; Milic-Emili, J. Caffeine effect on breathing pattern and vagal reflexes in newborn rabbits. *Respir. Physiol.* **1980**, *40*, 211–225. [CrossRef]

48. Julien, C.A.; Joseph, V.; Bairam, A. Caffeine reduces apnea frequency and enhances ventilatory long-term facilitation in rat pups raised in chronic intermittent hypoxia. *Pediatr. Res.* **2010**, *68*, 105–111. [CrossRef]

49. Montandon, G.; Bairam, A.; Kinkead, R. Long-term consequences of neonatal caffeine on ventilation, occurrence of apneas, and hypercapnic chemoreflex in male and female rats. *Pediatr. Res.* **2006**, *59*, 519–524. [CrossRef]

50. Montandon, G.; Kinkead, R.; Bairam, A. Disruption of adenosinergic modulation of ventilation at rest and during hypercapnia by neonatal caffeine in young rats: Role of adenosine A_1 and A_{2A} receptors. *Am. J. Physiol. Integr. Comp. Physiol.* **2007**, *292*, R1621–R1631. [CrossRef]

51. Montandon, G.; Bairam, A.; Kinkead, R. Neonatal caffeine induces sex-specific developmental plasticity of the hypoxic respiratory chemoreflex in adult rats. *Am. J. Physiol. Integr. Comp. Physiol.* **2008**, *295*, R922–R934. [CrossRef]

52. Henderson-Smart, D.J.; Davis, P.G. Prophylactic methylxanthines for endotracheal extubation in preterm infants. *Cochrane Database Syst. Rev.* **2010**, CD000139. [CrossRef]

53. Bhatia, J. Current options in the management of apnea of prematurity. *Clin. Pediatr.* **2000**, *39*, 327–336. [CrossRef]

54. Kreutzer, K.; Bassler, D. Caffeine for apnea of prematurity: A neonatal success story. *Neonatology* **2014**, *105*, 332–336. [CrossRef]

55. Skouroliakou, M.; Bacopoulou, F.; Markantonis, S.L. Caffeine versus theophylline for apnea of prematurity: A randomised controlled trial. *J. Paediatr. Child Health* **2009**, *45*, 587–592. [CrossRef]

56. Miao, Y.; Zhou, Y.; Zhao, S.; Liu, W.; Wang, A.; Zhang, Y.; Li, Y.; Jiang, H. Comparative efficacy and safety of caffeine citrate and aminophylline in treating apnea of prematurity: A systematic review and meta-analysis. *PLoS ONE* **2022**, *17*, e0274882. [CrossRef]

57. Teng, R.-J.; Jing, X.; Michalkiewicz, T.; Afolayan, A.J.; Wu, T.-J.; Konduri, G.G. Attenuation of endoplasmic reticulum stress by caffeine ameliorates hyperoxia-induced lung injury. *Am. J. Physiol. Cell. Mol. Physiol.* **2017**, *312*, L586–L598. [CrossRef]

58. Onland, W.; Cools, F.; Kroon, A.; Rademaker, K.; Merkus, M.P.; Dijk, P.H.; van Straaten, H.L.; Te Pas, A.B.; Mohns, T.; Bruneel, E.; et al. Effect of Hydrocortisone Therapy Initiated 7 to 14 Days After Birth on Mortality or Bronchopulmonary Dysplasia Among Very Preterm Infants Receiving Mechanical Ventilation. *JAMA* **2019**, *321*, 354. [CrossRef]

59. Tian, C.; Li, D.; Fu, J. Molecular Mechanism of Caffeine in Preventing Bronchopulmonary Dysplasia in Premature Infants. *Front. Pediatr.* **2022**, *10*, 902437. [CrossRef]

60. Balany, J.; Bhandari, V. Understanding the Impact of Infection, Inflammation, and Their Persistence in the Pathogenesis of Bronchopulmonary Dysplasia. *Front. Med.* **2015**, *2*, 90. [CrossRef]

61. Weichelt, U.; Cay, R.; Schmitz, T.; Strauss, E.; Sifringer, M.; Bührer, C.; Endesfelder, S. Prevention of hyperoxia-mediated pulmonary inflammation in neonatal rats by caffeine. *Eur. Respir. J.* **2013**, *41*, 966–973. [CrossRef] [PubMed]

62. Dayanim, S.; Lopez, B.; Maisonet, T.M.; Grewal, S.; Londhe, V.A. Caffeine induces alveolar apoptosis in the hyperoxia-exposed developing mouse lung. *Pediatr. Res.* **2014**, *75*, 395–402. [CrossRef] [PubMed]

63. Clyman, R.I.; Roman, C. The Effects of Caffeine on the Preterm Sheep Ductus Arteriosus. *Pediatr. Res.* **2007**, *62*, 167–169. [CrossRef] [PubMed]

64. Fredholm, B.B.; Bättig, K.; Holmén, J.; Nehlig, A.; Zvartau, E.E. Actions of caffeine in the brain with special reference to factors that contribute to its widespread use. *Pharmacol. Rev.* **1999**, *51*, 83–133. [PubMed]

65. Voskoboinik, A.; Kalman, J.M.; Kistler, P.M. Caffeine and arrhythmias: Time to grind the data. *JACC Clin. Electrophysiol.* **2018**, *4*, 425–432. [CrossRef] [PubMed]

66. Howlett, R.A.; Kelley, K.M.; Grassi, B.; Gladden, L.B.; Hogan, M.C. Caffeine administration results in greater tension development in previously fatigued canine muscle in situ. *Exp. Physiol.* **2005**, *90*, 873–879. [CrossRef]

67. Rasmussen, C.A.F.; Sutko, J.L.; Barry, W.H. Effects of ryanodine and caffeine on contractility, membrane voltage, and calcium exchange in cultured heart cells. *Circ. Res.* **1987**, *60*, 495–504. [CrossRef]

68. Miller, M.S.; Friedman, W.F.; Wetzel, G.T. Caffeine-induced contractions in developing rabbit heart. *Pediatr. Res.* **1997**, *42*, 287–292. [CrossRef]

69. Herrmann-Frank, A.; Lüttgau, H.; George Stephenson, D. Caffeine and excitation–contraction coupling in skeletal muscle: A stimulating story. *J. Muscle Res. Cell. Motil.* **1999**, *20*, 223–236. [CrossRef]

70. Villanueva-García, D.; Mota-Rojas, D.; Miranda-Cortés, A.; Ibarra-Ríos, D.; Casas-Alvarado, A.; Mora-Medina, P.; Martínez-Burnes, J.; Olmos-Hernández, A.; Hernández-Avalos, I. Caffeine: Cardiorespiratory effects and tissue protection in animal models. *Exp. Anim.* **2021**, *70*, 431–439. [CrossRef]

71. Robertson, S.M.; Friend, M.A.; Doran, G.S.; Edwards, S. Caffeine supplementation of ewes during lambing may increase lamb survival. *Animal* **2018**, *12*, 376–382. [CrossRef]

72. Ishida, S.; Ito, M.; Takahashi, N.; Fujino, T.; Akimitsu, T.; Saikawa, T. Caffeine Induces Ventricular Tachyarrhythmias Possibly Due to Triggered Activity in Rabbits in Vivo. *Jpn. Circ. J.* **1996**, *60*, 157–165. [CrossRef]

73. Mehta, A.; Jain, A.C.; Mehta, M.C.; Billie, M. Caffeine and cardiac arrhythmias. An experimental study in dogs with review of literature. *Acta Cardiol.* **1997**, *52*, 273–283.

74. Tofovic, S.P.; Kost, C.K.; Jackson, E.K.; Bastacky, S.I. Long-term caffeine consumption exacerbates renal failure in obese, diabetic, ZSF1 (fa-facp) rats. *Kidney Int.* **2002**, *61*, 1433–1444. [CrossRef]

75. Awaad, A.S.; Soliman, G.A.; Al-Outhman, M.R.; Al-Shdoukhi, I.F.; Al-Nafisah, R.S.; Al-Shamery, J.; Al-Samkhan, R.; Baqer, M.; Al-Jaber, N.A. The Effect of Four Coffee types on Normotensive rats and Normal/Hypertensive Human Volunteers. *Phyther. Res.* **2011**, *25*, 803–808. [CrossRef]

76. Rashid, A.; Hines, M.; Scherlag, B.J.; Yamanashi, W.S.; Lovallo, W. The effects of caffeine on the inducibility of atrial fibrillation. *J. Electrocardiol.* **2006**, *39*, 421–425. [CrossRef]

77. Ferraz, G.C.; Teixeira-Neto, A.R.; Mataqueiro, M.I.; Lacerda-Neto, J.C.; Queiroz-Neto, A. Effects of intravenous administration of caffeine on physiologic variables in exercising horses. *Am. J. Vet. Res.* **2008**, *69*, 1670–1675. [CrossRef]

78. DeGraves, F.J.; Ruffin, D.C.; Duran, S.H.; Spano, J.S.; Whatley, E.M.; Schumacher, J.; Riddell, M.G. Pharmacokinetics of caffeine in lactating dairy cows. *Am. J. Vet. Res.* **1995**, *56*, 619–622.

79. Van der Veeken, L.; Grönlund, S.; Gerdtsson, E.; Holmqvist, B.; Deprest, J.; Ley, D.; Bruschettini, M. Long-term neurological effects of neonatal caffeine treatment in a rabbit model of preterm birth. *Pediatr. Res.* **2020**, *87*, 1011–1018. [CrossRef]

80. Endesfelder, S.; Strauß, E.; Bendix, I.; Schmitz, T.; Bührer, C. Prevention of Oxygen-Induced Inflammatory Lung Injury by Caffeine in Neonatal Rats. *Oxid. Med. Cell. Longev.* **2020**, *2020*, 3840124. [CrossRef]

81. Winerdal, M.M.E.; Urmaliya, V.; Winerdal, M.M.E.; Fredholm, B.B.; Winqvist, O.; Ådén, U. Single dose caffeine protects the neonatal mouse brain against hypoxia ischemia. *PLoS ONE* **2017**, *12*, e0170545. [CrossRef] [PubMed]

82. Villanueva-García, D.; Mota-Rojas, D.; Miranda-Cortés, A.E.; Mora-Medina, P.; Hernández-Avalos, I.; Casas-Alvarado, A.; Olmos-Hernández, A.; Martínez-Burnes, J. Neurobehavioral and neuroprotector effects of caffeine in animal models. *J. Anim. Behav. Biometeorol.* **2020**, *8*, 298–307. [CrossRef]

83. Dobson, N.R.; Hunt, C.E. Pharmacology review. Caffeine use in neonates: Indications, pharmacokinetics, clinical effects, outcomes. *Neoreviews* **2013**, *14*, e540–e550. [CrossRef]

84. Alves-Martinez, P.; Atienza-Navarro, I.; Vargas-Soria, M.; Carranza-Naval, M.J.; Infante-Garcia, C.; Benavente-Fernandez, I.; Del Marco, A.; Lubian-Lopez, S.; Garcia-Alloza, M. Caffeine Restores Neuronal Damage and Inflammatory Response in a Model of Intraventricular Hemorrhage of the Preterm Newborn. *Front. Cell Dev. Biol.* **2022**, *10*, 908045. [CrossRef] [PubMed]

85. Yang, L.; Yu, X.; Zhang, Y.; Liu, N.; Xue, X.; Fu, J. Encephalopathy in preterm infants: Advances in neuroprotection with caffeine. *Front. Pediatr.* **2021**, *9*, 724161. [CrossRef]

86. Schmidt, B.; Roberts, R.S.; Davis, P.; Doyle, L.W.; Barrington, K.J.; Ohlsson, A.; Solimano, A.; Tin, W. Long-term effects of caffeine therapy for apnea of prematurity. *N. Engl. J. Med.* **2007**, *357*, 1893–1902. [CrossRef]

87. Bruschettini, M.; Moreira, A.; Pizarro, A.B.; Mustafa, S.; Romantsik, O. The effects of caffeine following hypoxic-ischemic encephalopathy: A systematic review of animal studies. *Brain Res.* **2022**, *1790*, 147990. [CrossRef]

88. Yang, L.; Yu, X.; Zhang, Y.; Liu, N.; Li, D.; Xue, X.; Fu, J. Proteomic analysis of the effects of caffeine in a neonatal rat model of hypoxic-ischemic white matter damage. *CNS Neurosci. Ther.* **2022**, *28*, 1019–1032. [CrossRef]

89. Yang, L.; Yu, X.; Zhang, Y.; Liu, N.; Xue, X.; Fu, J. Caffeine treatment started before injury reduces hypoxic–ischemic white-matter damage in neonatal rats by regulating phenotypic microglia polarization. *Pediatr. Res.* **2022**, *92*, 1543–1554. [CrossRef]

90. Sun, H.; Gonzalez, F.; McQuillen, P.S. Caffeine restores background EEG activity independent of infarct reduction after neonatal hypoxic ischemic brain injury. *Dev. Neurosci.* **2020**, *42*, 72–82. [CrossRef]

91. Nowland, T.L.; Kind, K.; Hebart, M.L.; van Wettere, W.H.E.J. Caffeine supplementation at birth, but not 8 to 12 h post-birth, increased 24 h pre-weaning mortality in piglets. *Animal* **2020**, *14*, 1529–1535. [CrossRef]

92. Jarratt, L.; James, S.E.; Kirkwood, R.N.; Nowland, T.L. Effects of caffeine and glucose supplementation at birth on piglet pre-weaning growth, thermoregulation, and survival. *Animals* **2023**, *13*, 435. [CrossRef]

93. Murdock, N.J.; Weaver, A.C.; Kelly, J.M.; Kleemann, D.O.; van Wettere, W.H.E.J.; Swinbourne, A.M. Supplementing pregnant Merino ewes with caffeine to improve neonatal lamb thermoregulation and viability. *Anim. Reprod. Sci.* **2021**, *226*, 106715. [CrossRef]

94. Dearlove, B.A.; Kind, K.L.; Gatford, K.L.; van Wettere, W.H.E.J. Oral caffeine administered during late gestation increases gestation length and piglet temperature in naturally farrowing sows. *Anim. Reprod. Sci.* **2018**, *198*, 160–166. [CrossRef]

95. Robertson, S.M.; Edwards, S.H.; Doran, G.S.; Friend, M.A. Maternal caffeine administration to ewes does not affect perinatal lamb survival. *Anim. Reprod. Sci.* **2021**, *231*, 106799. [CrossRef]

96. Christensen, Z.P.; Freedman, E.G.; Foxe, J.J. Caffeine exposure in utero is associated with structural brain alterations and deleterious neurocognitive outcomes in 9–10 year old children. *Neuropharmacology* **2021**, *186*, 108479. [CrossRef]

97. McLeod, R.; Rosenkrantz, T.; Fitch, R.H. Therapeutic Interventions in Rat Models of Preterm Hypoxic Ischemic Injury: Effects of Hypothermia, Caffeine, and the Influence of Sex. *Life* **2022**, *12*, 1514. [CrossRef]

98. Guo, H.-L.; Long, J.-Y.; Hu, Y.-H.; Liu, Y.; He, X.; Li, L.; Xia, Y.; Ding, X.-S.; Chen, F.; Xu, J.; et al. Caffeine Therapy for Apnea of Prematurity: Role of the Circadian CLOCK Gene Polymorphism. *Front. Pharmacol.* **2022**, *12*, 724145. [CrossRef]

99. Dai, H.-R.; Guo, H.-L.; Hu, Y.-H.; Xu, J.; Ding, X.-S.; Cheng, R.; Chen, F. Precision caffeine therapy for apnea of prematurity and circadian rhythms: New possibilities open up. *Front. Pharmacol.* **2022**, *13*, 1053210. [CrossRef]
100. Piccione, G.; Caola, G.; Refinetti, R. Annual rhythmicity and maturation of physiological parameters in goats. *Res. Vet. Sci.* **2007**, *83*, 239–243. [CrossRef]
101. Piccione, G.; Assenza, A.; Fazio, F.; Giannetto, C.; Caola, G. Chronobiologic blood pressure assessment: Maturation of the daily rhythm in newborn foals. *Biol. Res.* **2008**, *41*, 51–57. [CrossRef] [PubMed]

Disclaimer/Publisher's Note: The statements, opinions and data contained in all publications are solely those of the individual author(s) and contributor(s) and not of MDPI and/or the editor(s). MDPI and/or the editor(s) disclaim responsibility for any injury to people or property resulting from any ideas, methods, instructions or products referred to in the content.

Review

Neurobiology of Maternal Behavior in Nonhuman Mammals: Acceptance, Recognition, Motivation, and Rejection

Genaro A. Coria-Avila [1,*], Deissy Herrera-Covarrubias [1], Luis I. García [1], Rebeca Toledo [1], María Elena Hernández [1], Pedro Paredes-Ramos [2], Aleph A. Corona-Morales [3] and Jorge Manzo [1]

[1] Instituto de Investigaciones Cerebrales, Universidad Veracruzana, Avenida Luis Castelazo s/n Col. Industrial Ánimas, Xalapa 91190, Mexico
[2] Facultad de Medicina Veterinaria y Zootecnia, Universidad Veracruzana, Avenida Miguel Ángel de Quevedo s/n, esq. Yañez, Veracruz 91710, Mexico
[3] Facultad de Nutrición, Universidad Veracruzana, Avenida Médicos y Odontólogos s/n Col. Unidad del Bosque, Xalapa 91010, Mexico
* Correspondence: gcoria@uv.mx

Citation: Coria-Avila, G.A.; Herrera-Covarrubias, D.; García, L.I.; Toledo, R.; Hernández, M.E.; Paredes-Ramos, P.; Corona-Morales, A.A.; Manzo, J. Neurobiology of Maternal Behavior in Nonhuman Mammals: Acceptance, Recognition, Motivation, and Rejection. *Animals* 2022, 12, 3589. https://doi.org/10.3390/ani12243589

Academic Editor: Alan Tilbrook

Received: 18 November 2022
Accepted: 14 December 2022
Published: 19 December 2022

Publisher's Note: MDPI stays neutral with regard to jurisdictional claims in published maps and institutional affiliations.

Copyright: © 2022 by the authors. Licensee MDPI, Basel, Switzerland. This article is an open access article distributed under the terms and conditions of the Creative Commons Attribution (CC BY) license (https://creativecommons.org/licenses/by/4.0/).

Simple Summary: Maternal behavior involves active and passive responses associated with the willingness to nurse and protect the young. In some species, its expression is very selective toward individuals that are recognized as their own and may be long-lasting, whereas in other species the expression is not as selective or may be short-lasting. Brain processes of acceptance, social recognition, inhibition of rejection/fear, and increase in care motivation mediate its expression. The neurocircuitry of maternal behavior is activated upon exposure to the right natural stimuli, such as those that occur during pregnancy, parturition, and lactation. However, even virgin females and males can respond with maternal behaviors if they develop sensitization to the offspring via cohabitation or cross-sensitization via mating. Herein, we discuss behavioral expression in different species, the natural triggering stimuli, and the putative neurocircuitries of acceptance, social recognition, motivation, and rejection during maternal behavior.

Abstract: Among the different species of mammals, the expression of maternal behavior varies considerably, although the end points of nurturance and protection are the same. Females may display passive or active responses of acceptance, recognition, rejection/fear, or motivation to care for the offspring. Each type of response may indicate different levels of neural activation. Different natural stimuli can trigger the expression of maternal and paternal behavior in both pregnant or virgin females and males, such as hormone priming during pregnancy, vagino-cervical stimulation during parturition, mating, exposure to pups, previous experience, or environmental enrichment. Herein, we discuss how the olfactory pathways and the interconnections of the medial preoptic area (mPOA) with structures such as nucleus accumbens, ventral tegmental area, amygdala, and bed nucleus of stria terminalis mediate maternal behavior. We also discuss how the triggering stimuli activate oxytocin, vasopressin, dopamine, galanin, and opioids in neurocircuitries that mediate acceptance, recognition, maternal motivation, and rejection/fear.

Keywords: preoptic area; parturition; amygdala; oxytocin; dopamine; recognition; brain; motivation; bond

1. Introduction

Survival of the mammal offspring depends on the correct expression of maternal behaviors, particularly during the early postnatal period. Newborns must be a powerful source of incentive sensory stimulation to the dam, and in return, they must be capable of responding either actively or passively to such stimuli, expressing acceptance and motivation to invest energy and time and willingness to risk their physical safety. Hence, the capacity to express maternal behavior depends on the sensitivity to respond to the right stimuli under certain physiological, ontogenic, or cognitive conditions. Accordingly, to understand the neurobiology of this behavior, we must consider neural systems involved in acceptance, social recognition, motivation, and fear/rejection. This review aims to provide information on the neurobiology of such processes in nonhuman mammals. We begin by describing the objective measures of active and passive responses in different species. Then, we discuss the natural stimuli that facilitate the expression of maternal behavior and the putative neurocircuitries. Accordingly, scientific articles related to maternal behavior were analyzed. The electronic databases PubMed, GoogleScholar, and SciELO were searched using the following keywords in English: maternal behavior; animals; preoptic area; parturition; amygdala; oxytocin; dopamine; recognition; brain; motivation; bond. The exploration included studies on laboratory and domestic animals.

2. Active and Passive Maternal Behaviors

Maternal behavior involves the facilitation of acceptance, recognition, and motivation, along with the inhibition of rejection and fear toward offspring. Acceptance is inferred from behaviors that allow proximity to any newborn, whereas recognition involves selective acceptance of specific individuals. Thus, females may passively accept unfamiliar newborns (i.e., allowing nursing to any young) and recognize/accept only familiar ones. Similarly, rejection may involve active responses (i.e., aggression/infanticide) to discourage contact, whereas fear may be expressed via passive avoidance (i.e., not approaching them). Furthermore, care motivation involves active behaviors that indicate willingness to nurse and protect the young. So, although the expression of maternal behaviors varies considerably among species, the endpoints served are the same. For instance, some precocial species, such as sheep and horses, express very selective maternal behavior toward offspring they accept/recognize as theirs during the very first hours postpartum [1,2]. That kind of maternal selectivity requires strict mechanisms of acceptance/social recognition that occur via imprinting (i.e., associative learning) during a brief period following parturition. Disturbance of recognition between the dam and her offspring during the early imprinting period may result in rejection (perhaps fear), despite all the hormonal input or whelping experience. By contrast, altricial species with massive reproductive strategies, like rats, canids, or pigs, are considered less strict because they may accept alien offspring during extended periods [3–5]. In addition, some ungulates [6] and nonhuman primates [7,8] may display extensive maternal repertoires daily for weeks or months, whereas others, such as lagomorphs, will display only a few minutes of nursing once a day [9].

Some behaviors that start before parturition might be considered indirect maternal behaviors. For example, nest-building (e.g., digging, shredding paper, straw carrying, and hair pulling) and isolation from the pack or herd [10–13]. Other early behaviors, such as those observed in pregnant dogs, including restlessness, reduced appetite, lack of attention, drowsiness, aggression, anxiety, fickleness, capriciousness, irritation, and increase in attention request, may only reflect an imminent parturition [14,15]. Indeed, nest-building and isolation are associated with searching for and selecting the appropriate birthplace [16,17]. It is possible that nest-building is more likely observed in altricial species, whereas isolation from others may occur in precocial ones.

Direct maternal behaviors are observed after parturition, during the first contact with the newborns. Dogs will actively bite and tear the fetal membranes and cut the umbilical cord, which functions to prevent asphyxiation of the pups [18]. The dam actively licks the head and the mouth of the newborn to stimulate respiration and orient the pups toward the mammary gland [13,15]. They also lick the anogenital area to facilitate urination and excretion during the first 2 postnatal weeks [19]. Rats also express anogenital licking, especially toward males during the first 10 postnatal days [20]. Preference to lick males is evoked by attractive odors from preputial compounds, such as dodecyl propionate [21], which depend on the levels of systemic steroids. If female pups are treated with androgens (i.e., testosterone and dihydrotestosterone) on the day of birth, they receive an equivalent amount of active anogenital licking as males [22]. Enhanced anogenital stimulation appears to have positive long-term effects on male reproductive behavior [23]. Dams display other active responses as well, such as retrieving the pups, oral consumption of the placenta, and defense from predators and conspecifics, and passive behaviors such as huddling and crouching to regulate body temperature or allowing nursing. In the beginning, passive behaviors may depend on acceptance, whereas active behaviors may depend on enhanced motivation to care. Other species, such as cows, also express intense active licking for the first hour and will be very protective if someone approaches. Passive acceptance may occur within the first 30–120 min postpartum when the calf stands and searches for the udder [16] (Table 1).

Table 1. Examples of active and passive maternal behaviors that are associated with processes of acceptance, recognition, motivation, and rejection/fear.

Responses	Acceptance	Recognition	Motivation	Rejection/Fear
Passive	Allow proximity Nursing	Selective proximity Selective nursing	Staying in nest Crouching over	Avoidance
Active		Directed calls Directed attention	Licking Retrieving Placenta consumption Defense Nest-building	Aggression Infanticide

The lack of maternal behaviors represents a serious problem that jeopardizes not only the survival of the offspring but also a very important mechanism of early socialization, cognitive development, and epigenetic changes associated with resilience to stress [24,25]. Good maternal behavior is associated with the so-called stress hyporesponsive period [24,26], which refers to a delay of the timing of glucocorticoid elevation in infants, associated with reduced stress response in adulthood [26]. Inappropriate maternal behaviors may occur in 50% of primiparous dogs, especially following cesarean section [27] or as a result of early separation during the postpartum period [28].

3. Natural Stimuli That Facilitate the Expression of Maternal Behavior

3.1. Hormones

Gregarious species have a natural predisposition to care for the young. However, the capacity to express appropriate levels of maternal behavior develops gradually with hormonal changes that occur throughout pregnancy. Then, drastic changes during parturition are needed to trigger the expression of behavior. In rats, for example, concentrations of progesterone (P4) gradually start to increase from the very first day of pregnancy, reaching a peak at day 15, and are followed by a drastic reduction during the last 3 days before delivery. By contrast, the levels of estradiol (E2) and prolactin (PRL) stay relatively low at the beginning but increase dramatically during those last 3 days. The reduction of P4 and increase in hormones such as E2, PRL, oxytocin (OT), and corticosteroids, are the main hormonal drastic changes during parturition [29–31] and therefore are associated with sensitization of maternal behavior. For example, in the rabbit doe, digging is stimulated by changes in E2 and P4, while straw carrying and hair pulling are under the control of PRL. In the rat, the reduction in P4 and the increase in E2 and PRL levels facilitate active licking, retrieving, and gathering of pups. Pharmacological blockade of estrogen in the medial preoptic area (mPOA) and small interfering RNA silencing of estrogen receptors (ERα) disrupts maternal behavior in mice [32,33], whereas specific activation of ERα-positive mPOA neurons enhances pup retrieval [34,35]. Likewise, the blockade of PRL receptors within the mPOA in mice abolishes pup retrieval [36].

Table 2. *Cont.*

Stimuli	Effect	Behavioral Response	Representative References
Parturition VCS	⇑	Nursing Retrieving	[38,39] [40]
Mating	⇓	Infanticide	[54,55]
Exposure to pups	⇑	Nest-building, retrieving, licking, crouching posture	[48–50]
Experience Multiparous	⇑ ⇑ ⇑ ⇑ ⇓	Licking, nursing, contact Defense Suckling Following, grooming Rejection	[60] [61] [64]
Environmental enrichment	⇓ ⇑	Anxiety Licking, grooming, crouching posture, defense	[65] [66]

4. Neural Pathways Underlying Maternal Behavior

4.1. Areas Involved in Acceptance and Social Recognition

Maternal behavior requires at least three neural processes to mediate: (1) increase in acceptance/social recognition, (2) decrease in rejection/fear, and (3) increase in care motivation [67]. Acceptance/recognition occurs when a sensitized brain is exposed to the right stimulus. For example, during the postpartum period, bitches will accept any pup scented with amniotic fluid. Washing the pups immediately after birth results in a lack of acceptance. However, if pups are bathed in amniotic fluid, the process of acceptance is restored, even hours later [28]. Similarly, washing a newborn lamb impairs acceptance/recognition behaviors in sheep (low bleats, acceptance at the udder, nursing, and licking time) and increases rejection/fear behaviors (high-pitched bleats, rejection at the udder, and aggressive behavior). Preventing contact with the mother during the first 4 h postpartum worsens rejection. Interestingly, washing an alien lamb also improves its acceptance [68]. Indeed, both acceptance and rejection behaviors depend on the olfactory system because anosmic ewes (treated with intranasal zinc sulfate) fail to express any acceptance or rejection behavior [69]. However, if tested many hours after birth, anosmic sheep will be capable of accepting/recognizing their lamb through visual or auditory cues that have been learned during the first postpartum hours [70].

Olfactory social recognition is perhaps the most relevant sensory system in many mammals and depends on both main and accessory olfactory pathways. Nonvolatile odor molecules are detected by the accessory pathway via the vomeronasal organ (VNO), located in the soft tissue of the nasal septum. The VNO projects to the accessory olfactory bulb (AOB), which innervates both the bed nucleus of the stria terminalis (BNST) and medial amygdala (MeA). BNST and MeA project to the mPOA, which in turn projects to the lateral preoptic area (LPOA) and substantia innominata (SI). LPOA and SI also send efferents to the ventral tegmental area (VTA), which sends dopaminergic projections to the nucleus accumbens (NAcc) [71,72]. With regard to volatile odor molecules (those that have a low boiling point and evaporate easily at room temperature), they are detected in the roof of the nasal cavity by the main olfactory epithelium (MOE), where olfactory sensory neurons are located. These neurons project to the main olfactory bulb (MOB), and MOB has efferents to the piriform cortex (PirCtx), and from there to the NAcc and mPOA [71]. Thus, both volatile odors and nonvolatile odors not only access the social recognition system but also the mesolimbic pathway of motivation. Within these pathways, social recognition depends on the nonapeptides arginine vasopressin (AVP) and oxytocin (OT). Mice with depletion of AVP V1a receptor (V1aRKO) exhibit profound social recognition impairment. They can remember the odors of food but fail to remember the odors of individuals they just spent time with [73,74]. Likewise, OT knockout mice (OTKO) express impaired discrimination to odors of conspecifics [75]. Following a social encounter, OTKO mice express less activity in MeA, BNST, and mPOA, but an infusion of OT in the MeA is sufficient to restore social recognition [74,76,77]. Interestingly, during the postpartum period, OT receptor (OTR) binding increases not only in the MeA but also in the BNST, mPOA [78], and dopaminergic regions of the brain stem [74,79–83]. The increase in OTR is mainly the result of exposure to E2 [84]. Thus, OT and OTR start to play a gradual role in the acceptance and olfactory social recognition via MeA starting two days before parturition when E2 increases. Experiments in sheep have shown that infusion of local anesthesia in the MeA or cortical amygdala (CoA) impairs recognition of the lamb during the early postpartum period, but this was not found with infusions in the BLA region [85]. As we describe later, BLA mediates rejection/fear responses, whereas MeA and CoA mediate olfactory acceptance/recognition, mainly influenced by OT, AVP, and dopamine (DA) (Figure 1).

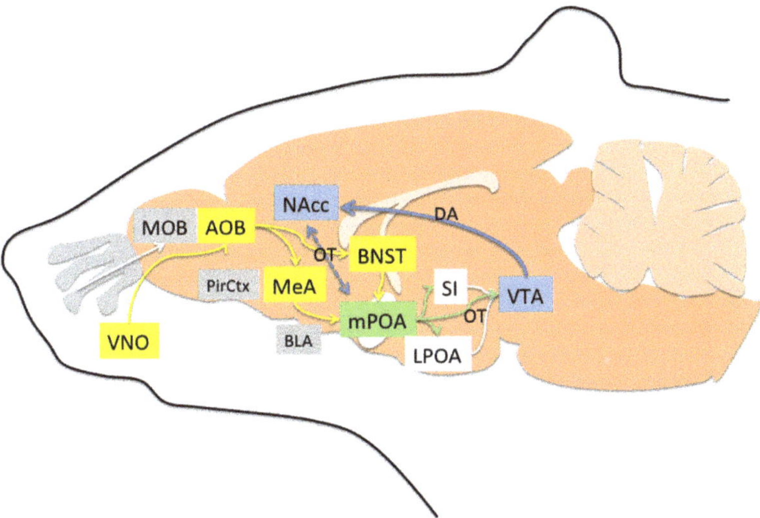

Figure 1. Sagittal drawing of brain olfactory pathways involved in acceptance, offspring social recognition, care motivation, and inhibition of rejection. Nonvolatile odor molecules are processed by the accessory pathway (yellow) via the vomeronasal organ (VNO), accessory olfactory bulb (AOB), bed nucleus of the stria terminalis (BNST), medial amygdala (MeA), and medial preoptic area (mPOA, green). The mPOA projects to the lateral preoptic area (LPOA), substantia innominata (SI), and ventral tegmental area (VTA, blue), which sends dopaminergic projections to the nucleus accumbens (NAcc, blue). Volatile odor molecules are processed by the main (gray) olfactory epithelium (MOE), main olfactory bulb (MOB), and piriform cortex (PirCtx), and from there to the NAcc and mPOA (which interact via oxytocin "OT" and dopamine "DA"). The basolateral (BLA) amygdala and prefrontal cortex (not shown) participate in rejection and fear.

In dogs, salivary OT (sOT) increases gradually during the postpartum period [86]. High levels of sOT are negatively correlated with the frequency of sniffing behavior toward the pups, perhaps because high sOT facilitates olfactory recognition, and recognized pups require less sniffing. Given that sniffing and time spent out of the whelping box have been positively correlated, low sOT is also a predictor of time away from the pups. Therefore, treatment with OT should improve maternal recognition and performance. In fact, rats that undergo cesarean section lack a naturally powerful stimulus to release OT, which results in reduced maternal behavior. However, intranasal OT in those rats restores pup retrieval and anogenital licking [87]. Similarly, intranasal OT facilitates paternal motivation, as observed in male marmosets that express shorter latency to respond with an approach to infant stimuli [88]. Dogs treated with intranasal OT turn more social toward humans [89] and more playful with other dogs [90]. In female rats, OT improves depression-like behaviors [91] and induces conditioned place preference (CPP) in the presence of a conspecific (social-CPP). Social-CPP reflects emotional memories of a place where cohabitation occurred, and it has been reported in rats that receive MDMA "Ecstasy" but not in those that receive AVP. Accordingly, both OT and AVP regulate olfactory recognition, but only OT augments the rewarding effects of social interaction [92].

In dogs, polymorphisms of the OTR gene have been associated with variations in the maternal behavior and with higher levels of sOT [93]. However, to date, no one has shown the potential good effects of intranasal OT in dogs with poor maternal behavior. Certainly, systemic injections of OT at birth are commonly indicated to facilitate uterine contractions, but only 1% of such systemic OT crosses the blood–brain barrier (BBB) [94]. Nevertheless, it has been demonstrated that fetal voles from mothers injected with OT at birth expressed increased methylation of OTR in the brain. In adulthood, OT-exposed voles are more gregarious, with increased alloparental caregiving toward pups and increased close social contact with other adults [95]. Thus, OT treatment at birth (even if injected) may facilitate both immediate and future social encounters.

4.2. Areas Mainly Involved in Increasing Motivation

The NAcc and mPOA are the main areas involved in maternal motivation [96]. In rats, broad lesions in the mPOA disrupted pup retrieval and reduced Fos-immunoreactivity in the NAcc evoked by exposure to pups [97]. However, specific lesions in the ventromedial region of the preoptic area (vmPOA) disrupt nest-building in mice but do not affect pup retrieval, whereas specific lesions in the central part of the mPOA (cmPOA) disrupt all maternal behaviors [98]. Lesions in the NAcc also decrease pup retrieval and bar pressing to gain access to pups [99], whereas lesions in the VTA (where NAcc DA afferents originate) disrupt the frequency of approaches and interaction with pups [100,101].

Between mPOA and NAcc, there is a synergistic role OT and DA. One study showed that female rats that display high levels of active maternal behavior (i.e., licking/grooming) express more OT-positive neurons in the mPOA and PVN, which in turn, project to the VTA. Thus, OT in the VTA facilitates DA release into the NAcc, which also enhances motivation [102]. The mPOA itself is a large and complex region formed by many subnuclei, such as the medial preoptic nucleus (MPN), median preoptic nucleus (MnPO), posterodorsal preoptic nucleus (PD), and ventrolateral preoptic nucleus (VLPO) in mice. Within all these nuclei, there is a vast diversity of neuron populations based on gene expression for transporters or neurotransmitters that are not found in the lateral preoptic area (LPOA). For instance, at least fifteen different markers of cell populations have been reported in the mice mPOA, including glutamic acid decarboxylase (GAD67), galanin, calbindin, a novel marker of sexually dimorphic nuclei (Moxd1), cocaine- and amphetamine-regulated transcript (CART), vesicular glutamate transporter (VGLUT2), vesicular GABA transporter (Vgat), proenkephalin (Penk), brain-derived neurotrophic factor (BDNF), leptin, cholecystokinin (CCK), neurotensin, neuropeptide tachykinin 2 (Tac2), prodynorphin (Pdyn), thyrotropin-releasing hormone (TRH), and oxytocin (reviewed in [103]). Many of these markers are known for controlling mechanisms of body temperature, wake–sleep cycle, or sexual behavior and appear not to have any direct role in maternal behavior. However, galaninergic neurons within the lateral part of the mouse mPOA (lmPOA) and MPN project toward MeA, VTA, PVN, and periaqueductal gray (PAG) and play a major role in motor coordination, motivation, and social recognition during maternal behavior by integrating inputs from many areas in the brain [104]. More than 70% of galanin-positive neurons in the mPOA also express alpha estrogen receptors (ERα) and androgen receptors [105] and therefore may become more active during the final phase of pregnancy. Optogenetic activation of mPOA galanin-positive neurons projecting to the PAG enhances pup grooming, whereas inhibition of those neurons impairs grooming. Likewise, activation of galanin neurons that project to VTA promotes interaction with pups but not pup retrieval, and their inhibition suppresses interaction [104]. The mPOA also receives input from hypocretin-1-containing neurons (HCRT-1, also known as orexin A) from the postero-lateral hypothalamus. Activation of HCRT-1 metabotropic receptors within the mPOA promotes active maternal behaviors such as licking and retrieving but also passive responses such as nursing [106]. The mPOA also receives inhibitory inputs via agouti-related neuropeptide (AGRP) neurons from the arcuate nucleus, which respond in conditions of caloric needs and normally produce hunger. Activation of AGRP inhibitory neurons or its projections to mPOA results in less maternal nest-building, without affecting pup retrieval, partly recapitulating suppression of maternal behaviors during food restriction [107]. Thus, retrieving pups and nest-building are associated with different neuronal activity patterns in the mPOA, such that neurons that are activated during pup retrieval tend to be inhibited during nest-building [34]. Some AGRP neurons that project directly to mPOA express Vgat. Optogenetic stimulation of Vgat neurons in the mPOA elicits both pup retrieval and nest-building, whereas inhibition of Vgat neurons results in decreased nest-building. Thus, GABA activation in mPOA can inhibit inhibitory AGRP neurons and facilitate all maternal behaviors. However, if stimulation is exclusively directed toward the subpopulation of Vgat neurons that express ERα, then only pup-retrieval is elicited, and not nest-building [107] (Table 3). In addition, ERα-positive neurons in the mouse mPOA project inhibitory efferents to nondopaminergic neurons in the VTA that promote maternal pup retrieval through disinhibition of dopaminergic neurons [34]. Accordingly, galanin-positive, ERα-positive, and Vgat-positive neurons may be involved in maternal motivation (pup retrieval) via mPOA and VTA [103] (Figure 2).

Like the mPOA, the NAcc also facilitates motivation toward recognized/accepted individuals. The NAcc is also important for the consolidation of social bonds. For example, neurons of the NAcc in monogamous voles express more OT receptors than in the polygamous voles [108]. Infusion of an OT antagonist into the NAcc of monogamous voles disrupts pair bonds facilitated by DA agonists, whereas OT antagonists disrupt bonds facilitated by DA agonists [109]. Pair bonds that develop after sex are disrupted by infusions of DA antagonists (i.e., haloperidol) in the NAcc, and by contrast, low doses of DA agonists (i.e., apomorphine) facilitate their development even in absence of sex [110,111]. Thus, having a working DA system is necessary, but not sufficient (OT is needed) to induce selective social bonds. During exposure to pups, maternal rats release more DA in the NAcc [112], as well as during licking and grooming of the pups [113]. Likewise, infusions of general DA antagonists such as flupenthixol into the NAcc of lactating rats impair maternal behavior (retrieval and licking) [114]. Specific injections of a D1-type antagonist (SCH23390) in the NAcc impairs pup retrieval, whereas a D2-type antagonist (eticlopride) does not impair such behavior [115]. Thus, maternal behavior is mainly facilitated by D1-type receptors and OT.

4.3. Areas Mainly Involved in Reducing Rejection/Fear

In addition to being motivated, individuals must reduce their rejection and fear of pups during the first postpartum encounters. This represents a dichotomy in excitatory/inhibitory neurocircuitries and appears to depend on the positive interaction of mPOA/BNST and a negative mechanism mediated by the anterior hypothalamic nucleus (AHN)/ventromedial nucleus (VMN)/PAG, amygdala, and some cortical areas [116]. Thus, activation of mPOA appears to facilitate maternal behavior upon its release from the inhibitory effects of the amygdala [117]. For instance, lesions of the vomeronasal inputs (amygdala) to the mPOA facilitate maternal behavior even in nulliparous female rats [118–120], mainly due to BLA projections that mediate fear responses [121,122]. BLA increases its expression of OTR during the postpartum period [108], and it can be inhibited by endogenous opioids as well [123]. Therefore, the elevation of OT and endogenous opioids during parturition (i.e., during VCS) may help reduce fear/rejection toward pups [124]. Opioids also regulate the release of both OT [125] and DA into the prefrontal cortex (PFC) [126]. PFC plays a key role in attention, memory, and negative emotions [127]. When activity in PFC is reduced, sheep express less aggressive behavior toward alien lambs without affecting the social recognition of their own lambs [128]. PFC also expresses more ERα and OTR in lactating rats, which appear to be associated with decreased anxiety [129].

The central region of mPOA and MPN also mediate the inhibition of some rejection responses. These regions express galanin, and direct optogenetic stimulation of galanin-positive neurons in male mice suppresses pup-directed aggression [130]. Within the vm-POA, some neurons mediate sickness symptoms [131]. However, it is unknown whether those neurons may induce rejection or fear toward pups.

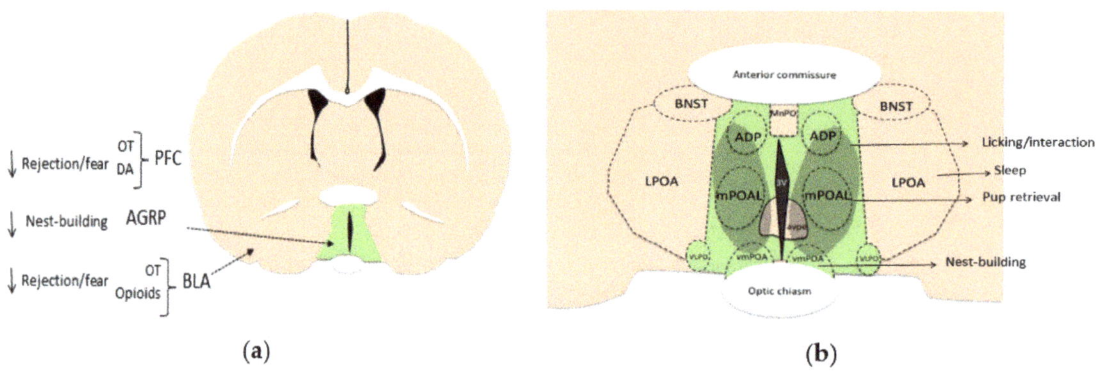

Figure 2. (**a**) Coronal drawing of a rat brain (modified from [132]). The green region represents the medial preoptic area (mPOA) as the main generator of maternal motivation via oxytocin (OT) and dopamine (DA) activity. Agouti-related peptide (AGRP) facilitates rejection in mPOA, but OT, DA, and opioids reduce rejection via other regions such as prefrontal cortex (PFC, not shown) and basolateral amygdala (BLA). (**b**) Amplification of the mPOA and subregions in the rat brain. Darker green spot represents putative galanin/ER+ neurons that mediate the expression of licking and pup retrieval in the lateral (mPOAL) and dorsal (ADP) parts of mPOA, where OT neurons are also found (modified from [103]). Nest-building depends on the activity of the ventral portion of the mPOA (vmPOA). Also shown are the bed nucleus of stria terminalis (BNST), median preoptic nucleus (MnPO), ventrolateral nucleus (vlPO), anteroventral periventricular nucleus (avpv), and lateral preoptic area (LPOA) that mediates sleep. Different mPOA subregions mediate diverse maternal behaviors interacting with mesolimbic regions.

Table 3. Brain areas and neurochemicals associated with maternal behaviors. Medial amygdala (MeA), bed nucleus of stria terminalis (BNST), cortical amygdala (CoA), medial preoptic area (mPOA), ventral tegmental area (VTA), basolateral amygdala (BLA), nucleus accumbens (NAc), prefrontal cortex (PFC), oxytocin (OT), estrogen (E2), prolactin (PRL), dopamine (DA), vesicular GABA transporter (Vgat), hypocretin-1 neurons (HCRT-1), agouti related protein (AGRP).

Brain Area	Neurochemical	Effect	Representative References
MeA BNST CoA	OT	Acceptance, social recognition	[74,79,83]

Table 3. *Cont.*

Brain Area	Neurochemical	Effect	Representative References
mPOA	E2 PRL OT DA Galanin	Pup retrieval Pup retrieval, nest-building Acceptance, licking/grooming Motivation, licking, grooming Motor coordination, motivation, recognition, grooming, inhibition of aggression	[32,33,40] [36] [102,103] [103,104,130]
	Vgat HCRT-1 AGRP	Nest-building Licking/retrieval Inhibition of nest-building	[130] [106] [107]
VTA	OT	Approach, interaction	[74,79,86]
BLA	OT opioids	Inhibit rejection/fear	[85,108,121,124]
NAc	DA OT	Motivation, approach, interaction, pup retrieval	[99–101,112]
PFC	OT DA	Inhibit rejection/fear	[125,126]

5. Conclusions

The capacity to express maternal behavior develops gradually as a consequence of hormonal priming during pregnancy. Around parturition time, sudden changes in hormones and physical stimuli trigger the expression of acceptance, recognition, and motivation at a time that reduces rejection and fear. The neurocircuitries of maternal behavior can be also sensitized by cohabitation, sex, and experience. Olfactory recognition depends mainly on the activity of OT in the MeA. Motivation is mainly mediated by OT and DA in NAcc and subregions of the mPOA, in which ventral parts mediate nest-building, and central and dorsal parts mediate licking retrieval and interaction with pups. Finally, BLA, PFC, and other hypothalamic regions (i.e., arcuate nucleus) mediate rejection and fear responses. These areas mediate discrimination of stimuli that might have an incentive value per se (i.e., amniotic fluid) or stimuli related to past experiences (i.e., memories of sexual reward, maternal reward, cohabitation, etc.).

Author Contributions: Conceptualization, G.A.C.-A.; investigation, G.A.C.-A., D.H.-C., L.I.G., R.T., M.E.H., P.P.-R., A.A.C.-M. and J.M.; writing—original draft preparation, G.A.C.-A.; writing—review and editing, G.A.C.-A., D.H.-C., L.I.G., R.T., M.E.H., P.P.-R., A.A.C.-M. and J.M. All authors have read and agreed to the published version of the manuscript.

Funding: This research received no external funding.

Institutional Review Board Statement: Not applicable.

Informed Consent Statement: Not applicable.

Data Availability Statement: Not applicable.

Conflicts of Interest: The authors declare no conflict of interest.

References

1. Kendrick, K.M.; Hinton, M.R.; Atkins, K.; Haupt, M.A.; Skinner, J.D. Mothers determine sexual preferences. *Nature* **1998**, *395*, 229–230. [CrossRef] [PubMed]
2. Crowell-Davis, S.L.; Houpt, K.A. Maternal Behavior. *Veter. Clin. N. Am. Equine Pract.* **1986**, *2*, 557–571. [CrossRef] [PubMed]
3. Zhang, X.; Wang, M.; He, T.; Long, S.; Guo, Y.; Chen, Z. Effect of Different Cross-Fostering Strategies on Growth Performance, Stress Status and Immunoglobulin of Piglets. *Animals* **2021**, *11*, 499. [CrossRef]
4. Grota, L.J. Effects of litter size, age of young, and parity on foster mother behaviour in *Rattus norvegicus*. *Anim. Behav.* **1973**, *21*, 78–82. [CrossRef] [PubMed]

5. Scharis, I.; Amundin, M. Cross-fostering in gray wolves (*Canis lupus lupus*). *Zoo Biol.* **2015**, *34*, 217–222. [CrossRef] [PubMed]
6. von Keyserlingk, M.A.; Weary, D.M. Maternal behavior in cattle. *Horm. Behav.* **2007**, *52*, 106–113. [CrossRef]
7. Dienske, H.; Van Vreeswijk, W. Regulation of nursing in chimpanzees. *Dev. Psychobiol.* **1987**, *20*, 71–83. [CrossRef]
8. Davenport, R.K., Jr.; Menzel, E.W., Jr.; Rogers, C.M. Maternal care during infancy: Its effect on weight gain and mortality in the chimpanzee. *Am. J. Orthopsychiatry* **1961**, *31*, 803–809. [CrossRef]
9. Schulte, I.; Hoy, S. Nursing and suckling behavior and mother-child contacts in domestic rabbits. *Berl. Munch. Tierarztl. Wochenschr.* **1997**, *110*, 134–138.
10. Kleiman, D. Reproduction in the Canidae. *Int. Zoo Yearb.* **1968**, *8*, 3–8. [CrossRef]
11. Denenberg, V.H.; Zarrow, M.; Taylor, R.E. Maternal Behavior in the Rat: An Investigation and Quantification of Nest Building. *Behaviour* **1969**, *34*, 1–16. [CrossRef]
12. Benedek, I.; Altbäcker, V.; Zsolnai, A.; Molnár, T. Exploring the Genetic Background of the Differences in Nest-Building Behavior in European Rabbit. *Animals* **2020**, *10*, 1579. [CrossRef] [PubMed]
13. Bleicher, N. Behavior of the bitch during parturition. *J. Am. Veter. Med. Assoc.* **1962**, *140*, 1076–1082.
14. Ferrari, J.; Monteiro-Filho, E.L.A. *Canid Familiaris-Comparative Analysis of pre and Postpartum Behavioral Patterns*; Universidade Federal do Paraná: Curitiba, Brazil, 2004.
15. Santos, N.R.; Beck, A.; Fontbonne, A. A review of maternal behaviour in dogs and potential areas for further research. *J. Small Anim. Pract.* **2019**, *61*, 85–92. [CrossRef] [PubMed]
16. Lidfors, L. Parental Behavior in Bovines. *Adv. Neurobiol.* **2022**, *27*, 177–212. [CrossRef] [PubMed]
17. Rørvang, M.V.; Nielsen, B.L.; Herskin, M.S.; Jensen, M.B. Prepartum Maternal Behavior of Domesticated Cattle: A Comparison with Managed, Feral, and Wild Ungulates. *Front. Veter. Sci.* **2018**, *5*, 45. [CrossRef]
18. Schweizer, C.M.; Meyers-Wallen, V.M. Medical managment of dystocia and indications of cesarean section in the bitch. In *Current Veterinary Therapy XIII*; Bonagura, W.B., Ed.; Saunders Co: Philadelphia, PA, USA, 2000; pp. 933–939.
19. Rheingold, H.L. Maternal behavior in the dog. In *Maternal Behaviors*; Mammals, I., Reingold, H.L., Eds.; John Wiley and Sons: New York, NY, USA, 1963; pp. 169–202.
20. Moore, C.L.; Morelli, G.A. Mother rats interact differently with male and female offspring. *J. Comp. Physiol. Psychol.* **1979**, *93*, 677–684. [CrossRef]
21. Lévy, F.; Keller, M.; Poindron, P. Olfactory regulation of maternal behavior in mammals. *Horm. Behav.* **2004**, *46*, 284–302. [CrossRef]
22. Moore, C. Maternal behavior of rats is affected by hormonal condition of pups. *J. Comp. Physiol. Psychol.* **1982**, *96*, 123–129. [CrossRef]
23. Moore, C.L. Maternal contributions to the development of masculine sexual behavior in laboratory rats. *Dev. Psychobiol.* **1984**, *17*, 347–356. [CrossRef]
24. Nagasawa, M.; Shibata, Y.; Yonezawa, A.; Takahashi, T.; Kanai, M.; Ohtsuka, H.; Suenaga, Y.; Yabana, Y.; Mogi, K.; Kikusui, T. Basal cortisol concentrations related to maternal behavior during puppy development predict post-growth resilience in dogs. *Horm. Behav.* **2021**, *136*, 105055. [CrossRef] [PubMed]
25. Weaver, I.C.; Cervoni, N.; Champagne, F.A.; D'Alessio, A.C.; Sharma, S.; Seckl, J.R.; Dymov, S.; Szyf, M.; Meaney, M.J. Epi-genetic programming by maternal behavior. *Nat. Neurosci.* **2004**, *7*, 847–854. [CrossRef] [PubMed]
26. Nagasawa, M.; Shibata, Y.; Yonezawa, A.; Morita, T.; Kanai, M.; Mogi, K.; Kikusui, T. The behavioral and endocrinological development of stress response in dogs. *Dev. Psychobiol.* **2013**, *56*, 726–733. [CrossRef] [PubMed]
27. Santos, N.R.; Beck, A.; Maenhoudt, C.; Billy, C.; Fontbonne, A. Profile of Dogs' Breeders and Their Considerations on Female Reproduction, Maternal Care and the Peripartum Stress—An International Survey. *Animals* **2021**, *11*, 2372. [CrossRef] [PubMed]
28. Abitbol, M.L.; Inglis, S.R. Role of amniotic fluid in newborn acceptance and bonding in canines. *J. Matern.-Fetal Med.* **1997**, *6*, 49–52. [CrossRef]
29. Rosenblatt, J.S.; Mayer, A.D.; Giordano, A.L. Hormonal basis during pregnancy for the onset of maternal behavior in the rat. *Psychoneuroendocrinology* **1988**, *13*, 29–46. [CrossRef] [PubMed]
30. Nelson, R.J. *An Introduction to Behavioral Endocrinology*, 2nd ed.; Sinauer Associates: Sunderland, MA, USA, 2000.
31. Bridges, R.S. The role of lactogenic hormones in maternal behavior in female rats. *Acta Paediatr. Suppl.* **1994**, *83*, 33–39. [CrossRef] [PubMed]
32. Ribeiro, A.C.; Musatov, S.; Shteyler, A.; Simanduyev, S.; Arrieta-Cruz, I.; Ogawa, S.; Pfaff, D.W. siRNA silencing of estrogen receptor-α expression specifically in medial preoptic area neurons abolishes maternal care in female mice. *Proc. Natl. Acad. Sci. USA* **2012**, *109*, 16324–16329. [CrossRef]
33. Catanese, M.; Vandenberg, L.N. Bisphenol S (BPS) alters maternal behavior and brain in mice exposed during pregnancy/lactation and their daughters. *Endocrinology* **2016**, *158*, 516–530. [CrossRef]
34. Fang, Y.-Y.; Yamaguchi, T.; Song, S.C.; Tritsch, N.X.; Lin, D. A Hypothalamic Midbrain Pathway Essential for Driving Maternal Behaviors. *Neuron* **2018**, *98*, 192–207.e10. [CrossRef]
35. Wei, Y.-C.; Wang, S.-R.; Jiao, Z.-L.; Zhang, W.; Lin, J.-K.; Li, X.-Y.; Li, S.-S.; Zhang, X.; Xu, X.-H. Medial preoptic area in mice is capable of mediating sexually dimorphic behaviors regardless of gender. *Nat. Commun.* **2018**, *9*, 279. [CrossRef] [PubMed]
36. Brown, R.S.E.; Aoki, M.; Ladyman, S.R.; Phillipps, H.R.; Wyatt, A.; Boehm, U.; Grattan, D.R. Prolactin action in the medial preoptic area is necessary for postpartum maternal nursing behavior. *Proc. Natl. Acad. Sci. USA* **2017**, *114*, 10779–10784. [CrossRef] [PubMed]

37. Lévy, F. Neuroendocrine control of maternal behavior in non-human and human mammals. *Ann. d'Endocrinol.* **2016**, *77*, 114–125. [CrossRef]

38. Kendrick, K.; Lévy, F.; Keverne, E. Importance of vaginocervical stimulation for the formation of maternal bonding in primiparous and multiparous parturient ewes. *Physiol. Behav.* **1991**, *50*, 595–600. [CrossRef]

39. Keverne, E.B.; Levy, F.; Poindron, P.; Lindsay, D.R. Vaginal Stimulation: An Important Determinant of Maternal Bonding in Sheep. *Science* **1983**, *219*, 81–83. [CrossRef] [PubMed]

40. Ahdieh, H.B.; Mayer, A.D.; Rosenblatt, J.S. Effects of Brain Antiestrogen Implants on Maternal Behavior and on Postpartum Estrus in Pregnant Rats. *Neuroendocrinology* **1987**, *46*, 522–531. [CrossRef]

41. Gandelman, R.; McDermott, N.J.; Kleinman, M.; DeJianne, D. Maternal nest building by pseudopregnant mice. *Reproduction* **1979**, *56*, 697–699. [CrossRef]

42. Steuer, M.A.; Thompson, A.C.; Doerr, J.C.; Youakim, M.; Kristal, M.B. Induction of maternal behavior in rats: Effects of pseudopregnancy termination and placenta-smeared pups. *Behav. Neurosci.* **1987**, *101*, 219–227. [CrossRef] [PubMed]

43. Misner, T.L.; A Houpt, K. Animal behavior case of the month. Aggression that began 4 days after ovariohysterectomy. *J. Am. Veter. Med. Assoc.* **1998**, *213*, 1260–1262.

44. Rosenblatt, J.S.; Siegel, H.I. Hysterectomy-induced maternal behavior during pregnancy in the rat. *J. Comp. Physiol. Psychol.* **1975**, *89*, 685–700. [CrossRef]

45. Rosenblatt, J.S.; Hazelwood, S.; Poole, J. Maternal Behavior in Male Rats: Effects of Medial Preoptic Area Lesions and Presence of Maternal Aggression. *Horm. Behav.* **1996**, *30*, 201–215. [CrossRef] [PubMed]

46. Perkinson, M.R.; Kim, J.S.; Iremonger, K.J.; Brown, C.H. Visualising oxytocin neurone activity in vivo: The key to unlocking central regulation of parturition and lactation. *J. Neuroendocr.* **2021**, *33*, e13012. [CrossRef] [PubMed]

47. Wang, P.; Wang, S.C.; Liu, X.; Jia, S.; Wang, X.; Li, T.; Yu, J.; Parpura, V.; Wang, Y.-F. Neural Functions of Hypothalamic Oxytocin and its Regulation. *ASN Neuro* **2022**, *14*, 17590914221100706. [CrossRef] [PubMed]

48. Terkel, J.; Rosenblatt, J.S. Humoral factors underlying maternal behavior at parturition: Corss transfusion between freely moving rats. *J. Comp. Physiol. Psychol.* **1972**, *80*, 365–371. [CrossRef] [PubMed]

49. Rosenblatt, J.S. Nonhormonal Basis of Maternal Behavior in the Rat. *Science* **1967**, *156*, 1512–1514. [CrossRef]

50. Jakubowski, M.; Terkel, J. Transition from pup killing to parental behavior in male and virgin female albino rats. *Physiol. Behav.* **1985**, *34*, 683–686. [CrossRef]

51. Harding, K.M.; Lonstein, J.S. Extensive juvenile "babysitting" facilitates later adult maternal responsiveness, decreases anxiety, and increases dorsal raphe tryptophan hydroxylase-2 expression in female laboratory rats. *Dev. Psychobiol.* **2016**, *58*, 492–508. [CrossRef]

52. Carcea, I.; Caraballo, N.L.; Marlin, B.J.; Ooyama, R.; Riceberg, J.S.; Navarro, J.M.M.; Opendak, M.; Diaz, V.E.; Schuster, L.; Torres, M.I.A.; et al. Oxytocin neurons enable social transmission of maternal behaviour. *Nature* **2021**, *596*, 553–557. [CrossRef]

53. Mennella, J.A.; Moltz, H. Infanticide in rats: Male strategy and female counter-strategy. *Physiol. Behav.* **1988**, *42*, 19–28. [CrossRef]

54. vom Saal, F.S. Time-contingent change in infanticide and parental behavior induced by ejaculation in male mice. *Physiol. Behav.* **1985**, *34*, 7–15. [CrossRef]

55. Dulac, C.; O'Connell, L.A.; Wu, Z. Neural control of maternal and paternal behaviors. *Science* **2014**, *345*, 765–770. [CrossRef] [PubMed]

56. Coolen, L.M.; Peters, H.J.; Veening, J.G. Fos immunoreactivity in the rat brain following consummatory elements of sexual behavior: A sex comparison. *Brain Res.* **1996**, *738*, 67–82. [CrossRef] [PubMed]

57. Malsbury, C.W. Facilitation of male rat copulatory behavior by electrical stimulation of the medial preoptic area. *Physiol. Behav.* **1971**, *7*, 797–805. [CrossRef] [PubMed]

58. Shimura, T.; Yamamoto, T.; Shimokochi, M. The medial preoptic area is involved in both sexual arousal and performance in male rats: Re-evaluation of neuron activity in freely moving animals. *Brain Res.* **1994**, *640*, 215–222. [CrossRef] [PubMed]

59. Arendash, G.W.; Gorski, R.A. Effects of discrete lesions of the sexually dimorphic nucleus of the preoptic area or other medial preoptic regions on the sexual behavior of male rats. *Brain Res. Bull.* **1983**, *10*, 147–154. [CrossRef]

60. Guardini, G.; Bowen, J.; Raviglione, S.; Farina, R.; Gazzano, A. Maternal behaviour in domestic dogs: A comparison between primiparous and multiparous dogs. *Dog Behav.* **2015**, *1*, 23–33. [CrossRef]

61. Vicentini, R.R.; El Faro, L.; Ujita, A.; Lima, M.L.P.; Oliveira, A.P.; Sant'Anna, A.C. Is maternal defensiveness of Gyr cows (*Bos taurus indicus*) related to parity and cows' behaviors during the peripartum period? *PLoS ONE* **2022**, *17*, e0274392. [CrossRef]

62. Jensen, M.B.; Webb, L.E.; Vaarst, M.; Bokkers, E. The effect of hides and parity on behavior of periparturient dairy cows at pasture. *J. Dairy Sci.* **2022**, *105*, 6196–6206. [CrossRef]

63. Aguggia, J.P.; Suárez, M.M.; Rivarola, M.A. Multiparity Dampened the Neurobehavioral Consequences of Mother–Pup Separation Stress in Dams. *Neuroscience* **2019**, *416*, 207–220. [CrossRef]

64. Lv, S.-J.; Yang, Y.; Li, F.-K. Parity and litter size effects on maternal behavior of Small Tail Han sheep in China. *Anim. Sci. J.* **2015**, *87*, 361–369. [CrossRef]

65. Zhang, Y.-M.; Cheng, Y.-Z.; Wang, Y.-T.; Wei, R.-M.; Ge, Y.-J.; Kong, X.-Y.; Li, X.-Y. Environmental Enrichment Reverses Maternal Sleep Deprivation-Induced Anxiety-Like Behavior and Cognitive Impairment in CD-1 Mice. *Front. Behav. Neurosci.* **2022**, *16*, 943900. [CrossRef] [PubMed]

66. Núñez-Murrieta, M.A.; Noguez, P.; Coria-Avila, G.A.; García-García, F.; Santiago-García, J.; Bolado-García, V.E.; Corona-Morales, A.A. Maternal behavior, novelty confrontation, and subcortical c-Fos expression during lactation period are shaped by gestational environment. *Behav. Brain Res.* **2021**, *412*, 113432. [CrossRef]

67. Coria-Avila, G.A.; Manzo, J.; Garcia, L.I.; Carrillo, P.; Miquel, M.; Pfaus, J.G. Neurobiology of social attachments. *Neurosci. Biobehav. Rev.* **2014**, *43*, 173–182. [CrossRef] [PubMed]

68. Poindron, P.; Otal, J.; Ferreira, G.; Keller, M.; Guesdon, V.; Nowak, R.; Lévy, F. Amniotic fluid is important for the maintenance of maternal responsiveness and the establishment of maternal selectivity in sheep. *Animal* **2010**, *4*, 2057–2064. [CrossRef] [PubMed]

69. Levy, F.; Poindron, P.; Le Neindre, P. Attraction and repulsion by amniotic fluids and their olfactory control in the ewe around parturition. *Physiol. Behav.* **1983**, *31*, 687–692. [CrossRef] [PubMed]

70. Ferreira, G.; Terrazas, A.; Poindron, P.; Nowak, R.; Orgeur, P.; Lévy, F. Learning of olfactory cues is not necessary for early lamb recognition by the mother. *Physiol. Behav.* **2000**, *69*, 405–412. [CrossRef] [PubMed]

71. Rymer, T.L. The Role of Olfactory Genes in the Expression of Rodent Paternal Care Behavior. *Genes* **2020**, *11*, 292. [CrossRef]

72. Fallon, J.H.; Moore, R.Y. Catecholamine innervation of the basal forebrain IV. Topography of the dopamine projection to the basal forebrain and neostriatum. *J. Comp. Neurol.* **1978**, *180*, 545–579. [CrossRef]

73. Bielsky, I.F.; Hu, S.-B.; Szegda, K.L.; Westphal, H.; Young, L.J. Profound Impairment in Social Recognition and Reduction in Anxiety-Like Behavior in Vasopressin V1a Receptor Knockout Mice. *Neuropsychopharmacology* **2003**, *29*, 483–493. [CrossRef]

74. Bielsky, I.F.; Young, L.J. Oxytocin, vasopressin, and social recognition in mammals. *Peptides* **2004**, *25*, 1565–1574. [CrossRef]

75. Kavaliers, M.; Colwell, D.; Choleris, E.; Ågmo, A.; Muglia, L.J.; Ogawa, S.; Pfaff, D.W. Impaired discrimination of and aversion to parasitized male odors by female oxytocin knockout mice. *Genes Brain Behav.* **2003**, *2*, 220–230. [CrossRef] [PubMed]

76. Ferguson, J.N.; Aldag, J.M.; Insel, T.R.; Young, L.J. Oxytocin in the Medial Amygdala is Essential for Social Recognition in the Mouse. *J. Neurosci.* **2001**, *21*, 8278–8285. [CrossRef] [PubMed]

77. Dumais, K.M.; Bredewold, R.; Mayer, T.E.; Veenema, A.H. Sex differences in oxytocin receptor binding in forebrain regions: Correlations with social interest in brain region- and sex- specific ways. *Horm. Behav.* **2013**, *64*, 693–701. [CrossRef] [PubMed]

78. Caughey, S.D.; Klampfl, S.M.; Bishop, V.R.; Pfoertsch, J.; Neumann, I.D.; Bosch, O.J.; Meddle, S.L. Changes in the Intensity of Maternal Aggression and Central Oxytocin and Vasopressin V1a Receptors Across the Peripartum Period in the Rat. *J. Neuroendocr.* **2011**, *23*, 1113–1124. [CrossRef]

79. Young, L.J.; Lim, M.M.; Gingrich, B.; Insel, T.R. Cellular Mechanisms of Social Attachment. *Horm. Behav.* **2001**, *40*, 133–138. [CrossRef]

80. Fuchs, A.-R.; Fuchs, F.; Husslein, P.; Soloff, M.S.; Fernström, M.J. Oxytocin Receptors and Human Parturition: A Dual Role for Oxytocin in the Initiation of Labor. *Science* **1982**, *215*, 1396–1398. [CrossRef]

81. Gimpl, G.; Fahrenholz, F. The Oxytocin Receptor System: Structure, Function, and Regulation. *Physiol. Rev.* **2001**, *81*, 629–683. [CrossRef]

82. Insel, T.R.; Gingrich, B.S.; Young, L.J. Oxytocin: Who needs it? *Prog. Brain Res.* **2001**, *133*, 59–66.

83. Grieb, Z.A.; Lonstein, J.S. Oxytocin receptor expression in the midbrain dorsal raphe is dynamic across female reproduction in rats. *J. Neuroendocr.* **2021**, *33*, e12926. [CrossRef]

84. Quiñones-Jenab, V.; Jenab, S.; Ogawa, S.; Adan, R.A.; Burbach, P.H.; Pfaff, D.W. Effects of Estrogen on Oxytocin Receptor Messenger Ribonucleic Acid Expression in the Uterus, Pituitary, and Forebrain of the Female Rat. *Neuroendocrinology* **1997**, *65*, 9–17. [CrossRef]

85. Keller, M.; Perrin, G.; Meurisse, M.; Ferreira, G.; Lévy, F. Cortical and medial amygdala are both involved in the formation of olfactory offspring memory in sheep. *Eur. J. Neurosci.* **2004**, *20*, 3433–3441. [CrossRef] [PubMed]

86. Ogi, A.; Mariti, C.; Pirrone, F.; Baragli, P.; Gazzano, A. The Influence of Oxytocin on Maternal Care in Lactating Dogs. *Animals* **2021**, *11*, 1130. [CrossRef] [PubMed]

87. Li, T.; Jia, S.-W.; Hou, D.; Liu, X.; Li, D.; Liu, Y.; Cui, D.; Wang, X.; Hou, C.; Brown, C.H.; et al. Intranasal Oxytocin Restores Maternal Behavior and Oxytocin Neuronal Activity in the Supraoptic Nucleus in Rat Dams with Cesarean Delivery. *Neuroscience* **2021**, *468*, 235–246. [CrossRef] [PubMed]

88. Taylor, J.H.; French, J.A. Oxytocin and vasopressin enhance responsiveness to infant stimuli in adult marmosets. *Horm. Behav.* **2015**, *75*, 154–159. [CrossRef] [PubMed]

89. Turcsán, B.; Román, V.; Lévay, G.; Lendvai, B.; Kedves, R.; Petró, E.; Topál, J. Intranasal Oxytocin Improves Social Behavior in Laboratory Beagle Dogs (*Canis familiaris*) Using a Custom-Made Social Test Battery. *Front. Veter. Sci.* **2022**, *9*, 785805. [CrossRef] [PubMed]

90. Romero, T.; Nagasawa, M.; Mogi, K.; Hasegawa, T.; Kikusui, T. Intranasal administration of oxytocin promotes social play in domestic dogs. *Commun. Integr. Biol.* **2015**, *8*, e1017157. [CrossRef] [PubMed]

91. Ji, H.; Su, W.; Zhou, R.; Feng, J.; Lin, Y.; Zhang, Y.; Wang, X.; Chen, X.; Li, J. Intranasal oxytocin administration improves depression-like behaviors in adult rats that experienced neonatal maternal deprivation. *Behav. Pharmacol.* **2016**, *27*, 689–696. [CrossRef]

92. Ramos, L.; Hicks, C.; Caminer, A.; Goodwin, J.; McGregor, I.S. Oxytocin and MDMA ('Ecstasy') enhance social reward in rats. *Psychopharmacology* **2015**, *232*, 2631–2641. [CrossRef]

93. Ogi, A.; Naef, V.; Santorelli, F.M.; Mariti, C.; Gazzano, A. Oxytocin Receptor Gene Polymorphism in Lactating Dogs. *Animals* **2021**, *11*, 3099. [CrossRef]

94. Ermisch, A.; Rühle, H.-J.; Landgraf, R.; Hess, J. Blood—Brain Barrier and Peptides. *J. Cereb. Blood Flow Metab.* **1985**, *5*, 350–357. [CrossRef]

95. Kenkel, W.M.; Perkeybile, A.-M.; Yee, J.R.; Pournajafi-Nazarloo, H.; Lillard, T.S.; Ferguson, E.F.; Wroblewski, K.L.; Ferris, C.F.; Carter, C.S.; Connelly, J.J. Behavioral and epigenetic consequences of oxytocin treatment at birth. *Sci. Adv.* **2019**, *5*, eaav2244. [CrossRef] [PubMed]

96. Walsh, C.J.; Fleming, A.S.; Lee, A.; Magnusson, J.E. The effects of olfactory and somatosensory desensitization on Fos-like immunoreactivity in the brains of pup-exposed postpartum rats. *Behav. Neurosci.* **1996**, *110*, 134–153. [CrossRef] [PubMed]

97. Stack, E.C.; Balakrishnan, R.; Numan, M.J.; Numan, M. A functional neuroanatomical investigation of the role of the medial preoptic area in neural circuits regulating maternal behavior. *Behav. Brain Res.* **2002**, *131*, 17–36. [CrossRef] [PubMed]

98. Tsuneoka, Y.; Maruyama, T.; Yoshida, S.; Nishimori, K.; Kato, T.; Numan, M.; Kuroda, K.O. Functional, anatomical, and neurochemical differentiation of medial preoptic area subregions in relation to maternal behavior in the mouse. *J. Comp. Neurol.* **2012**, *521*, 1633–1663. [CrossRef] [PubMed]

99. Lee, A.; Clancy, S.; Fleming, A.S. Mother rats bar-press for pups: Effects of lesions of the mpoa and limbic sites on maternal behavior and operant responding for pup-reinforcement. *Behav. Brain Res.* **1999**, *100*, 15–31. [CrossRef]

100. Lee, A.; Li, M.; Watchus, J.; Fleming, A.S. Neuroanatomical basis of maternal memory in postpartum rats: Selective role for the nucleus accumbens. *Behav. Neurosci.* **1999**, *113*, 523–538. [CrossRef]

101. Gaffori, O.; Le Moal, M. Disruption of maternal behavior and appearance of cannibalism after ventral mesencephalic tegmentum lesions. *Physiol. Behav.* **1979**, *23*, 317–323. [CrossRef]

102. Shahrokh, D.K.; Zhang, T.-Y.; Diorio, J.; Gratton, A.; Meaney, M.J. Oxytocin-Dopamine Interactions Mediate Variations in Maternal Behavior in the Rat. *Endocrinology* **2010**, *151*, 2276–2286. [CrossRef]

103. Tsuneoka, Y.; Funato, H. Cellular Composition of the Preoptic Area Regulating Sleep, Parental, and Sexual Behavior. *Front. Neurosci.* **2021**, *15*, 649159. [CrossRef]

104. Kohl, J.; Babayan, B.M.; Rubinstein, N.D.; Autry, A.E.; Marin-Rodriguez, B.; Kapoor, V.; Miyamishi, K.; Zweifel, L.S.; Luo, L.; Uchida, N.; et al. Functional circuit architecture underlying parental behaviour. *Nature* **2018**, *556*, 326–331. [CrossRef]

105. Tsuneoka, Y.; Yoshida, S.; Takase, K.; Oda, S.; Kuroda, M.; Funato, H. Neurotransmitters and neuropeptides in gonadal steroid receptor-expressing cells in medial preoptic area subregions of the male mouse. *Sci. Rep.* **2017**, *7*, 9809. [CrossRef] [PubMed]

106. Rivas, M.; Torterolo, P.; Ferreira, A.; Benedetto, L. Hypocretinergic system in the medial preoptic area promotes maternal behavior in lactating rats. *Peptides* **2016**, *81*, 9–14. [CrossRef] [PubMed]

107. Li, X.-Y.; Han, Y.; Zhang, W.; Wang, S.-R.; Wei, Y.-C.; Li, S.-S.; Lin, J.-K.; Yan, J.-J.; Chen, A.-X.; Zhang, X.; et al. AGRP Neurons Project to the Medial Preoptic Area and Modulate Maternal Nest-Building. *J. Neurosci.* **2018**, *39*, 456–471. [CrossRef]

108. Insel, T.R.; E Shapiro, L. Oxytocin receptor distribution reflects social organization in monogamous and polygamous voles. *Proc. Natl. Acad. Sci. USA* **1992**, *89*, 5981–5985. [CrossRef]

109. Liu, Y.; Wang, Z. Nucleus accumbens oxytocin and dopamine interact to regulate pair bond formation in female prairie voles. *Neuroscience* **2003**, *121*, 537–544. [CrossRef] [PubMed]

110. Aragona, B.J.; Liu, Y.; Curtis, J.T.; Stephan, F.K.; Wang, Z. A Critical Role for Nucleus Accumbens Dopamine in Partner-Preference Formation in Male Prairie Voles. *J. Neurosci.* **2003**, *23*, 3483–3490. [CrossRef] [PubMed]

111. Gingrich, B.; Liu, Y.; Cascio, C.; Wang, Z.; Insel, T.R. Dopamine D2 receptors in the nucleus accumbens are important for social attachment in female prairie voles (*Microtus ochrogaster*). *Behav. Neurosci.* **2000**, *114*, 173–183. [CrossRef]

112. Hansen, S.; Bergvall, A.H.; Nyiredi, S. Interaction with pups enhances dopamine release in the ventral striatum of maternal rats: A microdialysis study. *Pharmacol. Biochem. Behav.* **1993**, *45*, 673–676. [CrossRef]

113. Champagne, F.A.; Chretien, P.; Stevenson, C.W.; Zhang, T.Y.; Gratton, A.; Meaney, M.J. Variations in Nucleus Accumbens Dopamine Associated with Individual Differences in Maternal Behavior in the Rat. *J. Neurosci.* **2004**, *24*, 4113–4123. [CrossRef]

114. Keer, S.; Stern, J. Dopamine Receptor Blockade in the Nucleus Accumbens Inhibits Maternal Retrieval and Licking, but Enhances Nursing Behavior in Lactating Rats. *Physiol. Behav.* **1999**, *67*, 659–669. [CrossRef]

115. Numan, M.; Numan, M.J.; Pliakou, N.; Stolzenberg, D.S.; Mullins, O.J.; Murphy, J.M.; Smith, C.D. The effects of D1 or D2 dopamine receptor antagonism in the medial preoptic area, ventral pallidum, or nucleus accumbens on the maternal retrieval response and other aspects of maternal behavior in rats. *Behav. Neurosci.* **2005**, *119*, 1588–1604. [CrossRef] [PubMed]

116. Romero-Morales, L.; García-Saucedo, B.; Martínez-Torres, M.; Cárdenas, M.; Cárdenas-Vázquez, R.; Luis, J. Neural activation associated with maternal and aversive interactions with pups in the mongolian gerbil (*Meriones unguiculatus*). *Behav. Brain Res.* **2022**, *437*, 114153. [CrossRef] [PubMed]

117. Fleming, A.S.; Korsmit, M. Plasticity in the maternal circuit: Effects of maternal experience on Fos-Lir in hypothalamic, limbic, and cortical structures in the postpartum rat. *Behav. Neurosci.* **1996**, *110*, 567–582. [CrossRef] [PubMed]

118. Fleming, A.S.; Rosenblatt, J.S. Maternal behavior in the virgin and lactating rat. *J. Comp. Physiol. Psychol.* **1974**, *86*, 957–972. [CrossRef] [PubMed]

119. Fleming, A.S.; Rosenblatt, J.S. Olfactory regulation of maternal behavior in rats: I. Effects of olfactory bulb removal in experienced and inexperienced lactating and cycling females. *J. Comp. Physiol. Psychol.* **1974**, *86*, 221–232. [CrossRef]

120. Fleming, A.S.; Rosenblatt, J.S. Olfactory regulation of maternal behavior in rats: II. Effects of peripherally induced anosmia and lesions of the lateral olfactory tract in pup-induced virgins. *J. Comp. Physiol. Psychol.* **1974**, *86*, 233–246. [CrossRef]

121. Inoue, S.; Kamiyama, H.; Matsumoto, M.; Yanagawa, Y.; Hiraide, S.; Saito, Y.; Shimamura, K.-I.; Togashi, H. Synaptic Modulation via Basolateral Amygdala on the Rat Hippocampus–Medial Prefrontal Cortex Pathway in Fear Extinction. *J. Pharmacol. Sci.* **2013**, *123*, 267–278. [CrossRef]
122. Lahoud, N.; Maroun, M. Oxytocinergic manipulations in corticolimbic circuit differentially affect fear acquisition and extinction. *Psychoneuroendocrinology* **2013**, *38*, 2184–2195. [CrossRef]
123. Deji, C.; Yan, P.; Ji, Y.; Yan, X.; Feng, Y.; Liu, J.; Liu, Y.; Wei, S.; Zhu, Y.; Lai, J. The Basolateral Amygdala to Ventral Hippocampus Circuit Controls Anxiety-Like Behaviors Induced by Morphine Withdrawal. *Front. Cell. Neurosci.* **2022**, *16*, 894886. [CrossRef]
124. Wardlaw, S.L.; Frantz, A.G. Brainβ-Endorphin during Pregnancy, Parturition, and the Postpartum Period. *Endocrinology* **1983**, *113*, 1664–1668. [CrossRef]
125. Neumann, I.; Russell, J.; Landgraf, R. Chapter 8: Endogenous opioids regulate intracerebral oxytocin release during parturition in a region-specific manner. *Prog. Brain Res.* **1992**, *91*, 55–58. [CrossRef] [PubMed]
126. Kalivas, P.W.; Abhold, R. Enkephalin release into the ventral tegmental area in response to stress: Modulation of mesocorticolimbic dopamine. *Brain Res.* **1987**, *414*, 339–348. [CrossRef] [PubMed]
127. Beauregard, M.; Leroux, J.-M.; Bergman, S.; Arzoumanian, Y.; Beaudoin, G.; Bourgouin, P.; Stip, E. The functional neuroanatomy of major depression: An fMRI study using an emotional activation paradigm. *Neuroreport* **1998**, *9*, 3253–3258. [CrossRef] [PubMed]
128. Broad, K.; Hinton, M.; Keverne, E.; Kendrick, K. Involvement of the medial prefrontal cortex in mediating behavioural responses to odour cues rather than olfactory recognition memory. *Neuroscience* **2002**, *114*, 715–729. [CrossRef]
129. Stamatakis, A.; Kalpachidou, T.; Raftogianni, A.; Zografou, E.; Tzanou, A.; Pondiki, S.; Stylianopoulou, F. Rat dams exposed repeatedly to a daily brief separation from the pups exhibit increased maternal behavior, decreased anxiety and altered levels of receptors for estrogens (ERα, ERβ), oxytocin and serotonin (5-HT1A) in their brain. *Psychoneuroendocrinology* **2015**, *52*, 212–228. [CrossRef]
130. Wu, Z.; Autry, A.E.; Bergan, J.F.; Watabe-Uchida, M.; Dulac, C.G. Galanin neurons in the medial preoptic area govern parental behaviour. *Nature* **2014**, *509*, 325–330. [CrossRef]
131. Osterhout, J.A.; Kapoor, V.; Eichhorn, S.W.; Vaughn, E.; Moore, J.D.; Liu, D.; Lee, D.; DeNardo, L.A.; Luo, L.; Zhuang, X.; et al. A preoptic neuronal population controls fever and appetite during sickness. *Nature* **2022**, *606*, 937–944. [CrossRef]
132. Paxinos, G.; Watson, C. *The Rat Brain in Stereotaxic Coordinates*; Academic Press, Inc.: San Diego, CA, USA, 1998.

Review

Mother–Young Bonding: Neurobiological Aspects and Maternal Biochemical Signaling in Altricial Domesticated Mammals

Cécile Bienboire-Frosini [1], Míriam Marcet-Rius [2], Agustín Orihuela [3], Adriana Domínguez-Oliva [4], Patricia Mora-Medina [5], Adriana Olmos-Hernández [6], Alejandro Casas-Alvarado [4] and Daniel Mota-Rojas [4,*]

[1] Department of Molecular Biology and Chemical Communication, Research Institute in Semiochemistry and Applied Ethology (IRSEA), 84400 Apt, France

[2] Animal Behaviour and Welfare Department, Research Institute in Semiochemistry and Applied Ethology (IRSEA), 84400 Apt, France

[3] Facultad de Ciencias Agropecuarias, Universidad Autónoma del Estado de Morelos, Cuernavaca 62209, Mexico

[4] Neurophysiology, Behavior and Animal Welfare Assessment, DPAA, Universidad Autónoma Metropolitana, Xochimilco Campus, Mexico City 04960, Mexico

[5] Facultad de Estudios Superiores Cuautitlán, Universidad Nacional Autónoma de Mexico (UNAM), Cuautitlán Izcalli 54740, Mexico

[6] Division of Biotechnology—Bioterio and Experimental Surgery, Instituto Nacional de Rehabilitación-Luis Guillermo Ibarra Ibarra (INR-LGII), Tlalpan, Mexico City 14389, Mexico

* Correspondence: dmota@correo.xoc.uam.mx

Simple Summary: Mother–young bonding is an essential process that increases newborn survival through selective maternal care. In this mechanism, olfactory, visual, auditory, and tactile stimuli coming from the offspring activate specific brain structures to establish the affective recognition of the neonate. Immature species at birth, such as altricial animals, require extensive care, and their post-birth maturation influences the degree of maternal interaction. This review aims to discuss the neurobiological aspects of bonding processes in altricial mammals, with a focus on the brain structures and neurotransmitters involved and how these influence the signaling during the first days of the life of newborns.

Abstract: Mother–young bonding is a type of early learning where the female and their newborn recognize each other through a series of neurobiological mechanisms and neurotransmitters that establish a behavioral preference for filial individuals. This process is essential to promote their welfare by providing maternal care, particularly in altricial species, animals that require extended parental care due to their limited neurodevelopment at birth. Olfactory, auditory, tactile, and visual stimuli trigger the neural integration of multimodal sensory and conditioned affective associations in mammals. This review aims to discuss the neurobiological aspects of bonding processes in altricial mammals, with a focus on the brain structures and neurotransmitters involved and how these influence the signaling during the first days of the life of newborns.

Keywords: imprinting; bonding; maternal recognition; olfactory; maternal anogenital licking; vocalization

Citation: Bienboire-Frosini, C.; Marcet-Rius, M.; Orihuela, A.; Domínguez-Oliva, A.; Mora-Medina, P.; Olmos-Hernández, A.; Casas-Alvarado, A.; Mota-Rojas, D. Mother–Young Bonding: Neurobiological Aspects and Maternal Biochemical Signaling in Altricial Domesticated Mammals. *Animals* **2023**, *13*, 532. https://doi.org/10.3390/ani13030532

Academic Editor: Alan Tilbrook

Received: 15 December 2022
Revised: 27 January 2023
Accepted: 30 January 2023
Published: 2 February 2023

Copyright: © 2023 by the authors. Licensee MDPI, Basel, Switzerland. This article is an open access article distributed under the terms and conditions of the Creative Commons Attribution (CC BY) license (https://creativecommons.org/licenses/by/4.0/).

1. Introduction

Mother–young bonding refers to the early learning process by which the dam and offspring create a selective bond that allows both to identify each other, to enhance newborn survival, and to enable the acquisition of a preference for a particular type of stimulus [1–3]. At the time of parturition, the mother is the main source of protection, warmth, and initial postnatal feeding through lactation, as the young are unable to survive without milk, and adequate maternal care [4–6]. In addition, males and other filial members can also contribute [4,5].

The beneficial impact of maternal protection through social behaviors directed at newborns has been widely reported in animals, especially mammals [7–14]. The neural pathways activated to establish a mother–young preference [15] are important to avoid misdirected care, reduce energy outlays, and enhance reproductive success. The degree of interaction can also affect the offspring's social behavior, notably maternal behavior, when reaching adulthood [16–18], and the effect that maternal behavior can have on the stress response of neonates [19]. Finally, maternal behavior and the interest of the newborn in his/her mother decreases when the offspring are mature enough to explore and feed themselves outside of the nest [20]. The relationship between the mother and the offspring is extremely important to promote the welfare of both.

This process of mutual recognition demands a brain that can be able to respond to multimodal sensory signals [15]. In this sense, the perception of different signals during bonding depends on the activation of several structures such as the locus coeruleus, or the olfactory, auditory, and visual cortices [21]. These areas interconnect with other regions to promote the secretion of neurotransmitters associated with the behavioral responses observed during this process, promoting maternal behaviors [22]. Significant evolution of the neocortex has been documented in monkeys and apes. This increased complexity is associated with the level of social relationships and strategies that serve to establish links between individuals that live in groups [23].

Another relevant issue associated with the requirement of protection and care of offspring is related to the maturity it has at parturition (precocial vs. altricial young; selective vs. non-selective bond) [24]. According to this, altricial species such as canids, rodents, marsupials, cricetids, and felines [25], are those that are more immature at birth, unable to move without help, require continuous parental protection during the first weeks of life, and must be groomed and provided with food [25]. Canine puppies, for example, are typically born with non-functional ears and eyes and are severely limited in regards to mechanical movement [26], so they depend on the dam and on their maternal behavior to survive. In this species, some mothers form a nest in a safe place where the young (born with an immature musculoskeletal system) are safe from predators [14]. Sows have the peculiarity that, although the newborn piglet has one of the highest motor and sensory developments, the mother builds a nest and the litter remains there for approximately one week [25].

Due to the influence that several brain structures, neurotransmitters, and the species have in the development of mother–young recognition, this review aims to analyze the neurobiological aspects of bonding processes. To achieve that, we will focus on the neurophysiological mechanisms of the mothers in mammal altricial species, including the brain structures, neurotransmitters, and communication channels involved.

2. Classification of Altricials, Precocials and Semialtricials

Altricial and precocial species can be differentiated by the degree of the physical and behavioral developmental stage at birth (Figure 1) [25,27,28]. Altricial species (e.g., canids, felids, most rodents, and lagomorphs) are characterized by a lack of fur, impeding their adequate thermoregulation, and limited sensorial and musculoskeletal capabilities [24,29,30]. They must be fed and cared for over an extended period by their parents, and mothers tend to build a nest or seek a sheltered area to give birth to undeveloped offspring with closed eyes [14].

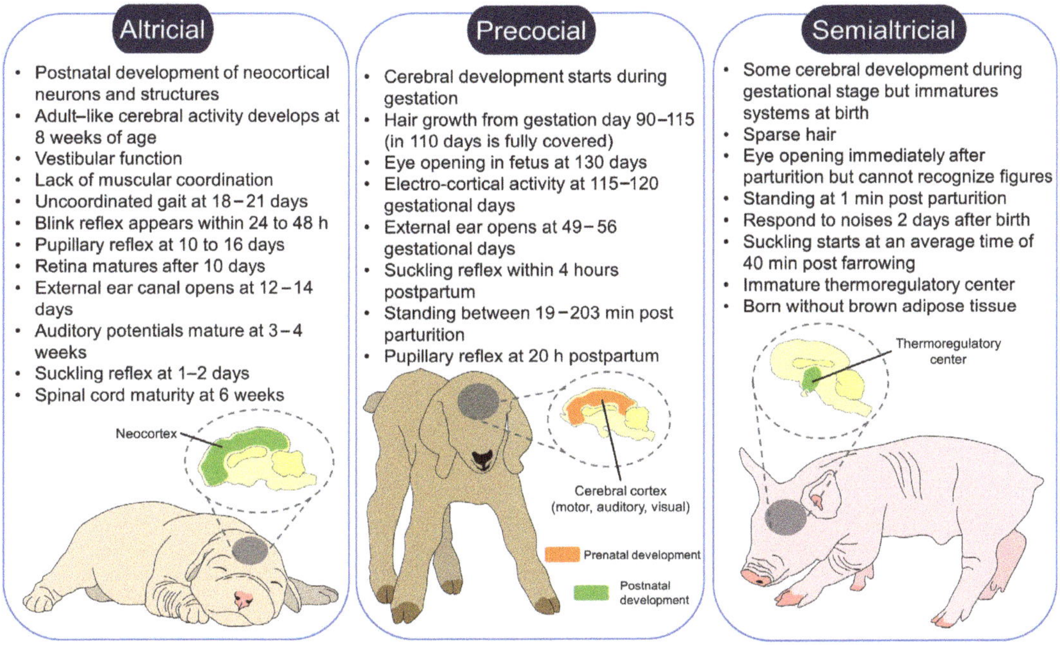

Figure 1. Neurodevelopmental differences between altricial, precocial, and semi-altricial species. According to the type of species, newborns from altricial, precocial, and semi-altricial females establish a different degree of bonding. In the case of precocial species, whose cerebral development starts during gestation, the time they require for maternal nursing is shorter since they can stand and move freely almost immediately after birth. In contrast, the maturation of sensorial systems in altricial species occurs during the postnatal period, where they are completely dependent on the mother to survive. Semi-altricial species are in between; while some of their senses are functional at birth, the mother nurses them until they are highly independent after several days after parturition.

For example, rat pups reach postnatal ambulation by Day 10, and just after leaving the nest (approximately on Days 17–21), they increase their locomotor activity [31]. Generally, motor development is measured by the righting reflex, vibrissae placing, and righting reflex in rat pups, as well as eye-opening and spontaneous locomotor activity. The presentation of these reflexes in newborns is altered by exposure to stressful situations in the dams, showing full development 14 days post-parturition. This shows how the prenatal stage influences not only the response from the mother but the offspring as well [32]. To compensate for the delayed maturation, murine mothers spend approximately a total of 10.6% of their time in the first week of life grooming and licking the pups [33]. In the case of dogs, although the vestibular function is present since birth, puppies lack muscular coordination and exploratory movements; they start using their limbs 7–10 days after parturition [34]. Marsupials are another important example of altricial species since the young are born in an embryonic state and require to be carried by their mothers at all times [35].

In contrast, offspring from precocial species (e.g., bovines, goats, sheep, and equines) can feed themselves, and follow their mother after birth, which is associated with a rapid recognition process between the two [36]. They have a completely functional vision and hearing since birth, and a locomotor system that is sufficiently well-developed to allow them to stand effectively and start suckling [37]. They also have efficient thermoregulatory systems and may be quite independent at an early age [38,39]. In some cases, precocial species hide and do not follow their mothers; the mothers move approximately 100 m away from the young to graze and return to nurse them. As these young do not move, their energy requirements are minimal, so all the food they consume translates into rapid growth [40,41].

Semi-altricial and semi-precocial species, such as primates and pigs, are those whose young have certain mobility or sensory independence, but still require prolonged maternal care [28]. The semi-altricial species have functional sensorial systems (e.g., auditory and visual), but deficient locomotor and thermoregulation. In primates, newborns are taken care of in a communal rearing system [42]. In domestic swine, nest-building is a natural behavior of the species since the offspring remains inside the nest for the first two weeks after farrowing [24,43].

Between types of species, there are some reported differences such as brain size and cognition, which may particularly influence the bonding systems established by altricial animals [25]. It is said that maternal investment and litter size influences brain size [44] and, at birth, altricial species have smaller brains [45] since most of the cerebral development occurs at the same time as eye-opening [46]. However, although their brain sizes might be smaller at birth when compared to precocial animals, their growth rate is higher during the first day post-birth [45].

An important fact to mention is that maturity at birth could also be associated with the status of the species being prey or predator. For example, precocial animals that stand up immediately after birth can avoid predators and escape [47], while altricial neonates require maternal protection, making them an easy prey when parents are not near the nest [48]. These differences are also related to the ability of the mother and the young to recognize each other in a very limited period where the brain and neuroendocrine pathways can form the bonding. Table 1 compares key factors associated to maternal–fetal bonding in altricial and precocial species.

Table 1. Summary and comparison of maternal bonding in altricial and precocial species.

Species	Pre-Natal Development	PostNatal Development	Functional Senses Immediately after Birth	Maternal Care	Bonding	Reference
Altricial	• Fetal response to umbilical cord occlusion	• Lung • Brown adipose tissue • Brain • Muscle	• Limited locomotion • Hairless • Limited thermoregulatory ability • Unable to hear and see immediately after birth	• Require nesting • Require constant parental care	• Nursing period within the nest • Mother–infant bonding in the first postnatal weeks	[49–51]
Precocial	• Organogenesis in utero (lung, liver, and brain) • Muscular development • Piloerection • Brown adipose tissue	• Coordinated locomotion	• Sight, hearing • Fully covered in fur • Locomotor capacity	• Do not require constant care. • Do not need nesting	• Bonding in the first postnatal hours	[52–55]

3. Sensitive Period

The process of filial bonding between the mother and the neonate is triggered during a decisive bounded interval called the sensitive (or critical) period [56–58]. In most animals, this period occurs during the first 4 to 12 h post-parturition [59]. Failure of maternal recognition can be due to interruptions caused by other animals or by human interventions. When these interruptions happen, reproductive, social, or behavioral disturbances, such as the mother's rejection of her offspring, can occur [60]. Specifically, in rats, odor-preference learning is achieved in the first 10 postnatal days and requires low levels of corticosterone in pups to maintain the normal length of the period [61]. In the case of rabbits, the first post-partum week is considered the sensitive period for pups to react to the mother or human intervention [62].

In the case of dogs, a sensitive period is considered to be established during the first two weeks of life, when they are highly dependent on their mother, and it is known that alteration in the maternal care or attachment during this period can negatively influence the behavior in adult dogs [63]. For kittens, a similar amount of time is considered the critical period, and during this time the mother is dedicated to nursing and feeding the newborns, only spending time away from the newborns when they become three weeks old [64].

Although the activities that occur during this period trigger behavioral responses [65], primarily, there exist neurophysiological phenomena in which the levels of neuronal plasticity in the offspring and the mother are more or less sensitive to experiences or environmental stimuli, depending on the species [66]. In the mother, the sensitive period is activated with the Fergusson reflex, a signal that is projected to the spinal cord, hypothalamus, and releases oxytocin (OXT) [67]. OXT stimulates uterine contractibility during parturition but also acts on the mother's olfactory bulb, and allows the secretion of dopamine (DA), thereby causing the mother to identify her offspring [68].

Modifications of neuronal plasticity may involve such mechanisms as synaptic consolidation through molecular adhesion cells, abolition of the activation of synapses because of the insertion of stabilizing molecules, and the construction of synapses through the growth of axons or dendrites. During this period, the influence of neurotransmitters and hormones is important for the establishment of the bond [69], but other factors can also alter mutual recognition at birth.

4. Factors That Influence the Bonding Process

The success or failure of dam–offspring bonding depends on the performance of certain behaviors, either by the mother toward her newborn, or vice versa. Primarily, communication pathways include tactile, visual [70,71], olfactory [72,73], and auditory stimuli [74], mainly related to the genetic background of each species and prior maternal experience.

4.1. Maternal Care

For proper maternal behavior to be carried out, various hormonal changes occur that trigger acceptance and interactions between the mother and the offspring, activating the attraction–acceptance circuit [75]. Maternal behavior includes feeding and defensive reactions such as increased vocalizations and locomotor activity when separated from their offspring [76]. The mother's brain is 'maternalized' by the hormones that provide emotional rewards for suckling, huddling, and grooming. At this point, the posterior pituitary gland releases (through neurological stimulation) the neuropeptide OXT [77–80], a nonapeptide that is critical for the neural processing of olfactory and social cues [81], increasing maternal interest in newborns because it helps reduce anxiety and stimulates maternal care [75,82]. This release of oxytocin occurs through two pathways: the Ferguson reflex, which consists of the pressure of the fetal head on the cervix at the time of parturition [83], and the stimulation of the newborns in the mammary gland at the time of suckling [77].

Studies of various mammal species have demonstrated that maternal care participates in the development of social skills, the brain, emotions, and certain behaviors [13]. For instance, in dogs, the amount of maternal care received during early life was associated with different patterns of behavioral responses and the coping strategies of puppies at two months of age [13]. Maternal care was also shown to influence hippocampal plasticity and cognitive functioning in rats, as well as decrease rodent pups' fear responses [10].

These long-term effects of maternal care on the offspring's behavior, emotional function, learning, memory, and neuroplasticity are notably mediated by epigenetic mechanisms that influence the transcriptional activity of many genes in the mammalian brain [84]. More precisely, some authors showed that early life experience (i.e., amount of parental care received) affected DNA methylation of oxytocin receptors (OXTR) in the nucleus accumbens of prairie voles offspring, hence altering the OXTR expression in this brain region [85,86].

Moreover, Champagne et al. [87] deciphered the mechanism of OXTR expression involvement in the "transmission of maternal behavior" between the mother and her offspring, which is mediated by the estrogen receptor genes' epigenetic modification in the MPOA in the hypothalamus; estrogen receptor activation acts to induce oxytocin receptor expression, hence, higher estrogen receptor expression causes higher oxytocin receptor expression. In turn, the oxytocin system in the MPOA is associated with the induction of maternal behavior. Thus, these epigenetic modulations might result in individual differences in maternal behavior [88].

The degree of maternal experience in females is also enhanced in multiparous animals, particularly in precocial species [89]. However, in virgin animals, such as rats, physiological and sensory signaling from their environment can induce maternal behavior. For example, low levels of corticosterone increase licking in virgin rats [90], and this maternal behavior is associated with the cell proliferation of the subventricular zones of the nulliparous rat brain [91]. Although maternal behavior can be induced with the use of hormones, such as estradiol or progesterone, maternal experience also depends on learning during the rearing of the pups, where offspring raised by their mothers become maternal more quickly and have high-licking behavior without the need of external hormones [92].

4.2. Importance of Nest-Building in Altricial Species

Nest-building provides shelter and comfort to the newborns and serves as a microclimate refuge station to prevent heat loss [93], to retrieve and transport the newborns there to have protection from predators and lick or nurse the offspring [29,30], particularly in altricial neonates that are born naked and with limited energy resources to thermoregulate [52]. Building the nest is an important action in most primates and rodents. Ungulate dams do not require building a nest since the newborns can actively move after the first hours of life and can respond to vocalizations calling them. Therefore, the predisposition of the mother to nest-building depends on the degree of the physical development of the offspring [24].

In rabbits, mice, and rats, nesting is the center where mother–young interactions take place [27]. For example, rats emerge from the nest by Days 19 and 20 [31], and mice spent at least 50% of their time with the mother in the nest within the first 21 days [94]. Long–Evans rat pups display motor activity to reach and huddle in the central positions of the nest at the first two postnatal days, an effect that is also influenced by the body mass of the animals, where light pups need to compensate for their high mass-to-volume ratio of heat loss [95]. This behavior and burrowing are considered indicators of well-being in rodents [96], but also a way to evaluate the influence of chemical signals that impede maternal care. Regarding this, mice lacking 2 adenylyl cyclase failed to construct a nest and did not retrieve pups during postpartum, which shows the importance of the main olfactory system to detect odorants or pheromones to start maternal behaviors [97].

For newborn rabbits, the nest is essential to survival during the first weeks of life due to the limited assistance of the mother at birth. The doe only feeds the pups once or twice per day for 4 to 5 weeks [98]. The suckling sessions have a duration of about 3 to 4 min. Therefore, the rest of the time the newborns huddle into the insulated nest material to save energy and prevent hypothermia [99].

Other species, such as marsupials, are born in an embryonic stage and immediately after birth reach the pouch and nipple of the mother to fully develop outside the uterus [27]. For these species, the role of a nest can seem less important; however, as Rowland et al. [100] studied in arboreal marsupials, tree hollows, instead of nest boxes, aid in the thermostability of neonate marsupials. According to the results, microclimates inside tree hollows provide a consistent thermal environment with temperatures no higher than 38 °C, in comparison to the nest boxes where the highest recorded temperature was 52 °C, a value that can trigger heat stress in the newborn.

In sows, nest-building is a natural behavior that permits piglets to be protected for long periods (up to seven days post-farrowing) and starts with foraging, rooting, and pawing [93]. This process starts three days before parturition, and studies conducted by Yun et al. [101] have demonstrated that high OXT concentrations (between 7.5 and 23.5 pg/mL) are associated with this behavior before farrowing, and that prolactin (between 23.9 and 27.0 pg/mL) and oxytocin participate in nesting after farrowing, on Days 1 to 7 of lactation. Domestication has altered the presentation of this behavior, and in intensive production systems, the animals are not able to perform it inside farrowing crates [102]. However, if given the required material, such as straw and non-confined environments, sows engage in this pre-farrowing behavior [103].

As has been mentioned, maternal behavior is triggered by the parturition process, the presence of the newborn, and by distinctive neural and endocrine signaling—being tactile, auditory, olfactory, and visual—that promotes the bonding and selective recognition of the filial offspring.

5. Neural Pathways Involved in the Bonding Process of Altricial Species

Maternal care and the attachment system in the mother–offspring binomial differ among species and depend on several factors: the newborn's condition at birth, postnatal development, and numerous neural circuits [104]. The cognitive, sensory, and locomotor capacities of both precocial and altricial species are associated with prenatal neurogenesis and degrees of brain maturation. Consequently, the maternal behavior towards the newborn (altricial and precocial) animals is an important way of promoting the survival of the offspring [14].

At birth, the neonate's mother is the first social contact and the main source of learning [28,39,104]. In altricial species, communication is practically unidirectional with multimodal signals being sent from the mother to her newborn but few going in the other direction [38].

Mother–young bonding involves structural changes in cortical regions and the continuous release of neurotransmitters [105] that respond to visual, auditory, tactile, and olfactory sensory stimuli [106]. Brain structures, such as the medial preoptic area (MPOA), ventral tegmental area (VTA), medial prefrontal cortex (MPC), and anterior cingulate subregions, dictate maternal motivation in the post-parturition period (on Day 8), while the medial prefrontal cortex prelimbic subregion contributes to late post-parturition maternal care (at day 16) [107].

The neuroendocrine response induces changes in the expression of a range of neuropeptides (e.g., b-endorphin, corticotrophin-releasing factor (CRF), OXT and arginine vasopressin (AVP)), substances that mediate maternal and other social behaviors according to the nature of the received stimuli [15,23]. For example, in the case of rodents, licking, nursing, nest-building, retrieving pups, and feeding the offspring are a group of behavioral responses triggered by a sensorial and endocrine stimulus. It requires the activation of maternal recognition systems and the suppression of responses such as defensive or aversive reactions to the newborn. To achieve this, the hormonal changes involving estradiol, OXT, prolactin, and progesterone during parturition act on specific receptors located in the hypothalamus (MPOA) and the bed nucleus of the stria terminalis (BNST) to activate maternal motivational systems. Simultaneously, the suppressed projections to the amygdala and other structures prevent fear and avoidance behaviors, resulting in proactive voluntary responses to nurse and care for the newborn (Figure 2) [63].

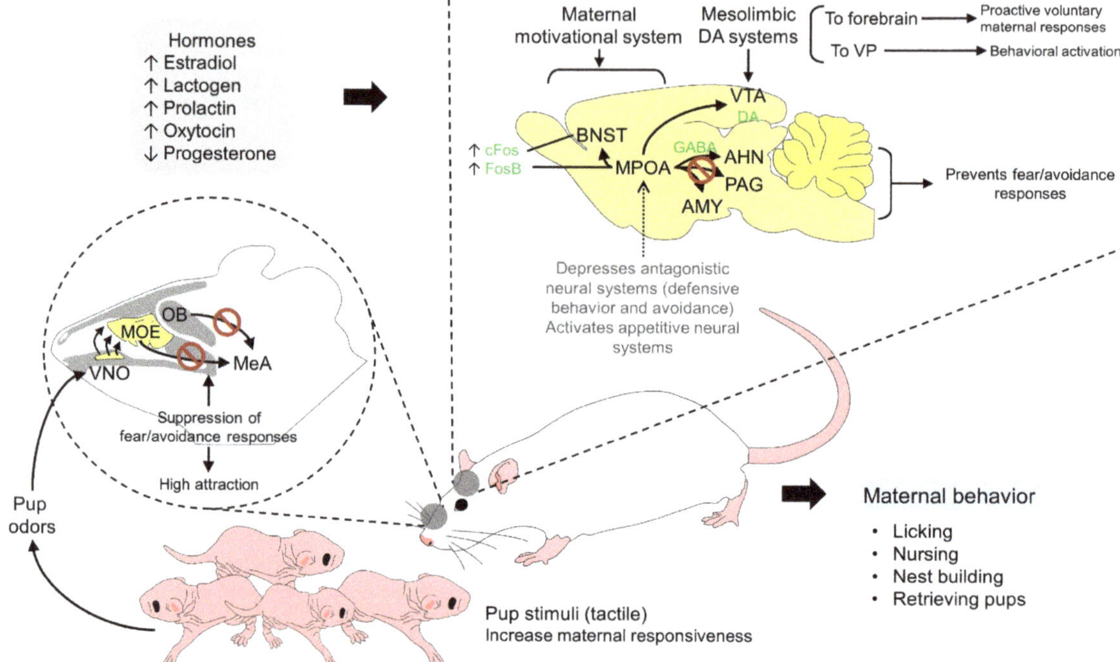

Figure 2. Neurobiology of maternal behavior in altricial species. To develop a maternal response in females, a series of neuroendocrine and sensorial stimuli need a dual interaction in cerebral structures, such as the MPOA, BNST, VTA, AMY, VNO, and OB, among others, to suppress aggressive and avoidance behaviors, while activating appetitive and motivational maternal pathways. AHN: anterior hypothalamic nucleus; BNST: bed nucleus of the stria terminalis; cFos: transcription factor and marker of neuronal activity; DA: dopamine; GABA: gamma-aminobutyric acid; FosB: Fos protein; MeA: medial amygdala; MOE: main olfactory epithelium; MPOA: Medial PreOptic Area; OB: olfactory bulb; PAG: periaqueductal gray; VNO: vomeronasal organ; VP: ventral pallidum.

5.1. Licking and Tactile Sensitivity of the Brain

Numerous studies have highlighted the importance of licking the newborn during parturition, not only because it provides benefits to the offspring but because it can also reduce the mother's stress [108–110].

In species such as horses, dogs, cats, and rats, among others, the hippocampus was observed while licking the newborn and an increase in glucocorticoid receptors was observed, thus conditioning the functioning of the hypothalamic–pituitary–adrenal (HHA) axis, a key system for the development of an autonomic/endocrine response, unconscious memory, and the regulation of emotional states in the face of external stressors [111,112]. For kittens, the mother is the one who initiates maternal nursing by tactile stimuli to the newborn, such as nuzzling and licking them, to guide them to the nipples to start suckling [64].

During parturition, the rabbit does constantly lick their hind limbs or the genital region of the recently born pup. However, contrary to rats, rabbits spend more time licking the blood, amniotic fluids, or eating the placental membranes than licking the newborn [113]. This reaction is associated with the limited interaction between the doe and the pups, with an average duration of 2 to 4 min [114]. Therefore, in these species, the thermal and olfactory cues emanating from the female are also the main signaling channel of the mother–young recognition [115]. Although it may be limited, rabbit pups can actively crawl and search for the nipple to suckle [114].

5.2. Olfactory Stimuli

Olfactory and odor stimuli are some of the most selective pathways for mother–young bonding [116]. The maternal brain undergoes neurogenesis, a type of adaptive plasticity of neurons, especially in olfactory structures, to facilitate olfactory memory [117]. Therefore, interneurons participate in the recognition of the offspring during the sensitive period [118].

Research has shown that non-lactating females are aversive to neonatal odors [119]. However, during postpartum, olfactory cues are partially inhibited to prevent neophobia [120], and these signals are a key stimulus to activate the maternal motivation system [119]. Mothers often recognize their offspring by smell, related to the release of OXT in the brain [121–123].

Odor stimuli from pups activate regions in the MPOA, BNST, or the lateral habenula and ventral tegmental area (VTA, through the action of OXT and prolactin [105]. Furthermore, norepinephrine in the brainstem projects on the paraventricular nucleus (PVN) in the hypothalamus and olfactory bulb. In turn, OXT neurons are activated in the PVN. They are responsible for promoting three mechanisms associated with maternal behavior: (1) OXT stimulation of the OB promotes maternal behaviors and reduces aggressiveness towards newborns; (2) Its action on the hypothalamus inhibits postpartum estrus; (3) It promotes behaviors associated with brood care in the VTA [105]. The birthing process triggers the Fergusson reflex through vaginocervical stimulation, increasing the oxytocin concentrations in the PVN.

In the case of rodents, the recognition of newborns and the development of social behaviors mainly depend on the detection of olfactory signals by the dam's main and accessory olfactory system [15]. In rats, the cortex receives olfactory inputs via the mediodorsal thalamic nucleus, and lesions to this system decrease the frequency of mother protection against cannibalism [124]. The amygdala also participates in the bonding process since lesions in that region alter the process of olfactory preference [125]. Detection of chemical signals induces maternal behavior in virgin females, such as retrieving isolated young [126]. Other studies have discovered that dodecyl propionate (DP), a pheromone released by the preputial gland of rodent pups, is associated with the mother's licking [9]. Contrarily, Morgan et al. [127] reported that maternal behavior is not affected if the vomeronasal organ is extirped in rats. Moreover, Brouette-Lahlou et al. [128] described that the pups' survival rate is not related to the vomeronasal activity.

In mice, the removal of the olfactory bulb eliminates maternal aspects in females [129]. A study by Liang et al. [130] concluded that the mother *Tylonycteris pachypus* bats recognize their young by scent. Although pheromones and olfactory stimuli are important for altricial species, in rabbit pups it has been reported that they search for the nipple with or without the association of suckling with odors [131].

5.3. Auditory Stimuli

Auditory cues have been described in bird species like penguins [132–134] and swallows [135], but are still little studied in mammals. In the case of dog puppies, maternal separation induces vocalizations in newborns from 6 to 8 weeks old, while high maternal care reduces whining and yelping emission, a sign of reduced distress [13]. In cats, vocalization from kittens elicits postural changes (e.g., lactation position) in the mother, and is also an exploratory approach to retrieving the offspring [136].

In rats and mice, pup vocalization highly influences the mother–pup bond. In this sense, the pups' ultrasonic vocalizations play an ethologically important role in mother–offspring communication by stimulating specific maternal behavioral responses [137]. Pups can emit vocalizations that reach intensities of 30–90 kHz. Ultrasonic vocalizations of rat pups have been recorded at 44 kHz in response to stressful environments such as maternal separation, unfamiliar odors, and thermal or tactile stimulus during birth, and their function is to call the mother to promote retrieval, grooming, and maternal care [138].

Maternal responsiveness can be modulated by the vocalization emitted by pups. For example, D'Amato et al. [139] reported that on the eighth postnatal day, female mice respond to calls from filial pups more frequently than those coming from alien individuals (latency time of 44.3 vs. 93.1 s, respectively). This response is also mediated by prolactin, according to Hashimoto et al. [140]; in 19 lactating rats, an increase in prolactin levels was registered after pup vocalization, eliciting retrieving and nest-building for the newborns. Conversely, in pups, maternal separation alters ultrasonic vocalization, producing low-frequency calls during isolation (around 40 kHz), but high maternal care was associated with more low-frequency vocalizations [141].

As observed in rodents—but also other species—maternal interaction and the responsiveness of the brain to neonatal cues highly depend on the participation of hormones and other neurochemicals that initiate and activate the cerebral structures involved in mother–young bonding.

5.4. Visual Stimuli

The cerebral hemispheres and roof of the forebrain are the main sites of visual recognition, as well as the prefrontal and cingulate areas of the cortex in mammals [142].

For canines and felines, the visual system, composed of the retina and axons from the optic nerve that project on the lateral geniculate nucleus, is essential for the perception of the environment [143] and, therefore, to recognize the offspring at birth. Although eyesight is important, in rabbits there was no behavioral difference in maternal care between normal and genetically blind individuals, showing that 98% and 97% of the animals in each group, respectively, developed expected newborn care, suggesting that other signaling pathways, such as auditory and olfaction, can compensate for this impairment [144].

Similarly, in mice, visual impairments due to albinism have not been associated with altered maternal behavior [145]. The same was reported by Sturman-Hulbe [146] in 15 female albino rats, for whom conditions such as blindness and anosmia did not result in altered maternal fitness, and other activities (e.g., nesting, suckling, and cleaning) were similar to those observed in healthy animals.

These findings show that, although visual signaling aids in the development of mother–young bonding, it does not directly trigger maternal behavior since animals require or use other neuroendocrine pathways to respond after parturition.

6. The Role of Neurobiological Systems for Mother–Young Recognition and Bonding

6.1. Neurotransmitters and Neurohormones

OXT is the main neurohormone associated with social recognition, maternal behavior, parturition, and milk ejection [14,81,88,104,147–149]. It is synthesized in the magnocellular neurons of the paraventricular (PVN) and supraoptic (SON) nuclei of the hypothalamus and is projected towards the posterior pituitary for further release into the peripheral circulation. Differences in individual maternal attachment are associated with the evolution of the dopaminergic and oxytocinergic neuroendocrine system, where OXT activates the mesocorticolimbic dopaminergic system in response to social signals. This is crucial for the expression of affiliative behaviors [148,150,151].

The perinatal role of the OXT system in animals is associated with dysfunctional maternal behaviors [152]. For example, in rats, OXT mediates the initiation of maternal behavior [153], even in virgin rats receiving intracerebroventricular doses [154], and OXT knockout mice have a low prevalence of pup retrieving and licking in comparison to nulliparous females [155]. Likewise, the administration of OXT antagonists inhibits maternal behaviors and interferes with the bonding process in the same species [152]. Aggression and maternal defensive behavior on Wistar rats also elicit a higher OXT release in the central nucleus of the amygdala and the PVN [156].

Maternal care such as licking is also related to OXT levels and OXT receptors in the central nervous system, with less expression of these in the hippocampus and OB in dams that lick the pups constantly [157], a trait considered to measure maternal care in this species. Likewise, the same kind of dam has high levels of dopamine (DA) receptors in the nucleus accumbens [158], estrogen receptors in the MPOA, and OXT receptors in the amygdala and the bed nucleus of the stria terminalis [157].

The interaction of OXT with other substances that might be secreted by the organism during the peripartum also dictates the maternal patterns that mothers elicit. In this sense, in rodents, prenatal stress activates the hypothalamic–pituitary–adrenocortical (HPA) axis, decreasing motor development and social and learning skills in the early stages, while OXT in the postpartum period can lower anxiety and depression [159]. It is important to decrease anxiety during the days after parturition [82] so the dam can easily accept the newborn, facilitating social bonding [160]. Specifically, it stimulates maternal behavior in dams and increases maternal care toward the offspring that decreases anxiety [75]. Additionally, Sohlstrom et al. [161] found that exogenous administration of OXT to rat pups in the first days of life results in lower corticosterone concentrations, low blood pressure, and greater weight gain in adults.

Prolactin is another important neuropeptide. Its main role is for milk production and ejections. However, it also participates during maternal care and parental behavior in birds and mammals [162–164]. In mice, prolactin promotes neurogenesis in the forebrain and the olfactory bulb, directly participating in the olfactory recognition and acceptance of the newborn [165]. Additionally, it regulates offspring-oriented care not only in females by promoting paternal behavior, shifting the endocrine response from one aimed at fertility to one focused on childcare [166]. Other substances secreted by neurons at this stage and that stimulate the limbic system are gamma-aminobutyric acid (GABA), glutamic acid (GLU), monoamines such as DA and serotonin (5-HT), as well as N-methyl-D-aspartate (NMDA) [167,168].

Additionally, the main function of GABA is to inhibit the signal transmission to cerebral structures such as the amygdala, thalamus, prefrontal cortex, and hippocampus [169], altering the affective aspect through inhibitory GABAergic connections [170]. In the MPOA and BNST, the administration of the GABA receptor agonist in lactating rats causes dose-dependent behavioral deficits such as lack of pup retrieving and maternal aggression, but does not affect licking [171]. In rat pups, catecholamines, such as DA and noradrenaline, have an influence on cardiac, respiratory, and motor activity in the first post-birth days [172]. The release of DA in nulliparous and multiparous female rats when exposed to pup stimuli was studied by Afonso et al. [173]. The researchers found that multiparous females had greater concentrations of DA in the nucleus accumbens, so it can be concluded that maternal experience is mediated by this neurotransmitter.

In the case of 5-HT, authors such as Angoa-Pérez et al. [174] found that female mice with mutations that cause 5-HT depletion have poor maternal performance (e.g., lack of pup retrieving, huddling, nest construction, and high-arched bac nursing), and also affected the survival rate of their offspring. The interaction of 5-HT with nitric oxide (NO) influences aggressive behavior toward the newborn due to diminished 5-HT concentration and lack of NO synthase [175]. Table 2 summarizes the mentioned neurotransmitters and their role in maternal recognition.

Table 2. Main neurotransmitters involved in mother–young recognition.

Neurotransmitter	Synthesis	Status	Role	Reference
OXT	PVN, SOP	↑	Maternal behavior	[153]
		↓	Lack of newborn retrieving and licking.	[155]
GABA	Presynaptic neuron	↑	Maternal defense	[176]

Table 2. *Cont.*

Neurotransmitter	Synthesis	Status	Role	Reference
GLU	Presynaptic neuron	↑	Long-term maternal experience.	[177]
DA	Dopaminergic neurons	↑	Maternal care, bonding, reward system.	[178]
		↓	Impaired maternal recognition.	
PRL	Lactotrophs in the anterior pituitary gland	↑	Maternal care.	[179]
		↓	Litter abandonment.	[180]
5-HT	Enteric nervous system, CNS, Merkel and pulmonary cells	↑	Modulates DA maternal effects.	[181]
		↓	Reduces pup survival and nursing behavior.	[182]
NMDA	Ionotropic neurons	↓	Impaired retrieval.	[183]

5-HT: serotonin; CNS: central nervous system; DA: dopamine; GABA: gamma-aminobutyric acid; GLU: glutamate; NMDA: N-methyl-D-aspartate; PVN: paraventricular nuclei; SON: supraoptic nuclei.

Interestingly, these neurohormonal/neurotransmitter parameters can be assessed to investigate the quality and development of mother–young bonding. In this context, the most studied neuromodulator is OXT [87,88,184–186]. Unfortunately, the methods used to measure OXT face several pitfalls, notably because there is no methodological consensus about the different OXT assays and pre-analytical treatments used: enzyme-linked immunosorbent assay, radioimmunoassay, liquid chromatography-based methods, with or without pre-analytical treatment steps, such as solid-phase extraction, reduction-alkylation, among others [187–191]. This leads to measurement discrepancies and inconsistent results [189,192,193] and raises the question of the existence of various molecular forms of OXT and their respective biological relevancy [192,194,195]. Additionally, another issue revolves around the matter of the sample volume and availability: OXT levels can be measured in excretory fluids (urine and saliva), blood, or cerebrospinal fluid, these two latter being more invasive but providing more consistent and robust results [189]. However, peripheral measures of OXT may not reflect the central nervous system activity. Cerebrospinal fluid, or even brain microdialysates, are the ideal samples to study the effects of OXT at the central level [196,197], hence, on the regulation of maternal behavior and young bonding. However, the sample collection methods are complicated and often provide only small volumes for the subsequent analyses; this constitutes substantial technical limitations, which could impact the quality of the final measures and the robustness of the findings [197,198].

6.2. Hypothalamic–Pituitary–Adrenal (HPA) and Hypothalamic –Pituitary–Thyroid (HPT) Axes

The endocrine response, because of bonding and maternal stimulation by the newborn and their signaling, is largely influenced by the hypothalamic–pituitary–adrenal axis (HPA) and the hypothalamic–pituitary–thyroid (HPT) axis.

The HPA axis is a complex system that regulates the body's response to stress. It involves interactions between the hypothalamus, pituitary gland, and adrenal glands, which together control the release of stress hormones such as glucocorticoids (GC: cortisol and/or corticosterone). During pregnancy and parturition, the HPA axis is activated, which leads to an increase in GC levels. This increase in glucocorticoids is thought to help the mother adapt to the physical and emotional demands of parturition and motherhood [199].

The activation of the HPA axis and the increase in GC also stimulates the release of OXT, a hormone that is important for maternal bonding and lactation as stated earlier, as a reaction to dampen the HPA axis and downregulate the GC stress response [200]. After parturition, GC levels typically decrease, and OXT levels increase, which is thought to promote maternal bonding and attachment to the young. The HPA axis also plays a role in regulating the mother's emotional state since it may be affected by dramatic hormonal shifts, which occur during pregnancy, parturition, and the postpartum period and influence her ability to bond with her offspring [201]. Additionally, studies have shown that exposure to stressors during pregnancy may negatively impact the HPA axis, maternal behavior, and mother–young bonding, in link with the oxytocinergic system [201,202]. In turn, maternal behavior serves to "program" HPA responses to stress in the offspring in rodents, the expression of GC receptors in the newborns' hippocampus is affected by dam licking [19] and pups receiving high tactile stimulation secrete low amounts of corticosterone, representing a least marked stress response [148]. In addition, it has been shown in rodent pups that a hypo-functioning HPA axis is necessary to prevent pups from learning an odor aversion to their mother in the nest, thus allowing the recognition of the mother and the bond formation with her [149].

Overall, the HPA axis is thought to play a role in mother–young recognition and bonding by regulating the release of hormones such as GC and OXT and influencing the young's brain activation and development.

The hypothalamic–pituitary–thyroid (HPT) axis plays a role in regulating the production and release of thyroid hormones, which exert broad effects on development and physiology. The two main hormones secreted by the thyroid gland are thyroxine (T4) and 3,5,3′-triiodothyronine (T3). While thyroid hormone status is associated with mood dis-orders, limited information is available about their involvement in social behaviors, including maternal behavior [203]. However, several studies have shown that thyroid hormones are implicated in the regulation of OXT and its signaling, acting on OXT plasma levels and OXT receptors' transcriptional regulation [204–206]. Hence, via their impact on OXT, maternal thyroid hormones could play a part in maternal behavior development and the establishment of mother–young bonds. Additionally, Stohn et al. [203] showed that an abnormal local regulation of thyroid hormone action in the mice brain due to type 3 deiodinase deficiency is related to abnormalities in the OXT system (low adult serum levels of OXT and an abnormal expression of the OXT gene and its receptor in the neonatal and adult hypothalamus). In addition, research on rodents has shown that maternal thyroid hormones influence the development of the brain regions that are necessary for offspring recognition, maternal behavior, and bonding, like the neurogenesis process in the maternal hippocampus [207,208].

Finally, it is noteworthy that the link between the HPT axis and mother–young recognition and bonding is not totally understood; more research is necessary to fully understand the relationship.

6.3. Brain–Gut–Microbiome Axis

The brain–gut axis refers to the communication pathway between the brain and the digestive system. This pathway is bidirectional, meaning that signals can travel from the brain to the gut, and from the gut to the brain. Additionally, the gut microbiome, which is the collection of microorganisms that live in the gut, also plays a part in the biochemical signaling events that take place between the gastrointestinal tract and the central nervous system [209,210].

Through the activation of the HPA axis, stress occurring at an early age and involving the bonding with the mother can impact the gut microbiota and the brain–gut axis function of the offspring [209,211]. It was shown that the intestinal permeability and the development of the HPA axis in the offspring were altered by limited nesting stress in Sprague Dawley rats [212]. In the same species, prevention of weaning results in changes in gut health and microbiome [213]. Early life stressful events such as maternal separation have also been implicated in the alteration of the intestinal microflora and the presentation of adult life disorders in the offspring (e.g., irritable bowel disease) [214,215].

The link between the gut–brain axis and the development of bonding between a mother and her offspring is also thought to be related to the role of certain hormones and neurotransmitters in both the gut and the brain. Indeed, the gut microbiome has been shown to play a role in regulating the production of certain neurotransmitters, such as serotonin, that are involved in maternal behavior [181,182,216]. These neurotransmitters and neurohormones can affect the mother's ability to bond with her child (see Table 2). Additionally, the mother's gut microbiome can influence her emotional state, which can benefit the bond development between the mother and her offspring [217,218].

7. Factors That Affect Maternal Recognition and Performance

The acceptance of the newborn in the first hours of life and the activation of the afore-mentioned neural pathways can be interrupted by a series of events that impede bonding. These events can be included in physiological and behavioral aspects. Regarding physiological elements, parturition pain is part of the normal process due to cervical dilatation and the activation of uterine pain receptors [219] (Figure 3). However, it is known to have a detrimental effect on the degree of maternal behavior [220], particularly during dystocia [221]. This has been reported in semi-altricial species such as pigs, whose peripartum pain causes consequences such as reduced food intake (hyporexia), reduced milk letdown, a lower ability to care for the newborn [222,223], and the use of anti-inflammatory drugs can decrease the presentation of peripartum fever, a disorder present in 40 to 100% of farrowing with a duration of 4 to 8 h [224]. For example, Kuller et al. [222] showed that pain and fever relief using paracetamol during peripartum improved the performance of the mother and reduced the variation of weight within litters and back fat loss during lactation. Pain can also be accompanied by cortisol and tumor necrosis factor α increase in multiparous sows, a stress-related response [223]. The use of analgesics has also shown a beneficial effect on mastitis-metritis-agalactia syndrome in sows, where Hirsch et al. [225] found that meloxicam and flunixin have similar efficiency when considering rectal temperature, inflammation of mammary glands, milk letdown, and nursing behavior ($p < 0.05$), but lower mortality rates were observed in litters from meloxicam group.

Another factor that can result in nest abandonment and even cannibalism is nervousness and fear-related responses in altricial species. Contrary to pregnant animals, whose fear and defensive behaviors are suppressed due to hormonal stimulation [226], in nulliparous or virgin rodents, avoidance is the first reaction to pups when exposed to newborns, and timidity and fearfulness are traits present in these animals [227]. However, maternal motivation can be induced in virgin rats and mice after only 5 to 21 days of exposure to pups [228,229], and even within the first 15 to 30 min [230]. The induced responsiveness, similar to parturient females retrieving pups and crouching over the offspring, shows that females can activate social mother–young bonding pathways even with foster offspring [88].

Nonetheless, when aversion to neonatal stimuli persists, nest abandonment and neonate mortality by a lack of nursing are observed in species such as altricial birds [231]. In them, the presence of predators such as mice can cause an up to a 10-fold increase in the rate of nest abandonment and, therefore, consequences to offspring survival [232]. In the case of mammals, together with the abandonment of the offspring, events such as cannibalism have been reported. Altered maternal responses such as cannibalism and rejection of puppies can be found in at least 10.7% of breeders [233]. In Kangal dogs, several studies reported maternal cannibalism occurring during the first 24 h after birth, and an association with high levels of adenosine deaminase and xanthine oxidase, metabolic enzymes that can be studied since cannibalism is associated with nutritional deficiencies and environmental stressors [234].

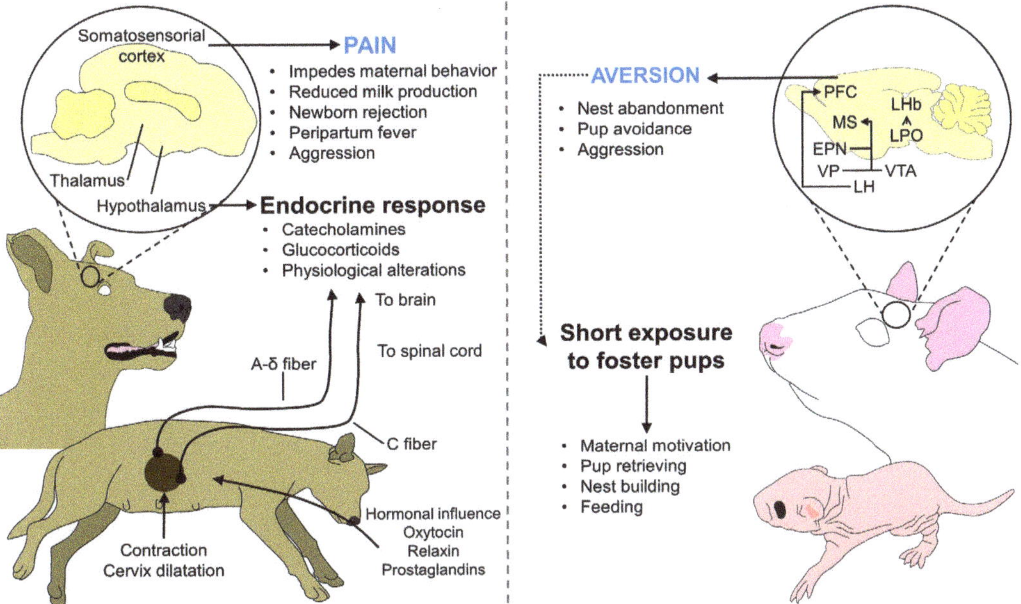

Figure 3. Influence of pain and aversion on maternal behavior in altricial species. Although pain is a physiological trait in the parturition process, if dystocia occurs and the pain is prolonged in the mother, the activation of nociceptors (A-delta and C fibers) activates the sympathetic nervous system, and the consequent release of catecholamines and glucocorticoids. Moreover, the conscious recognition of pain in the somatosensorial cortex alters maternal behavior and reduces milk production, resulting in a negative trait for both the mother and the offspring. On the other hand, the activation of aversion pathways in non-parturient females prevents them from maternal responsiveness. However, in species such as rodents, short periods of pup exposure elicit similar responses in multiparous, parturient, and lactating animals. EPN: entopeduncular nucleus; LH: lateral hypothalamus; LHb: lateral habenula; LPO: lateral preoptic area; MS: medial septal nucleus; PFC: prefrontal cortex; VP: ventral pallidum; VTA: ventral tegmental area.

Regarding the nutritional state of the mother before and during parturition, low lipid levels (high-density and low-density lipoprotein, as well as cholesterol) were found by Kockaya et al. [235] in Kangal dogs with a history of cannibalism ($p < 0.05$), as well as low OXT concentration. However, cannibalism is not only related to a detrimental mother–young interaction. It can also be caused by the high population density of a species within an area, as reported in dingoes in Australia [236].

Moreover, human–animal interaction can potentially alter the degree of acceptance of the newborn by the mother, as well as the social behavior and human-related response of the offspring when reaching adulthood [237]. Additionally, it is relevant to mention that performing ethological studies regarding maternal behavior requires special attention to sample size and reliable methods to obtain robust results. For example, it has been observed in ewes that human intervention alters the parturition process and, consequently, could affect the bonding period with bad-quality maternal care [238]. Therefore, the aspects that need to be considered during parturition and the sensitive period are the physiological and emotional states of the mother, as well as the intervention in the process.

8. Future Directions

Understanding the neurobiological pathways in mothers during the bonding process can help to avoid situations that cause discomfort or anxiety in females, and thereby improve their productivity. Many studies showed a correlation between the anxiety of primiparous mothers and the amount of maternal care in mammals [239–241], including dogs [242]. In the same way, by improving the facilities so that there is no stress on the mothers and the offspring, the imprinting process will develop without failure.

An important factor is recognizing the importance of nesting behavior in altricial species in commercial breeding units. In some cases, such as with fattening rabbits, not providing the mother with material to nest could affect the survival of the newborns, making them vulnerable to environmental factors [99]. To assess these behaviors, more studies focusing on precision livestock farming, such as Oczak et al.'s [243] work, could help evaluate the presence of this behavior with a sensitivity of 87% and specificity of 85%, an alternative that can be applied to other species to improve their welfare and the newborn survival.

Likewise, knowing the interaction between neurochemicals, brain structures, and behavior during and after parturition could help to propose protocols in cases of maternal cannibalism in species such as dogs [244] or piglets. This could be an approach that would also represent an advantage for the newborn and future gestation periods.

9. Conclusions

Bonding is an essential process for dams and newborns, where several factors play a key role in its development. Inherent characteristics of the animals such as the type of species (e.g., altricial and precocial) contribute to the degree of maternal care that the newborn requires during the first days of life. For example, altricial offspring born with limited motor capacities and immature sensorial systems often require prolonged nursing inside a nest or near the mother to obtain energy resources and protection.

The signaling of this interaction is triggered by brain pathways activated by external stimuli and neurotransmitters and hormones present at parturition. The presence of the newborn, their odor, vocalization, and tactile stimuli are processed by brain structures such as the MPOA, amygdala, visual and auditory cortex, as well as the olfactory main system, causing the release of dopamine, OXT, and prolactin, among others, to develop maternal motivation and a selective bond with the offspring.

A mother–young recognition represents a beneficial trait for the newborn and its survival, and contributes to the maternal experience for future litters. Thus, considering the necessary elements for this process to start (e.g., a sensitive period) is a way of avoiding elements that may interfere with bonding. Particularly, in productive units, strategies aimed to protect this period are relevant in the improvement of the welfare and commercial value of the animals, preventing neonatal losses, and gaining mothers with previous experience to raise their offspring.

Author Contributions: C.B.-F., M.M.-R., A.O., A.D.-O., P.M.-M., A.O.-H., A.C.-A. and D.M.-R. contributed to the conceptualization, writing, and reading. All authors have read and agreed to the published version of the manuscript.

Funding: This research received no external funding.

Institutional Review Board Statement: Not applicable.

Data Availability Statement: Not applicable.

Conflicts of Interest: The authors declare that they have no conflict of interest.

References

1. Fraser, A.F. *Comportamiento de Los Animales de Granja*; Acribia: Zaragoza, España, 1980; p. 292.
2. Maier, R. La Evolución Del Aprendizaje. In *Comportamiento Animal. Un Enfoque Evolutivo y Ecológico*; Maier, R., Ed.; McGrawHill Interamericana: Madrid, Spain, 2001; p. 541.
3. Mota-Rojas, D.; Bienboire-Frosini, C.; Marcet-Rius, M.; Domínguez-Oliva, A.; Mora-Medina, P.; Lezama-García, K.; Orihuela, A. Mother-Young Bond in Non-Human Mammals: Neonatal Communication Pathways and Neurobiological Basis. *Front. Psychol.* **2022**, *13*, 1064444. [CrossRef] [PubMed]

4. Sèbe, F.; Aubin, T.; Boué, A.; Poindron, P. Mother–young vocal communication and acoustic recognition promote preferential nursing in sheep. *J. Exp. Biol.* **2008**, *211*, 3554–3562. [CrossRef] [PubMed]
5. Clutton-Brock, T.H. *The Evolution of Parental Care*; Princenton University Press: Princenton, NJ, USA, 1991; p. 368.
6. Keverne, E.B. Neurobiological and molecular approaches to attachment and bonding. In *Attachment and Bonding. A New Synthesis*; Grossmann, K.E., Hrdy, S.B., Lamb, M.E., Porges, S.W., Sachser, N., Eds.; Dahlem Workshop: Berlin, Germany, 2005; pp. 101–117.
7. Tibbetts, E.A.; Dale, J. Individual recognition: It is good to be different. *Trends Ecol. Evol.* **2007**, *22*, 529–537. [CrossRef] [PubMed]
8. Francis, D.D.; Meaney, M.J. Maternal care and the development of stress responses. *Curr. Opin. Neurobiol.* **1999**, *9*, 128–134. [CrossRef] [PubMed]
9. Lévy, F.; Keller, M.; Poindron, P. Olfactory regulation of maternal behavior in mammals. *Horm. Behav.* **2004**, *46*, 284–302. [CrossRef]
10. Menard, J.L.; Champagne, D.L.; Meaney, M.J.P. Variations of maternal care differentially influence 'fear' reactivity and regional patterns of cfos immunoreactivity in response to the shock-probe burying test. *Neuroscience* **2004**, *129*, 297–308. [CrossRef]
11. D'Amato, F.R.; Zanettini, C.; Sgobio, C.; Sarli, C.; Carone, V.; Moles, A.; Ammassari-Teule, M. Intensification of maternal care by double-mothering boosts cognitive function and hippocampal morphology in the adult offspring. *Hippocampus* **2011**, *21*, 298–308. [CrossRef]
12. Lindeyer, C.M.; Meaney, M.J.; Reader, S.M. Early maternal care predicts reliance on social learning about food in adult rats. *Dev. Psychobiol.* **2013**, *55*, 168–175. [CrossRef]
13. Guardini, G.; Bowen, J.; Mariti, C.; Fatjó, J.; Sighieri, C.; Gazzano, A. Influence of maternal care on behavioural development of domestic dogs (*Canis familiaris*) living in a home environment. *Animals* **2017**, *7*, 93. [CrossRef]
14. Lezama-García, K.; Mariti, C.; Mota-Rojas, D.; Martínez-Burnes, J.; Barrios-García, H.; Gazzano, A. Maternal behaviour in domestic dogs. *Int. J. Vet. Sci. Med.* **2019**, *7*, 20–30. [CrossRef]
15. Broad, K.; Curley, J.; Keverne, E. Mother–infant bonding and the evolution of mammalian social relationships. *Philos. Trans. R. Soc. B Biol. Sci.* **2006**, *361*, 2199. [CrossRef]
16. Damián, J.P.; Hötzel, M.J.; Banchero, G.; Ungerfeld, R. Growing without a mother during rearing affects the response to stressors in rams. *Appl. Anim. Behav. Sci.* **2018**, *209*, 36–40. [CrossRef]
17. Toinon, C.; Waiblinger, S.; Rault, J. Maternal Deprivation Affects Goat Kids' Social Behavior before and after Weaning. *Dev. Psychobiol.* **2022**, *64*, e22269. [CrossRef]
18. Kikusui, T.; Isaka, Y.; Mori, Y. Early Weaning Deprives Mouse Pups of Maternal Care and Decreases Their Maternal Behavior in Adulthood. *Behav. Brain Res.* **2005**, *162*, 200–206. [CrossRef]
19. Liu, D.; Diorio, J.; Tannenbaum, B.; Caldji, C.; Francis, D.; Freedman, A.; Sharma, S.; Pearson, D.; Plotsky, P.M.; Meaney, M.J. Maternal Care, Hippocampal Glucocorticoid Receptors, and Hypothalamic-Pituitary-Adrenal Responses to Stress. *Science* **1997**, *277*, 1659–1662. [CrossRef] [PubMed]
20. Cramer, C.P.; Thiels, E.; Alberts, J.R. Weaning in rats: I. maternal behavior. *Dev. Psychobiol.* **1990**, *23*, 479–493. [CrossRef] [PubMed]
21. Moriceau, S.; Shionoya, K.; Jakubs, K.; Sullivan, R.M. Early-Life Stress Disrupts Attachment Learning: The Role of Amygdala Corticosterone, Locus Ceruleus Corticotropin Releasing Hormone, and Olfactory Bulb Norepinephrine. *J. Neurosci.* **2009**, *29*, 15745–15755. [CrossRef] [PubMed]
22. Ohki-Hamazaki, H. Neurobiology of Imprinting. *Brain Nerve* **2012**, *64*, 657–664.
23. Curley, J.P.; Keverne, E.B. Genes, brains and mammalian social bonds. *Trends Ecol. Evol.* **2005**, *20*, 561–567. [CrossRef]
24. Olazábal, D.E.; Pereira, M.; Agrati, D.; Ferreira, A.; Fleming, A.S.; González-Mariscal, G.; Lévy, F.; Lucion, A.B.; Morrell, J.I.; Numan, M.; et al. Flexibility and adaptation of the neural substrate that supports maternal behavior in mammals. *Neurosci. Biobehav. Rev.* **2013**, *37*, 1875–1892. [CrossRef]
25. Scheiber, I.B.R.; Weiß, B.M.; Kingma, S.A.; Komdeur, J. The importance of the altricial—precocial spectrum for social complexity in mammals and birds—A review. *Front. Zool.* **2017**, *14*, 1–20. [CrossRef] [PubMed]
26. Czerwinski, V.H.; Smith, B.P.; Hynd, P.I.; Hazel, S.J. The influence of maternal care on stress-related behaviors in domestic dogs: What can we learn from the rodent literature? *J. Vet. Behav.* **2016**, *14*, 52–59. [CrossRef]
27. Nowak, R.; Keller, M.; Lévy, F. Mother-young relationships in sheep: A model for a multidisciplinary approach of the study of attachment in mammals. *J. Neuroendocrinol.* **2011**, *23*, 1042–1053. [CrossRef]
28. Orihuela, A.; Mota-Rojas, D.; Strappini, A.; Serrapica, F.; Braghieri, A.; Mora-Medina, P.; Napolitano, F. Neurophysiological mechanisms of cow–calf bonding in buffalo and other farm animals. *Animals* **2021**, *11*, 1968. [CrossRef] [PubMed]
29. Numan, M. Hypothalamic neural circuits regulating maternal responsiveness toward infants. *Behav. Cogn. Neurosci. Rev.* **2006**, *5*, 163–190. [CrossRef] [PubMed]
30. González-Mariscal, G.; Poindron, P. Parental care in mammals. In Hormones, Brain and Behavior. Pfaff, D.W., Arnold, A.P., Fahrbach, S.E., Etgen, A.M., Rubin, R.T., Eds.; Elsevier: Amsterdam, The Netherlands, 2002; pp. 215–298.
31. Ruppert, P.H.; Dean, K.F.; Reiter, L.W. Development of locomotor activity of rat pups in figure-eight mazes. *Dev. Psychobiol.* **1985**, *18*, 247–260. [CrossRef]
32. Patin, V.; Vincent, A.; Lordi, B.; Caston, J. Does prenatal stress affect the motoric development of rat pups? *Dev. Brain Res.* **2004**, *149*, 85–92. [CrossRef]
33. Champagne, F.A.; Francis, D.D.; Mar, A.; Meaney, M.J. Variations in maternal care in the rat as a mediating influence for the effects of environment on development. *Physiol. Behav.* **2003**, *79*, 359–371. [CrossRef]
34. Lavely, J.A. Pediatric neurology of the dog and cat. *Vet. Clin. N. Am. Small Anim. Pract.* **2006**, *36*, 475–501. [CrossRef]

35. Kimble, D.P. Didelphid behavior. *Neurosci. Biobehav. Rev.* **1997**, *21*, 361–369. [CrossRef]
36. Dwyer, C. Behavioural development in the neonatal lamb: Effect of maternal and birth-related factors. *Theriogenology* **2003**, *59*, 1027–1050. [CrossRef] [PubMed]
37. Lévy, F.; Keller, M. Neurobiology of maternal behavior in sheep. In *Advances in the Study of Behavior*; Brockmann, J.H., Roper, T., Naguib, M., Edwards., W., Barnard, C., Mitani, J., Eds.; Academic Press: Cambridge, MA, USA, 2008; Volume 38, pp. 399–437.
38. Muir, G.D. Early ontogeny of locomotor behaviour: A comparison between altricial and precocial animals. *Brain Res. Bull.* **2000**, *53*, 719–726. [CrossRef] [PubMed]
39. Glatzle, M.; Hoops, M.; Kauffold, J.; Seeger, J.; Fietz, S.A. Development of deep and upper neuronal layers in the domestic cat, sheep and pig neocortex. *Anat. Histol. Embryol.* **2017**, *46*, 397–404. [CrossRef]
40. Fisher, D.O.; Blomberg, S.P.; Owens, I.P.F. Convergent maternal care strategies in ungulates and macropods. *Evolution* **2002**, *56*, 167–176. [CrossRef]
41. Jensen, P. Parental Behavior. In *Social Behavior in Farm Animals*; Keeling, L.J., Gonyou, H.W., Eds.; CABI Publishing: Wallingford, UK, 2001; p. 406. ISBN 9780851993973.
42. Nakamichi, M.; Yamada, K. Distribution of dorsal carriage among simians. *Primates* **2009**, *50*, 153–168. [CrossRef]
43. Algers, B.; Uvnäs-Moberg, K. Maternal Behavior in pigs. *Horm. Behav.* **2007**, *52*, 78–85. [CrossRef]
44. Isler, K. Energetic trade-offs between brain size and offspring production: Marsupials confirm a general mammalian pattern. *BioEssays* **2011**, *33*, 173–179. [CrossRef] [PubMed]
45. Bennett, P.M.; Harvey, P.H. Brain size, development and metabolism in birds and mammals. *J. Zool.* **1985**, *207*, 491–509. [CrossRef]
46. Halley, A.C. Brain at Birth. In *Encyclopedia of Evolutionary Psychological Science*; Shackelford, T., Weekes-Shackelford, V., Eds.; Springer International Publishing: Cham, Switzerland, 2018; pp. 1–8.
47. Gingerich, P.D.; Ul-Haq, M.; von Koenigswald, W.; Sanders, W.J.; Smith, B.H.; Zalmout, I.S. New protocetid whale from the middle eocene of pakistan: Birth on land, precocial development, and sexual dimorphism. *PLoS ONE* **2009**, *4*, e4366. [CrossRef]
48. Harris, M.A.; Murie, J.O. Inheritance of nest sites in female columbian ground squirrels. *Behav. Ecol. Sociobiol.* **1984**, *15*, 97–102. [CrossRef]
49. Geiser, F.; Wen, J.; Sukhchuluun, G.; Chi, Q.-S.; Wang, D.-H. Precocious Torpor in an Altricial Mammal and the Functional Implications of Heterothermy During Development. *Front. Physiol.* **2019**, *10*, 469. [CrossRef] [PubMed]
50. Canals, M.; Figueroa, D.P.; Miranda, J.P.; Sabat, P. Effect of gestational and postnatal environmental temperature on metabolic rate in the altricial rodent, Phyllotis darwini. *J. Therm. Biol.* **2009**, *34*, 310–314. [CrossRef]
51. Robinson, S.R. *A Comparative Study of Prenatal Behavioral Otogeny in Altricial and Precocial Murid Rodents*; Oregon State University: Corvallis, OR, USA, 1989.
52. Ferner, K.; Schultz, J.A.; Zeller, U. Comparative anatomy of neonates of the three major mammalian groups (monotremes, marsupials, placentals) and implications for the ancestral mammalian neonate morphotype. *J. Anat.* **2017**, *231*, 798–822. [CrossRef]
53. Grand, T.I. Altricial and precocial mammals: A model of neural and muscular development. *Zoo Biol.* **1992**, *11*, 3–15. [CrossRef]
54. Versace, E.; Vallortigara, G. Origins of Knowledge: Insights from Precocial Species. *Front. Behav. Neurosci.* **2015**, *9*, 338. [CrossRef]
55. Stroobants, S.; Creemers, J.; Bosmans, G.; D'Hooge, R. Post-weaning infant-to-mother bonding in nutritionally independent female mice. *PLoS ONE* **2020**, *15*, e0227034. [CrossRef]
56. Yamaguchi, S.; Aoki, N.; Kitajima, T.; Iikubo, E.; Katagiri, S.; Matsushima, T.; Homma, K.J. Thyroid Hormone Determines the Start of the Sensitive Period of Imprinting and Primes Later Learning. *Nat. Commun.* **2012**, *3*, 1081. [CrossRef] [PubMed]
57. Lemche, E. Research Evidence from Studies on Filial Imprinting, Attachment, and Early Life Stress: A New Route for Scientific Integration. *Acta Ethol.* **2020**, *23*, 127–133. [CrossRef] [PubMed]
58. Bolhuis, J.J. Early Learning and the Development of Filial Preferences in the Chick. *Behav. Brain Res.* **1999**, *98*, 245–252. [CrossRef] [PubMed]
59. Poindron, P.; Orgeur, P.; Le Neindre, P.; Kann, G.; Raksanyi, I. Influence of the Blood Concentration of Prolactin on the Length of the Sensitive Period for Establishing Maternal Behavior in Sheep at Parturition. *Horm. Behav.* **1980**, *14*, 173–177. [CrossRef] [PubMed]
60. Madigan, S.; Bakermans-Kranenburg, M.J.; Van Ijzendoorn, M.H.; Moran, G.; Pederson, D.R.; Benoit, D. Unresolved States of Mind, Anomalous Parental Behavior, and Disorganized Attachment: A Review and Meta-Analysis of a Transmission Gap. *Attach. Hum. Dev.* **2006**, *8*, 89–111. [CrossRef]
61. Upton, K.J.; Sullivan, R.M. Defining Age Limits of the Sensitive Period for Attachment Learning in Rat Pups. *Dev. Psychobiol.* **2010**, *52*, 453–464. [CrossRef] [PubMed]
62. Pongrácz, P.; Altbäcker, V.; Fenes, D. Human Handling Might Interfere with Conspecific Recognition in the European Rabbit (*Oryctolagus cuniculus*). *Dev. Psychobiol.* **2001**, *39*, 53–62. [CrossRef] [PubMed]
63. Dietz, L.; Arnold, A.-M.K.; Goerlich-Jansson, V.C.; Vinke, C.M. The Importance of Early Life Experiences for the Development of Behavioural Disorders in Domestic Dogs. *Behaviour* **2018**, *155*, 83–114. [CrossRef]
64. Rosenblatt, J.S. Learning in Newborn Kittens. *Sci. Am.* **1972**, *227*, 18–25. [CrossRef]
65. Manteca, X. *Etologia Veterinaria*, 1st ed.; Multimédica: Madrid, España, 2009.
66. Knudsen, E.I. Sensitive Periods in the Development of the Brain and Behavior. *J. Cogn. Neurosci.* **2004**, *16*, 1412–1425. [CrossRef]

67. Keverne, E.B.; Kendrick, K.M. Oxytocin Facilitation of Maternal Behavior in Sheepa. *Ann. N. Y. Acad. Sci.* **1992**, *652*, 83–101. [CrossRef]
68. Singh, P.K.; Kamboj, M.; Chandra, S.; Singh, R.K. Effect of Calf Suckling Dummy Calf Used and Weaning on Milk Ejection Stimuli and Milk Yield of Murrah Buffaloes (*Bubalus bubalis*). *J. Pharmacogn. Phytochem.* **2017**, *6*, 1012–1015.
69. Hübener, M.; Bonhoeffer, T. Neuronal Plasticity: Beyond the Critical Period. *Cell* **2014**, *159*, 727–737. [CrossRef]
70. Coulon, M.; Deputte, B.L.; Heyman, Y.; Delatouche, L.; Richard, C.; Baudoin, C. Visual Discrimination by Heifers (*Bos taurus*) of Their Own Species. *J. Comp. Psychol.* **2007**, *121*, 198–204. [CrossRef]
71. Coulon, M.; Deputte, B.L.; Heyman, Y.; Baudoin, C. Individual Recognition in Domestic Cattle (*Bos taurus*): Evidence from 2D-Images of Heads from Different Breeds. *PLoS ONE* **2009**, *4*, e4441. [CrossRef]
72. Alexander, G. Odour, and the Recognition of Lambs by Merino Ewes. *Appl. Anim. Ethol.* **1978**, *4*, 153–158. [CrossRef]
73. Booth, K.K.; Katz, L.S. Role of the Vomeronasal Organ in Neonatal Offspring Recognition in Sheep. *Biol. Reprod.* **2000**, *63*, 953–958. [CrossRef] [PubMed]
74. Alexander, G.; Shillito, E.E. The importance of odour, appearance and voice in maternal recognition of the young in merino sheep (*Ovis aries*). *Appl. Anim. Ethol.* **1977**, *3*, 127–135. [CrossRef]
75. Bridges, R.S. Neuroendocrine regulation of maternal behavior. *Front. Neuroendocrinol.* **2015**, *36*, 178–196. [CrossRef]
76. Solano, J.; Orihuela, A.; Galina, C.S.; Aguirre, V. A Note on Behavioral Responses to Brief Cow-Calf Separation and Reunion in Cattle (*Bos indicus*). *J. Vet. Behav.* **2007**, *2*, 10–14. [CrossRef]
77. Kustritz, M.V.R. Reproductive behavior of small animals. *Theriogenology* **2005**, *64*, 734–746. [CrossRef]
78. Pageat, P.; Gaultier, E. Current research in canine and feline pheromones. *Vet. Clin. N. Am. Small Anim. Pract.* **2003**, *33*, 187–211. [CrossRef]
79. Lim, M.M.; Young, L.J. neuropeptidergic regulation of affiliative behavior and social bonding in animals. *Horm. Behav.* **2006**, *50*, 506–517. [CrossRef]
80. Lee, H.-J.; Macbeth, A.H.; Pagani, J.; Young, W.S. Oxytocin: The great facilitator of life. *Prog. Neurobiol.* **2009**, *88*, 127–151. [CrossRef]
81. Bielsky, I.F.; Young, L.J. Oxytocin, Vasopressin, and Social Recognition in Mammals. *Peptides* **2004**, *25*, 1565–1574. [CrossRef] [PubMed]
82. Rich, M.E.; DeCárdenas, E.J.; Lee, H.-J.; Caldwell, H.K. Impairments in the Initiation of Maternal Behavior in Oxytocin Receptor Knockout Mice. *PLoS ONE* **2014**, *9*, e98839. [CrossRef]
83. Whitman, D.C.; Albers, H.E. Role of Oxytocin in the Hypothalamic Regulation of Sexual Receptivity in Hamsters. *Brain Res.* **1995**, *680*, 73–79. [CrossRef]
84. McGowan, P.O.; Suderman, M.; Sasaki, A.; Huang, T.C.T.; Hallett, M.; Meaney, M.J.; Szyf, M. Broad Epigenetic Signature of Maternal Care in the Brain of Adult Rats. *PLoS ONE* **2011**, *6*, e14739. [CrossRef] [PubMed]
85. Perkeybile, A.M.; Carter, C.S.; Wroblewski, K.L.; Puglia, M.H.; Kenkel, W.M.; Lillard, T.S.; Karaoli, T.; Gregory, S.G.; Mohammadi, N.; Epstein, L.; et al. Early Nurture Epigenetically Tunes the Oxytocin Receptor. *Psychoneuroendocrinology* **2019**, *99*, 128–136. [CrossRef]
86. Danoff, J.S.; Wroblewski, K.L.; Graves, A.J.; Quinn, G.C.; Perkeybile, A.M.; Kenkel, W.M.; Lillard, T.S.; Parikh, H.I.; Golino, H.F.; Gregory, S.G.; et al. Genetic, Epigenetic, and Environmental Factors Controlling Oxytocin Receptor Gene Expression. *Clin. Epigenetics* **2021**, *13*, 23. [CrossRef]
87. Champagne, F.; Diorio, J.; Sharma, S.; Meaney, M.J. Naturally occurring variations in maternal behavior in the rat are associated with differences in estrogen-inducible central oxytocin receptors. *Proc. Natl. Acad. Sci. USA* **2001**, *98*, 12736–12741. [CrossRef]
88. Nagasawa, M.; Okabe, S.; Mogi, K.; Kikusui, T. Oxytocin and mutual communication in mother-infant bonding. *Front. Hum. Neurosci.* **2012**, *6*, 31. [CrossRef]
89. Poindron, P. Mechanisms of Activation of Maternal Behaviour in Mammals. *Reprod. Nutr. Dev.* **2005**, *45*, 341–351. [CrossRef]
90. Rees, S.L.; Panesar, S.; Steiner, M.; Fleming, A.S. The Effects of Adrenalectomy and Corticosterone Replacement on Induction of Maternal Behavior in the Virgin Female Rat. *Horm. Behav.* **2006**, *49*, 337–345. [CrossRef]
91. Furuta, M.; Bridges, R.S. Effects of Maternal Behavior Induction and Pup Exposure on Neurogenesis in Adult, Virgin Female Rats. *Brain Res. Bull.* **2009**, *80*, 408–413. [CrossRef] [PubMed]
92. Novakov, M.; Fleming, A. The Effects of Early Rearing Environment on the Hormonal Induction of Maternal Behavior in Virgin Rats. *Horm. Behav.* **2005**, *48*, 528–536. [CrossRef] [PubMed]
93. Wischner, D.; Kemper, N.; Krieter, J. Nest-Building Behaviour in Sows and Consequences for Pig Husbandry. *Livest. Sci.* **2009**, *124*, 1–8. [CrossRef]
94. Bechard, A.; Mason, G. Leaving Home: A Study of Laboratory Mouse Pup Independence. *Appl. Anim. Behav. Sci.* **2010**, *125*, 181–188. [CrossRef]
95. Bautista, A.; García-Torres, E.; Prager, G.; Hudson, R.; Rödel, H.G. Development of Behavior in the Litter Huddle in Rat Pups: Within- and between-Litter Differences. *Dev. Psychobiol.* **2010**, *52*, 35–43. [CrossRef] [PubMed]
96. Jirkof, P. Burrowing and Nest Building Behavior as Indicators of Well-Being in Mice. *J. Neurosci. Methods* **2014**, *234*, 139–146. [CrossRef] [PubMed]
97. Wang, Z.; Storm, D.R. Maternal Behavior Is Impaired in Female Mice Lacking Type 3 Adenylyl Cyclase. *Neuropsychopharmacology* **2011**, *36*, 772–781. [CrossRef] [PubMed]

98. Castellini, C.; Dal Bosco, A.; Arias-Álvarez, M.; Lorenzo, P.L.; Cardinali, R.; Rebollar, P.G. The Main Factors Affecting the Reproductive Performance of Rabbit Does: A Review. *Anim. Reprod. Sci.* **2010**, *122*, 174–182. [CrossRef]

99. Baumann, P.; Oester, H.; Stauffacher, M. The Influence of Pup Odour on the Nest Related Behaviour of Rabbit Does (*Oryctolagus cuniculus*). *Appl. Anim. Behav. Sci.* **2005**, *93*, 123–133. [CrossRef]

100. Rowland, J.A.; Briscoe, N.J.; Handasyde, K.A. Comparing the Thermal Suitability of Nest-Boxes and Tree-Hollows for the Conservation-Management of Arboreal Marsupials. *Biol. Conserv.* **2017**, *209*, 341–348. [CrossRef]

101. Yun, J.; Swan, K.-M.; Vienola, K.; Farmer, C.; Oliviero, C.; Peltoniemi, O.; Valros, A. Nest-Building in Sows: Effects of Farrowing Housing on Hormonal Modulation of Maternal Characteristics. *Appl. Anim. Behav. Sci.* **2013**, *148*, 77–84. [CrossRef]

102. Damm, B.I.; Lisborg, L.; Vestergaard, K.S.; Vanicek, J. Nest-Building, Behavioural Disturbances and Heart Rate in Farrowing Sows Kept in Crates and Schmid Pens. *Livest. Prod. Sci.* **2003**, *80*, 175–187. [CrossRef]

103. Burri, M.; Wechsler, B.; Gygax, L.; Weber, R. Influence of Straw Length, Sow Behaviour and Room Temperature on the Incidence of Dangerous Situations for Piglets in a Loose Farrowing System. *Appl. Anim. Behav. Sci.* **2009**, *117*, 181–189. [CrossRef]

104. Mora-Medina, P.; Napolitano, F.; Mota-Rojas, D.; Berdugo-Gutiérrez, J.; Ruiz-Buitrago, J.; Guerrero-Legarreta, I. Imprinting, Sucking and Allosucking Behaviors in Buffalo Calves. *J. Buffalo Sci.* **2018**, *7*, 49–57. [CrossRef]

105. Insel, T.R.; Young, L.J. The Neurobiology of Attachment. *Nat. Rev. Neurosci.* **2001**, *2*, 129–136. [CrossRef]

106. Debiec, J.; Sullivan, R.M. The Neurobiology of Safety and Threat Learning in Infancy. *Neurobiol. Learn. Mem.* **2017**, *143*, 49–58. [CrossRef]

107. Pereira, M.; Morrell, J.I. Functional Mapping of the Neural Circuitry of Rat Maternal Motivation: Effects of Site-Specific Transient Neural Inactivation. *J. Neuroendocrinol.* **2011**, *23*, 1020–1035. [CrossRef]

108. Caldji, C.; Tannenbaum, B.; Sharma, S.; Francis, D.; Plotsky, P.M.; Meaney, M.J. Maternal Care during Infancy Regulates the Development of Neural Systems Mediating the Expression of Fearfulness in the Rat. *Proc. Natl. Acad. Sci. USA* **1998**, *95*, 5335–5340. [CrossRef]

109. Starr-Phillips, E.J.; Beery, A.K. Natural Variation in Maternal Care Shapes Adult Social Behavior in Rats. *Dev. Psychobiol.* **2014**, *56*, 1017–1026. [CrossRef]

110. Champagne, F.A.; Weaver, I.C.G.; Diorio, J.; Sharma, S.; Meaney, M.J. Natural Variations in Maternal Care Are Associated with Estrogen Receptor Alpha Expression and Estrogen Sensitivity in the Medial Preoptic Area. *Endocrinology* **2003**, *144*, 4720–4724. [CrossRef]

111. Bale, T.L.; Picetti, R.; Contarino, A.; Koob, G.F.; Vale, W.W.; Lee, K.-F. Mice Deficient for Both Corticotropin-Releasing Factor Receptor 1 (CRFR1) and CRFR2 Have an Impaired Stress Response and Display Sexually Dichotomous Anxiety-Like Behavior. *J. Neurosci.* **2002**, *22*, 193–199. [CrossRef] [PubMed]

112. Bale, T.L.; Vale, W.W. CRF and CRF Receptors: Role in Stress Responsivity and Other Behaviors. *Annu. Rev. Pharmacol. Toxicol.* **2004**, *44*, 525–557. [CrossRef] [PubMed]

113. Hudson, R.; Cruz, Y.; Lucio, R.A.; Ninomiya, J.; Martínez–Gómez, M. Temporal and Behavioral Patterning of Parturition in Rabbits and Rats. *Physiol. Behav.* **1999**, *66*, 599–604. [CrossRef]

114. Hudson, R.; Distel, H. Nipple Location By Newborn Rabbits: Behavioural Evidence for Pheromonal Guidance. *Behaviour* **1983**, *85*, 260–274. [CrossRef]

115. Val-Laillet, D.; Nowak, R. Early Discrimination of the Mother by Rabbit Pups. *Appl. Anim. Behav. Sci.* **2008**, *111*, 173–182. [CrossRef]

116. Sullivan, R.M.; Wilson, D.A. Molecular Biology Of Early Olfactory Memory. *Learn. Mem.* **2003**, *10*, 1–4. [CrossRef]

117. Corona, R.; Meurisse, M.; Cornilleau, F.; Moussu, C.; Keller, M.; Lévy, F. Disruption of Adult Olfactory Neurogenesis Induces Deficits in Maternal Behavior in Sheep. *Behav. Brain Res.* **2018**, *347*, 124–131. [CrossRef]

118. Brennan, P.A.; Kendrick, K.M. Mammalian Social Odours: Attraction and Individual Recognition. *Philos. Trans. R. Soc. B Biol. Sci.* **2006**, *361*, 2061–2078. [CrossRef]

119. Parrott, M.L.; Ward, S.J.; Temple-Smith, P.D. Olfactory Cues, Genetic Relatedness and Female Mate Choice in the Agile Antechinus (*Antechinus agilis*). *Behav. Ecol. Sociobiol.* **2007**, *61*, 1075–1079. [CrossRef]

120. Kinsley, C.H.; Bridges, R.S. Morphine Treatment and Reproductive Condition Alter Olfactory Preferences for Pup and Adult Male Odors in Female Rats. *Dev. Psychobiol.* **1990**, *23*, 331–347. [CrossRef]

121. Mota-Rojas, D.; Velarde, A.; Huertas, S.; Cajiao, M. (Eds.) *Animal welfare, A Global Vision in Ibero-America*, 3rd ed.; Elsevier: Barcelona, Spain, 2016; pp. 1–516.

122. Mota-Rojas, D.; Orihuela, A.; Napolitano, F.; Mora-Medina, P.; Alonso-Spilsbury, M. Olfaction in Animal Behaviour and Welfare. *CAB Rev. Perspect. Agric. Vet. Sci. Nutr. Nat. Resour.* **2018**, *13*, 1–13. [CrossRef]

123. Mota-Rojas, D.; Guerrrero-Legarreta, I.; de Rosa, G.; Mora-Medina, P.; Braghieri, A.; Fabio, N. Dairy Buffalo Behaviour and Welfare from Calving to Milking. *CAB Rev. Perspect. Agric. Vet. Sci. Nutr. Nat. Resour.* **2019**, *14*, 1–9. [CrossRef]

124. Ferreira, A.; Dahlöf, L.-G.; Hansen, S. Olfactory Mechanisms in the Control of Maternal Aggression, Appetite, and Fearfulness: Effects of Lesions to Olfactory Receptors, Mediodorsal Thalamic Nucleus, and Insular Prefrontal Cortex. *Behav. Neurosci.* **1987**, *101*, 709–717. [CrossRef]

125. Dhungel, S.; Masaoka, M.; Rai, D.; Kondo, Y.; Sakuma, Y. Both Olfactory Epithelial and Vomeronasal Inputs Are Essential for Activation of the Medial Amygdala and Preoptic Neurons of Male Rats. *Neuroscience* **2011**, *199*, 225–234. [CrossRef]

126. Trouillet, A.C.; Moussu, C.; Poissenot, K.; Keller, M.; Birnbaumer, L.; Leinders-Zufall, T.; Zufall, F.; Chamero, P. Sensory Detection by the Vomeronasal Organ Modulates Experience-Dependent Social Behaviors in Female Mice. *Front. Cell. Neurosci.* **2021**, *15*, 32. [CrossRef]
127. Morgan, H.D.; Fleming, A.S.; Stern, J.M. Somatosensory Control of the Onset and Retention of Maternal Responsiveness in Primiparous Sprague-Dawley Rats. *Physiol. Behav.* **1992**, *51*, 549–555. [CrossRef]
128. Brouette-Lahlou, I.; Godinot, F.; Vernet-Maury, E. The Mother Rat's Vomeronasal Organ Is Involved in Detection of Dodecyl Propionate, the Pup's Preputial Gland Pheromone. *Physiol. Behav.* **1999**, *66*, 427–436. [CrossRef]
129. Gandelman, R.; Zarrow, M.X.; Denenberg, V.H.; Myers, M. Olfactory Bulb Removal Eliminates Maternal Behavior in the Mouse. *Science* **1971**, *171*, 210–211. [CrossRef]
130. Liang, J.; Yang, J.; Chen, Y.; Zhang, L. The Role of Olfactory Cues in Mother-Pup, Groupmate, and Sex Recognition of Lesser Flat-Headed Bats, Tylonycteris Pachypus. *Ecol. Evol.* **2021**, *11*, 15792–15799. [CrossRef]
131. Hudson, R. Do Newborn Rabbits Learn the Odor Stimuli Releasing Nipple-Search Behavior? *Dev. Psychobiol.* **1985**, *18*, 575–585. [CrossRef]
132. Jouventin, P.; Aubin, T.; Lengagne, T. Finding a Parent in a King Penguin Colony: The Acoustic System of Individual Recognition. *Anim. Behav.* **1999**, *57*, 1175–1183. [CrossRef] [PubMed]
133. Aubin, T.; Jouventin, P. How to Vocally Identify Kin in a Crowd: The Penguin Model. *Adv. Study Behav.* **2002**, *31*, 243–277. [CrossRef]
134. Charrier, I.; Mathevon, N.; Jouventin, P. Mother's Voice Recognition by Seal Pups. *Nature* **2001**, *412*, 873. [CrossRef] [PubMed]
135. Beecher, M.D.; Beecher, I.M.; Hahn, S. Parent-Offspring Recognition in Bank Swallows (*Riparia riparia*): II. Development and Acoustic Basis. *Anim. Behav.* **1981**, *29*, 95–101. [CrossRef]
136. Haskins, R. Effect of Kitten Vocalizations on Maternal Behavior. *J. Comp. Physiol. Psychol.* **1977**, *91*, 830–838. [CrossRef]
137. Branchi, I.; Santucci, D.; Alleva, E. Ultrasonic Vocalisation Emitted by Infant Rodents: A Tool for Assessment of Neurobehavioural Development. *Behav. Brain Res.* **2001**, *125*, 49–56. [CrossRef]
138. Stark, R.A.; Harker, A.; Salamanca, S.; Pellis, S.M.; Li, F.; Gibb, R.L. Development of Ultrasonic Calls in Rat Pups Follows Similar Patterns Regardless of Isolation Distress. *Dev. Psychobiol.* **2020**, *62*, 617–630. [CrossRef]
139. D'Amato, F.R.; Scalera, E.; Sarli, C.; Moles, A. Pups Call, Mothers Rush: Does Maternal Responsiveness Affect the Amount of Ultrasonic Vocalizations in Mouse Pups? *Behav. Genet.* **2005**, *35*, 103–112. [CrossRef]
140. Hashimoto, H.; Saito, T.R.; Furudate, S.; Takahashi, K.W. Prolactin Levels and Maternal Behavior Induced by Ultrasonic Vocalizations of the Rat Pup. *Exp. Anim.* **2001**, *50*, 307–312. [CrossRef]
141. Kaidbey, J.H.; Ranger, M.; Myers, M.M.; Anwar, M.; Ludwig, R.J.; Schulz, A.M.; Barone, J.L.; Kolacz, J.; Welch, M.G. Early Life Maternal Separation and Maternal Behaviour Modulate Acoustic Characteristics of Rat Pup Ultrasonic Vocalizations. *Sci. Rep.* **2019**, *9*, 19012. [CrossRef]
142. Bradley, P. A Light and Electron Microscopic Study of the Development of Two Regions of the Chick Forebrain. *Dev. Brain Res.* **1985**, *20*, 83–88. [CrossRef] [PubMed]
143. Andrews, E.F.; Jacqmot, O.; Espinheira Gomes, F.N.C.M.; Sha, M.F.; Niogi, S.N.; Johnson, P.J. Characterizing the Canine and Feline Optic Pathways in Vivo with Diffusion MRI. *Vet. Ophthalmol.* **2022**, *25*, 60–71. [CrossRef] [PubMed]
144. El-Sabrout, K. Does the Blindness Affect the Behavioural Activities of Rabbit? *J. Anim. Behav. Biometeorol.* **2018**, *6*, 6–8. [CrossRef]
145. Zaharia, M.D.; Kulczycki, J.; Shanks, N.; Meaney, M.J.; Anisman, H. The Effects of Early Postnatal Stimulation on Morris Water-Maze Acquisition in Adult Mice: Genetic and Maternal Factors. *Psychopharmacology* **1996**, *128*, 227–239. [CrossRef]
146. Sturman-Hulbe, M.; Stone, C.P. Maternal Behavior in the Albino Rat. *J. Comp. Psychol.* **1929**, *9*, 203–237. [CrossRef]
147. Kenkel, W.M.; Yee, J.R.; Carter, C.S. Is Oxytocin a Maternal–Foetal Signalling Molecule at Birth? Implications for Development. *J. Neuroendocrinol.* **2014**, *26*, 739–749. [CrossRef]
148. Love, T.M. Oxytocin, Motivation and the Role of Dopamine. *Pharmacol. Biochem. Behav.* **2014**, *119*, 49–60. [CrossRef]
149. Sharpey-Schafer, E. Experimental Physiology. *Br. Med. J.* **1933**, *2*, 755. [CrossRef]
150. Orsucci, F.; Paoloni, G.; Conti, C.; Reda, M.; Fulcheri, M. The Role of Oxytocin in Plasticity, Memory and Attachment Dynamics. *J. Biol. Regul. Homeost Agents* **2013**, *27*, 947–954.
151. Strathearn, L. Maternal Neglect: Oxytocin, Dopamine and the Neurobiology of Attachment. *J. Neuroendocrinol.* **2011**, *23*, 1054–1065. [CrossRef]
152. Boccia, M.L.; Goursaud, A.-P.S.; Bachevalier, J.; Anderson, K.D.; Pedersen, C.A. Peripherally Administered Non-Peptide Oxytocin Antagonist, L368,899®, Accumulates in Limbic Brain Areas: A New Pharmacological Tool for the Study of Social Motivation in Non-Human Primates. *Horm. Behav.* **2007**, *52*, 344–351. [CrossRef] [PubMed]
153. Pedersen, C.A.; Caldwell, J.D.; Walker, C.; Ayers, G.; Mason, G.A. Oxytocin Activates the Postpartum Onset of Rat Maternal Behavior in the Ventral Tegmental and Medial Preoptic Areas. *Behav. Neurosci.* **1994**, *108*, 1163–1171. [CrossRef] [PubMed]
154. Pedersen, C.A.; Prange, A.J. Induction of maternal behavior in virgin rats after intracerebroventricular administration of oxytocin. *Proc. Natl. Acad. Sci. USA* **1979**, *76*, 6661–6665. [CrossRef]
155. Pedersen, C.A.; Vadlamudi, S.V.; Boccia, M.L.; Amico, J.A. Maternal Behavior Deficits in Nulliparous Oxytocin Knockout Mice. *Genes Brain Behav.* **2006**, *5*, 274–281. [CrossRef]
156. Bosch, O.J. Brain Oxytocin Correlates with Maternal Aggression: Link to Anxiety. *J. Neurosci.* **2005**, *25*, 6807–6815. [CrossRef]

157. Ruthschilling, C.A.; Albiero, G.; Lazzari, V.M.; Becker, R.O.; de Moura, A.C.; Lucion, A.B.; Almeida, S.; da Veiga, A.B.G.; Giovenardi, M. Analysis of Transcriptional Levels of the Oxytocin Receptor in Different Areas of the Central Nervous System and Behaviors in High and Low Licking Rats. *Behav. Brain Res.* **2012**, *228*, 176–184. [CrossRef]

158. Lauby, S.C.; Ashbrook, D.G.; Malik, H.R.; Chatterjee, D.; Pan, P.; Fleming, A.S.; McGowan, P.O. The Role of Interindividual Licking Received and Dopamine Genotype on Later–life Licking Provisioning in Female Rat Offspring. *Brain Behav.* **2021**, *11*, e02069. [CrossRef]

159. Ogi, A.; Naef, V.; Santorelli, F.M.; Mariti, C.; Gazzano, A. Oxytocin Receptor Gene Polymorphism in Lactating Dogs. *Animals* **2021**, *11*, 3099. [CrossRef] [PubMed]

160. Caldwell, H.K. Neurobiology of Sociability. *Sens. Nat.* **2012**, *1*, 187–205. [CrossRef]

161. Sohlström, A.; Carlsson, C.; Uvnäs-Moberg, K. Effects of Oxytocin Treatment in Early Life on Body Weight and Corticosterone in Adult Offspring from Ad Libitum-Fed and Food-Restricted Rats. *Neonatology* **2000**, *78*, 33–40. [CrossRef]

162. Costa-Brito, A.; Gonçalves, I.; Santos, C.A. The Brain as a Source and a Target of Prolactin in Mammals. *Neural Regen. Res.* **2022**, *17*, 1695. [CrossRef] [PubMed]

163. Bridges, R.S.; Numan, M.; Ronsheim, P.M.; Mann, P.E.; Lupini, C.E. Central Prolactin Infusions Stimulate Maternal Behavior in Steroid-Treated, Nulliparous Female Rats. *Proc. Natl. Acad. Sci. USA* **1990**, *87*, 8003–8007. [CrossRef] [PubMed]

164. Cabrera-Reyes, E.A.; Limón-Morales, O.; Rivero-Segura, N.A.; Camacho-Arroyo, I.; Cerbón, M. Prolactin Function and Putative Expression in the Brain. *Endocrine* **2017**, *57*, 199–213. [CrossRef]

165. Bridges, R.S.; Grattan, D.R. Prolactin-Induced Neurogenesis in the Maternal Brain. *Trends Endocrinol. Metab.* **2003**, *14*, 199–201. [CrossRef] [PubMed]

166. Hashemian, F.; Shafigh, F.; Roohi, E. Regulatory Role of Prolactin in Paternal Behavior in Male Parents: A Narrative Review. *J. Postgrad. Med.* **2016**, *62*, 182. [CrossRef]

167. Castro-Sierra, E.; Chico Ponce de León, F.; Portugal Rivera, A. Neurotransmisores Del Sistema Límbico. *Salud Ment.* **2005**, *28*, 27–32.

168. Angoa-Pérez, M.; Kuhn, D.M. Neuronal Serotonin in the Regulation of Maternal Behavior in Rodents. *Neurotransmitter* **2015**, *2*, e615.

169. Tortora, G.; Derrickson, B. *Principios de Anatomía y Fisiología*, 13th ed.; Tortora, G., Derrickson, B., Eds.; Editorial Médica Panamericana: Buenos Aires, Argentina, 2018; ISBN 9786078546114.

170. Nakazawa, K.; Zsiros, V.; Jiang, Z.; Nakao, K.; Kolata, S.; Zhang, S.; Belforte, J.E. GABAergic Interneuron Origin of Schizophrenia Pathophysiology. *Neuropharmacology* **2012**, *62*, 1574–1583. [CrossRef]

171. Arrati, P.; Carmona, C.; Dominguez, G.; Beyer, C.; Rosenblatt, J. GABA Receptor Agonists in the Medial Preoptic Area and Maternal Behavior in Lactating Rats. *Physiol. Behav.* **2006**, *87*, 51–65. [CrossRef]

172. Kuznetsov, S.V.; Dmitrieva, L.E.; Sizonov, V.A. Cardiac, Respiratory, and Motor Activity in Norm and after Activation of Catecholaminergic Systems in Newborn Rat Pups. *J. Evol. Biochem. Physiol.* **2012**, *48*, 430–444. [CrossRef]

173. Afonso, V.M.; Grella, S.L.; Chatterjee, D.; Fleming, A.S. Previous Maternal Experience Affects Accumbal Dopaminergic Responses to Pup-Stimuli. *Brain Res.* **2008**, *1198*, 115–123. [CrossRef] [PubMed]

174. Angoa-Pérez, M.; Kane, M.J.; Sykes, C.E.; Perrine, S.A.; Church, M.W.; Kuhn, D.M. Brain Serotonin Determines Maternal Behavior and Offspring Survival. *Genes Brain Behav.* **2014**, *13*, 579–591. [CrossRef] [PubMed]

175. Chiavegatto, S.; Nelson, R.J. Interaction of Nitric Oxide and Serotonin in Aggressive Behavior. *Horm. Behav.* **2003**, *44*, 233–241. [CrossRef] [PubMed]

176. Lee, G.; Gammie, S.C. GABA enhancement of maternal defense in mice: Possible neural correlates. *Pharmacol. Biochem. Behav.* **2007**, *86*, 176–187. [CrossRef]

177. Okino, E.; Morita, S.; Hoshikawa, Y.; Tsukahara, S. The glutamatergic system in the preoptic area is involved in the retention of maternal behavior in maternally experienced female rats. *Psychoneuroendocrinology* **2020**, *120*, 104792. [CrossRef]

178. Douglas, A.J. Baby Love? Oxytocin-Dopamine Interactions in Mother-Infant Bonding. *Endocrinology* **2010**, *151*, 1978–1980. [CrossRef]

179. Georgescu, T.; Swart, J.M.; Grattan, D.R.; Brown, R.S.E. The Prolactin Family of Hormones as Regulators of Maternal Mood and Behavior. *Front. Glob. Women's Health* **2021**, *2*, 767467. [CrossRef]

180. Brown, R.S.E.; Aoki, M.; Ladyman, S.R.; Phillipps, H.R.; Wyatt, A.; Boehm, U.; Grattan, D.R. Prolactin action in the medial preoptic area is necessary for postpartum maternal nursing behavior. *Proc. Natl. Acad. Sci. USA* **2017**, *114*, 10779–10784. [CrossRef]

181. Gao, J.; Chen, L.; Li, M. 5-HT2A receptors modulate dopamine D2-mediated maternal effects. *Pharmacol. Biochem. Behav.* **2019**, *180*, 32–43. [CrossRef]

182. Muzerelle, A.; Soiza-Reilly, M.; Hainer, C.; Ruet, P.-L.; Lesch, K.-P.; Bader, M.; Alenina, N.; Scotto-Lomassese, S.; Gaspar, P. Dorsal raphe serotonin neurotransmission is required for the expression of nursing behavior and for pup survival. *Sci. Rep.* **2021**, *11*, 6004. [CrossRef]

183. Zoicas, I.; Kornhuber, J. The Role of the N-Methyl-D-Aspartate Receptors in Social Behavior in Rodents. *Int. J. Mol. Sci.* **2019**, *20*, 5599. [CrossRef] [PubMed]

184. Francis, D.D.; Young, L.J.; Meaney, M.J.; Insel, T.R. Naturally Occurring Differences in Maternal Care are Associated with the Expression of Oxytocin and Vasopressin (V1a) Receptors: Gender Differences. *J. Neuroendocrinol.* **2002**, *14*, 349–353. [CrossRef] [PubMed]

185. Curley, J.P.; Jensen, C.L.; Franks, B.; Champagne, F.A. Variation in maternal and anxiety-like behavior associated with discrete patterns of oxytocin and vasopressin 1a receptor density in the lateral septum. *Horm. Behav.* **2012**, *61*, 454–461. [CrossRef]
186. Pearce, S.; Gabler, N.; Ross, J.; Escobar, J.; Patience, J.; Rhoads, R.; Baumgard, L. The effects of heat stress and plane of nutrition on metabolism in growing pigs. *J. Anim. Sci.* **2013**, *91*, 2108–2118. [CrossRef]
187. Szeto, A.; McCabe, P.M.; Nation, D.A.; Tabak, B.A.; Rossetti, M.A.; McCullough, M.E.; Schneiderman, N.; Mendez, A.J. Evaluation of Enzyme Immunoassay and Radioimmunoassay Methods for the Measurement of Plasma Oxytocin. *Psychosom. Med.* **2011**, *73*, 393–400. [CrossRef]
188. Brandtzaeg, O.K.; Johnsen, E.; Roberg-Larsen, H.; Seip, K.F.; MacLean, E.L.; Gesquiere, L.R.; Leknes, S.; Lundanes, E.; Wilson, S.R. Proteomics tools reveal startlingly high amounts of oxytocin in plasma and serum. *Sci. Rep.* **2016**, *6*, 31693. [CrossRef] [PubMed]
189. Bienboire-Frosini, C.; Chabaud, C.; Cozzi, A.; Codecasa, E.; Pageat, P. Validation of a Commercially Available Enzyme ImmunoAssay for the Determination of Oxytocin in Plasma Samples from Seven Domestic Animal Species. *Front. Neurosci.* **2017**, *11*, 524. [CrossRef] [PubMed]
190. Liu, D.; Han, X.; Liu, X.; Cheng, M.; He, M.; Rainer, G.; Gao, H.; Zhang, X. Measurement of ultra-trace level of intact oxytocin in plasma using SALLE combined with nano-LC–MS. *J. Pharm. Biomed. Anal.* **2019**, *173*, 62–67. [CrossRef] [PubMed]
191. Franke, A.A.; Li, X.; Menden, A.; Lee, M.R.; Lai, J.F. Oxytocin analysis from human serum, urine, and saliva by orbitrap liquid chromatography-mass spectrometry. *Drug Test. Anal.* **2019**, *11*, 119–128. [CrossRef]
192. MacLean, E.L.; Wilson, S.R.; Martin, W.L.; Davis, J.M.; Nazarloo, H.P.; Carter, C.S. Challenges for measuring oxytocin: The blind men and the elephant? *Psychoneuroendocrinology* **2019**, *107*, 225–231. [CrossRef]
193. Tabak, B.A.; Leng, G.; Szeto, A.; Parker, K.J.; Verbalis, J.G.; Ziegler, T.E.; Lee, M.R.; Neumann, I.D.; Mendez, A.J. Advances in human oxytocin measurement: Challenges and proposed solutions. *Mol. Psychiatry* **2023**, *28*, 127–140. [CrossRef] [PubMed]
194. Uvnäs Moberg, K.; Handlin, L.; Kendall-Tackett, K.; Petersson, M. Oxytocin is a principal hormone that exerts part of its effects by active fragments. *Med. Hypotheses* **2019**, *133*, 109394. [CrossRef] [PubMed]
195. Gnanadesikan, G.E.; Hammock, E.A.D.; Tecot, S.R.; Lewis, R.J.; Hart, R.; Carter, C.S.; MacLean, E.L. What are oxytocin assays measuring? Epitope mapping, metabolites, and comparisons of wildtype & knockout mouse urine. *Psychoneuroendocrinology* **2022**, *143*, 105827. [CrossRef]
196. Coplan, J.D.; Karim, A.; Chandra, P., St.; Germain, G.; Abdallah, C.G.; Altemus, M. Neurobiology of Maternal Stress: Role of Social Rank and Central Oxytocin in Hypothalamic–Pituitary Adrenal Axis Modulation. *Front. Psychiatry* **2015**, *6*, 100. [CrossRef] [PubMed]
197. Rault, J.-L. Effects of positive and negative human contacts and intranasal oxytocin on cerebrospinal fluid oxytocin. *Psychoneuroendocrinology* **2016**, *69*, 60–66. [CrossRef]
198. Freeman, S.M.; Samineni, S.; Allen, P.C.; Stockinger, D.; Bales, K.L.; Hwa, G.G.C.; Roberts, J.A. Plasma and CSF oxytocin levels after intranasal and intravenous oxytocin in awake macaques. *Psychoneuroendocrinology* **2016**, *66*, 185–194. [CrossRef]
199. Almanza-Sepulveda, M.L.; Fleming, A.S.; Jonas, W. Mothering revisited: A role for cortisol? *Horm. Behav.* **2020**, *121*, 104679. [CrossRef]
200. Brown, C.A.; Cardoso, C.; Ellenbogen, M.A. A meta-analytic review of the correlation between peripheral oxytocin and cortisol concentrations. *Front. Neuroendocrinol.* **2016**, *43*, 19–27. [CrossRef]
201. Carter, C. Biological perspectives on social attachment and bonding. In *Attachment and Bonding: A New Synthesis*; Carter, C., Ahnert, L., Grossmann, K., Hrday, S., Lamb, M., Porges, S., Sachser, N., Eds.; The Dahlemn Workshops: Berlin, Germany, 2005; pp. 85–100.
202. Gatta, E.; Mairesse, J.; Deruyter, L.; Marrocco, J.; Van Camp, G.; Bouwalerh, H.; Lo Guidice, J.-M.; Morley-Fletcher, S.; Nicoletti, F.; Maccari, S. Reduced maternal behavior caused by gestational stress is predictive of life span changes in risk-taking behavior and gene expression due to altering of the stress/anti-stress balance. *Neurotoxicology* **2018**, *66*, 138–149. [CrossRef]
203. Stohn, J.P.; Martinez, M.E.; Zafer, M.; López-Espíndola, D.; Keyes, L.M.; Hernandez, A. Increased aggression and lack of maternal behavior in Dio3-deficient mice are associated with abnormalities in oxytocin and vasopressin systems. *Genes Brain Behav.* **2018**, *17*, 23–35. [CrossRef]
204. Adan, R.A.; Cox, J.J.; van Kats, J.P.; Burbach, J.P. Thyroid hormone regulates the oxytocin gene. *J. Biol. Chem.* **1992**, *267*, 3771–3777. [CrossRef] [PubMed]
205. Vasudevan, N.; Davidkova, G.; Zhu, Y.-S.; Koibuchi, N.; Chin, W.W.; Pfaff, D. Differential Interaction of Estrogen Receptor and Thyroid Hormone Receptor Isoforms on the Rat Oxytocin Receptor Promoter Leads to Differences in Transcriptional Regulation. *Neuroendocrinology* **2001**, *74*, 309–324. [CrossRef] [PubMed]
206. Ciosek, J.; Drobnik, J. Vasopressin and oxytocin release and the thyroid function. *J. Physiol. Pharmacol.* **2004**, *55*, 423–441.
207. Desouza, L.A.; Ladiwala, U.; Daniel, S.M.; Agashe, S.; Vaidya, R.A.; Vaidya, V.A. Thyroid hormone regulates hippocampal neurogenesis in the adult rat brain. *Mol. Cell. Neurosci.* **2005**, *29*, 414–426. [CrossRef]
208. Noorjahan, N.; Cattini, P.A. Neurogenesis in the Maternal Rodent Brain: Impacts of Gestation-Related Hormonal Regulation, Stress, and Obesity. *Neuroendocrinology* **2022**, *112*, 702–722. [CrossRef]
209. Cryan, J.F.; Dinan, T.G. Mind-altering microorganisms: The impact of the gut microbiota on brain and behaviour. *Nat. Rev. Neurosci.* **2012**, *13*, 701–712. [CrossRef]
210. Bercik, P.; Collins, S.M.; Verdu, E.F. Microbes and the gut-brain axis. *Neurogastroenterol. Motil.* **2012**, *24*, 405–413. [CrossRef]

211. Sudo, N. Role of gut microbiota in brain function and stress-related pathology. *Biosci. Microbiota Food Heal.* **2019**, *38*, 75–80. [CrossRef]

212. Moussaoui, N.; Larauche, M.; Biraud, M.; Molet, J.; Million, M.; Mayer, E.; Taché, Y. Limited Nesting Stress Alters Maternal Behavior and In Vivo Intestinal Permeability in Male Wistar Pup Rats. *PLoS ONE* **2016**, *11*, e0155037. [CrossRef]

213. Farshim, P.; Walton, G.; Chakrabarti, B.; Givens, I.; Saddy, D.; Kitchen, I.; Swann, J.R.; Bailey, A. Maternal Weaning Modulates Emotional Behavior and Regulates the Gut-Brain Axis. *Sci. Rep.* **2016**, *6*, 21958. [CrossRef]

214. Bailey, M.T.; Coe, C.L. Maternal separation disrupts the integrity of the intestinal microflora in infant rhesus monkeys. *Dev. Psychobiol.* **1999**, *35*, 146–155. [CrossRef]

215. O'Mahony, S.M.; Marchesi, J.R.; Scully, P.; Codling, C.; Ceolho, A.-M.; Quigley, E.M.M.; Cryan, J.F.; Dinan, T.G. Early Life Stress Alters Behavior, Immunity, and Microbiota in Rats: Implications for Irritable Bowel Syndrome and Psychiatric Illnesses. *Biol. Psychiatry* **2009**, *65*, 263–267. [CrossRef] [PubMed]

216. O'Mahony, S.M.; Clarke, G.; Borre, Y.E.; Dinan, T.G.; Cryan, J.F. Serotonin, tryptophan metabolism and the brain-gut-microbiome axis. *Behav. Brain Res.* **2015**, *277*, 32–48. [CrossRef]

217. Rackers, H.S.; Thomas, S.; Williamson, K.; Posey, R.; Kimmel, M.C. Emerging literature in the Microbiota-Brain Axis and Perinatal Mood and Anxiety Disorders. *Psychoneuroendocrinology* **2018**, *95*, 86–96. [CrossRef]

218. Bernabé, B.P.; Tussing-Humphreys, L.; Rackers, H.S.; Welke, L.; Mantha, A.; Kimmel, M.C. Improving Mental Health for the Mother-Infant Dyad by Nutrition and the Maternal Gut Microbiome. *Gastroenterol. Clin. N. Am.* **2019**, *48*, 433–445. [CrossRef] [PubMed]

219. Mota-Rojas, D.; Velarde, A.; Marcet-Rius, M.; Orihuela, A.; Bragaglio, A.; Hernández-Ávalos, I.; Casas-Alvarado, A.; Domínguez-Oliva, A.; Whittaker, A.L. Analgesia during Parturition in Domestic Animals: Perspectives and Controversies on Its Use. *Animals* **2022**, *12*, 2686. [CrossRef]

220. Martínez-Burnes, J.; Muns, R.; Barrios-García, H.; Villanueva-García, D.; Domínguez-Oliva, A.; Mota-Rojas, D. Parturition in Mammals: Animal Models, Pain and Distress. *Animals* **2021**, *11*, 2960. [CrossRef] [PubMed]

221. Mota-Rojas, D.; Martínez-Burnes, J.; Napolitano, F.; Domínguez-Muñoz, M.; Guerrero-Legarreta, I.; Mora-Medina, P.; Ramírez-Necoechea, R.; Lezama-García, K.; González-Lozano, M. Dystocia: Factors Affecting Parturition in Domestic Animals. *CABI Rev.* **2020**, *2020*, 1–16. [CrossRef]

222. Kuller, W.; Sietsma, S.; Hendriksen, S.; Sperling, D. Use of Paracetamol in Sows around Farrowing: Effect on Health and Condition of the Sow, Piglet Mortality, Piglet Weight and Piglet Weight Gain. *Porc. Heal. Manag.* **2021**, *7*, 46. [CrossRef]

223. Ison, S.H.; Jarvis, S.; Hall, S.A.; Ashworth, C.J.; Rutherford, K.M.D. Periparturient Behavior and Physiology: Further Insight into the Farrowing Process for Primiparous and Multiparous Sows. *Front. Vet. Sci.* **2018**, *5*, 122. [CrossRef]

224. Tummaruk, P.; Sang-Gassanee, K. Effect of Farrowing Duration, Parity Number and the Type of Anti-Inflammatory Drug on Postparturient Disorders in Sows: A Clinical Study. *Trop. Anim. Health Prod.* **2013**, *45*, 1071–1077. [CrossRef] [PubMed]

225. Hirsch, A.C.; Philipp, H.; Kleemann, R. Investigation on the Efficacy of Meloxicam in Sows with Mastitis-Metritis-Agalactia Syndrome. *J. Vet. Pharmacol. Ther.* **2003**, *26*, 355–360. [CrossRef] [PubMed]

226. Stolzenberg, D.S.; Mayer, H.S. Experience-Dependent Mechanisms in the Regulation of Parental Care. *Front. Neuroendocrinol.* **2019**, *54*, 100745. [CrossRef] [PubMed]

227. Fleming, A.S.; Luebke, C. Timidity Prevents the Virgin Female Rat from Being a Good Mother: Emotionality Differences between Nulliparous and Parturient Females. *Physiol. Behav.* **1981**, *27*, 863–868. [CrossRef]

228. Martín-Sánchez, A.; Valera-Marín, G.; Hernández-Martínez, A.; Lanuza, E.; Martínez-García, F.; Agustín-Pavón, C. Wired for Motherhood: Induction of Maternal Care but Not Maternal Aggression in Virgin Female CD1 Mice. *Front. Behav. Neurosci.* **2015**, *9*, 197. [CrossRef]

229. Seip, K.M.; Morrell, J.I. Exposure to Pups Influences the Strength of Maternal Motivation in Virgin Female Rats. *Physiol. Behav.* **2008**, *95*, 599–608. [CrossRef]

230. Stolzenberg, D.S.; Rissman, E.F. Oestrogen-Independent, Experience-Induced Maternal Behaviour in Female Mice. *J. Neuroendocrinol.* **2011**, *23*, 345–354. [CrossRef]

231. Work, T.M.; Duhr, M.; Flint, B. Pathology of House Mouse (*Mus musculus*) Predation on Laysan Albatross (*Phoebastria aimutabilis*) on Midway Atoll National Wildlife Refuge. *J. Wildl. Dis.* **2021**, *57*, 125–131. [CrossRef]

232. Duhr, M.; Flint, E.N.; Hunter, S.A.; Taylor, R.V.; Flanders, B.; Howald, G.; Norwood, D. Control of House Mice Preying on Adult Albatrossess at Midway Atoll National Wildlife Refuge. In *Island Invasives: Scaling up to Meet the Challenge*; Veitch, C.R., Ciout, M.N., Russell, J.C., West, C.J., Eds.; IUCN: Gland, Switzerland, 2017; Volume 62, pp. 21–25.

233. Dos Santos, N.R.; Beck, A.; Fontbonne, A. The View of the French Dog Breeders in Relation to Female Reproduction, Maternal Care and Stress during the Peripartum Period. *Animals* **2020**, *10*, 159. [CrossRef]

234. Ercan, N.; Kockaya, M.; Kapancik, S.; Bakir, D. Determination of Serum Adenosine Deaminase and Xanthine Oxidase Activity in Kangal Dogs with Maternal Cannibalism. *Vet. World* **2017**, *10*, 1343–1346. [CrossRef]

235. Kockaya, M.; Ercan, N.; Demirbas, Y.S.; Da Graça Pereira, G. Serum Oxytocin and Lipid Levels of Dogs with Maternal Cannibalism. *J. Vet. Behav.* **2018**, *27*, 23–26. [CrossRef]

236. Meek, P.D.; Brown, S.C. It's a Dog Eat Dog World: Observations of Dingo (*Canis familiaris*) Cannibalism. *Aust. Mammal.* **2017**, *39*, 92. [CrossRef]

237. Goddard, P.; Waterhouse, T.; Dwyer, C.; Stott, A. The Perception of the Welfare of Sheep in Extensive Systems. *Small Rumin. Res.* **2006**, *62*, 215–225. [CrossRef]
238. Dwyer, C.M. Genetic and physiological determinants of maternal behavior and lamb survival: Implications for low-input sheep management. *J. Anim. Sci.* **2008**, *86*, E246–E258. [CrossRef] [PubMed]
239. Swanson, L.J.; Campbell, C.S. Maternal Behavior in the Primiparous and Multiparous Golden Hamster1. *Z. Tierpsychol.* **2010**, *50*, 96–104. [CrossRef]
240. Kemps, A.; Timmermans, P. Effects of Social Rearing Conditions and Partus Experience on Periparturitional Behaviour in Java-Macaques (*Macaca fascicularis*). *Behaviour* **1984**, *88*, 200–214.
241. Jacobs, B.S.; Moss, H.A. Birth Order and Sex of Sibling as Determinants of Mother-Infant Interaction. *Child Dev.* **1976**, *47*, 315–322. [CrossRef] [PubMed]
242. Guardini, G.; Bowen, J.; Raviglione, S.; Farina, R.; Gazzano, A. Maternal Behaviour in Domestic Dogs: A Comparison between Primiparous and Multiparous Dogs. *Behaviour* **2015**, *1*, 23–33. [CrossRef]
243. Oczak, M.; Maschat, K.; Berckmans, D.; Vranken, E.; Baumgartner, J. Classification of Nest-Building Behaviour in Non-Crated Farrowing Sows on the Basis of Accelerometer Data. *Biosyst. Eng.* **2015**, *140*, 48–58. [CrossRef]
244. Koçkaya, M.; Demirbas, Y.S. The Use of Carbetocin in the Treatment of Maternal Cannibalism in Dogs. *J. Vet. Behav.* **2020**, *40*, 98–102. [CrossRef]

Disclaimer/Publisher's Note: The statements, opinions and data contained in all publications are solely those of the individual author(s) and contributor(s) and not of MDPI and/or the editor(s). MDPI and/or the editor(s) disclaim responsibility for any injury to people or property resulting from any ideas, methods, instructions or products referred to in the content.

Review

The Effect of Maternal Behavior around Calving on Reproduction and Wellbeing of Zebu Type Cows and Calves

Agustín Orihuela [1,*] and Carlos S. Galina [2]

[1] Facultad de Ciencias Agropecuarias, Universidad Autónoma del Estado de Morelos, Cuernavaca 62209, Mexico

[2] Departamento de Reproducción, Facultad de Medicina Veterinaria y Zootecnia, Universidad Nacional Autónoma de México, Ciudad Universitaria, Ciudad de México 04510, Mexico; cgalina@unam.mx

* Correspondence: aorihuela@uaem.mx

Simple Summary: After a brief introduction establishing the importance of parturition and the establishment of the mother-young bond, this review summarizes knowledge on the maternal behavior of Zebu type cows, including maternal and young behavioral patterns, the protective behavior of the mother and how maternal behavior might be influenced by experience and weather conditions. In addition, the effect of some nutrition, heat stress, calf presence around parturition and its effects on cattle reproductive performance, welfare issues on reproductive behavior and calf performance are reviewed. Finally, some conclusions and practical implications are established.

Abstract: The behaviors associated with domestic cattle such as maternal care are quite similar to those behaviors observed in wild ungulates. These behaviors allow the cow to bond with her calf, protect and provide it with nourishment and ultimately reduce the bond at weaning. Although maternal behavior is an important factor influencing the survival and early development of the newborn calf, Zebu type cows around calving have not been studied extensively. Herein, we consider the four main aspects of maternal behavior in cattle and particularly *Bos indicus* cows and calves. Firstly, we provide a brief description of the behavior of cows around parturition and the behavior of the first stages of the calves' lives. In the second part, the protective behavior of the mother is analyzed. Subsequently, examples of animal welfare implications followed by an analysis of some factors that affect calf survival, including mother experience and weather conditions, are discussed, and in the last part, reproduction along with some peculiarities of reproductive behavior, and the wellbeing of mother and calves are examined. We concluded that knowledge of maternal behavior of Zebu type cows around calving and interactions with calves might contribute to an enhanced reproductive efficiency of the mother and the welfare of the calf.

Keywords: welfare; cow-calf bond; peripartum; *Bos indicus*; tropics

Citation: Orihuela, A.; Galina, C.S. The Effect of Maternal Behavior around Calving on Reproduction and Wellbeing of Zebu Type Cows and Calves. *Animals* **2021**, *11*, 3164. https://doi.org/10.3390/ani11113164

Academic Editor: David W. Miller

Received: 30 September 2021
Accepted: 2 November 2021
Published: 5 November 2021

Publisher's Note: MDPI stays neutral with regard to jurisdictional claims in published maps and institutional affiliations.

Copyright: © 2021 by the authors. Licensee MDPI, Basel, Switzerland. This article is an open access article distributed under the terms and conditions of the Creative Commons Attribution (CC BY) license (https://creativecommons.org/licenses/by/4.0/).

1. Introduction

Once a Zebu cow becomes pregnant, she carries, in most of the cases, a single calf for about 283 days. Male calves are carried for 4 to 5 days longer, resulting in a higher birth weight. It has been estimated that each day longer in gestation results in a 1.4 kg increase in the birth weight [1].

Calves stand and walk shortly after birth, and cows nurse their offspring for about six to eight months, with other members of the herd playing a critical role in the protection of the calves. However, the attention for the calves is provided primarily by their mothers [2]. Parental care entails how much energy is necessary to invest in a neonate at the cost of the reserves of the cow's resources for her calf [3], besides for her own reproductive performance. The cow after parturition is the most important social individual for the calf, providing protection against predators, food, warmth, shelter and immunological defense [4,5]; thus, important information concerning the physical and social environment is essential for the calf´s survival. At the same time, the presence and nursing behavior of the calf coupled with other environmental conditions including the body reserves of the mother, affect the interval from parturition to the resumption of the ovarian activity.

2. Maternal Behavior

2.1. Behavioral Patterns

When cattle are kept free under extensive conditions, it has been observed that there is a certain organization among the members of the herd according to their physiological state. Pregnant cows tend to form more cohesive groups that walk greater distances, away from calves and other cows [6].

Pregnant cows usually isolate themselves from the herd when near calving [7]. Arching of the back and an elevated tail occur for 1 to 3 h before the chorioallantoic membranes rupture. During this period, many cows licked or kicked their flanks. Some of the signs of imminent parturition are relaxation of the sacrosciatic ligament, slackening of the tissue of the perineum and vulva, distention of the udder and teats and mucous discharge from the vulva.

Labor onset occurs about 2.5 h before calving. During the birthing process, it is common for pregnant cows to lie down and stand frequently [8]. The identification of this type of behavior may have practical relevance, since the risk of a calf dying at birth is higher when the cow calves in a standing position (16%) compared to lying down (4.2%) [9]. This feature is particularly important in cattle calving at pasture. When membranes rupture, the cow often licks the fluid and tends to stay near the spot, now attractive to the cow, where the fluid fell. The discharge is followed by the appearance of the calf's feet. The calf is completely delivered during a period of 45 to 290 min, and time is often used as a measure of labor difficulty. Most calves are delivered in an extended posture, dorsal position and anterior presentation. Zebu calves weigh an average of about 25 kg, with variations between breeds.

Placenta expulsion can take place immediately after parturition or up to five hours later, with a tendency to involve more time in male calves. The placenta weights between 2 and 4.5 kg with 41 to 112 placentomes per placenta [1,10]. The placenta is eaten by about 82% of cows [11].

The newborn calf shakes its head, snuffles and sneezes. This behavior may begin during parturition as soon as the calf´s shoulders are free of the mother´s vulva. Some calves will remain motionless for up to 30 min after birth, but within an hour most calves can stand. It may take 30 min to an hour before the teats are located, and the cow´s conformation may not provide the higher recess that the calf appears to seek. Most calves suckle within 3 h (Figure 1).

Some previous studies described the occurrence of allosuckling in cattle, ranging from 3.0 [12] to 19.02% of the total suckling events [13], but it seems to be rare in Zebu cattle [14]. In general, ungulates have been classified as hiders and followers [15]. Bovines, including *Bos indicus* cows, are species where the young remain at or near the birth site for 3 to 7 days after parturition, whilst the mother may be away for long periods of time [16]. The young of this kind of animal can afford to have a low degree of social responsiveness, since the mothers actively seek them out at feeding times [17]. After this hiding period, Orihuela et al. [18] demonstrated that newborn Zebu calves frequently form groups with one or more cows. These researchers observed 142 groups, where most of them were formed by 1 to 3 cows and 2 to 32 calves between 4 and 32 days old. This kind of communal rearing may be a strategy developed to allow different cows to move for better access to feeding while their calves remain in larger groups and are cared by other adult cows [19]. Additionally, this feature indirectly improves maternal energy budgets [20]. This may also be an anti-predatory behavior [21,22], facilitating calves' socialization [23]. These communal groups of several calves with some cows might be favored by the fact that cows group according to their physiological state, affecting the way herd members interact with each other [6].

Similarly, Reinhardt and Reinhardt [24] studied the cohesive relationships in a *B. indicus* cattle herd over a 3–5-year period and concluded that the social structure is based on matriarchal families that are interconnected by means of friendship relationships between non-kin partners. In addition, Enriquez et al. [25] demonstrated that 33% of 25-day postpartum Brahman cows, during a maternal protective test, reacted when their calves or calves from other cows individually walked freely for 30 s in front of the continuous pen where the cows were held.

Figure 1. Postpartum cows allow their calves to suckle, which they identify through opposing signals favored by the calf's normal suckling position.

In cattle, the suckling bout lasts for some time and includes additional communication. Before milk ejection, there is about one minute of suckling and butting; afterwards, there is a period of several minutes of continued nursing from different teats and some butting, that might partly be a way for the calf to stimulate further milk production [26], and thereby, communicate nutritive needs to the dam [17]. Particularly in Zebu type cattle, it has been demonstrated that the tactile stimulation provided by the calf improves milk ejection and milk yield [27]. Some early studies showed that in Zebu breeds of cattle, milk ejection cannot be activated in the absence of a suckling calf [27] and, if the calf dies, blowing in the cow´s vagina [28] or presenting a dummy of a calf´s head [29] is necessary to obtain milk. For this reason, many herdsmen of Zebu cattle allow the calf to suckle first, to induce the milk let-down reflex [30].

2.2. Protective Behavior

In general, recently calved cows are considered dangerous and aggressive animals, especially *B. indicus* breeds [31,32], where the maternal aggression of mothers protecting their newborns is considered one of the main causes of accidents with cattle [33–35]. It has been proposed that this may be a consequence of an increase in their maternal protective behavior or a modification of their temperament triggering nervous responses, as well as the maintenance of an alertness attitude [4]. Orihuela et al. [36] found that the temperament of the cows did not affect their reaction to people handling their calves during the first days postpartum. This finding suggests that temperament measures cannot be used to predict the aggressiveness of the cows; however, there seems to be a genetic reason that explains the difference in aggressiveness at parturition within several beef cattle breeds [37].

Aggressive cows can be a high risk to handlers. Sometimes it is necessary to approach cows during parturition to intervene in some dystocia problems, and in general, newborn calves require care soon after delivery to treat navels and apply some identification procedure, which is essential to ensure good health and reduce the mortality rate of the newborns. It is during these last approaches that some cows, and in general those that give birth for the first time, might become nervous and sometimes show aggressive behavior towards the handlers. This reaction could be interpreted as a natural response to a potential predator that threatens her young. Aggressive cows can be a real problem. Costa [38] evaluated 4385 Nelore cows, finding this behavior in 13.2% of the cases, where 11.4% threatened and 1.8% attacked their handlers, which can be potentially risky in poor installations when working under extensive conditions. Negative interactions between animals and handlers lead to situations of stress and fear, increasing the risk of aggressions. Thus, de Oliveira Costa et al. [39] propose training handlers to improve the human-animal relationships, the use of adequate facilities and the slaughter of dangerous animals to solve this problem.

The possible presence of predators at the parturition place also affects the cow–calf relationship. Toledo [40] has given special attention to species such as the black vulture (*Coragyps atratus*) because in some countries the presence of these animals increases around the calving sites, interfering with the normal establishment of the cow–calf bond, in addition to direct attacks on the calves. The main effect that has been caused is a reduction in the time that the mother spends in contact with her offspring. This ends up negatively affecting the difficulties that newborns might have to stand up and suckle [40].

In cattle production systems, handling the calf at an early age is necessary for routine management procedures [41]. It is possible that cattle perceive humans as potential predators, and thus, cows may exacerbate their reactions [42].

2.3. Mother Experience and Weather Conditions

Some of the factors that affect the position of the cow at calving are the lack of calving experience, and the presence of potential predators around the calving site [7,40]. Episodes of aggressive behavior toward the newborn calf are also important. In accordance with Schmidek et al. [43], these events are more frequent in primiparous than multiparous cows (55.7 and 22.0%, respectively), resulting in a higher percentage of calves that do not achieve their first suckling within the first 3 h postpartum (15.7 and 5.7%, respectively) and in greater latencies for a first successful suckling (102.6 and 76.0 min, respectively). The risk of mortality in the neonate also increases with the delay of a first postpartum suckling for more than 3 h [43]. Weather conditions [44], calf weight at birth, udder shape and teat size [43] are other factors that could also result in a failure or delay of the first postpartum suckling. Up to a third of calves may not suckle within 6 h of birth. This is particularly apt to be the case when the cow has a pendulous udder or thicker teats, such as the Indobrasil breed [45,46].

In research carried out with 1527 calves born between 1992 and 2004, it was found that 279 (18.3%) failed to achieve their first suckling and required assistance to be able to suckle after 6 h of being born. Hanging udders and large teats, as well as calves weighing less than 25 kg at birth, were the two factors identified as the main causes of failure to achieve the first postpartum suckling [47]. Therefore, in practice, the conformation of the udder of the cows and the weight at birth of the calves become possible indicators that these animals will require help for the survival of the calf.

As mentioned above, Zebu calves congregate in newborn groups that also integrate at least one adult cow, preferably higher parity cows [18]. Older cows have more experience in calf care, which could influence some alloparent behavior, along with other physiological differences such as increased milk production and a higher number of antibodies in their colostrum than younger mothers [48].

In general, younger parents seem to have a higher residual fitness cost for reproduction and a poorer probability of raising young successfully and tend to show less parental care [49]. In a recent study, Enriquez et al. [25] found that during a maternal protective test, cows with slight follicular development reacted to a greater number of calves displaying a higher intensity of reaction. In addition, these cows did not display estrous behavior, suggesting a relationship between protective maternal behavior and some reproductive variables in Zebu-type cows (*B. indicus*). This seems to be in accord with the theory of terminal investment, where animals with a favorable potential to reproduce, would do so [49].

3. Reproductive Performance

Rust [50] suggested that extensive production systems will be under considerable pressure because of land availability and predicted climatic changes. Thus, there is a need to intensify production systems in the cattle industry [51]. This scenario has moved producers and scientists alike to find methods to change the traditional strategy, to leave the cows at pasture and collect the offspring at defined periods. Consequently, the economic and reproductive efficiency of the cows are influenced by the survival of their calves, as their livelihood and performance are the main source of income for beef cattle producers [52]. Knowledge about maternal and filial behavior is important to identify those situations that could increase the risk of weak, abandoned or stillborn newborns, since it offers the opportunity to reduce these types of problems, thereby reducing economic losses [47,53].

Orihuela and Galina [54] concluded that any method of calf separation should consider factors such as the time when the event is performed, the nutritional status of the cow and the effect of stress on the dam and the offspring. Breaking the cow–calf bond in the different methods utilized need to be reviewed, not only from the reproductive viewpoint but also from the welfare of the animals [55].

3.1. Nutrition and Onset of Ovarian Activity

Animals raised under pasture conditions depend on adequate fodder to gain weight after calving and have enough milk for their young [52]. This is not an easy feat considering that animals tend to calve in the spring, usually prior to the rainy season, where the fodder available in the last trimester of pregnancy could affect the onset of ovarian activity [56,57]. In effect, in a comparative study between two locations in the tropics of México and Costa Rica where the onset of the rainy season could vary, it was found that the pattern of body loss in the last trimester of gestation was directly related to the onset of the rainy season [58]. In fact, in a further experiment, when two groups of cows with different postpartum periods were subjected to a synchronization program, monitoring body fat change via ultrasound, it was found that the driving force for a prompt restoration of ovarian activity and pregnancy were the changes in body reserves rather than the time postpartum [56,57]. These examples illustrate the need for farmers and technicians alike to join resources to avoid body loss during the last trimester of gestation catering for the wellbeing of the dam and later for their offspring. This principle is hardly dependent on having an adequate use of the pasture available to avoid an unfair competition between animals of different weights and calving. First calf heifers are usually at a higher risk [59,60]. Burns et al. [61], in an extensive review about factors affecting the reproductive efficiency of beef cows in Northern Australia, underline the importance of nutrition as a key player for a prompt restoration of ovarian activity. A failure to meet nutritional requirements would cause, in those cows, calving intervals longer than 15 months.

There are alternatives to improve pasture availability and animal welfare in the tropics. For example, the use of native trees and shrubs [62] facilitating the shading area and soil restoration; therefore, pastures will be available and not dried out to direct solar radiation for longer periods of time [63].

Amendola et al. [64] performed an interesting experiment comparing a group of heifers grazing in a silvopastoral scheme as opposed to a monoculture system during two distinct seasons of the year in the dry tropics of Mexico, concluding that the hierarchy in the silvopastoral system was more constant and stable between seasons compared to the monoculture system. The same group [65] compared the effect of improved pastures and shaded areas in the hottest months of the year, as opposed to the traditional system of monoculture, suggesting that forage availability and access to shade allow cattle to have longer periods of rest, permitting a better rumination and general performance. These experiments were carried out in heifers; therefore, how such factors affect mother and offspring relationships remains to be elucidated.

This is particularly important in dams recently calving. The stress produced by parturition, suckling of the newborn, plus the need to search for a calving site and protect their calves from predators, are monumental tasks for the dam, still needing to search for adequate pastures to sustain their body metabolism.

A different story is the situation of cattle known as dual purpose, which is estimated to be present in 70% of the livestock living in the tropical and sub-tropical regions of the world [66]. This system started with crossbreeding Holstein or Brown Swiss with Zebu breeds, mainly Brahman, Guzerat and Gyr plus other local undefined Zebu type of cattle. This mélange of breeds has caused difficulties in characterizing their reproductive capacity, as the feeding regimens consisted of local or improved grasslands and various combinations of feeding supplementation [67].

This system is popular due to several factors related to their adaptation to the harsh environmental conditions in the tropics. There is less capital investment and technical support [66]; thus, it is not surprising that their level of technology adoption is so varied and heavily dependent on their location [68]. All these factors make it quite complicated to draw conclusions to the wellbeing of the calf and dam either by judging their nutritional requirements or reproductive efficiency.

3.2. Heat Stress

Breaking the mother–offspring bond can affect the welfare of both. Although, there is abundant information on the different methods to separate the calves from their dams (for an early review see [69]). In comparison, few control studies are available on the stress and their ultimate welfare. It has been shown that rising levels of cortisol have an important effect on the reproductive performance of the dams [70]. Additionally, raising animals under extreme heat environments affects their reproductive performance, influencing the wellbeing of livestock in general. Bernabucci et al. [71] have postulated that heat stress is caused when the body temperature of an animal is beyond its normal range, creating a difficulty in heat dissipation and finally reducing physiological as well as behavioral responses. Bearing in mind that beef cattle in the tropics are usually exposed to a high number of hours, and in many enterprises, without shelters for the animals, Diaz et al. [72] studied the effect of higher temperature indexes (THI) and found a negative effect on the resumption of ovarian activity, especially if these higher THI occurred during the last trimester of pregnancy. This study also underlined the importance of some animals being able to restore ovarian activity despite a hostile environment. How do high temperatures affect the performance of cattle during calf separation? A question in need of answers.

What is certainly known, is that *B. indicus* cattle are more resistant to high temperatures than *B. taurus*. Beatty et al. [73] suggested that *B. taurus* breeds have less resistance to heat stress, a subsequent reduced feed intake and raised plasma non-esterified fatty acids compared to *B. indicus*. The ability to tolerate heat in *B. indicus* is manifested in their reproductive ability to raise follicles capable of fertilization [74]. Even more, Blackshaw and Blakshaw [75], in a detailed review, point out that the advantages of *B. indicus* cattle over *B. taurus* are related to physiological events, such as the ability to disperse heat due to the coat color of their skin, food and water intake and sweating intensity. Most of these research communications are related to a comparison between *B. taurus* and *B. indicus*. It would be commendable to compare different breeds of *B. indicus* to find out if effectively all the breeds are similar in their behavior, particularly if the management system relies on allowing the calf to suckle at will.

3.3. Presence of the Calf

There seems to be a common agreement that suckling and the presence of the calf impaired the adequate growth of a follicle capable of producing a viable zygote [76]. This shortcoming is even more evident when the animals are under stress [77]. Nonetheless, calf separation remains the single most important tool to restore ovarian activity after calving [78] (Figure 2).

Figure 2. Two lactating cows display sexual behavior. The cow in the bottom shows sexual receptivity even while nursing. Reproductive activity in Zebu cows can be induced after a period of separation between the mother and her calf, inhibiting the negative effect of suckling and the presence of the offspring on the resumption of postpartum ovarian activity.

Whatever the method used to break the bond between the dam and the offspring, it can be strengthened with the use of hormonal therapies to promote follicular growth [79]. Mondragon et al. [80], in a detailed study of the follicular dynamics comparing animals under a regimen of restricted suckling (RS) with the continuous presence of the calf, observed that cows in the RS treatment had a larger follicle than the control cows. The authors concluded that interrupting suckling will favor the development of healthier follicles. Furthermore, Álvarez-Rodriguez et al. [81] found that the management practices limiting suckling must also avoid a close cow–calf association to reduce long postpartum intervals to first ovulation.

A different issue in need of research relates to the diverse systems utilized in dual purpose cattle, starting from having the calf present, allowing the milk let down from the cow, to allowing the calf to suckle one teat or other combinations. In recent years, the introduction of more elaborate procedures, such as extracting the milk without the presence of the calf either by hand or machine are more common. No matter what system is used, it is not surprising to find calving intervals wider than 14 months in the literature [82,83]. More research is obviously in demand for human interventions [84]. One of these procedures is related to the beneficial effect of calf separation. Recent evidence suggests that neither calf separation nor hormonal therapies are strictly necessary to restore ovarian activity, since techniques such as restricted suckling have proven to be effective [85].

3.4. Sexual Behavior

The sexual behavior of Zebu cows offers certain specific characteristics, which must be taken into account when conducting heat detection programs, since on many occasions the manifestation of external signs of estrus can be affected by many factors [86,87]. Some of these factors are the complex social order and dominance characteristics in cows. For example, not all heifers in a herd mount cows in estrus [88]. Orihuela et al. [89] found that 60% of the mounts among females were performed by a heavier and larger cow than the one being mounted. Similarly, when evaluating a herd composed of Charolais and Brahman cows, it was observed that most of the mounts were carried out by Charolais cows on Brahman cows, which were comparatively larger and heavier than the latter, while Brahman cows rarely rode Charolais cows [90]. On the other hand, the larger, dominant cows inhibit the riding activity of the smaller, subordinate ones, sometimes even of young or inexperienced males [91]. However, a correlation between the dominance rank of the females in a herd and the number of mounts given or received, has not been demonstrated [92]. Other factors that affect the manifestation of estrus are breed, age, weight, presence of horns and the number of cows in the sexually active group [86,93]. Sexual receptivity, or allowing oneself to ride, remains the most unequivocal sign of estrus, and homosexual interactions are very common in Zebu cattle, coupled with the fact that mounting between Zebu females occurs almost exclusively when both animals are in estrus [94].

When the estrus of Zebu cows is synchronized artificially, either through the use of drugs or by separating the calf from her mother, or both, a high percentage of animals participate in sexually active groups, leading to a greater number of animals in estrus, with longer and more intense estruses, due to social facilitation and sexual stimulation. Orihuela et al. [95] found that by staying together, about 80% of the unsynchronized cattle showed signs of heat similar to those of their progestin-treated herd mates. Similar studies have programmed the induction of estrus in Zebu cows so that they display behavioral signs of estrus, one after another, every third day. In response, the cows grouped their riding behavior on the days for which they were scheduled [96]. Natural heat synchronization appears to be common in Zebu cattle against artificially synchronized herd mates [97]. Medrano et al. [96] found that 85% of the mounts detected after an estrous synchronization program are received and given by cows in estrus, although, due to imitation behavior, some of them exhibit active mounts in the absence of follicles capable of ovulating, affecting the final results of artificial insemination. The type of heat could be affecting the duration of estrus. However, researchers generally agree that the duration of estrus in *B. indicus* cattle is shorter than in *B. taurus* cattle [98,99].

Handling can also affect when estrus is displayed. Vaca et al. [100] kept Zebu cattle reared under extensive conditions in a pen to facilitate the detection of estrus and found that 50% of the animals did not show estrous signs until they were released to the pasture again. Likewise, during strong tropical storms, the Zebu cow can suspend the manifestation of sexual behavior.

Environmental temperature is also a factor, and in general, *B. indicus* cattle show sexual behavior more frequently during the hot summer months [101], compared to European cattle that manifest more in cold climates [102]. However, although Zebu cattle are better adapted to the heat, most of the sexual activity seems to take place during the cooler hours of the night [89,95].

The presence of the male is another factor that affects the sexual behavior of the Zebu cow. The bull must be dominant enough to prevent cow–cow riding as well as younger bulls mounting. When a bull is used in breeding cows with synchronized estrus, the sexual behavior is grouped in a shorter period and the proportion of estruses with a shorter duration increases than in groups without a bull [95]. In addition, the male affects the start time of estrus [103].

4. Conclusions

Parturition is a very important moment, a determinant in the young´s life through the establishment of the cow–calf bond, which allows food and protection from the mother. In addition, the reproduction performance of the cow is also impacted at this time, as the period between parturition and the restart of ovarian activity is affected by the presence of the calf, the suckling events and the reserves of the cow, among others. Both the life of the calves and the reproductive activity of the cows are factors that can mean considerable economic losses for the producer. The knowledge of how the behavior and physiology of the animals affect these aspects is very important, without neglecting the aspects of wellbeing, both in the mother and in the offspring.

Author Contributions: A.O. and C.S.G. contributed equally to this article. All authors have read and agreed to the published version of the manuscript.

Funding: This research received no external funding.

Institutional Review Board Statement: Not applicable.

Data Availability Statement: Not applicable.

Conflicts of Interest: The authors declare no conflict of interest.

References

1. Mukasa-Mugerwa, E.; Mattoni, M. Parturient behaviour and placental characteristics of *Bos indicus* cows. *Rev. d'Élevage Médecine Vétérinaire Pays Trop.* **1990**, *43*, 105–109.
2. Zilkha, N.; Sofer, Y.; Beny, Y.; Kimchi, T. From classic ethology to modern neuroethology: Overcoming the three biases in social behavior research. *Curr. Opin. Neurobiol.* **2016**, *38*, 96–108. [PubMed]
3. Almansa, J. *La Evolución y la Conservación de la Biodiversidad*; Universidad de Extremadura: Extremadura, Spain, 2012.
4. McGee, M.; Earley, B. Review: Passive immunity in beef-suckler calves. *Animal* **2019**, *13*, 810–825. [CrossRef] [PubMed]
5. Kour, H.; Corbet, N.J.; Patison, K.P.; Swain, D.L. Changes in the suckling behaviour of beef calves at 1 month and 4 months of age and effect on cow production variables. *Appl. Anim. Behav. Sci.* **2021**, *236*, 105219.
6. Pérez-Torres, L.I.; Averós, X.; Esteves, I.; Rubio, I.; Corro, M.; Clemente, N.; Orihuela, A. Herd dynamics and distribution of Zebu cows and calves (*Bos indicus*) are affected by state before, during and after calving. *Appl. Anim. Behav. Sci.* **2020**, *229*, 105013.
7. Paranhos da Costa, M.J.R.; Schmidek, A.; Toledo, L.M. Mother-offspring interactions in Zebu cattle. *Reprod. Dom. Anim.* **2008**, *43*, 213–216. [CrossRef]
8. Paranhos da Costa, M.J.R.; Cromberg, V.U. Relações materno-filiais em bovinos de corte nas primeiras horas após o parto. In *Comportamento Materno em Mamíferos: Bases Teóricas e Aplicações aos Ruminantes Domésticos*; Sociedade Brasileira de Etologia: São Paulo, Brazil, 1998.
9. Paranhos da Costa, M.J.R.; Schmidek, A.; Toledo, L.M. *Boas Práticas de Manejo: Bezerros ao Nascimento*; Editora Funep: Jaboticabal, SP, Brazil, 2006.
10. Rao, C.H.C.; Rao, A.R. Foetal membranes of Ongole and crossbred cows. *Indian J. Anim. Sci.* **1980**, *50*, 953–956.
11. Houpt, K.A. *Domestic Animal Behavior for Veterinarians and Animal Scientists*; Iowa State University Press: Ames IW, USA, 1998.
12. Das, S.M.; Redbo, I.; Wiktorsson, H. Effect of age of calf on suckling behaviour and other behavioural activities of Zebu and crossbred calves during restricted suckling periods. *Appl. Anim. Behav. Sci.* **2000**, *67*, 47–57. [CrossRef]
13. Víchová, J.; Bartoš, L. Allosuckling in cattle: Gain or compensation? *Appl. Anim. Behav. Sci.* **2005**, *94*, 223–235. [CrossRef]
14. Paranhos da Costa, M.J.R.; Albuquerque, L.G.; Eler, J.P.; de Silva, J.Á.V., II. Suckling behaviour of Nelore, Gir and Caracu calves and their crosses. *Appl. Anim. Behav. Sci.* **2006**, *101*, 276–287.
15. Lent, P.C. Mother-infant relationships in ungulates. In *The Behaviour of Ungulates and its Relationship to Management*; New Series, 24; Geist, U., Walther, F., Eds.; IUCN Publisher: Morges, Switzerland, 1974; pp. 14–55.

16. Reinhardt, V. Social behaviour of *Bos indicus*. *Rev. Rural Sci.* **1980**, *4*, 143–146.

17. Jensen, P. Parental Behavior. In *Social Behaviour in Farm Animals*; Keeling, L.J., Gonyou, H.W., Eds.; CABI Publishing: Saskatoon, SK, Canada, 2001.

18. Orihuela, A.; Pérez-Torres, L.; Ungerfeld, R. Evidence of cooperative calves´ care and providers characteristics in Zebu cattle (*Bos indicus*) raised under extensive conditions. *Trop. Anim. Health Prod.* **2021**, *53*, 143–148. [PubMed]

19. Riedman, M.L. The evolution of alloparental care and adoption in mammals and birds. *Q. Rev. Biol.* **1982**, *57*, 405–435. [CrossRef]

20. Rosenbaum, S.; Gettler, L.T. With a little help from her friends (and family) part II: Non-maternal caregiving behavior and physiology in mammals. *Physiol. Behav.* **2018**, *193*, 12–24.

21. Caro, T.M. *Antipredator Defences in Birds and Mammals*; University of Chicago Press: Chicago, IL, USA, 2005.

22. Von Keyserlingk, M.A.G.; Weary, D.M. Maternal behavior in cattle. *Horm. Behav.* **2007**, *52*, 106–113. [CrossRef]

23. Bouissou, M.F.; Boissy, A.; Le Neindre, P.; Veissier, I. The social behavior of domestic species. In *Social Behaviour in Farm Animals*; Keeling, L.J., Gonyou, H.W., Eds.; CABI Publishing: Skara, Sweeden, 2001; pp. 113–145.

24. Reinhardt, V.; Reinhardt, A. Cohesive relationships in a cattle herd (*Bos indicus*). *Behaviour* **1981**, *77*, 121–151. [CrossRef]

25. Enríquez, M.F.; Pérez-Torres, L.I.; Orihuela, A.; Rubio, I.; Corro, M.; Galina, C.S. Relationship between protective maternal behavior and some reproductive variables in Zebu-type cows (*Bos indicus*). *J. Anim. Behav. Biometeorol.* **2021**, *9*, 2124.

26. Lidfords, L.; Jensen, P.; Algers, B. Suckling in free-ranging beef cattle -temporal patterning of suckling bouts and effects of age and sex. *Ethology* **1994**, *98*, 321–332. [CrossRef]

27. Orihuela, A. Effect of calf stimulus on the milk yield of Zebu type cattle. *Appl. Anim. Bahav. Sci.* **1990**, *26*, 187–190. [CrossRef]

28. Debackere, M.; Peters, G. The influence on vaginal distension in milk ejection diuresis in the lactating cow. *Arch. Int. Pharmacodyn.* **1960**, *123*, 462–471.

29. Folley, S. The milk ejection reflex. A neuroendocrine theme in biology, myth, and art. *Proc. Soc. Endocrinol.* **1969**, *44*, 476–490. [CrossRef]

30. Zamora, S.F. *Estudio de Factores de Manejo que Afectan la Producción de Leche en el Trópico*; Facultad de Medicina Veterinaria y Zootecnia de la Universidad Nacional Autónoma de México: Mexico City, Mexico, 1981.

31. Hearnshaw, H.; Morris, C.A. Genetic and environmental effects of temperament score in beef cattle. *Aust. J. Agric. Res.* **1984**, *35*, 723–733. [CrossRef]

32. Fordyce, G.; Goddard, M.E.; Seifert, G.W. The measurement of temperament in cattle and the effect of experience and genotype. *Proc. Aust. Soc. Anim. Prod.* **1982**, *14*, 329–333.

33. Grandin, T. Safe handling of large animals. *Occup. Med.* **1999**, *14*, 195–212. [PubMed]

34. Le Neindre, P.; Gringnard, L.; Trillat, G.; Boissy, A.; Ménissier, F.; Sapa, F.; Boivin, X. Docile Limousin cows are not poor mothers. In Proceedings of the 7th World Congress on Genetics Applied to Livestock Production, Montpellier, France, 19–23 August 2002; pp. 59–62.

35. De Oliveira, C.F.; Valente, T.; Paranhos da Costa, M.J.R.; del Campo, M. Expression of maternal protective behavior in cattle: A review. *Rev. Acad. Cienc. Anim.* **2018**, *16*, e161106.

36. Orihuela, A.; Pérez-Torres, L.I.; Ungerfeld, R. The time relative to parturition does not affect the behavioral or aggressive reactions in Zebu cows (*Bos indicus*). *Livest. Sci.* **2020**, *234*, 103978. [CrossRef]

37. Buddenberg, B.J.; Brown, C.J.; Johnson, Z.B.; Honea, R.S. Maternal behavior of beef cows at parturition. *J. Anim. Sci.* **1986**, *62*, 42–46. [CrossRef]

38. Costa, F.O. Avaliaçao do Comportamento de Proteçao Materna de vacas Nelore e Hereford e seus Potenciais Efeitos no Ganho de Peso dos Bezerros. Ph.D. Thesis, Programa de Pos-Graduaçao em Zootecnia, Faculdade de Ciencias Agrarias e Veterinarias—UNESP, Jaboticabal, SP, Brazil, 2017; p. 111.

39. De Oliveira Costa, F.; Valente, T.S.; de Toledo, L.M.; Ambrósio, L.A.; del Campo, M.; Paranhos da Costa, M.J.R. A conceptual model of the human-animal relationships dynamics during newborn handling on cow-calf operation farms. *Livest. Sci.* **2021**, *246*, 104462. [CrossRef]

40. Toledo, L.M. Fatores Intervenientes no Comportamento de Vacas e Bezerros do Parto até a Primeira Mamada. Ph.D. Thesis, Faculdade de Ciências Agrárias e Veterinárias, Universidade Estadual Paulista, Jaboticabal, SP, Brazil, 2005.

41. Turner, S.P.; Lawrence, A.B. Relationship between maternal defensive aggression, fear of handling and other maternal care traits in beef cows. *Livest. Sci.* **2007**, *106*, 182–188. [CrossRef]

42. Pérez-Torres, L.; Orihuela, A.; Corro, M.; Rubio, I.; Cohen, A.; Galina, C.S. Maternal protective behavior of Zebu type cattle (*Bos indicus*) and its association with temperament. *J. Anim. Sci.* **2014**, *92*, 4694–4700. [CrossRef]

43. Schmidek, A.; Paranhos da Costa, M.J.R.; Mercadante, M.E.Z.; Toledo, L.M. The effect of newborn calves´ vigor in their mortality probability. In *Proceedings of the 40th International Congress of the International Society for Applied Ethology*; University of Bristol: Bristol, UK, 2006; p. 297.

44. Toledo, L.M.; Paranhos da Costa, M.J.R.; Titto, E.A.L.; Figueiredo, L.A.; Ablas, D.S. Impactos de variáveis climáticas na agilidade de bezerros Nelore neonatos. *Ciênc. Rural* **2007**, *37*, 1399–1404. [CrossRef]

45. Edwards, S.A. Factors affecting time to first suckling in dairy calves. *Anim. Prod.* **1982**, *34*, 339–346.

46. Ventrop, M.; Michanek, P. The importance of udder and teat conformation for teat seeking by the newborn calf. *J. Dairy Sci.* **1992**, *75*, 262–268. [CrossRef]

47. Schmidek, A.; Mercadante, M.E.Z.; Paranhos da Costa, M.J.R.; Figueiredo, L.A. Falhas na primeira mamada em um rebanho da raça Guzerá: Fatores predisponentes, reflexos na sobrevivência do bezerro e parâmetros genéticos. *Revista Brasileira de Zootecnia* **2008**, *37*, 998–1004. [CrossRef]

48. Allbright, J.L.; Arave, C.W. *The Behaviour of Cattle*; CAN International: New York, NY, USA, 1997.

49. Clutton-Brock, T.H. *The Evolution of Parental Care*; Princeton University Press: Princeton, NJ, USA, 1991.

50. Rust, J. The impact of climate change on extensive and intensive livestock production systems. *Anim. Front.* **2019**, *9*, 20–25. [CrossRef]

51. Seré, C.; Steinfeld, H. *World Livestock Production Systems. Current Status Issues and Trends*; FAO Animal Production and Health Paper; Food and Agriculture Organization (FAO): Rome, Italy, 1996; p. 127.

52. Diskin, M.G.; Kenny, D.A. Optimizing reproductive performance of beef cows and replacements heifers. *Animal* **2014**, *8*, 27–39. [CrossRef]

53. Cromberg, V.U.; Paranhos da Costa, M.J.R. Mamando logo, para crescer a receita. *Anualpec* **1997**, *97*, 215–217.

54. Orihuela, A.; Galina, C.S. Effects of separation of cows and calves on reproductive performance and animal welfare in tropical beef cattle. *Animals* **2019**, *9*, 223–236.

55. Orihuela, A. Animal welfare and sustainable animal production. *Adv. Anim. Biosci.* **2016**, *2*, 214–216.

56. Díaz, R.; Galina, C.S.; Rubio, I.; Corro, M.; Pablos, J.L.; Orihuela, A. Monitoring changes in back fat thickness and its effect on the restoration of ovarian activity and fertility in *Bos indicus* cows. *Reprod. Domest. Anim.* **2018**, *53*, 495–501. [CrossRef]

57. Díaz, R.F.; Galina, C.S.; Estrada, S.; Montiel, F.; Castillo, G.; Romero-Zúñiga, J.J. Variations in the temperature-humidity index and dorsal fat thickness during the last trimester of gestation and early postpartum periods affect fertility of *Bos indicus* cows in the tropics. *Vet. Med. Int.* **2018**, *2018*, 1–8. [CrossRef]

58. Díaz, R.; Galina, C.S.; Rubio, I.; Corro, M.; Pablos, J.L.; Rodríguez, A.; Orihuela, A. Resumption of ovarian function, the metabolic profile and body condition in Brahman cows (*Bos indicus*) is not affected by the combination of calf separation and progestogen treatment. *Anim. Reprod. Sci.* **2017**, *185*, 181–187. [PubMed]

59. Maquivar, M.; Galina, C.S.; Verduzco, A.; Galindo, J.; Molina, R.; Estrada, S.; Mendoza, M.G. Reproductive response in supplemented heifers in the humid tropics of Costa Rica. *Anim. Reprod. Sci.* **2006**, *93*, 16–23. [CrossRef] [PubMed]

60. Maquivar, M.; Galina, C.S. Factors Affecting the Readiness and Preparation of Replacement Heifers in Tropical Breeding Environments. *Reprod. Dom. Anim.* **2010**, *45*, 937–942. [CrossRef]

61. Burns, B.M.; Fordyce, G.; Holroyd, R.G. A review of factors that impact on the capacity of beef cattle females to conceive, maintain a pregnancy and wean a calf-Implications for reproductive efficiency in northern Australia. *Anim. Reprod. Sci.* **2010**, *122*, 1–22.

62. Murgueitio, E.; Calle, Z.; Uribe, F.; Calle, A.; Solorio, B. Native trees and shrubs for the productive rehabilitation of tropical cattle ranching lands. *For. Ecol. Manag.* **2011**, *261*, 1654–1663. [CrossRef]

63. Ainsworth, J.A.; Moe, S.R.; Skarpe, C. Pasture shade and farm management effects on cow productivity in the tropics. *Agric. Ecosyst. Environ.* **2012**, *155*, 105–110. [CrossRef]

64. Améndola, L.; Solorio, F.J.; Ku-Vera, J.C.; Améndola-Massiotti, R.D.; Zarza, H.; Galindo, F. Social behaviour of cattle in tropical silvopastoral and monoculture systems. *Animal* **2016**, *10*, 863–867. [CrossRef]

65. Améndola, L.; Solorio, F.J.; Ku-Vera, J.C.; Améndola-Massioti, R.D.; Zarza, H.; Mancera, H.F.; Galindo, F. A pilot study on the foraging behaviour of heifers in intensive silvopastoral and monoculture systems in the tropics. *Animal* **2018**, *13*, 606–616. [CrossRef] [PubMed]

66. Rangel, J.; Perea, J.; De-Pablos-Heredero, C.; Espinosa-García, J.A.; Mujica, P.T.; Feijoo, M.; Barba, C.; García, A. Structural and technological characterization of tropical smallholder farms of dual-purpose cattle in Mexico. *Animals* **2020**, *10*, 86. [CrossRef]

67. Cunningham, E.P.; Syrstad, O. *Cross-Breeding B. Indicus and B. taurus for Milk Production in the Tropics*; FAO Animal Production and Health Papers; Food and Agriculture Organization; United Nations: Rome, Italy, 1987; Volume 68.

68. Villarroel-Molina, O.; De-Pablos-Heredero, C.; Rangel, J.; Vitale, M.P.; García, A. Usefulness of Network Analysis to Characterize Technology Leaders in Small Dual-Purpose Cattle Farms in Mexico. *Sustainability* **2021**, *13*, 2291.

69. Galina, C.S.; Rubio, I.; Basurto, H.; Orihuela, A. Consequences of different suckling systems for reproductive activity and productivity of cattle in tropical conditions. *Appl. Anim. Behav. Sci.* **2001**, *72*, 255–262. [CrossRef]

70. Fernández-Novo, A.; Pérez-Garnelo, S.S.; Villagrán, A.; Pérez-Villalobos, N.; Astiz, A. The Effect of Stress on Reproduction and Reproductive Technologies in Beef Cattle—A Review. *Animals* **2020**, *10*, 2096.

71. Bernabucci, U.; Lacetera, N.; Baumgard, L.H.; Rhoads, R.P.; Ronchi, B.; Nardone, A. Metabolic and hormonal acclimation to heat stress in domesticated ruminants. *J. Anim. Sci.* **2010**, *4*, 1167–1183. [CrossRef] [PubMed]

72. Díaz, R.F.; Galina, C.S.; Aranda, E.M.; Aceves, L.A.; Gallegos, J.S.; Pablos, J.L. Effect of temperature—Humidity index on the onset of post- partum ovarian activity and reproductive behavior in *Bos indicus* cows. *Anim. Reprod.* **2020**, *17*, 1–11. [CrossRef]

73. Beatty, D.; Barnes, A.; Pethick, D.W.; Taylor, E.; Dunshea, F.R. *Bos indicus* cattle can maintain feed intake and fat reserves in response to heat stress better than *Bos taurus* cattle. *J. Anim. Feed Sci.* **2014**, *13*, 619–622. [CrossRef]

74. Torres-Júnior, J.R.S.; Pires, M.F.A.; De Sá, W.F.; Ferreira, A.M.; Viana, J.H.M.; Camargo, L.S.D.A.; Ramos, A.A.; Folhadella, I.M.; Polisseni, J.; De Freitas, C.; et al. Effect of maternal heat-stress on follicular growth and oocyte competence in *Bos indicus* cattle. *Theriogenology* **2008**, *69*, 155–166.

75. Blackshaw, J.K.; Blackshaw, A.W. Heat stress in cattle and the effect of shade on production and behaviour: A review. *Aust. J. Exp. Agric.* **1994**, *34*, 285–295. [CrossRef]

76. Xavier, E.; Galina, C.S.; Pimentel, C.; Fiala, S.; Maquivar, M. Calf presence and estrous response, ovarian follicular activity and the pattern of luteinizing hormone in postpartum *Bos indicus* cows. *Anim. Reprod.* **2018**, *15*, 1208–1213. [CrossRef]
77. Wolferson, D.; Roth, Z. Impaired reproduction in heat-stressed cattle: Basic and applied aspects. *Anim. Reprod. Sci.* **2000**, *60–61*, 535–547. [CrossRef]
78. Sanders, J.O.; Cartwright, T.C. A general cattle production systems model. I. Structure of the model. *Agric. Syst.* **1979**, *4*, 217–227. [CrossRef]
79. Bó, G.A.; Baruselli, P.S.; Mapletoft, R.J. Synchronization techniques to increase the utilization of artificial insemination in beef and dairy cattle. *Anim. Reprod.* **2013**, *10*, 137–142.
80. Mondragón, V.; Galina, C.S.; Rubio, I.; Corro, M.; Salmerón, F. Effect of restricted suckling on the onset of follicular dynamics and body condition score in Brahman cattle raised under tropical conditions. *Anim. Reprod. Sci.* **2016**, *167*, 89–95. [PubMed]
81. Álvarez-Rodríguez, J.; Revilla, R.; Sanz, A. Relationships between cow-calf sensory cues and the postpartum interval in suckler cows. *Red. Vet.* **2009**, *10*, 1–16.
82. Galina, C.S.; Arthur, G.H. Review on cattle reproduction in the tropics. Part 2. Parturition and calving intervals. *Anim. Breed. Abstr.* **1989**, *58*, 697–707.
83. Arroyo-Rebollar, R.; López-Villalobos, N.; García-Martínez, A.; Arriaga-Jordán, C.M.; Albarrán-Portillo, B. Relationship between calving interval and profitability of Brown Swiss cows in a subtropical region of Mexico. *Trop. Anim. Health Prod.* **2021**, *53*, 1–8. [CrossRef] [PubMed]
84. Placzek, M.; Christoph-Schulz, I.; Barth, K. Publica attitude towards cow-calf separation and other common practices of calf rearing in dairy farming—A review. *Org. Agric.* **2021**, *11*, 41–50.
85. Pérez-Torres, L.I.; Ortiz, F.; Martínez, J.F.; Orihuela, A.; Rubio, I.; Corro, M.; Galina, C.S.; Ungerfeld, R. Short and long-term effects of temporary early cow-calf separation or restricted suckling on well-being and performance in Zebu cattle. *Animal* **2021**, *15*, 100132.
86. Orihuela, A. Some factors affecting the behavioural manifestation of oestrus in cattle: A review. *Appl. Anim. Behav. Sci.* **2000**, *70*, 1–16. [CrossRef]
87. González-Tous, M.; Ruiz-Valencia, E.; Hoyos-García, L.A.; Prieto-Manrique, E.; Betancur-Hurtado, C. Sexual behavior of Brahan cows in a region of the colombian caribbean. *Orinoquia* **2010**, *14*, 168–177.
88. Hafez, E.S.E.; Bouissou, M.F. The behavior of cattle. In *The Behaviour of Domestic Animals*, 3rd ed.; Hafez, E.S.E., Ed.; Williams and Wilkins: Baltimore, MD, USA, 1975; pp. 203–245.
89. Orihuela, A.; Galina, C.S.; Duchateau, A. Behavioural patterns of Zebu bulls toward cows previously synchronized with PGF2α. *Appl. Anim. Behav. Sci.* **1988**, *21*, 267–276.
90. Galina, C.S.; Calderón, A.; Mcclosky, M. Detection of signs of oestrus in the Charolais cow and its Brahman cross under continuous observation. *Theriogenology* **1982**, *17*, 485–498. [PubMed]
91. Orihuela, A.; Polanco, A. El comportamiento sexual de las especies de interés zootécnico. *Rev. Chapingo* **1980**, *25*, 60–65.
92. Orihuela, A.; Galina, C.S. Social order measured in field and pen conditions and its relation with sexual behaviour in Brahman (*Bos indicus*) cows. *Appl. Anim. Behav. Sci.* **1997**, *52*, 3–11. [CrossRef]
93. Galina, C.S.; Orihuela, A.; Rubio, I. Behavioural trends affecting oestrus detection in Zebu cattle. *Anim. Reprod. Sci.* **1996**, *42*, 465–470. [CrossRef]
94. Galina, C.S.; Arthur, G.H. Review of cattle reproduction in the tropics. Part 4. Oestrous cycles. *Anim. Breed. Abstr.* **1990**, *58*, 697–707.
95. Orihuela, A.; Galina, C.S.; Escobar, J.; Riquelme, E. Oestrous behaviour following prostaglandin F2α injection in Zebu cattle under continuous observation. *Theriogenology* **1983**, *9*, 795–809. [CrossRef]
96. Medrano, E.A.; Hernández, O.; Lamothe, C.; Galina, C.S. Evidence of asynchrony in the onset of signs of oestrus in Zebu cattle treated with a progestagen ear implant. *Res. Vet. Sci.* **1996**, *60*, 51–54.
97. Lamonthe, C.; Montiel, F.; Fredriksson, G.; Galina, C.S. Reproductive performance of Zebu cattle in Mexico. 3. Influence of season and social interaction on the timing of expressed oestrus. *J. Trop. Agric.* **1995**, *72*, 319–323.
98. Mattoni, M.; Mukasa-Mugerwa, E.; Cecchini, G.; Sovani, S. The reproductive performance of east African (*Bos indicus*) Zebu cattle in Ethiopia. 1. Estrous cycle length, duration, behavior and ovulation time. *Theriogenology* **1988**, *30*, 961–971. [CrossRef]
99. Plasse, D.; Warnick, A.C.; Koger, M. Reproductive behaviour of *Bos indicus* females in a subtropical environment. IV. Length of estrus cycle, duration of estrus, time of ovulation, fertilization and embryo survival in grade Brahman heifers. *J. Anim. Sci.* **1970**, *30*, 63–72. [PubMed]
100. Vaca, L.A.; Galina, C.S.; Fernandez-Baca, S.; Escobar, F.J.; Ramirez, B. Oestrus cycles, oestrus and ovulation of the Zebu in the Mexican tropics. *Vet. Rec.* **1985**, *117*, 434–437.
101. Kumar, D. A study on the breeding behaviour of Sahiwal cows and their breeding efficiency through artificial insemination under farm conditions. *Indian Vet. J.* **1978**, *2*, 34–40.
102. De Silva, A.W.M.V.; Anderson, G.W.; Gwazdauskas, F.C.; McGilliard, M.L.; Lineweaver, J.A. Interrelationships with oestrus behaviour and conception in cattle. *J. Dairy Sci.* **1981**, *64*, 2409–2418. [CrossRef]
103. Espinoza, J.L.V.; López, R.A.; Palacios, E.A.; Ortega, R.P.; Avila, S.; Murillo, B.A. Efecto del toro sobre el comportamiento estral de vacas Chinampas (*Bos taurus*) en una región tropical seca. *Zootec. Trop.* **2007**, *25*, 19–28.

Article

Time Spent in a Maternity Pen during Winter Influences Cow and Calf Behavior in Pasture-Based Dairy Systems

Fabiola Matamala [1], Helen Martínez [2], Claudio Henríquez [3] and Pilar Sepúlveda-Varas [4,*]

1 Escuela de Graduados, Facultad de Ciencias Veterinarias, Universidad Austral de Chile, Valdivia 5090000, Chile; fabiolamatamalahernandez@gmail.com
2 Escuela de Graduados, Facultad de Ciencias Agrarias y Alimentarias, Universidad Austral de Chile, Valdivia 5090000, Chile; helenmichellemartinez@gmail.com
3 Instituto de Farmacología y Morfofisiología, Facultad de Ciencias Veterinarias, Universidad Austral de Chile, Valdivia 5090000, Chile; claudio.henriquez@uach.cl
4 Instituto de Ciencias Clínicas Veterinarias, Facultad de Ciencias Veterinarias, Universidad Austral de Chile, Valdivia 5090000, Chile
* Correspondence: pilar.sepulveda@uach.cl; Tel.: +56-63-293700

Citation: Matamala, F.; Martínez, H.; Henríquez, C.; Sepúlveda-Varas, P. Time Spent in a Maternity Pen during Winter Influences Cow and Calf Behavior in Pasture-Based Dairy Systems. *Animals* **2022**, *12*, 1506. https://doi.org/10.3390/ani12121506

Academic Editors: Daniel Mota-Rojas, Julio Martínez-Burnes and Agustín Orihuela

Received: 4 April 2022
Accepted: 7 June 2022
Published: 9 June 2022

Publisher's Note: MDPI stays neutral with regard to jurisdictional claims in published maps and institutional affiliations.

Copyright: © 2022 by the authors. Licensee MDPI, Basel, Switzerland. This article is an open access article distributed under the terms and conditions of the Creative Commons Attribution (CC BY) license (https://creativecommons.org/licenses/by/4.0/).

Simple Summary: One of the major challenges of spring calving pasture-based systems in temperate regions is the exposure of periparturient dairy cows and their newborn calves to cold and wet winter conditions. We investigated whether moving precalving cows from an outdoor paddock to an indoor maternity pen affects the behavior of the cow and her newborn during winter and if this was influenced by the timing relative to calving. Our results indicated that cows housed in a maternity pen spent more time lying and ruminating compared with cows kept in an outdoor paddock. Moreover, newborn calf vitality improved when cows were moved to a maternity pen 3 weeks before calving when compared with those moved during the week before calving or those that remained in an outdoor paddock precalving. This information may aid in the design of calving management systems that can minimize any negative effects of inclement winter weather on cow and calf welfare in seasonal, year-round pasture-based systems.

Abstract: Our study compared the behavior of prepartum dairy cows that either remained in an outdoor paddock until calving (OP) during winter or were moved to an indoor maternity pen either early (EM) or late (LM) relative to calving. Forty-two multiparous Holstein cows were divided into three treatments (OP, EM, or LM) and monitored from 3 weeks before to 1.5 h after calving. Cows in EM and LM were moved to a maternity pen starting at week three and week one before the expected calving date, respectively. We assessed the cleanliness of the cows at calving, immunoglobulin G concentration in colostrum, and the behavior and vitality of calves across treatments. Cows spent more time lying in EM compared to OP and LM during the weeks −3 and −2 relative to calving, but lying time was increased in LM cows compared with OP cows during the week −1 relative to calving. Prepartum rumination time was lowest in OP cows but not different between EM or LM. Calves from OP cows spent more time lying and had lower vitality after calving than those from LM and EM cows, respectively; calves from EM and LM cows were intermediate for lying and vitality, respectively, but did not differ from either group. The cleanliness was greatest in cows that calved indoors (EM or LM); nevertheless, precalving management did not affect the IgG concentration in colostrum. Our study demonstrates that, in comparison with OP, EM and LM have positive implications for the welfare of the dam and its newborn calf during winter.

Keywords: calving management; welfare; cattle; pasture-based systems; winter

1. Introduction

The time around calving has been recognized as a critical period for the life and welfare of the dam [1,2] and its newborn calf [3,4]. In pasture-based systems located in temperate regions, such as southern Chile or New Zealand, parturition management aims to ensure that most cows calve during the spring calving season (from late winter to early spring) to allow for maximum pasture utilization to support lactation and thus minimize feeding costs [5]. In this type of dairy production system, during the prepartum period (i.e., three weeks before the expected calving date), cows are commonly located in outdoor paddocks with or without access to pasture, where they remain until calving [6,7]. Because the calving season typically coincides with cold and wet weather conditions [8], it has been recommended that outdoor calving paddocks should provide dry ground, shelter, and protection to dairy cattle during periods of winter weather [9]. However, despite these recommendations, shelter (natural or artificial) is rarely provided in practice [10]. For instance, a survey restricted to pasture-based dairy farms in southern Chile found that about half of the farms did not have indoor maternity areas and left the prepartum cows in small "sacrifice" outdoor paddocks without shelter during winter months [7].

The accumulated body of evidence indicates that clean, dry, well-bedded, and comfortable resting surfaces are important to dairy cows, particularly during the prepartum and calving periods, as they increase lying time, ensure hygiene, facilitate the calving process, and decrease the risk of illness after calving for both the cow and calf [11]. Thus, the use of maternity facilities in pasture-based systems could be an attractive option for farmers because it provides a clean and dry environment that minimizes stress and ensures the comfort of the cow and its newborn calf [12,13]. Moreover, it could promote better calving supervision to ensure assistance at calving if needed [14,15]. This could be particularly important because problems related to parturition may increase the risk of stillbirth, low calf vitality, and lower passive transfer of immunity [16,17]. However, the use of calving facilities has received less research in pasture-based systems than in confinement systems.

It has been recommended that cows should be moved to a calving facility within 1 or 2 days before calving to allow animals to adapt to their new environment [14,18]. Because environmental winter factors in temperate areas, such as rain and low temperatures, can reduce lying time [8,19–21], it may be beneficial to move prepartum cows from an outdoor paddock to an indoor calving area before calving to ensure the protection from adverse weather. A previous study has shown that when managed in outdoor paddocks without access to shelter during winter, cows spend less time lying, body cleanliness is impacted, and their blood NEFA concentrations increase prepartum [22]; these results are of interest provided that decreased lying times in prepartum dairy cows is associated with lower calf survival at parturition [23]. Moreover, elevated prepartum serum NEFA concentrations are associated with dystocia [24] and stillborn calf [23]. To our knowledge, there have been no studies directly evaluating the timing of access to a maternity pen on the behavior of dairy cows and their offspring exposed to winter weather in a temperate climate.

This study has two main objectives: (1) to investigate differences in lying, rumination, and activity behavior of prepartum dairy cows that remain in an outdoor paddock until calving (OP) or were moved indoors to a maternity pen either 3 (EM) or 1 week (LM) before calving during winter, and (2) to assess the effect of these different treatments on both dam and calf behavior immediately after calving. In addition, we assessed the cleanliness of the cows at calving and IgG concentration in colostrum.

2. Materials and Methods

2.1. Animals, Management, and Experimental Design

The study was carried out at the Austral Agricultural Experimental Station of the Universidad Austral de Chile in Valdivia, Chile (39°47′46″ S, 73°13′13″ W) between June to September 2019 (Southern Hemisphere winter). Valdivia has a Temperate Oceanic Climate (Cfb: temperate rainy climate in winter) according to Köppen–Geiger classification, as described by Sarricolea et al. [25].

A total of 42 clinically healthy multiparous late pregnant Holstein cows (mean ± SD parity = 3.1 ± 1.6; BW = 624 ± 72 kg; BCS = 2.7 ± 0.3 (5—point scale, [26]) were used. Cows were selected from a group of dry cows based on expected calving dates. Cows were managed according to the standard operating procedures for this facility, which included a typical herd of grazing cows under a seasonal spring calving system. Cows were dried-off approximately 8 weeks before their expected calving date. At the beginning of the dry-off period, as a preventive measure for lameness, cow's feet were trimmed by a trained veterinarian. From the beginning of the dry period until approximately three weeks before the expected calving date (far-off period), cows were maintained in outdoor paddocks with pasture (mixture of grasses and legumes) and fresh water ad libitum. The stocking rate in these outdoor paddocks was maintained at approximately 20 to 25 cows/ha, but this was dynamic as cows came and left the outdoor paddock depending on their expected calving date. One week before enrolment, a trained veterinarian performed a general clinical examination of all cows. Cows clinically healthy pre-, during, and post-calving were used in this study, whereby no clinical signs indicative of illness or lameness were observed, and no cows required assisted calving. The cows selected for the study were marked with numbers using hair dye (White Bleach and Oxi-Cream, Elgon, Italy). After calving, cows remained with their newborn calf for at least 1.5 h (range: 1.5 to 12 h). Then, the calf was moved to the calf barn by the farm personnel, and the cow was integrated into the lactating group.

Cows were enrolled in the study three weeks before their expected calving date and allocated to 1 of 3 treatments based on the expected calving date. Provided the limited availability of maternity pens, the sample size was restricted to a total of 14 cows per treatment.

The treatments were as follows (Figure 1):

1. Outdoor paddock (OP): Cows were individually kept in an outdoor paddock for the last three weeks before their expected calving date until calving day and exposed to natural winter weather conditions (i.e., rain, wind, and cold).
2. Early Movement (EM): Cows were individually housed in a maternity pen for the last three weeks before their expected calving date until calving day.
3. Late Movement (LM): Cows were individually kept in an outdoor paddock during weeks −3 and −2 and moved to individual indoor maternity pens during the week −1 relative to calving until the day of calving.

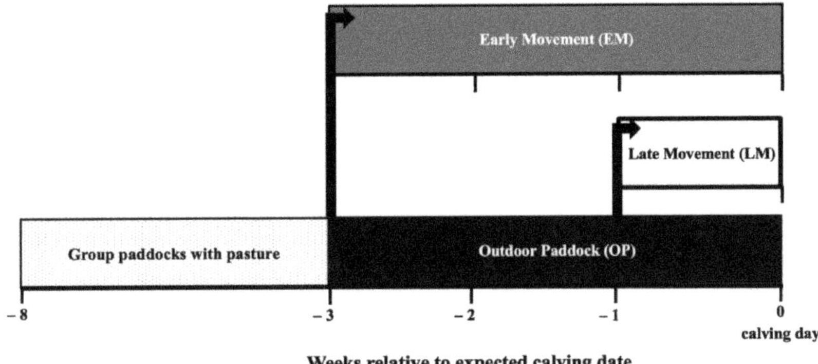

Figure 1. Diagram of the study design. During the far-off period, cows were grouped together in an outdoor paddock with pasture. Three weeks before the expected calving date, cows were individually allocated to 1 of 3 treatments: Outdoor paddock (OP; black bar), where cows were individually kept in an outdoor paddock from 3 weeks before until the day of calving; Early movement (EM; grey bar), where cows were individually housed in a maternity pen from 3 weeks before until the day of calving; and, Late movement (LM; white bar), where cows were individually kept in an outdoor paddock during weeks −3 and −2 and moved to individual indoor maternity pens during week −1 relative to calving until the day of calving.

Each outdoor paddock measured 3.5 × 5 m and had a bare soil surface with no grass cover, reduced water infiltration, and moderate mud content. Outdoor paddock set-up allowed for visual, auditory, olfactory, and limited tactile contact between cows because an electric fence separated the experimental outdoor paddocks. The amount of mud was controlled through a boot test used previously by our research group [22] and adapted from Chen et al. [27]. To do this, the researcher stands on 2 spots inside of each outdoor paddock, and the marks left on the boots indicate how far mud comes up. The boot test consists of a three-point scale: 1 = dry soil (mud does not cover the boots); 2 = muddy soil (boots covered with mud below ankle level); and 3 = very muddy soil (mud-covered boots above ankle level).

Each indoor maternity pen used measured 3.5 × 5 m and consisted of an aluminum roof, wood walls, and a thick layer of sawdust (10 cm deep approximately) on concrete flooring. In this area, manure and urine were raked and removed daily after the morning feeding. Sawdust was changed or added daily or when necessary to ensure that each lying space was clean and dry.

Cows were individually fed twice daily at approximately 09:00 and 15:00 h; the rations were provided in a feed bin placed in the corner of the outdoor paddock or maternity pen. The ration was formulated following NRC [28] guidelines and consisted of approximately 26 kg of grass silage per day on an as-fed basis (27% DM, 14% CP, 49% NDF, and 2.4 Mcal/kg; reported on a DM basis) and approximately 3 kg of commercial concentrates per day on an as-fed basis (87% DM, 22.5% CP, 12.0% NDF, and 11.56 Mcal/kg; reported on a DM basis) with anionic mineral mix (Mg 4.0%, Cl 33.0%, S 2.6%, Ca 0.7%, K 0.3%, 2 mg/kg Na, 1050 mg/kg Cu, 2100 mg/kg Mn, 3500 mg/kg Zn, 140 mg/kg I, 13 mg/kg Co, 10 mg/kg Se). Clean, fresh water was provided *ad libitum* in a water trough (600 L water trough).

2.2. Weather Conditions Measurements

The daily measurements of rainfall (mm), air temperature (°C), relative humidity (%), and wind speed (m/s) were obtained from a weather station (A720, ADCON Telemetry GMBH, Klosterneuburg, Austria) located 1 km from the research location. The daily measurements of ambient temperature and relative humidity inside the maternity pen were recorded using electronic data loggers (RC-4HC, Elitech Technology, Inc., Milpitas, San Jose, CA, USA). All measures of weather conditions were included to describe the weather conditions in this study.

2.3. Behavioral Measurements

Prepartum lying behavior was recorded using electronic data loggers (HOBO Pendant G Acceleration Data Logger; Onset Computer Corp, Bourne, MA, USA) attached to the hind leg of each cow using a flexible bandage. The loggers were removed weekly from the animals for data download and then reattached to the leg. The data logger was set to record the y-axis at 1 min intervals for consecutive hours, and lying data were processed using the cutoff point validated by Ledgerwood et al. [29]. This information was used to determine whether the cow was standing or lying and subsequently was used to calculate daily lying time, number of lying bouts (i.e., frequency of transitions from lying to standing positions), and duration of lying bouts (minutes lying per day/the number of lying bouts per day). Lying down events shorter than two minutes were removed from the data following the recommendation of Mattachini et al. [30].

Prepartum rumination and neck activity behavior were measured using the Hr-Tag collars (SCR Engineers Ltd., Netanya, Israel) as described and validated by Schirmann et al. [31]. These collars consisted of a microphone to monitor rumination time and an accelerometer that quantifies neck activity. Rumination time was recorded in minutes per 2 h interval, and neck activity data were also recorded every 2 h as an arbitrary number. Data were transferred and stored in the control unit via radio frequency and downloaded daily to the database. This information was used to determine total rumination time and neck activity per cow per day.

Calving time was defined as the time when the calf was fully expelled from the cow [32]. The individual behavior of the cows and their calves during the first 1.5 h after calving were video recorded. Outdoor paddocks and maternity pens were equipped with one camera each (Ezviz; Model CS-CV310-A0-1B2WFR, City of Industry, CA, USA) that were mounted on wooden poles 2.2 m above the ground to visualize the complete individual area in the outdoor paddock or pen. The area under observation was naturally lit during daylight hours (08:00 to 17:59 h), and infrared lighting was used for night-time recording (18:00 to 07:59 h). The signal from the cameras went through an NVR recorder (Ezviz; Model Wi-Fi EZVIZ X5C). The behaviors posture, maternal behavior, comfort-related behavior, and calf vitality are described in Table 1 were continuously monitored at one-second intervals by one trained observer using the Behavioral Observation Research Interactive Software (BORIS, http://www.boris.unito.it/; access date: October 2019 [33]). Inter-observer agreement was calculated through the intraclass correlation coefficient (ICC = 0.96).

Table 1. Description of dam and calf recorded behaviors during the first 1.5 h after calving in all treatments.

Behavior	Classification	Definition
Dam		
Standing	Posture	Standing with all four limbs fully extended and perpendicular to the ground [34].
Lying down	Posture	Lying on sternal and/or lateral recumbence [34].
Grooming her calf	Maternal behavior	Standing dam's muzzle or tongue is in physical contact with, or in close proximity of the calf's body [34].
Self-grooming	Comfort related behavior	Dam licking herself [16].
Calf		
Standing	Posture	Standing with all four limbs fully extended and perpendicular to the ground [34].
Lying	Posture	Lying on sternal or lateral recumbence [34].
Lying bout	Posture	Transitions from lying to standing position [29].
Standing attempt	Vitality behavior	Calf is partially standing upright with its four limbs placed under its body, with the ventral part not touching the ground. Calf does not fully extends its limbs [32].
Successful standing	Vitality behavior	Calf is standing upright with all limbs fully extended for longer than 5 s [32].
Suckle attempt	Vitality behavior	Standing calf is positioned below standing dam with its head located under the front of the dam's udder [32].
Successful suckling	Vitality behavior	Standing calf is positioned below standing cow with head located at the udder for more than 5 s [32].

2.4. Cow Cleanliness and Immunoglobulin G Colostrum Concentration

Cow cleanliness was evaluated by a single observer at enrollment in the study (3 weeks before expected calving date) and immediately after calving. Cleanliness scores of the tailhead, upper leg (thigh), ventral abdomen, udder, and lower hind leg were evaluated using a 5—point scale, where 1 = clean to 5 = dirty [35].

Colostrum samples of 50 mL were taken from each cow at first milking (between 3 to 24 h after calving) and immediately frozen at $-20\ ^\circ$C until analysis. The IgG colostrum concentration was measured using a commercially available bovine IgG ELISA kit according to the manufacturer's instructions (ab205078, Abcam, Cambridge, UK). In brief, after thawing at room temperature, colostrum samples were serially diluted in ELISA wash buffer to final dilutions of 1:1,000,000.

2.5. Data Handling and Statistical Analyses

Statistical analyses were performed through R (version 4.0.3; https://www.r-project.org/; access date: October 2020) with linear mixed-effects models using R packages lme4 [36] and using the cow as the experimental unit. Models were checked to ensure normality of residuals, and the appropriate covariance structures were selected based on the lowest Akaike information criterion (AIC). Significant effects were defined as $p < 0.05$, $p < 0.001$ and tendencies were considered $p < 0.10$.

Descriptive statistical analyses (median and range) were performed to summarize the weather conditions. Three cows were excluded from the analysis of prepartum behavior, cleanliness score, and IgG colostrum concentrations because they calved before their expected calving date (missing more than 1 week), and thus, they did not complete the 3 weeks prepartum period in their respective treatment (OP, $n = 1$; EM, $n = 1$; LM, $n = 1$). Therefore, the final analysis of prepartum behavior, cleanliness score, and immunoglobulin G concentration in colostrum included 39 cows (13 cows per treatment).

In order to evaluate prepartum behavior, the lying, rumination, and activity behavior data were summarized into 3 periods based on week relative to calving: week -3 (days -21 to -15), week -2 (days -14 to -8), and week -1 (days -7 to -2). The data were analyzed using linear mixed-effects models, considering the treatment (OP, EM, and LM), and period (week -3, week -2, and week -1) as fixed effects and the cow as random effect. To evaluate cow cleanliness, the model considered the treatment (OP, EM, and LM), and observation period (at enrollment and after calving) as fixed effects, while the cow was considered as random effect. A linear mixed effects model was also used for evaluating IgG concentration in colostrum (mg/mL), considering the treatment (OP, EM, and LM) as fixed effect, whereas the calving time (day: from 08:00 to 17:59 h; night: from 18:00 to 07:59 h), and colostrum sampling time (>12 h and <12 h) were considered as random effects.

An additional 10 cows (and their newborn calves) were excluded from the analysis of behavior after calving due to technical problems with the video recordings (OP, $n = 4$; EM, $n = 3$; LM, $n = 1$) or having a stillborn calf (EM, $n = 2$). Hence, 29 cows and their calves were included (OP, $n = 9$; EM, $n = 8$; LM, $n = 12$). Because all calves performed a standing attempt, this behavior was calculated as the time from birth to the onset of this behavior within the first 1.5 h after calving (latency). However, because other vitality behaviors (e.g., successful standing, suckle attempt, or successful suckling) were not performed by all calves, we decided to analyze them as behavior performed (yes/no).

To examine the effect of treatment (OP, EM, and LM) on both dam (latency to stand after calving, latency to groom her calf, and time spent grooming her calf) and calf behaviors (lying time, frequency lying bouts, and latency to stand attempt) linear mixed-effects models were used. Characteristics of the calving (parity of the cow, cow body condition score at calving, and sex of the calf) were considered random effects.

We analyzed the association between treatment (OP, EM, and LM) and performed behaviors (dam: lying after calving and self-grooming; calf: successful standing, suckle attempt, and successful suckling) using a Fisher's exact test R package RVAideMemoire [34].

3. Results

3.1. Weather Conditions

During the study period, daily median precipitation was 2.4 mm (range: 0 to 62.8 mm), the daily temperature reached 7.6 (range: 3.4 to 12.8 °C), and daily relative humidity was 83.4% (range: 60.8 to 96.9%), and daily wind speed was 2.9 m/s (range: 0 to 17.6 m/s). Inside the maternity pen, daily temperature and relative humidity averaged 9.5 °C (range: 4.2 to 13.3 °C) and 85.6 ± 0.7% (range: 61.3 to 92.4%), respectively.

3.2. Prepartum Behavior

The lying time differed among treatments (Figure 2A). During week -3 and -2 before calving, OP and LM cows spent approximately 3 h/d lesser time lying than EM cows (week $-3 = 9.4 \pm 0.7$ h/d, 10.3 ± 0.8, and 13.2 ± 0.8; week $-2 = 9.8 \pm 0.5$ h/d, 9.1 ± 0.6, and 13.3 ± 0.6, respectively; $p < 0.001$). During the week before calving, cows in the OP treatment showed a decreased lying time versus LM (9.6 ± 0.5 vs. 11.9 ± 0.6 h/d; $p < 0.001$), but there was no difference between EM and LM cows (12.9 ± 0.6 vs 11.9 ± 0.6 h/d; $p > 0.1$).

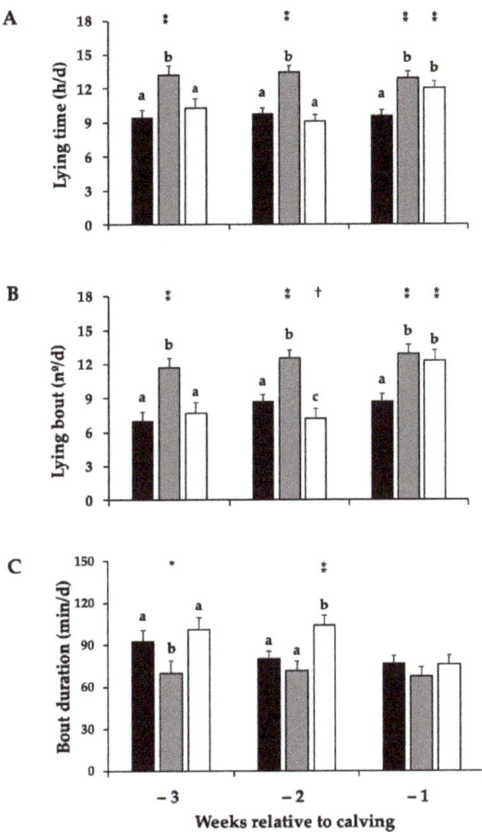

Figure 2. Lying time (**A**), lying bouts (**B**) and lying bout duration (**C**) of dairy cows kept in an outdoor paddock from 3 weeks before calving until calving day (black bar), or subjected to an early (3 weeks before calving; gray bar) or late movement (1 week before calving; white bar) to an indoor maternity pen where they remained until calving day (*n*: 13 cows per treatment). Least squares means and SE are reported. Different letters (a, b, c) indicate a statistical difference in the same week. † $p < 0.10$, * $p < 0.05$, ** $p < 0.001$.

Cows in OP and LM had lesser lying bouts compared with EM during week −3 and −2 before calving (week −3 = 7.0 ± 0.8, 7.7 ± 0.9, and 11.7 ± 0.8; week −2 = 8.7 ± 0.6, 7.2 ± 0.9, and 12.5 ± 0.7 bouts/d, respectively; $p < 0.001$; Figure 2B). We observed that for OP cows the number of lying bouts decreased in week −1 compared to LM (8.7 ± 0.7 vs. 12.3 ± 0.9 bouts/d; $p < 0.001$); this difference was no longer evident with EM cows (12.3 ± 0.9 vs. 12.9 ± 0.8 bouts/d; $p > 0.1$).

Mean lying bout duration was longer in OP and LM cows than EM cows during week −3 and −2 (week −3 = 93.0 ± 7.4, 101.3 ± 8.5, and 69.9 ± 9.0; week −2 = 80.3 ± 5.4, 104.0 ± 7.1, and 71.8 ± 6.4 min/d, respectively; $p < 0.05$; Figure 2C). During the week before calving, we did not observed differences among treatments (OP = 76.8 ± 5.5, EM = 67.7 ± 6.6, and LM = 76.0 ± 6.7; $p > 0.1$).

Cows in OP had a lower daily rumination time compared to EM cows on week −2 (546 ± 16 min/d vs. 584 ± 18; $p < 0.05$) and to LM cows on week −1 (536 ± 16 min/d vs. 575 ± 19; $p < 0.05$; Figure 3A).

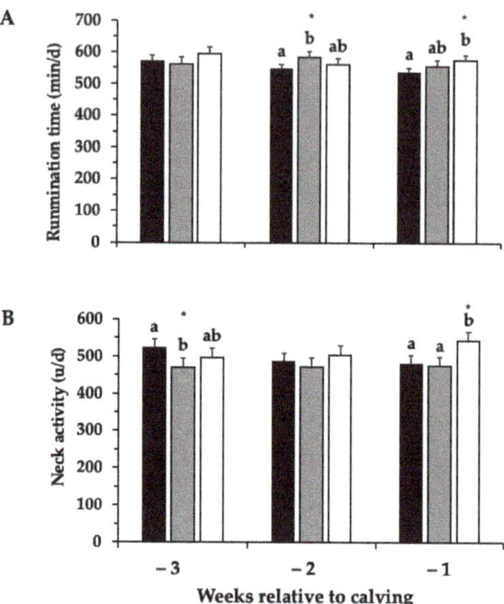

Figure 3. Rumination time (**A**) and neck activity (**B**) of dairy cows kept in an outdoor paddock from 3 weeks before calving until calving day (black bar), or subjected to an early (3 weeks before calving; gray bar) or late movement (1 week before calving; white bar) to an indoor maternity pen were they remained until calving day (*n*: 13 cows per treatment). Least squares means and SE are reported. Different letters (a, b) indicate a statistical difference in the same week. * $p < 0.05$.

For daily neck activity, we found the main effects of treatment on specific weeks before calving (Figure 3B). Cows in OP had more neck activity compared to EM cows on week -3 (525 ± 22 vs. 469 ± 25 units/d; $p < 0.05$), without differences among these treatments in any other week. During week -1, LM cows showed an increase in neck activity compared to OP and EM cows (543 ± 25, 481 ± 22, and 476 ± 24 units/d, respectively; $p < 0.05$).

3.3. Dam and Calf Behavior

After parturition, the latency of the dam to stand after calving or to groom her calf did not differ among treatments ($p > 0.1$; Table 2). Further, there was no effect of treatment on the time that the cow spent grooming her calf ($p > 0.1$; Table 2), but self-grooming behavior was performed lesser frequently in OP dams compared to LM ($p < 0.05$; Table 3). There was no difference in self-grooming behavior between OP and EM cows ($p > 0.1$; Table 3). No differences in the number of cows that lay down after calving was found, regardless of treatment ($p > 0.1$; Table 3).

Calves from OP cows spent more time lying during the first 1.5 h after calving compared with calves from LM cows ($p < 0.05$; Table 2), but there was no difference between OP and EM treatments ($p > 0.1$; Table 2).

The number of lying bouts and latency to stand attempt was similar among treatments (Table 2; $p > 0.1$). We did not find differences in the success of calves to stand among treatments ($p > 0.1$; Table 3). However, all calves (100%) in the EM treatment and most (83%) in the LM treatment performed a suckle attempt or a successful suckling event, compared with less than a half in OP treatment (44%; $p < 0.05$; Table 3).

Table 2. Effect of housing treatment on dam and calf behaviors during the first 1.5 h after calving (LSM ± SE).

	Housing Treatment		
Behavior	**OP** [1]	**EM** [2]	**LM** [3]
Dam			
Latency to stand after calving (min)	1.4 ± 1.7 [a]	1.0 ± 2.3 [a]	2.6 ± 2.1 [a]
Latency to groom her calf (min)	1.5 ± 1.7 [a]	1.1 ± 2.3 [a]	2.7 ± 2.1 [a]
Duration to groom her calf (min)	47.1 ± 5.1 [a]	53.5 ± 5.8 [a]	50.7 ± 5.2 [a]
Calf			
Lying time (min)	72.0 ± 8.8 [a]	57.7 ± 11.8 [a,b]	45.5 ± 10.4 [b]
Lying bouts (No.)	2.4 ± 0.9 [a]	3.4 ± 1.3 [a]	3.1 ± 1.2 [a]
Latency to stand attempt (min)	19.8 ± 4.4 [a]	10.8 ± 6.2 [a]	16.9 ± 5.4 [a]

[1] Outdoor Paddock (OP): Cows kept individually in an outdoor paddock for the last 3 weeks before their expected calving date until calving day (*n*: 9). [2] Early Movement (EM): Cows housed individually in an indoor maternity pen for the last 3 weeks before their expected calving date until calving day (*n*: 8). [3] Late Movement (LM): Cows were individually kept in an outdoor paddock during weeks −3 and −2 and moved to individual indoor maternity pens during week −1 relative to calving until the day of calving (*n*: 12). [a,b] Indicates a statistical difference in the same row with *p* < 0.05.

Table 3. Association of housing treatment on the number (and percentage) of dams and calves that performed different behaviors during the first 1.5 h after calving.

	Housing Treatment		
Behavior	**OP** [1]	**EM** [2]	**LM** [3]
Dam			
Lying after calving	3/9 (33) [a]	0/8 [a]	4/12 (33) [a]
Self-grooming	1/9 (11) [a]	4/8 (50) [a,b]	9/12 (75) [b]
Calf			
Successful standing	7/9 (78) [a]	8/8 (100) [a]	11/12 (92) [a]
Suckle attempt	4/9 (44) [a]	8/8 (100) [b]	10/12 (83) [a,b]
Successful suckling	4/9 (44) [a]	8/8 (100) [b]	10/12 (83) [a,b]

[1] Outdoor Paddock (OP): Cows kept individually in an outdoor paddock for the last 3 weeks before their expected calving date until calving day (*n*: 9). [2] Early Movement (EM): Cows housed individually in an indoor maternity pen for the last 3 weeks before their expected calving date until calving day (*n*: 8). [3] Late Movement (LM): Cows were individually kept in an outdoor paddock during weeks −3 and −2 and moved to individual indoor maternity pens during week −1 relative to calving until the day of calving (*n*: 12). [a,b] Indicates a statistical difference in the same row with *p* < 0.05.

3.4. Cleanliness and Immunoglobulin G Colostrum Concentration

At enrollment, cows scored lower than three on the cleanliness score scale regardless of the treatment (*p* > 0.1; Table 4). After calving, OP cows were dirtier than EM and LM cows for all cleanliness scores (*p* < 0.05; Table 4). Furthermore, EM cows were cleaner on their upper legs than LM cows (*p* < 0.05; Table 4).

Table 4. Effect of housing treatment on cow cleanliness scores (LSM and SE) at enrollment of the study (3 weeks before to calving) and after calving (day 0) (*n*: 13 per treatment).

	At Enrollment			After Calving		
Cleanliness Score	**OP** [1]	**EM** [2]	**LM** [3]	**OP** [1]	**EM** [2]	**LM** [3]
Tail head	1.1 ± 0.2 [a]	1.1 ± 0.2 [a]	1.1 ± 0.2 [a]	2.6 ± 0.6 [b]	1.3 ± 0.2 [a]	1.3 ± 0.2 [a]
Upper leg	1.5 ± 0.2 [a,c]	2.1 ± 0.3 [a,c]	2.1 ± 0.3 [a,c]	4.3 ± 0.2 [b]	1.5 ± 0.3 [c]	2.7 ± 0.3 [a]
Ventral abdomen	1.8 ± 0.2 [a]	1.9 ± 0.3 [a]	2.3 ± 0.3 [a]	4.0 ± 0.2 [b]	1.2 ± 0.3 [a]	1.7 ± 0.3 [a]
Udder	1.3 ± 0.2 [a]	1.7 ± 0.3 [a]	1.8 ± 0.3 [a]	3.7 ± 0.2 [b]	1.4 ± 0.3 [a]	1.4 ± 0.3 [a]
Lower leg	1.9 ± 0.3 [a]	2.5 ± 0.4 [a]	2.6 ± 0.4 [a]	4.8 ± 0.3 [b]	2.5 ± 0.4 [a]	3.0 ± 0.4 [a]

[1] Outdoor Paddock (OP): Cows kept individually in an outdoor paddock for the last 3 weeks before their expected calving date until calving day. [2] Early Movement (EM): Cows housed individually in an indoor maternity pen for the last 3 weeks before their expected calving date until calving day. [3] Late Movement (LM): Cows were individually kept in an outdoor paddock during weeks −3 and −2 and moved to individual indoor maternity pens during week −1 relative to calving until the day of calving. [a,b,c] Indicates a statistical difference in the same row with *p* < 0.05.

The mean IgG colostrum concentration was 22.0 ± 1.8 mg/mL (±SD; range: 1.1 to 41.4 mg/mL). The IgG colostrum concentrations were similar between OP and EM cows (25.4 ± 3.8 mg/mL vs. 20.9 ± 3.8 mg/mL; $p > 0.1$), and higher than LM cows (15.4 ± 3.8; $p < 0.05$).

4. Discussion

Although much research has focused on understanding the effect of the type of housing and time spent in a maternity pen in confinement-managed housed dairy cows, far less research has focused on dairy cows on pasture. In seasonal pasture dairy systems in temperate regions precalving cows might be especially affected by environmental factors such as rain, wind, or mud. For instance, in southern Chile, many farmers kept prepartum cows during late winter in outdoor paddocks with little or no opportunity to shelter, and stand-off surfaces are usually uncovered and can easily become wet and muddy if they are not well managed [7]. The objective of this study was to understand the impact of moving prepartum cows exposed to winter weather to a maternity facility at different times before calving. In addition, we investigated the effect of prepartum housing on the behavior of the dam and its offspring in the first hour after calving. As expected, we found numerous behavioral differences between cows that remained in outdoor paddocks until calving or were moved to a maternity pen in the weeks before calving. An important feature of the current study is that we also examined the effect of housing treatments on cow cleanliness levels, which was highest in the cows that calved indoors in the maternity pen. This research supports a new strategy to design calving management practices in seasonal pasture-based systems in temperate climates that minimize the effect of winter weather on cow lying behavior and calf vitality.

4.1. Prepartum Behavior

It is well-documented that dairy cattle spend more time lying down on dry surfaces compared with wet [37,38] or muddy ones [27]. We found that prepartum cows spent more time lying when they were housed in an indoor maternity pen (EM: 13.1 h/d and LM: 12.0 h/d) compared to when they were kept in the outdoor paddock (OP: 9.6 h/d). These results are similar to findings by Black and Krawczel (2016) [39], who observed that prepartum dairy cows spent around 10 and 13 h/d lying down when they were kept in pasture or free stall facilities, respectively. These authors [39] argued that the lower daily lying time of cows kept on pasture was due to the motivation of cows to graze. However, in our study, the cows kept in outdoor paddocks were on a surface without pasture. It is possible that the continued exposition of cows to wet, muddy and cold conditions could explain the shorter lying time. Hendriks et al. (2019) [8] observed that prepartum dairy cows managed in a pasture-based system in New Zealand spent 9.8 h/d lying down when they were exposed to winter weather. An adequate amount of time lying down is often used as an indicator of dairy cow welfare [40], and it has been described as particularly important during the latter part of pregnancy [41]. Moreover, one study found that reduced lying time caused by a lack of suitable resting areas with dry surfaces may also affect the quality of the rest [38].

In this study, we observed that access to a maternity area (both EM and LM) increased the number of transitions between standing and lying compared to the outdoor paddock (OP). Previous work reported that the dryness of the surfaces also has a marked effect on the postural changes; cows kept on wet and muddy surfaces had fewer lying bouts compared to drier ones [27,42], probably caused by discomfort in the process of lying down. Campler et al. [41] reported that soft bedding—as in our study, the sawdust bedding inside a maternity pen—improves the cow traction and facilitates her postural change from lying to standing, or vice versa.

It is worth noting that our management proposal of moving the pregnant cow from an outdoor paddock to an indoor facility the last week before calving (LM) provided an increase in lying time of 31% compared to cows that remained in the outdoor paddock

(OP). One of the greatest barriers for a grazing dairy producer, in order to consider moving a precalving cow to a maternity area during the calving season, is their limited indoor accommodation and human resources [41]. An advantage of the LM management in the present study is that it can be implemented with higher practicability than the EM strategy. Further studies are needed to evaluate other management strategies (e.g., access to shelter and dry surfaces to lay down) for calving cows managed in pasture-based spring calving systems, where they are typically exposed to inclement weather.

Overall, the cows kept in outdoor paddocks (OP) ruminated ~5% less than the cows that were moved to a calving facility (EM and LM). Rumination is a health status indicator in dairy cows [43], and several studies have shown deviations in rumination time related to health disorders [44,45]. Soriani et al. [44] described, in confinement systems, that cows with reduced rumination time during the last week of pregnancy maintained a reduced rumination time after calving and presented a higher frequency of diseases. We recommend that our results should be interpreted with caution as this decrease in rumination time (~5%) may not be biologically relevant.

We also found higher neck activity in cows who were moved to maternity areas the last week before calving (LM) compared with cows that remained in their housing treatments (OP or EM); this could be a response to new housing surroundings. However, we do not know whether this behavioral change can be interpreted as something positive for the cow or not.

4.2. Dam and Calf Behavior

To our knowledge, this is the first study to assess the maternal behavior of dairy cows managed in a pasture-based system. We assessed the maternal behavior through latency and the amount of time that the dam spent grooming her calf. We did not find differences, among housing treatments, both in the latency of the dam to groom her calf and the time that they spent grooming it. When considering the three treatments, dams spent an average of ~56% of the time of study grooming their calves, a result similar to that described by Jensen [34] in cows managed in a confinement system, suggesting that this is a behavior of high priority for them. An early and long expression of grooming after parturition benefits both dam and calf [46]. During the grooming, the dam obtains an important analgesic effect with the ingestion of amniotic fluid [47], while that calf activity is stimulated [48]. More studies are required to understand the importance of the expression of maternal behavior in dairy cows managed in intensive production systems.

Although our study does not provide evidence to declare an association between housing treatment and lying after calving behavior, we considered it important to highlight that none of the cows kept in the maternity pen throughout the prepartum period (EM) laid down after calving. It is well known that when the dam is in a standing position during the few hours after calving, it facilitates the teat-seeking behavior of the newborn calf and, therefore, the opportune ingestion of colostrum [49]. However, the effects on the dam still are unknown.

Unfortunately, due to difficulties in continuously observing self-grooming behavior, we were unable to obtain the duration of this behavior. However, we demonstrate that self-grooming was associated with housing treatment. Only one cow that calved in the outdoor paddock (OP) expressed self-grooming, while ≥50% of cows that calved in maternity pens (EM and LM) expressed this behavior. Research in dairy cows has suggested that the expression of grooming behavior is an indicator of a positive cow affective state [50,51]. In addition, the deprivation of self-grooming after the calf is delivered has been described in dams with assisted calving, and it has been suggested to utilize this deprivation as an indicator of discomfort or pain [16]. This result highlights the relevance of providing calving facilities for the improvement of the affective state of cows when calving. We encourage more research in this area.

Calves born from cows kept in outdoor paddocks (OP) spent 80% of the time of the study lying down after calving, whereas those born from cows housed in maternity pens spent 64% (EM) and 51% (LM) of time lying down. Newborn calves exposed to cold conditions are more prone to heat loss than adult cows [52]. Likely as a thermoregulatory response, the OP calves found more protection in lying than standing posture, which is a situation of concern related to poor calf welfare. Although in our study all calves tried to stand up, only the calves born from cows moved to a maternity area at the beginning of the prepartum period (EM) achieved the behaviors "successful standing", "suckle attempt", and "successful suckling". Likewise, Campler et al. [32] found that calves with an early start of standing behavior were the same as those with early suckling. The adequate ingestion of colostrum (i.e., passive transfer of immunity) is essential for calf survival. Recommendations suggest that the first ingestion of colostrum should occur within the first 4 h of life [53]. To safeguard the welfare of the calf newborn exposed to winter conditions, we decided to remove the calves from the outdoor paddock at 1.5 h after calving. The impact of early exposure to winter conditions could lead to negative effects on the calves' health and performance. Although we did not investigate this topic, we suggest that future studies do so.

Our study has some limitations. Our results are specific to cows that were kept individually housed in indoor maternity pens or outdoor paddocks, which may not represent the conditions in commercial grazing dairy herds. In addition, due to the reduced numbers of indoor maternity pens, we used a small sample size which could have prevented us from finding differences in variables such as latency to stand up after calving, latency and duration of grooming her calf, and lying behavior in the dam and her calf. A larger sample size decreases the type II error, and, therefore, we would have a higher probability of finding differences in the variables that are more difficult to detect. Furthermore, a larger sample size would allow analyzing whether any of the housing treatments is a risk factor for the expression behavior of dam and calf. In order to evaluate the applicability of our suggestion, further studies are needed to investigate the timing to move the cow to a maternity pen in large commercial pasture-based systems.

4.3. Cow Cleanliness and Immunoglobulin G Colostrum Concentration

As we expected, cows moved early (EM) or late to a maternity area (LM) were cleaner after calving than those kept in the outdoor paddocks (OP). The interest in promoting good hygiene conditions aims to reduce the risk of exposure to pathogens that could cause mastitis [54] and foot diseases [55]. Additionally, cleanliness conditions have been demonstrated to be an indicator of comfort [56] and welfare in dairy cows [57]. In our study, the cows came from outdoor paddocks provided with pasture. Investigations suggest that cows with access to pasture have little dirt on their bodies [58,59]. This should explain the low level of dirtiness in the cow at enrollment of study.

Beyond the effect of housing treatment, it was a matter of concern that all the cows in our study had concentrations of IgG lower than the minimum satisfactory threshold (50 mg/mL; [60]). Therefore, none of the cows in our study provided immunologically satisfactory colostrum to the calves. In an observational study in pasture-based dairy systems, Dunn et al. [61] reported that 42% of farms produced an average of <50 mg/mL IgG; thus, we are facing a disturbing scene regarding the feeding of the newborn calf. Calves achieve adequate immunocompetence through passive transfer of immunoglobulins obtained from colostrum [62]. Immunoglobulin G is the most abundant isotype found in colostrum, and it has been determined as a marker of colostrum quality [60]. Undoubtedly, the quality of colostrum in pasture-based systems sparks many questions and challenges. We encourage further studies aimed at investigating this topic.

5. Conclusions

We found that cows moved to a maternity pen at the beginning of the prepartum period (EM) spent more time lying and had a higher number of lying bouts than cows kept in outdoor paddocks (OP). However—and favorably—our proposal of moving cows to a maternity pen one week before calving (LM) resulted in these cows promptly increasing their lying time and lying bouts, even resembling the cows that came from three weeks housed in maternity pens (EM). In addition, in the LM treatment, more cows performed self-grooming after calving. On the other hand, we observed that calves born from cows moved early to a maternity pen (EM) had higher vitality than those calves from cows kept in outdoor paddocks (OP). Finally, we also found that cleanliness was highest in cows that calved in the maternity pens (EM or LM). Our findings demonstrate that moving the cow, early or late, from a winter outdoor paddock to a maternity pen had positive effects on the behavioral response of the cow and her newborn calf.

Author Contributions: Conceptualization, F.M. and P.S.-V.; methodology, F.M., H.M., C.H., P.S.-V.; formal analysis, F.M.; investigation, F.M. and P.S.-V.; data curation, F.M.; writing—original draft preparation, F.M.; writing—review and editing, P.S.-V.; supervision, P.S.-V.; project administration, P.S.-V.; funding acquisition, P.S.-V. All authors have read and agreed to the published version of the manuscript.

Funding: This research was funded by Fondo Nacional de Desarrollo Científico y Tecnológico (Grant FONDECYT No: 11170820). F.M. was supported by Scholarship "Agencia Nacional de Investigation y Desarrollo (ANID, Chile)" Doctorado Nacional (Folio number 21170749).

Institutional Review Board Statement: The experiment was approved by The Animal Care Ethics Committee of the Universidad Austral de Chile (Protocol N° 361/ 2019), and all procedures were performed according to the management of the farm.

Data Availability Statement: The data can be found at https://doi.org/10.6084/m9.figshare.19119389.

Acknowledgments: The authors thank the dairy farm workers of the Austral Agricultural Experimental Station and to Salvador Cruz (student at the Universidad Nacional de México), Paulina Sánchez, Gabriela Tagliaferro, Francisca Rocha (students at the Universidad Andrés Bello de Chile), Alvaro Navarro and Consuelo Zumelzu (students at the Universidad Austral de Chile), especially for their help with data recording.

Conflicts of Interest: The authors declare no conflict of interest.

References

1. Grummer, R.R. Impact of changes in organic nutrient metabolism on feeding the transition dairy cow. *J. Anim. Sci.* **1995**, *73*, 2820. [CrossRef] [PubMed]
2. von Keyserlingk, M.A.G.; Rushen, J.; de Passillé, A.M.; Weary, D.M. Invited review: The welfare of dairy cattle—Key concepts and the role of science. *J. Dairy Sci.* **2009**, *92*, 4101–4111. [CrossRef] [PubMed]
3. Arnott, G.; Roberts, D.; Rooke, J.A.; Turner, S.P.; Lawrence, A.B.; Rutherford, K.M.D. Board Invited Review: The importance of the gestation period for welfare of calves: Maternal stressors and difficult births1. *J. Anim. Sci.* **2012**, *90*, 5021–5034. [CrossRef] [PubMed]
4. Mellor, D.J.; Stafford, K.J. Animal welfare implications of neonatal mortality and morbidity in farm animals. *Vet. J.* **2004**, *168*, 118–133. [CrossRef]
5. Pérez-Prieto, L.A.; Delagarde, R. Meta-analysis of the effect of pregrazing pasture mass on pasture intake, milk production, and grazing behavior of dairy cows strip-grazing temperate grasslands. *J. Dairy Sci.* **2012**, *95*, 5317–5330. [CrossRef]
6. Zobel, G.; Proudfoot, K.; Cave, V.; Huddart, F.; Webster, J. The Use of Hides during and after Calving in New Zealand Dairy Cows. *Animals* **2020**, *10*, 2255. [CrossRef]
7. Calderón-Amor, J.; Hernández-Gotelli, C.; Strappini, A.; Wittwer, F.; Sepúlveda-Varas, P. Prepartum factors associated with postpartum diseases in pasture-based dairy cows. *Prev. Vet. Med.* **2021**, *196*, 105475. [CrossRef]
8. Hendriks, S.J.; Phyn, C.V.C.; Turner, S.-A.; Mueller, K.R.; Kuhn-Sherlock, B.; Donaghy, D.J.; Huzzey, J.M.; Roche, J.R. Effect of weather on activity and lying behaviour in clinically healthy grazing dairy cows during the transition period. *Anim. Prod. Sci.* **2020**, *60*, 148. [CrossRef]
9. New Zealand National Animal Welfare Advisory Committee. *Code of Welfare: Dairy Cattle*; New Zealand National Animal Welfare Advisory Committee: Wellington, New Zealand, 2019.
10. Mee, J.; Boyle, L. Assessing whether dairy cow welfare is "better" in pasture-based than in confinement-based management systems. *N. Z. Vet. J.* **2020**, *68*, 168–177. [CrossRef]
11. Proudfoot, K. Maternal Behavior and Design of the Maternity Pen. *Vet. Clin. N. Am. Food Anim. Pract.* **2019**, *35*, 111–124. [CrossRef]
12. Cook, N.B.; Nordlund, K.V. Behavioral needs of the transition cow and considerations for special needs facility design. *Vet. Clin. N. Am. Food Anim. Pract.* **2004**, *20*, 495–520. [CrossRef] [PubMed]

13. Vasseur, E.; Rushen, J.; de Passillé, A.M.; Lefebvre, D.; Pellerin, D. An advisory tool to improve management practices affecting calf and heifer welfare on dairy farms. *J. Dairy Sci.* **2010**, *93*, 4414–4426. [CrossRef] [PubMed]

14. Mee, J.F. Managing the dairy cow at calving time. *Vet. Clin. N. Am. Food Anim. Pract.* **2004**, *20*, 521–546. [CrossRef] [PubMed]

15. Relić, R.; Starič, J.; Ježek, J. Management practices that influence the welfare of calves on small family farms. *J. Dairy Res.* **2020**, *87*, 93–98. [CrossRef] [PubMed]

16. Barrier, A.C.; Ruelle, E.; Haskell, M.J.; Dwyer, C.M. Effect of a difficult calving on the vigour of the calf, the onset of maternal behaviour, and some behavioural indicators of pain in the dam. *Prev. Vet. Med.* **2012**, *103*, 248–256. [CrossRef] [PubMed]

17. Barrier, A.C.; Haskell, M.J.; Birch, S.; Bagnall, A.; Bell, D.J.; Dickinson, J.; Macrae, A.I.; Dwyer, C.M. The impact of dystocia on dairy calf health, welfare, performance and survival. *Vet. J.* **2013**, *195*, 86–90. [CrossRef] [PubMed]

18. Proudfoot, K.L.; Jensen, M.B.; Heegaard, P.M.H.; von Keyserlingk, M.A.G. Effect of moving dairy cows at different stages of labor on behavior during parturition. *J. Dairy Sci.* **2013**, *96*, 1638–1646. [CrossRef]

19. Tucker, C.B.; Rogers, A.R.; Verkerk, G.A.; Kendall, P.E.; Webster, J.R.; Matthews, L.R. Effects of shelter and body condition on the behaviour and physiology of dairy cattle in winter. *Appl. Anim. Behav. Sci.* **2007**, *105*, 1–13. [CrossRef]

20. Webster, J.R.; Stewart, M.; Rogers, A.R.; Verkerk, G.A. Assessment of welfare from physiological and behavioural responses of New Zealand dairy cows exposed to cold and wet conditions. *Anim. Welf.* **2008**, *17*, 19–26.

21. Schütz, K.E.; Clark, K.V.; Cox, N.R.; Matthews, L.; Tucker, C.B. Responses to short-term exposure to simulated rain and wind by dairy cattle: Time budgets, shelter use, body temperature and feed intake. *Anim. Welf.* **2010**, *19*, 375–383.

22. Cartes, D.; Strappini, A.; Sepúlveda-Varas, P. Provision of shelter during the prepartum period: Effects on behavior, blood analytes, and health status in dairy cows in winter. *J. Dairy Sci.* **2021**, *104*, 3508–3521. [CrossRef] [PubMed]

23. Menichetti, B.T.; Piñeiro, J.M.; Barragan, A.A.; Relling, A.E.; Garcia-Guerra, A.; Schuenemann, G.M. Association of prepartum lying time with nonesterified fatty acids and stillbirth in prepartum dairy heifers and cows. *J. Dairy Sci.* **2020**, *103*, 11782–11794. [CrossRef] [PubMed]

24. Dyk, P. The Association of Prepartum Non-Esterified Fatty Acids and Body Condition with Peripartum Health Problems on 95 Michigan Dairy Farms. Mater's Thesis, Michigan State University, East Lansing, MI, USA, 1995.

25. Sarricolea, P.; Herrera-Ossandon, M.; Meseguer-Ruiz, Ó. Climatic regionalisation of continental Chile. *J. Maps* **2017**, *13*, 66–73. [CrossRef]

26. Ferguson, J.D.; Galligan, D.T.; Thomsen, N. Principal Descriptors of Body Condition Score in Holstein Cows. *J. Dairy Sci.* **1994**, *77*, 2695–2703. [CrossRef]

27. Chen, J.M.; Stull, C.L.; Ledgerwood, D.N.; Tucker, C.B. Muddy conditions reduce hygiene and lying time in dairy cattle and increase time spent on concrete. *J. Dairy Sci.* **2017**, *100*, 2090–2103. [CrossRef]

28. NRC. *Nutrient Requirements of Dairy Cattle*, 7th rev. ed.; The National Academies Press: Washington, DC, USA, 2001.

29. Ledgerwood, D.N.; Winckler, C.; Tucker, C.B. Evaluation of data loggers, sampling intervals, and editing techniques for measuring the lying behavior of dairy cattle. *J. Dairy Sci.* **2010**, *93*, 5129–5139. [CrossRef]

30. Mattachini, G.; Antler, A.; Riva, E.; Arbel, A.; Provolo, G. Automated measurement of lying behavior for monitoring the comfort and welfare of lactating dairy cows. *Livest. Sci.* **2013**, *158*, 145–150. [CrossRef]

31. Schirmann, K.; von Keyserlingk, M.A.G.; Weary, D.M.; Veira, D.M.; Heuwieser, W. Technical note: Validation of a system for monitoring rumination in dairy cows. *J. Dairy Sci.* **2009**, *92*, 6052–6055. [CrossRef]

32. Campler, M.; Munksgaard, L.; Jensen, M.B. The effect of housing on calving behavior and calf vitality in Holstein and Jersey dairy cows. *J. Dairy Sci.* **2015**, *98*, 1797–1804. [CrossRef]

33. Friard, O.; Gamba, M. BORIS: A free, versatile open-source event-logging software for video/audio coding and live observations. *Methods Ecol. Evol.* **2016**, *7*, 1325–1330. [CrossRef]

34. Jensen, M.B. Behaviour around the time of calving in dairy cows. *Appl. Anim. Behav. Sci.* **2012**, *139*, 195–202. [CrossRef]

35. Reneau, J.K.; Seykora, A.J.; Heins, B.J.; Endres, M.I.; Farnsworth, R.J.; Bey, R. Association between hygiene scores and somatic cell scores in dairy cattle. *J. Am. Vet. Med. Assoc.* **2005**, *227*, 1297–1301. [CrossRef] [PubMed]

36. Bates, D.; Mächler, M.; Bolker, B.; Walker, S. Fitting Linear Mixed-Effects Models Using lme4. *J. Stat. Softw.* **2015**, *67*. [CrossRef]

37. Fregonesi, J.A.; Veira, D.M.; von Keyserlingk, M.A.G.; Weary, D.M. Effects of Bedding Quality on Lying Behavior of Dairy Cows. *J. Dairy Sci.* **2007**, *90*, 5468–5472. [CrossRef]

38. Schütz, K.E.; Cave, V.M.; Cox, N.R.; Huddart, F.J.; Tucker, C.B. Effects of 3 surface types on dairy cattle behavior, preference, and hygiene. *J. Dairy Sci.* **2019**, *102*, 1530–1541. [CrossRef]

39. Black, R.; Krawczel, P. A Case Study of Behaviour and Performance of Confined or Pastured Cows During the Dry Period. *Animals* **2016**, *6*, 41. [CrossRef]

40. Tucker, C.B.; Jensen, M.B.; de Passillé, A.M.; Hänninen, L.; Rushen, J. Invited review: Lying time and the welfare of dairy cows. *J. Dairy Sci.* **2021**, *104*, 20–46. [CrossRef]

41. Campler, M.R.; Munksgaard, L.; Jensen, M.B. The effect of transition cow housing on lying and feeding behavior in Holstein dairy cows. *J. Dairy Sci.* **2019**, *102*, 7398–7407. [CrossRef]

42. Fisher, A.; Stewart, M.; Verkerk, G.; Morrow, C.; Matthews, L. The effects of surface type on lying behaviour and stress responses of dairy cows during periodic weather-induced removal from pasture. *Appl. Anim. Behav. Sci.* **2003**, *81*, 1–11. [CrossRef]

43. Radostits, O.M.; Gay, C.C.; Hinchcliff, K.W.; Constable, P.D. *Veterinary Medicine: A Textbook of the Diseases of Cattle, Horses, Sheep, Pigs and Goats*, 10th ed.; Saunders Ltd: Philadelphia, PA, USA, 2007.

44. Soriani, N.; Trevisi, E.; Calamari, L. Relationships between rumination time, metabolic conditions, and health status in dairy cows during the transition period1. *J. Anim. Sci.* **2012**, *90*, 4544–4554. [CrossRef]
45. Calamari, L.; Soriani, N.; Panella, G.; Petrera, F.; Minuti, A.; Trevisi, E. Rumination time around calving: An early signal to detect cows at greater risk of disease. *J. Dairy Sci.* **2014**, *97*, 3635–3647. [CrossRef]
46. von Keyserlingk, M.A.G.; Weary, D.M. Maternal behavior in cattle. *Horm. Behav.* **2007**, *52*, 106–113. [CrossRef]
47. Pinheiro Machado, L.C.; Hurnik, J.F.; Burton, J.H. The effect of amniotic fluid ingestion on the nociception of cows. *Physiol. Behav.* **1997**, *62*, 1339–1344. [CrossRef]
48. Metz, J.; Metz, J.H.M. Maternal influence on defecation and urination in the newborn calf. *Appl. Anim. Behav. Sci.* **1986**, *16*, 325–333. [CrossRef]
49. Selman, I.E.; McEwan, A.D.; Fisher, E.W. Studies on natural suckling in cattle during the first eight hours post partum II. Behavioural studies (calves). *Anim. Behav.* **1970**, *18*, 284–289. [CrossRef]
50. Boissy, A.; Manteuffel, G.; Jensen, M.B.; Moe, R.O.; Spruijt, B.; Keeling, L.J.; Winckler, C.; Forkman, B.; Dimitrov, I.; Langbein, J.; et al. Assessment of positive emotions in animals to improve their welfare. *Physiol. Behav.* **2007**, *92*, 375–397. [CrossRef] [PubMed]
51. Newby, N.C.; Duffield, T.F.; Pearl, D.L.; Leslie, K.E.; LeBlanc, S.J.; von Keyserlingk, M.A.G. Short communication: Use of a mechanical brush by Holstein dairy cattle around parturition. *J. Dairy Sci.* **2013**, *96*, 2339–2344. [CrossRef] [PubMed]
52. Van laer, E.; Moons, C.P.H.; Sonck, B.; Tuyttens, F.A.M. Importance of outdoor shelter for cattle in temperate climates. *Livest. Sci.* **2014**, *159*, 87–101. [CrossRef]
53. Weaver, D.M.; Tyler, J.W.; VanMetre, D.C.; Hostetler, D.E.; Barrington, G.M. Passive Transfer of Colostral Immunoglobulins in Calves. *J. Vet. Intern. Med.* **2000**, *14*, 569–577. [CrossRef]
54. Rowe, S.; Tranter, W.; Laven, R. Longitudinal study of herd udder hygiene and its association with clinical mastitis in pasture-based dairy cows. *J. Dairy Sci.* **2021**, *104*, 6051–6060. [CrossRef]
55. Moreira, T.F.; Nicolino, R.R.; Meneses, R.M.; Fonseca, G.V.; Rodrigues, L.M.; Facury Filho, E.J.; Carvalho, A.U. Risk factors associated with lameness and hoof lesions in pasture-based dairy cattle systems in southeast Brazil. *J. Dairy Sci.* **2019**, *102*, 10369–10378. [CrossRef] [PubMed]
56. Robles, I.; Zambelis, A.; Kelton, D.F.; Barkema, H.W.; Keefe, G.P.; Roy, J.P.; von Keyserlingk, M.A.G.; DeVries, T.J. Associations of freestall design and cleanliness with cow lying behavior, hygiene, lameness, and risk of high somatic cell count. *J. Dairy Sci.* **2021**, *104*, 2231–2242. [CrossRef] [PubMed]
57. Bowell, V.; Rennie, L.J.; Tierney, G.; Lawrence, A.; Haskell, M.J. Relationship Between Building Desing, Management System And Dairy Cow Welfare. *Anim. Welf.* **2003**, *12*, 547–552.
58. Ellis, K.A.; Innocent, G.T.; Mihm, M.; Cripps, P.; McLean, W.G.; Howard, C.V.; Grove-White, D. Dairy cow cleanliness and milk quality on organic and conventional farms in the UK. *J. Dairy Res.* **2007**, *74*, 302–310. [CrossRef] [PubMed]
59. Nielsen, B.H.; Thomsen, P.T.; Sørensen, J.T. Identifying risk factors for poor hind limb cleanliness in Danish loose-housed dairy cows. *Animals* **2011**, *5*, 1613–1619. [CrossRef]
60. McGuirk, S.M.; Collins, M. Managing the production, storage, and delivery of colostrum. *Vet. Clin. N. Am. Food Anim. Pract.* **2004**, *20*, 593–603. [CrossRef]
61. Dunn, A.; Ashfield, A.; Earley, B.; Welsh, M.; Gordon, A.; Morrison, S.J. Evaluation of factors associated with immunoglobulin G, fat, protein, and lactose concentrations in bovine colostrum and colostrum management practices in grassland-based dairy systems in Northern Ireland. *J. Dairy Sci.* **2017**, *100*, 2068–2079. [CrossRef]
62. Godden, S. Colostrum Management for Dairy Calves. *Vet. Clin. N. Am. Food Anim. Pract.* **2008**, *24*, 19–39. [CrossRef]

 *animals*

Review

Allonursing in Wild and Farm Animals: Biological and Physiological Foundations and Explanatory Hypotheses

Daniel Mota-Rojas [1,*], Míriam Marcet-Rius [2], Aline Freitas-de-Melo [3], Ramon Muns [4], Patricia Mora-Medina [5], Adriana Domínguez-Oliva [1] and Agustín Orihuela [6,*]

[1] Neurophysiology, Behavior and Animal Welfare Assessment, DPAA, Universidad Autónoma Metropolitana, Unidad Xochimilco, Mexico City 04960, Mexico; mvz.freena@gmail.com

[2] Animal Behaviour and Welfare Department, IRSEA (Research Institute in Semiochemistry and Applied Ethology), Quartier Salignan, 84400 Apt, France; m.marcet@group-irsea.com

[3] Departamento de Biociencias Veterinarias, Facultad de Veterinaria, Universidad de la República, Montevideo 11600, Uruguay; alinefreitasdemelo@hotmail.com

[4] Agri-Food and Biosciences Institute, Hillsborough, Co Down BT 26 6DR, Northern Ireland, UK; Ramon.Muns@afbini.gov.uk

[5] Facultad de Estudios Superiores Cuautitlán, Universidad Nacional Autónoma de México, Cuautitlán Izcalli 54714, Mexico; mormed2001@yahoo.com.mx

[6] Facultad de Ciencias Agropecuarias, Universidad Autónoma del Estado de Morelos, Cuernavaca 62209, Mexico

* Correspondence: dmota100@yahoo.com.mx (D.M.-R.); aorihuela@uaem.mx (A.O.)

Citation: Mota-Rojas, D.; Marcet-Rius, M.; Freitas-de-Melo, A.; Muns, R.; Mora-Medina, P.; Domínguez-Oliva, A.; Orihuela, A. Allonursing in Wild and Farm Animals: Biological and Physiological Foundations and Explanatory Hypotheses. *Animals* **2021**, *11*, 3092. https://doi.org/10.3390/ani11113092

Academic Editor: Cecilie Marie Mejdell

Received: 14 September 2021
Accepted: 27 October 2021
Published: 29 October 2021

Publisher's Note: MDPI stays neutral with regard to jurisdictional claims in published maps and institutional affiliations.

Copyright: © 2021 by the authors. Licensee MDPI, Basel, Switzerland. This article is an open access article distributed under the terms and conditions of the Creative Commons Attribution (CC BY) license (https://creativecommons.org/licenses/by/4.0/).

Simple Summary: Allonursing and allosuckling are behaviors displayed by some females, characterized by nursing and feeding non-filial offspring. Although both are costly behaviors, this type of communal parenting is widespread in various species; however, not all animals display this behavior, and even among species, some differences can be observed. This review aims to analyze the biological and physiological foundations of allonursing and allosuckling in wild and farm animals. It also summarizes some current hypotheses to explain these behaviors as a strategic approach for the mother or the offspring, describing the individual and collective advantages and disadvantages and their implications on an animal.

Abstract: The dams of gregarious animals must develop a close bond with their newborns to provide them with maternal care, including protection against predators, immunological transference, and nutrition. Even though lactation demands high energy expenditures, behaviors known as allonursing (the nursing of non-descendant infants) and allosuckling (suckling from any female other than the mother) have been reported in various species of wild or domestic, and terrestrial or aquatic animals. These behaviors seem to be elements of a multifactorial strategy, since reports suggest that they depend on the following: species, living conditions, social stability, and kinship relations, among other group factors. Despite their potential benefits, allonursing and allosuckling can place the health and welfare of both non-filial dams and alien offspring at risk, as it augments the probability of pathogen transmission. This review aims to analyze the biological and physiological foundations and bioenergetic costs of these behaviors, analyzing the individual and collective advantages and disadvantages for the dams' own offspring(s) and alien neonate(s). We also include information on the animal species in which these behaviors occur and their implications on animal welfare.

Keywords: animal perinatology; non-offspring nursing; fostering; mismothering; lactation

1. Introduction

In most mammal species, attention to newborns is provided primarily by the mother [1]. Parental care entails decisions that consider the number and size of the offspring and how much energy to invest in a neonate at the cost of the reserves of parental resources for present or future offspring [2]. The mother is essential for the newborn, providing protection against predators, food, warmth, shelter, and immunological defense [3,4]. In addition, mothers aid the neonates in acquiring important information concerning their physical and social environment. During the first minutes after birth, an exchange of sensory signals occurs between the dam and their offspring(s) (sight, touch, smell, and hearing), allowing the recognition of and attachment to each other [4–6]. In precocial species, including most ungulates, ensuring the offspring's survival requires establishing a mother–young bond as soon as possible after birth [7,8]. Parturient females of precocial species deliver one or more fully developed neonates that can stand and follow the mother soon after birth (around 30 min in sheep), and feeding begins as soon as this mutual recognition and attachment takes place [9]. In contrast, in altricial species, locomotor activity, sight, audition, and thermoregulation are restricted at birth [10], and the offspring depends entirely on parental care for their nutrition and growth, and, in the case of rodents, calls and ultrasonic sounds facilitate maternal bonding [11]. In these species, the newborns require constant care to nurse, feed, and provide a warm environment, representing a maternal bond formed over prolonged periods [12]. On the other hand, precocial animals (e.g., ungulates) achieve rapid inter-individual recognition, thanks to the neurochemical signaling that has a central role in the selective attachment within the first two hours after birth, to discriminate and reject any non-filial suckling after this stage [13].

The offspring of terrestrial mammals find the mother's udder by exploring her underside from chest to teats, guided by various signals emanating from her body. Dams usually help the newborns by arching their backs and flexing their hindlimbs to facilitate access to their teats. Newborns quickly learn the physical characteristics of the udder by visual, olfactory, and thermal cues [9,14]. The neonatal behavior also depends on the type of maternal behavior. One example comes from Surti buffaloes, where neonates whose mothers were categorized as 'highly aggressive' or 'attentive' to protecting their calves reached the udder faster and fed longer than calves from females classified as 'indifferent' or 'apathetic' [15].

However, caring for the offspring is not an activity that is exclusive to the biological parents [16]. Among humans and non-human animals, there are several practices in which members of the same nest or brood care for non-filial newborns of conspecifics, either sharing care and provisioning (communal breeders), or with assistance in protection and feeding by a nonbreeding helper [17,18]. In these social systems, a pair of animals perform parent-like behaviors in non-filial young [19,20]. These include feeding, grooming, nursing, and allosuckling [20]. In perching birds, they are reported in 13% of the species and 9% in all species of birds, mammals, and fish [17]. Because they entail consequences for the fitness of animals, these behaviors can be observed in populations with stable or unfavorable conditions [17]. Some of the advantages of breeding communally at the social and group level are direct benefits, such as efficient foraging and the cooperative detection of predators [20], while the benefits towards the mother and the newborn include better reproductive performance and inclusive fitness of the mothers, weight gain, and a higher survival rate in non-filial newborns, as well as thermoregulation in the critical stage of birth [21]. Recently, Orihuela et al. [22] found some evidence of alloparental care in zebu cattle (*Bos indicus*) raised under extensive conditions (Figure 1). Similarly, Pérez-Torres et al. [23] found that zebu cows allonurse and protect non-filial calves in the first 120 days postpartum, regardless of the animal's temperament. However, some behavioral components of protective behavior might be related with other reproductive variables, as Enríquez et al. [24] found an inverse association between the number of cows reacting to more calves and the presence of follicles, and cows displaying a more intense reaction towards their calf and estrous display.

Figure 1. In extensive livestock production, it has recently been shown that specific cows of the herd take care of groups of zebu (*Bos indicus*) calves while their mothers graze, adepted from [22].

In gregarious animals, the establishment of the mother–young bond allows the newborn(s) to be cared for by the dam, and to establish the nursing–suckling relation [25,26]. Several studies have determined that one of the most acute senses in post-parturient females is smell (olfactory) because in species such as buffaloes and sheep, dams are especially receptive to their offspring's odor [3,27]. The maternal responsiveness facilitates acceptance of their own young or even of alien young impregnated with her amniotic fluid. In sheep, recognition of the chemical signals emitted by the lamb occurs within the first four hours postpartum, approximately, based on the dam's detection of the specific olfactory signals emitted by her newborn [28]. Those signals are mainly produced by the wool around the anal region, the area that the dam licks most often, and where she obtains the largest amounts of chemical substances [28,29]. During the mother–young interaction, at the cerebral level, plasticity occurs in specific areas of the mother's brain, such as the principal olfactory bulb [30].

In most mammals, the development of the maternal bond requires recognition of the offspring through olfactory, visual, and hormonal signals that culminate in said behavior [12]. Despite the need for such specific bonding mechanisms, during alloparental care, it is hypothesized that, somehow, species ignore the cost of raising a non-filial offspring and lean towards the benefits, in terms of lactation, milk synthesis, and the nutrient density of milk, to improve the development of the newborn [31].

The vomeronasal organ (VNO) is the main structure involved in maternal behavior during the olfactory recognition of the offspring, although it participates in sexual, social, and aggressive behaviors [32–35]. After birth, the body coat of the newborn is covered by the amniotic fluid [36,37], and pheromones are present in the anal region of the young [38,39]. These factors are detected by the VNO and integrated into superior and cortical structures of the brain to promote maternal bonding [40–42], social interactions [43–46], and sexual [47–56] or aggressive behaviors [48]. The VNO consists of sensory olfactory epithelial cells that communicate the oral and nasal cavities through the incisive duct in the roof of the nasal passage and the incisive papillae, respectively [57,58]. The neuronal axons in the VNO project dorsally to the margin of the olfactory bulb [59]. Subsequently, the signal is transmitted to the medial amygdala and hypothalamic centers to generate the aforementioned behaviors [60]. Therefore, through this olfactory system, the chemosensory mother–offspring communication is developed [61], and mammals can identify kin or conspecifics [59]. This may facilitate the development of an olfactory memory that persists, favoring attention and the survival of the offspring. In most ungulates, mothers develop exclusive care towards the young that they recognize during the first hours postpartum [62,63]. The exclusive mother–young bond allows the other to ration a valuable resource—the mother's milk—and ensure it is consumed only by her own offspring [64], since nursing non-filial newborns would constitute a costly behavior for the female [65].

Colostrum is the first food that newborns receive, providing passive immunity, nutrition, and thermoregulation, and enhances offspring survival for several weeks or months [66,67]. The mother's milk is the food that fosters optimal growth and development of the offspring. Interestingly, recent studies of diverse species of wild and farm animals have documented different strategies of care and attention reserved exclusively for a dam's own offspring. One of these behaviors is denominated allonursing, which is characterized by the nursing of another female's offspring. The term allosuckling is used to refer to the suckling of offspring from lactating females that are not the offspring's mother [18,64,68–72]. These behaviors have been reported in animals such as ungulates, in swine, for example, where females deliver large litters, and in species where only one neonate is born, except for goats [73]. Allonursing is less common in monotocous (primates, cetaceans, all ungulates, except swine) than polytocous species (swine). However, in the latter, where multiple offspring are born at the same time, the energetic demand is higher than in females with a single newborn (monotocous) [74]. In the primate taxa, including prosimians (*Propithecus candidus*), allonursing has been reported in 17 of over 620 species, and some authors suggest that kin selection and nulliparity have a key role in non-filial nursing [75]. Regarding this, Dušek [76] studied the differences between both types of species and found that in monotocous animals, such as the red deer (*Cervus elaphus*), allonursing behaviors seem to benefit the offspring's fitness, while in politocous animals (*Mus musculus*), alloparental care is performed to maximize the number of newborns [77,78]. In light of those findings, the aim of this review is to analyze the biological foundations and bioenergetic costs of allonursing and allosuckling behaviors, analyzing the individual and collective advantages and disadvantages for the dams' own offspring and alien neonate(s), while also identifying the animal species in which they occur and their implications for the welfare of the young. In addition, some current hypotheses attempting to explain these behaviors are also reviewed.

2. Lactation: An Energetic Costly Period

The present work provides new findings on the costs and benefits of allonursing/allosuckling during lactation; for example, in the association between reduced maternal lactation effort and faster weaning [31], compensatory growth of the offspring, the risks of pathogen transmission or improvement of the immune response [79], and milk production increments in the cows [80], among others.

Progenitors invest in their offspring regardless of the energetical cost that could affect their own survival or reproduction [81]. This is related to the principle of "assignation", which states that organisms have finite energy and nutrient resources to maintain all their body functions, especially growth, maintenance, and reproduction. Because every activity that an animal performs consumes energy, the process of sustaining life entails costs, since circulation, respiration, excretion, and muscular contraction never cease, not even during absolute repose [82].

An animal's energy requirements depend on intrinsic factors, such as its basal metabolism, activity, physiological state, age, and sex [83]. Environmental features also influence animals' energy requirements, such as the environmental temperature, humidity, precipitation, shelter, and protection from severe weather. Furthermore, the amount and quality of food available, and water consumption, affect feed intake and the possibility of animals meeting their energy requirements. Lactation is related to an animal's reproductive success, being a process that demands particularly high energy expenditure and constitutes the most energetic aspect of mammalian biology [84]. The nutritional requirements of lactating females increase, and they maximize their food ingestion to provide the energy for satisfying their own energy needs and those of their offspring. Females in lactation produce a nutritious liquid that is well adapted to promote the growth and development of their young. The components of the energy expenditure of lactating females include the resting metabolic rate and milk energy output, both of which increase markedly during lactation [85]. This means that energy ingestion must also increase to compensate for the energy spent [86]. Marotta and Lagreca [87] observed that the energy requirements of lactating sows raised in fields correspond to the sum of the maintenance needs and milk production plus the effect of climate and physical activity. They calculated the energy required for maintenance as 110 kcal of metabolizable energy/kg$^{0.75}$/day. Additionally, the amount of energy required for milk production during the growth of the young varied according to external environmental factors, such as confinement (19.0) Mcal/digestible energy (DE/day) vs. open-air conditions (20.7 Mcal/DE/day for autumn–winter vs. 19.8 Mcal/DE/day for spring–summer). In polygynous mammals, the energy expenditure during lactation also varies with sex, with males being more energetically demanding than females [88,89]. Furthermore, Moen [90] found that the female white-tailed deer's largest energy outlay occurred while lactating to nurse two fawns. The lowest cost (1.5 times her basal metabolism) occurred during the reproductive period in winter.

Despite evidence that supports this asseveration, the gestation and lactation expenditure of the mother does not always follow the same pattern, since lactation is commonly consistent with intervals of food security [91]. In fact, after controlling for individual variation in a study of wild deer (*Cervus elaphus* L., an ungulate), Clutton-Brock et al. [91] demonstrated that the cost of lactation demand is higher than that for reproduction or survival.

Nursing can also generate conditions of physiological stress, reflected in weight loss, despite greater food consumption, which could lead to susceptibility to parasitic action, reduce fertility indices, and increase mortality rates compared to non-lactating females [92]. Considering these costs, several questions have been raised regarding allonursing or allosuckling behaviors, which are erratic in some species; for example, in wild mammals, allonursing may increase maternal mortality due to the physiological and energetic cost that lactation requires to feed not only their offspring, but a non-biological individual [91]. On the contrary, in species such as chimpanzees (*Pan troglodytes schweinfurthii*), Bădescu et al. [31] have reported that cooperative breeding involves some benefits to alloparents, such as an improvement in their reproduction, and direct and indirect fitness. This also correlates to the growth, weight gain, and early weaning of the offspring. Some benefits, costs, and causes are included in Figure 2 [64,77,79,80,84,93–100]. However, there have been reports in bat roosts, animal groups that practice communal reproduction—such as lions—and in species such as seals that do not develop strong mother–infant bonds (Figure 2). Is this a strategy that females adopt to reduce energy expenditures by not nursing their own neonates? Is it an option that newborns seek when their nutritional needs are unsatisfied? Or is it a group strategy that distributes energy outlays to ensure greater success in neonate survival?

Some stressors might come at a cost in lactation; for example, high ambient temperatures can decrease food consumption, milk production, reproductive performance, and growth of the young in farm animals [101]. In pigs, Black et al. [102] showed that when the environmental temperature increased from 18 to 28 °C, milk production decreased by 25%, consumption by 40%, and oxygen uptake decreased from 523 to 411 mL/min. Animals might succumb to hyperthermia if they cannot maintain thermoneutrality, affecting not only the energy balance, but also water, Na, K and Cl metabolism, which are important constituents of sweat, the most important thermoregulatory mechanism used to dissipate excess body heat [103]. In addition, it is possible that even some stressful situations during pregnancy might influence postpartum milk production [104] and the adaptability of the offspring [105].

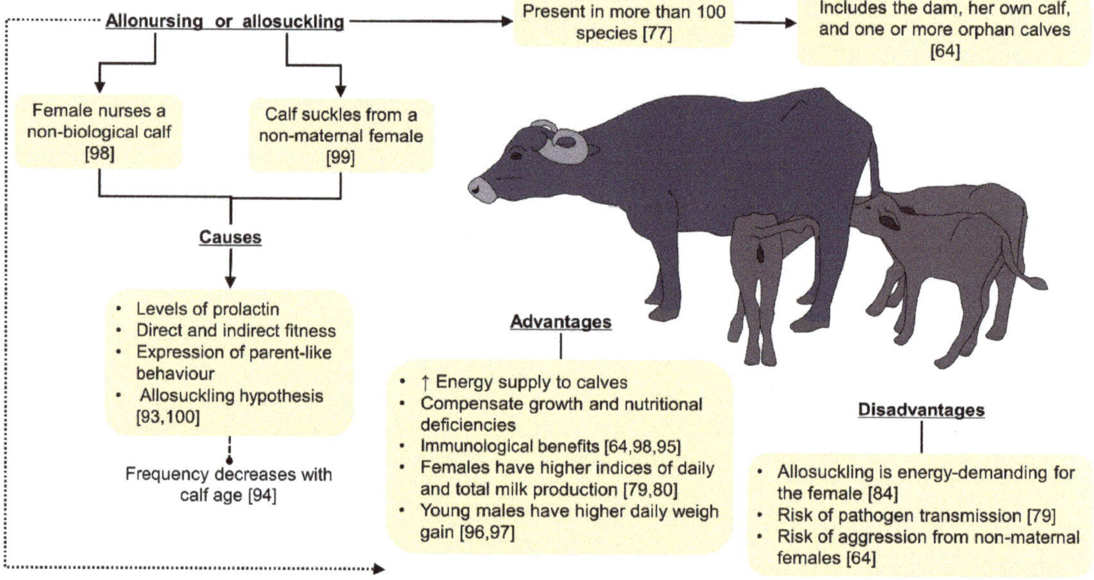

Figure 2. Allonursing and allosuckling causes, disadvantages and advantages to the mother and the offspring. While allonursing and allosuckling respond to endocrine and social contexts, they involve difficulties to the mother, such as the high energy demand, and to the young because of the risk of aggression of the foster mother, and the risk of pathogen transmission to the calves. However, some benefits include the compensation of nutritional deficiencies in the calves, and the acquisition of a wider range of immunoglobulins to enhance the immune system, as well as an enhancement in productivity parameters, such as daily weight gain.

3. The Neurophysiology of Suckling

The mother's milk is produced in the mammary gland, by extracting several chemical compounds from the blood. At the histological level, the alveolae produce the milk that accumulates and is stored in the excretory ducts and lactiferous sinuses before nursing begins. In terms of neurophysiology, the neurohormonal stimuli generated by palpation and suction of the teat—or any other stimulus that a dam associates with milking—are governed by the somatic nerves of the central nervous system that pass through the spinal cord [106]. These signals reach the hypothalamus, which releases oxytocin, principally, though a whole cascade of hormones is involved, including vasopressin from the posterior lobe of the hypophysis [107,108], which travels through the bloodstream to the mammary gland. Oxytocin is directly responsible for the myoepithelial contraction that releases the milk from the lactiferous ducts into the cistern of the gland and, from there, to the teat or nipple where it is ingested by the infant [109]. The neurophysiology of suckling or allosuckling is summarized in Figure 3.

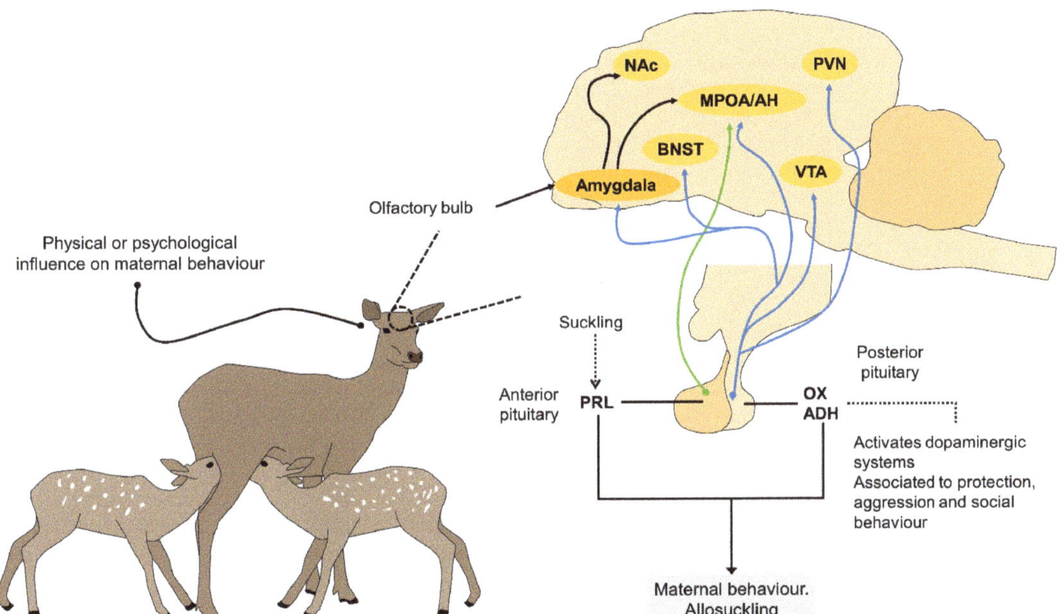

Figure 3. The neurophysiology of maternal behavior, milk let down, suckling and allosuckling. The main hormones associated with these processes (PRL, OX and ADH) act on different cerebral structures. The OX response is mediated by the VTA, PVN, amygdala, BNST and MPOA/AH. PRL site of action is in the MPOA/AH. The amygdala, NAc and BDNST also have a role in the reward system and maternal memory, and the former in the olfactory recognition of the offspring. ADH: antidiuretic hormone; BNST: bed nucleus of the stria terminalis; MPOA/AH: medial preoptic area/anterior hypothalamus; NAc: nucleus accumbens; OX: oxytocin; PRL: prolactin; PVN: paraventricular nucleus; VTA: ventral tegmental area, adepted from [110–114].

The mechanism of suction, or suckling, has been associated with inhibition of the releasing of the luteinizing hormone (LH) and the return to ovarian cyclicity after birth. In beef cows, the episodic secretion of the gonadotropin-releasing hormone (GnRH) by the hypothalamus modulates the hypophyseal pulsatile release of LH. Although the hypothalamic content of GnRH is not affected by the condition of lactation, the concentrations of GnRH in the pituitary portal system are suppressed by the act of suckling. During the postpartum period, the patterns of LH secretion remain at sub-optimal levels during the development of the preovulatory follicles. The renewal of cyclicity and ovulation is delayed until the frequency of the LH pulses increases to the threshold observed during proestrus. Once the offspring is removed, there is a delay of 24–48 h for the frequency of the LH pulse to increase, but inhibition is often reestablished quickly after the return of the calf in the early puerperium period [115]. Silveira et al. [115] explored the possible role of maternal behavior in anovulation mediated by lactation, hypothesizing that identifying the suckling calf is a critical determinant of anovulation induced by nursing. Those researchers randomly assigned 27 crossbreed meat cows to one of the following three study groups: dams that nursed an alien calf ($n = 11$); dams that nursed their own calf ($n = 8$); dams whose calves were separated for 6 days ($n = 8$). They observed that forced suckling by alien calves did not attenuate the release of LH in the cows after extraction by their own calves. In addition, the anovulatory intervals of the allonursing cows were similar to those of the weaned, non-nursing cows (study group 3). Moreover, neither the suckling posture nor the stimuli that resulted from contact with the udder affected LH secretion. The authors concluded that the maternal bond is important, but not essential, for anovulation mediated by lactation.

On the other hand, stimulation of the sow's teat by both filial and non-filial offspring benefits milk production. In pigs, constant teat stimulation by the piglet has been shown to maximize colostrum production [116,117], enhance mammary gland growth [118], and promote better local blood circulation [119]. Therefore, although the effect of nursing frequency, littler size, and weight, among others, can affect milk production, massaging of the teat, and the consequent hormonal and circulatory changes increase milk production in females [120]. This enhanced mammary gland development has also been reported in meerkats (*Suricata suricatta*), in whom the parity (primiparous and multiparous) is associated with greater tolerance to allosuckling from alien pups [121].

4. The Main Hypotheses Explaining Allonursing and Allosuckling

There are some theories that attempt to explain why females accept non-filial offspring and why young allosuckle [80]. It may be intentional (i.e., when a dam is aware that the newborn she nurses is not her own, but allows it to accede to her udder anyway), or due to errors in breeding, where the dam is unaware that she is nursing an alien newborn, perhaps because she does not recognize it or fails to identify it while nursing it together with her own newborn [73]. Evolutionary theories on the origins of allonursing behaviors have proposed various phenomena, as follow: reciprocal altruism, selective parenting, poor targeting of parental care, a resource-optimizing strategy (care, food) [122], and maintaining social stability [65]. To date, the literature includes eight hypotheses regarding this kind of cooperative breeding, where the cost–benefit relation has a relevant role in the decision [92]. These include genetic bases as well as social and immunological benefits for both the offspring and the foster mother [123]. In addition, it is important to emphasize that all hypotheses are usually related to allosuckling and allonursing, so they are not mutually exclusive [92], with the exception of the improved nutrition and the compensation hypotheses. The main allonursing and allosuckling hypotheses are summarized in Figure 4 [19,22,77,80,95,124–128].

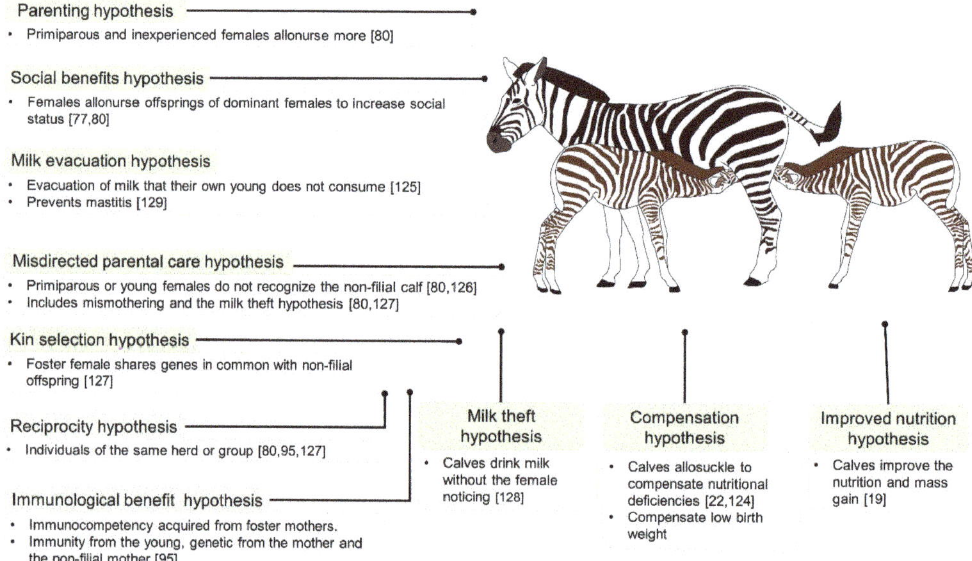

Figure 4. The main hypotheses explaining allonursing and allosuckling. Some of the hypotheses of these two cooperative maternal behaviors include genetic origins of the foster mother and the non-filial young, social and health benefits for the female, and advantages to the offspring to compensate and improve nutrition and immunocompetency.

4.1. Allonursing: A Strategy Adopted by Dams?

4.1.1. Kin Selection or Selective Parenting

This hypothesis describes the relationship that is established between females and offspring that share kinship, or between those who are close relatives, where dams only allow alien offspring to suckle if they share genes of common ascendence, as this ensures that the infants will survive to disseminate those genes in the group [73,77]. Analyses of the association between allonursing and litter or group size suggest that this is more common in primiparous females than multiparous species [122]. Another observation is that allonursing occurs more often in reduced groupings with close kinship [122]. In monotocous species, such as African elephants (*Loxodonta africana*), the prevalence of calves attempting to feed on individuals within their families, or closely related individuals, is approximately 78.9% [129]. However, this hypothesis does not apply to all species. Banded mongoose (*Mungos mungo*) is a species where the collective care of the offspring is performed by an adult who is responsible for feeding, protecting, and transmitting skills to single young. It has been reported that allonursing in these animals does not follow a genetic distinction between young. However, the durations of care and interaction are longer for those with genetic heritage, and sex-specific behavior has been observed in mongooses, where females care more for same-sex, non-filial newborns [130]. In water buffaloes, this kinship relationship did not show influence in a herd of 30 buffaloes. Of 570 allosuckling events, 351 were from alien calves; moreover, the calves of sisters and half-sisters had a higher success rate when requesting allosuckling from genetically unrelated mothers [90]. In otariids, such as *Arctocephalus australis* that practice philopatry shared breeding areas, allonursing has been observed to improve reproductive success [92].

In animals in which dams group according to kinship, alloparental care does not have an association with a greater variation in kinship, but is more frequent among species with litters [78]. In this case, the relative inversion by descendance is likely reduced, together with the additional costs of lactation, by dividing them among various newborns. The findings reported by MacLeod and Lukas [78] suggest that feeding alien offspring can progress rapidly when the additional costs for all dams are reduced in relation to the benefits it represents for their offspring. However, this distinction that some animals can make about kin selection among offspring has also been related to a high risk of aggression by unrelated herd members [131]. While the kin selection hypothesis can be prevalent in rodents, in the Sinai spiny mice (*Acomys dimidiatus*), allonursing was attributed to the maternal experience of the animals and to misdirected parental care [132]. Similarly, in the domestic reindeer (*Rangifer tarandus*), Engelhardt et al. [133] found no support for this hypothesis in 25 pairs of animals (mothers and calves). From 5176 successful allosuckling events, no correlation was found between the relatedness of the reindeers and the acceptance of non-filial calves.

The inclusive fitness/kin selection theory cannot explain how helping behavior between non-kin, and direct and indirect fitness can interact [134,135]. Alloparents are not always related to the offspring they help [136,137]. The direct fitness of helping is likely also important between kin [135]. Variation in relatedness explains approximately 10% of the variation in provisioning behavior in avian and mammalian cooperative breeding social systems [138,139]. The importance of indirect fitness is overestimated, whereas the importance of direct fitness is underestimated [140].

4.1.2. Reciprocity Hypothesis

Reciprocity is observed when females allonurse the non-filial young of mothers who have previously fed their offspring, which some authors considered to be some kind of reciprocal altruism, where the animal nurses or helps other individuals at the expense of their wellbeing [141]. This relation is proposed when free-roaming or captive dams spend substantial amounts of time separated from their offspring, though this does not seem to be an essential pre-condition for allonursing, since it occurs in otariids when the mother is absent, feeding in the sea, while her newborn is being fed on land by an alien female. Under those circumstances, mothers cannot be sure which female is reciprocating [92].

Tendencies in reciprocity imply a benefit of group rearing, since it promotes an improvement in the fitness of both females who assist in any milk deficiency of the biological female, and contributes to reinforcing the calf's immunity, which are aspects that involve other hypotheses that will be addressed later [80]. In a study with reindeers (*Rangifer tarandus*), by Engelhardt et al. [127], with 25 does, it was reported that females allonurse by reciprocity among lactating females from the group. Lactating wild sows in social groups also tend to accept suckling from alien piglets. This behavior is considered as inclusive fitness, where the piglet benefits from the alloparental care and the sow can forage without exposing its offspring to environmental and predator dangers [131]. Similarly, in wild Yellowstone bison (*Bison bison*), Jones and Treanor [142] observed cooperative behaviors, such as cleaning of the newborns, and even mutual consumption of the placentas, which are activities that could be a maternal temporary relief and increase the chances of the young's survival. In contrast, in water buffaloes, the reciprocity hypothesis has not been linked to allomaternal care [79].

There are other types of reciprocity, including generalized and indirect. In the first type, there is an effect that affects the entire population and increases the probability that all individuals will cooperate after social interaction, regardless of the recipient. In contrast, individuals practicing indirect reciprocity decide to help and cooperate with others despite whether they have received help from them or not [143]. The frequency of interaction is important in this type of reciprocity and involves a certain degree of evolutionary cooperation in the population.

4.1.3. Parenting Hypothesis

Among some terrestrial species, nursing the young of other dams in a proportion similar to their own newborns—collective lactation—may allow mothers to improve their reproductive performance compared to dams that do not share their milk with other offspring in the group [73], particularly in primiparous females [80,92]. An example of this is observed in female water buffalo, in which 97% of 30 dams allonursed alien offspring due to the mother's experience [90]. Additionally, studies show that dams who feed their own calves and non-filial ones have an increased quantity of produced milk [79,80]. Furthermore, studies have determined that dams nurse alien offspring to improve their maternal abilities. This may help explain why inexperienced dams, or females that have yet to reproduce, perform allonursing behavior more often than experienced ones, to learn to parent [73]. Spontaneous allonursing events were reported in dwarf mongooses (*Helogale parvula*) in the work of Creel et al. [144]. In this study, females with spontaneous lactation could nurse and suckle non-filial young. In the case of otariids, however, juvenile or inexperienced dams have not been observed to participate in allonursing, perhaps because they only give birth once a year, so it would be difficult for them to obtain maternal experience by nursing or caring for two offspring at the same time [92].

This theory also includes non-breeding females or those with pseudopregnancy and spontaneous lactation, such as carnivores, particularly canids, in whom alloparental care has also been reported [145]. Additionally, in these taxa, allonursing is correlated with high concentrations of prolactin and oxytocin, and low concentrations of testosterone and glucocorticoids [145]. High concentrations of these neurochemicals have been associated with an increase in cooperative and maternal behaviors [146], as well as milk ejection. They bind to their respective receptors in the hypothalamus, the posterior pituitary gland, and other structures, such as the mammary gland [145]. When a filial or alien young suckles a female, this stimulates the receptors located in the teats and induces the secretion of prolactin and oxytocin [19]. On the other hand, testosterone induces paternal behaviors [145], while the reduction in glucocorticoids is associated with improved fitness in communally breeding species, such as rodents [147].

4.1.4. Social Benefits Hypothesis

Cooperative nursing and feeding of non-filial offspring are also associated with social benefits within the group, such as a reduction in aggression by other members, a reduction in infanticide, and the maintenance of social rank [69]. Some reports describe allonursing as an extreme communal form of raising offspring in mammals, a phenomenon that may be due to a herd's social stability, since it rarely appears in unstable groups; for example, this behavior is often observed in artiodactyls, but rarely in equines [65]. Pluhácek and Bartosová [148] reported, for the first time, allosuckling in a common hippopotamus herd (*Hippopotamus amphibius*), composed of one adult male, one male offspring, and two multiparous females, who were the mother and grandmother of the offspring. Although these animals were in captivity, this type of allosuckling event could be related to kin selection and a certain type of social benefit, where the daughters preferred to stay and rear their offspring in the same group, to protect and increase the survival rate of the young. Similarly, in species with a strong social group structure, such as meerkats (*Suricatta suricatta*), allonursing practices are common, mainly in mothers that have lost their young. These animals do not only allosuckle non-filial young, but also engage in anti-predatory activities, such as staying alert to protect the young, involving the entire community [149]. A similar situation was observed in wild cavies (*Cavia aperea*). Cavies usually live in groups of one to four adults, where the mother–offspring bond is strong. In these animals, alloparental care is unusual. However, it was observed in a group of one male, two females, and four young, in which social structure, food availability, calving time, group adaptation, and cohesiveness may be factors that contribute to the presentation of this behavior. In contrast, in reindeers, cooperative care of animals has not been shown to provide a social benefit for the mother [133].

Another benefit that is associated with social improvement includes the group augmentation hypothesis, which is considered to be a way to enhance indirect fitness [150]. This hypothesis refers to a reproductive improvement in breeders, survival of the herd, and keeping a large population of individuals within a group that can promote long-term benefits [151]. It is important to consider that the group augmentation hypothesis can also enhance direct fitness [151–153].

4.1.5. Milk Evacuation (Milk Dumping)

The milk evacuation hypothesis is related to the excess of milk and the inflammatory consequences that this causes in the dam. This hypothesis states that allonursing helps to evacuate milk that the dam's own infant does not consume by offering it to alien offspring. Among wild animals, this practice could have the advantage of reducing the female's weight, improving their agility when being hunted [73]. This pattern is observed more frequently in polytocous species, due to the large litter size, where it also constitutes a benefit by reducing the cost of milk production, since they are species with large litters by nature [104]. In the evening bat (*Nycticeius humeralis*), Wilkinson [125] stated that mothers who produce large amounts of milk allonursed alien pups to prevent mastitis development and promote milk evacuation.

In some species, milk production has also been associated with allonursing. Allosuckling also increases teat stimulation, with consequent oxytocin release and increased milk production in some cases. This observation is similar to that found in a group of 35 buffaloes of the Murrah, Jaffarabadi, and Mediterranean breeds. The buffalo females that fed alien offspring showed higher levels of daily milk production and total production (peak of 6.96 ± 0.3 kg/day, and 1853.0 ± 76.2 kg, respectively), especially in those who allonursed male calves [97]. In contrast, as mentioned by Paranhos da Costa et al. [96], the daily milk production of buffaloes suckled by female calves was higher than in those suckled by males (4.56 ± 1.01 vs. 4.181 ± 0.60, respectively), although the bull-calves obtained greater weight gains (0.49 ± 0.13) compared to the heifers (0.39 ± 0.11). Something remarkable regarding this hypothesis is that sharing milk with non-filial young does not imply a nutritional imbalance for the filial calf, since the milk offered represents an amount that the filial calf cannot consume from its mother [78]. Similarly, in females who have lost their young, due to miscarriages or other events, alloparental care benefits the health of the udder by promoting the expulsion of milk and preventing mastitis [129].

4.1.6. Misdirected Parental Care

In contrast to these proposals, but keeping in mind that the priority for the females of gregarious animals is to optimize the care and feeding of their own offspring, allonursing has also been conceived—negatively—as an erratic or misdirected behavior, attributed to dams who fail to perceive that they are nursing an alien newborn. This hypothesis includes the allomaternal care events in which the female nurses and feeds non-filial offspring without recognizing it as alien young, or perhaps she is aware that an alien newborn is stealing her milk, but does not reject it. Some authors infer that in this case, the benefit of group rearing is greater than the cost of being vigilant and preventing an alien calf from stealing the milk [98]. Within this hypothesis, mismothering and milk theft are also included. Both are considered to be maladaptive processes [133]; however, although some authors use both without distinction, the hypothesis of milk theft is often associated with a trait linked to the newborn rather than to the mother [154].

Zapata et al. [99] reported such apparent unawareness, or lapses in attention, among wild guanaco dams in the Parque Nacional Torres del Paine in Southern Chile. They observed two guanaco calves, about one month in age, being nursed simultaneously by a female. One was her own young, and the other an alien. Both newborns appeared to be healthy and in a good nutritional condition. During the periods of allosuckling, the dam apparently did not realize that the alien calf was positioned behind her own offspring. Even after moving her head to look behind, she showed no active signs of rejection, such as walking away, kicking, or spitting; she did not reject the alien offspring. The authors hypothesized that her behavior might be based on a cost–benefit balance, in that the milk thieving produced less pathological risk than the energy expenditure required to remain sufficiently alert to detect, recognize, and reject the alien calf [73].

In pinnipeds, such as the South American fur seal (*Arctocephalus australis*), recognition of the young, after foraging trips to the sea, is performed through vocal and olfactory signals. The pup does not usually set connections with members other than its biological mother. However, low percentages of allosuckling have been observed (3.37%), and are attributed to misdirected parental care when the mother returns from the sea and cannot recognize the pup. It can also be motivated by the hunger of the offspring, who approaches the non-filial mother to suckle, with the risk of suffering aggression from them, while the mother returns. Another likely reason could be the individual characteristics, such as behavioral syndromes and the personality of the offspring that performs the mentioned behaviors [92]. This result was similar to that reported in Steller sea lions (*Eumetopias jubatus*), species in which the period required to raise a single pup is long. Maniscalco et al. [126] observed only 28 cooperative breeding events, where eight primiparous females devoted more time to non-filial care (median of 359.1 s), due to maternal inexperience, in comparison to the same number of multiparous females (29.8 s), who also rejected any approach unless they were found sleeping. Furthermore, adoption was described in a mother who had lost her pup after a few days of birth. Although another 19 pups died, no other adoption case was reported, so the authors did not consider it as misdirected parenting, since the adoptive mother had previous maternal experience. Lastly, even though females are considered to give allomaternal care because they are not capable of identifying filial from non-filial young, selectivity in nursing certain offspring has also been reported, which can be interpreted as the mother understands that it is not her young, but the benefit of nursing the alien offspring is greater [125].

The studies mentioned above show that the maternal strategy of allonursing has significant benefits for the newborn fitness that can also strengthen the immune system. Examples of these benefits are accelerated growth, better nutrition, and survival of the newborn among the group [155]. This is due to the protection of the young by other members of the herd and the reduction in time between meals when several mothers feed non-offspring [78]. Similarly, this could be due to licking, huddling, thermoregulation of the newborn, and the improvement in their immunocompetence by the transmission of immunoglobulins and lymphocytes through the milk of different mothers [156]. For the mothers, the improved reproductive capacity of females has also been reported [156]. In addition, there are some limited benefits to the mother when parenting is directed at close relatives [78], although in house mice this is mentioned as a direct adaptive benefit [157]. These benefits, however, entail a cost for the female, since lactation requires the availability of a greater energy load [78]. Various explanations have been proposed to elucidate the benefits of this costly behavior for dams, most of which are focused on adaptive aspects derived from maternal care. One suggests that females nurse to acquire experience in maternal care (as mentioned in the misdirected parental care hypothesis), while others posit allonursing as an effective means of evacuating excess milk that can accumulate and cause pain [78], foster systemic or mammary gland infections [92], or impede mobility. Still, others propose that it increases the probability of reciprocal behaviors, or provides indirect benefits when dams nurse the offspring of females in the group who are related by kinship ties (as stated in the kinship hypothesis) [78]. In pigs, the teat stimulation of filial and non-filial offspring has been shown to increase milk production in the current or subsequent nursing and allonursing behaviors [158]. However, there are some exceptions, and additional scientific studies must be conducted to clarify the precise causes of this maternal behavior and the circumstances in which it is performed in a wider range of species [122]. Future studies must consider the non-nutritional aspects of allonursing, such as when the newborns do not consume the mother's milk, but the mother accepts the interaction. In this situation, why are mothers giving offspring access to their udders? Understanding whether this effect is due to a benefit to the mother, in terms of milk production by stimulating receptors in the teats, or whether the mother simply responds to a biological need of the calf, could expand the current understanding of community care of the young. Similarly, the immunological benefits that can be obtained from communal breeding is another field that needs in-depth study to discern the physiological benefits of allosuckling.

4.2. Allosuckling: A Strategy for Offspring?

Allosuckling is often correlated with milk thieving by alien offspring and in females who allonurse after losing an infant [90]. This behavior provides advantages for newborns by allowing them to satisfy their nutritional needs by stealing milk from an alien dam [77,122].

4.2.1. Milk Theft

Milk theft constitutes one of the most studied hypotheses, in which non-filial offspring steal milk from non-filial mothers by placing themselves in positions where the female cannot see them, or by doing so in conjunction with filial calves to increase the probability of feeding success, as in the case of giraffes [123]. If the infant's own mother cannot satisfy its needs—because she died, is primiparous or inexperienced, or has inadequate milk production—it will turn to an alternative source and begin to steal milk from a substitute female [73]. This condition has been reported in domestic animals, such as bovines of the Curraleiro Pé breed in Brazil, where allosuckling occurred in the presence of the filial newborn and the non-filial calf. In those cases, fattening cows apparently did not discriminate between grooming and nursing alien offspring [159]. These cows, with low selectivity for their calves, have some practical uses within some cattle farms (Figure 5). In the river buffalo (*Bubalus bubalis*), allosuckling has been linked with maternal inexperience in young females, whose hungry calves may begin to steal milk from alien mothers [73] to satisfy their nutritional needs and ensure their survival. Among wild animals, reports on Iberian red deer (*Cervus elaphus hispanicus*) indicate that if the mother has low milk production, the offspring tend to suckle more from other females. Studies of this species thus propose allosuckling as a compensatory response to a reduced supply of the mother's milk [73].

Figure 5. Even though, in most ungulates, cows develop exclusive care towards the young that they recognize in the first hours postpartum, adepted from [56], some cows display no rejection to calves from other cows. In practice, these cows may be selected for the rearing of orphaned calves or for those whose mothers do not produce enough milk.

Regarding the observations of attempts by newborns to perform allosuckling, recent studies that compare the results of research from the 1990s show an increase in the percentage of time that offspring perform this behavior, over the 10% of total time that offspring devoted to suckling from alien dams in the 1990s. Attempts to explain this difference cite the controls imposed in earlier research, advances in the tools provided by the science of etiology, and increases in the frequency of animals being held in captivity, in intensive production systems that may trigger this behavior as part of a pattern to protect the offspring. The observations, in this regard, show that both domesticated and wild animals kept in captivity show increased allosuckling behavior by the offspring, reaching levels as high as 50 and 43% in river buffaloes and fallow deer (*Dama dama*), respectively, as well as in wild mouflon (*Ovis musimon*). Studies of captive Iberian deer found that milk ingestion by allosuckling reached levels as high as 37.8% of all nursing events when performed under group conditions [73]. Zapata et al. [99] analyzed 123 h of video recordings of wild guanacos, but observed only one case of allosuckling. They described the event as follows: While the mother was caring for her own calf in an antiparallel position, an alien young accessed the female in a parallel position behind the filial offspring. When the non-filial calf began to suckle, the mother smelled and recognized her own young. The dam showed a passive attitude after smelling her calf and remained still for around 30 s. The alien calf took advantage of that time to suckle, separating from the female when the group began to move. The researchers concluded that this behavior is uncommon among free-roaming guanacos. This agrees with Murphey et al. [79], who mentioned that the acceptance rate of an alien water buffalo calf increases when filial calves are feeding.

Concerning the position adopted in cases of successful allosuckling, Olléová et al. [65] found that the most common orientation of the Grevy's zebra foals was inverse parallel, followed by the parallel position, with very few cases of the perpendicular posture. Their study documented allonursing in zebra species housed at the Dvur Králové Zoo in the Czech Republic; there were 29 foals and 23 mares of plains zebras in three herds, 8 foals and 18 mares of Grevy's zebras (*Equus grevyi*) in one herd, and 6 foals and 9 mares of mountain zebras. They observed the following four specific events: successful nursing (uninterrupted from 5 to 60 s); attempts by foals to suckle at their mothers' udder (<5 s or when the mother did not permit contact with the teats); attempts to allosuckle; successful allosuckling. They also noted the nursing foal's position relative to the mare (inverse parallel, perpendicular, or parallel). Their findings for the Grevy's zebras revealed a relation of 1729 successful filial nursing attempts vs. 824 unsuccessful attempts, with 13 successful allosuckling events vs. 117 failed attempts. Among the plains zebra foals, both successful and unsuccessful attempts at nursing from their own mothers were more frequent (4614 successes vs. 3192 failures). No successful attempts at allosuckling were observed, and only one failed attempt was recorded. The observations of the mountain zebras showed greater success in nursing behavior with the foals' own mothers (843) compared to failed attempts (296), but only one, unsuccessful, attempt at allosuckling. The authors concluded that Grevy's zebra mares show greater tolerance towards alien foals than those of the other two zebra species, perhaps due to their social bonds and the lack of a hierarchy in this species compared to the other two. In addition, allosuckling may occasionally occur among Grevy's zebras, only during a specific period, as has been observed in captivity. Finally, the foal's perpendicular position means that it is easily detectable by the alien dam, but even so, she allowed it to nurse. This led the authors to suggest that the dams' position in the group's social system could influence the frequency of allosuckling in this species.

In semi-domestic reindeers (*Rangifer tarandus*), milk theft and antiparallel positions were attributed to allosuckling events [127], in which females recognize non-filial offspring (misdirection does not occur) and kin selection has no relevance in their care. A similar case was observed in Bactrian camels (*Camelus bactrianus*); this theory has also been reported as the reason for allosuckling, a prevalent behavior of older calves allosuckling, motivated by the weaning of the biological mother [146]. Regarding Bactrian camels (*Camelus bactrianus*), Miková and Sovják [73] observed that animals held in captivity seemed to exhibit allosuckling behavior similar to that observed in other animals. Similarly to other ungulates in the wild, this species lives in social groups, which is a fact that seems to favor allosuckling behavior by offspring. These groups of camels are made up of mostly females with their young. During the reproductive period, they are joined by males that fight to maintain their harem. Groups maintained in captivity are similar—females and offspring, adults, and males—but males are allowed to mate with the females during the estrus period, and they remain together year-round.

There is no scientific evidence suggesting that captivity generates allosuckling in these offspring, which remain with their mother for 1–2 years, a period of virtually exclusive care, though the dams do not perform active maternal behaviors, such as cleaning the calf or bringing it close to the udder [73]. This is in contrast with other domesticated animals' characteristic behaviors, such as buffaloes, cows, and sheep [64]. In contrast, there are reports that allosuckling in guanacos held in captivity ranges from 4.1–40% of all nursing episodes. This suggests that the indices of allosuckling in one species can vary considerably, suggesting the need to contemplate environmental conditions as well [99]. In a study of Pampas deer (*Ozotoceros bezoarticus*, Linnaeus 1758), Villagrán [160] mentions that, under conditions of semi-captivity, the duration of the mother–infant bond may be prolonged naturally, due to space limitations and the increased frequency of contact among individuals. This may also facilitate allosuckling behavior.

When alien offspring allosuckle, there are certain modifications in the signals of the filial behavior of both the mother and the young. Silveira et al. [115] studied 24 primiparous crossbreed heifers and 11 multiparous beef cows (parity 2–9) and observed that after the implementation of a regimen of controlled nursing, the dams initially smelled the alien calf, often thoroughly, but ignored it later when it attempted to nurse. When the calves born to the beef cows were exposed to the alien mother, they attempted—almost without exception—to avoid face-to-face interaction, instead of concentrating their efforts to bypass a board placed to block access to the udder. In contrast, the dams of that group generally responded to their offspring by exchanging vocalizations, smelling, licking, and staying close during periods of face-to-face interaction. Another behavior observed frequently was pseudo-suckling of the dam's chest area and neck. The filial calves, in comparison, sought face-to-face interaction and waited, showing some patience, for the board to be removed, so they could begin to suckle. The filial calves of the cows were nursed completely and adequately shortly before the 10 min period elapsed. They were then removed easily from the corral where the mother was held. The alien calves nursed more vigorously, more often, and apparently touched the teats with greater force. Invariably, the alien calves had to be forcibly separated from the teats after 10 min and then expelled from the pen. The typical nursing posture adopted by all calves was either the classic inverse parallel position at a 45° angle, or perpendicular to the mother's body [115].

In general, this hypothesis applies better for monotocous females, for whom milk production means a high energy demand, so they do not usually share their production voluntarily [78]; it also applies for species whose litter size is large, so it is more difficult for the mother to identify non-filial offspring [77]. Although, König [157] considers that in polytocous species, the milk theft hypothesis is less probable. In house mice, communal nursing represents a benefit for the pup, regarding thermoregulation, defense against predators, and feeding. In the African lion, the females may recognize the alien newborn, but perform allonursing nonetheless.

4.2.2. Compensation

In primiparous females that usually give birth to young with low birth weights, or whose amount of milk is not enough to cover the needs of the calf [88,129], compensation by allosuckling has been reported in cattle and buffaloes [161]. Zapata et al. [124] have reported that, in guanacos, the mothers of calves that request allosuckling have a low body condition ($p = 0.02$) and nutritional deficiencies that can lead the young to feed on other dams, who accept the interaction 57% of the time. On the other hand, female water buffalo that suckle other calves tend to restrict the amount of milk they consume, so the calf prolongs the sucking time and performs allosuckling to meet their nutritional requirements [93]. It is important to note that allosuckling should not be seen as a negative aspect, or isolated from the characteristics of productive units and conditions of captivity, because it impacts the offspring's productive performance, just as birth weight, birth order, sex, and age do [73]. Víchová and Bartoš [94] reported that female offspring suckle more from alien mothers than male calves, and in the young of fattening cattle than those of crossed dairy breeds. Paranhos Da Costa et al. [96] found that male buffaloes feed more from their own mother (2.25 times) and non-filial females (2.4 times), with greater daily weight gain (0.490 ± 0.13 kg/day), than females, who spend more time, on average, suckling from their own mothers (two times) [96,97].

Engelhardt et al. [19] compared the compensation and improved nutrition hypotheses in reindeer (*Rangifer tarandus*). Regarding compensation, they evaluated 25 animals with their calves to quantify whether newborns benefit when having a low birth mass, insufficient maternal milk supply, or delayed growth. The results showed that, although the number of allosuckling events was not influenced by low birth weight, the same animals increased their mass gain while being more involved in allosuckling bouts (0.46% increase from birth to 67 days old); however, no data were found to support the compensation hypothesis. In contrast, in other species, allosuckling has been reported to respond to deficiencies in newborns, such as domestic cattle (*Bos taurus*). In these, the incidence of allo-suckling was higher in calves that suckled from alien mothers, which was associated with low birth weight or an insufficient supply of maternal milk (GEE, X2 (1) = 3:73, *p* = 0.05) [94]. In contrast, in 33 fallow deer (*Dama dama*), despite a high incidence of allosuckling (73%), no data were found that could confirm that fawns allosuckle to compensate for their milk requirements not being met [162].

Another example is found in the work of Réale et al. [163], with mouflon (*Ovis gmelini musimon*) during the lambing season. In this study, the authors found that the lambs performed allosuckling in seasons in which the food resources were limited, and had lower growth rates, which is considered to be a constraint in maternal expenditure. In 10 cattle dams, with 20 twin calves from several-bred crosses who showed a high frequency of allosuckling (in 42% of the bouts), the compensation hypothesis was associated with mothers who did not have enough milk for their young [164].

4.2.3. Improved Nutrition

In river buffalo, calves consume the milk of a non-filial female, despite feeding on their mother, perhaps due to nutritional deficiencies that are compensated by allosuckling [64], although a low growth rate of the donors' calves has also been reported [79]. On the other hand, there are also situations in which allosuckling does not imply any advantage to the calf, as mentioned by Lee [129] in African elephants (*Loxodonta africana*), where allosuckling events represented only 3.7% (of a total of 1865 events) and were performed on nulliparous females. Therefore, they did not provide milk to the alien offspring, and nutritional improvement was absent. The effect that allosuckling exerts on foster mothers was also investigated in a group of meerkats (*Suricatta suricatta*), where allosuckling females lose around 1.43 ± 44.07 g of weight overnight, representing a non-significant change in their weight. Contrarily, non-allonursing mothers had an average weight gain of 20.07 ± 56.14 g, which implies that there is a very low degree of energy stress that does not represent a risk in these species, where the maintenance of an adequate body condition is important for their survival, fertility, and dominance [149].

A greater frequency of allosuckling is also sex-related, as male and female calves, respectively, may perform this behavior occasionally or frequently. On this topic, Drábková et al. [165] studied a group of 28 farmed red deer over two seasons, recording 1730 episodes of suction during 1696 episodes of lactation in 38 fawns born to 23 does. They classified the fawns as frequent or sporadic allosucklers. They observed at least one allosuckling event in 26 of the 38 fawns. Allosucking behavior began during the first week of the fawn's life. The male fawns performed more allosuckling than females and their duration of allosuckling was longer. The authors concluded that allosuckling fawns have different nursing behaviors, and that recurrent and sporadic allosucklers should be considered when analyzing this behavior.

Effects of Sex and Age of Offspring on Allonursing

Another feature observed is that as the offspring's exclusive alimentation with the mother's milk decreases with greater age, the frequency of allosuckling also diminishes. Víchová and Bartoš [94] found, in beef cattle, that as the age of the calves increased, allosuckling tended to decrease. Naturally, this tendency towards a reduced intake of the mother's milk with greater age has also been observed in red deer in the wild, where nursing ends after six to seven months, and on commercial farms, where weaning may occur between 3 and 12 months of life. This act invariably reduces the duration of the mother–infant bond and fosters the offspring's independence [160]. Hence, allosuckling also decreases as the age of the alien offspring increases, as Drábková et al. [165] reported for farmed red deer (*Cervus elaphus*).

4.2.4. Immunological Function or Benefit

One of the main benefits of allosuckling for the calves is the possibility to obtain a wide variety of antibodies by feeding on different females, to enhance their immune system and resistance to diseases [80]. The dam's own offspring and an alien newborn could obtain cross-transfer antibodies that are acquired passively by newborns in the group.

When the dam is exposed to pathogenic microorganisms, the mammary gland produces IgA immunoglobulins through plasmatic cells derived from B lymphocytes in the intestine. These immunoglobulins migrate with the plasmatic cells into the milk 2 to 4 weeks before farrowing, then the transfer of IgG and IgM from the blood to the mammary tissue takes place and peaks at parturition. Therefore, premature births, excessively short drying periods, or induced births can produce low amounts of colostrum in immunoglobins, such as IgG. This is important because colostrum is the fluid that, on average, contains the highest concentration of immunoglobulins; it contains 6 g in 100 g, ranging from 2 to 23%, compared to just 0.1% in milk [109]. In foster cows, suckling by alien calves in the first hours after birth is important for colostrum intake [166]. In contrast, wild pigs seem to prevent suckling from non-filial offspring by separating from the group before farrowing and rejoining 10 days after birth [158]. In another aspect, the mother–young binomial (her own or alien) benefits from allonursing because the newborn's oral cavity contains a variety of microorganisms, many of which could be pathogenic. Contact with the dam's tissues can cause cell lysis that activates the production of antibodies, which are transported in the bloodstream to the antigens contained on the cell surface of the microorganisms. As a result, ingesting maternal antibodies through the milk of the real or alien mother protects the young from possible infection. The allosuckling infant receives greater protection because it obtains higher antibody levels from the milk of different dams, especially when we consider that the females may be in distinct stages of antibody production for a certain pathogen, and implies genetic advantages in which some females may have specific alleles that allow them to produce distinct or more effective antibodies than those of the biological mother [95]. The immunological advantages not only refer to those destined for the calf, as Roulin [80] has described an endocrine regulation of allosuckling, where the maintenance of high concentrations of prolactin contributes to the immune reinforcement of the mother.

In rodents (*Octodon degus*), the relationship between communal rearing and the transmission of passive immunity, measured in the ratio of white cells and IgG in the mother and newborns, has been studied. However, none had an association between immunocompetence and communal rearing [167]. In the same species, Becker et al. [168] found that communal breeding offers immunological properties, in which feeding on more than one female allows them to obtain antibodies and defense cells, such as IgG and IgA. Although it was shown that mothers can transfer antibodies to offspring and non-offspring during lactation, the data did not show an immunological improvement to support the hypothesis of immunocompetence.

Finally, it is important to highlight that allosuckling may entail disadvantages for the alien offspring, such as a higher probability of suffering lesions by the responses of particularly aggressive alien dams [124], and a high risk of the transmission of pathogens through the milk [64]. Dalto et al. [169] suggest that allosuckling is a risk factor for contracting Johne's disease or paratuberculosis in buffalo, while milk production can also be affected in the dam-rearing systems of cattle and buffaloes [170].

4.3. Altruism in Allomaternal Care?

Although the mentioned hypotheses seek to explain why females engage in allomaternal behaviors, the probability that non-filial mothers accept alien offspring has also been associated with an act of altruism [64]. In these circumstances, the non-filial offspring succeeds in allosuckling without the presence of the biological calf; for example, Brandlová et al. [171] observed five events of allosuckling in camels (*Camelus bactrianus*) and concluded that there is a probability that females recognize the young as a non-filial calf, but tolerate the approach because the benefit is higher or does not represent a considerable cost. These reasons continue to be research elements for further work in all species in which alloparental care predominates.

The effect of stress deserves a final consideration. In several studies [172,173], suckling is associated with significant alterations in the reactivity of the hypothalamic–pituitary–adrenal (HPA) axis during the postpartum period. During lactation, suckling increases both oxytocin and prolactin release, and decreases the plasma levels of ACTH and cortisol, suggesting an inhibitory influence of lactogenic peptides on the HPA axis [174], which could add positive mental states, reducing stress levels during allonursing/allosuckling.

In general, nowadays, different benefits from allonursing/allosuckling have been suggested; for example, individuals can gain experience by caring for other offspring, which can translate into better reproductive performance when they become mothers [100]. However, in the future, it is necessary to evaluate this and other types of possible advantages, and carry out experiments that allow for controlling the effect that variables such as parity, age, and the female's own experience may have on those individuals. In addition to the above, it is necessary to increase not only the evolutionary approaches, but also to deepen the physiological mechanisms involved in allonursing/allosuckling. Similarly, within livestock production, it is necessary to evaluate the economic costs or productive and reproductive benefits that this type of behavior can bring.

5. Conclusions

The phenomena of allonursing and allosuckling are currently being documented in an ever-greater number of domesticated and wild animal species, including both terrestrial and aquatic mammals. Because allonursing forces alien dams to increase their energy expenditures, by allowing alien offspring to feed at their udders, it is a strategy that most commonly develops within social groups where the group members do not change frequently, and the social structure is maintained for long periods. During allonursing, both the alien offspring and the non-filial mothers obtain the following benefits: improved welfare for the former (by improving immunity, satiating hunger, and satisfying behavioral needs, such as sucking, among others) and better maternal performance for the latter (i.e., increasing experience). However, we cannot lose sight of the other side of the coin, since allonursing also has the potential to compromise animal health by transmitting disease-causing microorganisms. Nevertheless, even this negative aspect has a positive side, since increased antibody production and transport through lactation may benefit both the dam and the offspring. Further studies will be required to more precisely determine the specific benefits that allosuckling and allonursing could provide to dams and offspring, both their own and alien.

Author Contributions: Conceptualization, D.M.-R., A.O. and M.M.-R.; investigation, D.M.-R., M.M.-R., A.F.-d.-M., R.M., P.M.-M., A.D.-O. and A.O.; writing—original draft preparation, A.O. and D.M.-R., writing—review and editing, D.M.-R., M.M.-R., A.F.-d.-M., R.M., P.M.-M., A.D.-O. and A.O.; project administration, D.M.-R., A.O., R.M. and M.M.-R. All authors have read and agreed to the published version of the manuscript.

Funding: This research received no external funding.

Data Availability Statement: Not applicable.

Conflicts of Interest: The authors declare no conflict of interest.

References

1. Zilkha, N.; Sofer, Y.; Beny, Y.; Kimchi, T. From classic ethology to modern neuroethology: Overcoming the three biases in social behavior research. *Curr. Opin. Neurobiol.* **2016**, *38*, 38–96. [CrossRef] [PubMed]
2. Martínez, J.G. La evolución y la conservación de la biodiversidad. In *Evolución: La Base de la Biología*; Sur, P., Ed.; Universidad de Extremadura: Extremadura, Spain, 2012; pp. 407–416.
3. Lévy, F.; Keller, M. Neurobiology of maternal behavior in sheep. *Adv. Study Behav.* **2008**, *38*, 399–437. [CrossRef]

4. Mora-Medina, P.; Orihuela-Trujillo, A.; Santiago, R.; Arch-Tirado, E.; Vázquez-Cruz, C.; Mota-Rojas, D. Metabolic changes during brief periods of ewe-lamb separation at different ages. *Anim. Prod. Sci.* **2018**, *58*, 1297–1306. [CrossRef]

5. Mandujano, C. Ecología y sociobiología de la impronta: Perspectivas para su estudio en los Crocodylia. *Cienc. Mar.* **2010**, *14*, 49–54.

6. Yamaguchi, S.; Aoki, N.; Kitajima, T.; Iikubo, E.; Katagiri, S.; Matsushima, T.; Homma, K.J. Thyroid hormone determines the start of the sensitive period of imprinting and primes later learning. *Nat. Commun.* **2012**, *3*, 1081. [CrossRef]

7. Poindron, P. Mechanisms of activation of maternal behaviour in mammals. *Reprod. Nutr. Dev.* **2005**, *45*, 341–351. [CrossRef]

8. Val-Laillet, D.; Nowak, R. Socio-spatial criteria are important for the establishment of maternal preference in lambs. *Appl. Anim. Behav. Sci.* **2006**, *96*, 269–280. [CrossRef]

9. Nowak, R.; Keller, M.; Val-Laillet, D.; Lévy, F. Perinatal visceral events and brain mechanisms involved in the development of mother-young bonding in sheep. *Horm. Behav.* **2007**, *52*, 92–98. [CrossRef]

10. Muir, G.D. Early ontogeny of locomotor behaviour: A comparison between altricial and precocial animals. *Brain Res. Bull.* **2000**, *53*, 719–726. [CrossRef]

11. Wöhr, M.; Oddi, D.; D'Amato, F.R. Effect of altricial pup ultrasonic vocalization on maternal behavior. *Handb. Behav. Neurosci.* **2010**, *19*, 159–166. [CrossRef]

12. Broad, K.D.; Curley, J.P.; Keverne, E.B. Mother–infant bonding and the evolution of mammalian social relationships. *Philos. Trans. R. Soc. B Biol. Sci.* **2006**, *361*, 2199. [CrossRef] [PubMed]

13. Hernández, H.; Terrazas, A.; Poindron, P.; Ramírez-Vera, S.; Flores, J.A.; Delgadillo, J.A.; Vielma, J.; Duarte, G.; Fernández, I.G.; Fitz-Rodríguez, G.; et al. Sensorial and physiological control of maternal behavior in small ruminants: Sheep and goats. *Trop. Subtrop. Agroecosyst.* **2012**, *15*, S91–S102.

14. Nowak, R.; Porter, R.H.; Lévy, F.; Orgeur, P.; Schaal, B. Role of mother-young interactions in the survival of offspring in domestic mammals. *Rev. Reprod.* **2000**, *5*, 153–163. [CrossRef]

15. Dubey, P.; Singh, R.R.; Choudhary, S.S.; Verma, K.K.; Kumar, A.; Gamit, P.M.; Dubey, S.; Prajapati, K. Post parturient neonatal behaviour and their relationship with maternal behaviour score, parity and sex in surti buffaloes. *J. Appl. Anim. Res.* **2018**, *46*, 360–364. [CrossRef]

16. Glasper, E.R.; Kenkel, W.M.; Bick, J.; Rilling, J.K. More than just mothers: The neurobiological and neuroendocrine underpinnings of allomaternal caregiving. *Front. Neuroendocrinol.* **2019**, *53*, 100741. [CrossRef] [PubMed]

17. Koenig, W.D. What drives cooperative breeding? *PLoS Biol.* **2017**, *15*, e2002965. [CrossRef]

18. Lukas, D.; Clutton-Brock, T. Cooperative breeding and monogamy in mammalian societies. *Proc. R. Soc. B Biol. Sci.* **2012**, *279*, 2151–2156. [CrossRef] [PubMed]

19. Engelhardt, S.C.; Weladji, R.B.; Holand, Ø.; Nieminen, M. Allosuckling in reindeer (*Rangifer tarandus*): A test of the improved nutrition and compensation hypotheses. *Mamm. Biol.* **2016**, *81*, 146–152. [CrossRef]

20. Jennions, M.D.; Macdonald, D.W. Cooperative breeding in mammals. *Trends Ecol. Evol.* **1994**, *9*, 89–93. [CrossRef]

21. Schubert, M.; Pillay, N.; Schradin, C. Parental and alloparental care in a polygynous mammal. *J. Mammal.* **2009**, *90*, 724–731. [CrossRef]

22. Orihuela, A.; Pérez-Torres, L.I.; Ungerfeld, R. Evidence of cooperative calves' care and providers' characteristics in zebu cattle (*Bos indicus*) raised under extensive conditions. *Trop. Anim. Health Prod.* **2021**, *53*, 143. [CrossRef] [PubMed]

23. Pérez-Torres, L.; Orihuela, A.; Corro, M.; Rubio, I.; Cohen, A.; Galina, C.S. Maternal protective behavior of zebu type cattle (*Bos indicus*) and its association with temperament. *J. Anim. Sci.* **2014**, *92*, 4694–4700. [CrossRef] [PubMed]

24. Enriquez, M.F.; Pérez-Torres, L.; Orihuela, A.; Rubio, I.; Corro, M.; Galina, C.S. Relationship between protective maternal behavior and some reproductive variables in zebu-type cows (*Bos indicus*). *J. Anim. Behav. Biometeorol.* **2021**, *9*, 2124. [CrossRef]

25. Špinka, M.; Illmann, G. Nursing behavior. In *The Gestating and Lactating Sow*; Farmer, C., Ed.; Wageningen Academic Publishers: Wageningen, The Netherlands, 2015; pp. 297–318.

26. Alonso-Spilsbury, M.; Ramirez-Necoechea, R.; Gonzalez-Lozano, M.; Mota-Rojas, D.; Trujillo-Ortega, M. Piglet survival in early lactation: A review. *J. Anim. Vet. Adv.* **2007**, *6*, 76–86.

27. Sánchez-Andrade, G.; James, B.M.; Kendrick, K.M. Neural encoding of olfactory recognition memory. *J. Reprod. Dev.* **2005**, *51*, 547558. [CrossRef] [PubMed]

28. Mora-Medina, P.; Orihuela-Trujillo, A.; Arch-Tirado, E.; Roldan, S.; Terrazas, A.; Mota-Rojas, D. Sensory factors involved in mother-young bonding in sheep: A review. *Vet. Med.* **2016**, *61*, 595–611. [CrossRef]

29. Ramírez, M.; Soto, R.; Poindron, P.; Alvarez, L.; Valencia, J.; Gonzalez, F.; Terrazas, A. Maternal behaviour around birth and mother-young recognition in Pelibuey sheep. *Vet. Mex.* **2011**, *42*, 27–46.

30. Brus, M.; Meurisse, M.; Keller, M.; Lévy, F. Interactions with the young down-regulate adult olfactory neurogenesis and enhance the maturation of olfactory neuroblasts in sheep mothers. *Front. Behav. Neurosci.* **2014**, *8*, 53. [CrossRef]

31. Bădescu, I.; Watts, D.P.; Katzenberg, M.A.; Sellen, D.W. Alloparenting is associated with reduced maternal lactation effort and faster weaning in wild chimpanzees. *R. Soc. Open Sci.* **2016**, *3*, 1–14. [CrossRef] [PubMed]

32. Mota-Rojas, D.; Orihuela, A.; Napolitano, F.; Mora-Medina, P.; Gregorio, H.; Alonso-Spilsbury, M. Olfaction in animal behaviour and welfare. *CAB Rev. Perspect. Agric. Vet. Sci. Nutr. Nat. Resour.* **2018**, *13*, 1–13. [CrossRef]

33. Poindron, P.; Lévy, F.; Krehbiel, D. Genital, olfactory, and endocrine interactions in the development of maternal behaviour in the parturient ewe. *Psychoneuroendocrinology* **1988**, *13*, 99–125. [CrossRef]

34. Poindron, P.; Lévy, F. Physiological, sensory, and experiential determinants of maternal behavior in sheep. In *Mammalian Parenting: Biochemical; Neurobiological; and Behavioral Determinants*; Krasnegor, N.A., Bridges, R.S., Eds.; Oxford University Press: Oxford, UK, 1990; pp. 133–156.

35. Porter, R.H.; Lévy, F.; Poindron, P.; Litterio, M.; Schaal, B.; Beyer, C. Individual olfactory signatures as major determinants of early maternal discrimination in sheep. *Dev. Psychobiol.* **1991**, *24*, 151–158. [CrossRef]

36. Price, E.O.; Dunn, G.C.; Talbot, J.A.; Dally, M.R. Fostering lambs by odor transfer: The substitution experiment. *J. Anim. Sci.* **1984**, *59*, 301–307. [CrossRef] [PubMed]

37. Alexander, G.; Stevens, D. Odour cues to maternal recognition of lambs: An investigation of some possible sources. *Appl. Anim. Ethol.* **1982**, *9*, 165–175. [CrossRef]

38. Booth, K.; Katz, L.S. Role of the vomeronasal organ in neonatal offspring recognition in sheep. *Biol. Reprod.* **2000**, *63*, 953–958. [CrossRef]

39. Alexander, G.; Stevens, D.; Bradley, L.R. Washing lambs and confinement as aids to fostering. *Appl. Anim. Ethol.* **1983**, *10*, 251–261. [CrossRef]

40. Fraser, E.J.; Shah, N.M. Complex chemosensory control of female reproductive behaviors. *PLoS ONE* **2014**, *9*, e90368. [CrossRef]

41. Kratzing, J. The structure of the vomeronasal organ in the sheep. *J. Anat.* **1971**, *108*, 247–260.

42. Fleming, A.; Vaccarino, F.; Tambosso, L.; Chee, P. Vomeronasal and olfactory system modulation of maternal behavior in the rat. *Science* **1979**, *203*, 372–374. [CrossRef] [PubMed]

43. Wysocki, C.J.; Beauchamp, G.K.; Reidinger, R.R.; Wellington, J.L. Access of large and nonvolatile molecules to the vomeronasal organ of mammals during social and feeding behaviors. *J. Chem. Ecol.* **1985**, *11*, 1147–1159. [CrossRef]

44. Bean, N.J. Modulation of agonistic behavior by the dual olfactory system in male mice. *Physiol. Behav.* **1982**, *29*, 433–437. [CrossRef]

45. Bean, N.J. Olfactory and vomeronasal mediation of ultrasonic vocalizations in male mice. *Physiol. Behav.* **1982**, *28*, 31–37. [CrossRef]

46. Winans, S.S.; Powers, J.B. Olfactory and vomeronasal deafferentation of male hamsters: Histological and behavioral analyses. *Brain Res.* **1977**, *126*, 325–344. [CrossRef]

47. Bellringer, J.F.; Hester Pratt, P.M.; Keverne, E.B. Involvement of the vomeronasal organ and prolactin in pheromonal induction of delayed implantation in mice. *J. Reprod. Fertil.* **1980**, *59*, 223–228. [CrossRef]

48. Clancy, A.N.; Macrides, F.; Singer, A.G.; Agosta, W.C. Male hamster copulatory responses to a high molecular weight fraction of vaginal discharge: Effects of vomeronasal organ removal. *Physiol. Behav.* **1984**, *33*, 653–660. [CrossRef]

49. Coquelin, A.; Clancy, A.N.; Macrides, F.; Noble, E.P.; Gorski, R.A. Pheromonally induced release of luteinizing hormone in male mice: Involvement of the vomeronasal system. *J. Neurosci.* **1984**, *4*, 2230–2236. [CrossRef]

50. Johns, M.A.; Feder, H.H.; Komisaruk, B.R.; Mayer, A.D. Urine-induced reflex ovulation in anovulatory rats may be a vomeronasal effect. *Nature* **1978**, *272*, 446–448. [CrossRef] [PubMed]

51. Lehman, M.N.; Winans, S.S.; Powers, J.B. Medial nucleus of the amygdala mediates chemosensory control of male hamster sexual behavior. *Science* **1980**, *210*, 557–560. [CrossRef] [PubMed]

52. Powers, J.B.; Winans, S.S. Vomeronasal organ: Critical role in mediating sexual behavior of the male hamster. *Science* **1975**, *187*, 961–963. [CrossRef] [PubMed]

53. Reynolds, J.; Keverne, E.B. The accessory olfactory system and its role in the pheromonally mediated suppression of oestrus in grouped mice. *J. Reprod. Fertil.* **1979**, *57*, 31–35. [CrossRef]

54. Wysocki, C.J.; Katz, Y.; Bernhard, R. Male vomeronasal organ mediates female-induced testosterone surges in mice. *Biol. Reprod.* **1983**, *28*, 917–922. [CrossRef]

55. Wysocki, C.J.; Nyby, J.; Whitney, G.; Beauchamp, G.K.; Katz, Y. The vomeronasal organ: Primary role in mouse chemosensory gender recognition. *Physiol. Behav.* **1982**, *29*, 315–327. [CrossRef]

56. Estes, R.D. Mammalia: The role of the vomeronasal organ in mammalian reproduction. *Mammalia* **1972**, *36*, 315–341. [CrossRef]

57. McCotter, R.E. The connection of the vomeronasal nerves with the accessory olfactory bulb in the opossum and other mammals. *Anat. Rec.* **1912**, *6*, 299–318. [CrossRef]

58. Karimi, H.; Mansoori Ale Hashem, R.; Ardalani, G.; Sadrkhanloo, R.; Hayatgheibi, H. Structure of vomeronasal organ (jacobson organ) in male *camelus domesticus var. dromedaris persica*. *Anat. Histol. Embryol.* **2014**, *43*, 423–428. [CrossRef] [PubMed]

59. Hovis, K.R.; Ramnath, R.; Dahlen, J.E.; Romanova, A.L.; LaRocca, G.; Bier, M.E.; Urban, N.N. Activity regulates functional connectivity from the vomeronasal organ to the accessory olfactory bulb. *J. Neurosci.* **2012**, *32*, 7907–7916. [CrossRef]

60. Bergan, J.F.; Ben-Shaul, Y.; Dulac, C. Sex-specific processing of social cues in the medial amygdala. *eLife* **2014**, *3*, e02743. [CrossRef]

61. Charra, R.; Datiche, F.; Casthano, A.; Gigot, V.; Schaal, B.; Coureaud, G. Brain processing of the mammary pheromone in newborn rabbits. *Behav. Brain Res.* **2012**, *226*, 179–188. [CrossRef]

62. Poindron, P.; Otal, J.; Ferreira, G.; Keller, M.; Guesdon, V.; Nowak, R.; Lévy, F. Amniotic fluid is important for the maintenance of maternal responsiveness and the establishment of maternal selectivity in sheep. *Animal* **2010**, *4*, 2057–2064. [CrossRef]

63. Poindron, P.; Lévy, F.; Keller, M. Maternal responsiveness and maternal selectivity in domestic sheep and goats: The two facets of maternal attachment. *Dev. Psychobiol.* **2007**, *49*, 54–70. [CrossRef]

64. Mora-Medina, P.; Napolitano, F.; Mota-Rojas, D.; Berdugo-Gutiérrez, J.; Ruiz-Buitrago, J.; Guerrero-Legarreta, I. Imprinting, sucking and allosucking behaviors in buffalo calves. *J. Buffalo Sci.* **2018**, *7*, 49–57. [CrossRef]

65. Olléová, M.; Pluháček, J.; King, S.R.B. Effect of social system on allosuckling and adoption in zebras. *J. Zool.* **2012**, *288*, 127–134. [CrossRef]
66. Quigley, J.D.; Strohbehn, R.E.; Kost, C.J.; O'Brien, M.M. Formulation of colostrum supplements, colostrum replacers and acquisition of passive immunity in neonatal calves. *J. Dairy Sci.* **2001**, *84*, 2059–2065. [CrossRef]
67. Silva, F.L.M.; Miqueo, E.; da Silva, M.D.; Torrezan, T.M.; Rocha, N.B.; Salles, M.S.V.; Bittar, C.M.M. Thermoregulatory responses and performance of dairy calves fed different amounts of colostrum. *Animals* **2021**, *11*, 703. [CrossRef]
68. Solomon, N.G.; French, J.A. The study of mammalian cooperative breeding. In *Cooperative Breeding in Mammals*; Solomon, N., French, J., Eds.; Cambridge University Press: Cambridge, UK, 1997; pp. 1–10.
69. Riedman, M.L. The evolution of alloparental care and adoption in mammals and birds. *Q. Rev. Biol.* **1982**, *57*, 405–435. [CrossRef]
70. Komdeur, J. The effect of kinship on helping in the cooperative breeding Seychelles warbler (*Acrocephalus sechellensis*). *Proc. R. Soc. B Biol. Sci.* **1994**, *256*, 47–52. [CrossRef]
71. Bon, R.; Campan, R. Unexplained sexual segregation in polygamous ungulates: A defense of an ontogenetic approach. *Behav. Processes* **1996**, *38*, 131–154. [CrossRef]
72. Doolan, S.P.; Macdonald, D.W. Co-operative rearing by slender-tailed meerkats (*Suricata suricatta*) in the southern Kalahari. *Ethology* **2001**, *105*, 851–866. [CrossRef]
73. Miková, K.; Sovják, R. Agricultura tropica et subtropica a review: Possibilities of allosuckling occurrence in camels (*Camelus bactrianus*). *Agric. Trop. Subtrop.* **2005**, *38*, 91–94.
74. Lukas, D.; Clutton-Brock, T. Monotocy and the evolution of plural breeding in mammals. *Behav. Ecol.* **2020**, *31*, 943. [CrossRef]
75. Patel, E. Non-maternal infant care in wild silky sifakas non-maternal infant care in wild silky sifakas (*Propithecus candidus*). *Lemur News* **2007**, *12*, 39–42.
76. Dušek, A. Maternal Investment Strategy in Model Monotocous and Polytocous Mammals: A Life-History Perspective. Ph.D. Thesis, Charles University, Prague, Czech Republic, 12 May 2011.
77. Packer, C.; Lewis, S.; Pusey, A. A comparative analysis of non-offspring nursing. *Anim. Behav.* **1992**, *43*, 265–281. [CrossRef]
78. MacLeod, K.J.; Lukas, D. Revisiting non-offspring nursing: Allonursing evolves when the costs are low. *Biol. Lett.* **2014**, *10*, 20140378. [CrossRef]
79. Murphey, R.M.; Paranhos da Costa, M.J.R.; Da Silva, R.G.; De Souza, R.C. Allonursing in river buffalo, *Bubalus bubalis*: Nepotism, incompetence, or thievery? *Anim. Behav.* **1995**, *49*, 1611–1616. [CrossRef]
80. Roulin, A. Why do lactating females nurse alien offspring? A review of hypotheses and empirical evidence. *Anim. Behav.* **2002**, *63*, 201–208. [CrossRef]
81. Royle, N.J.; Smiseth, P.T.; Kölliker, M. *The Evolution of Parental Care*; Oxford University Press: Oxford, UK, 2013; pp. 1–377.
82. Davyt, P.D.; Talmón, T.D.; Villanueva, C.Á. Efecto de la Estrategia de Alimentación sobre el Gasto Energético en Vacas Lecheras. Bachelor's Thesis, Universidad de la República, Montevideo, Uruguay, 4 September 2017.
83. Gallina, S.; Bello, J. El gasto energético del venado cola blanca (*Odocoileus virginianus texanus*) en relación a la precipitación en una zona semiárida de México. *Therya* **2010**, *1*, 9–22. [CrossRef]
84. Gittleman, J.L.; Thompson, S.D. Energy allocation in mammalian reproduction. *Integr. Comp. Biol.* **1988**, *28*, 863–875. [CrossRef]
85. Freer, M.; Dove, H.; Nolan, J. *Nutrient Requirements of Domesticated Ruminants*; Csiro Publishing: Collingwood, ON, Canada, 2007; pp. 50–61. [CrossRef]
86. Zhao, Z.J.; Hambly, C.; Shi, L.L.; Bi, Z.Q.; Cao, J.; Speakman, J.R. Late lactation in small mammals is a critically sensitive window of vulnerability to elevated ambient temperature. *Proc. Natl. Acad. Sci. USA* **2020**, *117*, 24352–24358. [CrossRef] [PubMed]
87. Marotta, E. Determinación del requerimiento energético de la cerda reproductora mantenida a campo en base al clima y la etología. *Analecta Vet.* **2003**, *23*, 28–35.
88. Clutton-Brock, T.H.; Albon, S.D.; Guinness, F.E. Parental investment in male and female offspring in polygynous mammals. *Nature* **1981**, *289*, 487–489. [CrossRef]
89. Bérubé, C.H.; Festa-Bianchet, M.; Jorgenson, J.T. Reproductive costs of sons and daughters in Rocky Mountain bighorn sheep. *Behav. Ecol.* **1996**, *7*, 60–68. [CrossRef]
90. Moen, A.N. Seasonal changes in heart rates, activity, metabolism, and forage intake of white-tailed deer. *J. Wildl. Manag.* **1978**, *42*, 715–738. [CrossRef]
91. Clutton-Brock, T.H.; Albon, S.D.; Guinness, F.E. Fitness costs of gestation and lactation in wild mammals. *Nature* **1989**, *337*, 260–262. [CrossRef]
92. Stábile, F. Aloamamantamiento en *Arctocephalus australis* en Isla de Lobos (Uruguay): Estrategia individual de las crías? Bachelor's Thesis, Universidad de la República de Uruguay, Montevideo, Uruguay, February 2013.
93. Andriolo, A.; Paranhos, M.J.R.; Costa, D.A.; Schmidek, W.R.; Costa, M.J.R.P.; De Estudos, E.-G. Suckling behaviour in water buffalo suckling behaviour in water buffalo (*Bubalus bubalis*): Development and individual differences. *Rev. Etol.* **2001**, *3*, 129–136.
94. Víchová, J.; Bartoš, L. Allosuckling in cattle: Gain or compensation? *Appl. Anim. Behav. Sci.* **2005**, *94*, 223–235. [CrossRef]
95. Roulin, A.; Heeb, P. The immunological function of allosuckling. *Ecol. Lett.* **1999**, *2*, 319–324. [CrossRef]
96. Paranhos Da Costa, M.J.R.; Andriolo, A.; Simplício De Oliveira, J.F.; Schmidek, W.R. Suckling and allosuckling in river buffalo calves and its relation with weight gain. *Appl. Anim. Behav. Sci.* **2000**, *66*, 1–10. [CrossRef]
97. Oliveira, A.D.F.M.; Quirino, C.R.; Bastos, R. Effect of nursing behaviour, sex of the calf, and parity order on milk production of buffaloes. *Rev. Colomb. De Cienc. Pecu.* **2017**, *30*, 30–38. [CrossRef]

98. Bartoš, L.; Vaňková, D.; Šiler, J.; Illmann, G. Adoption, allonursing and allosucking in farmed red deer (*Cervus elaphus*). *Anim. Sci.* **2001**, *72*, 483–492. [CrossRef]

99. Zapata, B.; Gaete, G.; Correa, L.A.; González, B.A.; Ebensperger, L.A. A case of allosuckling in wild guanacos (*Lama guanicoe*). *J. Ethol.* **2009**, *27*, 295–297. [CrossRef]

100. Mumme, R.L. A bird's-eye view of mammalian cooperative breeding. In *Cooperative Breeding in Mammals*; Solomon, N., French, J., Eds.; Cambridge University Press: Omaha, NE, USA, 2009; pp. 364–388.

101. Fuquay, J.W. Heat stress as it affects animal production. *J. Anim. Sci.* **1981**, *52*, 164–174. [CrossRef]

102. Black, J.L.; Mullan, B.P.; Lorschy, M.L.; Giles, L.R. Lactation in the sow during heat stress. *Livest. Prod. Sci.* **1993**, *35*, 153–170. [CrossRef]

103. Kadzere, C.T.; Murphy, M.R.; Silanikove, N.; Maltz, E. Heat stress in lactating dairy cows: A review. *Livest. Prod. Sci.* **2002**, *77*, 59–91. [CrossRef]

104. Herrenkohl, L.R.; Whitney, J.B. Effects of prepartal stress on postpartal nursing behavior, litter development and adult sexual behavior. *Physiol. Behav.* **1976**, *17*, 1019–1021. [CrossRef]

105. Lay, D.C.; Randel, R.D.; Friend, T.H.; Jenkins, O.C.; Neuendorff, D.A.; Bushong, D.M.; Lanier, E.K.; Bjorge, M.K. Effects of prenatal stress on suckling calves. *J. Anim. Sci.* **1997**, *75*, 3143–3151. [CrossRef]

106. Benson, G.K.; Fitzpatrick, R.J. The Neurohypophysis and the Mammary Gland. In *Pars Intermedia and Neurohypophysis*; Butterworthpp & Co.: London, UK, 2020; pp. 414–443.

107. Bisset, G.W. The Milk-Ejection Reflex and the Actions of Oxytocin, Vasopressin and Synthetic Analogues on the Mammary Gland. In *Neurohypophysial Hormones and Similar Polypeptides*; Berde, B., Ed.; Springer: Berlin, Germany, 1968.

108. Ni, Y.; Chen, Q.; Cai, J.; Xiao, L.; Zhang, J. Three lactation-related hormones: Regulation of hypothalamus-pituitary axis and function on lactation. *Mol. Cell. Endocrinol.* **2021**, *520*, 111084. [CrossRef]

109. Arizala, J.A.; Olivera, M. Fisiología de la Producción Láctea en Bovinos: Involución de la Glándula Mamaria, Lactogénesis, Galactopoyesis, y Eyección de la Leche. In *Buenas Prácticas de Producción Primaria de Leche*; Olivera, M.A., Ed.; Fondo Editorial Biogénesis: Medellín, Colombia, 2007; pp. 143–151.

110. Svennersten-Sjaunja, K.; Olsson, K. Endocrinology of milk production. *Domest. Anim. Endocrinol.* **2005**, *29*, 241–258. [CrossRef] [PubMed]

111. Perez-Marquez, H.J.; Ambrose, D.J.; Schaefer, A.L.; Cook, N.J.; Bench, C.J. Infrared thermography and behavioral biometrics associated with estrus indicators and ovulation in estrus-synchronized dairy cows housed in tiestalls. *J. Dairy Sci.* **2019**, *102*, 4427–4440. [CrossRef] [PubMed]

112. Farmer, C.; Robert, S. Hormonal, behavioural and performance characteristics of Meishan sows during pregnancy and lactation. *Can. J. Anim. Sci.* **2003**, *83*, 1–12. [CrossRef]

113. Brogan, R.S.; Mitchell, S.E.; Trayhurn, P.; Smith, M.N. Suppression of leptin during lactation: Contribution of the suckling stimulus versus milk production. *Endocrinology* **1999**, *140*, 2621–2627. [CrossRef]

114. Bridges, R.S. The behavioral neuroendocrinology of maternal behavior: Past accomplishments and future directions. *Horm. Behav.* **2020**, *120*, 104662. [CrossRef] [PubMed]

115. Silveira, P.A.; Spoon, R.A.; Ryan, D.P.; Williams, G.L. Evidence for maternal behavior as a requisite link in suckling-mediated anovulation in cows. *Biol. Reprod.* **1993**, *49*, 1338–1346. [CrossRef] [PubMed]

116. Fraser, D. Some factors influencing the availability of colostrum to piglets. *Anim. Sci.* **1984**, *39*, 115–123. [CrossRef]

117. Muns, R.; Nuntapaitoon, M.; Tummaruk, P. Non-infectious causes of pre-weaning mortality in piglets. *Livest. Sci.* **2016**, *184*, 46–57. [CrossRef]

118. Muns, V. Welfare and Management Strategies to Reduce Pre-Weaning Mortality in Piglets. Ph.D. Thesis, Universitat Autonoma de Barcelona, Barcelona, Spain, 20 June 2013.

119. Muns, V.R.; Tummaruk, P. Management strategies in farrowing house to improve piglet pre-weaning survival and growth. *Thai J. Vet. Med.* **2016**, *46*, 347–354.

120. Algers, B. Nursing in pigs: Communicating needs and distributing resources. *J. Anim. Sci.* **1993**, *71*, 2826–2831. [CrossRef]

121. Brachet, M.A.A.; Vullioud, P.; Ganswindt, A.; Manser, M.B.; Keller, M.; Clutton-Brock, T.H. Parity predicts allonursing in a cooperative breeder. *J. Mammal.* **2021**, *1*, gyab084. [CrossRef]

122. Monticelli, P.F.; Tokumaru, R.S.; Ades, C. Allosuckling in a captive group of wild cavies *Cavia aperea*. *Mammalia* **2018**, *82*, 355–359. [CrossRef]

123. Gloneková, M.; Brandlová, K.; Pluháček, J. Stealing milk by young and reciprocal mothers: High incidence of allonursing in giraffes, *Giraffa camelopardalis*. *Anim. Behav.* **2016**, *113*, 113–123. [CrossRef]

124. Zapata, B.; Correa, L.; Soto-Gamboa, M.; Latorre, E.; González, B.A.; Ebensperger, L.A. Allosuckling allows growing offspring to compensate for insufficient maternal milk in farmed guanacos (*Lama guanicoe*). *Appl. Anim. Behav. Sci.* **2010**, *122*, 119–126. [CrossRef]

125. Wilkinson, G.S. Communal nursing in the evening bat, *Nycticeius humeralis*. *Behav. Ecol. Sociobiol.* **1992**, *31*, 225–235. [CrossRef]

126. Maniscalco, J.M.; Harris, K.R.; Atkinson, S.; Parker, P. Alloparenting in Steller sea lions (*Eumetopias jubatus*): Correlations with misdirected care and other observations. *J. Ethol.* **2006**, *25*, 125–131. [CrossRef]

127. Engelhardt, S.C.; Weladji, R.B.; Holand, Ø.; Røed, K.H.; Nieminen, M. Evidence of reciprocal allonursing in reindeer, *Rangifer tarandus*. *Ethology* **2015**, *121*, 245–259. [CrossRef]

128. Zapata, B.; González, B.A.; Ebensperger, L.A. Allonursing in captive guanacos, lama guanicoe: Milk theft or misdirected parental care? *Ethology* **2009**, *115*, 731–737. [CrossRef]

129. Lee, P.C. Allomothering among African elephants. *Anim. Behav.* **1987**, *35*, 278–291. [CrossRef]

130. Vitikainen, E.I.K.; Marshall, H.H.; Thompson, F.J.; Sanderson, J.L.; Bell, M.B.V.; Gilchrist, J.S.; Hodge, S.J.; Nichols, H.J.; Cant, M.A. Biased escorts: Offspring sex, not relatedness explains alloparental care patterns in a cooperative breeder. *Proc. R. Soc. B Biol. Sci.* **2017**, *284*, 20162384. [CrossRef]

131. Lukas, D.; Huchard, E. The evolution of infanticide by females in mammals. *Philos. Trans. R. Soc. B* **2019**, *374*, 20180075. [CrossRef]

132. Tučková, V.; Šumbera, R.; Čížková, B. Alloparental behaviour in Sinai spiny mice *Acomys dimidiatus*: A case of misdirected parental care? *Behav. Ecol. Sociobiol.* **2016**, *70*, 437–447. [CrossRef]

133. Engelhardt, S.C.; Weladji, R.B.; Holand, Ø.; De Rioja, C.M.; Ehmann, R.K.; Nieminen, M. Allosuckling in reindeer (*Rangifer tarandus*): Milk-theft, mismothering or kin selection? *Behav. Process.* **2014**, *107*, 133–141. [CrossRef]

134. Queller, D.C. Kinship, reciprocity and synergism in the evolution of social behaviour. *Nature* **1985**, *318*, 366–367. [CrossRef]

135. Taborsky, M. Cichlid Fishes: A Model for the Integrative Study of Social Behavior. In *Cooperative Breeding in Vertebrates: Studies of Ecology, Evolution, and Behavior*; Koenig, W.D., Dickinson, J.L., Eds.; Cambridge University Press: Cambridge, UK, 2016; pp. 272–292.

136. Magrath, R.D.; Whittingham, L.A. Subordinate males are more likely to help if unrelated to the breeding female in cooperatively breeding white-browed scrubwrens. *Behav. Ecol. Sociobiol.* **1997**, *41*, 185–192. [CrossRef]

137. Riehl, C. Living with strangers: Direct benefits favour non-kin cooperation in a communally nesting bird. *Proc. R. Soc. B Biol. Sci.* **2011**, *278*, 1728–1735. [CrossRef]

138. Kingma, S.A. Direct benefits explain interspecific variation in helping behaviour among cooperatively breeding birds. *Nat. Commun.* **2017**, *8*, 1094. [CrossRef]

139. Griffin, A.S.; West, S.A. Kin discrimination and the benefit of helping in cooperatively breeding vertebrates. *Science* **2003**, *302*, 634–636. [CrossRef] [PubMed]

140. Clutton-Brock, T. Breeding together: Kin selection and mutualism in cooperative vertebrates. *Science* **2002**, *296*, 69–72. [CrossRef] [PubMed]

141. Trivers, R.L. The Evolution of reciprocal altruism. *Q. Rev. Biol.* **1971**, *46*, 35–57. [CrossRef]

142. Jones, J.D.; Treanor, J.J. Allonursing and cooperative birthing behavior in Yellowstone Bison, Bison bison. *Can. Field-Nat.* **2008**, *122*, 2008. [CrossRef]

143. Taborsky, M.; Frommen, J.G.; Riehl, C. Correlated pay-offs are key to cooperation. *Philos. Trans. R. Soc. B Biol. Sci.* **2016**, *371*, 20150084. [CrossRef]

144. Creel, S.R.; Monfort, S.L.; Wildt, D.E.; Waser, P.M. Spontaneous lactation is an adaptive result of pseudopregnancy. *Nature* **1991**, *351*, 660–662. [CrossRef]

145. Montgomery, T.M.; Pendleton, E.L.; Smith, J.E. Physiological mechanisms mediating patterns of reproductive suppression and alloparental care in cooperatively breeding carnivores. *Physiol. Behav.* **2018**, *193*, 167–178. [CrossRef] [PubMed]

146. Madden, J.R.; Clutton-Brock, T.H. Experimental peripheral administration of oxytocin elevates a suite of cooperative behaviours in a wild social mammal. *Proc. Biol. Sci.* **2011**, *278*, 1189–1194. [CrossRef] [PubMed]

147. Ebensperger, L.A.; Ramírez-Estrada, J.; León, C.; Castro, R.A.; Tolhuysen, L.O.; Sobrero, R.; Quirici, V.; Burger, J.R.; Soto-Gamboa, M.; Hayes, L.D. Sociality, glucocorticoids and direct fitness in the communally rearing rodent, *Octodon degus*. *Horm. Behav.* **2011**, *60*, 346–352. [CrossRef]

148. Pluháček, J.; Bartošová, J. A case of suckling and allosuckling behaviour in captive common hippopotamus. *Mamm. Biol.* **2011**, *76*, 380–383. [CrossRef]

149. MacLeod, K.J.; Clutton-Brock, T.H. Low costs of allonursing in meerkats: Mitigation by behavioral change? *Behav. Ecol.* **2015**, *26*, 697–705. [CrossRef]

150. Ogino, M.; Maldonado-Chaparro, A.A.; Farine, D.R. Drivers of alloparental provisioning of fledglings in a colonially breeding bird. *Behav. Ecol.* **2021**, *32*, 316–326. [CrossRef]

151. Kingma, S.A.; Santema, P.; Taborsky, M.; Komdeur, J. Group augmentation and the evolution of cooperation. *Trends Ecol. Evol.* **2014**, *29*, 476–484. [CrossRef]

152. Kokko, H.; Johnstone, R.A.; Clutton-Brock, T.H. The evolution of cooperative breeding through group augmentation. *Proc. R. Soc. B Biol. Sci.* **2001**, *268*, 187–196. [CrossRef] [PubMed]

153. Teunissen, N.; Kingma, S.A.; Fan, M.; Roast, M.J.; Peters, A. Context-dependent social benefits drive cooperative predator defense in a bird. *Curr. Biol.* **2021**, *31*, 4120–4126. [CrossRef] [PubMed]

154. Gloneková, M.; Brandlová, K.; Pluháček, J. Further behavioural parameters support reciprocity and milk theft as explanations for giraffe allonursing. *Sci. Rep.* **2021**, *11*, 1–9. [CrossRef]

155. Engelhardt, S.C.; Weladji, R.B.; Holand, Ø.; Røed, K.H.; Nieminen, M. Allonursing in reindeer, *Rangifer tarandus*: A test of the kin-selection hypothesis. *J. Mammal.* **2016**, *97*, 689–700. [CrossRef]

156. König, B. Cooperative care of young in mammals. *Naturwissenschaften* **1997**, *84*, 95–104. [CrossRef] [PubMed]

157. König, B. Non-Offspring Nursing in Mammals: General Implications from a Case Study on House Mice. In *Cooperation in Primates and Humans. Mechanism and Evolution*; Kappeler, P.P., Van Schaik, C.P., Eds.; Springer: Berlin, Germany, 2009; pp. 191–205. [CrossRef]

158. Drake, A.; Fraser, D.; Weary, D.M. Parent–offspring resource allocation in domestic pigs. *Behav. Ecol. Sociobiol.* **2007**, *62*, 309–319. [CrossRef]
159. Castanheira, M.; McManus, C.M.; De Paula Neto, J.B.; Da Costa, M.J.R.P.; Mendes, F.D.C.; Sereno, J.R.B.; Bértoli, C.D.; Fioravanti, M.C.S. Maternal offspring behaviour in curraleiro pé duro naturalized cattle in brazil. *Rev. Bras. Zootec.* **2013**, *42*, 584–591. [CrossRef]
160. Villagrán, B.M. Evolución del Comportamiento Alimenticio del Venado de Campo (*Ozotoceros bezoarticus, Linnaeus* 1758) en Semicautiverio Durante las 12 Primeras Semanas de Vida. Ph.D. Thesis, Universidad de la República, Montevideo, Uruguay, 17 April 2009.
161. Bertoni, A.; Napolitano, F.; Mota-Rojas, D.; Sabia, E.; Álvarez-Macías, A.; Mora-Medina, P.; Morales-Canela, A.; Berdugo-Gutiérrez, J.; Guerrero-Legarreta, I. Similarities and differences between river buffaloes and cattle: Health, physiological, behavioral and productivity aspects. *J. Buffalo Sci.* **2020**, *9*, 92–109. [CrossRef]
162. Pélabon, C.; Yoccoz, N.G.; Ropert-Coudert, Y.; Caron, M.; Peirera, V. Suckling and allosuckling in captive fallow deer (*Dama dama, Cervidae*). *Ethology* **1998**, *104*, 75–86. [CrossRef]
163. Réale, D.; Boussès, P.; Chapuis, J.L. Nursing behaviour and mother–lamb relationships in mouflon under fluctuating population densities. *Behav. Process.* **1999**, *47*, 81–94. [CrossRef]
164. Waltl, B.; Appleby, M.C.; Sölkner, J. Effects of relatedness on the suckling behaviour of calves in a herd of beef cattle rearing twins. *Appl. Anim. Behav. Sci.* **1995**, *45*, 1–9. [CrossRef]
165. Drábková, J.; Bartošová, J.; Bartoš, L.; Kotrba, R.; Pluháček, J.; Švecová, L.; Dušek, A.; Kott, T. Sucking and allosucking duration in farmed red deer (*Cervus elaphus*). *Appl. Anim. Behav. Sci.* **2008**, *113*, 215–223. [CrossRef]
166. Loberg, J. Behaviour of Foster Cows and Calves in Dairy Production Acceptaof Calves, Cow-Calf Interactions and Weaning. Ph.D. Thesis, Universitatis Agriculturae Sueciae, Uppsala, Sweden, 23 November 2007.
167. Ebensperger, L.A.; León, C.; Ramírez-Estrada, J.; Abades, S.; Hayes, L.D.; Nova, E.; Salazar, F.; Bhattacharjee, J.; Becker, M.I. Immunocompetence of breeding females is sensitive to cortisol levels but not to communal rearing in the degu (*Octodon degus*). *Physiol. Behav.* **2015**, *140*, 61–70. [CrossRef]
168. Becker, M.I.; De Ioannes, A.E.; León, C.; Ebensperger, L.A. Females of the communally breeding rodent, Octodon degus, transfer antibodies to their offspring during pregnancy and lactation. *J. Reprod. Immunol.* **2007**, *74*, 68–77. [CrossRef]
169. Dalto, A.C.; Bandarra, P.M.; Pavarini, S.P.; Boabaid, F.M.; De Bitencourt, A.P.G.; Gomes, M.P.; Chies, J.; Driemeier, D.; Da Cruz, C.E.F. Clinical and pathological insights into Johne's disease in buffaloes. *Trop. Anim. Health Prod.* **2012**, *44*, 1899–1904. [CrossRef]
170. Mota-Rojas, D.; De Rosa, G.; Mora-Medina, P.; Braghieri, A.; Guerrero-Legarreta, I.; Napolitano, F. Dairy buffalo behaviour and welfare from calving to milking. *CAB Rev. Perspect. Agric. Vet. Sci. Nutr. Nat. Resour.* **2019**, *14*, 1–14. [CrossRef]
171. Brandlová, K.; Bartoš, L.; Haberová, T. Camel calves as opportunistic milk thefts? the first description of allosuckling in domestic bactrian camel (*Camelus bactrianus*). *PLoS ONE* **2013**, *8*, e53052. [CrossRef] [PubMed]
172. Carter, C.S.; Altemus, M. Integrative functions of lactational hormones in social behavior and stress management. *Ann. N. Y. Acad. Sci.* **1977**, *807*, 164–174. [CrossRef] [PubMed]
173. Lightman, S.L. Alterations in hypothalamic-pituitary responsiveness during lactation. *Ann. N. Y. Acad. Sci.* **1992**, *652*, 340–346. [CrossRef] [PubMed]
174. Schlein, P.A.; Zarrow, M.X.; Denenberg, V.H. The role of prolactin in the depressed or 'buffered' adrenocorticosteroid response of the rat. *J. Endocrinol.* **1974**, *62*, 93–99. [CrossRef]

MDPI
St. Alban-Anlage 66
4052 Basel
Switzerland
www.mdpi.com

Animals Editorial Office
E-mail: animals@mdpi.com
www.mdpi.com/journal/animals

Disclaimer/Publisher's Note: The statements, opinions and data contained in all publications are solely those of the individual author(s) and contributor(s) and not of MDPI and/or the editor(s). MDPI and/or the editor(s) disclaim responsibility for any injury to people or property resulting from any ideas, methods, instructions or products referred to in the content.